ORGANIC CHEMISTRY

SECOND EDITION

Frank C. Whitmore

Late Research Professor of Organic Chemistry
The Pennsylvania State College

With the Assistance of a Committee of Colleagues

VOLUME ONE

Part I: Aliphatic Compounds
Part II: Alicyclic Compounds

Dover Publications Inc.
Mineola, New York

Bibliographical Note

This Dover edition, first published in 1961 and reissued in 2011, is an unabridged and corrected republication of the second edition of the work originally published in 1951 by the D. Van Nostrand Company, Inc., New York.

Library of Congress Cataloging-in-Publication Data

Whitmore, Frank C. (Frank Clifford). 1887-1947.
    Organic chemistry / Frank C. Whitmore. — Dover ed.
      p. cm.
    Originally published: 2nd ed. New York : D. Van Nostrand, 1951.
    Includes bibliographical references and index.
    ISBN-13: 978-0-486-60700-9 (pbk.)
    ISBN-10: 0-486-60700-3 (pbk.)
    1. Chemistry, organic. 2. Aliphatic compounds. 3. Alicylclic compounds.
I. Title.

QD251.W5 2011
547—dc23

20011018514

Manufactured in the United States by Courier Corporation
60700342
www.doverpublications.com

THE COMMITTEE

Dr. N. C. COOK, *General Electric Company, Schenectady, N. Y.*

Dr. J. A. DIXON, *The California Research Corporation, Richmond, Calif.*

Dr. M. R. FENSKE, *The Pennsylvania State College*

Dr. G. H. FLEMING, *The Pennsylvania State College*

Dr. R. S. GEORGE, *Hercules Powder Company, Wilmington, Del.*

Dr. A. H. HOMEYER, *Mallinckrodt Chemical Company, St. Louis, Mo.*

Dr. J. H. JONES, *The Pennsylvania State College*

Dr. J. A. KRIMMEL, *Industrial Research Institute, University of Denver, Denver, Colo.*

Dr. J. F. LAUCIUS, *The Du Pont Company, Wilmington, Del.*

Dr. A. R. LUX, *The Du Pont Company, Wilmington, Del.*

Dr. H. S. MOSHER, *Stanford University*

Dr. W. A. MOSHER, *University of Delaware*

Dr. C. I. NOLL, *The Pennsylvania State College*

Dr. T. S. OAKWOOD, *The Pennsylvania State College*

Dr. R. W. SCHIESSLER, *The Pennsylvania State College*

Dr. L. H. SOMMER, *The Pennsylvania State College*

Dr. R. B. WAGNER, *The Pennsylvania State College*

Dr. H. D. ZOOK, *The Pennsylvania State College*

This Dover Edition is dedicated to the memory of

FRANK C. WHITMORE

1887 - 1947

for his contribution to chemical research and education

PREFACE TO SECOND EDITION

Only a few hours before his death in June 1947, Dr. Whitmore completed the revision of the aliphatic section of his Organic Chemistry. The remaining sections were partly revised and extensive material for their completion was left in the form of notes and rough draft. Many people felt that the book should be finished since it served so many needs of chemists. Accordingly, former students and friends have helped finish this revision as a token of gratitude to a man who gave himself freely to science and scientists.

In this edition, several changes have been made in the organization of the book. The sections treating metal alkyls, phosphorous compounds, and organo-metallic compounds have been transferred to the end of the book.

The abbreviated references used in the first edition have been replaced with complete references.

The Index has been changed to a style similar to the style of Chemical Abstracts.

Major changes and additions have been made in the material of the aliphatic and heterocyclic sections to keep pace with the rapid advances in these fields. Additions necessary to bring Dr. Whitmore's revision of the aliphatic section up to date since his death have been made. Treatment of the terpenes, alkaloids and dyes have received special attention.

I am deeply obligated to the committee of colleagues who assisted in the completion of this revision.

I am indebted to Dr. M. L. Wolfrom for reading and suggesting changes in the carbohydrate section.

I am grateful to Dr. F. E. Cislak for checking the phenol section.

I wish to thank Dr. J. G. Aston for his valuable criticism and advice.

It is impossible to thank adequately the graduate students, friends and secretaries in the School of Chemistry and Physics of the Pennsylvania State College, for their help in referencing and indexing.

My thanks are due to Dean George L. Haller who generously provided stenographic and clerical help.

Dr. Whitmore was particularly grateful to Dr. H. B. Hass, Dr. Edward Lyons, and many others too numerous to name for their corrections of the first edition. I will consider it a favor if Dr. Whitmore's friends will also advise me concerning errors in this edition.

MRS. FRANK C. WHITMORE

STATE COLLEGE, PENNSYLVANIA,
March 28, 1951

PREFACE

This work is a one-volume reference-text designed for practising organic chemists, for others who have to take occasional cognizance of organic compounds and their reactions, and for students who have pursued organic chemistry for at least a year with the aid of the many excellent elementary and intermediate textbooks now available. All efforts have been directed to making it a text of advanced character designed for those already possessing reasonable knowledge and experience in organic chemistry.

In the compilation of this work practically all available sources of information have been drawn upon. Special use has been made of all the existing advanced one-volume texts in German, French and Italian. In cases where English translations of such texts contain material not in the originals they also have been used.

The author's thanks are due to his many friends in industry who have helped freely with facts and comments on industrial processes to the full extent compatable with the ordinary practices of trade secrecy. During the eight years in which the book has been written Marion Mason Whitmore has helped in all stages of the work. To Ann Reimel Young and Margaret Zerbey, Librarians of the School of Chemistry and Physics, go thanks for extensive help in literature searches. Beulah Mattern Knop and Alice Long have carried on the difficult typing year after year. Hearty thanks go to the score of organic chemists in Pond Laboratory who speeded the preparation of the Index. No thanks could be adequate for the painstaking and faithful work of Delcena Crabtree and Mary Louise Shaner to whom fell the arduous task of sorting the forty-nine thousand index slips.

The author will be grateful to his colleagues in the industrial and academic fields for advice on errors both of commission and omission which they note in the work.

FRANK C. WHITMORE

STATE COLLEGE, PENNSYLVANIA,
February 1, 1937

CONTENTS

VOLUME ONE

PART I. ALIPHATIC COMPOUNDS

PART II. ALICYCLIC COMPOUNDS

CONTENTS

INTRODUCTION

In keeping with the present trend toward aliphatic chemistry, especially in British and American industry, nearly three-fourths of the work is devoted to aliphatic and alicyclic chemistry. The section on aromatic chemistry is shorter than in most volumes of this type while that on heterocyclic compounds is relatively larger. The wide occurrence of aromatic properties is emphasized. The complex alkaloids are presented in an orderly arrangement based on an analysis and classification of possible combinations of nitrogen-ring systems.

Other works and the literature should be consulted for the application of analytical and physical principles to organic chemistry and for many details in the development of the science. Thus, the reader will fail to find in this work many of the historically interesting formulas which have been proposed for benzene.

Another type of omission is that of details about the distillation and utilization of coal tar. This is because of the lack of any uniformity at the present time in the working up of this important material. A still different type of omission is that of the details of the work on the sex hormones. In this, as in many similar cases throughout the work, references are given to sources of as detailed information as the reader can wish.

No attempt has been made to recognize priority among workers. In fact, in many cases the name cited is that of a recent worker in whose articles can be found summaries of earlier work.

General principles have been stressed throughout the work. Many of these such as the initiation of reactions by preliminary addition and the tendency for ring closure appear repeatedly.

A deliberate attempt has been made to explode what might be called the fallacy of homologous series in which it is often assumed that a knowledge of the first two or three members of a series furnishes a satisfactory knowledge of the series itself. Thus, in the alcohol series it has been necessary to go at least to the seven carbon member before distinct novelties in properties and reactions cease to appear.

The use of electronic conceptions has been definitely limited to those cases in which ordinary structural formulas fail. In most processes in organic chemistry the bond corresponds exactly to the effect of an electron pair and nothing is achieved by substituting two dots for the conventional dash.

British Annual Reports and Organic Syntheses are constantly referred to because they offer, respectively, excellent summaries and detailed preparative directions. Unfortunately, in neither case are the references as complete as might be possible.

INTRODUCTION

The explanation of all references and abbreviations is included in the Index.

Entirely aside from its use in locating specific material, the Index will be helpful, especially to advanced students, in selecting the important compounds and processes of organic chemistry and then following them through a range of examples covering the entire science.

In treating the whole of organic chemistry in a single volume a decidedly condensed style has been necessary. Thus the two chief users of this work, the practising chemist and the advanced student, will find the use of paper and pencil helpful in expanding many of the formulas and equations.

PART I

ALIPHATIC COMPOUNDS

I. HYDROCARBONS

A. Saturated Hydrocarbons, C_nH_{2n+2}

Paraffins, alkanes, homologs of methane.

This simplest homologous series of organic chemistry shows the gradation in physical properties characteristic of such series. From a gas, only slightly less volatile than oxygen, the repeated increase of CH_2 in the compounds produces volatile liquids at C_5 and low-melting solids at C_{16}. The increase in boiling point for an increase of CH_2 decreases with the higher members (p. 4). Of isomers, the normal (n-) (straight chain) member has the highest boiling point. In the series to C_8 the n-hydrocarbons boil lower than the lowest boiling isomer of the next homolog. Thus all the octanes boil higher than n-heptane. At that point in the series, however, the spread between two successive n-hydrocarbons becomes so small and the possibility of branching, with accompanying lowering of the b.p., so great that two of the highly branched nonanes boil lower than n-octane. These are 2,2,5-trimethylhexane and 2,2,4,4-tetramethylpentane.[1] The densities of the n-alkanes increase from 0.4 to a limiting value of about 0.78. The value 0.77 is reached by the C_{11} member.

The index of refraction (n^{20}D) for the liquid n-alkanes ranges from 1.3577 for n-pentane to 1.4270 for n-pentadecane. A rise of 1° *decreases* the n_D by 0.00055 for n-pentane and 0.00044 for n-dodecane. The use of the α line of the hydrogen spectrum instead of the D line *decreases* the n^{20} for n-pentane by 0.0019 and for n-dodecane by 0.0022 while the use of the β line in place of the D line gives *increases* of 0.0044 and 0.0053 respectively.

The alkanes are practically insoluble in water but soluble in most organic liquids. In aniline their solubility is limited at ordinary temperature. The Critical Solution Temperatures (C.S.T.) in aniline and in liquid sulfur dioxide are characteristic of the individual hydrocarbons both in this and other series. The C.S.T. in aniline for some of the *normal* alkanes in °C. follow: C_5, 71.4; C_6, 69.0; C_7, 69.9; C_8, 71.8; C_9, 74.4; C_{10}, 77.5; C_{11}, 80.6; C_{12}, 83.7. The C.S.T. in liquid SO_2 are as follows: C_6, 10.2; C_8, 26.9; C_{10}, 37.3; C_{12}, 47.3; C_{14}, 55.5; C_{32}, 110.0°. The values for furfural are: C_6, 92; C_7, 95; C_{12}, 112.5; C_{13}, 115.9; C_{14}, 119.6; C_{15}, 122.7; C_{16}, 125.9; C_{17}, 129.3; C_{18}, 132.0; C_{20}, 138.1; C_{24}, 147.2; C_{26}, 150.3.

---

[1] Doss. "Physical Constants of the Principal Hydrocarbons," 4th Ed. The Texas Co.

Because of their inability to *add* reagents the alkanes are called *saturated hydrocarbons*. Thus they can react with halogens only by *substitution*, a hydrogen being removed for each halogen which enters the molecule. They do not ordinarily react with hydrogen. Under extreme conditions, *hydrogenolysis* occurs with splitting of the C–C linkage. Ethane, under such conditions, gives methane. An important application is the conversion of the easily obtainable alkylation product, 2,2,3-Me_3-pentane, to Me_3-butane (Triptane) by hydrogenolysis.

A mixture of the higher members of the series, paraffin wax, received its name because of its inertness to acids and oxidizing agents (from *parum affinis*). Because methane is inert to most reagents and because the next few *normal* homologs are rather inert, the name paraffin hydrocarbons has given the impression that the entire series is very inactive chemically. This is not true. Even paraffin wax is fairly reactive with oxygen at slightly elevated temperatures (preparation of acids).[2] While reagents such as nitric acid, sulfuric acid, chromic acid mixture or potassium permanganate do not act *readily* with the lower *normal* members, some of them act with the higher members and with the branched members containing a *tertiary hydrogen*, R_3CH. This hydrogen can be replaced by $-NO_2$, $-SO_3H$ or $-OH$ with nitric acid, sulfuric acid or oxidizing agents respectively.

Paraffins react readily with chlorine in light or at slightly elevated temperature to give *substitution* of H by Cl (chlorination). Polychlorides are readily obtained. The reaction may become dangerously explosive if not controlled.

Vapor phase nitration of paraffins, replacement of H or alkyl by NO_2, is increasingly important commercially.[3] The high temperature necessary for nitration favors splitting of C–C. Thus the nitration of propane gives not only 1- and 2-nitropropane but also nitromethane and nitroethane.

Paraffin hydrocarbons react with SO_2 and Cl_2 (Reed Reaction) in the presence of actinic light to give alkyl sulfonyl chlorides, RSO_2Cl.[4]

In the first part of the series the increase of CH_2 makes a marked difference in the percentage composition. Successive additions of CH_2 have a decreasing effect as the composition approaches that of $(CH_2)_n$. Thus an ordinarily accurate C and H determination would barely distinguish C_{20} from C_{30}.

Possible and Known Isomers. Using only the conception of the tetravalence of carbon the following numbers of *structural isomers* are predicted for the alkanes: 1 each for C_1, C_2 and C_3, 2 for C_4, 3 for C_5, 5 for C_6, 9 for C_7, 18 for C_8, 35 for C_9 and 75 for C_{10}. Methods of calculating the number of theoretically possible isomers have been developed.[5, 6, 7] The numbers in-

[2] Ellis. "Chemistry of Petroleum Derivatives." Reinhold, 1934. p. 959.
[3] Hass et al. *Ind. Eng. Chem.* **28**, 339 (1936); **35**, 1146 (1943); **39**, 817 (1947). *Chem. Rev.* **32**, 373 (1943).
[4] Lockwood. *Chem. Inds.* **62**, 760 (1948).
[5] Henze, Blair. *J. Am. Chem. Soc.* **53**, 3077 (1931).
[6] Blair, Henze. *J. Am. Chem. Soc.* **54**, 1538 (1932).
[7] Perry. *J. Am. Chem. Soc.* **54**, 2918 (1932).

dicated are 366,319 for C_{20} and over 4 billion for C_{30}. Many of the structural isomers contain asymmetric carbon atoms and can give rise to stereoisomers. Thus, of the 18 structurally isomeric octanes, 3-Me-heptane, 2,3-Me$_2$-hexane, 2,4-Me$_2$-hexane and 2,2,3-Me$_3$-pentane each contains an asymmetric carbon and could exist in *dextro* and *levo* optically active forms. A fifth octane, 3,4-Me$_2$-hexane, contains two similar asymmetric carbons and could exist in *d*-, *l*- and *meso*-forms. Thus the total number of isomers of the octanes becomes 24, of which 11 are stereoisomers and 13 are non-stereoisomers. Similarly for C_{10}, the 75 structure isomers give rise to 101 stereoisomers and 35 non-stereoisomers. Soon the numbers predicted lose all physical significance, there being a total of 3,395,964 "possible" isomeric eicosanes (C_{20}).

Turning from the predicted to the known we find that all the predicted structural isomers have been prepared for the first nine members of the alkane series. Of the 75 possible structurally isomeric decanes, about half have been prepared. Many optically active hydrocarbons have also been prepared (p. 22).[8]

The preparation of the higher alkanes involves many difficulties among which are (1) the decreased activity of the larger molecules, (2) the failure of many reactions when extreme branching of the carbon chain occurs,[9] (3) rearrangement due to branched chains,[10] and (4) the difficulty of separating and purifying the intermediates and products. The distillation methods used have been improved remarkably.[11,12,13,14] The newer combination of distillation with solvent extraction has been studied by many workers (*Distex Process*).[15]

The effect of branching in isomeric paraffins may be seen from the melting and boiling points, °C., and refractive indices, n^{20}D, of *n*-octane, 4-Me-heptane, 2,2,4-Me$_3$-pentane ("iso-octane"), and 2,2,3,3-Me$_4$-butane which are respectively: -56.8, 125.6, 1.3976; -121.3, 117.5, 1.3980; -107.3, 99.2, 1.3914; $+101.6$, 106.5. The effect of symmetry in raising the m.p. is notable in the last.

Occurrence of the Alkanes. The alkanes are widely distributed in nature. Methane occurs in natural gas from 75 to nearly 100%, in "fire damp" in coal mines and as "marsh gas" formed by the decay of vegetable matter. The higher homologs are found to a decreasing extent in natural gas and to an

[8] Levene, Marker. *J. Biol. Chem.* **91**, 405 (1931); **91**, 761 (1931); **92**, 455 (1931); **95**, 1 (1932).

[9] Conant, Blatt. *J. Am. Chem. Soc.* **51**, 1227 (1929).

[10] Whitmore. *J. Am. Chem. Soc.* **54**, 3274 (1932); *Chem. Eng. News* **26**, 668 (1948).

[11] Fenske et al. *Ind. Eng. Chem.* **28**, 644 (1936); **24**, 408 (1932); **26**, 1164 (1934); **30**, 297 (1938).

[12] Podbielniak. *Ind. Eng. Chem., Anal. Ed.* **5**, 119 (1933); **13**, 639 (1941); *C.A.* **36**, 3989 (1942); **38**, 1400 (1944).

[13] Stedman. *Can. Chem. Met.* **21**, 214 (1937); *Trans. Am. Inst. Chem. Eng.* **33**, 153 (1937); *Can. J. Research.* **15**, B, 383 (1937).

[14] Ewell et al. *Ind. Eng. Chem., Anal. Ed.* **12**, 544 (1940); *Ind. Eng. Chem.* **36**, 871 (1944).

[15] Griswold, Van Berg. *Ind. Eng. Chem.* **38**, 170 (1946).

C atoms	Name	m. °C.	b. °C. (Doss)
1	Methane	−182.5	−161.4
2	Ethane	−183.2	− 89.0
3	Propane	−187.7	− 42.1
4	Butane	−138.3	− 0.6
5	Pentane	−129.7	36.0
6	Hexane	− 95.4	68.7
7	Heptane	− 90.6	98.4
8	Octane	− 56.8	125.6
9	Nonane	− 53.7	150.7
10	Decane	− 29.7	174.0
11	Undecane	− 25.6	195.8
12	Dodecane	− 9.6	216.3
13	Tridecane	− 6.2	236.5
14	Tetradecane	5.5	253.5
15	Pentadecane	10.0	272.7
16	Hexadecane	18.1	286.5
17	Heptadecane	21.0	305.8
18	Octadecane	28.	317.9
19	Nonadecane	31.4	336.2
20	Eicosane	36.7	205*
21	Heneicosane	40.4	215*
22	Docosane	45.7	230*
23	Tricosane	47.5	320.7
24	Tetracosane	49.4	250*
25	Pentacosane	53.3	259*
26	Hexacosane	56.4	268*
27	Heptacosane	59.	277*
28	Octacosane	60.3	286*
29	Nonacosane	62.3	295*
30	Triacontane	64.7	304*
31	Hentriacontane	65.3	312*
32	Dotriacontane (dicetyl)	69.6	320*
33	Tritriacontane	71.1	328*
34	Tetratriacontane	72.3	336*
35	Pentatriacontane	74.5	341*
36	Hexatriacontane	76.	265**
37	Heptatriacontane	77.	
38	Octatriacontane	77.	
39	Nonatriacontane	79.	
40	Tetracontane	81.	
41	Hentetracontane	81.	
42	Dotetracontane	83.	
43	Tritetracontane	85.	336**
44	Tetratetracontane	86.	
45	Pentatetracontane	86.	354**
50	Pentacontane	92.	200***
52	Dopentacontane	94.	
54	Tetrapentacontane	95.	
60	Hexacontane	99.	250***
62	Dohexacontane	100.	
64	Tetrahexacontane	102.	
66	Hexahexacontane	104.	
67	Heptahexacontane	105.	300***
70	Heptacontane	105.	300***

* at 15 mm. Hg.   ** at 1.5 mm.   *** at 0.041 mm.

increasing extent in petroleum. A typical analysis of a natural gas from a large high pressure line supplied from many wells of various ages and from different sands gave the following percentages: methane 78, ethane 13, propane 6, butanes 1.7, pentanes .6, hexanes .3, heptanes and above .4. Gas from the Lower Oriskany Sand of Pennsylvania has 98.8% methane while a gas from Glasgow, Kentucky has been found with only 23%.[16]

Analysis of natural gas.[17]

The most important occurrence of the alkanes is in petroleum. Probably all petroleums contain at least some of this series. The proportions of hydrocarbons of other series and of non-hydrocarbon constituents vary over wide ranges. Pennsylvania grade petroleum probably contains the largest amount of paraffin hydrocarbons, although there are indications that some Michigan crudes contain a still larger proportion, especially of the normal hydrocarbons.

Recent studies indicate that all living organisms may form hydrocarbons as by-products of their metabolism. Whenever the non-saponifiable portion of a plant or animal product is freed from sterols and related products, the residue is practically sure to contain hydrocarbons. The normal paraffins containing 7, 9, 11, 15, 19, 21, 22, 23, 25, 27, 28, 29, 30, and 31 carbon atoms have been reported, mainly since 1935. The identification of hydrocarbons beyond C_{20} should be supplemented by X-Ray studies.

Formation of Alkanes. There is no agreement as to the probable mode of formation of natural gas and petroleum.[18,19,20] The destructive distillation of vegetable materials, such as wood and the various forms of coal, as well as certain bituminous shales gives varying amounts of the alkanes. "Low temperature tar" obtained by heating soft coal at about 600° contains alkanes which, at higher temperatures, are converted to aromatic substances. The cracking of cotton seed oil gives a gas containing 35% methane, 12% ethane and 5% propane and a liquid distillate containing 37% of higher alkanes.

The formation of hydrocarbons from fatty acids by alpha-particle bombardment has been demonstrated.[21]

If all of the hydrocarbons formed by plants and animals since life appeared had survived, the total amount would probably be thousands of times the total amount of petroleum and natural gas. Most of such hydrocarbons have apparently been destroyed by micro-organisms which can utilize them in the absence of more reactive sources of energy.

The methods of preparation of the alkanes will be given under the individual members of the series.

Petroleums consist mainly of mixtures of hydrocarbons with admixtures of compounds of oxygen, nitrogen, and sulfur varying from traces to 10% or

[16] Ellis. "Chemistry of Petroleum Derivatives." Reinhold, 1934. p. 13.
[17] Ellis, *ibid.* pp. 1092–1124.
[18] Engler. *Chem. Zt.* **30**, 711 (1906).
[19] Brooks. *Bull. Am. Assoc. Petroleum Geol.* **15**, 611 (1931); **20**, 280 (1936).
[20] Ellis. "Chemistry of Petroleum Derivatives." Reinhold, 1934. p. 35.
[21] Sheppard, Burton. *J. Am. Chem. Soc.* **68**, 1636 (1946).

more in various crudes. Definite knowledge on the compounds in petroleums
is very limited. It is fairly certain that all petroleums contain members of the
methane series, the polymethylene (alicyclic) series, and the benzene (aromatic)
series of hydrocarbons. All petroleums contain optically active material of
MW about 400. The amount of this material is lowest in petroleum from the
Pennsylvania area. The general composition of crude petroleums from
different sources may be indicated by the following chart.[22]

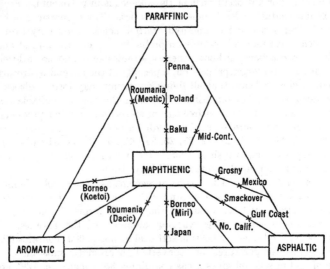

Many petroleum fractions contain hydrocarbons more deficient in hydrogen
than any series of known structure. These extend to C_nH_{2n-20}.[23] There is no
conclusive evidence that any natural petroleum contains members of the
olefin series. The difference between petroleums of various sources is in the
proportions of the different types of hydrocarbons and in the nature and
amounts of impurities.[24] Thus Pennsylvania crude oil contains a large proportion of methane hydrocarbons and practically no impurities of sulfur or
nitrogen compounds. Mid-Continent (Oklahoma and Texas) crudes contain
larger proportions of aromatic and polymethylene compounds and larger
amounts of sulfur compounds.

The present ignorance of the actual compounds in petroleum is colossal.
Less than 100 definite hydrocarbons have been isolated and certainly identified.[25, 26] Most of these come from less than 30% of a single Mid-Continent

[22] Gruse. "Petroleum and Its Products." McGraw-Hill Book Co., 1928.
[23] Ellis. "Chemistry of Petroleum Derivatives." Reinhold, 1934. p. 28.
[24] Ellis, *ibid*. p. 19.
[25] Ellis, *ibid*. p. 27.
[26] Rossini. *Refiner Natural Gasoline Mfr.* 20, 494 (1941); *Chem. Eng. News* 25, 230 (1947).

crude. No other crude has been studied even to that extent. Many hydrocarbons believed to be present in petroleums have been reported on insufficient evidence. Extreme care is necessary in purifying and identifying even relatively simple paraffin hydrocarbons.[27] California crudes contain larger amounts of sulfur compounds, as well as some nitrogen compounds.[28, 29] Mexican, Venezuelan and Colombian crudes contain still larger amounts of sulfur compounds. The nature of the sulfur compounds in crude petroleum is little known.[30] Other important petroleum fields are those of Russia, Persia, Rumania, Borneo and Canada. Smaller fields are found in many parts of the world including even France, England, Germany and Italy. The United States in recent years has produced over 70% of the world's crude oil. The American Petroleum Institute, 250 Park Avenue, New York City, issues frequent bulletins covering world petroleum statistics. The extent and amount of crude petroleum deposits is now known to be many times what was suspected a few years ago.

The basis of petroleum technology was laid by a report by Benjamin Silliman, Jr., of Yale, made in 1855 [31] on a sample of surface petroleum from Titusville, Pennsylvania. In 1859 the first oil well was sunk near Titusville by E. L. Drake. Its production was 25 barrels per day.

The refining of petroleum has grown into a most complex chemical engineering industry. Distillation is the chief method used in separating crude petroleum into useful products. At the present time the distilled fractions from crude petroleum are casinghead gasoline, gasoline, kerosine, gas oil and, in some cases, lighter lubricating oils ("neutrals"). The residues from distillation supply most of the lubricating oils ("bright stock"), petrolatum (vaseline) and either paraffin wax or petroleum pitch, depending on the nature of the crude. Fractions of the distillate boiling about 0° and about 20° have been called cymogene and rhigoline respectively. Higher boiling portions of the volatile part of petroleum are called *petroleum ether* and *ligroin.* Such names should always be accompanied by boiling ranges to avoid confusion. It is also helpful to know the type of crude used so as to have at least an approximate idea of the amount of aromatic materials present since these change the solvent properties markedly. *Benzine* is an indefinite term roughly corresponding to a volatile gasoline. Gasoline (petrol) is any mixture of hydrocarbons which can be used in spark ignition internal combustion engines. Too much low boiling material will prevent the fuel from being sucked as liquid into the carburetor thus causing "vapor lock," while too much high boiling material results in imperfect combustion and excessive carbon deposition in the engine cylinders. Formerly the only requirements for gasoline

[27] Washburn. *Ind. Eng. Chem.* 22, 985 (1930).
[28] Bailey et al. *J. Am. Chem. Soc.* 52, 1239 (1930).
[29] Ellis. "Chemistry of Petroleum Derivatives." Reinhold, 1934. p. 819.
[30] Ellis, *ibid.* pp. 421–463.
[31] Johns. *Ind. Eng Chem.* 15, 446 (1923).

were that it should be neither too volatile nor too non-volatile and should not contain enough sulfur compounds to cause corrosion. The end point for gasoline is 400° F. These simple requirements were changed by the high compression motor, introduced to increase the power for a given weight of engine. Increased compression tends to give detonation or *knock* instead of rapid smooth combustion of the fuel. Gasolines from different crudes and from different processes vary widely in knock characteristics. In general, straight chain paraffin hydrocarbons knock worse and aromatic hydrocarbons, olefins, and branched hydrocarbons knock less (p. 24). The discovery of catalysts which decreased knock, notably *tetraethyllead*,[32] has revolutionized the gasoline industry. At first used only in one special gasoline (*Ethyl Gas*), its use soon spread to other grades. Probably the annual consumption of this organometallic compound approaches a quarter of a billion pounds per year. Moreover, it has catalyzed the large scale production of compounds of high octane number (pp. 23, 24, 51).

Kerosine is any mixture of hydrocarbons which is not volatile enough for use as gasoline but which can be burned in lamps and similar devices. Since gasoline is subject to tax in many cases and kerosine is not, it has become necessary to have a legal definition. U. S. Government specifications for kerosine to be used as a burning oil require a distillation end point of not over 625° F. and a flash point of not less than 115° F.

Analysis of petroleum distillates.[33]

Until the end of the Nineteenth Century, the most important product from petroleum was kerosine. The lower boiling products were of little use. To prevent the inclusion of them in kerosine, stringent regulations were made as to the "flash point" and "fire point" of kerosine in order to insure its safe use. The first "cracking" or thermal decomposition of the higher fractions of petroleum was for the purpose of increasing the yield of kerosine above that obtainable by straight distillation. With the development of the internal combustion engine, gasoline has become the important product, with lubricating oil a close second. Kerosine is only a by-product. In order to increase the yield of gasoline, many different cracking processes have been perfected. These operate under a wide range of conditions of temperature and pressure and on materials such as gas oil, kerosine and even the crude petroleum itself. Yields of gasoline as high as 70% of the crude have been obtained. The cracking of kerosine or gasoline to obtain more gasoline or gasoline of higher anti-knock quality is called "re-forming."

The peculiar features of the more important thermal cracking processes have now been combined with others in most large installations, so that the classification of cracking processes by name is frequently not possible or desirable.[34, 35]

[32] Midgley. *C. A.* **20**, 1514 (1926).
[33] Ellis. "Chemistry of Petroleum Derivatives." Reinhold, 1934. p. 1125.
[34] Ellis. "Chemistry of Petroleum Derivatives." Reinhold, 1934. p. 91.
[35] Waverly Handbook, 1941.

The increased demand for high octane aviation fuel in World War II speeded the development of the Houdry catalytic cracking process in which various catalytic clays are used at high temperatures to give more cracking with less carbonization. Many "fluid catalyst" processes have been developed. In these the finely powdered catalyst is circulated with the hot vapors to be cracked. The four principal catalytic processes are Houdry, Thermofor, Fluid and Cycloversion. In 1945 these processes accounted for nearly 1,000,000 barrels of charging stock daily, 29% of the total cracking in the United States. A "Cat Cracker" is to be found in practically every modern refinery.

Cracked gasolines are rich in olefins and diolefins. This has an advantage due to the anti-knock properties of these unsaturated compounds,[36] but the disadvantage that the unsaturated compounds, especially the diolefins, tend to polymerize and form "gums" which clog the carburetor.[37] The tendency of higher olefins and diolefins from cracking operations to form complex products is being utilized in the manufacture of resins from petroleum.

Sulfuric acid is used in refining petroleum fractions to remove various objectionable materials including sulfur compounds. Unfortunately this treatment also removes some of the aromatic compounds and the olefins and diolefins (from cracked gasoline). This is undesirable since these materials have great anti-knock value. The present tendency is to add small amounts of stabilizers of a wide variety to cracked gasoline to delay polymerization. These stabilizers are usually compounds of the anti-oxidant type such as naphthylamines, p-aminophenol and diphenylhydrazine.[38]

The use of sulfuric acid in petroleum refining is decreasing.

The refining of lubricants has been largely revolutionized by the use of extraction by partially miscible solvents such as phenol, cresylic acid (mixture of cresols, xylenols and higher phenols), nitrobenzene, dichlorodiethyl ether ("chlorex") and furfural. These extraction processes are natural outgrowths of one long used in the industry, the extraction of petroleum fractions with liquid sulfur dioxide (Edeleanu process) for the removal of sulfur compounds.

Paraffin wax was originally obtained from tars from the distillation of wood, peat, and lignite, but is now obtained from petroleum, especially from paraffin-base oils such as Pennsylvania grade crude. Little is known about the composition of paraffin wax except that it consists mainly of higher alkanes and probably contains very large amounts of the normal compounds. Wax may separate from lubricating oils at low temperatures and thus decrease their rate of flow (decrease the "pour point"). To avoid this, the lubricant is sometimes "dewaxed" by dilution with a low boiling petroleum fraction, or better with propane under pressure. The solution is then refrigerated to separate the wax which is removed by means of filter presses or centrifuges.

[36] Ellis. "Chemistry of Petroleum Derivatives." Reinhold, 1934. p. 961.
[37] Ellis, *ibid.* pp. 881, 889.
[38] Ellis, *ibid.* p. 889.

The solvent is then removed by distillation. Propane dewaxing is now combined with a de-asphaltizing step, based on the insolubility of asphalts in propane at high temperatures. The lubricating cut is heated with propane under pressure and the precipitated asphalts are separated hot. The solution is then cooled to remove the wax. Even a *paraffin base* oil like Pennsylvania grade oil yields a small amount of asphaltic material by this process. Dewaxing is not only costly but there is some possibility that the wax is an advantage in the lubricant except for the danger of its solidifying. Hence substances have been introduced to delay or inhibit wax crystallization (Paraflow, Santopour). Such substances consist of complex mixtures of complicated molecules incapable of crystallizing. Perhaps they are adsorbed on the first microcrystals and prevent their growth as nuclei for the crystallization of the main mass of wax. A typical crystallization inhibitor is obtained by heavily chlorinating paraffin and treating the product with naphthalene and aluminum chloride.

Petrolatum (vaseline) is a buttery mixture of hydrocarbons similar to paraffin.

Liquid petrolatum (sometimes called Russian oil, white oil or "Nujol") is a high boiling petroleum distillate which has been treated with fuming sulfuric acid until no further reaction takes place. It is practically odorless and tasteless and is used as a laxative.

Heavier grades of petroleum oils which are not suitable for other purposes are now being burned in Diesel engines. This involves their being sprayed into the cylinders and ignited by compression without the use of spark plugs. The present rapid increase in the use of Diesel engines is leading to a definite effort to standardize and find the best Diesel fuels. Cetane and cetene (p. 46) and methylnaphthalene are used as low knock and high knock standard fuels for rating Diesel fuels. It is to be noted that the knock qualities of a fuel are exactly opposite for a Diesel motor and for a spark ignition motor. Thus an aromatic hydrocarbon like methylnaphthalene is a good anti-knock material for the latter while a long straight chain compound like cetane or cetene (n-C_{16}) gives extreme knock (p. 46).

Ozokerite, earthwax, is a natural paraffin wax found in Galicia and near Baku. When bleached it is used as "ceresin." It is harder and has a higher melting point than ordinary paraffin wax.

Asphalt is found in large deposits on the island of Trinidad and in smaller amounts in many other places. It is a complex oxidation and polymerization product of hydrocarbons, probably from crude petroleum. Large amounts of petroleum pitch are now used in place of natural asphalt. Gilsonite is a high grade asphalt found in Utah.

Carbon black is obtained by the partial combustion of natural gas.[39] Thermatomic carbon (p. 14).

[39] Ellis. "Chemistry of Petroleum Derivatives." Reinhold, 1934. p. 234.

Artificial Petroleum from Coal. Coals, especially those of the lower ranks such as bituminous and brown coals, contain considerable amounts of hydrogen but less than that contained in the heaviest petroleums. These coals can be hydrogenated under high pressure with suitable catalysts [40] to give a material essentially like petroleum. Methane, which is a considerable by-product, is used with steam as the source of the hydrogen used in the process. The hydrogenation of low grade coal to give liquid fuels and lubricants is at present assuming industrial importance in countries which have coal but no petroleum. In the future it will be important for the whole world because the coal reserve is undoubtedly many times that of petroleum. It is interesting to note in passing that the first hydrogenation of coal was accomplished by Berthelot by means of hydriodic acid.

A better solution involves treatment of hot coal or coke with steam to give water gas (CO and H_2) which is then passed over suitable catalysts to give complex liquid fuels (*Synthol*, Franz Fischer, Fischer-Tropsch) (p. 420).[41, 42, 43, 44]

The petroleum chemicals industry has been expanding at a rapid rate. It was estimated in 1944 that over 3.5 billion pounds of "Petro chemicals" were produced annually. Important raw materials include natural gas and refinery gas, the latter produced chiefly in the cracking operation used for the production of gasoline. Typical products manufactured in large quantities include ethyl and isopropyl alcohol, acetone, ammonia, synthetic glycerol, "Isooctane" (2,2,4-trimethylpentane), butadiene, styrene, methylethyl ketone, tertiary butyl alcohol, toluene, detergents, and various solvents. In smaller volumes aliphatic sulfur compounds, fungicides, insecticides, oxidation inhibitors, high molecular weight polymers, methanol, formaldehyde, orthoxylene, resins and other special chemicals are produced. Altogether several thousand finished products are now made from petroleum.

INDIVIDUAL PARAFFIN HYDROCARBONS

Methane, CH_4

Commercial Sources:

The total annual tonnage production of methane is about half that of petroleum.

1. Natural gas consists largely of methane (75–100%).

2. Gas formed by heating soft coal contains 30–40% methane.

3. Anaerobic bacterial decomposition of vegetable matter (mainly cellulose) gives gases rich in methane. The gas from the activated sludge process of

[40] Bergius. *Ind. Eng. Chem., News Ed.* 4, No. 23, 9 (1926).
[41] Fischer. *Ber.* 71A, 56 (1938).
[42] *Gas J.* 216, 278 (1936).
[43] *Petroleum Times* 36, 613 (1936).
[44] Fischer, Tropsch. *Ber.* 59B, 830, 923 (1926).

sewage disposal contains as much as 80% methane. A process has been proposed for converting farm waste, such as cornstalks, into a fuel gas containing about 50% methane.[1]

4. The hydrogenation of coal, petroleum and similar products forms large amounts of methane gas.[2]

Preparation.

   A. General Methods (applicable to higher hydrocarbons).

   1. Hydrolysis of the Grignard reagent.

$$CH_3MgX + H_2O \rightarrow CH_4 + MgX(OH)$$

Dilute acid or a solution of an ammonium salt is usually used to dissolve the basic magnesium salt formed.

   2. Purified dimethylmercury or methylmercuric sulfate treated with concentrated sulfuric acid gives pure dry methane. The mercury dimethyl (very toxic) is a high boiling liquid while the methylmercuric sulfate is a slightly volatile easily crystallized solid.

   3. By reduction of alkyl halides by metallic sodium in liquid ammonia.

$$CH_3I + 2 Na + NH_3 \rightarrow CH_4 + NaI + NaNH_2$$

The escaping methane can be freed from ammonia by washing with acid.

   B. Special Methods for Methane.

   1. By removing the higher hydrocarbons from natural gas by activated carbon.[3]

   2. By heating anhydrous sodium acetate with soda lime (a mixture of the hydroxides of sodium and calcium obtained by slaking quick lime in concentrated sodium hydroxide solution).

$$CH_3CO_2Na + NaOH \rightarrow CH_4 + Na_2CO_3$$

Although this method is often assumed to be suitable for the homologs of methane it is entirely unsatisfactory. Thus Na propionate, butyrate, and caproate, RCOONa, give approximately the following percentages of *RH*, CH_4, and H_2 under conditions which give a 98% yield of CH_4 from CH_3COONa: *40*, 20, 30; *20*, 40, 30; *10*, 40, 40.[4] The results are even less satisfactory with branched compounds.

   [1] Buswell, Boruff. *Cellulose* 1, 162 (1930). *Ind. Eng Chem.* 21, 1181 (1929); 25, 147 (1933).
   [2] Bergius. *Ind Eng. Chem., News Ed.* 4, No. 23, 9 (1926); *C. A.* 16, 261 (1922).
   [3] Storch, Golden. *J. Am. Chem. Soc.* 54, 4662 (1932).
   [4] Oakwood, Miller. *J. Am. Chem. Soc.* 72, 1849 (1950).

3. By the hydrolysis of aluminum carbide.[5]

$$Al_4C_3 + 12\ H_2O \rightarrow 3\ CH_4 + 4\ Al(OH)_3$$

Aluminum carbide probably has the molecular structure.

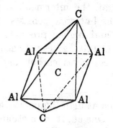

Impurities usually give traces of H_2 and C_2H_2.

4. By passing hydrogen at 1200° over carbon,[6] or at 500–600° over carbon with catalysts of Ni, Co, or Fe.[7]

5. By the action of hydrogen sulfide and carbon disulfide with hot copper, the first synthesis of methane.[8]

6. By passing CO or CO_2 over hot calcium hydride, CaH_2.

Physical Properties. Methane is a colorless gas with a faint odor. It is slightly soluble in water but more so in alcohol. Its critical temperature and pressure are −82.5° and 45.7 atm. Its m.p. and b.p. are −183° and −161.4°. It can be liquefied by liquid air. The best method of determining methane in the presence of other hydrocarbons is by low temperature distillation (ethane, b. −88.3°, propane, b. −42.2°).[9–12]

Tetrahedral Structure. The suggestion that methane has a pyramidal structure[13] rather than the usually accepted tetrahedral structure is based on an incorrect interpretation of X-ray and macrocrystalline data.

Reactions of Methane. The surprising inactivity of methane may be due to its having an outer shell of eight electrons as in the rare gases, the four hydrogen atoms being inside this outer shell. The ionization potential of methane is of the same order of magnitude as that of argon.

1. When heated at about 1000°, methane gives a small yield of benzene (0.2 gal. per 1000 cu. ft.) and other aromatic compounds. Careful studies of the pyrolysis of methane indicate that the first step is its conversion to hydrogen

[5] *Ann. Rep. Chem. Soc.* (London) **1913**, 56.
[6] Mayer, Altmayer. *Ber.* **40**, 2134 (1907).
[7] Bone, Jerdan. *J. Chem. Soc.* **71**, 41 (1897).
[8] Berthelot. *Compt. rend.* **43**, 236 (1856).
[9] Rosen, Robertson. *Ind. Eng. Chem., Anal Ed.* **3**, 284 (1931).
[10] Podbielniak. *Ind. Eng. Chem., Anal. Ed.* **3**, 177 (1931).
[11] Rose. *Ind. Eng. Chem., Anal Ed.* **8**, 478 (1936).
[12] Hicks-Bruun, Brunn. *J. Am. Chem. Soc.* **58**, 810 (1936).
[13] Henri. *Chem. Rev.* **4**, 189 (1927).

and free methylene,[14] (CH_2). The latter reacts with a molecule of methane to give ethane. Under properly controlled conditions as high as 95% yields of ethane can be obtained.[15] The ethane loses hydrogen to form ethylene and then further to form acetylene. The latter polymerizes to form benzene. Under ordinary conditions all the intermediate products are less stable and the mixture issuing from the hot zone consists of unchanged methane and a small amount of benzene and higher products.[16-20] In the absence of all possible catalysts, methane at 1000° gives only acetylene and hydrogen.[21] The Thermatomic Carbon Process gives carbon and hydrogen at high temperatures.[22]

2. Oxidation.[23, 24]

(a) With excess of oxygen at high temperatures complete combustion gives carbon dioxide and steam. One cu. ft. of methane gives 1000 B.T.U. whereas the same amounts of coal gas and ordinary water gas give only 500 and 300 B.T.U. respectively.

(b) With insufficient oxygen it is possible to limit the combustion and obtain carbon monoxide and water as the chief products. There has been much speculation as to the mechanism of the combustion of methane. The first product is probably methanol formed by the direct introduction of an atom of oxygen into a molecule of methane. Methanol is readily dehydrogenated or oxidized to formaldehyde and hydrogen or steam. Formaldehyde is decomposed by heat to carbon monoxide and hydrogen or is readily oxidized to formic acid which decomposes to form carbon monoxide and water. Since all of these possible intermediate products are more sensitive to heat and oxidation than are methane, carbon monoxide and water, they do not survive among the reaction products.

In spite of enormous amounts of work on the oxidation of methane, meagre practical results have been obtained. Even under the best conditions the yields of methanol and formaldehyde amount to only a few per cent.[25] The formaldehyde obtained by oxidizing natural gas probably comes from the higher homologs present (p. 19).

(c) Nitrogen peroxide oxidizes methane to give a small yield of formaldehyde.[26]

[14] Kassel. *J. Am. Chem. Soc.* **54**, 3949 (1932).
[15] Storch. *J. Am. Chem. Soc.* **54**, 4188 (1932).
[16] Ellis. "Chemistry of Petroleum Derivatives." Reinhold, 1934. pp. 37–90.
[17] *Ann. Rep. Chem. Soc.* (London) 1930, 82.
[18] Hague, Wheeler. *J. Chem. Soc.* 1929, 378.
[19] Schneider, Frolich. *Ind. Eng. Chem.* **23**, 1405 (1931).
[20] Hessels, vanKrevelen, Watermann. *J. Soc. Chem. Ind.* **58**, 323 (1939).
[21] Holliday, Gooderham. *J. Chem. Soc.* 1931, 1594.
[22] Moore. *Ind. Eng. Chem.* **24**, 21 (1932).
[23] Egloff, Schaad. *Chem. Revs.* **6**, 91.
[24] Ellis. "Chemistry of Petroleum Derivatives." Reinhold, 1934. p. 846.
[25] Boomer, Thomas. *Can. J. Research* **15**, B, 401 (1937).
[26] Frolich, Harrington, Waitt. *J. Am. Chem. Soc.* **50**, 3216 (1928).

3. With steam. The most important practical reaction of methane next to its complete combustion for the production of heat is its reaction with steam at high temperatures (800–1000° C.) in the presence of catalysts such as nickel activated by promoters such as alumina or thoria to give carbon monoxide and hydrogen. This is the chief source of hydrogen for commercial hydrogenation of coal and petroleum and for the synthesis of ammonia. The carbon monoxide formed will further react with steam at about 500 °C. with an iron oxide catalyst promoted with chromium oxide to give carbon dioxide and hydrogen. Thus four molecules of hydrogen are obtained from one molecule of methane and two molecules of steam.[27]

4. With chlorine. The reaction of methane with chlorine is likely to be explosive. The products are hydrogen chloride, carbon, and the four possible chlorinated methanes. If the violence of the reaction is abated by bringing the gases together in a reactor filled with sand, the formation of carbon is practically eliminated. By modern methods of distillation it is possible to separate the methyl chloride, methylene chloride, chloroform and carbon tetrachloride formed in the reaction.[28, 29] By using volume ratios of N_2, CH_4 and Cl_2 of 80:8:1 over a catalyst of cupric chloride on pumice at 450°, it is possible to convert 45% of the chlorine to methyl chloride.

The conversion of methane to its chlorination products may become commercially important in spite of the following complications: (a) There are cheap methods for making pure methyl chloride, chloroform and carbon tetrachloride. (b) Most of the possible uses for methylene chloride are better served by formaldehyde. (c) Cheap methane in the form of natural gas and cheap chlorine from cheap hydroelectric power are usually not located near each other geographically. One or the other has to be transported. (Compare p. 81)

Bromine would presumably act with methane much like chlorine. Iodine is much less reactive and, moreover, the hydrogen iodide formed would tend to reverse any substitution, thus giving the original hydrocarbon. Fluorine reacts violently with methane giving almost entirely carbon and hydrogen fluoride.

5. With nitric acid. Using small diameter Pyrex tubing at 444° and 100 psi, with a 10:1 ratio of methane to nitric acid, it is possible to obtain 27% conversion per pass.[30] Recycling of the methane gives a 90% yield.

6. Methane under the influence of alpha particles or of an electrical discharge is condensed to a mixture of complex hydrocarbons.[31, 32]

7. Methane can be converted to HCN in 10% yield by passing it with

[27] Ellis. "Chemistry of Petroleum Derivatives." Reinhold, 1934. p. 276.
[28] *Ann. Rep. Chem. Soc.* (London) 1919, 69; 1923, 74.
[29] Ellis. "Chemistry of Petroleum Derivatives." Reinhold, 1934. p. 686.
[30] Hass et al. *Ind. Eng. Chem.* 39, 919 (1947).
[31] Lind, Glockler. *J. Am. Chem. Soc.* 52, 4450 (1930).
[32] Ellis. "Chemistry of Petroleum Derivatives." Reinhold, 1934. p. 264.

NH_3 over Al_2O_3 at 1000°. An electric discharge through CH_4 and N_2 converts the former nearly completely to HCN.

8. Methane does *not react* with the following: (*a*) ordinary oxidizing agents, (*b*) hydrogen, (*c*) reducing agents, (*d*) acids, (*e*) bases, (*f*) metals.

Methane, Storch, U.S.B. of Mines, Information Circular 6549 (1932).

Ethane, CH_3CH_3, occurs in varying amounts in natural gas. The statement that it forms the chief gas of a well near Pittsburgh is erroneous. It can be prepared by the general methods from ethyl Grignard reagents, ethyl halides, and ethylmercury compounds. Heating sodium propionate with soda lime gives a poor yield of ethane along with ethylene and hydrogen.

In common with the other homologs of methane it can be made by the Wurtz reaction from the proper alkyl halide and metallic sodium.

$$2\ CH_3I + 2\ Na \rightarrow CH_3CH_3 + 2\ NaI$$

The yield is poor.

Ethane is also obtained during the electrolysis of sodium acetate solution.[33]

$$2\ CH_3CO_2Na + 2\ H_2O \rightarrow C_2H_6 + 2\ CO_2 + 2\ NaOH + H_2$$

The ethane and carbon dioxide are liberated at the positive electrode (anode). With higher sodium salts, $R\text{-}CO_2Na$, this reaction gives very poor yields of R-R.

The best way of making *pure ethane* is by the catalytic hydrogenation of ethylene prepared from ethanol.

According to its ordinary physical and chemical properties the single bond between the two methyl groups in ethane allows free rotation. Its thermodynamic properties at low temperatures show that there is a mutual repulsion between the H atoms in the two methyls.[34]

Reactions of Ethane. Ethane is much more reactive than methane.

1. On *heating* it gives hydrogen and ethylene which then undergoes a very complex series of decompositions and polymerizations.[35]

2. When treated with insufficient *oxygen*, especially in the presence of platinum as a catalyst, ethane gives a considerable amount of formaldehyde, much more than can be obtained from methane by any oxidation method.[36] Presumably the process involves the intermediate formation of ethylene and ethylene glycol.

3. Chlorine and bromine react readily with ethane giving a complex mixture of halogenated products.[37] These reactions have never been studied thoroughly because of their complexity, the lack of a cheap source of ethane, and the fact that the desired derivatives of ethane are readily available by other means.

[33] Kolbe. *Ann.* **69**, 257 (1849).
[34] Kemp, Pitzer. *J. Chem. Phys.* **4**, 749 (1936).
[35] Pease. *J. Am. Chem. Soc.* **50**, 1779 (1928).
[36] Ellis. "Chemistry of Petroleum Derivatives." Reinhold, 1934. p. 850.
[37] Ellis, *ibid.* p. 712.

4. The silent electric discharge converts ethane to hydrocarbon mixtures of M.W. 100–500.[38]

5. Hydrogenolysis at high temperature and pressure gives methane.

6. Ethane is inert under ordinary conditions to oxidizing and reducing agents, and to acids, bases, and metals.

Propane, $CH_3CH_2CH_3$. It is not necessary to prepare propane because it is available at less than a cent a pound in tank car lots. It is separated from "natural gasoline" or "Casinghead gas" by distillation under pressure or by selective adsorption on activated carbon.

Rather impure propane is sold in tanks under moderate pressure for use as a household fuel gas (Pyrofax, "Bottle gas"). Propane has recently found wide use in the petroleum industry as a combined solvent and refrigerant for simultaneously extracting and dewaxing a lubricating fraction (Mueller process) (Duo-Sol process).[39] It is also used to de-asphaltize oils (p. 19).

Propane is more reactive than ethane. Thermal decomposition gives mainly ethylene and methane with smaller amounts of propylene and hydrogen.[40] The ethylene is used to make ethylene glycol and related products. The propylene is changed to isopropyl alcohol and acetone.

In the thermal rupture of the $C^{12}-C^{12}$ and $C^{12}-C^{13}$ bond of propane, the first was found to occur 8% greater than the latter.[41]

Chlorination at 300° gives nearly equal amounts of 1- and 2-chloropropanes.[42] Dichlorination gives all theoretically possible dichlorides including geminal and vicinal.

A carbon tetrachloride solution of propane when treated with sulfur dioxide and chlorine with ultraviolet radiation at 25° gives equal amounts of the 1- and 2-isomers of $C_3H_7SO_2Cl$ (sulfochlorination).[43] Disubstitution gives only 1,3-propane disulfonyl chloride with no geminal or vicinal product.

The vapor-phase nitration of propane gives 1- and 2-nitropropanes and smaller amounts of nitroethane and nitromethane.[44]

Propane with O_2 below the ignition point gives propylene, acetaldehyde and CO_2.[45]

Butanes, C_4H_{10}. The two isomers, normal butane and isobutane are available in any desired quantity and degree of purity by separation from natural gas by means of selective adsorption or fractional distillation under pressure.

1. *n-Butane,* $CH_3CH_2CH_2CH_3$, b. − 0.6°, shows practically the same reactions as propane or ethane but is more reactive.

On pyrolysis it gives all the theoretically possible products, namely,

[38] Lind, Glockler. *J. Am. Chem. Soc.* **50**, 1767 (1928).
[39] Wilson. *C. A.* **30**, 1223 (1936).
[40] Pease. *J. Am. Chem. Soc.* **50**, 1779 (1928).
[41] Stevenson, Wagner, Bieck. *J. Chem. Phys.* **16**, 993 (1948).
[42] Hass, McBee, Weber *Ind. Eng. Chem.* **28**, 333 (1936).
[43] Asinger, Schmidt, Ebeneder. *Ber.* **75B**, 34 (1942).
[44] Hass, Hodge, Vanderbilt. *Ind. Eng. Chem.* **28**, 339 (1936).
[45] Pease. *J. Am. Chem. Soc.* **51**, 1839 (1929).

methane, ethane, ethylene, propylene, 1-butene, 2-butene and hydrogen.   The predominant reaction gives methane and propylene.

The most important reaction of *n*-butane is its catalytic dehydrogenation to give butenes and then butadiene for synthetic rubber.   A reaction of nearly as great importance is the isomerization of *n*-butane to isobutane by means of catalysts such as aluminum chloride or bromide or alumina with certain promoters.

The non-catalytic high temperature oxidation of butane has been highly developed (Celanese, Bludworth).   The chief products are acetaldehyde, formaldehyde, propionaldehyde, acetone, methanol, tetramethylene oxide (H_4-furan), and methyl ethyl ketone.   Minor products include acrolein and the following alcohols: ethyl, propyl, isopropyl, allyl, butyl, and methallyl.

Chlorination at 300° gives the 1- and 2-chlorobutanes in about 1:2 ratio.[46]   Chlorination in light at 50° to introduce about 7 Cl, followed by high temperature chlorination gives a good yield of perchlorobutadiene, $Cl_2C = CCl - CCl = CCl_2$, a very inert chlorocarbon.[47]   Sulfochlorination by means of SO_2 and Cl_2 gives the same ratio of substitution by $-SO_2Cl$.   Introduction of a second group gives the 1,4- and 1,3-isomers in 1:4 ratio.   This is in sharp contrast of the dichlorination of butane which gives all possible dichlorides including the geminal and vicinal compounds.

n-Butane and sulfur at 600° react to give *n*-butylenes, butadiene, and thiophene.[48]   Yields of 30-50% of either of the last two can be obtained by suitable recycling.

An electrical discharge converts *n*-butane to a mixture of hydrocarbons of somewhat the same complexity as crude petroleum.   One of the resulting fractions contains a mixture of octanes and octenes.[49]

Pure *n*-butane, mixed with a definite amount of dry air, is used as a standard fuel for the accurate heat treatment of special steels.

2. *Isobutane*, trimethylmethane, $(CH_3)_3CH$, b. −11.7°.   On heating at 600° it gives more isobutylene and hydrogen than methane and propylene while at 700° the reverse is true.[50]   The most important reaction of isobutane is in the alkylation reaction with olefins (p. 51).   Thus, with propylene and conc. sulfuric acid it gives mainly 2,3- and 2,4-Me_2-pentanes.

Catalytic dehydrogenation over alumina activated by chromic oxide (best), vanadyl sulfate, ferric oxide or zinc oxide gives isobutylene for polymerization and alkylation reactions (p. 46).   A trace of water is absolutely necessary for this reaction.[51]

Chlorination at 300° gives isobutyl and t-butyl chlorides in 2:1 ratio.   Dichlorination gives all possible products.

[46] Hass, McBee, Weber.   *Ind. Eng. Chem.* **28**, 333 (1936).
[47] McBee, Hatton.   *Abstracts 109 Meeting Am. Chem. Soc.* (1946).
[48] Rasmussen, Hansford, Sachanen.   *Ind. Eng. Chem.* **38**, 376 (1946).
[49] Lind, Glockler.   *J. Am. Chem. Soc.* **51**, 3655 (1929).
[50] Hurd, Spence.   *J. Am. Chem. Soc.* **51**, 3353 (1929).
[51] Burgin, Groll, Roberts.   *Oil Gas J.* **37**, 48 (1938).

In the direct chlorinations of paraffins the hydrogen atoms are always substituted at rates which are in the order primary > secondary > tertiary. At 300° C. in the vapor phase the relative rates are 1.00:3.25:4.43.

At increasing temperatures the relative rates approach 1:1:1 in both liquid and vapor phase.[52]

Sulfochlorination gives only $Me_2CHCH_2SO_2Cl$. Introduction of a second group gives only $MeCH(CH_2SO_2Cl)_2$.

In addition to the expected reactions, isobutane gives others on account of its *tertiary* hydrogen atom. Mild oxidation with oxygen or an oxidizing agent like potassium permanganate converts this hydrogen to a hydroxyl, forming tertiary butyl alcohol.[53] Concentrated sulfuric acid and fuming nitric acid replace the tertiary hydrogen by the sulfonic acid group (SO_3H) and the nitro group (NO_2) respectively. With various catalysts it adds to olefins like isobutylene to give complex mixtures containing 2,2,4-Me_3-pentane.[54] A similar reaction takes place in the presence of conc. sulfuric acid.[55] Anhydrous hydrogen fluoride also catalyzes a similar change. These *alkylation* processes were of the utmost importance in World War II in supplying the enormous amounts of 100-octane gasoline needed for aviation fuel (p. 51).

Isobutylene adds to ethylene at high pressure and temperature without a catalyst to form neohexane, 2,2-Me_2-butane.[56]

Mixed butanes are used for "carburetting" water gas to render its flame luminous.

Pentanes, C_5H_{12}. *Normal pentane*, $CH_3(CH_2)_3CH_3$, b. 36°, and *isopentane*, dimethylethylmethane, 2-Me-butane, $(CH_3)_2CHCH_2CH_3$, b. 28°, are readily obtained from natural gas. A name such as 2-methylbutane, in which the name and position of a substituent group in a parent molecule are given, is known as a Geneva name from a congress of organic chemists at Geneva in 1892 which greatly expanded the system of nomenclature for complex organic compounds devised by Hofmann in 1865.[57] The prefix *iso-* is properly used only for compounds containing Me_2CH—attached to a normal or straight chain of atoms. The casinghead or natural gasoline made from natural gas by compression or adsorption methods contains 15–35% of a pentane mixture boiling 27–40°.[58]

The reactions of normal and isopentane are like those of the butanes. Less usual reactions are the following: *n*-Pentane with acetyl chloride and $AlCl_3$ gives 2-acetylpentane and with CO, HCl, and $AlCl_3$ gives ethyl isopropyl ketone.[59] Isopentane with CO and HCl in the presence of $AlBr_3$ and Cu_2Cl_2

[52] Hass, McBee, Weber. *Ind. Eng. Chem.* **28**, 333 (1936).
[53] Meyer. *Ann.* **220**, 1 (1883).
[54] Ipatieff, Komarewsky, Grosse. *J. Am. Chem. Soc.* **57**, 1722 (1935).
[55] Birch et al. *Ind. Eng. Chem.* **31**, 884 (1939).
[56] Frey, Hepp. *C. A.* **35**, 611 (1941).
[57] Hofmann. *Ber.* **26**, 1595 (1865).
[58] Ellis. "Chemistry of Petroleum Derivatives" Reinhold, 1934. p. 18.
[59] Hopff. *Ber.* **64B**, 2739 (1931)

gives, among other products, 2,2-dimethylbutanol. Isopentane in vapor phase oxidation gives isovaleraldehyde, then isobutyraldehyde and finally acetone, at which point the oxidation becomes slower. Isopentane with dil. nitric acid gives 2-nitro-2-Me-butane. With more concentrated acid it gives di- and trinitro compounds.[60]

The chlorination of mixed n- and iso-pentane has been developed commercially.[61-63] A mixture of chlorine with a large excess of hydrocarbon is passed very rapidly through iron tubes at about 300°. The chlorine reacts completely with the formation of all the theoretically possible monochlorides and smaller amounts of higher chlorides. These chlorides are used as such or are converted to alcohols, olefins and acetates (Pentasol, Pentacetate, Sharples).

Pentane is used in thermometers for low temperatures.

Neopentane, m. −16.8°; b. 9.5° (the prefix *neo-* indicates the presence of a C attached by all four valences to other C atoms), tetramethylmethane, 2,2-Me_2-propane, $(CH_3)_4C$, is obtained by the action of dimethylzinc on acetone dichloride, 2,2-dichloropropane, prepared from acetone and phosphorus pentachloride, or on a tertiary butyl halide. The best preparation is from t-butyl chloride and methylmagnesium chloride.[64] Vapor phase nitration proceeds normally.[65]

Neopentane, on pyrolysis, gives methane and isobutylene.[66] On chlorination it gives neopentyl chloride, $(CH_3)_3CCH_2Cl$, without any rearrangement.[67] This chloride cannot be made from the corresponding alcohol (pp. 75, 120).

Neopentyl deuteride (neopentane with one atom of hydrogen replaced by deuterium) is obtained by the action of heavy water on neopentylmagnesium chloride.

This removal of the symmetry of the molecule destroys the simplicity of its Raman spectrum.[68]

Pure normal alkanes have a faint but decidedly pleasant odor entirely different from that of the impure materials which resemble the characteristic odor of petroleum fractions.

The sizes and shapes of the molecules of n-heptane, n-octane, and n-nonane in the gaseous state have been determined.[69] They exist as loose helixes.

Hexanes.[70] All five of the hexanes have been isolated from petroleum and have been repeatedly synthesized.

2,3-Me_2-butane, diisopropyl, in common with all paraffins having tertiary

[60] Ellis. "Chemistry of Petroleum Derivatives." Reinhold, 1934. p. 1042.
[61] Ayres. *Ind. Eng. Chem.* **21**, 899 (1929).
[62] Clark. *Ind. Eng. Chem.* **22**, 439 (1930).
[63] Ellis. "Chemistry of Petroleum Derivatives." Reinhold, 1934. p. 726.
[64] Whitmore, Fleming. *J. Am. Chem. Soc.* **55**, 3803 (1933).
[65] Howe, Hass. *Ind. Eng. Chem.* **38**, 251 (1946).
[66] Frey. *Ind. Eng. Chem.* **25**, 441 (1933).
[67] Whitmore, Fleming. *J. Am. Chem. Soc.* **55**, 4161 (1933).
[68] Whitmore, Rank. *J. Am. Chem. Soc* **56**, 749 (1934).
[69] Melaven, Mack. *J. Am. Chem. Soc.* **54**, 888 (1932).
[70] Maman. *Compt. rend.* **198**, 1323 (1934).

H atoms, reacts readily with dilute nitric acid to give a tertiary nitro compound. More vigorous treatment also replaces the other tertiary H.[71]

2,2-Me$_2$-butane, trimethylethylmethane, neohexane, is made commercially from ethylene and isobutane (p. 18). It reacts readily with nitric acid to give a 3-nitro compound.[72, 73] The structure is proved by reduction with alkaline stannite solution to give the oxime of methyl t-butyl ketone.

Heptanes. Normal heptane is obtained by distillation and chemical treatment of the oil obtained from the resin of the Jeffrey pine.[74] In common with other normal paraffins it exhibits a bad "knock" in an internal combustion engine. n-Heptane is used as a standard in knock rating.[75] For this purpose it was originally made from the Jeffrey pine but is now prepared synthetically. Several processes are available.

n-Butyraldehyde condenses readily with acetone. The following steps are then easy.

$$PrCHO + CH_3COCH_3 \rightarrow PrCHOHCH_2COCH_3$$
$$\rightarrow PrCH=CHCOCH_3 \rightarrow PrCH_2CH_2COCH_3 \rightarrow CH_3(CH_2)_5CH_3$$

n-Heptane can also be made by the reduction of heptaldehyde, (oenanthol) obtained in the destructive distillation of castor oil (mainly glyceryl ricinoleate).

n-Heptane has been chlorinated by means of sulfuryl chloride, SO$_2$Cl$_2$, to give a 90% yield of chlorides consisting of 15% 1-chloroheptane and 85% of 2-chloroheptanes.[76]

n-Heptane, 2,2-Me$_2$-pentane and 2- and 3-Me-hexane have been found in petroleum.[77]

All nine of the possible isomers, C$_7$H$_{16}$, have been prepared in a high state of purity.[78] A variety of physical properties such as b.p., index of refraction, density, f.p., C.S.T. in aniline, molecular refractivity and dispersion, molecular volume, compressibility, coefficient of expansion, surface tension, viscosity and dielectric properties have been measured for these pure materials.

Six of the isomeric heptanes were made by the following steps:

1. Grignard reagent + ketone → tertiary alcohol.
2. Alcohol → olefin mixture.
3. Olefin mixture → heptane.

Thus 2-methylhexane was made as follows:

1. n-Butylmagnesium bromide in anhydrous ether with acetone followed by treatment with acid gave 2-methyl-2-hexanol (dimethyl-n-butylcarbinol).

[71] Ellis. "Chemistry of Petroleum Derivatives." Reinhold, 1934. p. 1043.
[72] Markownikoff. Chem. Zentr. 1899 II, 472.
[73] Howe, Hass. Ind. Eng. Chem. 38, 251 (1946).
[74] Kremers. J. Am. Pharm. Assoc. 6, 11 (1917).
[75] Edgar. Ind. Eng. Chem. 19, 145 (1927).
[76] Kharasch, Brown. J. Am. Chem. Soc. 61, 2142 (1939).
[77] Ellis. "Chemistry of Petroleum Derivatives." Reinhold, 1934. p. 27.
[78] Edgar, Calingaert, Marker. J. Am. Chem. Soc. 51, 1483 (1929).

2. Dehydration of the alcohol by refluxing with a trace of iodine (applicable to higher tertiary alcohols, Hibbert) gave mainly 2-methyl-2-hexene with a small amount of the isomeric 2-methyl-1-hexene.

3. Passage of the mixed olefins with purified hydrogen over hot finely divided nickel gave 2-methylhexane, isoheptane.

Similar methods gave 3-methylhexane, 3-ethylpentane, 2,3-dimethylpentane, 2,4-dimethylpentane and 2,2,3-trimethylbutane.

2,2-Dimethylpentane and 3,3-dimethylpentane could not be prepared by this series of reactions because the corresponding alcohols, 2,2-dimethyl-3-pentanol and 3,3-dimethyl-2-pentanol, contain the "neopentyl alcohol" system (p. 128) and undergo *rearrangements* on dehydration, giving chiefly 2,3-dimethyl-2-pentene. These two heptanes were made by a modified Wurtz reaction involving the reaction of a suitable Grignard reagent with an alkyl halide in the presence of dry ether and mercuric chloride. The yield was poor but better than that obtainable by any other known method.

One of the simplest optically active hydrocarbons, *l*-2,3-dimethylpentane (methylethylisopropylmethane), has been made in the following steps.[79]

Optically active material is obtained by resolving *dl*-2-butanol into its *d*- and *l*-forms.[80] The alcohol is heated with phthalic anhydride to give hydrogen *sec*-butyl phthalate. This is then combined with the optically active base, brucine, to give brucine *dextro-sec*-butyl and brucine *levo-sec*-butyl phthalates. Recrystallization separates these, the former being less soluble. Acidification precipitates the optically active hydrogen *sec*-butyl phthalate which is then saponified to give the optically active 2-butanol.

The conversion of dextrorotatory 2-butanol involves the steps: Conversion to the levorotatory bromide by hydrogen bromide, reaction of the latter with the sodium derivative of ethyl methylmalonate, followed by saponification and acidification to give the methyl *sec*-butyl malonic acid which loses carbon dioxide at 190° to give dextrorotatory methyl-*sec*-butyl-acetic acid (2,3-dimethyl-*n*-valeric acid). This is converted to the dextrorotatory ethyl ester which is reduced by sodium and absolute alcohol to the primary alcohol, levorotatory 2,3-dimethyl-1-pentanol. This is converted to the dextrorotatory bromide by PBr₃. Conversion of the bromide to the Grignard reagent and treatment of the latter with acid gives levorotatory 2,3-dimethylpentane. The specific rotations $[\alpha]_D^{25}$ (p. 115) of the optically active substances involved in this series of changes are approximately as follows: 2-Butanol, +8.4; 2-Bromobutane, −13.8; 2,3-Dimethyl-valeric acid, +1.3; Ethyl ester, +1.9; 2,3-Dimethyl-1-pentanol, −0.9; 1-Bromo-2,3-dimethylpentane, +2.9; 2,3-Dimethylpentane, −9.4.

The most important of the heptanes is *Triptane*, 2,2,3-Me₃-butane, because of its unusual octane rating (120–150?). It can be made in a variety of ways. *Triptanol* (p. 130), 2,3,3-Me₃-2-butanol, can be dehydrated to *triptene*, 2,3,3-

[79] Levene, Marker. *J. Biol. Chem.* 91, 405 (1931).
[80] Pickard, Kenyon. *J. Chem. Soc.* 99, 45 (1911).

Me₃-1-butene, which is readily hydrogenated to triptane. Me₃-ethylene from t-AmOH can be methylated by MeCl and lime under pressure to give Me₄-ethylene and then triptane.[81] Triptane can also be made by the *hydrogenolysis* of 2,2,3-Me₃-pentane obtained either by the alkylation of 2-butene with isobutane or by the hydrogenation of the co-dimer from isobutylene and 2-butene. Byproducts of the hydrogenolysis process are 2,3-Me₂-butane and 2,3-Me₂-pentane, also valuable in aviation fuels. Other methods are available.

Octanes. The eighteen possible octanes have been prepared by methods similar to those described for the heptanes.[82–84]

The Wurtz reaction was first used (1855) in the action of isobutyl iodide and sodium to give 2,5-Me₂-hexane. n-Octane is obtained by the Wurtz reaction in 65% yield.[85] As always, the reaction gives considerable amounts of by-products.

The reaction may be assumed to involve the following steps:

The halide reacts with the metallic sodium (a free radical with an odd electron) to give NaX and a free organic radical with an odd electron. This cannot exist as such. The most likely change is for it to react with another atom of sodium to form a sodium alkyl. This is a strong hydrocarbon base which can react with the alkyl halide by metathesis to form the higher hydrocarbon or by elimination of HX to form a 1-olefin, NaX, and the saturated hydrocarbon corresponding to the halide.

$$BuX + Na. \rightarrow Bu. + NaX$$
$$Bu. + Na. \rightarrow BuNa$$
$$BuNa + BuX \rightarrow NaX + Bu_2$$
$$BuNa + CH_3CH_2CH_2CH_2X \rightarrow NaX + BuH + CH_3CH_2CH = CH_2$$

The assumption that the butane and 1-butene are formed by disproportionation of the free organic radicals is probably unsound.

n-Hexane, n-octane, n-decane and n-tetradecane can be prepared by the Wurtz reaction by the action of Na on the normal halides containing 3, 4, 5 and 7 carbon atoms which are readily available. A trace of acetonitrile acts catalytically. It must be remembered that numerous by-products are formed and must be removed from the product. This is easy because of wide differences in the boiling points (The Wurtz Reaction).[86]

The preparation of a normal hydrocarbon from a mixture of two alkyl halides of nearly the same carbon content with sodium is seldom practical. There are *three* main products with boiling points rather close together and a correspondingly large number of by-products.

[81] Miller, Lovell. *Ind. Eng. Chem.* **40**, 1138 (1948).
[82] Noller. *J. Am. Chem. Soc.* **51**, 594 (1929).
[83] Pope, Dykstra, Edgar. *J. Am. Chem. Soc.* **51**, 2203 (1929).
[84] Whitmore, Laughlin. *J. Am. Chem. Soc.* **55**, 5056 (1933).
[85] Lewis, Hendricks, Yohe. *J. Am. Chem. Soc.* **50**, 1993 (1928).
[86] Wooster. *Chem. Rev.* **11**, 1 (1932).

n-Octane can also be obtained readily by chilling the proper fraction from petroleum with solid CO_2 and acetone.

n-Octane with anhydrous $AlCl_3$ at 150° gives *n*-butane and polymeric olefins.[87]

The preparation of hexamethylethane, 2,2,3,3-Me_4-butane, requires special processes. Of those, perhaps the best is the peculiar reaction of silver bromide with *t*-butylmagnesium chloride to form the hydrocarbon, MgX_2 and metallic Ag.

2,2,4-Trimethylpentane is made by the catalytic hydrogenation of the diisobutylenes obtained from *t*-butyl alcohol or isobutylene and sulfuric acid. This substance, incorrectly called "iso-octane," is used as a standard in determining detonation properties, "knock rating" or "octane number" of gasoline.[88] The "octane number" (O.N.) is the per cent of 2,2,4-Me_3-pentane in a mixture of that substance and *n*-heptane which has the same "knock" as the gasoline being tested. An isomer, 2,2,3-Me_3-pentane, has an even higher knock rating but there is no feasible way to prepare it. The limiting compression ratios for those two octanes are 7 to 1 and 12 to 1 for the 2,2,4- and the 2,2,3-isomers respectively.

The vapor phase oxidation of *n*-octane has been carefully studied.[89] The first product is octanal. This then goes to heptanal and CO and CO_2. The change continues step by step through the lower aldehydes. This process apparently involves "chain reactions" which give detonation or "knock." These processes are inhibited by tetraethyl lead, a good "anti-knock." When branched chain octanes are used the aldehyde formation and degradation take place until a branch in the chain is reached, when the reaction is distinctly retarded. This is related to the low knock properties of highly branched hydrocarbons such as 2,2,4-Me_3-pentane.[90]

Higher Alkanes. The normal paraffins C_5 to C_{18} and C_{20}, C_{24}, and C_{26} have been prepared and their physical constants determined.[91-93]

Hundreds of other alkanes have been prepared by methods analogous to those described. With some, even less satisfactory methods must be used. Thus 2,2,4,4-Me_4-pentane is obtained in poor yield by the action of dimethylzinc on 2,2,4-Me_3-4-Br-pentane made by adding HBr to the diisobutylenes. The boiling point of this nonane is 122°, thus being lower than that of *n*-octane, b. 125°.

3,3,4,4-Me_4-hexane, di-*t*-amyl, b. 169.7°/744, $n^{20}D$ 1.4376, is obtained in 5% yield in the preparation of *t*-AmMgX. The compactness and symmetry raise the index of refraction but do not lower the b.p. as much as would be

[87] Grignard, Stratford. *Compt. rend.* **178**, 2149 (1924).
[88] Edgar. *Ind. Eng. Chem.* **19**, 145 (1927).
[89] Pope, Dykstra, Edgar. *J. Am. Chem. Soc.* **51**, 1875 (1929).
[90] Ellis. "Chemistry of Petroleum Derivatives." Reinhold, 1934. pp. 9, 61.
[91] Shepard, Henne, Midgley. *J. Am. Chem. Soc.* **53**, 1948 (1931).
[92] Mair. *Bur. Standards J. Research.* **9**, 457 (1932).
[93] Schiessler et al. *Proc. A.P.I.*, III (1946).

expected. Thus n-decane and 2,2,5,5-Me_4-hexane have b.p. and $n^{20}D$ respectively as follows: 174.0°, 1.4114; 135.0°/736, 1.4057. The effect of the two ethyl groups is not understood.

Triisopropylmethane, $(Me_2CH)_3CH$, b. 157.04/760, $n^{20}D$ 1.4265 is another decane of interesting properties. It has been prepared by the addition of diisopropyl ketone to isopropyl lithium. The resulting carbinol is converted to the alkane by dehydration and hydrogenation.[94] It is also made starting with Ac_2O, $ZnCl_2$, and 2,4-Me_2-2-pentene. The resulting unsaturated ketone reacts smoothly with MeMgBr to give the unsaturated t-alcohol which on dehydration over alumina to the diene and hydrogenation gives the desired alkane.[95]

2,2,3,3,6,6,7,7-Me_8-octane, m. 74.2°, is obtained in small yield in the action of Mg with the monochloride of Me_6-ethane. The isomeric n-hexadecane has m. 18°.

Because of the danger of fires from ordinary aviation fuels on crash landings, much effort is being made to produce *safety fuels* of high flash points but with good knock characteristics. This is opening up the chemistry of the higher highly branched paraffins and liquid aromatic compounds much as did the work of Edgar, Calingaert and Marker on the isomeric heptanes and octanes after World War I.

n-Decane is readily separated from petroleum by careful distillation and freezing.[96]

n-Undecane has been made from undecoic acid and HI[97] and from n-butyl toluene sulfonate and n-heptylmagnesium bromide (14% yield).[98]

2,11-Me_2-dodecane is obtained by hydrogenating the olefins from the ditertiary glycol made from ethyl sebacate and excess methylmagnesium bromide.

n-Hexadecane is the standard for the *cetane number* of Diesel fuels. The cetane number of a fuel is about 0.88 of the formerly used and less satisfactory cetene number. Typical cetane numbers are approximately as follows: n-hexadecane 100, 1-Me-naphthalene 0, n-heptane 55, 2,2,4-Me_3-pentane (*Isooctane*) 23, tetraisobutylene 5, benzene −10, and toluene −20. It is notable that highly branched paraffins and aromatic compounds which give high octane ratings give low cetane numbers.

The normal C_{20}, C_{30}, C_{40}, C_{50}, C_{60}, C_{70} and about 25% of higher unidentified hydrocarbons, C_{10n}, have been obtained by the Wurtz reaction with sodium and decamethylene bromide, $Br(CH_2)_{10}Br$, in absolute ether.[99] The separation was accomplished by a molecular still.[100] The reduction involved in the

[94] Howard et al. *J. Research. Natl. Bur. Standards* **38**, 365 (1947).
[95] Cook, Krimmel, Whitmore. *Abstracts 112th Meeting Am. Chem. Soc.* **28L** (1947).
[96] Bruun, Hicks-Bruun. *J. Research Natl. Bur. Standards* **8**, 583 (1932).
[97] Krafft. *Ber.* **15**, 1687 (1882).
[98] Gilman, Beaber. *J. Am. Chem. Soc.* **47**, 518 (1925).
[99] Carothers et al. *J. Am. Chem. Soc.* **52**, 5279 (1930).
[100] Washburn. *Bur. Standards J. Research* **2**, 476 (1929).

process is explained as follows: (Q = an organic group.)

$$QBr + 2 Na \rightarrow QNa + NaBr$$
$$QNa + Et_2O \rightarrow QH + EtONa + C_2H_4$$

Perhydrolycopene, $C_{40}H_{82}$, 2,6,10,14,19,23,27,31-Me_8-dotriacontane, is obtained by catalytic hydrogenation of lycopene.

The largest alkane prepared is $C^{94}H^{190}$ − 2,5,9,13,17,24,28,32,36,41,45,49,-53,60,64,68,72,75-octadecamethylhexaheptacontane.[101]

The most important chemical reaction of higher hydrocarbons is their instability to heat, their ability to be "cracked" at about 500° into smaller more volatile molecules suitable for gasoline (p. 9).[102] The detailed chemistry of this important industrial process is little known. All cracking processes applied to all petroleums give in varying amounts alkanes, alkenes, naphthenes and aromatic compounds. The last two are probably formed from the alkanes. The initial step in the pyrolysis of an alkane may be

$$RCH_2CH_2CH_2R' \rightarrow RCH = CH_2 + CH_3R'$$

This is probably preceded by the formation of free radicals.[103] The conversion of a naphthene to an aromatic compound involves the removal of hydrogen. This may react with olefins or be evolved. Some hydrogen is practically always found in cracked gases. The lower olefins formed tend to polymerize less and thus survive to a larger extent in the gases which are always byproducts of cracked gasoline. These form the basis for the important industrial processes involving ethylene, propylene and the butylenes. Mixed or partially separated cracked gases are utilized in dimerization or co-dimerization processes, followed by hydrogenation to give additional highly branched octanes, nonanes, and decanes for high octane gasoline (p. 46). Similarly, properly modified mixtures of cracked gases are used in alkylation processes to give similar products without the necessity of hydrogenation (p. 46).

A constant by-product of cracking is petroleum coke.[104] This is the ultimate result of continued polymerization and removal of hydrogen.

The chlorination of paraffin wax is important because of the use of the chlorinated products in making wax crystallization inhibitors (Paraflow) and extreme pressure lubricants.

Treatment of high alkanes, either pure or as mixtures with SO_2 and Cl_2, followed by NaOH gives sulfonates, RSO_3Na, with important surface active properties.[105]

A reaction of the higher alkanes which will undoubtedly be of great practical importance in the future is their oxidation by means of air or oxygen to form

[101] Karrer, Stoll, Stevens. *Helv. Chim. Acta* 14, 1194 (1931).
[102] Ellis. "Chemistry of Petroleum Derivatives." Reinhold, 1934. p. 91.
[103] Rice. *J. Am. Chem. Soc.* 55, 3035 (1933).
[104] Ellis. "Chemistry of Petroleum Derivatives." Reinhold, 1934. p. 225.
[105] Reed. *C. A.* 30, 5593 (1936); 34, 554 (1940).

a great variety of products. Thus the oxidation of n-heptane with insufficient air gives a mixture from which about 25 pure compounds have been isolated and identified. In addition to degradation products, these include all of the possible tetrahydrofurans. Evidently the initial attack by O on the first or second C atom is followed by closure of the 5-ring.

Blowing heavy oils with air in presence of catalysts such as Mn oleate gives oil additives of varying content of higher alcohols, acids and the like (*Alox*).

B. UNSATURATED HYDROCARBONS, C_nH_{2n}

Olefins, Alkylenes or Alkenes, Ethylene and its Homologs

An olefin contains two less H atoms than the corresponding paraffin and is characterized by an unsaturation which makes possible the *addition* of two univalent groups to two adjacent carbon atoms. This condition of unsaturation is indicated by a *double bond*, $C = C$. Such a double bond is merely an empirical indication of a point of unsaturation and activity in the molecule. Even less is known about its true nature than about that of the single bond between two carbon atoms. Electronically a double bond is represented by a "sharing" of two *electron pairs* by two adjacent atoms, $R_2C::CR_2$ or $R_2C \vdots CR_2$ in which R is H or an organic group. Addition to the double bond may be due to an "activation" or polarization (complete or partial).[1-7]

$$\begin{array}{ccc} \text{H} \quad \text{H} & \text{H} \quad \text{H} & \text{H} \quad \text{H} \\ \text{H} : \ddot{\text{C}} : \ddot{\text{C}} : \text{H} \rightleftharpoons & \text{H} : \ddot{\text{C}} :: \ddot{\text{C}} : \text{H} \rightleftharpoons & \text{H} : \ddot{\text{C}} : \ddot{\text{C}} : \text{H} \end{array}$$

In most cases, this conception adds little to the cruder one that the double bond "opens" and adds a univalent group at each end. The three formulas written above represent some of the possible members of the "resonance hybrid" which corresponds more accurately to the properties of ethylene than can any individual structure. The olefinic bond has been subjected to a quantum mechanics analysis.[8]

Olefins exhibit both chain and position isomerism. Thus the possible numbers of *structural isomers* are: C_4, 3; C_5, 5; C_6, 13; C_7, 27; C_8, 66. All of these have been made through the octenes.[9,10]

In addition to structural isomerism, *stereoisomerism* is possible in olefins in which neither unsaturated carbon has two identical groups or atoms attached to it (p. 39). Very few of such stereomers have been isolated.

[1] Lewis. *J. Am. Chem. Soc.* **38**, 762 (1916).
[2] Lowry. *J. Chem. Soc.* **123**, 822 (1923).
[3] Carothers. *J. Am. Chem. Soc.* **46**, 2226 (1924).
[4] Kharasch, Reinmuth. *J. Chem. Education* **5**, 404 (1928).
[5] Prevost, Kirrmann. *Bull. soc. chim.* (4) **49**, 194 (1931).
[6] Pauling, Brockway, Beach. *J. Am. Chem. Soc.* **57**, 2705 (1935).
[7] Harman, Eyring. *J. Chem. Phys.* **10**, 557 (1942).
[8] Mulliken. *Phys. Rev.* **41**, 751 (1932).
[9] Schurman, Boord. *J. Am. Chem. Soc.* **55**, 4930 (1933).
[10] Whitmore, Whitmore, Cook. *J. Am. Chem. Soc.* **72**, 51 (1950).

Occurrence and Formation. Because of their great reactivity, olefins are not found in natural products. They are formed in the thermal decomposition of almost all carbon compounds. Thus the cracking of cottonseed oil gives about 15% unsaturated compounds.[11] The total amount of olefins obtained as by-products in cracking processes for making gasoline is enormous. In 1936 the amount of the first three olefins formed in this way was about 5×10^{10} cu. ft.

Pintsch gas obtained by cracking the gas oil fraction of petroleum contains 35–40% olefins.

The *physical properties* of the alkenes closely resemble those of the alkanes. Ethylene boils some 15° below ethane but the corresponding higher members in the two series boil at nearly the same points. In general, the 1-olefin is lower boiling than the isomers with the double bond farther from the end of the chain. Thus the boiling points of *n*-heptane, 1-heptene, and 2-heptene are 98.4°, 93.5°, and 98.3° respectively. The melting points of the olefins are somewhat lower than those of the corresponding paraffins. The presence of a double bond increases the density about 0.015 for straight chain compounds. For branched chain compounds the increase varies widely. Thus the values of d^{20} for 2-Me-pentane, 2-Me-1-pentene and 2-Me-4-pentene are 0.652, 0.682

ALPHA OLEFINS

	m. °C.	b. °C. (Doss)
Ethylene	− 169.4	− 103.9
Propylene	− 185.2	− 47.6
1-Butene	− 190.	− 6.3
1-Pentene	− 138.	+ 30.1
1-Hexene	− 141.	63.7
1-Heptene	− 120.	92.8
1-Octene	− 102.1	121.6
1-Nonene	− 88.	145.3
1-Decene	− 66.	172.
1-Undecene	− 49.	189.
1-Dodecene	− 33.6	213.
1-Tridecene	− 22.	233
1-Tetradecene	− 12.	127/15 mm.
1-Pentadecene	− 4.	247
1-Hexadecene (cetene)	+ 4.	158/11.5
1-Heptadecene	11.	156/10
1-Octadecene	18.	180/15
1-Nonadecene	21.7	177/10
1-Eicosene (C$_{20}$)		315
1-Docosene (C$_{22}$)	41.	178/0.6
1-Tricosene		259/50
1-Tetracosene		273/50
1-Hexacosene (*Cerotene*) (C$_{26}$)	51.5	200/2
1-Octacosene	56.5	222/1.5
?-Triacontene (*Melene*) (C$_{30}$)	62.	218/0.5
1-Hentriacontene (C$_{31}$)	64.	295/15

[11] Egloff, Morrell. *Ind. Eng. Chem.* **24**, 1426 (1932).

and 0.665. In the 1-olefin the branching of the chain and the double bond each strengthens the effect of the other while in the 4-olefin they are apparently independent. The density of 1-pentene is 0.641 while that of 1-nonene is 0.731. The double bond increases the index of refraction (n^{20}D) about 0.011 for 1-olefins. The increase is greater when the double bond is nearer the middle of the molecule and also in branched molecules.

The *chemical properties* of the alkenes are in strong contrast to those of the alkanes. Not only do they *add* a great variety of reagents with the greatest ease but they also react with themselves to form *polymers*.

The presence and position of a double bond has a distinct effect on the properties such as boiling point and refractive index. Thus these values for *n*-octane and the 1-, 2-, 3-, and 4-octenes are respectively in °C and n^{20}D: 125.6, 1.3976; 121.6, 1.4090; 125.2, 1.4130; 123.2, 1.4138; 122.3, 1.4119.[12] The combination of branching and shifting of the double bond gives more profound effects. Thus the properties of the above compounds can be compared with those of 2,3-Me$_2$-hexane and the four olefins of the same carbon skeleton, 2,3-Me$_2$-1- and 2-hexenes and 4,5-Me$_2$-1- and 2-hexenes which show the following b. and n^{20}D: 115.8, 1.4013; 110.5, 1.4110; 121.8, 1.4267; 108, 1.4152; 110, 1.4132 (Doss).[13] The effect of substituting all of the H atoms in ethylene is especially striking.

The study of structure and the identification of olefins are greatly aided by *infrared* spectra and *Raman* spectra.[14,15] The rapid methods thus made possible were valuable in World War II in aviation gasoline production.[16]

Heats of formation, of combustion and of hydrogenation have been assembled for ethylene, propylene, the 4 butenes, the 6 pentenes, the 13 hexenes and higher 1-olefins.[17]

Individual Olefins

Methylene, (CH$_2$) is too reactive to be isolated.[18] Attempts to make it have yielded ethylene or higher polymers (CH$_2$)$_n$.[19] The free methylene radical has been obtained by heating iodomethylmagnesium iodide in nitrogen. It reacts with O$_2$ to give formaldehyde and with CO to give ketene. The pyrolysis of diazomethane, CH$_2$N$_2$, gives the methylene radical.[20]

Ethylene, ethene, CH$_2$=CH$_2$, was discovered in 1795 by the four Dutch chemists Deimann, Van Troostwick, Bondt, and Louwrenburgh. It occurs in the gases obtained by *cracking* almost any organic material. It is thus found in illuminating gas and in the gases from the manufacture of cracked gasoline.

[12] Doss. "Physical Constants of the Principal Hydrocarbons." The Texas Co.
[13] Doss. "Physical Constants of the Principal Hydrocarbons." The Texas Co.
[14] Barnes, Liddell, Williams. *Ind. Eng. Chem., Anal. Ed.* 15, 659 (1943).
[15] Stamm. *Ind. Eng. Chem., Anal. Ed.* 17, 318 (1945).
[16] Demmerle. *Chem. Eng. News.* 24, 2020 (1946).
[17] Prosen, Rossini. *J. Research. Natl. Bur. Standards.* 36, 269 (1946).
[18] Nef. *Ann.* 270, 267 (1892).
[19] Butlerow. *Ann.* 120, 356 (1861).
[20] Rice, Glasebrook. *J. Am. Chem. Soc.* 55, 4329 (1933).

Its older name "olefiant gas" comes from its action with chlorine to form an oil (ethylene dichloride).

Ethylene may be prepared by a number of methods which are also general for its homologs.

1. By dehydration of ethanol above 150° by means of sulfuric acid, phosphoric acid, benzene sulfonic acid, etc.; by catalytic action of hot aluminum oxide, etc. This last method, using a mixture of superheated steam and ethanol passed over kaolin was employed for making ethylene for the preparation of "mustard gas" ($\beta\beta'$-dichlorodiethylsulfide) during World War I.

2. By removal of halogen acid from an ethyl halide by means of an alcoholic solution of a base, usually potassium hydroxide. The yield of olefin is poor because of the simultaneous formation of an *ether* corresponding to the alkyl halide and the alcohol used. The complexity of the process can be seen from the fact that neopentyl iodide with alcoholic KOH gives mainly neopentane (p. 75).

3. By removal of two halogen atoms from an ethylene halide, XCH_2CH_2X by means of zinc or magnesium. A modification of this method depends on the fact that ethylene iodide loses its iodine on heating and forms ethylene. Thus if ethylene bromide is heated with an alcoholic solution of potassium iodide the products are potassium bromide, iodine and ethylene.

4. By electrolysis of sodium succinate solution (Kolbe).[21]

5. Oxidation of ethane. This process involves the dehydrogenation of ethane in the presence of oxygen in the ratio of 3:1 at 600° C.

6. The pyrolysis of ethane involves the conversion of ethane into ethylene in 80% yield by a cracking treatment at about 900° C.

7. Exposure of waste hydrocarbon gases to an electric arc is used primarily to produce acetylene, but gives a 10% yield of ethylene which is recovered by fractionation after removal of acetylene and carbon black. Waste gas from hydrogenation of coal, natural gas, and coke-oven residues are used as inexpensive raw materials for this method.

8. The partial hydrogenation of acetylene at 200° C. using a palladium catalyst gives an 85% yield of ethylene.

Processes (5–8) have received commercial application, especially in Europe.

9. The large amount of ethylene used for making ethylene glycol and the many related commercial products (some 200 in 1946) was formerly prepared by cracking fairly pure propane from natural gas.[22] The accompanying methane was used to supply part of the heat for the cracking process. Modern plants are located near large petroleum refineries from which enormous amounts of ethylene are available in the cracked gases.[23]

The reactions of ethylene are characteristic of compounds containing a double bond between carbon atoms.

[21] Petersen. *Chem. Zentr.* **1897** II, 518.
[22] Cambron. *Can. J. Research* **7**, 646 (1932).
[23] Ellis. "Chemistry of Petroleum Derivatives." Reinhold, 1934. p. 120.

1. Hydrogen adds with catalysts such as platinum black or colloidal palladium in water or alcohol suspension or, best, finely divided nickel[24] at temperatures around 300° (Raney Ni, UOP Ni catalyst). The discovery of this process laid the basis for the countless hydrogenations of theoretical and practical importance which are now conducted in nearly every field of organic chemistry. A copper catalyst can be used in place of nickel. The combination of ethylene and hydrogen is a homogeneous second-order reaction.[25] Ethylene may be activated by the action of alpha particles[26] and by excited Hg atoms.[27]

A mechanism has been suggested for hydrogenation involving the steps: (1) Addition of a proton to an activated ethylene molecule (p. 27) (2) addition of two electrons (from the metal catalyst), and (3) addition of a second proton.[28]

Hydrogenation of ethylene with a Pt catalyst is used as an accurate method of its determination in presence of ethane and methane. The volume of hydrogen consumed equals the volume of ethylene present.

2. Halogens add to ethylene, chlorine most readily and iodine least. Besides ethylene dichloride, chlorine gives higher chlorides such as 1,1,2-trichloroethane.[29] This is ascribed to activation of the molecules during the reaction. This activation may well take the form of the addition of a positive Cl and the loss of a proton followed by the further addition of Cl_2. Similar changes are common with olefins having the grouping $RR'C=C$ (p. 39).

Ethylene with a mixture of chlorine and bromine (bromine chloride) gives ethylene chlorobromide (1-chloro-2-bromoethane). No ethylene chloride is obtained.

The reaction of ethylene with bromine has been studied extensively.[30] In the gaseous state there is no reaction.[31] Glass, moisture and light favor the reaction. A lining of paraffin in the reaction vessel practically stops the reaction between ethylene and gaseous bromine while a lining of stearic acid catalyzes it better than glass.[32] Ethylene reacts very slowly in the dark with a solution of bromine in dry CCl_4. Strangely, it acts more rapidly at 0° than at 25°. The reaction is accelerated by traces of moisture or by light.[33]

The reaction of ethylene with iodine is incomplete and reversible.[34] Iodine chloride and iodine bromide also add to ethylene.

3. Halogens in presence of reactive solvents or ionized salts give products containing halogen and a radical from the salt or solvent.

[24] Sabatier, Senderens. *Compt. rend.* 124, 1358 (1897).
[25] Pease. *J. Am. Chem. Soc.* 54, 1876 (1932).
[26] Lind, Bardwell, Perry. *J. Am. Chem. Soc.* 48, 1556 (1926).
[27] Taylor, Hill. *J. Am. Chem. Soc.* 51, 2922 (1929).
[28] Burton, Ingold. *J. Chem. Soc.* 1929, 2022.
[29] Stewart, Smith. *J. Am. Chem. Soc.* 51, 3082 (1929).
[30] Ingold, Ingold. *J. Chem. Soc.* 1931, 2354.
[31] Stewart, Edlund. *J. Am. Chem. Soc.* 45, 1014 (1923).
[32] Norrish. *J. Chem. Soc.* 123, 3006 (1923).
[33] Davis. *Ind. Eng. Chem.* 20, 1055 (1928).
[34] Mooney, Reid. *J. Chem. Soc.* 1931, 2597.

(a) Water. Chlorine or bromine in water with ethylene give the halo-hydrins, XCH_2CH_2OH, which are important in commercial syntheses.[35] Their formation is usually ascribed to the addition of hypochlorous or hypo-bromous acid formed from the halogen and water. Only small amounts of ethylene dihalides are formed.

(b) Aqueous salt solutions. Bromine and NaCl solution convert ethylene to the chlorobromide. With sodium nitrate solution the product is the nitrate of ethylene bromohydrin, $BrCH_2CH_2ONO_2$.

(c) Alcohols. Ethers of the halohydrins are obtained. Thus a methanol solution of bromine with ethylene gives β-bromoethyl methyl ether.

These reactions seem to favor the conception that the initial step is the addition of a positive halogen to the activated ethylene molecule to form a carbonium ion (A) which then combines with an anion or coordinates with a molecule of solvent to give the observed product.

$$: \overset{..}{\underset{..}{X}} \ + \ : CH_2 : \underset{+}{CH_2} \ \rightarrow \ : \overset{..}{\underset{..}{X}} : CH_2 : \underset{+}{CH_2}$$
$$(A)$$

$$(A) \ + \ : \overset{..}{\underset{..}{Y}} : \ \rightarrow \ : \overset{..}{\underset{..}{X}} : CH_2 : CH_2 : \overset{..}{\underset{..}{Y}} :$$

$$(A) \ + \ : \overset{..}{\underset{H}{O}} : R \ \rightarrow \ : \overset{..}{\underset{..}{X}} : CH_2 : CH_2 : \overset{..}{\underset{H}{O}} : R$$

The coordination complex loses a proton to the solution leaving the beta-halogen ether. If water is involved in the last equation, $R = H$ and the result is a halohydrin. The ease of formation of a coordination complex between a carbonium ion and a solvent like HOH or ROH explains the strong tendency of hydroxyl and alkoxyl to add in these reactions. This tendency is out of all proportion to any possible "ionization" of HOH or ROH.

4. Halide acids add HI most readily and HCl least. Traces of ether assist the addition, probably through the formation of oxonium salts.[36] The addition is also catalyzed by carbon and silica. The technique of the addition of HX to ethylenic linkages has been improved.[37] Pure ethylene and hydrogen chloride do not react in the absence of the liquid phase. The presence of $AlCl_3$ catalyzes the addition effectively.[38] HF can be added to double bonds.[39,40]

[35] Curme, Young. *Chem. Met. Eng.* 25, 1091 (1921).
[36] *Ann. Rep. Chem. Soc.* (London) 1920, 55.
[37] Kharasch et al. *J. Am. Chem. Soc.* 55, 2468, 2521, 2531 (1933).
[38] Tulleners, Tuyn, Waterman. *Rec. trav. chim.* 53, 544 (1934).
[39] Grosse, Linn. *J. Org. Chem.* 3, 26 (1938).
[40] Henne, Waalkes. *J. Am. Chem. Soc.* 67, 1639 (1945).

5. Sulfuric acid adds H to one carbon and $-OSO_3H$ to the other. Accompanying this addition reaction, is a small amount of polymerization in which two or more molecules of ethylene add to each other. The absorption of ethylene by sulfuric acid is much slower than that of the higher olefins. Large quantities of acid are required, nearly 30 parts of concentrated acid for 1 part of ethylene. This method is used, however, for the commercial production of ethyl alcohol.[41] Sulfuric acid saturated with salts such as silver and nickel sulfates absorbs ethylene more rapidly than the pure acid. Ethylene is absorbed by silver nitrate solution.[42] Oleum, $H_2S_2O_7$, forms ethionic acid, $HO_3SCH_2CH_2OSO_3H$, the acid sulfate of isethionic acid, β-hydroxyethyl sulfonic acid, $HOCH_2CH_2SO_3H$. Sulfur trioxide gives carbyl sulfate, the anhydride of ethionic acid.

$$\begin{array}{ccc} CH_2 & & CH_2 - SO_2 - O \\ \| & + \ 2\,SO_3 \rightarrow & | \qquad\qquad / \\ CH_2 & & CH_2 - O - SO_2 \end{array}$$

Chlorosulfonic acid, $ClSO_3H$, absorbs ethylene much more rapidly than does sulfuric acid.

6. Nitric acid gives β-nitroethyl nitrate, $O_2NCH_2CH_2ONO_2$.

7. The chlorides of sulfur add to ethylene. This method is the best preparation for mustard gas, the most important poison gas of World War I. The reaction was first noted by the Scotchman, Guthrie.[43]

$$2\ C_2H_4 + S_2Cl_2 \rightarrow S + (ClCH_2CH_2)_2S$$

Similarly, selenium monochloride adds but the latter also acts as a chlorinating agent on the selenide formed.

$$2\ C_2H_4 + 2\ Se_2Cl_2 \rightarrow 3\ Se + (ClCH_2CH_2)_2SeCl_2$$

Selenium oxychloride, $SeOCl_2$, gives the same compound and SeO_2.[44]

8. Mercuric salts which hydrolyze, such as the sulfate, nitrate and acetate, add to ethylene in alkaline water or alcohol solution to give compounds of the types

$$YHgCH_2CH_2OH$$
$$(YHgCH_2CH_2)_2O \qquad\qquad YHgCH_2CH_2OR$$

in which Y is the acid radical.[45] Probably the reason that non-hydrolyzable Hg salts like the chloride do not add to ethylene is that they are not sufficiently ionized to give a positive addend to combine with the free electron pair of an activated ethylene molecule (p. 27). A Hg salt which can add in this way

[41] Ellis. "Chemistry of Petroleum Derivatives." Reinhold, 1934. p. 300.
[42] Morris. *J. Am. Chem. Soc.* **51**, 1460 (1929).
[43] Guthrie. *Ann.* **113**, 266 (1860).
[44] Ellis. "Chemistry of Petroleum Derivatives." Reinhold, 1934. p. 577.
[45] Hofmann, Sand. *Ber.* **33**, 1340 (1900).

gives a carbonium ion $YHgCH_2CH_2^+$ which then coordinates with HOH or ROH to give the observed products (p. 33). The mercurated ether indicated above would be formed by action of the carbonium ion with the hydroxyl of the mercurated alcohol. The halides of these compounds are less soluble than the other salts and are obtained by simply adding the calculated amount of NaX to the reaction mixture. These Hg compounds react with iodine to give ethylene iodohydrin, diiodo ethyl ether, and iodoethyl alkyl ethers thus proving the position of the C—Hg linkage. With acids they would be expected to give the corresponding hydrogen substitution products, namely, EtOH, Et_2O and EtOR. Instead, all three types of compounds react with mineral acids to give quantitative yields of ethylene. Because of this surprising behavior, these compounds were formerly regarded as mere molecular compounds, $C_2H_4.Hg(OH)Y$, $2 C_2H_4.Hg_2OY_2$ and $C_2H_4.Hg(OR)Y$.[46] The question has been settled in favor of the *structural* formulas by the preparation in optically active form of such a compound with an asymmetric carbon.[47] The peculiar action of the grouping —Hg—C—C—O— with acid to give ethylene is merely an especially easy case of beta-fission.[48] The proton from the acid coordinates with the oxygen to give an oxonium ion which loses HOH or ROH leaving a carbonium ion $YHg:CH_2:CH_2^+$. The electronically deficient carbon attracts the electron pair which holds the Hg, this forming $CH_2::CH_2$ and YHg^+. The simultaneous effect of the anion from the acid to combine to form the non-ionized or slightly ionized Hg salt forces the process in the same direction.

9. Ethylene adds benzene and other aromatic compounds under the influence of anhydrous aluminum chloride.[49-51] Thus benzene and ethylene give all the possible ethyl substituted benzenes.

10. Dilute solutions of oxidizing agents such as potassium permanganate or barium chlorate add two hydroxyl groups to the adjacent carbons to form glycol, $HOCH_2CH_2OH$. With more drastic conditions, the glycol first formed is further oxidized with the separation of the two carbons.

11. Incomplete oxidation of ethylene with air gives formaldehyde.[52] Ethylene with ozone gives an ozonide which is hydrolyzed to formaldehyde.

$$\begin{matrix} CH_2 & \\ | & \diagdown \\ O & \quad O_2 + H_2O \rightarrow 2\ HCHO + H_2O_2 \\ | & \diagup \\ CH_2 & \end{matrix}$$

Ethylene can be oxidized catalytically to ethylene oxide (p. 310).

[46] Manchot. *Ann.* **420**, 170 (1920).
[47] Marvel et al. *J. Am. Chem. Soc.* **48**, 1409 (1926); **53**, 789 (1931).
[48] Whitmore. *Ind. Eng. Chem. News* 26, 668 (1948).
[49] Friedel, Crafts. *Ann. chim. phys.* (6) **14**, 433 (1888).
[50] Thomas. "Anhydrous Aluminum Chloride in Organic Chemistry." Reinhold, 1941.
[51] Ellis. "Chemistry of Petroleum Derivatives." Reinhold, 1934. p. 559.
[52] Ellis. "Chemistry of Petroleum Derivatives." Reinhold, 1934. p. 874.

12. Ethylene, in common with other olefins, burns with a luminous flame. This property is utilized in making carburetted water gas by adding the cracked products from gas oil and similar materials. Ethylene is used with oxygen for welding and cutting metals (Calorene). While its flame is not as hot as that obtained with acetylene, it has the advantage of being stable under high pressures in ordinary steel cylinders.

13. Ethylene can be polymerized difficultly by heat alone or catalytically by means of complex silicates such as floridin,[53] but only with difficulty by acids. When ethylene is thus polymerized the products are extremely complex.[54] Polymerization by the silent electric discharge yields 1-butene and 1-hexene.[55]

High pressure polymerization of ethylene gives *Polythene*, a very inert resin of about 20,000 M.W. having low electrical losses.[56] Chemically pure polyethylene is used as a substitute for cartilege and bone in plastic surgery.

14. The pyrolytic reactions of ethylene are most complex, giving among other products, carbon, hydrogen, methane, ethane, propylene, butylene, butadiene and aromatic products.[57]

15. Ethylene reacts with aqueous alkalies at 400° and 50 atm. pressure to give acetates and hydrogen. Similarly propylene gives propionic acid.

16. With CO_2 at high temperature and pressure in the presence of catalysts and ultraviolet light, acrylic acid, $CH_2 = CHCO_2H$, and its polymers are obtained in low yields.[58]

SUMMARY

The chemical versatility of ethylene is shown by more than two hundred commercial products available from it in 1946. One company alone markets more than one hundred chemicals derived from this basic raw material. Included are acetaldehyde, acetic acid and anhydride, ethylene chlorohydrin, ethylene glycol and its mono- and di-esters, ethylene chloride, ethylene oxide, ethanol, 1-butanol, 2-Et-1-hexanol, higher alcohols and their esters, ethyl ether, vinyl chloride and acetate (basis of *Vinylite* plastics), ethylene diamine, diethylene triamine, triethylene tetramine ($NH_2CH_2CH_2NHCH_2CH_2NHCH_2$-$CH_2NH_2$) and higher analogs, ethyl sulfate, $\beta\beta'$-dichloroethyl ether (*Chlorex*), di- and tri-ethylene glycols ($\beta\beta'$-dihydroxyethyl ether and 1,2-bis-β-hydroxy-ethoxy-ethane, $HO(CH_2CH_2O)_3H$) and higher liquid and solid polyethylene glycols (*Carbowax*), 2-ethyl-1,3-hexandiol, 2-ethylbutyrates and 2-ethylhexoates of polyethylene glycols (*Flexol* plasticizers 3GH,360 and 460), mono methyl, ethyl, *n*-butyl, 2-ethylbutyl and phenyl ethers of ethylene glycol (*Cellosolves*), diethyl ether of ethylene glycol and higher polyethylene glycol diethers, mono

[53] Ellis, *ibid.* p. 588.
[54] Montmollin. *Bull. soc. chim.* 19, 242 (1916).
[55] Mignonac, Vanier de Saint-Aunay. *Compt. rend.* 189, 106 (1929).
[56] Hahn, Macht, Fletcher. *Ind. Eng. Chem.* 37, 526 (1945).
[57] Egloff et al. *Ind. Eng. Chem.* 27, 917 (1935); *J. Phys. Chem.* 35, 1825 (1931).
[58] Ellis. "Chemistry of Petroleum Derivatives." Reinhold, 1934. p. 586.

methyl, ethyl and n-butyl ethers of diethylene glycol (Carbitols), mono-, di- and tri-ethanol amines ($HOCH_2CH_2NH_2$ etc.), N,N-Et$_2$-ethanolamine ($Et_2NCH_2CH_2OH$), phenylethanolamine and other aliphatic and aromatic N-substituted alkanolamines, morpholine, $O(CH_2CH_2)_2NH$ and various derivatives, acetates of the cellosolves and carbitols, 1,4-dioxan (diethylene oxide), $O(CH_2CH_2)_2O$, glyoxal, butyraldehyde, ethyl orthosilicate, ethyl acetoacetate, 2-Et-butyric acid, 2-Et-hexoic acid, styrene, polyethylene plastics, mercapto-ethanol, sodium alkyl sulfates (*Tergitol* penetrants 08,4 and 7), water-soluble hydroxyethyl cellulose (*Cellosize WS*), 2-Et-hexyl phthalate, butadiene, acrylonitrile, methyl acrylate and higher ester, ethylene glycol dinitrate, butyl carbitol thiocyanate, phthalates and higher fatty esters of cellosolves and carbitols, ethyl bromide, ethyl chloride, ethylamines and many synthetic plasticizers and surface-active agents.

Ethylene is used to hasten the development of bananas, citrous fruits, bulbs and potatoes. It is especially valuable in ripening fruit which is picked green for transportation. Treatment with ethylene stimulates the proteolytic enzymes of pineapples.[59] Many other chemicals have similar effects.[60, 61] Among them are ethylene chlorohydrin, sodium thiocyanate, thiourea, ethylene dichloride, furfural, ethyl bromide and vinyl chloride.

Ethylene can be determined by absorption in a solution of 22% $HgSO_4$ dissolved in 22% H_2SO_4 after removal of CO_2 and higher olefins.[62, 63]

In addition to its ordinary industrial significance, the military importance of a well-developed ethylene industry cannot be overemphasized.

Propylene, propene, methylethylene, $CH_3CH=CH_2$, can be made by the standard methods from n-propyl alcohol, isopropyl alcohol, n-propyl halides, isopropyl halides and from propylene dihalides. It is also formed by heating glycerol (1,2,3-trihydroxypropane) with zinc dust.

Studies of the dehydration of propanol containing radioactive carbon show that meta phosphoric acid and alumina cause complete or nearly complete isomerization, while pyrolysis of the trimethyl-n-propyl ammonium hydroxide gives little or no rearrangement.[64]

The propylene used commercially for the preparation of isopropyl alcohol (Petrohol), propylene glycol and acetone is obtained as a by-product in the cracking of petroleum fractions to make gasoline and of propane to make ethylene.[65] Pure propylene is best made from the propyl alcohols passed in the vapor phase over heated Al_2O_3 in copper tubes.[66]

The reactions of propylene are like those of ethylene with certain added

[59] Harvey. *The Minnesota Horticulturist* **54**, 140 (1926).
[60] Denny. *C. A.* **18**, 557 (1924).
[61] Denny. *Contrib. Boyce Thompson Inst.* **5**, 441 (1933).
[62] Francis, Lukasiewicz. *Ind. Eng. Chem., Anal. Ed.* **17**, 703 (1945).
[63] Brooks, Benjamin, Zahn. *Ind. Eng. Chem., Anal. Ed.* **18**, 339 (1946).
[64] Fries, Calvin. *J. Am. Chem. Soc.* **70**, 2235 (1948).
[65] Ellis. "Chemistry of Petroleum Derivatives." Reinhold, 1934. p. 127.
[66] Ipatieff. *Ber.* **36**, 1990 (1903); **67B**, 1061 (1934).

complications. Thus propylene with pure bromine gives a 90% yield of propylene dibromide and 4% of 1,2,3-tribromopropane. With bromine and iron it gives 40% of the 1,2,3- and 17% of the 1,1,2-tribromo compound. With ferric bromide or aluminum as catalyst the yields of the tribromides are 8% and 33% respectively.[67]

It is possible to chlorinate propylene directly to allyl chloride (p. 96).

Oxidation of propene by permanganate has been shown, using radioactive carbon, to occur at both the double bond and the CH_3-CH bond.[68]

When the addend has two unlike parts, two different addition products are possible since the double bond is not symmetrical. Both possible products are formed but one usually predominates to the extent of 90% or more. This chief product is the one in which the hydrogen or the "positive" part adds to the end carbon, Markownikoff's Rule (1870). Thus propylene gives mainly isopropyl compounds with halide acids and sulfuric acid.[69] With hypochlorous acid it gives mainly 1-chloro-2-propanol.[70] The direction of addition depends on the polarity of the parts of the addend. With hypohalous acids the OH is "negative" and the halogen "positive" although the ionization of these substances gives H^+ and XO^- ions.

The phenomena expressed in Markownikoff's Rule may be pictured as follows:

1. The inactive double bond is activated in some way. The extra electron pair instead of being shared by the two olefinic carbons becomes more closely associated with one of them. In an unsymmetrical olefin this can take place in two ways:

$$\text{(A)} \quad \overset{\displaystyle H \quad H}{\underset{\displaystyle H}{Me:C:C}} \quad \leftrightarrow \quad \overset{\displaystyle H \quad H}{\underset{\displaystyle H}{Me:C::C}} \quad \leftrightarrow \quad \overset{\displaystyle H \quad H}{\underset{\displaystyle H}{Me:C:C:}} \quad \text{(B)}$$

2. The H ion or the positive halogen adds to the free electron pair of the activated ethylene linkage. Then the negative part of the addend adds to the electronically deficient carbon.[70a, 70b] Thus the reason for Markownikoff's Rule may be that form B is predominant.

The whole subject of the addition of an unsymmetrical molecule XY to an unsymmetrical olefin is still unsettled.[71–73] The influence of the solvent used is important. Thus glacial acetic acid favors the addition of HBr to an alpha

[67] *Ann. Rep. Chem. Soc.* (London) 1923, 74.
[68] Fries, Calvin. *J. Am. Chem. Soc.* 70, 2235 (1948).
[69] Ellis. "Chemistry of Petroleum Derivatives." Reinhold, 1934. p. 320.
[70] Ellis, *ibid.* p. 499.
[70a] Francis. *J. Am. Chem. Soc.* 47, 2314 (1925).
[70b] Terry, Eichelberger. *J. Am. Chem. Soc.* 47, 1067 (1925).
[71] Kharasch, Darkis. *Chem. Revs.* 5, 571 (1928).
[72] *Ann. Rep. Chem. Soc.* (London) 1919, 89; 1923, 105.
[73] Ingold et al. *J. Chem. Soc.* 1931, 2742, 2746.

olefin to give mainly the primary bromide.[74, 75] Neopentylethylene,[76] 1-pentene and 1-heptene[77] have been found to add HBr gas to give *only* the *primary* bromides. Much of the confusion was eliminated by the discovery of the *peroxide effect*.[78] Thus, propylene in the presence of peroxides adds HBr contrary to Markownikoff's Rule to give *n*-propyl bromide.

While olefins do not add iodine readily they add iodine monochloride, ICl. In the case of unsymmetrical olefins both possible addition products are obtained but the predominant one has the I on the carbon with the most H atoms (Markownikoff's Rule). Iodine in presence of silver salts causes the addition of I and the radical of the salt to the double bond. Thus the nitrite and acetate with iodine add a nitro and an acetate group respectively (p. 31).

Propylene and H_2S over a NiS catalyst at 250° give the *n*- and iso-propyl mercaptans in 2:1 ratio.[79]

Propylene is much more reactive than ethylene. For instance, propylene is absorbed completely by 85% sulfuric acid whereas ethylene is absorbed rapidly only by fuming sulfuric acid. Propylene is also oxidized more rapidly than ethylene. With oxygen at 500°, propylene gives acetaldehyde, formaldehyde, oxides of carbon, butylenes, amylenes, and hexylenes.[80]

The pyrolysis of propylene at 600° gives a complex mixture including H_2, methane, ethylene, acetylene, ethane and about 25% of aromatic hydrocarbons.[81]

Propylene is polymerized more readily by acids than ethylene. It is rapidly polymerized at 250° by various silicates such as floridin and "catasil" (Al_2O_3 adsorbed on silica gel).[82] The volatile products contain 6,7,5,9 and 8 carbon atoms in decreasing abundance. Obviously the reactions involved are complex.

The following products are commercially available from propylene as a raw material: isopropyl alcohol, isopropyl acetate, isopropyl ether, acetone, allyl chloride, glycerol, diacetone alcohol, mesityl oxide, methyl isobutyl ketone, methylisobutylcarbinol, 4-Me-1,3-pentadiene, propylene chlorohydrin, propylene glycol, propylene oxide, propylene dichloride, $\beta\beta'$-dichloro isopropyl ether, and methyl isobutyl ketone.

The magnitude of propylene production is shown by the fact that one medium refinery annually produces enough of the gas to make about 80 million pounds of glycerol if it were so converted.

[74] Ipatieff, Dechanoff. *Chem. Zentr.* 1904, II, 691.
[75] Delacre. *Bull. sci. acad. roy. Belg.* 1906, 7.
[76] Whitmore, Homeyer. *J. Am. Chem. Soc.* 55, 4555 (1933).
[77] Sherrill, Mayer, Walter. *J. Am. Chem. Soc.* 56, 926 (1934).
[78] Kharasch et al. *J. Am. Chem. Soc.* 55, 2468, 2521, 2531 (1933).
[79] Barr, Keyes. *Ind. Eng. Chem.* 26, 1111 (1934).
[80] Lenher. *J. Am. Chem. Soc.* 54, 1830 (1932).
[81] Hurd, Meinert. *J. Am. Chem. Soc.* 52, 4978 (1930).
[82] Gayer. *Ind. Eng. Chem.* 25, 1122 (1933).

Butylenes, C_4H_8. The isomers are:

1-Butene, alpha-butylene, ethylethylene, $CH_3CH_2CH = CH_2$, b. $-6.3°$.

2-Butene, beta-butylene, symmetrical Me_2-ethylene (pseudo butylene), $CH_3CH = CHCH_3$, *cis* form, b. $3.6°$; *trans* form, b. $0.9°$.

Isobutylene, gamma-butylene, methylpropene, unsymmetrical Me_2-ethylene, $(CH_3)_2C = CH_2$, b. $-7.0°$.

With increasing complexity, new complications appear.

Preparation. 1. Dehydration of 1-butanol by acids gives 1-butene and varying amounts of the 2-butenes. Dehydration over hot Al_2O_3 *free from all acid* impurities gives pure 1-butene.[83]

2. Pure 1-butene can be made from allyl bromide or chloride and a methyl Grignard reagent.[84] This is a good method for making pure 1-olefins.

3. 1-Butene from cracked gases contains some 2-butene. This does no harm because both give 2-butanol on hydration with sulfuric acid. The C_4 fraction of cracked gases also contains some isobutylene. Commercially, this is absorbed by more dilute sulfuric acid than will react with the *n*-butylenes. These are then absorbed in stronger acid. The first acid sulfate solution on dilution and hydrolysis gives *t*-butyl alcohol while the second gives *s*-butyl alcohol.

4. Perhaps the strangest formation of 1-butene is by the high temperature acid catalyzed dehydration of isobutyl alcohol which gives about equal amounts of the expected isobutylene and of the *n*-butylenes (p. 114).

5. 2-Butene exists in stereoisomeric forms due to the restricted rotation of the doubly bonded carbon atoms. The more symmetrical *trans* form is the more volatile.

$$
\begin{array}{cc}
\text{Me} - \text{C} - \text{H} & \text{Me} - \text{C} - \text{H} \\
\| & \| \\
\text{Me} - \text{C} - \text{H} & \text{H} - \text{C} - \text{Me} \\
cis & trans
\end{array}
$$

These have been prepared from angelic and tiglic acids, stereoisomeric alpha methylcrotonic acids, $MeCH = CMeCO_2H$, respectively.

6. Isobutylene is best prepared in the laboratory by dehydrating *t*-butyl alcohol. No rearrangement has been observed in this process. It can also be prepared by careful high temperature dehydration of isobutyl alcohol over acid-free alumina.[85, 86]

Reactions. The butylenes resemble propylene in their reactions. Thus 1-butene and isobutylene add HBr according to Markownikoff's Rule to give 2-Br-butane and *t*-butyl bromide respectively. This addition can be reversed in the presence of peroxides to give at least 80% of 1-Br-butane and isobutyl bromide.[87]

[83] Pines. *J. Am. Chem. Soc.* **55**, 3892 (1933).
[84] Lucas, Dillon. *J. Am. Chem. Soc.* **50**, 1460 (1928).
[85] Ipatiev. *Ber.* **36**, 2011 (1903).
[86] Pines. *J. Am. Chem. Soc.* **55**, 3892 (1933).
[87] Kharasch, Hinckley. *J. Am. Chem. Soc.* **56**, 1212, 1243 (1934).

The *n*-butylenes react with SO_2 to give resins whereas isobutylene does not react. Thus 50 lbs. of *n*-butenes heated with 750 lbs. of liquid SO_2 and 13 g. of $LiNO_3$ give an insoluble resin.

Isobutylene is much more stable to heat than isobutane.[88] At 700° it gives methane, hydrogen, propylene and a complex mixture of aromatic hydrocarbons including benzene and toluene and probably xylenes, naphthalene, phenanthrene and anthracene. At 630−885° and very short contact times, 0.053−2.1 sec, it gives as much as 60% $CH_3C \equiv CH$ and $CH_2 = C = CH_2$ with little tar or carbonaceous material.[89]

Isobutylene and chlorine fail to give the dichloride under any known conditions. Instead, methallyl and isocrotyl chlorides are formed (p. 98). The formation of unsaturated chlorides by the action of olefins with chlorine has long been known. Olefins which add HX with great ease react with chlorine and bromine to give unsaturated halides instead of the expected dihalides.[90]

Isobutylene with bromine gives 75% isobutylene dibromide, 17% 1,2,3-tribromo-2-Me-propane and 7% *t*-butyl bromide. This is the same type of change observed in the action of chlorine by Kondakoff on isobutylene and by Stewart on ethylene. The unsaturated bromide adds another molecule of bromine.

Isobutylene with aqueous KI_3 solution gives *t*-butyl alcohol and isobutylene oxide,

$$Me_2C - CH_2$$
$$\diagdown \diagup$$
$$O$$

This is probably due to (*a*) the instability of a 1,2-diiodide, (*b*) the ease of addition of HI and HIO to isobutylene, (*c*) the ease of hydrolysis of a tertiary iodide, (*d*) the reducing action of HI on an iodide, (*e*) the ready loss of HI by an iodohydrin.

Isobutylene with acetyl chloride and $AlCl_3$ gives the HCl addition product of mesityl oxide, Me_2CClCH_2COMe.

Isobutylene adds nitrogen trichloride according to Markownikoff's Rule (positive chlorine as in HOCl) to give the remarkable compound, 2-dichloro-amino-2-chloromethylpropane.[91]

$$\begin{array}{c} CH_3 \\ | \\ ClCH_2-C-NCl_2 \\ | \\ CH_3 \end{array}$$

This is stable and forms a hydrochloride whereas most NCl compounds react with HCl to form Cl_2.

[88] Hurd, Spence. *J. Am. Chem. Soc.* 51, 3561 (1929).
[89] Rice, Haymes. *J. Am. Chem. Soc.* 70, 964 (1948).
[90] Kondakoff. *Ber.* 24, 929 (1891).
[91] Coleman, Mullins, Pickering. *J. Am. Chem. Soc.* 50, 2739 (1928).

The butylenes polymerize much more readily than the lower olefins. Isobutylene polymerizes most readily. This is characteristic of olefins containing the grouping, $RR'C = C$, related to tertiary alcohols. Various acid and silicate catalysts convert isobutylene to polymers in which n is 2–7 or even higher.[92–94] Boron trifluoride converts isobutylene to a high molecular weight rubbery solid (*Vistanex*). Co-polymerization of isobutylene and isoprene (3%) by aluminum chloride at $-140°$ F. gives *Butyl Rubber* which is especially valuable for tire inner tubes.

Isobutylene unites with tertiary hydrocarbons such as isobutane (Alkylation) (p. 19).

Thioglycolic acid adds to isobutylene contrary to Markownikoff's Rule[95] to give $Me_2CHCH_2SCH_2CO_2H$.

Amylenes, C_5H_{10}. All the possible structural isomers have been prepared.

1. 1-Pentene, *n*-propylethylene, $CH_3CH_2CH_2CH = CH_2$, b. 30°.

2. 2-Pentene, *sym*-methylethylethylene, $CH_3CH_2CH = CHCH_3$, b. 36.4°.

3. 2-Me-1-butene, *unsym*-methylethylethylene, $CH_3CH_2C(CH_3) = CH_2$, b. 32°.

4. 2-Me-2-butene, trimethylethylene, $(CH_3)_2C = CHCH_3$, b. 38.4°.

5. 3-Me-1-butene, isopropylethylene, $(CH_3)_2CHCH = CH_2$, b. 20.1°.

These can be prepared like the butylenes but with correspondingly greater complications due to rearrangements. Thus *besides* the expected dehydration products, 1-pentanol gives 2-pentene; *sec*-butylcarbinol gives also 2-pentene in small amounts; isoamyl alcohol gives mainly 2-methyl-2-butene, etc.

Commercial "amylene" was formerly made by dehydrating commercial amyl alcohol, itself a mixture. Now 2-pentene and trimethylethylene are available (Sharples). 2-Chloro-2-methylbutane and 3-chloro-2-methylbutane formed during the chlorination of isopentane lose HCl during the subsequent treatment with alkali, giving mainly trimethylethylene.

The reactions of the amylenes present few peculiarities. At high temperatures and in the presence of catalysts, methylbutenes suffer shifts of the double bond.[96] Isopropylethylene and *unsym*-methylethylethylene are converted mainly to trimethylethylene. The change may be regarded as the addition of H^+ to one carbon and a loss of H^+ from a different carbon. This is the equivalent of the older conception that water was added one way and split off in a different way (p. 43).

1-Pentene in solution in hexane, glacial acetic acid or CCl_4 adds HBr contrary to Markownikoff's Rule to give only *n*-amyl bromide.[97]

[92] Butlerow. *Ber.* 9, 1687 (1876).

[93] Lebedev, Koblyanskii. *Ber.* **63B**, 103 (1930).

[94] Flory. *J. Am. Chem. Soc.* **65**, 372 (1943).

[95] Kharasch, Read, Mayo. *Chemistry & Industry* 752 (1938).

[96] Ipatieff. *Ber.* 36, 2003 (1903).

[97] Sherrill, Mayer, Walter. *J. Am. Chem. Soc.* 54, 926 (1934).

Apparently the addition of HBr to an unsymmetrical olefin compound of the type RCH=CHR' is not influenced by peroxides.[98]

The amylenes give complex addition products with N_2O_3, N_2O_4 and NOCl.[99, 100]

Trimethylethylene with nitrosyl chloride gives the blue nitrosochloride, $Me_2CClCH(NO)Me$. On standing or on heating this changes to the oxime, $Me_2CClC(=NOH)Me$. The bimolecular nitrosochloride is obtained directly from trimethylethylene, an alkyl nitrite and fuming hydrochloric acid.

Nitrous anhydride and nitrogen tetroxide give a nitrosite and a nitrosate respectively of which the blue nitroso forms are as follows:

$$Me_2C(ONO)CH(NO)Me \quad \text{and} \quad Me_2C(ONO_2)CH(NO)Me$$

These are convertible to the hydroximino and bis- forms.

Me_3-ethylene reacts with Ac_2O and $ZnCl_2$ to form 3,4-Me_2-3-penten-2-one.[101] The rate of hydrogenation of the isomeric pentenes has been studied.[102]

The amylenes differ widely in their activity with acids. Thus isopropylethylene does not act with 5N HCl in months whereas trimethylethylene acts in 5 hrs. 2-Me-1-butene acts even more rapidly.

The amylenes related to t-amyl alcohol react with phenols in the presence of acid catalysts to give phenol ethers and, on rearrangement, tertiary amyl phenols.[103] This reaction is general for olefins, $R_2C=C$.

The amylenes polymerize with different degrees of ease, trimethylethylene forming "di-isoamylenes" and higher polymers most readily.[104, 105] The dimers from the polymerization of t-amyl alcohol have been shown to consist mainly of 2,3,4,4-Me_4-1-hexene, 3,4,5-5-Me_4-2-hexene, 3,4,4,5-Me_4-2-hexene and 3,5,5-Me_3-2-heptene.[106, 107]

Hexylenes, C_6H_{12}. All thirteen possible structural isomers have been prepared. Most of these offer no peculiarities not observed with the amylenes.

In the case of 3-Me-2-pentene, the cis and trans stereomers boil far enough apart (5°) so that they can be separated by careful fractional distillation.[108]

Tetramethylethylene, 2,3-Me_2-2-butene, $(CH_3)_2C=C(CH_3)_2$, b. 72°.

Preparation. 1. The best large scale laboratory method is by dehydration of pinacolyl alcohol, $Me_3CCHOHMe$, prepared from t-BuMgCl and MeCHO. Acid dehydration gives the following products: Tetramethylethylene, b. 72°, 61% yield; unsym-methylisopropylethylene, b. 54°, 31%; t-butylethylene, b.p.

[98] Smith, Harris. *Nature* 135, 187 (1935).
[99] Schmidt. *Ber.* 35, 2323 (1902); 36, 1775 (1903).
[100] Wallach. *Ann.* 241, 288 (1887).
[101] Kondakow. *Ber.* 27R, 309, 941 (1894).
[102] Davis, Thomson, Crandall. *J. Am. Chem. Soc.* 54, 2340 (1932).
[103] Niederl, Natelson. *J. Am. Chem. Soc.* 53, 272 (1931).
[104] Brooks, Humphrey. *J. Am. Chem. Soc.* 40, 822 (1918).
[105] Norris, Joubert. *J. Am. Chem. Soc.* 49, 873 (1927).
[106] Drake, Kline, Rose. *J. Am. Chem. Soc.* 56, 2076 (1934).
[107] Stehman, Cook, Whitmore. *J. Am. Chem. Soc.* 72, 4163 (1950).
[108] van Rissegheim. *Bull. soc. chim. Belg.* 31, 213 (1922).

41°, 3%. Thus the "normal" dehydration involving the methyl group becomes only a side reaction and rearrangement is the principal process.[109]   The Me_4-ethylene is readily separated by a distilling column of at least 20 equivalent plates.   The low boiling material can be isomerized to the equilibrium mixture by warming with acid.

2. Commercially, by the alkylation of Me_3-ethylene with MeCl and lime.[110]

3. By dehydration of dimethylisopropylcarbinol made from isopropyl Grignard reagent and acetone or from MeMgX and an isobutyric ester or methyl isopropyl ketone.

4. By removal of Br_2 from its dibromide obtained by the action of HBr on pinacol (tetramethylethylene glycol) prepared by the reduction of acetone with amalgamated magnesium.

The *reactions* of tetramethylethylene are typical olefin reactions.   In many ways they are simplified by the symmetry of the molecule and the absence of H from the ethylene carbons.   Thus it gives with NOCl, N_2O_3 and N_2O_4 simple addition compounds of the blue true nitroso type without any complication of tautomerism or dimerization,

$Me_2CClC(NO)Me_2$, $Me_2C(ONO)C(NO)Me_2$,
Nitrosochloride   Nitrosite

$Me_2C(ONO_2)C(NO)Me_2$
Nitrosate

In the presence of acid catalysts Me_4-ethylene gives an equilibrium mixture of the three olefins listed above (1).[111]   Thus the shift of a double bond involving only the rearrangement of H atoms[112] can include shifts of a double bond which involve the rearrangement of alkyl groups as well.   This phenomenon is of the greatest importance in determining the final olefin mixture obtained from an alcohol by dehydration with rearrangement.   Because of this and its relation to the changes in structure involved in polymerization, copolymerization, and alkylation, it is worth while to consider it in detail.

$$Me_2C=CMe_2 + H^+ \rightleftharpoons Me_2\overset{+}{C}-CHMe_2$$

$$\diagdown \ : Me \ \text{rearrange}$$

$$CH_2=C-CHMe_2 + H^+ \qquad Me_3C-\overset{+}{C}HMe$$
$$\quad Me$$

$$Me_3C-CH=CH_2 + H^+$$

It should be noted that the intermediate carbonium ions or fragments of very short life are formed by *addition* of protons from the catalyst.   All members of the above system are in equilibrium.   Energy relationships control the proportions of the three olefins in the equilibrium mixture of approximately

[109] Whitmore, Rothrock.   *J. Am. Chem. Soc.* **55**, 1106 (1933).
[110] Miller, Lovell.   *Ind. Eng. Chem.* **40**, 1138 (1948).
[111] Laughlin, Nash, Whitmore.   *J. Am. Chem. Soc.* **56**, 1395 (1934).
[112] Ipatieff.   *Ber.* **36**, 2003 (1903).

60:30:5 reading down from left to right.[113] Under suitable conditions, the carbonium ions can add to the olefins to give polymers. It is interesting that Me_4-ethylene gives about 20 dimers, no one of which is directly related to the monomer. In other words, the above equilibrium is established before dimerization takes place. Then the other olefins and the carbonium ions react more readily than does the original olefin even though it is present in largest amount.

Me_4-ethylene reacts with acetyl chloride or acetic anhydride in presence of $ZnCl_2$ to give a good yield of 3,3,4-Me_3-4-penten-2-one. This is a valuable intermediate for higher highly branched hydrocarbons.

t-Butylethylene, 3,3-Me_2-1-butene, $(CH_3)_3CCH=CH_2$, b. 41°. Ordinary dehydration of the corresponding alcohol gives mainly rearranged products. This is avoided by pyrolysis of pinacolyl acetate.[114,115] A still better method is the high temperature dehydration of pinacolyl alcohol over a relatively inactive alumina catalyst free from acid impurities.[116] This olefin was first obtained by the Tschugaeff Reaction (1899) which is useful in obtaining unrearranged dehydration products of alcohols. It consists in decomposing the methyl xanthate of the alcohol related to the desired olefin. Pinacolyl alcohol is converted to the sodium alcoholate and treated with carbon disulfide to give the sodium xanthate. This is converted to the methyl compound by MeI or Me_2SO_4. This is decomposed by heat to give the desired olefin, carbon oxysulfide and methyl mercaptan.[117,118]

$$Me_3CCH(ONa)Me \xrightarrow{CS_2} Me_3CCH(Me)OCS_2Na \xrightarrow{MeI} Me_3CCH(Me)OCS_2Me$$
$$\xrightarrow{\Delta} Me_3CCH=CH_2 + COS + MeSH$$

t-Bu-ethylene adds HBr contrary to Markownikoff's Rule to give the primary bromide. This is probably due to peroxides.[119] Acid catalysts convert the olefin almost completely to a 2:1 mixture of Me_4-ethylene and 2,3-Me_2-1-butene (p. 43).

Heptylenes, C_7H_{14}. All twenty-seven possible structural isomers have been prepared, largely by methods devised by Boord.[120] Most heptyl alcohols give difficultly separable mixtures of heptylenes. The methods of Boord lead to pure individual olefins. They depend on the following reactions:

a. An aldehyde with an alcohol and HCl to give an alpha chloroether.

$$RCH_2CHO + HCl + EtOH \rightarrow RCH_2CHCl(OEt)$$

R may be H or an organic group.

[113] Laughlin, Nash, Whitmore. *J. Am. Chem. Soc.* **56**, 1395 (1934).
[114] Whitmore, Rothrock. *J. Am. Chem. Soc.* **55**, 1106 (1933).
[115] Dolliver et al. *J. Am. Chem. Soc.* **59**, 831 (1937).
[116] Cramer, Glasebrook. *J. Am. Chem. Soc.* **61**, 230 (1939).
[117] Fomin, Sochanski. *Ber.* **46**, 246 (1913).
[118] Schurman, Boord. *J. Am. Chem. Soc.* **55**, 4930 (1933).
[119] Kharasch et al. *J. Org. Chem.* **4**, 428 (1939).
[120] Boord et al. *J. Am. Chem. Soc.* **52**, 3396 (1930); **55**, 4930 (1933).

b. The action of the chloroether with bromine to give an alpha, beta dibromoether, RCHBrCHBr(OEt).

c. The action of a Grignard reagent to replace the alpha bromine by an alkyl group, RCHBrCHR'(OEt).

d. The resulting beta bromoether reacts with zinc with removal of the bromine and the alkoxyl group. This is like the action of Zn with a 1,2-dibromide. It leaves an olefin, determined by the aldehyde and Grignard reagent used, RCH=CHR'.

Examples of syntheses of heptylenes[121] illustrating the five possible types of substituted olefins are given below. The aldehyde used in *Step a* of the Boord Synthesis, and the Grignard Reagent required in *Step c* are indicated.

Type I.

	Aldehyde	Grignard R.
1-Heptene, AmCH=CH$_2$, b. 94.9°	MeCHO	Am

A Type I olefin containing a tertiary group such as 3,3-Me$_2$-1-pentene, Me$_2$EtCCH=CH$_2$, b. 76.9°, cannot be made by this method but has been made by Tschugaeff's xanthate method from 3,3-Me$_2$-2-pentanol.[122]

Type II.

2-Heptene, BuCH=CHMe, b. 98.3°	EtCHO	Bu
	AmCHO	Me

A tertiary group causes trouble due to rearrangement during dehydration, as in the preparation of 4,4-Me$_2$-2-pentene, Me$_3$CCH=CHMe, b. 76°, which is made from 2,2-Me$_2$-3-pentanol through the xanthate.

Type III.

2-Me-1-hexene, MeBuC=CH$_2$, b. 91.3°	MeCHO	Me and Bu

For this type, the Grignard Reagent must be used twice. The beta bromoether from *Step c* is treated with KOH to give an alpha alkyl-vinyl ether, CH$_2$=C(R)OEt which is then converted to the dibromide, BrCH$_2$CBr(R)OEt. This is then treated with the second Grignard reagent to form

$$BrCH_2CR'R(OEt)$$

The removal of the Br and OEt takes place by Zn in *Step d.*

2,3,3-Me$_3$-1-butene, Triptene, (Me$_3$C)MeC=CH$_2$, b. 80°, is formed by dehydrating Me$_5$-ethanol. No rearrangement is possible. It is also obtained by alkylating Me$_3$- or Me$_4$-ethylene with MeCl and lime.[123]

Type IV.

2-Me-2-hexene, Me$_2$C=CHPr, b. 94.5°	BuCHO	Me

[121] Soday, Boord. *J. Am. Chem. Soc.* **55**, 3293 (1933).
[122] Schurman, Boord. *J. Am. Chem. Soc.* **55**, 4930 (1933).
[123] Miller, Lovell. *Ind. Eng. Chem.* **40**, 1138 (1948).

In this type, the higher beta bromoether, obtained in *Step c*, is converted to the $\alpha\beta$-unsaturated ether. A small by-product of $\beta\gamma$-unsaturated ether can be removed by distillation. The $\alpha\beta$-product is then converted to the dibromide which is treated with a second mol of Grignard reagent. Zinc then gives the olefin.

Type V.

2,3-Me₂-2-pentene, MeEtC=CMe₂, b. 91°, can be made by dehydrating dimethyl-*s*-butylcarbinol.

Octylenes, C_8H_{16}. All of the possible 66 structural isomers have been prepared. The last three were 5,5-Me₂-2-hexene, 2,4-Me₂-3-hexene, and 4-Me-3-Et-1-pentene.[124]

The best known octylenes are the two diisobutylenes, b. 101.2° and 104.5°. They are 2,4,4-Me₃-1-pentene and 2,4,4-Me₃-2-pentene.[125] As usual, the 1-isomer has the lower boiling point. They are formed in 4:1 ratio from isobutylene or *t*-BuOH. Either isomer is rapidly converted to the equilibrium mixture by acids, and by silica at 25°, but not at −20.[126] Their hydrogenation gives 2,2,4-Me₃-pentane ("iso-octane"), the anti-knock standard used with *n*-heptane in rating gasolines (p. 24). HCl and HBr add readily to give 2,2,4-Me₃-4-halogen-pentanes. Chlorine and bromine give unsaturated halides instead of the dihalides. Oxidation of mixed diisobutylenes with dichromate and acid gives a fair yield of methyl neopentyl ketone, MeCOCH₂CMe₃. Other products are Me₃-acetic acid and methylneopentylacetic acid. The primary oxidation products of the diisobutylenes with CrO_3 and Ac₂O are the corresponding epoxides, accompanied by smaller amounts of methyl neopentyl ketone, methyl neopentylacetaldehyde and the glycols. The glycol of the 2-isomer undergoes the pinacol rearrangement thus:[127]

Similar type products are formed with perbenzoic and peracetic acid.[128]

In the Oxo Process the diisobutylenes, unlike most olefins, react with carbon monoxide to produce only one isomer, 3,5,5-trimethylhexanal. From this reaction a number of pure highly branched aliphatic compounds have become commercially available[129] (Oxo Process, p. 420).

The xanthate reaction gives two of the otherwise difficultly obtainable octylenes, 3,3-Me₂-4-hexene, b. 106°; and 2,2-Me₂-3-hexene, b. 100.1°.[130]

[124] Whitmore, Cook, Whitmore. *J. Am. Chem. Soc.* 72, 51 (1950).
[125] Whitmore et al. *J. Am. Chem. Soc.* 54, 3706 (1932).
[126] Gallaway, Murray. *J. Am. Chem. Soc.* 70, 2584 (1948).
[127] Byers, Hickinbottom. *Nature* 160, 402 (1947).
[128] Byers, Hickinbottom. *J. Chem. Soc.* 1948, 284.
[129] *Chem. Eng. News* 27, 245 (1948).
[130] Schurman, Boord. *J. Am. Chem. Soc.* 55, 4930 (1933).

Highly branched octylenes: 1,1,-diisopropylethylene, b. 103°; 2,3,3-Me$_3$-1-pentene, b. 108°; 3,4,4-Me$_3$-2-pentene, b. 112°; and 2,3,4-Me$_3$-2-pentene, b. 115°. The last two are obtained from the co-polymerization of the olefins formed by treating a mixture of *t-* and *s*-butyl alcohols with excess of 75% sulfuric acid.[131] They are also important constituents of the "co-dimer" obtained by polymerization of cracked gases from petroleum. The last octylene is probably formed by successive 1,2-rearrangements.

Dodecylenes and Higher Olefins. The best known are the *triisobutylenes* which consist mainly of 2,2,4,6,6-Me$_5$-3-heptene and 2-neopentyl-4,4-Me$_2$-1-pentene. The latter can be obtained pure by oxidizing the mixture with insufficient chromic acid.[132] The 1-olefin is less readily oxidized. About 10% of the triisobutylene mixture consists of 2,4,4,6,6-Me$_5$-1-heptene and its 2-isomer.[133]

Of the multitude of known higher olefins, two pairs of symmetrical and unsymmetrical substituted ethylenes show interesting effects of changes in structure. The 1,1- and 1,2-di-*t*-Bu-ethylenes have respectively b. p. and n^{20}D of: 149.5°, 1.4364; 123.6°/734, 1.4114. The latter decylene is to be compared with the highest boiling octylene, *cis*-2-octene with b. 124.6°/750 and n^{20}D 1.4139. The effect of symmetry, compactness and "exposure" of the double bond in the two decylenes is worth study. The comparable pair of dodecylenes having neopentyl in place of *t*-butyl show no such effects. 1,1- and 1,2-dineopentylethylenes have respectively: b. 178° and 179.6°/732; and n^{20}D 1.4293 and 1.4265 (supercooled liquid, m. 23.1°). The high melting point of the symmetrical compound is notable for an olefin of this MW.

2,2,3,5,5,6,6-Me$_7$-heptane, *ditriptene*, b. 132°/80, n^{20}D 1.4510, is obtained in good yield by the dimerization of triptene.[134]

Using cobalt catalysts, olefins can be made to react with carbon monoxide and hydrogen to produce aldehydes and alcohols. This is the so-called *Oxo* reaction (p. 420) and is of considerable commercial importance.

Olefins can be made by heating the palmitic or stearic esters of the higher alcohols under suitably diminished pressure.[135]

Higher normal olefins, cetene (C$_{16}$), cerotene (C$_{26}$) and melene (C$_{30}$) have been made from cetyl alcohol, chinese wax and beeswax respectively. Cetene has been used as a standard for Diesel fuel. In this type of engine it gives no "knock" (pp. 10, 25). It has now been replaced by the more stable cetane.

Mixed higher alkenes obtained by cracking a petroleum rich in alkanes have been polymerized to give an artificial lubricating oil (Syntholube).

Olefin mixtures obtained by cracking processes may be converted to special resins.

[131] Whitmore, Laughlin et al. *J. Am. Chem. Soc.* **63**, 756 (1941).
[132] Whitmore, Surmatis. *J. Am. Chem. Soc.* **63**, 2200 (1941).
[133] Whitmore et al. *J. Am. Chem. Soc.* **63**, 2035 (1941).
[134] Whitmore, Ropp, Cook. *J. Am. Chem. Soc.* **72**, 1507 (1950).
[135] Krafft. *Ber.* **16**, 3018 (1883).

Determination of the Structure of Olefins. This may be done by the action of ozone to give ozonides. This process is little understood. The decomposition of the ozonides (often dangerously explosive) gives carbonyl compounds (ketones, aldehydes, acids) related to the two parts of the molecule held together by the double bond.[136-137] As applied to a pure olefin, ozonolysis may prove a good qualitative test provided reaction takes place under the conditions used. With mixtures of olefins, it is likely to be treacherous in the extreme since different olefins react with widely different speeds and some are inactive by any ordinary treatment. The recovery of the products is rarely as much as 50%. The only satisfactory method for identifying the members of a mixture of olefins involves the following steps: 1. Separate the mixture by fractionation in an inert atmosphere through distilling columns of at least 100 theoretical plates using a reflux ratio of the same order (100:1) until fractions are obtained which consist mainly of one olefin with admixtures of not over two others; 2. Synthesize and purify each of the suspected olefins; 3. Take the infrared (or/and the Raman) spectra of these pure known olefins; 4. Study the spectra of the distillation fractions in comparison with the known spectra. An example of this method is the identification of the six octylenes from the dehydration of t-butylisopropylcarbinol (p. 131).

The auto-oxidation of olefins is of great practical importance because of gum formation from olefins in cracked gasolines.[138]

Mechanism of Olefin Polymerization

The increasing theoretical and practical importance of polymerization warrants a discussion of the probable course of the conversion of isobutylene to its dimers and trimers. This is a good example because the ease of the reaction and the absence of rearrangement make it simpler than the polymerization of less reactive monomers. It should be emphasized that the following processes are reversible to varied degrees. The final products are those less reactive under the experimental conditions.

$$Me_2C=CH_2 + H^+ \rightleftharpoons Me_3C^+ \qquad (A)$$

$$Me_2C=CH_2 + (A) \rightleftharpoons Me_2\overset{+}{C}-CH_2CMe_3 \qquad (B)$$

A proton from the acid catalyst adds to the extra electron pair of the double bond in isobutylene according to Markownikoff's Rule. This gives the t-butyl carbonium ion (A). This may unite more or less firmly with the catalyst, may revert to isobutylene, or may add to a molecule of isobutylene as indicated to give the larger carbonium ion (B). It is important to note that this addition is like that of the proton in the first step.

The carbonium ion (B) cannot exist as such but may depolymerize, may unite with the catalyst, or may stabilize itself by the attraction of an electron

[136] Harries. *Ber.* **37**, 839 (1904).
[137] Long. *Chem. Rev.* **27**, 437 (1940).
[138] Milas. *Chem. Rev.* **10**, 295 (1932).

pair from a carbon atom adjacent to the electronically deficient carbon (C⁺) *without its proton*. This establishes a double bond involving the formerly deficient atom. Thus a proton is expelled to the catalyst or attracted by the catalyst. If this takes place with one of the methyl groups, the product is 2,4,4-Me₃-1-pentene (C), the low-boiling diisobutylene. If the methylene group is involved, the product is 2,4,4-Me₃-2-pentene (D).

$$H^+ + CH_2=C-CH_2CMe_3 \rightleftharpoons (B) \rightleftharpoons Me_2C=CHMe_3 + H^+$$

Me

(C)   (D)

At this point the mixture contains all of the above olefins and carbonium ions, the latter probably in loose combination with the catalyst. Reactions, similar to those outlined, continue but less rapidly because of the greater size and consequent inertness of the molecules and carbonium ions involved. The addition of protons to isobutylene or to the diisobutylenes (C) and (D) give the carbonium ions (A) and (B). A readily predictable reaction would involve the addition of a *t*-Bu carbonium ion to (C) which resembles isobutylene in its reactive double bond structure.

$$(A) + (C) \rightleftharpoons Me_3CCH_2 - \overset{+}{C} - CH_2CMe_3 \quad (E)$$

Me

The new carbonium ion (E) could stabilize its electronically deficient carbon by the expulsion of a proton from the lone Me group or from one of the adjacent methylene groups. The first would give 1,1-dineopentylethylene, and the second would give 2,2,4,6,6-Me₅-3-heptene. These two dodecylenes account for about 90% of the triisobutylenes. The addition of a *t*-Bu to (D) is understandably difficult because of the "crowded" nature of the double bond in the latter. No product of such an addition has been found. Another reason for the absence of this addition is that (C) and (D) occur in the ratio 4:1. Moreover (D) is convertible to (C) through the carbonium ion (B).

The other 10% of the triisobutylenes is accounted for by the addition of the large carbonium ion (B) to isobutylene.

Me   Me

$$Me_2C=CH_2 \;+\; {}^+C-CH_2CMe_3 \rightleftharpoons Me_2\overset{+}{C}-CH_2-C-CH_2CMe_3$$

Me   Me

This carbonium ion expels a proton from one of the adjacent methyl groups to give

Me

$$CH_2=C-CH_2-C-CH_2CMe_3$$

Me   Me

Loss of a proton from the adjacent methylene group gives the 2-isomer. These two dodecylenes constitute the 10% high boiling material in the triisobutylenes.

The *dodecylenes* formed by dimerizing Me₄-ethylene illustrate the intimate relation between rearrangement and polymerization by acid catalysts or their equivalent. They also show the powerful effect of even slight changes of conditions in such processes. Thus, with 84% sulfuric acid at 0° a 70% yield of dimers is obtained. These all involve rearrangement either before or during dimerization or both. The products identified include seven Me₅-3-heptenes with the methyl groups in the positions 22366, 22466, 23466, 22456, 22556, 22356, and 22566. The first forms 30% of the dimer. On the other hand 77–80% acid gives mainly the 22356 compound which is formed only in small amount with the slightly stronger acid.[139]

Co-polymerization of two different olefins gives added complications but apparently follows similar courses. Thus the addition of *t*-butyl alcohol to a solution of *s*-butyl alcohol in 75% sulfuric acid gives a 75% yield of nearly equal amounts of 3,4-Me₃-2-pentene and 2,3,4-Me₃-2-pentene (p. 47). The former is the expected product from the addition of a *t*-butyl carbonium ion to 2-butene followed by loss of a proton to give the olefin. The latter appears to have been formed by a 1,3-shift of a methyl group, but probably is the result of successive 1,2-shifts.

Mechanism of Alkylation

In the petroleum industry alkylation may be defined as the union of a tertiary hydrocarbon with a reactive olefin to form a saturated paraffin either without a catalyst or with a catalyst such as conc. sulfuric acid. Without a catalyst, temperatures are necessary which are high enough to give free radicals which may add to the olefin in the ordinary way. (Free radical polymerization, p. 52). With conc. sulfuric acid and similar catalysts, the process bears certain similarities to the dimerization of olefins. The same type of rearrangements take place. It is more than probable that carbonium ions are involved. A key to the problem may be the long observed but usually neglected fact that dimerization of olefins always gives some saturated paraffins corresponding to the olefins used. It thus appears possible for the following exchange to occur between carbonium ions and paraffins:

$$R^+ + R'H \rightleftharpoons RH + R'^+$$

A tertiary H takes part in this reaction probably because a tertiary carbonium ion is more stable than the secondary and primary structures. The change

---

[139] Whitmore. *Chem. Eng. News* **26**, 668 (1948).

actually involves an *intermolecular* form of the common intramolecular change which takes place during rearrangements, in that a carbon having only 6 electrons appropriates a needed electron pair with its attached proton.[140] An electron pair with its proton $(:H)$ is most readily removed from a tertiary grouping as is shown by the ready formation of the tertiary chlorides from the grouping $R_2CHCHOH-$ instead of the expected secondary chlorides (pp. 83, 123, 129). There is evidence that an H which is alpha to a double bond may also change R^+ to RH. The olefin system thereby becomes

$$-C-\overset{+}{C}-C=C-$$

which can lose a proton to form the conjugated diene $-C=C-C=C-$. This would account for the accumulation of highly unsaturated material in the acid layer during alkylations.

A typical alkylation is the action of a large local excess of isobutane with isobutylene in presence of conc. sulfuric acid. The first step may be the formation of a *t*-Bu carbonium ion from isobutylene and the catalyst. This is surrounded by perhaps 1000 isobutane molecules. Its reaction with these is futile because R and R' are identical. When it meets an isobutylene molecule, however, it can unite to give the larger carbonium ion, $Me_3CCH_2\overset{+}{C}Me_2$, which can react with one of the surrounding isobutane molecules to form 2,2,4-Me$_3$-pentane and a *t*-Bu carbonium ion which continues the chain. Before the octyl carbonium ion undergoes this change it may undergo internal rearrangement. One resulting carbonium ion would be $Me_2\overset{+}{C}CHMeCHMe_2$, which would react with isobutane to form 2,3,4-Me$_3$-pentane and a *t*-Bu ion to carry on the chain. When mixtures of C$_4$ and C$_5$ olefins and paraffins are used, the predicted C$_8$, C$_9$, and C$_{10}$ paraffins correspond to those obtained insofar as these have been identified.

Sometimes rearrangement is necessary *before* alkylation can take place. Thus *n*-butane or cyclohexane will alkylate olefins in presence of AlCl$_3$ which is capable of isomerizing them to isobutane and methylcyclopentane respectively but not with H_2SO_4 or BF$_3$ which cannot bring about this preliminary isomerization to a tertiary RH. Cyclopentane will not alkylate olefins even with AlCl$_3$, presumably because it cannot be isomerized to a tertiary hydrocarbon.[141-147]

[140] Whitmore. *J. Am. Chem. Soc.* **54,** 3274 (1932).
[141] Birch et al. *Ind. Eng. Chem.* **31,** 884 (1939).
[142] Caesar, Francis. *Ind. Eng. Chem.* **33,** 1426 (1941).
[143] Bartlett, Condon, Schneider. *J. Am. Chem. Soc.* **66,** 1531 (1944).
[144] Egloff, Hulla. *Chem. Rev.* **37,** 323 (1945).
[145] Schmerling. *J. Am. Chem. Soc.* **67,** 1778 (1945).
[146] Ciapetta. *Ind. Eng. Chem.* **37,** 1210 (1945).
[147] Gorin, Kuhn, Miles. *Ind. Eng. Chem.* **38,** 795 (1946).

Free Radical Polymerization

Free radical polymerization offers some interesting similarities and differences with acid or carbonium ion polymerization. The driving force of these reactions lies in the *single* electron deficiency of an atom of the free radical. An initiating radical or atom attaches itself to one of the olefinic carbons, usually the more exposed one, of an unsaturated molecule by capturing one of the π electrons. The odd electron left on the adjacent carbon creates a new deficiency which reacts similarly with another unsaturated molecule to give a still larger free radical. The molecule continues to increase by this chain type reaction until it finds some way of satisfying its deficiency without creating another in the same molecule.

The principal steps in radical polymerization are initiation, propagation and cessation.

$$R \cdot + M \rightarrow RM \cdot \underset{\longrightarrow}{\rightleftharpoons} RM_i \rightarrow RM_n$$

(Initiation)   (Propagation)   (Cessation)

Initiation: Some of the compounds which enter into radical polymerization are styrene, butadiene, acrylates, vinyl chloride and acetate, vinylidine chloride, allyl esters and vinyl pyridine. Catalysts for the initiation reactions include acyl peroxide, diazo compounds, alkyl peroxides, radicals from photolysis of aldehydes and ketones, free metals and metal from decomposition of metal alkyls.

By labeling the initiating radicals it has been possible to verify their presence in many polymer samples and thus prove their role as initiators. Selective reactivity of radicals can be studied in this same manner. Nitrosoanilides, which presumably generate an acyl and acyloxy radical simultaneously always introduce only the acyl fragment into the polymer molecules indicating that the acyloxy radical is far less effective as an initiating radical.

It has been almost universally observed that the reactions involving catalysts of the above type proceed at a rate proportional to the square root of the catalyst concentration. This is in agreement with the observed first order reaction rate for the decomposition of acyl peroxides and similar radical catalysts.

Propagation: The most important factor to be considered in the growth of polymer chains is the mode of addition. There is little evidence for any other type of addition except head to tail, which may be expected for several reasons: (1) The stability of free radicals is in the order tertiary > secondary > primary. (2) Sterically a deficient carbon can attack a terminal position more easily than an inside position. (3) The electrons are more apt to be in the vicinity of the terminal carbon than the inside carbon.

The symmetrical growth of the molecule in free radical polymerization is not disturbed by intramolecular rearrangement (as in ionic polymerization) but may be affected by further reaction of functional groups along the chain, either within the molecule or with another polymer molecule. Since polystyrene and other polymers can easily be deprived of their tertiary hydrogen atoms by other radicals, branched polymers result from chain transfer reactions:

$$-CH_2-\underset{\underset{\phi}{|}}{\overset{\overset{H}{|}}{C}}-CH_2- \ + \ R\cdot \ \longrightarrow \ R{:}H \ + \ -CH_2\underset{\underset{\phi}{|}}{\overset{\cdot}{C}}-CH_2- \ \xrightarrow{\phi CH=CH_2}$$

$$\phi-\overset{\cdot}{CH}$$
$$|$$
$$CH_2$$
$$|$$
$$-CH_2-\underset{\underset{\phi}{|}}{C}-CH_2-$$

In contrast to most ionic reactions, free radical reactions are seldom reversible. This might be the result of some of the following factors: (1) Exhaustion of the catalyst due to its entering into the reaction. (2) Frequent absence of weak spots for attack in products of the reaction. (3) The inability of a carbon, deficient in only one electron, to satisfy its needs by getting them from an adjacent carbon to carbon linkage. This of course means that the molecule has easier ways of getting an electron, such as disproportionation or further propagation.

Under drastic conditions however some polymers can be forced into equilibrium reactions. Polystyrene at high temperature and with free radical catalysts reaches an equilibrium viscosity. Methyl methacrylate and methacrylate polymers readily yield monomer on thermal cracking.

Cessation: There are a number of ways in which an individual polymer chain may become stabilized. Reaction with another radical to couple or disproportinate may be regarded as normal cessation.

$$2\,R\cdot \ \rightarrow \ R{:}R \quad (coupling)$$

$$2\,RCH_2\underset{\underset{X}{|}}{CH}\cdot \ \rightarrow \ RCH_2\underset{\underset{X}{|}}{CH_2} + RCH=\underset{\underset{X}{|}}{CH} \quad (disproportionation)$$

The latter seems to be more favored when it can occur, i.e., when the radical has an active hydrogen atom on the beta carbon. Disproportionation of free radicals is similar to olefin formation in ionic polymerization. The ionic polymer loses a proton to its environment (acid medium) and produces a double bond, whereas the free radical loses a hydrogen atom to its environment (other free radicals) and likewise produces a double bond.

It is possible to control free radical polymerization to an unusual extent by limiting the number of initiating reactions, selecting molecules which do not readily undergo side reactions such as disproportionation, and making proper choice of concentration, solvent, inhibitors, chain stoppers and transfer reagents. The use of monomer mixtures adds still another variation to these reactions, which with their products have received a tremendous amount of investigation and commercial application in the last twenty years. They form the basis of the plastic industry.[148]

C. Unsaturated Hydrocarbons, C_nH_{2n-2}

1. Diolefins, Dienes

These may be classified according to the relative position of the two olefinic linkages: (a) Allene, $CH_2=C=CH_2$, and its homologs, (b) 1,3-butadiene, $CH_2=CH-CH=CH_2$, in which the double bonds are conjugated, and its homologs,[1] and (c) compounds in which the double bonds are separated by at least two single linkages as in 1,4-pentadiene, $CH_2=CHCH_2CH=CH_2$.

Allene, propadiene, dimethylenemethane, $CH_2=C=CH_2$, b. $-32°$, is best made by the action of zinc dust on 2,3-dichloropropene, obtained by the action of KOH and a trace of water on 1,2,3-Cl_3-propane from the addition of chlorine to allyl chloride (Shell). This is better than the ordinary preparation from the corresponding bromo compounds. It is also among the products of the electrolysis of salts of itaconic acid. Allene polymerizes readily at 400–600° to give complex liquid mixtures containing aromatic compounds.[2] Sulfuric acid converts allene to acetone, the addition of a molecule of water producing the enol form of acetone, $CH_2=C(OH)CH_3$. Allene with excess bromine gives $BrCH_2CBr_2CH_2Br$, a liquid of camphor odor. Hypochlorous acid gives $ClCH_2COCH_2Cl$, $ClCH_2COCH_2OH$, and $CH_2=CClCH_2OH$. Alcoholic KOH gives $CH_2=C(CH_3)OC_2H_5$, ethyl isopropenyl ether. Such an addition does not ordinarily take place with a $C=C$ bond unless it is alpha-beta to a $C=O$ bond. Thus the grouping $C=C-C=O$ bears a certain resemblance to the $C=C=C$ grouping. Allene reacts with Na in ether to give sodium methylacetylide, $CH_3C\equiv CNa$.

A consideration of the spatial relations of an allene of the type

$$RR'C=C=CRR'$$

will show that it could exist in enantiomorphic and, therefore, optically active forms. Mills has obtained optically active allenes of this type by dehydrating the related tertiary alcohols by d-camphorsulfonic acid. The levo acid gives the enantiomer.

Homologs of allene of the type $RCH=C=CH_2$ are prepared as follows:[3] acrolein, $CH_2=CHCHO$, is treated with RMgX to give the substituted allyl alcohol, $RCHOHCH=CH_2$; with PBr_3 this largely undergoes an allylic rearrangement to give $RCH=CH-CH_2Br$; addition of bromine followed by

[148] Price. "Reactions at the Carbon Carbon Double Bond." Interscience Publishers, 1947.

[1] *Ann. Rep. Chem. Soc.* (London) 1916, 79; 1917, 130; 1922, 102.

[2] Meinert, Hurd. *J. Am. Chem. Soc.* 52, 4540 (1930).

[3] Bouis. *Ann. chim.* 9, 402 (1928).

fusion with 75% aqueous KOH gives $RCHBrCBr=CH_2$, which reacts with zinc dust to give the allene. 1,2-Pentadiene can be made by these five steps with an overall yield of 21% from acrolein. The simpler members of the series boil as follows: Me-allene (1,2-butadiene) 19°, Et 45°, n-Pr 78°, isoPr-allene (4-Me-1,2-pentadiene) 70°, n-Bu 106°, isoBu 96°, t-Bu 82°. Addition of bromine (1 mol) is mainly in the 2,3 position. Cold sulfuric acid gives methyl ketones.

Allene homologs of the type $RR'C=C=CH_2$ can be made from ketones, $RR'CHCOMe$, by treatment with PCl_5 and then with alcoholic KOH.

sym-Me₂-allene, 2,3-pentadiene, b. 49°/735, is made from butyl chloral with MeMgX to give $3,3,4-Cl_3$-2-pentanol which is treated with PCl_5 to give $2,3,3,4-Cl_4$-pentane. With zinc this gives the allene in 15% yield for the three steps.

Me₄-allene, b. 70°, is made by KOH with $2,4-Me_2-3,3-Cl_2$-pentane.

All except the tetrasubstituted allenes react on heating with sodamide or sodium to form $RC\equiv CNa$ by migration of the unsaturation to the end of the chain.[4]

1,3-Butadiene, divinyl, vinylethylene, erythrene, pyrrolylene,

$$CH_2=CH-CH=CH_2, \text{ b. } 1°.$$

Beginning in 1941 butadiene became very important in the making of synthetic rubber. Two chief sources were developed. The cracked gases of the petroleum industry contain in the C_4 fraction a small amount of butadiene which can be separated either by means of the solid compound with cuprous chloride or by solvent extraction with furfural or other suitable solvents. The n-butenes and n-butane in the gases can be cracked to give more butadiene. From ethyl alcohol, two routes are available. A direct catalytic process gives a fair yield of butadiene as originally found by Ipatieff.[5] Acetaldehyde from ethyl alcohol or from acetylene can be condensed to acetaldol which can be reduced catalytically to the 1,3-glycol ("1,3-butylene glycol"). Dehydration gives a good yield of butadiene. A third process involves a special fermentation to give 2,3-butylene glycol. Dehydration produces methyl ethyl ketone by a rearrangement like that of pinacol (p. 210). Butadiene can be obtained by cracking the diacetate of the glycol much as t-butylethylene is made from the acetate of pinacolyl alcohol (p. 43).

In German wartime synthetic rubber manufacture, the aldol process was used starting with acetylene from carbide or methane (p. 65). A new process involved the reaction of acetylene with 15% formaldehyde in presence of a Cu-Bi-silica catalyst at 5 atm. and 100° to give butynediol in good yield. Hydrogenation and dehydration gave butadiene.

[4] Faworsky. *J. prakt. Chem.* (2) **37**, 417 (1888).
[5] Ipatieff. *J. prakt. Chem.* (2), **67**, 420 (1903).

Butadiene.[6–9]

Chlorine adds to butadiene to give a 1,2- and a 1,4-dichloride and the two stereoisomeric tetrachlorides. The boiling points at 40 mm. are 45°, 76°, 111° and 132° (m. 72°) respectively. The 1,2-dichloride does not rearrange to the 1,4-dichloride as is possible with the dibromide. Thus the formation of the 1,4-compound does not take place by rearrangement of the 1,2-compound but is due to an allylic shift of the double bond during the reaction.[10]

Butadiene gives a 1,2-dibromide, $CH_2=CHCHBrCH_2Br$, b. 52°/10 and two isomeric 1,4-dibromides, $BrCH_2CH=CHCH_2Br$, one m. 54° and the other a liquid b. 59°/13. The solid is the *trans* form since it gives racemic erythritol.[11] This 1,4- addition was explained by Thiele (1899) on the basis of his famous theory of *partial valences*. This assumed the unsaturation of the olefinic linkage to be made up of partial valences. With double bonds in the 1,3 position the neighboring partial valences could satisfy each other, thus leaving the partial valences in the 1,4 positions as the most unsaturated points in the molecule.

$$CH_2 \dashrm{---} CH \dashrm{---} CH \dashrm{---} CH_2$$

The present tendency is to explain the reactions of conjugated dienes by the mechanism involving stepwise addition with or without allylic rearrangement of the intermediate carbonium ion.

$$CH_2=CH-CH=CH_2 \rightleftharpoons XCH_2-\overset{+}{C}H-CH=CH_2 \leftrightarrow XCH_2-CH=CH-\overset{+}{C}H_2$$

$$XCH_2-CHX-CH=CH_2 \qquad XCH_2-CH=CH-CH_2X$$

1,3-Butadiene with excess bromine gives two tetrabromides, one difficultly soluble in petroleum ether and melting at 119° and the other more soluble, m. 39°. These are stereoisomers of the same type as the tartaric acids, one being meso and the other racemic. In this case it is not possible to resolve the latter into optically active forms because of the lack of a suitable group for compound formation with an optically active reagent.

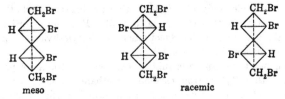

meso   racemic

[6] Grosse, Morrell, Mavity. *Ind. Eng. Chem.* **32**, 309 (1940).
[7] Lebedev et al. *Oil Gas J.* **41** No. 42, 76 (1943).
[8] Frolich, Morrell. *Chem. Eng. News* **21**, 1138 (1943).
[9] Egloff, Hulla. *Chem. Rev.* **35**, 279 (1944).
[10] Muskat, Northrup. *J. Am. Chem. Soc.* **52**, 4043 (1930).
[11] Ellis. "Chemistry of Petroleum Derivatives." Reinhold, 1934. p. 649.

Butadiene adds ICl in CH_2Cl_2 at $-35°$ to give the 1,4- and 1,2-products in the ratio 4:1.[12] It adds HCl gas at $-80°$ to give 1,2- and 1,4-addition in about 8:2 ratio.[13] Treatment of butadiene with conc. HCl and $ZnCl_2$ gives a 75% yield of addition products with the 1,4-predominating about 3:2.[14]

It adds ammonia, 1:4, to give tributenylamine,

$$(MeCH = CHCH_2)_3N.$$

Like other conjugated dienes it reacts with sulfur dioxide to give a cyclic sulfone,

It adds one or two mols of diazomethane, CH_2N_2, to give 5-vinylpyrazoline and 5,5'-dipyrazoline. In this case 1,2-addition gives 5-membered rings whereas the more usual 1,4-addition would give a 7-membered ring.

1-Bromobutadiene adds Br_2, HBr, and HOBr in the 1,4-position.[15]

An interesting 1,4-addition of 1,3-butadiene and other conjugated diolefins is the *Diels-Alder reaction*[16,17] with maleic anhydride and other conjugated ($\alpha\beta$) unsaturated carbonyl compounds.

Anhydride of 1,2,3,6-tetrahydro-
phthalic acid

The three simplest conjugated dienes can be identified by the Diels-Alder reaction with maleic anhydride to give anhydrides of tetrahydrophthalic acid and its homologs.[18] The mps. of the anhydrides from butadiene, isoprene, and piperylene are respectively 104°, 64°, 62°.

Chloromaleic anhydride and conjugated dienes form an adduct that can be used for diene analysis. The highly reactive chlorine of the adduct can be titrated without affecting the vinyl chlorine of the reagent.[19]

[12] Ingold, Smith. *J. Chem. Soc.* **1931**, 2752.
[13] Kharasch, Kritchevsky, Mayo. *J. Org. Chem.* **2**, 489 (1937).
[14] Stehman, Cook, Whitmore. *J. Am. Chem. Soc.* **72** (1950).
[15] Muskat, Grimsley. *J. Am. Chem. Soc.* **55**, 2140 (1933).
[16] Diels, Adler. *Ann.* **460**, 98 (1928).
[17] Norton. *Chem. Rev.* **31**, 319 (1942).
[18] Birch, Scott. *Ind. Eng. Chem.* **24**, 49 (1932).
[19] Putnam, Moss, Hall. *Ind. Eng. Chem., Anal. Ed.* **18**, 628 (1946).

Butadiene can also add to certain cyclic olefinic bonds which are not $\alpha\beta$ to carbonyl groups. Thus, with 3,6-endomethylene-1,2,3,6-tetrahydrophthalic anhydride obtained from cyclopentadiene and maleic anhydride, it gives the anhydride of 1,4-endomethylene-Δ6,7-octahydronaphthalene-2,3-dicarboxylic acid.

Similar reactions take place with such widely different compounds as acrolein and benzoquinone giving tetrahydrobenzaldehyde, tetrahydro-α-naphthoquinone and octahydroanthraquinone depending on whether one or two mols of the diene react with the quinone.[20-22]

"Die Methoden der Dien-Synthese," Alder, Berlin, 1933.

The Diels-Alder reaction gives a good method for separating dienes from other materials and identifying them since the compounds with maleic anhydride are well crystallized and have characteristic melting points.[23]

1,3-Butadiene polymerizes to a rubber-like material. Before 1939 German chemists had developed two successful synthetic rubbers by co-polymerization of butadiene with styrene (Buna-S) and with acrylonitrile (Buna-N, Perbunan). After 1941 these and related synthetic rubbers became of world-wide importance. For the production of synthetic rubber, butadiene of very high purity is necessary. Minute traces of certain impurities such as acetylenic compounds inhibit the polymerization. The co-polymerization is carried out under moderate pressure in glass lined equipment by agitating an emulsion of the components prepared in a soap or other suitable medium with catalysts such as organic peroxides. The result is an artificial latex which is then treated much like *hevea latex.*

The American rubber industry has made important advances in these types of rubber and have added new types. The chemical literature from 1940 to 1946 is literally full of articles on synthetic rubber and related *elastomers.*

The sodium salts of methyl *n*-alkylcarbinols and sodium alkenes form a complex called *alfin* catalysts that are extremely active polymerization reagents for dienes.[24]

[20] Ellis. "Chemistry of Petroleum Derivatives." Reinhold, 1934. p. 650.
[21] *Ann. Rep. Chem. Soc.* (London) 1930, 88.
[22] Allen. *J. Chem. Education* 10, 494 (1933).
[23] Ellis. "Chemistry of Petroleum Derivatives." Reinhold, 1934. p. 660.
[24] Morton, Magat, Letsinger. *J. Am. Chem. Soc.* 69, 950 (1947).

Traces of butadiene in gasoline cause gum formation which clogs the carburetor.

One type of polymerization of dienes results in the formation of cyclic compounds and probably resembles the Diels-Alder reaction in being a 1,4-addition of the diene to a double bond in another molecule.[25]

$$CH-CH \qquad\qquad CH=CH$$
$$CH_2 \diagup\diagdown CH_2 \qquad\rightarrow\qquad CH_2 \diagup\diagdown CH_2$$
$$CH_2=CH-CH=CH_2 \qquad\qquad CH_2-CH-CH=CH_2$$

Butadiene slowly forms dangerously explosive peroxides. These can be destroyed safely by strong NaOH.[26]

Isoprene, 2-Methyl-1,3-butadiene, b. 37°, until the development of butadiene for synthetic rubber, was the best known diene.[27] It can be made by the destructive distillation of natural rubber (caoutchouc). Because it can be polymerized readily to rubber, many processes have been evolved for its preparation. The synthesis of rubber by these processes did not become commercial partly because their threat to the natural rubber industry stimulated the improvement in the growing of plantation rubber to the point where it has sold as low as five cents per pound. During World War II the success of the copolymers of butadiene relegated isoprene to a position as an additive in such products. For such use it is made by catalytic dehydrogenation of isopentane and isoamylenes. It can also be obtained in about 60% yield by cracking *dl*-limonene (dipentene) from turpentine.[28]

During the years of interest in isoprene as a possible source of synthetic rubber, its synthesis was worked out from a variety of materials such as beta-methylpyrrolidine (by exhaustive methylation), Me_3-ethylene (by dehydrohalogenation of its dichloride,[29] isoamyl chloride (by chlorination and removal of 2 HCl), and methyl ethyl ketone (by the Mannich reaction with CH_2O and Me_2NH, followed by exhaustive methylation).

Isoprene adds bromine to give a 1,4-dibromide and a tetrabromide. With 2 HBr it gives 2-Me-2,4-dibromobutane.[30] The addition of 1 HBr gives an unstable tertiary bromide (1,2-addition) which changes on standing to a primary bromide (the 1,4-product), 2-Me-4-bromo-2-butene. On hydrolysis, this undergoes an allylic rearrangement to dimethylvinylcarbinol ("isoprene alcohol").[31] With bromine water the chief product is 1-bromo-2-Me-2-buten-4-ol.

[25] Vaughan. *J. Am. Chem. Soc.* **54,** 3863 (1932).
[26] Scott. *Am. Chem. Soc., News Ed.* **18,** 404 (1940).
[27] Ellis. "Chemistry of Petroleum Derivatives." Reinhold, 1934. pp. 153–635.
[28] Davis, Goldblatt, Palkin. *Ind. Eng. Chem.* **38,** 53 (1946).
[29] Friedländer. "Fortschritte der Teerfarbenfabrikation," Vol. II. Springer, Berlin. p. 804.
[30] Ellis. "Chemistry of Petroleum Derivatives." Reinhold, 1934. p. 647.
[31] *Ann. Rep. Chem. Soc* (London) **1923,** 59.

Bromine, in aqueous EtOH, gives about 45% $BrCH_2CMe=CHCH_2OH$, 15% $BrCH_2CMe(OH)CH=CH_2$ and a little $HOCH_2CMe=CHCH_2Br$, thus showing that the attack on the conjugated system starts mainly at the 1-carbon.[32]

If Thiele's conception of partial valences gives the true explanation of 1,4-addition the reduction of isoprene would be expected to give mainly trimethylethylene and isopentane. Reduction with a platinum catalyst using equimolar proportions of hydrogen and isoprene gave isopentane, 30%, Me_3-ethylene, 15%, *unsym*-MeEt-ethylene, 13%, isopropylethylene, 12% and unchanged isoprene, 30%. Similar results were obtained with 1,3-butadiene, 1,3-pentadiene and 2,3-Me$_2$-1,3-butadiene. Diisobutenyl,

$$Me_2C=CHCH=CMe_2$$

on the other hand gave 1,4-addition as the sole primary process.[33]

Isoprene with Na or K and EtBr gives 1,4-addition of the free ethyl radical to give 4-Me-4-octene.[34, 35] This is analogous to the addition of free triphenyl-methyl radicals to form

$$(C_6H_5)_3CCH_2CMe=CHCH_2C(C_6H_5)_3.[36]$$

Isoprene, in glacial acetic acid, treated with NaSCN and bromine gives a dithiocyano addition product,[37] m. 77°, probably

$$NCSCH_2CH=C(Me)CH_2SCN.$$

Isoprene adds SO_2 to form the cyclic sulfone, 2-Me-2-buten-1,4-sulfone, m. 63°.[38, 39] At 120–135° the sulfone decomposes and the isoprene is regenerated.[40]

Isoprene and other substituted dienes react with CO under high pressure to give alkyl cyclopentanones.[41]

Isoprene was known to react at 180° with benzoquinone to give a product m. 234° containing two molecules of the diene for one of the quinone[42] for some time before the discovery of the general reaction of conjugated dienes with $\alpha\beta$-unsaturated carbonyl compounds.

The *polymerization* of isoprene is a most complex subject. In nature it may give rise to the terpenes and rubber and related vegetable products. In the laboratory it gives products at least as complicated.[43]

[32] Ingold, Smith. *J. Chem. Soc.* 1931, 2752.
[33] Lebedev, Yakubchik. *J. Chem. Soc.* 1928, 2190.
[34] *Ann. Rep. Chem. Soc.* (London) 1930, 88.
[35] Midgley, Henne. *J. Am. Chem. Soc.* 52, 2075 (1930).
[36] Conant et al. *J. Am. Chem. Soc.* 53, 1941 (1931); 55, 3475 (1933).
[37] Bruson, Calvert. *J. Am. Chem. Soc.* 50, 1735 (1928).
[38] *Ann. Rep. Chem. Soc.* (London) 1930, 90.
[39] Ellis. "Chemistry of Petroleum Derivatives." Reinhold, 1934. p. 655.
[40] Menshik. *C. A.* 3771 (1948).
[41] Raasch, Theobald. *C. A.* 42, 2987 (1948).
[42] v. Euler, Josephson. *Ber.* 53B, 822 (1920).
[43] Conant, Peterson. *J. Am. Chem. Soc.* 54, 628 (1932).

The dimerization of isoprene was formerly thought to give a dimethylcyclooctatriene (Harries) but the process is actually an internal dimerization resulting in menthadienes.[44] If, instead of ring closure, the polymerization produces long chains of the type,

$$Me_2C = CHCH_2(CH_2CMe = CHCH_2)_nCH = CMeCH = CH_2$$

the product is caoutchouc. Natural acylic terpene hydrocarbons, dimers of isoprene, are ocimene and myrcene.

```
C = C - C - C - C = C - C = C          C - C = C - C - C - C = C
    |           |                           |               ||
    C           C                           C               C
       Ocimene                                   Myrcene
```

Isoprene, with glacial acetic and sulfuric acids, gives geraniol, cyclogeraniol, linalool, α-terpineol and 1,4- and 1,8-cineole.[45] This complex result may be analyzed into simple reactions involving the addition of a proton to an olefinic linkage to give a carbonium ion which can then undergo the ordinary reactions of allylic shift, reaction with anions from the medium, and dimerization by addition to an olefinic linkage. The new carbonium ion can go through the same changes and also internal dimerization (cyclization). These reactions are so important that they may well be discussed in detail.

1. H[+] adds to the more active double bond of an isoprene molecule leaving the second carbon with only 6 electrons (indicated as [+])

$$CH_2 = C(Me)CH = CH_2 + H^+ \rightarrow Me_2\overset{+}{C}CH = CH_2$$

2. This carbon relieves its deficiency by attracting an electron pair from the double bond of the next carbon, leaving the fourth carbon deficient and giving $Me_2C = CH - \overset{+}{C}H_2$.

3. The rearranged carbonium ion reacts with another isoprene molecule exactly as the H[+].

$$Me_2C = CH\overset{+}{C}H_2 + CH_2 = C(Me) - CH = CH_2$$
$$\rightarrow Me_2C = CHCH_2CH_2\overset{+}{C}(Me)CH = CH_2$$

4. This undergoes a similar electron shift to that in 2, giving

$$Me_2C = CHCH_2CH_2C(Me) = CH\overset{+}{C}H_2$$

This fragment can undergo changes listed in 5-10.

5. It can coordinate with a molecule of water from the medium.

$$R^+ + \ :\overset{..}{O}:H \rightarrow \left[R:\overset{..}{O}.H \right]^+ \rightarrow H^+ + ROH$$
$$\quad\quad\quad\ \overset{..}{\underset{H}{}}\quad\quad\quad \overset{..}{\underset{H}{}}$$

[44] *Ann. Rep. Chem. Soc.* (London) 1931, 87.
[45] ibid. 1932, 151.

The loss of the proton may be regarded as an expulsion to give a more stable molecule, ROH, or as a removal by water to form $(H_3O)^+$. In this case ROH is geraniol, $Me_2C=CHCH_2CH_2C(Me)=CHCH_2OH$. Union of the carbonium ion with acetate ion or bisulfate ion would give the same product on hydrolysis.

6. The C^+ can approach the 6-C in space and can add to its double bond to form a 6-ring.

$$Me_2C=CH \begin{array}{c} CH_2-CH_2 \\ \diagup \qquad \diagdown \\ \qquad\qquad C-Me \rightarrow \\ \diagup \qquad \\ {}^+CH_2-CH \end{array}$$

$$Me_2\overset{+}{C}-CH \begin{array}{c} CH_2-CH_2 \\ \diagup \qquad \diagdown \\ \qquad\qquad C-Me \quad (A) \\ \diagdown \qquad \diagup \\ CH_2-CH \end{array}$$

$$Me_2CH-C^+ \begin{array}{c} CH_2-CH_2 \\ \diagup \qquad \diagdown \\ \qquad\qquad C-Me \quad (B) \\ \diagdown \qquad \diagup \\ CH_2-CH \end{array}$$

7. (A) can undergo changes including coordination with H_2O as in 5. Loss of a proton then gives *α-terpineol* having OH in place of the $^+$.

8. The C^+ can attract an electron pair from the adjacent tertiary C together with the attached H to give (B). This could give 4-terpineol with OH in place of $^+$.

9. Addition of H^+ to α-terpineol from (7) would give

$$Me_2C(OH)-CH(CH_2CH_2)_2\overset{+}{C}Me$$

In this the C^+ and the OH can approach each other in space. Coordination as in 5, followed by expulsion of a proton, gives ordinary 1,8-cineol, an inner ether with a 6-ring.

10. Entirely similar processes with the product of (8) would give 1,4-cineol, an internal ether containing a 5-ring.

The products actually isolated are evidently those most stable under the experimental conditions.

Methylisoprene, 2,3-Me_2-1,3-butadiene, diisopropenyl,

$$CH_2=C(CH_3)C(CH_3)=CH_2,$$

b. 70°, is made by catalytic dehydration of pinacol. Its polymerization by aniline hydrobromide as a catalyst gives *methyl rubber* used as a rubber substitute during World War I. It is inferior to natural rubber.

The chief product with HBr is $Me_2C = CMeCH_2Br$, formed by 1,4-addition. The influence of methyl groups on the polymerization of substituted butadienes has been studied.[46]

1,3-Pentadiene, piperylene, $CH_3CH = CH - CH = CH_2$, b. 43°, has been obtained by the exhaustive methylation of piperidine. This might be expected to give the 1,4-diene, but rearrangement takes place to give the conjugated diene. It is also obtainable in excellent yields from the pyrolysis of the acetate of 2,4-pentanediol and in small amounts from cracking operations.[47]

2,4-Hexadiene adds HBr to give

$$MeCH_2CHBrCH = CHMe, \quad MeCH_2CH = CHCHBrMe$$

in 9:1 ratio. Thus the chief addition is 1,2. This may be interpreted as indicating that there is less tendency for the group

$$Me - CH_2 - \overset{+}{C}H - CH = CH - Me$$

to undergo the allylic rearrangement to form $Me - CH_2 - CH = CH - \overset{+}{C}H - Me$ than for $Me - \overset{+}{C}H - CH = CH_2$ to change to $Me - CH = CH - \overset{+}{C}H_2$ as in the 1,4-addition to butadiene. This appears to be a case of "hyperconjugation" of a methyl group with an unsaturated linkage.[48] Apparently the grouping $H_3 \equiv C - CH = CH - CH_2 -$ is more stable than $H_3 \equiv C - CH_2 - CH = CH -$ in the same way that

$$CH_2 = CH - CH = CH - CH_2 -$$

is more stable than the non conjugated form $CH_2 = CH - CH_2 - CH = CH -$. Organic chemistry is full of examples of the ready conversion of a non conjugated system to the isomeric conjugated form. This is due to the greater resonance energy of the latter.[49,50]

Piperylene, has been prepared in a variety of ways in stereoisomeric forms, *cis* b. 44°, *trans* b. 42°.[51] *cis*- and *trans*-Piperylene react with SO_2 to form the same sulfone which on decomposition gives only the trans isomer. *trans*-Piperylene reacts rapidly and completely with maleic anhydride, affording a method for the isolation of pure *cis*-piperylene. Refluxing with iodine gives a 14:86 *cis-trans* ratio; thermal pyrolysis results in a 3:4 *cis-trans* mixture.[52, 53]

4-Me-1,3-pentadiene is readily obtained by careful catalytic reduction of diacetone alcohol and dehydration of the resulting glycol.

Higher alkylated dienes of the type, $Me_2C = CHC(Me) = CHR$, can be made by dehydrating the product of mesityl oxide with RCH_2MgX.

[46] Whitby. *Ann. Rep. Chem. Soc.* (London) **1932,** 113.
[47] Ellis. "Chemistry of Petroleum Derivatives." Reinhold, 1934. p. 637.
[48] Mulliken, Rieke, Brown. *J. Am. Chem. Soc.* **63,** 41 (1941).
[49] Pauling. "The Nature of the Chemical Bond." Cornell University Press, 1940.
[50] Wheland. "The Theory of Resonance." John Wiley and Sons, Inc., 1944.
[51] Berkenheim, Dankova. *J. Gen. Chem.* (U.S.S.R.) **9,** 924 (1939).
[52] Craig. *J. Am. Chem. Soc.* **65,** 1007 (1943).
[53] Frank. *J. Am. Chem. Soc.* **69,** 2313 (1947).

2,4-Dimethyl-3-isopropyl-1,3-pentadiene, b. 141°, $n^{20}D$ 1.4390, is readily obtained by dehydration of the corresponding tertiary alcohol 2,4-dimethyl-3-isopropyl-1-penten-4-ol over $CuSO_4$.

Dienes with Independent Double Bonds. Of these the most readily available is diallyl, 1,5-hexadiene, $CH_2 = CHCH_2CH_2CH = CH_2$, b. 61°, obtainable in good yield by the action of sodium on allyl bromide or iodide.[54] This type of diolefin shows the ordinary reactions of the double bond with complications due to the possibility of ring formation. Diallyl with benzoyl peroxide[55] gives a furan derivative instead of the expected tetrahydroxy compound. Hydration with sulfuric acid gives 2,5-epoxyhexane, the inner ether of the expected glycol.

1,4-Diolefins have been obtained by Boord's bromoether synthesis.[56]

Diisobutenyl, $CH_2 = CMeCH_2CH_2C(Me) = CH_2$, b. 114°. The effect of 1-double bonds may be seen by comparing with *diisocrotyl,*

$$Me_2C = CHCH = CMe_2,$$

b. 128°. The latter also has conjugated double bonds and a very symmetrical molecule which would ordinarily be expected to lower the boiling point. Under the influence of alcoholic KOH or silica catalysts at 200° diisobutenyl changes to the extent of 80% into the conjugated compound, diisocrotyl.[57]

2,3,3,4-Me₄-1,4-pentadiene, b. 127.7°, $n^{20}D$ 1.4403, d^{20} 0.7712, is obtainable starting with Me_4-ethylene. The latter is treated with acetic anhydride and $ZnCl_2$ to give the unsaturated ketone which reacts with MeMgX to give 2,3,3,4-Me_4-4-penten-2-ol which is then dehydrated to the desired diene with $CuSO_4$. More vigorous dehydrating agents, even pure alumina, give mainly fission to Me_4-ethylene and acetone showing that the proton attack is on the olefinic carbon rather than on the hydroxyl oxygen.

1,3,5-Hexatriene, b. 82°, can be made from allyl chloride with sodamide and NH_3.[58]

2. ACETYLENES, ALKINES OR ALKYNES

The acetylenes resemble the olefins in being unsaturated with the unsaturation localized at two adjacent carbon atoms. Each of these atoms is capable of *adding* two univalent groups or their equivalent. This unsaturation is indicated by a *triple bond* between the two carbons. Although an acetylene can add twice as much as an olefin, the *ease of addition* to a triple bond may be less than that to a double bond. This lowered activity of the triple bond is perhaps due to an electronic structure resembling those of nitrogen, carbon monoxide and hydrocyanic acid (Lewis). In each case the *pair* of atoms has a

[54] Cortese. *J. Am. Chem. Soc.* **51**, 2266 (1929).
[55] *Ann. Rep. Chem. Soc.* (London) 1927, 87.
[56] Shoemaker, Boord. *J. Am. Chem. Soc.* **53**, 1505 (1931).
[57] Lebedef. *J. Gen. Chem.* (U.S.S.R.) **4**, 23 (1934).
[58] Kharasch, Sternfeld. *J. Am. Chem. Soc.* **61**, 2318 (1939).

complete octet.

$$: N : N : \qquad : C : O : \qquad H : C : N : \qquad H : C : C : H$$

The physical properties of the acetylenes resemble closely those of the paraffins and olefins.

Acetylene, ethine, ethyne, $HC \equiv CH$

Acetylene b. $-84°$ (sublimes) m. $-81°$ is formed during many thermal decompositions of organic material. Thus it is found in illuminating gas. It is also formed in partial combustion of hydrocarbons as in the flame of a Bunsen burner which is burning at the base. The synthesis of acetylene from carbon and hydrogen in the electric arc by Berthelot in 1862 was of great theoretical interest because it opened up the synthesis of nearly all types of organic compounds from the elements and definitely overthrew any remaining idea that organic and inorganic materials obeyed different laws. Acetylene is also formed during the electrolysis of solutions of sodium fumarate and sodium maleate. The treatment of bromoform, or iodoform, CHX_3, with metals such as Zn or Ag forms acetylene.

Preparation. 1. Since 1892 when Willson made calcium carbide, CaC_2, available commercially, acetylene has been made by the action of water on that substance.[59]

2. Acetylene can be made from natural gas by compressing it sufficiently to raise the temperature to $1370°$ C. and then allowing it to expand rapidly, thus cooling it to a point at which acetylene is stable.[60, 61]

3. The arc process for making acetylene which has been experimented with so long was finally reduced to large scale practice in Germany in connection with part of the wartime synthetic rubber program. A direct current of 7000 volts at 7000 kilowatts per reactor using 2800 cu.m. per hour of mixed methane and ethane from coal hydrogenation was employed. The conversion is about 50% per pass with weight yields of about 45% C_2H_2 and 10% C_2H_4. By-products are carbon black, hydrogen and smaller amounts of higher acetylenes and aromatic compounds.

4. It was formerly made in the laboratory by the action of alcoholic KOH on ethylene bromide to remove 2 HBr. The reaction proceeds in two steps, vinyl bromide being formed first. Since vinyl bromide is volatile, the yield is cut down correspondingly. Moreover, considerable vinyl ethyl ether, $CH_2 = CHOEt$, is formed.

Properties. Acetylene is a gas with a very slight odor when pure. The odor of the impure gas is caused by traces of H_2S and PH_3 due to impurities of

[59] Ahrens. *Sammlung Chemische-Technischen Vorträge* **1896**, 1–46.
[60] Wulff. *C. A.* **25**, 751, 962 (1931); **26**, 2040 (1932); **28**, 5640 (1934); **35**, 4577 (1941); **36**, 6784 (1942).
[61] Hasche. *Chem. & Met. Eng.* **49**, No. 7, 78 (1942).

sulfates and phosphates in the lime used for making commercial calcium carbide. It dissolves at ordinary temperature in water, benzene, ethanol and acetone to the extent of 1, 4, 6, and 25 volumes respectively. Since pure compressed acetylene is dangerously explosive, it is handled commercially in solution at 10–12 atmospheres in acetone absorbed in some porous material (Prestolite tanks). A better solvent is tetrahydrofuran (Reppe). The b.p. of acetylene is −84° while the m.p. is −81°. The liquid and solid are violently explosive, as it is a highly endothermic compound.

Studies of the specific heat of acetylene indicate that acetylene exists in equilibrium with isoacetylene (acetylidene, Nef),[62] ($H_2C = C$).[63]

The normal vibrations of acetylene have been calculated from its infra-red spectrum.[64]

Contrary to earlier work, acetylene forms no compound with hemoglobin In small amounts it has a narcotic effect (*Narcylene*).

Acetylene became very important in the German technology and economy during World War II.[65]

Reactions. 1. With oxygen. Since acetylene is an endothermic compound, its combustion gives more heat than would be obtained by burning the same weight of pure carbon and hydrogen. The acetylene-oxygen flame is also superior to the oxy-hydrogen flame because the temperature of dissociation of carbon monoxide is far above that of steam. Consequently the hottest part of the acetylene flame is correspondingly hotter than the hydrogen flame. The brilliant white flame of acetylene burning in a special burner in air is due to the incandescent particles of carbon formed during the combustion. This gave acetylene its former wide use as an illuminant. At present the chief use of acetylene is for cutting and welding metals.

Acetylene forms explosive mixtures with air over wide limits from 3 to 82% acetylene. This behavior is unusual. Thus methane and ethylene form explosive mixtures with air only between the limits 5 and 13% and 4 and 22% respectively. The kinetics of the slow oxidation of acetylene by oxygen are very complex.[66]

Controlled partial oxidation gives *acetylene black* which is especially valuable in dry batteries.

2. With hydrogen. One or two molecules of hydrogen can be added to acetylene catalytically. In Germany during World War II ethylene was made in this way for the production of its many useful derivatives (p. 35).

Even styrene for Buna-S was made by acetylene to ethylene to Et-benzene to styrene.

Chromous chloride solution, $CrCl_2$, will reduce $-C \equiv C-$ to $-CH = CH-$.

[62] Nef. *Ann.* 298, 202 (1897).
[63] Usherwood. *Chemistry & Industry* 42, 1246 (1923).
[64] Olson, Kramers. *J. Am. Chem. Soc.* 54, 136 (1932).
[65] Monrad. *Chem. & Met. Eng.* 53, No. 7, 120 (1946).
[66] Kistiakowsky, Lenher. *J. Am. Chem. Soc.* 52, 3785 (1930).

3. While acetylene and chlorine react explosively to give carbon and HCl, the reaction can be moderated by the use of a catalyst such as $SbCl_5$. The chief products are acetylene tetrachloride and dichloride, $Cl_2CHCHCl_2$ and $ClCH=CHCl$. Smaller amounts of pentachloro- and hexachloroethane are obtained.

Bromine acts similarly.

Iodine at 150° gives solid *trans* and liquid *cis* acetylene diiodide.[67] The configurations have been determined by conversion to fumaric and maleic acids respectively. The addition of two univalent groups to a triple bond to give a *trans* olefin has been observed in many cases but has not been satisfactorily explained.

4. Halogen acids including HF add to form vinyl halides and ethylidene halides, $CH_2=CHX$ and CH_3-CHX_2. The latter addition is in accord with Markownikoff's Rule. Traces of peroxides cause the reverse addition to give ethylene bromide, $BrCH_2CH_2Br$.[68]

Charcoal activated by HBr at 450° causes the addition of HBr to acetylene at 200° to give mainly 1,2-Br_2-ethane.[69]

Vinyl chloride for polymers and co-polymers can be made by acetylene and HCl over $HgCl_2$ on activated carbon. The use of acetic acid and zinc acetate gives vinyl acetate for similar uses.

5. Acetylene adds HCN in presence of a catalyst of barium cyanide and carbon to give vinyl cyanide, *acrylonitrile*, $CH_2=CHCN$.

6. Water can be added to acetylene in presence of mercuric sulfate and sulfuric acid to form acetaldehyde (b.p. 20°) which distills out of the reaction mixture and can be used for many important syntheses. Another conversion of acetylene to acetaldehyde is by a B_2O_3 and H_3PO_4 catalyst at 360°.[70] A better preparation of acetaldehyde may be by way of the vinyl alkyl ethers (p. 194). At 450° C. using $ZnO \cdot Cr_2O_3$ acetylene reacts with water as follows:

$$2C_2H_2 + 3H_2O \rightarrow CH_3COCH_3 + CO_2 + 2H_2$$

7. Acetic acid can also be added in the presence of mercuric salts to form vinyl acetate, $CH_2=CHOCOCH_3$, and ethylidene diacetate,[71]

$$CH_3CH(OCOCH_3)_2$$

Zn naphthenate causes the addition of organic acids to C_2H_2 to form vinyl esters (Reppe). Even the complex higher acids from *tall oil* from sulfate paper pulp plants form such esters. These are good drying oils.

8. Acetylene with oleum gives $(HO)_2CHCH(SO_3H)_2$ which, with barium hydroxide, forms $CH_2(SO_3)_2Ba$ and Ba formate.

[67] *Am. Rep. Chem. Soc.* (London) 1911, 97.
[68] Kharasch, McNab, Mayo. *J. Am. Chem. Soc.* 55, 2521 (1933).
[69] Wilson, Wylie. *J. Chem. Soc.* 1941, 596.
[70] Schaad, Ipatieff. *J. Am. Chem. Soc.* 62, 178 (1940).
[71] *Ann. Rep. Chem. Soc.* (London) 1921, 63.

9. Alcohols in the presence of BF_3 and mercuric compounds give acetals, $CH_3CH(OR)_2$.[72] With a KOH catalyst, alcohols add to acetylene to give vinyl ethers $CH_2 = CHOR$.

10. Aromatic compounds add to acetylene in presence of cuprous catalysts according to Markownikoff's Rule. Thus phenol gives $(HOC_6H_4)_2CHCH_3$. Zn naphthenate similarly introduces vinyl groups into phenols. The products polymerize. *Koresin*, a tackifier for synthetic rubber, is such a product from *t*-Bu-phenol.

Acetylene with benzene and $AlCl_3$ gives a complex mixture containing 1,1-diphenylethane and 9,10-dimethyldihydroanthracene and only traces of styrene.[73]

While it is possible to make styrene directly from acetylene and benzene, the process is so likely to give explosions that even in Germany during World War II acetylene was first hydrogenated to ethylene in producing styrene.

11. Acetylene adds to itself (polymerizes) under the influence of cuprous catalysts to give vinylacetylene and divinylacetylene, $CH_2 = CH - C \equiv CH$ and $CH_2 = CH - C \equiv C - CH = CH_2$, and higher polymers.[74]

When heated, acetylene polymerizes to a great variety of products including benzene and a complex mixture of hydrocarbons called cuprene.[75] The silent electric discharge and ultraviolet light produce similar results.[76, 77]

Acetylene is not polymerized by sulfuric acid. Different concentrations of acid give acetaldehyde and crotonaldehyde, $CH_3CH = CHCHO$.

It can be polymerized to cyclopolyolefins (p. 604).

The effect of heat on acetylene has been extensively studied.[78, 79]

12. Acetylene reacts when heated under the proper conditions with S or NH_3 to give thiophene or pyrrole.[80] Under other conditions NH_3 gives acetonitrile.

13. In a silent electric discharge, acetylene and nitrogen form HCN.

14. Acetylene reacts with $AsCl_3$ in presence of $AlCl_3$ (Nieuwland) to form chlorovinyl dichloroarsine, $ClCH = CHAsCl_2$ (the war gas, Lewisite),[81] and the more highly substituted arsines, $(ClCH = CH)_2AsCl$ and $(ClCH = CH)_3As$.

15. The hydrogens in acetylene have acidic properties. Treatment of acetylene with ammonical cuprous or ammonical silver solutions precipitates cuprous and silver acetylides. These are explosive when dry. Treatment with acids regenerates acetylene. One or both hydrogens of acetylene can be

[72] Nieuwland, Vogt, Foohey. *J. Am. Chem. Soc.* 52, 1018 (1930).
[73] *Ann. Rep. Chem. Soc.* (London) 1921, 81.
[74] Nieuwland et al. *J. Am. Chem. Soc.* 53, 4197 (1931).
[75] Ellis. "Chemistry of Petroleum Derivatives." Reinhold, 1934. p. 194.
[76] Lind, Livingston. *J. Am. Chem. Soc.* 54, 94 (1932).
[77] Glockler, Walz. *Trans. Electrochem. Soc.* 88, 10 (1945).
[78] *Ann. Rep. Chem. Soc.* (London) 1912, 74; 1919, 57; 1920, 53; 1928, 68; 1932, 112.
[79] Egloff, Lowry, Schaad. *J. Phys. Chem.* 36, 1457 (1932).
[80] *Ann. Rep. Chem. Soc.* (London) 1922, 61.
[81] Lewis, Perkins. *Ind. Eng. Chem.* 15, 290 (1923).

replaced by heating with metallic sodium[82] or potassium or by passing acetylene into a solution of the metal or metal amide in liquid NH_3. These compounds differ from the heavy metal acetylides in being stable when dry. They are completely hydrolyzed by water. Acetylene reacts like an acid with Grignard reagents forming the corresponding hydrocarbon and $HC \equiv CMgX$ or $XMgC \equiv CMgX$. These acetylene Grignard reagents can be used in the usual Grignard syntheses. Acetylene reacts with dilute alkaline solutions of organo-mercuric halides $RHgX$, to give precipitates of organomercuric acetylides, $RHgC \equiv CHgR$, of characteristic melting points.[83] The melting points (°C.) for the derivatives of the n-alkyls from methyl to decyl are 233, 196, 151, 126, 92, 105, 96, 108, 98, and 111. Those of the derivatives of isopropyl, sec-butyl and isoamyl are 111, 106 and 107. Acetylene is regenerated by the action of acid.

Many $C-H$ compounds other than acetylenic compounds of the type $C \equiv CH$ show feebly acidic properties.

16. The hydrogen atoms in acetylene can be *replaced* by iodine by the action of an alkaline solution of iodine. The product, diiodoacetylene, is a very insoluble solid of peculiar odor. The mono and disubstituted acetylenes containing chlorine and bromine are spontaneously inflammable.

17. Acetylene with nitric acid gives isoxazole-α-carboxylic acid.[84]

$$\begin{array}{c} N\text{------}O \\ \| \qquad \diagdown \\ \qquad \qquad C\text{---}CO_2H \\ \| \qquad \diagup \\ CH\text{---}CH \end{array}$$

18. Acetylene reacts with ketones in presence of sodamide to give ethynyl dialkyl carbinols, $RR'C(OH)C \equiv CH$.[85]

With KOH and more vigorous conditions, each of the H atoms of acetylene takes part in such an "aldol" or "Grignard" type of addition to give butyne-diols, $R_2C(OH)C \equiv C(OH)CR_2$, in which the R's may be almost any combination of H or alkyl groups. Thus formaldehyde gives the first member (p. 55). This type of reaction has been widely exploited by Reppe and his co-workers as part of the German wartime chemical development of acetylene.[86]

19. In general, acetylene in the presence of alkaline catalysts will add active H compounds (H attached to O, S, or N) in such a way as to attach a vinyl group in place of the H.

$$H-Q + HC \equiv C H \rightarrow CH_2 = CH-Q$$

Many examples of this type appear above. Others include the conversion of $R-SH$ to $R-SCH=CH_2$, and the NH of primary or secondary amines to

[82] Guernsey, Sherman. *J. Am. Chem. Soc.* **48**, 141 (1926).
[83] Spahr, Vogt, Nieuwland. *J. Am. Chem. Soc.* **55**, 2465 (1933).
[84] Ellis "Chemistry of Petroleum Derivatives." Reinhold, 1934. p. 683.
[85] Carothers, Jacobson. *J. Am. Chem. Soc.* **55**, 1097 (1933).
[86] Monrad. *Chem. & Met. Eng.* **53**, No. 7, 120 (1946).

$N-CH=CH_2$. Such compounds can be polymerized. N-Vinylcarbazole obtained in this way, using Zn naphthenate as catalyst, gives a useful fibrous and thermoplastic resin.

20. Carbon monoxide and water add to C_2H_2 in presence of nickel carbonyl to give acrylic acid. When alcohols are used in place of water, the result is the acrylic ester $CH_2=CHCO_2R$.

21. Chlorine water converts acetylene to Cl_2-acetaldehyde.

As indicated above the entire acetylene industry boomed in Germany during World War II.[87]

Allylene, methylacetylene, propyne, $CH_3C\equiv CH_2$, b. $-23.3°$, can be made by the action of alcoholic potash on propylene dibromide (1,2-dibromopropane) or on acetone dichloride (2,2-dichloropropane). In each case allene

$$(CH_2=C=CH_2)$$

is also formed. This can be converted to $MeC\equiv CNa$ by treatment with Na in ether. Allylene can also be made by a modified Wurtz reaction by methyl iodide or methyl sulfate with monosodium acetylide in liquid NH_3 or the mono Grignard reagent of acetylene, $HC\equiv CMgX$.

Allylene gives all the addition reactions of acetylene. These follow Markownikoff's Rule. Thus mercuric salts cause the addition of water to form acetone.

Allylene also shows the acid H reactions of acetylene. With alkaline $Hg(CN)_2$ solution it gives $(MeC\equiv C)_2Hg$, m. 204°. This is a general reaction valuable for identifying α-acetylenes.[88, 89] The corresponding derivatives of $RC\equiv CH$ in which R corresponds to the n-alkyl groups ethyl to decyl have melting points of 163, 118, 96, 61, 80, 68, 80, 79, and 85.

$MeC\equiv CNa$ reacts with acetaldehyde, ethyl formate, acetyl chloride and CO_2 to give pentyn-3-ol-2, butyn-2-al, pentyn-3-one-2 and methylpropiolic acid (1-carboxy-1-propyne). The sodium derivatives of monosubstituted acetylenes in general form the best sources of $\alpha\beta$-acetylenic alcohols, aldehydes, ketones and acids.[90] They act like Grignard reagents.

The polymerization of allylene by sulfuric acid yields mesitylene (*sym*-trimethylbenzene). Heat converts it to allene which then polymerizes.

Ethylacetylene, 1-butyne, $C_2H_5C\equiv CH$, b. 8°, is made like its lower homolog or from 1,1-Cl_2-butane and sodamide. The latter exemplifies a good general preparation for higher 1-acetylenes starting with aldehydes.[91]

Long heating with alcoholic KOH at 150° converts 1-butyne to 2-butyne. This is a general reaction of alpha acetylenes of the type $RCH_2C\equiv CH$. Those

[87] Hasche. *Chem. & Met. Eng.* **52**, No. 10, 116 (1945).
[88] Johnson, McEwen. *J. Am. Chem. Soc.* **48**, 469 (1926).
[89] Vaughn. *J. Am. Chem. Soc.* **55**, 3453 (1933).
[90] Moureu, Delange. *Ann. chim.* (7) **25**, 239 (1902).
[91] Bourguel. *Ann. chim.* **3**, 191, 325 (1925).

of type, $RR'CHC \equiv CH$ are converted to allenes, $RR'C = C = CH_2$, while those of type $R_3CC \equiv CH$ are stable to alcoholic KOH at 200°.[92]

The production of higher 1-acetylenes from a 1-olefin through its dibromide is unsatisfactory. Thus 1,2-dibromobutane with alcoholic KOH gives 1- and 2-bromo-1-butene, 1- and 2-ethoxy-1-butene, 1,2-butadiene and 1- and 2-butyne in varying amounts.

The addition and polymerization reactions of 1-butyne are like those of propyne.

Crotonylene, dimethylacetylene, 2-butyne, $CH_3C \equiv CCH_3$, b. 29°, can be made in the usual ways. Since it contains no $C \equiv CH$ group it does not react with ammonical solutions of silver and cuprous salts. Heating with metallic sodium or sodamide, however, isomerizes it with the formation of the sodium derivative of 1-butyne. Thus acetylenes of the 1- and 2- types are interconvertible.

t-Butylacetylene, Me₃-allylene, $(CH_3)_3CC \equiv CH$, b. 39°, is the lowest boiling C_6 compound. It is obtained from pinacolone by treatment with PCl₅ followed by alcoholic KOH. No rearrangement takes place. It is unchanged by alcoholic KOH at 200°.

Disubstituted acetylenes give no reaction with the Grignard reagent.[93] They are hydrated by sulfuric acid to give ketones, RCH_2COR'. With chlorine water they form the corresponding dichloroketones.

Higher Acetylenes. The boiling points of 1-pentyne (alpha-valerylene), 2-pentyne, and 3-Me-1-butyne are 40°, 55°, and 35°, illustrating the effect of position of the triple bond and of branching of the chain. Those of 1- and 2-hexyne are 71° and 84°, of 1-heptyne (oenanthylidene), and its 2- and 3-isomers are 110°, 113°, and 106°; and of 1-octyne (caprylidene) and its 2-isomer are 125° and 133°.

Because of the ready availability of heptaldehyde by cracking castor oil, 1-heptyne is the commonest higher acetylene. Its sodium derivative, in common with all sodium acetylides, reacts with CO_2 to give an acetylene acid. The methyl ester is a perfume, "methyl heptine carbonate."

1-Heptyne with an alcoholic KOH solution of isopropylmercuric bromide forms an oil, $AmC \equiv CHgCHMe_2$, d^{20} 1.628.[94]

Higher alpha acetylenes containing an even number of carbons can be made from an alkyl halide and disodium acetylide in liquid ammonia to give $RC \equiv CR$ which can be rearranged by $NaNH_2$ to $R'C \equiv CH$.[95] 1-Decyne and 1-dodecyne are thus obtained from n-Bu and n-Am halides.

Acetylenes are more active than olefins with acid oxidizing agents. They split readily at the triple bond giving organic acids.

[92] Favorsky. *J. Chem. Soc. Abs.* **54**, 798 (1888).
[93] Gilman, Shumaker. *J. Am. Chem. Soc.* **47**, 514 (1925).
[94] Spahr, Vogt, Nieuwland. *J. Am. Chem. Soc.* **55**, 3728 (1933).
[95] Vaughn, Nieuwland. *J. Am. Chem. Soc.* **55**, 2150 (1933).

Vinylacetylene, butenyne, $CH_2=CHC\equiv CH$, is obtained from acetylene and a cuprous chloride catalyst (DuPont). It adds HCl to give an unstable 1,4-addition product $ClCH_2-CH=C=CH_2$, which rearranges spontaneously, or more rapidly in presence of excess HCl, to give *chloroprene.* It dimerizes in the presence of $CuCl \cdot NH_4Cl$ to $CH_2=CH-CH=C-C\equiv C-CH=CH_2$.[96]

Diacetylene, butadiyne, $HC\equiv C-C\equiv CH$, is obtained as the cuprous derivative by heating ammonium diacetylene dicarboxylate with ammoniacal cuprous solution. The cuprous compound can also be obtained by heating cuprous acetylide with $CuCl_2$ solution.[97] Treatment with KCN gives the hydrocarbon. The Cu compound is violet red and the Ag compound is yellow. The latter is violently explosive even when wet.

Dipropargyl, 1,5-hexadiyne, is made from the tetrabromide of diallyl with alcoholic KOH at 110°. It forms a liquid tetrabromide and an octabromide, m. 141°. It polymerizes readily. Heating with alcoholic KOH in a sealed tube rearranges the triple bonds in the usual way.

$$HC\equiv CCH_2CH_2C\equiv CH \rightarrow CH_3C\equiv C-C\equiv CCH_3$$
$$\text{m. } -6°, \text{ b. } 87° \qquad\qquad \text{m. } 64°, \text{ b. } 130°.$$

The substances are isomeric with benzene, m. 5°, b. 80°. The second is **Me₂-diacetylene,** 2,4-hexadiyne. It can also be made by oxidation of cuprous methylacetylide with ferricyanide. In contrast to its isomer it does not polymerize.

The only important **dienyne** is divinylacetylene,

$$CH_2=CH-C\equiv C-CH=CH_2,$$

obtained as a byproduct in the manufacture of vinylacetylene for Neoprene rubber. On standing it polymerizes to a dangerously explosive solid resin. When partly polymerized and dissolved in a suitable solvent it forms a useful synthetic drying oil (S.D.O., DuPont).

Polyenes or poly-olefins occur in carotenoids. The only aliphatic examples are lycopene and squalene (p. 588–9).

II. HALOGEN DERIVATIVES OF THE PARAFFINS

A. **Alkyl halides,** $C_nH_{2n+1}X,RX$

Methyl fluoride, chloride and bromide and ethyl fluoride and chloride are gases. The other alkyl halides to about C_{16} are liquids with sweetish odors. The freezing points of the chlorides and iodides of methyl and n-butyl are respectively $-103°$, $-66°$ and $-123°$, $-103°$. The densities for the same pairs are respectively 0.95, 2.29 and 0.91 and 1.64.

[96] Dolgapal'skii. *C. A.* **42**, 4517 (1948).
[97] Straus, Kollek. *Ber.* **59B**, 1664 (1926).

BOILING POINTS OF ALKYL HALIDES, °C.

	Fluorides	Chlorides	Bromides	Iodides
Methyl................	− 78.5	− 24.	+ 4.6	42.6
Ethyl.................	− 37.1	+ 12.2	38.4	72.2
n-Propyl..............	− 2.5	46.6	70.9	102.4
iso-Propyl.............	− 9.4	36.5	59.6	89.5
n-Butyl...............	32.5	77.9	101.6	127.
sec-Butyl.............	25.1	66.5	91.	120.
iso-Butyl.............		68.5	92.	119.
ter-Butyl.............	12.1	51.5	72.	100.
n-Amyl...............	62.8	106.6	127.9	156.
ter-Amyl.............	44.8	85.	108.	127.
Neopentyl............		84.4	105.	133.
n-Hexyl..............	85	134.	156.	177.
n-Heptyl.............	119	159.	178.	204.
n-Octyl..............	142	180.	202.	221.

The alkyl halides are nearly insoluble in water (MeI soluble 1.4 in 100 ml.), soluble in ether, in hydrocarbons and in alcohols. They are insoluble in cold concentrated sulfuric acid. This fact is used in removing impurities of alcohols from the halides. Doubt has been cast on this method of purification.[1] Like other organic halogen compounds, alkyl halides give the *Beilstein Test*, a green flame when heated on a clean copper wire in a Bunsen flame.

While the alkyl halides have a halogen atom in place of one hydrogen of a paraffin they are not usually prepared by direct halogenation. Fluorine reacts violently with paraffins giving hydrogen fluoride and carbon. The chlorination of hydrocarbons usually gives mixtures. By means of an enormous amount of work on the chlorination of methane the process has been developed for the preparation of methyl chloride. The chlorination of *n*- and *iso*-pentane has been developed commercially to give monohalides which are important intermediates for the amyl alcohols (Sharples). Similar processes are being studied for the butanes.

The chlorination of paraffins or of paraffinic side chains in aromatic compounds can effectively be brought about by treatment with sulfuryl chloride, SO_2Cl_2, and a peroxide.[2]

While the bromination of paraffins is possible, the process has received little study. The iodination of paraffins is not possible, the reverse action of an alkyl iodide and hydrogen iodide to give a paraffin and I_2 taking place too readily.

The numbers of alkyl halides theoretically possible have been calculated.[3] Thus 1553 formulas can be written for the decyl chlorides.

The preparations and chemical reactions of alkyl halides, RX, cannot be safely generalized except for the first two members. This is because of the

[1] McCullough, Cortese. *J. Am. Chem. Soc.* 51, 225 (1929).
[2] Kharasch, Brown. *J. Am. Chem. Soc.* 61, 2142 (1939).
[3] Blair, Henze. *J. Am. Chem. Soc.* 54, 1098 (1932).

frequent occurrence of *molecular rearrangements* both in the preparation and the *reactions* of the higher members and because of the great differences in *primary, secondary* and *tertiary halides*, RCH_2X, $RR'CHX$, and $RR'R''CX$.

When it is possible, addition of a halogen acid, HX, to an olefin gives the best method of preparing an alkyl halide. Even this is occasionally troublesome as in the case of the addition of HCl to isopropylethylene which gives *t*-AmCl as well as the desired product. HBr can be added to many 1-olefins in presence of peroxides to give 1-bromo compounds.[4]

The halides may be divided into several classes within each of which generalization is fairly safe except that our knowledge is still inadequate.

1. **The primary halides,** MeX, EtX and the general type RCH_2CH_2X can be prepared from the corresponding alcohols in good yields as follows: (a) Chlorides by means of HCl and pressure, HCl and $ZnCl_2$ heated,[5] $HCl + H_2SO_4$, $NaCl + H_2SO_4$, PCl_3 (alone or with $ZnCl_2$), PCl_5, $SOCl_2$ (with pyridine);[6] (b) Bromides by the corresponding bromine compounds; (c) Iodides by I_2 and P,[7] or from the corresponding chloride and KI in methanol or acetone. The reactions of the alcohols are slow especially with the halide acids. The first action is undoubtedly the formation of an oxonium salt, $ROH \cdot HX$ or $(ROH)_2HX$, which reacts gradually to give RX. Thus there is no real resemblance to the action of an acid with a base in spite of the apparent similarity of the two equations.

$$ROH + HX \rightarrow RX + H_2O$$
$$MOH + HX \rightarrow MX + H_2O$$

In the latter case, the base first dissociates into ions which then react. In the organic reaction the *first process is one of addition followed by slower decomposition*. This difference between inorganic and organic reactions is common.

Formerly the only primary halides which could be made by adding halide acid to olefins were the ethyl halides. It is now possible to make *primary bromides* by adding HBr to 1-olefins in presence of a peroxide (p. 38).

The halides of Class 1 give ordinary metathetical reactions, without rearrangement, with AgOH ($Ag_2O + H_2O$), Ag salts, inorganic cyanides, etc. The iodides are most reactive and the chlorides least. The bromides are usually preferred to the chlorides because they are made more easily and are more reactive without too great cost.

Type 1 halides include those of the normal alkyl groups, $Me(CH_2)_n-$, and others such as isoamyl, $Me_2CH(CH_2)_2-$, and neopentylcarbinyl (β-*t*-butylethyl), $Me_3C(CH_2)_2-$. In other words, *if the halogen is on the No. 1 carbon and there is no branch in the chain nearer than the No. 3 carbon rearrangements are unlikely both in the preparation and reactions of the halides.*

---

[4] Kharasch et al. *J. Org. Chem.* 2, 195 (1937).
[5] "Org. Syntheses."
[6] Darzens. *Compt. rend.* 152, 1314 (1911).
[7] "Org. Syntheses."

2. **Primary halides of the type** $RR'CHCH_2X$ can be obtained in only fair yields from the corresponding alcohols. Examples are isobutyl, Me_2CHCH_2- and "active amyl," sec-butylcarbinyl, $MeEtCHCH_2-$. The best yields are obtained by $SOCl_2$ or HBr under carefully controlled conditions. The halides of Type 2, in metathetical reactions, give mixtures of the expected product with the corresponding tertiary compound, depending on the conditions of the experiment. Thus isobutyl bromide with silver acetate, without a solvent, gives mainly isobutyl acetate whereas, in the presence of glacial acetic acid, the chief product is tertiary butyl acetate. Isobutyl iodide with AgNCO gives the tertiary and iso derivatives in the ratio 2:1. Isobutyl iodide with alcoholic $AgNO_3$ gives ethyl t-butyl ether.

3. **Primary halides of the type** $RR'R''CCH_2X$ cannot be made from the corresponding alcohols except in very small yields, the main product being a tertiary halide formed by the shift of one of the alkyl groups. Two chlorides of this type have been obtained by chlorination of neopentane and neohexane respectively (p. 83). These halides are very inert except toward Mg and Na.

4. **Secondary halides of propane and butane** can be made from isopropyl alcohol and sec-butyl alcohol without complications. The chlorides can be made by treating the alcohols with HCl and $ZnCl_2$ in the cold. The reactions of these halides are entirely normal.

5. **Secondary halides of the normal paraffins beginning with pentane** are little known. Only with the greatest difficulty were the 2- and 3-halogen derivatives of n-pentane finally prepared pure. Either secondary alcohol gives a mixture of the two halides.[8,9]

6. **Secondary halides having a branch on the carbon next to the carbinyl group,** $RR'CHCHXR''$, cannot be made from the corresponding alcohol even under the mildest conditions. Thus methylisopropylcarbinol,

$$MeCHOHCHMe_2,$$

under all known conditions gives the t-amyl halide exclusively.[10] The desired secondary chloride can be obtained only by adding HX to isopropylethylene (p. 83). It is present to about 10% in commercial mixed amyl chlorides from which it can be readily separated by columns of about 100 plates.

7. **Secondary chlorides having two branches on** a carbon atom adjacent to the carbinyl carbon, $RR'R''CCHXR'''$ cannot be made from the corresponding alcohols since the neopentyl alcohol system in them readily undergoes rearrangement during substitution. Pinacolyl alcohol, $Me_3CCHOHMe$, gives the tertiary halides, $Me_2CXCHMe_2$. The addition of HCl to t-butylethylene under pressure gives 35–40% pinacolyl chloride and 60–65% rearranged chloride. The addition of HBr to t-butylethylene, $Me_3CCH=CH_2$, might be expected to give the desired product, but this addition proceeds contrary to

[8] Lauer, Stodola *J. Am. Chem. Soc.* 56, 1215 (1934).
[9] Hass, Weber. *Ind. Eng. Chem., Anal. Ed.* 7, 231 (1935).
[10] Whitmore, Johnston. *J. Am. Chem. Soc.* 60, 2265 (1938).

Markownikoff's Rule and gives mainly the primary halide, $Me_3CCH_2CH_2Br$. Hydrogen iodide reacts with t-butylethylene to give only 10% rearranged iodide and 90% pinacolyl iodide.[11]

Pinacolyl chloride can also be made by the chlorination of neohexane, the use of sulfuryl chloride being the most convenient. It may be readily separated from the tertiary chloride, which boils only 1° away, by washing with aqueous silver nitrate.

The pinacolyl halides are the only known halides of this type.

8. **Tertiary halides,** $RR'R''CX$, are formed with the greatest ease by the action of the tertiary alcohols with HX either in concentrated solution or as gas. Thus t-butyl alcohol reacts almost instantaneously with hydrochloric acid.

GENERAL REACTIONS OF ALKYL HALIDES

The many reactions given by these substances are far from the simple metathetical reactions ordinarily pictured. They are usually far from complete and are almost always accompanied by side reactions including rearrangements. Their mechanism and kinetics have been studied intensively.[12–18]

1. Effect of Heat.

Primary halides of Class 1 decompose only at high temperatures. The pyrolysis of n-butyl chloride without a catalyst gives 1-butene. With $CaCl_2$ as catalyst, the chief products are *trans* and *cis* 2-butenes.[19] Without a catalyst, s-BuCl gives 1-butene and *trans* and *cis* 2-butenes in yields of 33, 44 and 22%. With $CaCl_2$, the yield of 1-butene falls to 15%. If the primary carbon bearing the halogen has a tertiary hydrogen next to it (Class 2) the temperature at which a change takes place is lowered.

Secondary halides lose HX more readily than their primary isomers and tertiary halides do so even more readily. Stability in this respect decreases from chlorides to iodides.

2. Reducing Agents.

The halogen can be replaced by H by means of powerful reducing agents such as sodium amalgam and alcohol, hydriodic acid, and sodium in alcohol or liquid ammonia, and metals with acids. Alkaline reagents tend to give olefins with tertiary halides.

3. Bases and Hydrolysis.[20]

[11] Ecke, Cook and Whitmore. J. Am. Chem. Soc. 72, 1511 (1950).
[12] Steigman, Hammett. *J. Am. Chem. Soc.* 59, 2536 (1937).
[13] Ingold et al. *J. Chem. Soc.* 1937, 1196, 1252.
[14] Bartlett. *J. Am. Chem. Soc.* 61, 1630 (1939).
[15] Winstein. *J. Am. Chem. Soc.* 61, 1635 (1939).
[16] Ogg. *J. Am. Chem. Soc.* 61, 1946 (1939).
[17] Elliott, Sugden. *J. Chem. Soc.* 1939, 1836.
[18] Dostrovsky, Hughes, Ingold. *J. Chem. Soc.* 1946, 146 ff.
[19] Weston, Hass. *J. Am. Chem. Soc.* 54, 3337 (1932).
[20] Ellis. "Chemistry of Petroleum Derivatives." Reinhold, 1934. p. 781.

(a) Aqueous bases convert primary halides mainly to alcohols but always give some olefin. With secondary halides the proportion of olefin increases at the expense of the alcohol formed. With tertiary halides practically no alcohol is formed unless the base is very dilute and the temperature is low.

(b) Alcoholic bases give olefins with by-products of the corresponding ethyl ethers in the case of the primary and secondary halides. Often the ether is formed in as large amount as the olefin (p. 82). Neopentyl iodide is abnormal as it gives mainly neopentane with alcoholic KOH (p. 75). In the case of the tertiary halides, olefin formation is rapid and there is practically no ether formation. Hydrolysis by water in alcohol solution is more rapid with secondary halides than with primary.[21]

4. Ammonia adds alkyl halides, especially those of Class 1, giving substituted ammonium halides. A complicated equilibrium system is set up.

$$RX + NH_3 \rightarrow RNH_3X$$
$$RNH_3X + NH_3 \rightleftharpoons NH_4X + RNH_2$$
$$RNH_2 + \cdot RX \rightarrow R_2NH_2X$$
$$R_2NH_2X + NH_3 \rightleftharpoons NH_4X + R_2NH$$
$$R_2NH + RX \rightarrow R_3NHX$$
$$R_3NHX + NH_3 \rightleftharpoons NH_4X + R_3N$$
$$R_3N + RX \rightarrow R_4NX$$

The addition of alkyl halides to amines follows the same course. It is important in exhaustive methylation (p. 174-5).

The amine or NH_3 molecule may be assumed to attack the rear of the carbon bearing the halogen by means of its free pair of electrons, thus causing the "ionization" of the halogen from the carbon without having left it with only 6 electrons.

$$R : \overset{\cdot\cdot}{\underset{\cdot\cdot}{N}} : \ + \ R : \overset{\cdot\cdot}{\underset{\cdot\cdot}{X}} : \ \rightarrow \ R : \overset{\cdot\cdot}{\underset{\cdot\cdot}{N}} : R \ + \ : \overset{\cdot\cdot}{\underset{\cdot\cdot}{X}} :$$
$$\ \ R \qquad\qquad\qquad R$$

(with R above and below each N)

In addition to these reactions, the ammonia or amine can act also as a base removing HX from the halide, slightly in the case of the primary halides, appreciably with the secondary halides and practically completely with the tertiary. Thus amines of tertiary alkyl groups are not obtainable from the halide and NH_3.

5. Metals.

(a) Sodium reacts with most primary halides either in anhydrous ether or without a solvent to give fair yields of hydrocarbons (Wurtz).

$$2\,RX + 2\,Na \rightarrow R-R + 2\,NaX$$

The reaction is best with iodides and poorest with chlorides. Sodium vapor

[21] Nicolet, Stevens. *J. Am. Chem. Soc.* **50,** 135 (1928).

with gaseous MeBr or EtBr gives NaBr and the free radicals. These react in
various ways.[22] Two ethyl radicals can form butane or ethylene and ethane
or ethylene and hydrogen. The reaction in the liquid phase has long been
assumed to take place through free radicals. This assumption involves certain
difficulties and has been proved to be unnecessary.[23,24] The first step is
apparently the formation of an alkyl sodium compound RNa, which then
reacts with another molecule of RX in the manner characteristic of alkyl
halides. If metathesis takes place, the products are NaX and R−R and the
Wurtz reaction has been "successful." The RNa may also act as a hydro-
carbo base,[25] and remove the elements of HBr from the halide to form an
olefin and a paraffin corresponding to R. The reaction in a solvent may be
even more complex. Thus if a higher olefin is used as a solvent it takes part in
the reaction. When the Wurtz reaction is carried out in ether solution the
sodium alkyl, RNa, can react with the ether to form EtONa and RH.[26,27]
With secondary halides the amount of unsaturated by-products increases
enormously and with tertiary halides they form the main product.

The Wurtz reaction "fails" almost completely with neopentyl chloride, giving 1,1-Me$_2$-cyclopropane but oddly goes quite well with pinacolyl chloride.

(b) The one reaction which apparently takes place with all types of halides
is that with magnesium and ether to give the Grignard reagent, RMgX. With
the simpler alkyl halides of all types it is possible to approach 100% yields
by the use of proper precautions. Higher tertiary halides as ordinarily
prepared are likely to give poor yields of Grignard reagents because of the loss
of HX to give olefins. This difficulty is overcome if very pure halides are used.
The simple Grignard reagents are commercially available (Arapahoe).

(c) Zinc-copper alloy reacts with the lower primary iodides difficultly to
give zinc dialkyls which are volatile and spontaneously inflammable.[28] Secondary and tertiary halides yield mainly olefins.

(d) Sodium amalgam in the presence of a trace of ethyl acetate as a catalyst
reacts with primary bromides and iodides to form mercury dialkyls.[29] These
are high boiling very toxic liquids.

6. Salts and Related Compounds.

(a) Cyanides and sodioesters. The reactions of alkyl halides most frequently mentioned in textbooks are those of synthetic importance like the
action with KCN or NaCN and with sodiomalonic ester, sodioacetoacetic ester
and sodiocyanoacetic ester. These reactions work well only with halides of
Class 1 (p. 74). The yields with those of Class 2 are much poorer. With

[22] Rice, Herzfeld. J. Am. Chem. Soc. 56, 284 (1934).
[23] Morton et al. J. Am. Chem. Soc. 64, 2240, 2242 (1942).
[24] Whitmore, Zook. J. Am. Chem. Soc. 64, 1783 (1942).
[25] Jones, Werner. J. Am. Chem. Soc. 40, 1257 (1918).
[26] Schorigin. Ber. 41, 2723 (1908).
[27] Carothers et al. J. Am. Chem. Soc. 52, 5279 (1930).
[28] Frankland. Ann. 71, 213 (1849).
[29] Frankland, Duppa. J. Chem. Soc. 16, 415 (1863).

halides of Classes 3 to 6 the reactions either fail completely, give poor yields or result in olefins or rearranged products. With tertiary halides, olefins are the main products. Small yields of cyanides, $RR'R''CCN$, are reported from the action of the tertiary iodides with the double cyanide of Hg and K and with silver cyanide, mercuric cyanide or *cuprous cyanide* (available for use in antifouling paint for boat bottoms).

The preparation of a higher halide, RCH_2X, from RX is often very important. The steps commonly given are

$$RX \rightarrow RCN \rightarrow RCH_2NH_2 \rightarrow RCH_2OH \rightarrow RCH_2X$$

The first step is likely to give trouble unless R is of Class 1. The second step gives relatively little trouble although there is a tendency to form some secondary amine, $(RCH_2)_2NH$. The third step spoils the entire process since the yield of normal product is poor even when R belongs to Class 1. A practical process for converting the primary amine, RCH_2NH_2 to RCH_2Cl consists in converting it to the benzoyl derivative which reacts with PCl_5 to give RCH_2Cl, C_6H_5CN and $POCl_3$.[30]

Another process for going up the series involves the action of monochloromethyl ether (from methanol, formaldehyde and HCl) with the Grignard reagent followed by the splitting of the resulting ether with HBr.[31]

$$RX \rightarrow RMgX \rightarrow RCH_2OMe \rightarrow RCH_2Br$$

The advantages of this method are (1) the formation of the Grignard reagent offers fewer complications than any other reaction of alkyl halides, (2) alpha halogenated ethers are easily prepared and react nearly quantitatively with Grignard reagents, and (3) the splitting of methyl ethers by HBr offers no complications unless R in the above compounds is tertiary.

In some cases it is possible to convert the halide to the Grignard reagent and treat the latter with formaldehyde to give the next higher primary alcohol which may be converted to the corresponding halide.

$$RMgX \rightarrow RCH_2OMgX \rightarrow RCH_2OH \rightarrow RCH_2X$$

Of course rearrangements are to be expected in the last step if R is secondary or tertiary.

The *addition of two carbon atoms* can be accomplished by the use of sodiomalonic ester in the following steps:

$$RX \rightarrow RCH(CO_2Et)_2 \rightarrow RCH_2CO_2H \rightarrow RCH_2CO_2Et$$
$$\rightarrow RCH_2CH_2OH \rightarrow RCH_2CH_2X$$

Once again the first step is likely to fail unless RX belongs to Class 1.

Another way of adding two carbon atoms involves the action of ethylene oxide with Grignard reagents of primary and secondary halides (p. 311).

[30] *Ann. Rep. Chem. Soc.* (London) 1911, 87.
[31] *ibid.* 1926, 101.

An alkyl halide may be lengthened by *three carbon atoms* by treating its Grignard reagent with γ-chloropropyl *p*-toluenesulfonate.[32]

$$RMgX + Tol\text{-}SO_3(CH_2)_3Cl \rightarrow R(CH_2)_3Cl + Tol\text{-}SO_3MgX$$

The preparation of a *lower alkyl halide from a higher member* of the series is also often important. The classical series of steps is as follows:

$$RCH_2X \rightarrow RCH_2OH \rightarrow RCHO \rightarrow RCO_2H \rightarrow RCOCl$$
$$\rightarrow RCONH_2 \rightarrow RNH_2 \rightarrow ROH \rightarrow RX$$

The first step will involve serious trouble unless the halide is of Class 1. The steps from the alcohol to the primary amine all *give excellent yields*. It is interesting that these are *not metathetical* processes. Even the action of the acid chloride with NH_3 probably involves an addition to the carbonyl group rather than a simple metathesis. Again, the conversion of the primary amine to the alcohol presents serious difficulty but the amine can be converted to the halide through von Braun's benzoyl method (p. 79). A useful modification of the above set of steps is the treatment of the acid chloride with sodium azide (NaN_3) to give the amine[33] which can then be treated by von Braun's method.

$$RCOCl \rightarrow RCON_3 \rightarrow RNH_2 \rightarrow RNHCOC_6H_5 \rightarrow RCl + C_6H_5CN$$

(b) *Silver salts.* Esters not readily available in other ways may sometimes be made from alkyl halides and the silver salts of the desired acids. Even this reaction works well only with halides of Class 1.

The action of alkyl halides with $AgNO_2$ has been very thoroughly studied.[34] Both nitro compound and nitrite are obtained in all cases, the combined yield being 60–90%.

(c) *Other salts.* Higher boiling halides react on heating with anhydrous sodium acetate ("fused") to give the corresponding acetates. As would be expected, the yields from secondary chlorides are poor and tertiary chlorides are converted entirely into olefins.

A special method for replacing primary and secondary halogen by hydroxyl is to reflux with anhydrous potassium formate in absolute methyl alcohol. The first step probably converts the halide to the formate which then undergoes "alcoholysis" $HCO_2R + MeOH \rightarrow HCO_2Me + ROH$. An easy way to be sure of having anhydrous potassium formate is to add thoroughly dried potassium hydroxide to a mixture of anhydrous methyl formate and methanol and reflux before adding the halide.[35]

The replacement of one halogen by another is often possible. Thus chlorides with $AlBr_3$ give bromides, with NaI, KI or CaI_2 give iodides;[36]

[32] Rossander, Marvel. *J. Am. Chem. Soc.* **50**, 1491 (1928).
[33] Naegeli, Lendorff. *Helv. Chim. Acta* **15**, 49 (1932).
[34] Reynolds, Adkins. *J. Am. Chem. Soc.* **51**, 279 (1929).
[35] "Org. Syntheses."
[36] Conant, Hussey. *J. Am. Chem. Soc.* **47**, 476 (1925).

bromides with NaI etc. give iodides less readily, with $SbCl_5$ give chlorides; iodides with $CuBr_2$ give bromides, with $HgCl_2$ give chlorides and with AgF or HgF_2 give fluorides. As usual these replacement reactions give good results only with primary halides of Class 1. Thus isobutyl iodide with ICl gives I_2 and *tertiary* butyl chloride exlusively. Replacement reactions of alkyl halides have been studied in great detail by the English organic chemists.[37]

(*d*) Anhydrous aluminum chloride reacts vigorously with alkyl halides giving HX and rearrangement and polymerization products. This type of reaction has not been adequately studied.

(*e*) Dialkylzinc compounds, Grignard reagents and organolithium compounds, with alkyl halides, give higher hydrocarbons in poor yields. These reactions are used only as a last resort.[38-40]

INDIVIDUAL ALKYL HALIDES

1. **Methyl fluoride** is prepared (1) by heating KF with $MeKSO_4$ and washing the products with conc. H_2SO_4 to remove ethylene and methyl ether; (2) by heating Me_4NF.

Methyl chloride is available liquefied in tanks. It is prepared from methanol with NaCl and H_2SO_4 or with hydrochloric acid and zinc chloride in autoclaves. It has also been made by heating trimethylamine with HCl, the former being obtained by the decomposition of betaine during the destructive distillation of sugar beet residues. It is prepared commercially by the direct chlorination of methane. It is used as a refrigerant. Because of its toxic character and very slight odor, it may be mixed with a small amount of a warning agent (ethyl mercaptan) for use in refrigeration. It is easily converted to the Grignard reagent, MeMgCl.

Methyl bromide is available in cylinders (Dow). It can also be generated by warming a mixture of NaBr, MeOH and sulfuric acid. It is useful as a methylating agent[41] and a fumigant. It should be handled with extreme caution as it is surprisingly toxic.[42] For large laboratory reactions involving methyl, it is convenient because of the greater solubility of $MeMgBr$ as compared with MeMgCl.

Methyl iodide is best made from methanol, iodine and phosphorus.[43] One possible equation for this complex reaction is

$$10\ CH_3OH + 5\ I_2 + 2\ P \rightarrow 10\ CH_3I + 2\ H_3PO_4 + 2\ H_2O$$

It can also be prepared by heating dimethyl sulfate (poisonous, odorless) with sodium iodide. Methyl iodide has been much used as a methylating

[37] Dostrovsky, Hughes, Ingold. *J. Chem. Soc.* 1946, 146 ff.
[38] Marvel, Hager, Coffman. *J. Am. Chem. Soc.* 49, 2323 (1927).
[39] Noller. *J. Am. Chem. Soc.* 51, 594 (1929).
[40] Edgar, Calingaert, Marker. *J. Am Chem. Soc.* 51, 1483 (1929).
[41] Lucas, Young. *J. Am. Chem. Soc.* 51, 2535 (1929).
[42] Watrous. *Brit. J. Ind. Med.* 4, 111 (1947).
[43] "Org. Syntheses."

agent. Thus methyl glucoside with MeI and Ag_2O gives polymethyl ethers. Ketones, with sodamide and MeI, have alpha H atoms replaced by Me groups. MeI reacts with finely divided Hg formed by photochemical decomposition of mercurous iodide to give crystalline MeHgI.[44]

2. **Ethyl chloride** is prepared like methyl chloride and is available in tanks. Small amounts of ethyl chloride can be made conveniently by heating an alcoholic solution of $CaCl_2$ with Et_2SO_4. Recently the preparation from ethylene and HCl has become important. Ethyl chloride has replaced the bromide used for the preparation of lead tetraethyl (anti-knock). It is also used for making ethyl mercaptan, EtSH, for the preparation of sulfonal. It is used as a refrigerant, in thermo-regulators for house heating and as a local freezing anesthetic for minor operations.

Ethyl bromide is prepared from ethanol, sodium bromide and sulfuric acid or from ethanol and a mixture of HBr and H_2SO_4 obtained from bromine, a limited amount of water and sulfur dioxide.[45] Recently the tendency in commercial preparation is to use the action of HBr with ethylene. The conversion of ethyl bromide to HBr and ethylene has been studied as a typical monomolecular reaction.[46]

Ethyl iodide is prepared like methyl iodide.

Infra-red spectrum studies have been made on the ethyl halides.[47]

3. **n-Propyl chloride** is best prepared from 1-propanol (from fusel oil or methanol synthesis) by heating with concentrated hydrochloric acid and zinc chloride.[48] It can also be prepared by means of PCl_3, PCl_5 or $SOCl_2$. In addition to the normal reactions giving the alkyl chloride there is always some loss due to the formation of the ester, R_3PO_3, R_3PO_4 and R_2SO_3.

n-Propyl bromide is prepared like ethyl bromide. It can also be prepared with PBr_3. The pentabromide is less suitable because it readily dissociates into the tribromide and free bromine and thus has an oxidizing effect on the alcohol. *n-Propyl iodide* is made like methyl iodide. **Isopropyl chloride** is made from isopropyl alcohol (Petrohol) by hydrochloric acid and zinc chloride. *Isopropyl bromide* and *iodide* are made like the corresponding normal compounds.

4. **n-Butyl halides** are readily prepared from 1-butanol by the same methods as the ethyl halides. **sec-Butyl halides** are prepared from 2-butanol (from cracked gases) like the isopropyl halides. **Isobutyl halides** are obtained from isobutyl alcohol (from fusel oil or from methanol synthesis) in only fair yields. This is due to the presence of the tertiary H atom next to the carbinol group. Considerable amounts of isobutylene and tertiary butyl halides are formed. This is a good example of rearrangement during a replacement

[44] Maynard. *J. Am. Chem. Soc.* 54, 2108 (1932).
[45] "Org. Syntheses."
[46] Vernon, Daniels. *J. Am. Chem. Soc.* 54, 2563 (1932).
[47] Cross, Daniels. *J. Chem. Physics* 1, 48 (1933).
[48] Norris, Taylor. *J. Am. Chem. Soc.* 46, 753 (1924).

reaction. The isobutyl carbonium ion formed by the decomposition of the initial oxonium compound may unite with a halide ion or may stabilize itself by loss of a proton to give isobutylene or may change to the more stable t-Bu ion before combining with halide ion. The isobutyl halides are also prepared by thionyl chloride, either alone or with pyridine, by phosphorus tribromide and by phosphorus and iodine. In such processes at least two types of reaction can occur.

$$3 \text{ ROH} + PX_3 \rightarrow 3 \text{ RX} + H_3PO_3$$
$$3 \text{ ROH} + PX_3 \rightarrow R_3PO_3 + 3 \text{ HX}$$

The factors controlling these have never been studied adequately.

Tertiary butyl halides, like tertiary halides in general, are prepared with extreme ease by merely treating tertiary butyl alcohol (from cracked gases) with the proper halide acid. The reaction is almost as rapid and complete as between a base and an acid. The only precaution which has to be used with the tertiary halides is due to the ease with which they lose halide acid to give olefins at higher temperatures. t-Butyl iodide is very unstable.

Chlorination of n-BuCl by SO_2Cl_2 and Bz_2O_2 gives the 1,2-, the 1,3-, and the 1,4-Cl_2 compounds in the ratio $1:2:1$.[49]

5. Amyl halides. The commercial chlorination of n- and iso pentane (Sharples) gives all the possible amyl chlorides except neopentyl chloride. The amounts in which the individual chlorides appear is now settled.[50] The preparation of the *pure* amyl halides may be discussed as typical of higher halides of the corresponding types.

n-Amyl halides, $Me(CH_2)_4X$, are now readily available from n-amyl alcohol (Sharples). Even with such a Class 1 alcohol, rearrangement takes place. Thus HCl and $ZnCl_2$ give about a 50% yield of the 1-chloride and about 10% of the 2- and 3-compounds. Thionyl chloride and pyridine give an 80% yield of the 1-chloride with no detectable rearrangement products.[51]

Iso-amyl halides, $Me_2CHCH_2CH_2X$, have long been available by the usual methods from isoamyl alcohol from fusel oil. The amount of rearrangement is probably little more than with n-amyl alcohol.

Active amyl halides, $MeEtCHCH_2X$, are obtained from active amyl alcohol from fusel oil. The chloride, bromide, and iodide have $[\alpha]D$ of about 1.7, 3.8, and 6.1 respectively.[52] The yields are poor because of the nearby branch in the chain.

Neopentyl halides (Me_3CCH_2X) are available only by special reactions.[53,54] The chloride is obtained by the chlorination of neopentane. The fact that this gives no rearrangement products whereas the replacement of OH in the

[49] Kharasch, Brown. *J. Am. Chem. Soc.* **61**, 2142 (1939).
[50] Hass, Weber. *Ind. Eng. Chem., Anal. Ed.* **7**, 231 (1935).
[51] Whitmore, Karnatz, Popkin. *J. Am. Chem. Soc.* **60**, 2540 (1938).
[52] Whitmore, Olewine. *J. Am. Chem. Soc.* **60**, 2570 (1938).
[53] Whitmore, Fleming. *J. Am. Chem. Soc.* **55**, 4161 (1933).
[54] Whitmore, Wittle, Popkin. *J. Am. Chem. Soc.* **61**, 1585 (1939).

corresponding alcohols gives practically entirely rearranged products indicates that in chlorination the unstable neopentyl carbonium ion is not formed at any stage in the process. The chlorination may go through a complex in which a neopentane molecule is held by a hydrogen bond to a chlorine molecule.

$$Me_3CCH_2- \quad = \quad Np-$$

$$Np:H \ + \ :\overset{..}{\underset{..}{Cl}}:\overset{..}{\underset{..}{Cl}}: \ \rightarrow \ Np:H:\overset{..}{\underset{..}{Cl}}:\overset{..}{\underset{..}{Cl}}:$$

As the greater attraction of the Cl for the H pulls it away from the C of the neopentyl group, the latter is never deprived of its electron pair but attracts the "positive" Cl with its sextet of electrons. There result the more stable compounds HCl and NpCl.

These chlorides react normally with Mg to give Grignard reagents which give RHgCl with HgCl₂. The Hg compounds react with Br₂ and I₂ to give the corresponding bromides and iodides which are not obtainable by ordinary reactions because of the tendency of the neopentyl system to rearrange during replacement reactions. Halides of this type react with Na surprisingly to give cyclopropane derivatives.[55] With other reagents they react slowly by way of an ionization step.[56] The action of neopentyl iodide with alcoholic KOH is remarkable in giving neopentane as the chief product.[57]

Secondary *n*-amyl halides (2- and 3-) are difficult to prepare pure. Probably the best methods are the use of the alcohols with thionyl chloride or phosphorus tribromide with pyridine in each case.

Secondary isoamyl halides, $Me_2CHCHXMe$, are available only by the addition of HX to isopropyl ethylene. Even this reaction gives an equal amount of tertiary halide.[58] This rearrangement goes through a carbonium ion formed by *addition*. The active form of the olefin may add a molecule of HCl

The addition complex (A) may give a Cl⁻ ion and the carbonium ion (B) which is in equilibrium with (C). Union of Cl⁻ with the last two would give the observed products. On the other hand, (A) may undergo an intra-molecular shift of the chlorine with its complete octet of electrons to the

[55] Whitmore et al. *J. Am. Chem. Soc.* **63**, 124 (1941); **65**, 1469 (1943).
[56] Dostrovsky, Hughes, Ingold. *J. Chem. Soc.* **1946** 146 ff.
[57] Whitmore, Wittle, Popkin. *J. Am. Chem. Soc.* **61**, 1586 (1939).
[58] Whitmore, Johnston. *J. Am. Chem. Soc.* **55**, 5020 (1933).

secondary C^+, thus satisfying its deficiency and forming the normal addition product. If this is the mechanism, the C^+ in (A) also satisfies itself by attracting an electron pair and its attached proton from the t-C. Then an intramolecular shift of the Cl with its octet gives the rearranged chloride. The formation of s-isoamyl chloride and t-amyl chloride by the addition of dry HCl to isopropylethylene is most significant for the theory of rearrangements. It is interesting that the secondary isoamyl chloride when once formed is unusually unreactive and stable, showing no tendency to rearrange to the tertiary chloride. Thus the rearrangement takes place during the change of the alcohol to the chloride. This tendency to rearrangement can be made useful in synthesis. Thus diisopropylcarbinol, which is available as a by-product of the synthesis of methanol from CO, reacts smoothly with HCl to give dimethylisobutylcarbinyl chloride $Me_2CHCH_2C(Cl)Me_2$.

Tertiary amyl halides, Me_2EtCX, are readily available by the action of HX with commercial t-amyl alcohol.

Hexyl halides, as would be expected offer still more complications. A good example is the result of warming 2-Et-1-butanol with $ZnCl_2$ and conc. HCl for eight hours. The product contains at least the following substituted pentanes: 3-Me-3-Cl-, 2-Me-2-Cl-, 3-Me-2-Cl, 4-Me-2-Cl, and 3-Cl- and 2-Cl-hexanes, listed in order of decreasing formation, together with a smaller amount of the "normal" product 2-Et-1-Cl-butane. The only predicted rearrangement product *not* found was 2-Me-3-Cl-pentane. On the other hand, treatment of 2-Et-1-butanol with $SOCl_2$ and pyridine gave an 80% yield of unrearranged chloride.[59]

The most difficultly obtainable hexyl chlorides are those containing the neopentyl chloride system, 1-Cl- and 3-Cl-2,2-Me_2-butanes, which are formed in about 20% and 10% yields in the chlorination of neohexane.[60] The latter can also be made by the addition of HCl to t-butylethylene (p. 76). The isomeric 1-Cl-3,3-Me_2-butane is also formed but is more readily obtained by passing ethylene into t-BuCl and $AlCl_3$ at $-30°$.[61] This probably involves an *ionization* to give $AlCl_4^-$ and t-Bu carbonium ions, the latter adding to the activated ethylene molecules (p. 27) to give $Me_3CCH_2CH_2$ carbonium ions which unite with the $AlCl_4^-$ ions. The product is decomposed by water to give the primary chloride and aluminum hydroxide.

sec-Hexyl iodide obtained by the vigorous action of HI on the hexahydric alcohols, mannitol and dulcitol, is the 2-compound as shown by conversion to the 1-olefin and the hydration of the latter to an alcohol which gave only n-butyric and acetic acids on oxidation.

Higher alkyl halides have been made in great number. In most cases their preparation depends on the preparation of the higher alcohol. A method of adding several carbons to a given halide consists in treating its Grignard

[59] Whitmore, Karnatz. *J. Am. Chem. Soc.* **60,** 2533 (1938).
[60] Whitmore, Bernstein, Mixon. *J. Am. Chem. Soc.* **60,** 2539 (1938).
[61] Schmerling. *J. Am. Chem. Soc.* **67,** 1152 (1945).

compound with the tolyl sulfonate of a 1-chloro-ω-hydroxy compound obtained from a polymethylene glycol.[62]

$$C_7H_7SO_3(CH_2)_nCl + RMgX \rightarrow R(CH_2)_nCl + C_7H_7SO_3MgBr$$

The chlorination of n-heptane with sulfuryl chloride and benzoyl peroxide gives 90% of monochlorides consisting about one fifth of the 1-chloride and the rest of mixed secondary chlorides.[63]

Identification of Alkyl Halides. This is possible by conversion to a variety of crystalline compounds. Among these are the acyl-anilides formed from phenyl isocyanate (C_6H_5NCO) and the Grignard reagent.

$$RMgX + C_6H_5NCO \rightarrow C_6H_5NHCOR$$

The m. (°C.) for the acyl-anilides from the simpler alkyl groups follow (Schwartz 1931): Me 114, Et 105, n-Pr 92, iso-Pr 103, n-Bu 63, iso-Bu 114, sec-Bu 111, t-Bu 132, n-Am 96, 2-sec-Am 95, 3-sec-Am 124, iso-Am 112, sec-iso-Am 78, $active$-Am 88, t-Am 92, neo-Am 132, n-Hex 69, n-Hep 57, 2-sec-Octyl 73. Correspondingly α-naphthyl isocyanate gives naphthalides, $RCONHC_{10}H_7$, m. (°C.): Me 160, Et 126, n-Pr 121, n-Bu 112.[64]

Another method of identification is the reaction with potassium 3-nitrophthalimide to give $(NO_2)C_6H_3(CO)_2N-R$. The m. (°C.) of typical derivatives follow: Me 113, Et 106, n-Pr 85, n-Bu 72, iso-Am 94. Obviously this method can give best results only with halides of Class 1. A similar method uses Na o-benzoic sulfimide.[65]

Still another means is the conversion to the Grignard reagent, treatment with oxygen and identification of the resulting alcohol in one of the usual ways (p. 133).

Alkyl halides have been identified by conversion to RHgX through the Grignard reagents.[66-68] The resulting alkylmercuric halides are readily crystallized (poison) and have definite melting points. This method fails with tertiary halides and is only fair with secondary. The mercury compounds can be treated in alkaline solution with acetylene to give

$$RHgC \equiv CHgR,$$

compounds of high melting points (p. 69).[69]

[62] Rossander, Marvel. *J. Am. Chem. Soc.* **50**, 1491 (1928).

[63] Kharasch, Brown. *J. Am. Chem. Soc.* **61**, 2142 (1939).

[64] Gilman, Furry. *J. Am. Chem. Soc.* **50**, 1214 (1928).

[65] Merritt, Levey, Cutter. *J. Am. Chem. Soc.* **61**, 15 (1939).

[66] Marvel, Calvery. *J. Am. Chem. Soc.* **45**, 820 (1923).

[67] Hill. *J. Am. Chem. Soc.* **50**, 167 (1928).

[68] Vaughn. *J. Am. Chem. Soc.* **55**, 3453 (1933).

[69] Spahr, Vogt, Nieuwland. *J. Am. Chem. Soc.* **55**, 2465 (1933).

B. Saturated Dihalides, $C_nH_{2n}X_2$

Methylene Halides, CH_2X_2

Methylene chloride, b. 39.8°, is obtained as a by-product in the commercial reduction of carbon tetrachloride to chloroform. It is remarkably stable and unreactive.[70] It can be used as a refrigerant. It is a useful solvent for nitrocellulose, cellulose acetate, polyvinyl acetate, waxes, fats, rubber and the like.

Methylene bromide, b. 99°, and *methylene iodide,* b. 181°, are obtained by reducing bromoform and iodoform with alkaline arsenite solution.[71] The iodide is used as a heavy liquid (d. 3.3) in separating heavy minerals such as sulfides from materials such as silicates which have densities below 3. Lighter minerals can be separated by diluting the iodide with lighter liquids to give a mixture in which one mineral will float and others will sink. The iodide exists in two crystalline forms, m. 5.5° and 6.0° (stable form).[72]

The halogen atoms in the methylene halides are less active than those in methyl halides. They react readily only with strongly alkaline reagents.

Ethylidene Halides, CH_3CHX_2. Ethylidene chloride, b. 58°, is made from acetaldehyde with phosphorus pentachloride or phosgene, the other product being $POCl_3$ or CO_2. It can be made by passing ethylene chloride over Al_2O_3 at 400° to give vinyl chloride and HCl which combine in presence of $AlCl_3$ at 125° to give ethylidene chloride. The bromide is more difficult to prepare from acetaldehyde and PBr_5, because the ready dissociation to PBr_3 and Br_2 causes bromination of the alpha carbon. The best preparation of ethylidene bromide or iodide is from HBr or HI and acetylene. Hydrolysis of the ethylidene halides gives acetaldehyde.

Ethylene Halides, XCH_2CH_2X, are obtained by direct addition of the halogens to ethylene.

Ethylene chloride, b. 84°, is available as a very cheap by-product of the production of ethylene chlorohydrin from ethylene and chlorine water.[73] It is "the oil of the Dutch chemists" (p. 29). It can also be made from liquid chlorine and dry ethylene.[74] It is characterized by great chemical stability and high solvent power. Like all polyhalides it burns difficultly. Under pressure it gives ordinary halide reactions. Thus, with NaCN, it gives ethylene dicyanide, succinonitrile. With SO_3 it gives $ClCH_2CH_2SO_2Cl$, beta-chloroethane sulfonyl chloride, which reacts with CaO to give ethylene oxide. A rubber substitute, Thiokol, which is resistant to solvents which attack ordinary rubber, has been made from ethylene chloride and Na_2S_x. This consists of high polymers containing alternate ethylene and polysulfide residues.

[70] Carlise, Levine. *Ind. Eng. Chem.* **24,** 146 (1932).
[71] "Org. Syntheses."
[72] Stone. *J. Am. Chem. Soc.* **54,** 112 (1932).
[73] Ellis. "Chemistry of Petroleum Derivatives." Reinhold, 1934. p. 468.
[74] Curme. *Chem. Met. Eng.* **25,** 999 (1921).

Ethylene diamine is also made from the chloride. Treatment of the chloride with bases under pressure produces vinyl chloride, $CH_2 = CHCl$. When sodium acetate is used vinyl chloride and acetate are obtained. These readily polymerize (Vinylite resins).

The U. S. Dept. of Agriculture has a publication giving a complete summary of the chemistry and uses of ethylene chloride with 470 references to the literature. Misc. Publication No. 117 (1932).

Ethylene bromide, b. 131°, m. 9°, is made in enormous amounts (over one million pounds per month in 1936) for use with tetraethyl lead in anti-knock fuel. It supplies bromine to convert the lead to lead bromide which escapes as such from the exhaust of the engine. The amount of ethylene bromide used for this purpose is so great that processes have been developed for extracting the bromine from sea water.[75] Sea water contains 65–70 parts per million of bromine whereas it contains less than 5 parts per *billion* of gold.[76]

As would be expected, ethylene bromide is more reactive than the chloride. In addition to the reactions of primary alkyl bromides it reacts with zinc dust to give the olefin, a reaction typical of 1,2-dibromides.

Ethylene iodide, m. 81°, is formed from ethylene and iodine wet with alcohol. It reverts to ethylene and iodine on heating.[77] The instability of 1,2-diiodides is utilized for making olefins by heating the corresponding dichloride or dibromide with NaI.

Ethylene chlorobromide, $ClCH_2CH_2Br$, b. 107°, is obtained by passing ethylene into bromine containing an equivalent amount of chlorine ("bromine chloride"). For a time it was used in Ethyl fluid in place of the bromide until it was found that only the bromine in the compound was effective. Its reactions are practically identical with those of the bromide or chloride. There is little difference in the reactivity of the two different halogen atoms in the molecule. One exception is the reaction with sodium sulfite which takes place practically entirely with the bromine giving sodium β-chloroethyl sulfonate, $ClCH_2CH_2SO_3Na$.

Ethylene chloroiodide, $ClCH_2CH_2I$, b. 140°, is prepared from ethylene and iodine monochloride, ICl. Similarly *ethylene bromoiodide*, b. 163°, has been prepared.

Ethylidene and ethylene halides react with an excess of alcoholic potash to give acetylene.

Propylidene Halides, 1,1- or αα-, $CH_3CH_2CHX_2$. These are made from propionaldehyde similarly to the ethylidene halides. $EtCHCl_2$, b. 86°; $EtCHBr_2$, b. 130°.

Acetone Dihalides, 2,2- or ββ-, $CH_3CX_2CH_3$. The chloride is prepared from acetone and phosphorus pentachloride in poor yield. A considerable amount of $CH_3CCl = CH_2$ is formed. Phosphorus pentabromide acts as a

[75] Stewart. *Ind. Eng. Chem.* **26**, 361 (1934).
[76] Gurévich. *Chimie & industrie* **29**, 284 (1933).
[77] Arnold, Kistiakowsky. *J. Chem. Physics* **1**, 166 (1933).

brominating agent, introducing bromine into the methyl groups. The bromide and iodide are obtained by adding HX to methylacetylene.

Propylene Halides, 1,2- or $\alpha\beta$-, CH_3CHXCH_2X. These are made by adding halogens to propylene. In the addition of iodine chloride to propylene the chief product is 1-iodo-2-chloropropane. Thus the iodine is the positive part of the addend.

The propylidene, propylene and acetone dihalides give methylacetylene with reagents like alcoholic potash or sodamide. With metallic zinc the propylene halides give propylene.

Trimethylene Halides, 1,3- or $\alpha\gamma$-$XCH_2CH_2CH_2X$. These are obtained from trimethylene glycol (1,3-dihydroxypropane), a by-product of glycerol manufacture, by the ordinary methods for making alkyl halides. The addition of HBr to allyl bromide or allyl chloride in the presence of peroxides takes place rapidly to give the 1,3-dihalide.[78] In the presence of antioxidants the addition is slow and produces the 1,2-compound. The two halogen atoms in the 1,3-compound act independently and almost exactly like other primary halides. It is possible to replace one or both of them under suitable experimental conditions. The dichloride is important as an intermediate for the general anesthetic, cyclopropane. For this purpose it is made by chlorinating propane. Heating with Zn gives cyclopropane.[79] (Mallinckrodt).

Dihalides of butane and isobutane are available by the methods already outlined. Special methods may be named for the following:

a.) Tetramethylene dichloride is available from the 1,4-addition of chlorine to butadiene, followed by hydrogenation (DuPont).

b.) Tetramethylene dibromide, $Br(CH_2)_4Br$, has been made from trimethylene dibromide[80] and from tetramethylenediamine.[81] The reactions follow:

$$Br(CH_2)_3Br + PhONa \rightarrow PhO(CH_2)_3Br \xrightarrow[H_2SO_4]{KCN} \xrightarrow{EtOH} \xrightarrow{Na} \xrightarrow[EtOH]{HBr} Br(CH_2)_4Br$$

$$H_2N(CH_2)_4NH_2 \rightarrow PhCONH(CH_2)_4NHCOPh \xrightarrow{PBr_5} PhCN + Br(CH_2)_4Br$$

The hydrolysis of dihalides containing a tertiary halogen gives aldehydes or ketones by rearrangement (p. 219).[82] Thus isobutylene dibromide and Me_3-ethylene dibromide give isobutyraldehyde and methyl isopropyl ketone respectively.

Tetramethylethylene dibromide reacts with sodio-organic compounds to give NaBr, tetramethylethylene and free radicals or their products.

[78] Kharasch, Mayo. *J. Am. Chem. Soc.* **55**, 2468 (1933).
[79] Hass et al. *Ind. Eng. Chem.* **28**, 1178 (1936).
[80] Marvel, Tanenbaum. *J. Am. Chem. Soc.* **44**, 2645 (1922).
[81] v. Braun, Kemke. *Ber.* **55B**, 3526 (1922).
[82] Evers. "Org. Syntheses."

Many *higher dihalides* have been prepared. In general the aldehydes react with phosphorus pentachloride to give $RCHCl_2$. These substances react readily with sodamide to give monosubstituted acetylenes. Ketones with the same reagent give dichlorides, $RCCl_2R'$, but in even poorer yields than with acetone. The dihalides, $RCHXCHXR'$, in which R' is either an alkyl group or hydrogen, are obtained in the usual way from halogens and the corresponding olefins. Amylene dichlorides are available commercially (Sharples). A special method is available for preparing dibromides of the type $RR'CBr-CHBrR''$ in which R'' is an alkyl group or hydrogen. Tertiary alcohols heated to about 80° are treated with bromine as rapidly as the latter is decolorized. The process may be illustrated by means of tertiary amyl alcohol.

$$(CH_3)_2C(OH)CH_2CH_3 + Br_2 \rightarrow (CH_3)_2CBrCHBrCH_3 + H_2O$$

This process is simpler than the preparation of the olefin and addition of bromine to it.

The higher $\alpha\omega$-dihalides, $X(CH_2)_nX$, can be prepared by the ordinary methods from the corresponding glycols obtained by the reduction of dibasic esters by means of a large excess of sodium and absolute alcohol[83] or by hydrogenation with Cu chromite type catalysts.[84] The reactions of these dihalides are like those of the corresponding propane derivatives. They give the ordinary reactions of Class 1 alkyl halides (p. 74). Decamethylene bromide with sodium and ether gives higher paraffins, $H-[(CH_2)_{10}]_n-H$, in which n is 2 to 7 and higher.[85]

C. TRIHALIDES

The most important of these are the *haloforms*, CHX_3.

Fluoroform, CHF_3, b. −82°, is made from bromoform and the fluorides of antimony and mercury.[86] SbF_3 with $CHCl_3$ gives chlorodifluoromethane, $CHClF_2$, a colorless gas of faint sweet odor, b. −40°, very soluble in water but not hydrolyzed. It is non-toxic.[87] Dichlorofluoromethane, $CHCl_2F$, b. 15°, has similar properties but is more reactive and is toxic.

Chloroform, $CHCl_3$, b. 61°, is now prepared by the reduction of carbon tetrachloride with moist iron. An older method is the action of bleaching powder with ethanol or acetone. This is an example of the *Haloform Reaction* which is given by primary and secondary carbinols containing at least one methyl group and by the corresponding carbonyl compounds. The first products are probably substances like Cl_3CCHO or Cl_3CCOCH_3 which then react with the lime to give chloroform and formate or acetate. The grouping X_3CCO- acts with bases to give CHX_3 and organic salts. The first step

[83] Bouveault, Blanc. *Bull. soc. chim.* (3) 31, 666 (1904).
[84] Adkins. The University of Wisconsin Press, 1937.
[85] Carothers et al. *J. Am. Chem. Soc.* 52, 5279 (1930).
[86] Henne. *J. Am. Chem. Soc.* 59, 1400 (1937).
[87] Booth, Bixby. *Ind. Eng. Chem.* 24, 637 (1932).

apparently involves addition of the base to the carbonyl group. When addition is blocked, the haloform reaction fails.[88]

Chloroform was one of the first anesthetics. It acts very rapidly. For this use it has to be protected from oxygen which reacts with it to give phosgene, $COCl_2$, and HCl, both very toxic.[89] Chloroform U.S.P. (United States Pharmacopeia) and D.A.B. (Deutches Arzneibuch) contains a trace of alcohol to convert any phosgene to harmless diethyl carbonate.

Chloroform, with anhydrous alcohol and sodium, gives ethyl orthoformate, $HC(OEt)_3$. With sodium ethylate it gives the same product but also CO, sodium formate, ethyl ether and some ethylene. Alcoholic KOH converts chloroform to potassium formate.

With primary amines, in presence of a base, chloroform gives an isocyanide (carbylamine) of intensely disagreeable odor (p. 419).

Chloroform adds to acetone in the presence of a solid base (aldol condensation) to give trichloro-t-butyl alcohol, $(Cl_3C)Me_2COH$, which is used as an antispasmodic (chloretone, Mothersills remedy for seasickness). Chloroform will add similarly to methyl ethyl ketone but not to aldehydes nor to higher ketones.[90]

Chloroform can be added to 1-olefins in presence of acetyl peroxide.[91]

$$RCH = CH_2 \rightarrow RCH_2CH_2CCl_3$$

Chloroform has a pleasant odor and a sweet burning taste. It is slightly soluble in water but completely miscible with most organic liquids. It is a good solvent for fats, waxes and hydrocarbons. It is used as an antiseptic to prevent fermentation.

Bromoform, $CHBr_3$, b. 151°, m. 7.7°, d. 2.9, is most conveniently made by treating acetone with sodium hydroxide and bromine.

Iodoform, CHI_3, m. 120°, is made by the carefully controlled electrolysis of an aqueous solution of alcohol, an inorganic iodide and sodium carbonate. In this way, practically all the iodide is converted to iodoform. Substances which give the haloform reaction (p. 90) can be recognized by treatment with iodine and a base to form a yellowish precipitate of iodoform of characteristic odor (*Iodoform Test*).[92]

The older use of iodoform as a dusting powder for wounds has been superseded by the use of the sulfa drugs. Modern medical practice makes little use of dusting powders for wounds, however.

Fluorochlorobromomethane, CHFClBr, b. 36°, has not yet been resolved into its optically active enantiomers.[93]

[88] Fuson, Tullock. *J. Am. Chem. Soc.* **56**, 1638 (1934).
[89] Hill. *J. Am. Chem. Soc.* **54**, 32 (1932).
[90] Howard. *J. Am. Chem. Soc.* **47**, 455 (1925).
[91] Kharasch, Jensen, Urry. *Science* **102**, 128 (1945).
[92] Lieben. *Ann. Spl. Bd.* **7**, 218 (1870).
[93] Berry, Sturtevant. *J. Am. Chem. Soc.* **64**, 1599 (1942).

Methyl chloroform, CH_3CCl_3, b. 74°, is made by chlorinating ethylidene chloride or from acetyl chloride and phosphorus pentachloride.

1,1,2-Trichloroethane, vinyl trichloride, Cl_2CHCH_2Cl, b. 113°, is made by chlorinating ethylene chloride.

The only readily obtainable *trihalides of propane* are of the 1,2,3 type.

Glyceryl chloride, trichlorohydrin, $CH_2ClCHClCH_2Cl$, b. 158°, is readily obtained from glycerol and PCl_5 It is available in large quantities as a by-product from the synthesis of glycerine. The corresponding bromide is readily obtained by adding bromine to allyl bromide, $CH_2 = CH - CH_2Br$. The corresponding iodide loses two atoms of iodine spontaneously giving allyl iodide, the product of the action of glycerol with phosphorus and iodine in the absence of water.

D. TETRAHALIDES

Carbon tetrafluoride, b. −130°, is obtained by electrolyzing fused KHF_2 using a soft carbon anode.[94] It is very inert.

Dichlorodifluoromethane, CCl_2F_2, b. −30°, is made from CCl_4 and SbF_3 as an important refrigerant[95] (*Freon*, F-12, Kinetic No. 12), is non-toxic, non-corrosive and non-inflammable. Its thermodynamic properties are ideal.[96] Freon is important as the solvent for pyrethrum and other insecticides in the Aerosol Bombs used in controlling malarial mosquitoes in long range planes, and for domestic purposes.

Carbon tetrachloride, b. 76.7°,[97] is the compound most readily obtained by the chlorination of methane because the intermediate products are more easily chlorinated than methane itself. This process is not yet commercial. The commercial preparation of carbon tetrachloride is from CS_2 with sulfur monochloride in the presence of iron as a catalyst (p. 457).

Carbon tetrachloride is rather inactive. With alcoholic KOH at 100° it gives CO and formates but no K_2CO_3 and no orthocarbonate. This peculiar behavior is probably due to the primary replacement of one chlorine by hydrogen under the influence of the base. This difference in one of the four halogen atoms is even more noticeable with the bromide and iodide.

Carbon tetrachloride reacts with fuming sulfuric acid to give phosgene, $COCl_2$, and pyrosulfuryl chloride, $S_2O_5Cl_2$. This reaction is convenient for generating small amounts of phosgene (poison).

Reduction of CCl_4 with moist iron gives chloroform. Methylene chloride, hexachloroethane and tetrachloroethylene are obtained as by-products.

Carbon tetrachloride is used in fire extinguishers because of its high volatility and the incombustibility of its heavy vapors (Pyrene). It is effective

[94] Lebeau, Damiens. *Compt. rend.* 182, 1340 (1926).

[95] Ellis. "Chemistry of Petroleum Derivatives." Reinhold, 1934. pp. 707–10.

[96] Thompson. *Ind. Eng. Chem.* 24, 620 (1932).

[97] Ahrens. *Sammling Chemische-Technischen Vorträge* 1905, 116.

in fires near electrical equipment where water would cause short circuits. Its use may involve danger because it reacts with water vapor at flame temperature to give phosgene which is dangerously toxic. Thus a space in which a fire has been extinguished by carbon tetrachloride should be ventilated at once.

Carbon tetrachloride is an excellent solvent. Alone and mixed with hydrocarbons it is widely used in dry cleaning (Carbona).

Carbon tetrachloride was formerly used to eliminate certain internal parasites (hook worm). For this purpose it had to be carefully freed from traces of impurities such as CS_2.

Carbon tetrabromide, b. 189°, m. 94°, is obtained from bromoform and sodium hypobromite solution on long standing. Boiling with alkali readily gives bromoform and a hypobromite. Compare the results of Kharasch. With alcoholic KOH, carbon tetrabromide gives mainly CO and ethylene but no carbonate. CBr_4 has been used as a brominating agent.[98]

Carbon tetraiodide (solid) is obtained from CCl_4 and AlI_3. It is very unstable toward oxygen. The reported hydrolysis to hypoiodite is erroneous. There is no basis for the assumption of positive iodine in the compound.[99] (See p. 425.)

Carbon tetrahalides can be obtained by the pyrolysis of X_3C-COX in which the halogen atoms may be varied at will. Thus $CBrCl_3$, b. 104°, and $ClCl_3$, b. 142°, have been made from trichloroacetic halides.[100]

Sym-**Tetrachloroethane,** acetylene tetrachloride, $Cl_2CHCHCl_2$, b. 146°, is obtained by passing acetylene and chlorine into antimony pentachloride. This moderates the explosive reaction which would take place between chlorine and acetylene gases if mixed directly. Another device for moderating this violent reaction is to pass the gases into a chamber filled with fine sand. Tetrachlorethane is also prepared by *the liquid phase chlorination* of acetylene in a lead lined reactor using iron filings as catalyst and tetrachlorethane as the reaction medium 70–90° C. It reacts readily with bases to give trichloroethylene, another valuable solvent. With air and steam in ultraviolet light it gives mainly dichloroacetic acid.

Tetrachloroethane is the most powerful solvent in the chlorinated series. It is an excellent solvent for cellulose acetate, varnishes, paints and general organic compounds. It is more toxic than the unsaturated chlorides and must be used in special apparatus to avoid industrial hazards. In the presence of moisture it attacks metals with the production of unsaturated chlorides.

The tetrabromide is made similarly. The tetraiodide does not exist.

Unsym-**Tetrachloroethane,** $ClCH_2CCl_3$, b. 130°, is obtained by chlorinating ethylene chloride.

---

[98] Hunter, Edgar. *J. Am Chem. Soc.* **34**, 2025 (1932).

[99] Kharasch, Mayo. *J. Org. Chem.* **2**, 76 (1937).

[100] Simons, Sloat, Meunier. *J. Am. Chem. Soc.* **61**, 435 (1939).

E. Higher Halides

A large number of fluorocarbons have been prepared.[101]

FLUOROCARBONS

	FC	HC
CF_4	b. $-128°$	b. -161
C_2F_4	$-78°$	
C_2F_6	$-79°$	-89
C_3F_8	$-36°$	-42
C_4F_{10}	$4°$	$-.5$
C_5F_{12}	$30°$	36
C_6F_{14}	$51°(60°)$	$68°$

Three general methods have been developed for the production of fluoro-carbons: (1) fluorination with gaseous fluorine; (2) fluorination with CoF_3 and similar agents; and (3) electrolysis of organic substances in liquid hydrogen fluoride.[102] The latter process promises to become the most important. The unusual stability of the fluorocarbons makes them of tremendous importance.[103]

The reaction of chlorine with acetylene gives as by-products **pentachloro-ethane,** b. 162°, m. $-29°$, and **hexachloroethane,** a white crystalline solid of camphor odor which melts and sublimes at 186°. C_2Cl_6 is also obtained as a by-product in making chloroform from CCl_4. Vigorous chlorination of perchloroethane gives two molecules of carbon tetrachloride. Similarly after most hydrocarbons are exhaustively chlorinated, further chlorination breaks a carbon to carbon linkage giving smaller completely chlorinated molecules. Pentachloroethane is a valuable high boiling solvent. It is toxic and must be used with proper care. Hexachloroethane is used in safety explosives, smoke screens and insecticides.

Treatment of a normal hydrocarbon with bromine and ferric bromide results in the introduction of one bromine on each carbon. This is probably because ferric bromide catalyzes the removal of HBr leaving a double bond which adds bromine. A more practical method of making 1,2,3,4-tetrabromo-butane is from butadiene.[104]

It was formerly believed that chlorine acted differently from bromine with aliphatic hydrocarbons, in that the chlorine tended to accumulate on the carbon where chlorination started. This is due to the fact that no catalyst is needed with chlorine. In the presence of ferric chloride the result would probably be the same as with bromine.

Vigorous chlorination, first at moderate temperatures and then, destruc-tively at high temperature, converts ethane to Cl_4-ethylene; propane to the latter substance and CCl_4; butane to Cl_6-butadiene; and n- or iso-pentane to

[101] Henne. "Org. Reactions," II.

[102] Simons et al. *Abstracts of papers, 114th Meeting, Am. Chem. Soc. Sept. 13–17, 1948* (Portland).

[103] Finger, Reed. *Chem. Ind.* **64**, 51 (1949).

[104] Jacobson. *J. Am. Chem. Soc.* **54**, 1545 (1932).

Cl_6-cyclopentadiene. As usual, the chlorine atoms attached to unsaturated carbon atoms are inactive.

1,1,1,2,3-Cl_5-propane, m. 178°, subl., is available commercially.

Asym-**Heptachloropropane,** $Cl_2CHCCl_2CCl_3$, b. 248°, m. 30°, is readily obtained by the action of chloroform and tetrachloroethylene in presence of aluminum chloride.

Highly chlorinated hydrocarbons and their action with $AlCl_3$ have been studied.[105] Chlorinated paraffin wax is used with aromatic hydrocarbons and $AlCl^3$ to give complex materials such as Paraflow which inhibit the separation of wax from lubricating oils on cooling (pour-point depressers). Chlorinated wax with phenols and naphthols in presence of $AlCl_3$ give similar compounds which can be esterfied with dibasic acids such as phthalic acid (Santopour).[106]

Many chlorides of carbon are known. They are perchlorohydrocarbons. CCl_4, b. 77°; C_2Cl_4, b. 121°; C_2Cl_6, b. 187°; C_3Cl_6, b. 210°; C_3Cl_8, m. 160°, b. 269°; C_4Cl_6, perchlorobutadiene, m. 32°, b. 269°; C_4Cl_6, perchlorobutyne, m. 39°, b. 284°; C_4Cl_8, b. 275°; C_6Cl_6, hexachlorobenzene, m. 226°, b. 326°.

F. UNSATURATED HALIDES, $C_nH_{2n-1}X$

1. **Vinyl halides,** $CH_2 = CHX$, are obtained by the action of alcoholic KOH on the corresponding ethylene and ethylidene dihalides. They are also formed by the addition of HX to acetylene and by the action of HX with CaC_2.[107] As is general for halogens attached to unsaturated carbon atoms, those in the vinyl halides are very unreactive. Almost the only ˜eaction which they give readily is with an excess of alkali to form acetylene. The vinyl halides add HX to form ethylidene halides. In the presence of peroxides, vinyl bromide adds HBr contrary to Markownikoff's Rule to give ethylene bromide.[108] Vinyl halides polymerize readily[109] (Vinylite resins Geon, Koroseal). **Vinyl fluoride** b. −51°, **chloride** b. −13.9°, **bromide** b. 16°, **iodide** b. 56°.

2. *a.* The **allyl halides,** $CH_2 = CH - CH_2X$, are made by the usual methods for alkyl halides from allyl alcohol obtained from glycerol. The iodide can also be made from glycerol, phosphorus and iodine under anhydrous conditions. Perhaps 1,2,3-triiodo-propane is formed and loses two iodine atoms from adjacent carbons.

The allyl halides act like alkyl halides of Class 1 (p. 74) but are more reactive. The freedom from side reactions is due to lack of tendency to form the related unsaturated compound, allene, $CH_2 = C = CH_2$, and to the fact that a 1:3 shift of the double bond would give an identical product.

$$CH_2 = CHCH_2X \rightarrow XCH_2CH = CH_2$$

[105] *Prins. Rec. trav. chim.* 51, 1065 (1932).
[106] Rieff. *C. A.* 31, 848 (1937).
[107] *Ann. Rep. Chem. Soc.* (London) 1922, 73.
[108] Kharasch, McNab, Mayo. *J. Am. Chem. Soc.* 55, 2521 (1933).
[109] *Ann. Rep. Chem. Soc.* (London) 1930, 93.

When HBr is added to an allyl halide in the absence of peroxides (presence of antioxidants) the product is a propylene halide (1,2) while in the presence of peroxides a trimethylene dihalide (1,3) is formed.[110] The addition in the presence of peroxides is more rapid in their absence. Addition to allyl halides without any special precautions gives varying proportions of the two dihalides depending on the amount of oxidation which the double bond has undergone. The addition of HI follows Markownikoff's Rule since it effectively reduces peroxides.

Allyl chloride, b. 44°, is made by the direct chlorination of propylene (Shell). The mechanism is not definitely known.[111] It seems unlikely, however, that the methyl group is chlorinated in the ordinary way while the olefinic linkage is left intact. The following changes seem more probable

$$H_2C:CH:CH_2 + \left[\ddot{C}l: \right]^+ \longrightarrow \left[H_2C:CH:CH_2 \atop H \quad +:\ddot{C}l: \right]^+$$
$$\underset{H}{\overset{\cdot\cdot}{}}$$

$$HCl \xleftarrow{\ :\ddot{C}l:\ } H^+ + H_2C::CH:CH_2Cl$$

A positive chlorine adds to the free electron pair of an activated propylene leaving the middle carbon with only six electrons. This condition is relieved by the attraction of an electron pair from the methyl group with the expulsion of a proton which unites with a negative chlorine. This possibly occurs by a chain mechanism

$$H^+ + :\overset{\cdot\cdot}{Cl}:\overset{\cdot\cdot}{Cl}: \rightarrow H:\overset{\cdot\cdot}{Cl}:\overset{\cdot\cdot}{Cl}: \rightarrow H:\overset{\cdot\cdot}{Cl}: + \overset{\cdot\cdot}{Cl}:$$

The entire change could be written without indicating the electrons as follows:

$$H_3C-CH-CH_2 + Cl^+ \rightarrow [H_3C-CH-CH_2Cl]^+$$
$$\overset{+}{} \quad \overset{-}{}$$
$$H^+ + H_2C=CH-CH_2Cl$$

$$H^+ + Cl_2 \rightarrow [HCl_2]^+ \rightarrow HCl + Cl^+$$

Allyl chloride, made from propylene, is used for the commercial preparation of allyl alcohol and of glycerol (Shell).

Allyl chloride, on standing, changes slowly to polymers with n as 9, 12, 5, 25, 11, and 7 in order of decreasing amount.[112] With $NaNH_2$ in liquid ammonia,[113] allyl chloride gives complex reactions. The simplest product is *1,3,5-hexatriene*, b. 81°, in 30% yield. A larger yield of 3-vinyl-4-chloromethylcyclohexene was formed, apparently by a Diels-Alder reaction of the first product with allyl chloride. Use of excess $NaNH_2$ gives a 40% yield of

[110] Kharasch, Mayo. *J. Am. Chem. Soc.* **55**, 2468 (1933).
[111] Groll, Hearne. *Ind. Eng. Chem.* **31**, 1530 (1939).
[112] *Ann. Rep. Chem. Soc.* (London) 1930, 93.
[113] Kharasch, Sternfeld. *J. Am. Chem. Soc.* **61**, 2318 (1939).

3-vinyl-4-α-butadienylcyclohexene, together with higher polymers of hexadiene. The non-conjugated position of the ring double bond in the cyclic compounds is strange. Allyl halides react with conjugated dienes to give substituted H_4-benzyl halides (Diels-Alder).

Allyl bromide, b. 71°; **iodide,** b. 102°. Allyl iodide reacts 60 times as rapidly with sodium ethylate as does n-propyl iodide.

Treatment of allyl bromide with Mg ordinarily gives diallyl. Good yields of the Grignard reagent can be obtained by using large amounts of ether.[114]

Allyl Grignard reagents give higher yields of addition products with carbonyl compounds than do the corresponding n-Pr reagents.[115] The same is true of higher homologs.

b. The **alpha** and **beta halogen propylenes,** $XCH = CHCH_3$, and $CH_2 = CXCH_3$, are obtained by removing one HX from propylidene and acetone dihalides respectively. The halogen atoms attached to unsaturated carbon atoms are inactive except to alcoholic potash which removes HX. The compounds react with dilute acids to give propionaldehyde and acetone respectively. This is because they add water under the influence of the acid and form compounds containing the grouping, C(OH)X, which loses HX spontaneously to give a carbonyl group.

$$ClCH = CHMe, \text{ b. } 36°; \qquad CH_2 = CClMe, \text{ b. } 23°.$$

3. *a.* **Crotyl halides,** $CH_3CH = CHCH_2X$, present interesting examples of 1:3 or allylic rearrangements. Crotyl alcohol, $CH_3CH = CHCH_2OH$, obtained by the reduction of crotonic aldehyde, reacts with PBr_3 and pyridine to give crotyl bromide and methylvinylcarbinyl bromide.[116] The same mixture is obtained from HBr and methylvinylcarbinol, $CH_3CHOHCH = CH_2$, formed from acrolein and MeMgX. When the halogen in a crotyl halide is replaced by hydroxyl the product is not pure crotyl alcohol but a mixture with methylvinylcarbinol (p. 136). This type of change is very common.[117] As in the case of 1:2 rearrangements, it is readily pictured on a simple electronic basis. The removal by the reagent of the OH from crotyl alcohol and from methylvinylcarbinol give the following two forms of the carbonium ion which are in resonance

$$CH_3CH = CH\overset{+}{C}H_2 \leftrightarrow CH_3\overset{+}{C}HCH = CH_2$$

Union with halide ions thus gives a mixture of the halides. Each of the latter can give the same resonance hybrid carbonium ion and consequently the mixture of the two alcohols.

Another preparation of these halides is by addition of HCl or HBr to butadiene (p. 57). 3-Cl-1-butene, b. 84°, $n^{20}D$ 1.4350: 1-Cl-2-butene, b. 64°,

[114] Gilman, McGlumphy. *Bull. soc. chim.* **43,** 1322 (1928).
[115] Young, Roberts. *J. Am. Chem. Soc.* **67,** 319 (1945).
[116] Young et al. *J. Am. Chem. Soc.* **58,** 104 (1936); **59,** 2051 (1937); **60,** 847 (1938).
[117] *Ann. Rep. Chem. Soc.* (London) 1923.

n^{20}D 1.4150. Slow distillation of the high boiling isomer through an efficient column give the other isomer.

b. **1-Halogen-1-butene,** EtCH = CHX, can be made by removing HX from a butylidene halide. Its halogen is inert.

c. **4-Halogen-1-butene,** $XCH_2CH_2CH = CH_2$, can be made from a tetramethylene dihalide. The reactions are those of a primary bromide and of a 1-olefin, uninfluenced by each other. Contrast the behaviour of ally halides.

d. **Methallyl bromide,** CH_2CMeCH_2Br, and isocrotyl bromide, Me_2CCHX, are obtained by removing HBr from isobutylene dibromide, the latter predominating. **Methallyl chloride,** b. 72°, is available commercially (Shell) from the action of chlorine and isobutylene. Even at 0° no dichloride is obtained. This is an example of Kondakoff's Rule (p. 40). The older assumption that the dichloride is first formed and then loses HCl is untenable. Methallyl chloride at 100° unites with fuming HCl to form the dichloride. The more modern conception that isobutylene undergoes ordinary chlorination of one of the Me groups and of the CH_2 group without attack on the double bond is likewise untenable. The action is probably like the high temperature chlorination of propylene (p. 95). A positive Cl adds to the terminal unsaturated carbon thus leaving the central carbon electronically deficient. This is overcome by expulsion of a proton from one of the Me groups to give methallyl chloride or from the $-CH_2Cl$ to give isocrotyl chloride. The proton combines with Cl^- to form HCl. The formation of HCl may be the controlling factor. In the case of the action of bromine with isobutylene, the dibromide is formed, possibly because of the lesser tendency to form HBr.

Many other cases of the formation of unsaturated halides from the action of halogens on olefins are known. For instance, the diisobutylenes (p. 40) fail to give dichlorides and dibromides, in both cases giving unsaturated halides and HX. The processes involved are probably like that involved in the ready halogenation of an enol (pp. 202, 216).

Methallyl halides react like allyl halides being very reactive and giving normal replacement reactions.

e. **Isocrotyl chloride,** $Me_2C = CHCl$, b. 68°, is formed by the action of bases with isobutylidene chloride, $Me_2CHCHCl_2$ and from isobutylene and chlorine. Its halogen is inactive. Treatment with acid gives isobutyraldehyde. This involves addition of water contrary to Markownikoff's Rule or, more probably, a rearrangement like that in the hydrolysis of isobutylene dibromide to give isobutyraldehyde (p. 117). Isocrotyl bromide reacts with sodium alcoholates at 140° to give the unsaturated ethers, $Me_2C = CHOR$. These react with dilute acids in sealed tubes forming isobutryaldehyde.

4. a. **n-Pentenyl halides** resemble the other unsaturated halides in their preparation and reactions depending on the relation of the halogen and the double bond. Those with the halogen on an unsaturated carbon are inactive. With the halogen and double bond in the allyl arrangement, $C = C - CX$, the

halogen is very reactive and allylic rearrangements take place. With farther separation of the halogen and the double bond they act independently.

b. **3,3-Dimethylallyl halides,** $Me_2C=CHCH_2X$, offer an example of complete allylic rearrangement. Treatment of dimethylvinylcarbinol with PX_3 gives only the primary halide while replacement of the primary halogen gives only the tertiary alcohol. Isoprene with HBr gives $Me_2C=CHCH_2Br$.[118]

G. Unsaturated Halides, $C_nH_{2n-3}X$

1. **Chloroprene,** 2-chloro-1,3-butadiene, is made from vinylacetylene and HCl (p. 72).[119] It is important as the monomer from which Neoprene rubber (formerly Duprene) is made.[120,121] It was the first practical synthetic rubber. For special uses it is superior to natural or other synthetic rubbers because of its low solubility in petroleum hydrocarbons.

Chloroprene heated to 80° for several days in the presence of charcoal and a polymerization inhibitor gives greater than 20% conversion to 1,5-dichloro-1,5-cyclooctadiene. Butadiene and 2,3-dichlorobutadiene likewise give cyclooctadienes but in lower yields.[122]

2. **Monohalogen acetylenes,** $HC\equiv CX$, are obtained from sym-dihalogenated ethylenes and bases. Both the bromo and chloro compounds have been made and are violently explosive, spontaneously inflammable, have strong odors and are toxic. Nef explained these peculiar properties on the basis that the halogenated acetylenes contain bivalent carbon much as do the isocyanides.[123]

$$XHC=C \qquad X_2C=C \qquad RN=C$$

On the other hand, compounds $R-C\equiv C-X$ are stable non-toxic unobjectionable substances. The explosive nature of monohalogen acetylenes suggests that of the halides of nitrogen in which the halogen atoms are regarded as positive. The following analogy may be significant,

$$:\!Cl\;[\;:\!\overset{..}{\underset{..}{N}}\!:\;]\;Cl:\qquad\qquad H\;[\;:\!\overset{..}{C}:\overset{..}{C}:\;]\;Cl:$$

$$:Cl:$$

$RNCl_2$ and $RC\equiv CCl$ have none of the properties of NCl_3 and C_2HCl except that hydrolysis gives HOCl. In the absence of water they are stable.

3. *a.* **Propargyl halides,** propiolic halides, $HC\equiv C-CH_2X$, can be made from propargyl alcohol by the usual reagents. They give normal reactions much like those of the allyl halides.

[118] Ellis. "Chemistry of Petroleum Derivatives." Reinhold, 1934. p. 646.
[119] *Ann. Rep. Chem. Soc.* (London) 1932, 108, 125.
[120] Nieuwland et al. *J. Am. Chem. Soc.* 53, 4197 (1931).
[121] Carothers et al. *J. Am. Chem. Soc.* 53, 4203 (1931).
[122] Foster, Schreiber. *J. Am. Chem. Soc.* 70, 2303 (1948).
[123] Nef. *Ann.* 298, 202 (1897).

b. **Allylene halides,** $CH_3C \equiv CX$, are obtained from allylene and hypohalite solutions. The iodo compounds can best be made from the acetylene and iodine in liquid NH_3. They react with mercuric cyanide solution to give $(RC \equiv C)_2Hg$, crystalline compounds of high melting points.[124] The halogen in $R - C \equiv CX$ is reactive but does not behave like that in an alkyl halide but rather like halogen in combination with O or N. Thus NaOEt gives the acetylene and a hypohalite.

c. Many higher acetylenic halides have been prepared. They involve rearrangements similar to those of the substituted allyl halides.

Unsaturated polyhalides are obtained by removing HX or X_2 from higher halides.

1. **1,1-Dichloroethene,** acetylidene dichloride, $CH_2 = CCl_2$, b. 37°; *Dibromo*, b. 92°. The former polymerizes to Saran plastics (Dow).

2. **1,2-Dihalogen ethenes,** acetylene dihalides, exist in *cis* and *trans* forms

$$\begin{array}{ccc} \text{HCCl} & & \text{HCCl} \\ \| & & \| \\ \text{HCCl} & & \text{ClCH} \\ cis \quad \text{b. } 48° & & trans \quad \text{b. } 60° \end{array}$$

Direct addition of Cl_2 to excess acetylene at 200° gives both forms. The commercial product is a mixture, b. 52°±, obtained by the reduction of acetylene tetrachloride with iron. The conversion of the two forms has been studied.[125] In common with other halogenated solvents, *dichloroethylene* is non-inflammable. Its halogen atoms are very unreactive. Consequently there is no danger of corrosion. It is an excellent solvent for the extraction of materials ranging from plant perfumes to rubber. It can be used in place of ordinary ether for general extractions.

Unsaturated iodides form polyvalent iodine derivatives having a fair degree of stability. They thus resemble the corresponding aromatic compounds. Thus Cl_2 in an inert solvent will react with the iodine atom of 1-Cl-2-I-ethylene. The dichloride reacts with sodium carbonate to give the iodoso compound and the latter disproportionates on warming to give the iodo and iodoxy compounds.

$$ClCH = CHI \rightarrow ClCH = CHICl_2 \rightarrow ClCH = CHIO \rightarrow ClCH = CHIO_2$$

3. **1,3-Dichloro-2-butene,** b. 128°, is available commercially. As would be expected, only the 1-Cl is reactive (DuPont).

4. **Trichloroethene,** $ClCH = CCl_2$, b. 87°, is made from acetylene tetrachloride and bases. It is stable and inactive except on exposure to both air and light. It is the most important of the chlorinated solvents. It is a powerful solvent for fats, resins, bitumens, rubber, sulfur and phosphorus. It is a valuable noninflammable substitute for benzene and petroleum hydrocarbons in dry cleaning. It is used for extracting the residual oil from vegetable oil cakes left after pressing out most of the oil, for degreasing leather and,

[124] Vaughn. *J. Am. Chem. Soc.* **55,** 3453 (1933).
[125] Wood, Dickinson. *J. Am. Chem. Soc.* **61,** 3259 (1939).

textiles and as an assistant in soaps for scouring textiles. It is used with stabilizers or inhibitors for degreasing metals.

The reactions of trichloroethylene are initiated at the unsaturated linkage rather than at a chlorine atom. Thus, dilute acids convert it to chloroacetic acid by hydration of the double bond followed by loss of HCl from the system $-CCl_2(OH)$ and hydrolysis of the resulting acid chloride. It reacts violently with conc. HNO_3 to give dichlorodinitromethane, $Cl_2C(NO_2)_2$. With formaldehyde and H_2SO_4 it gives $O(CH_2CHClCO_2H)_2$ by a peculiar combination of condensation, hydrolysis and dehydration.

5. **Tetrachloroethene, perchloroethylene,** $Cl_2C=CCl_2$, b. 120°, is made from pentachloroethane obtained as a by-product in the action of chlorine with acetylene and from hexachlorethane obtained in the same way and as a by-product in the reduction of carbon tetrachloride. With conc. HNO_3 it gives mainly CO_2. With formaldehyde and H_2SO_4 it forms $HOCH_2CCl_2CO_2H$. Tetrachloroethylene is more expensive than the other chlorinated solvents. It has been used in place of CCl_4 for treatment of hook worms, and similar parasites.

6. **Tetrafluoroethylene,** $F_2C=CF_2$, and trifluorochloroethylene,

$$F_2C=CFCl,$$

give the heat resistant chemically inert "noble plastics" known respectively as Teflon (DuPont) and Kel$-$F (Kellogg).[126,127]

7. Bases, with trichloropropane from glycerol, give **γ-chloroallyl chloride,** $ClCH=CHCH_2Cl$, which contains an active and an inactive chlorine. It and the dibromo compound, with Grignard reagents, give $RCH_2CH=CHX$ from which $NaNH_2$ gives 1-acetylenes.

The important soil amendment and disinfectant *DD* is a 1:1 mixture of 1,2-Cl_2-propane and 1,3-Cl_2-propene obtained as a by-product in the manufacture of allyl chloride from propylene. It is especially valuable in pineapple culture.

8. A special method for making unsaturated triiodo derivatives is the treatment of cuprous acetylides with iodine to give $RCI=CI_2$.[128]

9. **1,3-Dichloro-2,4-hexadiene** occurs as *cis* and *trans* isomers of b. 81° and 61° at 17 mm.

10. **Dichloroacetylene,** b. 32°, and **dibromoacetylene,** b. 77°, are toxic and have odors like isocyanides. The older conception of these as derivatives of acetylidene, $H_2C=C$ (p. 99),[129,130] is disproved by electron diffraction and interference studies.[131,132]

Strangely diiodoacetylene, C_2I_2, is an inert solid, m. 80°.

[126] Renfrew, Lewis. *Ind. Eng. Chem.* **38**, 870 (1946).
[127] *Chem. Ind.* **63**, 586 (1948).
[128] Lespieau. *Bull. soc. chim.* (4) **3**, 638 (1908).
[129] Nef. *Ann.* **298**, 202 (1897).
[130] Ellis. "Chemistry of Petroleum Derivatives." Reinhold, 1934. p. 675.
[131] de Laszlo. *Nature* **135**, 474 (1935).
[132] Finbak, Hassel. *Arch. Math. Naturvidenskab.* **45**, 38 (1941).

III. ALCOHOLS

A. Saturated Alcohols, $C_nH_{2n+1}OH$

These are mono hydroxyl derivatives of the paraffin hydrocarbons. The reactions of the OH group are like those of water but are modified by the length and structure of the attached carbon skeleton. The electronic similarity between water and the alcohols is worth emphasizing.

$$: \overset{..}{\underset{..}{O}} : H \qquad\qquad\qquad : \overset{..}{\underset{..}{O}} : H$$
$$H \qquad\qquad\qquad\qquad\qquad R$$

In each case the ionization of H is enough to give reaction with metals of the Na and Ca families. In both cases, the attack of acids is by coordination of a proton with one of the free electron pairs of the oxygen atom. The carbon attached to the hydroxyl is rendered more reactive than in the paraffins. The reactions of both the C and O in the system COH are profoundly influenced by the number of alkyl groups attached to the C. Hence the important classification into *primary*, *secondary* and *tertiary alcohols* or *mono-*, *di-* and *tri-substituted carbinols*, RCH_2OH, $RR'CHOH$ and $RR'R''COH$ in which the alkyl groups may be alike or different. This system of classification and nomenclature of alcohols depends on regarding them as substitution products of the simplest alcohol, CARBINOL (methanol CH_3OH), in which one, two or three of the methyl H atoms are replaced by organic groups. In the simple alcohols, these groups contain only C and H.

The alcohols range from non-viscous liquids of b.p. 66° and density 0.8 to solids (*n*-decyl alcohol, m. 7°, b. 230°, d. 0.84). Alcohols higher than C_{10} are waxy solids increasingly like the higher hydrocarbons as the carbon content rises. The boiling points of the alcohols are much higher than those of the corresponding hydrocarbons because of the *association* of the liquids. This constitutes another resemblance to water. These differences in boiling points are 230°, 171°, 102°, 58°, and 57° for the C_1, C_2, C_5, C_{10} and C_{16} members. Evidently the association effect decreases as the alcohols lose their resemblance to water.

The first three alcohols are completely soluble in water. The higher ones become less soluble as the carbon content increases.

Many primary alcohols occur in natural products, usually in the form of esters.

In contrast to the paraffins, the alcohols are very reactive with a great variety of chemical reagents. Apparently the introduction of the oxygen atom into the non-polar hydrocarbon molecule decreases the stability and inactivity of the latter. Moreover, the oxygen atom through its ability to form oxonium compounds offers an active spot in the molecule.

It is impossible to generalize satisfactorily on either the preparations or reactions of alcohols, ROH, since the size and especially the branching of the alkyl group have such profound effects. Consequently the individual alcohols will be considered until enough members of the series have been presented to show the many peculiarities produced by changes in the alkyl group attached to the hydroxyl. It is not possible to generalize accurately even according to the three classes of alcohols. The alcohols form the best refutation to the older idea that knowledge of the first two or three members of an homologous series makes possible the prediction of the reactions of the higher members.

Individual Alcohols

Methanol, methyl alcohol, wood alcohol, wood spirit, carbinol, CH_3OH, b. 66°, occurs in nature in a few esters such as methyl salicylate, in oil of wintergreen and as complex methyl ethers in the lignin portion of wood and many alkaloids and natural dyes. Pine lignin contains 15% methoxyl.[1]

The crude pyroligneous acid which separates from the tar obtained in wood distillation contains up to 3% methanol along with a smaller amount of acetone and a larger amount of acetic acid. The latter is neutralized with lime and the methanol and acetone are removed and separated by distillation. The methanol first obtained contains impurities which render it specially suitable as a denaturant for ethanol. One of these impurities which is now being separated commercially is diacetyl to the extent of about 3% of the crude wood alcohol.

The modern method of preparing methanol is from carbon monoxide and hydrogen. It is produced in important quantities from the Fischer Tropsch synthesis of gasoline (p. 420).

Methanol has also been made from hydrogen and fermentation carbon dioxide (CSC).

Although methanol is assumed to be an intermediate in the oxidation of methane[2] it is not obtained in that way because it is so much more easily oxidized than CH_4. With a 9:1 mixture of methane and oxygen at 360° and 100 atm. 17% of the methane oxidized was recovered as methanol. Less than 1% of formaldehyde was obtained. The other products were CO, CO_2 and H_2O. No H_2 was found.[3]

Methanol forms azeotropic mixtures (constant b.p.) with many liquids. Such a mixture with $CHCl_3$ (b. 61°) boiling at 53° is useful in separating methanol (b. 66°) from acetone (b. 56°).[4] The per cent of the second liquid and the boiling points of several such mixtures follow: acetone, 86, 56°; carbon disulfide, 86, 38°; benzene, 60, 58°; carbon tetrachloride, 79, 56°; chloroform, 87, 53°; methyl iodide, 93, 39°; n-hexane, 73, 50°. A commercial azeotrope of methanol, acetone and methyl acetate (24:48:28) is sold as "Methyl Acetone."

Dynax (DuPont) is a methanol-benzene blended fuel for speed boats. Some mixtures of substances which cannot be separated by simple distillation

[1] Hibbert et al. *J. Am. Chem. Soc.* **61**, 509, 516, 725 (1939).
[2] Bone, Allum. *Proc. Roy. Soc.* (London) **A134**, 578 (1932).
[3] Newitt, Haffner. *Proc. Roy. Soc.* (London) **A134**, 591 (1932).
[4] *Ann. Rep. Chem. Soc.* (London) **1921**, p. 61.

can be mixed with methanol and fractionated because one forms an azeotrope and the other does not. If the liquid desired is insoluble in water the methanol can then be washed out.

Since it forms no azeotrope with water it can be separated from the latter by careful fractionation (difference from ethanol).

The purest methanol can be obtained:

1. By conversion to the solid compound $CaCl_2 \cdot 4$ MeOH which can be heated to 100° to remove all impurities not removable from the methanol by distillation. Treatment of the compound with water and distillation gives pure methanol.

2. By conversion to the oxalate, $(CO_2Me)_2$, and purification by crystallization, m. 54°, and treatment with a base, followed by distillation.

As a matter of fact the best grade of commercial synthetic methanol is practically as pure as any ever obtained.

Acetone in wood distillation methanol can be detected as iodoform (odor, yellow crystals) by treating with a base and iodine.

Methanol, in common with all compounds containing the group $MeO-$, can be determined by the *Zeisel methoxyl method* which consists in treating with constant boiling HI, distilling MeI formed, reacting it with $AgNO_3$ and determining the AgI formed.

Methanol dissolves solid carbon dioxide.

The most important reactions of methanol are:

1. Mild oxidation or dehydrogenation to give formaldehyde, H_2CO, important as a disinfectant and as an intermediate in synthetic resins. This is accomplished at 500–600° in contact with silver or copper gauze either with or without the admission of air. More vigorous oxidation of methanol gives formic acid and finally CO_2.

Most methods for the qualitative and quantitative determination of methanol depend on its conversion to formaldehyde.

2. With sulfuric acid to form dimethyl ether, Me_2O. Conditions which might be expected to dehydrate the methanol to give methylene, $CH_2=$, give ethylene and higher polymers instead. Under other conditions, the acid sulfate can be formed. This on distillation gives the volatile dimethyl sulfate (*toxic, odorless*). $2 \text{ MeHSO}_4 \rightarrow \text{Me}_2\text{SO}_4 + \text{H}_2\text{SO}_4$.

3. Halide acids yield methyl halides. The chloride is best formed by HCl and $ZnCl_2$ under pressure. The bromide is formed readily from a mixture of methanol, sodium bromide and concentrated H_2SO_4.

The action of alcohols with halide acids is very different from that of a base and an acid. It undoubtedly takes place through the formation of an *oxonium salt* which then decomposes to give the halide.

$$\text{MeOH} + \text{HX} \rightleftharpoons \begin{bmatrix} \text{H---O---H} \\ | \\ \text{Me} \end{bmatrix}^+ \text{X}^- \rightleftharpoons \text{H}_2\text{O} + \text{MeX}$$

Both these reactions are reversible but the second is less easily reversed. The fact that HI acts much more readily than HCl with methanol and most other alcohols is due to the greater ease with which it forms oxonium salts. The *addition* reaction to form oxonium ions is probably the first step in all reactions of alcohols with acidic reagents and catalysts.

In the case of some of the higher alcohols crystalline oxonium salts with halide acids are obtained (p. 129).

4. With halides of phosphorus to give methyl halides. The only reaction actually used is that with phosphorus and iodine to give methyl iodide (p. 81).

5. With metals to give methylates (methoxides). Sodium and potassium react violently giving MeONa and MeOK which crystallize with MeOH. Dry sodium methylate is available commercially (Mathieson). Aluminum (amalgamated to give a clean surface) reacts readily. Even magnesium reacts with no difficulty (difference from higher alcohols). The easiest way to obtain *absolute methanol* is to make a suspension of magnesium methylate in a portion of the alcohol and add it to the rest. Any trace of water reacts to give insoluble MgO.

$$(MeO)_2Mg + H_2O \rightarrow 2\ MeOH + MgO$$

Magnesium methylate is valuable in removing the last traces of water (0.1–0.2%) from ordinary "absolute" ethyl alcohol for use as a solvent in malonic ester and acetoacetic ester syntheses and similar reactions in which a small trace of moisture is very harmful and a little methanol is harmless.

6. With organic acids to give esters (p. 120).

$$HCO_2H + CH_3OH \rightleftharpoons HCO_2CH_3 + H_2O$$

In this case the formic acid is a strong enough acid to require no other catalyst. In the case of higher acids, it is usual to add a trace of sulfuric acid or to saturate the mixture of acid and alcohol with dry hydrogen chloride gas. Since methanol is cheap the reversible reaction can be forced more nearly to completion by using a large excess of the alcohol.

7. When esters of more expensive acids are needed it is best first to convert the acid to its chloride by PCl₅ or SOCl₂

$$\underset{RC-Cl}{\overset{O}{\parallel}} + CH_3OH \rightarrow \underset{RC-OCH_3}{\overset{O}{\parallel}} + HCl$$

This reaction is not reversible.

8. Methanol reacts with fused alkalies to give hydrogen, formates and finally carbonates.[5]

$$CH_3OH \rightarrow HCO_2K + 2\ H_2 \rightarrow K_2CO_3 + H_2$$

9. Methanol combines with many substances as alcohol of crystallization: CaCl₂·4 MeOH, MgCl₂·6 MeOH, CuSO₄·2 MeOH, BaO·2 MeOH,

[5] Fry, Otto. *J. Am. Chem. Soc.* **50**, 1122 (1928).

MeOK·MeOH etc. Because alcohols readily form such compounds, especially with $CaCl_2$, the latter is not used to dry them.

10. Boron trifluoride with methyl and ethyl alcohols gives highly conducting solutions.[6] This is another example of the tendency for completion of an octet.

$$
\begin{array}{c}
\text{F} \\
\overset{\cdot\cdot}{\text{B}} : \text{F} \\
\text{F}
\end{array}
+ \text{R} : \overset{\cdot\cdot}{\underset{\cdot\cdot}{\text{O}}} : \text{H} \rightarrow
\left[
\begin{array}{c}
\text{F} \\
\text{R} : \overset{\cdot\cdot}{\underset{\cdot\cdot}{\text{O}}} : \overset{\cdot\cdot}{\underset{\cdot\cdot}{\text{B}}} : \text{F} \\
\text{F}
\end{array}
\right]^{-}
\text{H}^+
$$

The linkage between the oxygen and boron is a coordinate link and may also be represented as follows:

$$
\begin{array}{c}
\text{H} \\
\searrow \\
\overset{+}{\text{O}} - \overset{-}{\text{B}}\text{F}_3 \qquad \text{H}^+[\text{RO} \rightarrow \text{BF}_3]^- \\
\nearrow \\
\text{R}
\end{array}
$$

The chief uses of methanol are:

1. Denaturing ethyl alcohol to free it from beverage taxes. The methanol is much more toxic than ethyl alcohol. Sub-lethal doses are likely to cause blindness. When prepared by wood distillation it contains impurities which give the resulting denatured alcohol a bad taste and odor. During the prohibition era in the United States many tragedies resulted from the removal of the disagreeable materials from denatured alcohol without the removal of the methanol before the mixture was diverted to beverage uses. Ethyl alcohol completely denatured with methanol (10%) can be used successfully even for making ethyl esters.

2. Solvent for shellac and varnishes.

3. Antifreeze for engine radiators. Its low MW is an advantage, its relatively high volatility a disadvantage.

4. Production of formaldehyde for synthetic resins.

5. Methylation of OH and NH_2 groups of intermediates and dyes.

6. Production of MeCl as a refrigerant and an intermediate.

Analytical reactions of methanol.[7]

Ethanol, ethyl alcohol, alcohol, grain alcohol, methylcarbinol, spirits of wine, CH_3CH_2OH, b. 78.3°.

The oldest preparation of ethyl alcohol is still the most important, namely the fermentation of glucose and materials related to it.

Before World War II, practically all ethanol in the U.S.A. was made from blackstrap molasses, mainly from Cuba. Since then surplus grain has been used as available. An increasing amount of ethanol has been made from

[6] *Ann. Rep. Chem. Soc.* (London) 1932, 104.

[7] Ahrens. *Sammlung Chemische-Technischen Vorträge* 1913, 20.

ethylene by sulfuric acid hydration. Direct catalytic hydration has been achieved[8] (Shell). In other countries potatoes and various grains are used as the source of glucose. During the prohibition era in the United States enormous amounts of crude "yellow chip" glucose from corn starch was used in the illegal production of alcohol.

The net reaction for the production of ethanol by the enzymes of yeast is

$$C_6H_{12}O_6 \rightarrow 2\ C_2H_5OH + 2\ CO_2.$$

Much study of this important process and its by-products has led to the conclusion that the steps involved include the formation of hexose diphosphate and its oxidative fission to an equilibrium mixture of phospho-dihydroxy-acetone and phospho-glyceraldehyde. These then go through pyruvic acid and acetaldehyde to ethanol.[9–13]

The fermented liquid contains about 10% ethanol which is concentrated by careful fractional distillation. The fore-run always contains acetaldehyde. The alcohol is obtained in various concentrations depending on its ultimate use. The maximum obtainable is an azeotropic mixture with water containing 95.6% alcohol by weight and boiling at 78.2° whereas absolute ethanol boils at 78.4°.

Protein impurities in the starting material give fusel oil containing n-propyl, isobutyl, isoamyl and active amyl alcohols together with smaller amounts of a very complex mixture of higher alcohols and oily compounds.[14] The amount of fusel oil is about 3–11 parts per 1000 of ethanol produced. Part of the fusel oil is fractionated to give n-propyl, isobutyl and mixed amyl alcohols. The latter cannot be separated by distillation in any ordinary equipment (p. 118).

The residue ("slops") from the fermentation mixture is worked up for glycerol and potash salts. The final step is to burn the residue giving an ash containing about 30% K_2O.

During the fermentation, the weight of CO_2 produced nearly equals that of the ethanol. This is nearly pure and free from air. It is compressed and sold as liquid in tanks or as solid Dry Ice for special refrigeration purposes.

Fermentation.[15,16]

Another preparation for ethyl alcohol which is assuming increasing commercial importance is the hydration of ethylene from cracked gases.[17] The

[8] *Chem. Ind.* **61**, 787 (1947).
[9] Michaelis. *Ind. Eng. Chem.* **27**, 1037 (1935).
[10] Embden. *Z. physiol. Chem.* **230**, 1 (1934).
[11] Meyerhof, Kiessling. *Biochem. Z.* **276**, 239 (1935).
[12] Nord. *Chem. Rev.* **26**, 423 (1940).
[13] Fulmer. *Brewer's Digest* **1943**.
[14] Schorigin et al. *Ber.* **66B**, 1087 (1933).
[15] Ahrens. *Sammlung Chemische-Technischen Vorträge* **1902**, 50; **1914**, 46; **1924**, 48.
[16] *Chem. Rev.* **3**, 41 (1926).
[17] Ellis. "Chemistry of Petroleum Derivatives." Reinhold, 1934. p. 332.

ethylene is absorbed in sulfuric acid to form ethyl hydrogen sulfate. This is diluted and heated to hydrolyze the ester to ethyl alcohol. The dilute sulfuric acid is then concentrated and used again. The cost of concentration has been reduced to a small fraction of the cost of fresh acid. Ethyl ether is obtained very cheaply as a by-product of this process. A possible alternative to the concentrating of the dilute acid is its conversion to ammonium sulfate for fertilizer.

In 1942 the British were hydrogenating coal to oil, cracking the oil to ethylene and hydrating the latter to about 100 tons of ethanol per day.

A practical process was developed in Germany for the direct hydration of ethylene to ethanol. The catalyst consists of WO_2 and WO_3 from wolframite, promoted with 5% ZnO. Ethylene and water are introduced into the top of a tower reactor packed with catalyst and the reaction proceeds at 200–300 atmospheres and 300° C. A 20% solution of ethanol is obtained from the bottom of the tower. This continuous process represents an improvement over the batch hydration of ethylene using sulfuric acid.

This process is claimed to be applicable for *isopropanol* and *secondary butyl alcohol* from propylene and butylene, respectively. The shortage of molasses during World War II and the exhaustion of the surplus grain supply increased the production of ethanol from ethylene.

It is produced in important quantities from the Fischer Tropsch synthesis of gasoline (p. 420–1).

A practically unlimited source of glucose for alcohol production exists in the hydrolysis of cellulose (from sawdust or any cheap vegetable waste) by HCl under pressure.[18,19]

Absolute ethyl alcohol is usually made by treating the 96% alcohol with fresh quicklime, CaO. Anhydrous alcohol is made commercially by distilling the 96% alcohol with dry benzene. A ternary mixture of benzene, water and alcohol distills at 64.8°, followed by a binary mixture of benzene and alcohol, b. 68.2°, and finally by absolute alcohol at 78.3°.[20,21] The benzene can be recovered and used again. This method of drying by distillation with dry benzene is very useful with a variety of organic materials.

Anhydrous ethanol is also prepared commercially by heating the 96% alcohol with a fused mixture of anhydrous potassium and sodium acetates (CSC).

Ethanol of 99.5% can be obtained by a special use of calcium chloride.[22]

BaO is soluble in absolute alcohol. When such a solution is added to alcohol containing even a trace of water, insoluble $Ba(OH)_2$ is formed.

The density of absolute alcohol at 20° is 0.789 while that of 96% alcohol is 0.801.

[18] Bergius. *Chemistry & Industry* **1933**, 1045.
[19] Johnson. *Chem. Inds.* **56**, 226 (1944).
[20] Keyes. *Ind. Eng. Chem.* **21**, 998 (1929).
[21] Othmer, Wentworth. *Ind. Eng. Chem.* **32**, 1588 (1940).
[22] Noyes. *J. Am. Chem. Soc.* **56**, 226 (1923).

Highly purified ethanol lacks the burning taste of the ordinary substance. Proof spirit (U.S.) contains 50% ethanol by volume. Thus 90% alcohol (by volume) is 180 proof and absolute alcohol is 200 proof. In Great Britain proof spirit is 49.3% ethanol by weight or 57% by volume. A distilled liquor was originally defined as proof spirit if when poured over gun powder and lighted, it would burn and finally ignite the powder. If the ethanol content is less than 49.3% by weight the water left after the burning of the alcohol is sufficient to prevent the powder from burning.

Denatured alcohol is ethanol containing materials which prevent its use as a beverage. It is exempt from the tax on beverage alcohol. The nature of the denaturants used varies in different countries and at different times. The commonest are crude wood alcohol, benzene, and pyridine bases.

Alcohol finds very wide and increasing uses in industry. About 40% of the production in the United States is used in antifreeze for automobiles. It is a valuable solvent and chemical intermediate. Ethanol has been widely used outside the United States as a motor fuel both alone and mixed with gasoline (petrol). Absolute alcohol is fairly soluble in gasoline. The absorption of a small amount of water causes a separation of aqueous alcohol. This can be prevented by adding to the mixture benzene and higher alcohols.

Ethanol is used for preserving biological specimens. Because of its affinity for water (48 cc. H_2O + 52 cc. abs. EtOH give only 96.3 cc. of mixture, all measured at 20°) it is used for dehydrating such materials. For this purpose it should be free from traces of aldehyde. This is best removed by adding a small amount of potassium hydroxide and zinc dust or sodium amalgam, refluxing and then distilling the alcohol.[23]

In pharmaceutical chemistry, solutions in ethanol are known as *tinctures*.

Ethyl alcohol has been isolated in minute amounts from brain, blood and liver of non-alcoholic humans, dogs and pigs.[24]

Ethanol unites with $CaCl_2$ and other salts as methanol does, forming $CaCl_2.4$ EtOH etc.

Ethanol gives *reactions* similar to those of methanol:

1. Dehydrogenation and oxidation give acetaldehyde, CH_3CHO. The easiest small scale preparation is by the oxidation of ethyl alcohol with "chromic acid mixture" (a dichromate and sulfuric acid). Although acetaldehyde is much more sensitive to oxidation than ethanol, this process is effective because of the greater volatility of acetaldehyde, b. 20°. More vigorous oxidation of ethanol gives acetic acid which is stable to oxidation. Oxidation with half the theoretical amount of chromic acid mixture gives ethyl acetate.

Oxidation with nitric acid involves the methyl as well as the carbinol group, the products including glycolic acid, glyoxal, glyoxylic acid and oxalic acid. In the presence of bases alcohol is readily oxidized even by air. The aldehyde

[23] Stout, Schuette. *Ind. Eng. Chem., Anal. Ed.* 5, 100 (1933).
[24] Gettler et al. *Mikrochemie* 11, 167 (1932).

formed is changed to resinous materials by the base. Alcoholic potash (KOH) is a reducing agent especially for nitro compounds.

2. Treatment with sulfuric acid gives diethyl ether or ethylene according to the conditions. p-Toluene sulfonic acid (a by-product of saccharin manufacture) is sometimes used in place of sulfuric acid because it gives fewer side reactions. The best way to make pure ethylene from ethanol is by passing its vapor with superheated steam over a dehydrating catalyst such as Al_2O_3 at about 350°. At about 300° alumina gives a good yield of ether and very little ethylene.[25]

Ethanol with sulfur trioxide gives beta hydroxyethane sulfonic acid, *isethionic acid.* There is also formed the acid sulfate of this acid,

$$HOSO_2 - OCH_2CH_2SO_3H,$$

ethionic acid.

3. The reactions with halide acids and halides of phosphorus are the same as with methanol.

4. The reactions with metals are the same as with methanol except that magnesium does not react readily. Aluminum gives $Al(OEt)_3$, m. 130°, which can be distilled under reduced pressure (b. 205°/14 mm.) and serves as a useful catalyst and reagent. Aluminum alkoxides unite with alcohols to form *alkoxyacids* (Meerwein). $Al(OR)_3 + ROH \rightarrow H^+[Al(OR)_4]^-$. These are strong acids which can be titrated with indicators. The acid radical bears a striking resemblance electronically to those of sulfuric, phosphoric and perchloric acids, each involving a system of a central atom, four oxygen atoms, and 32 electrons. Sodium ethylate or ethoxide, $EtONa \cdot 2\ EtOH$, is converted to $EtONa$ at 200°.

The treatment of nearly anhydrous ethanol with metallic sodium will not remove the last trace of water because of the equilibrium

$$C_2H_5OH + NaOH \rightleftharpoons C_2H_5ONa + H_2O$$

If $Mg\ (OMe)_2$ is used, the last traces of water are removed because no soluble base is formed. Metallic calcium can be used because the $Ca(OH)_2$ and CaO formed are insoluble. Boiling absolute ethanol with BaO gives a precipitate of $Ba(OEt)_2$.

5. Ethanol reacts with acids to form esters slightly less rapidly than methanol.

6. It reacts with acid chlorides and acid anhydrides to give esters. Ethyl benzoate, $C_6H_5CO_2Et$, has a characteristic odor. The p-nitro compound, $NO_2C_6H_4CO_2Et$, m. 57° whereas the Me ester m. 97°. The Me and Et esters of 3,5-dinitrobenzoic acid m. 107.8° and 92.7° respectively.

7. With fused alkalies ethanol at 500° reacts to give hydrogen and acetates and then carbonates and methane[1] (Methanol, p. 105).

8. Ethanol reacts with chlorine to give chloral, Cl_3CCHO.

[25] Alvarado. *J. Am. Chem. Soc.* **50**, 790 (1928).

9. The formation of iodoform by treatment with a base and iodine is not a test for ethanol unless substances such as acetaldehyde, acetone and isopropyl alcohol are absent (p. 91).

Reactions of Ethyl Alcohol.[26]

Propyl Alcohols

1. **Normal propyl alcohol,** 1-propanol, ethylcarbinol, $CH_3CH_2CH_2OH$, b. 97°, constitutes 3–7% of the fusel oil from ordinary alcoholic fermentation. The amount is so small that most distillers do not separate it. The fusel oil from the Nipa palm of the Philippines is said to contain more propyl and butyl alcohols than other fusel oil.

n-Propyl alcohol is now available in considerable amounts as a by-product of the action of CO and H_2 (p. 420).

The *reactions* of n-propyl alcohol, a typical primary alcohol, RCH_2OH, are like those of ethyl alcohol. For instance oxidation, as with all primary alcohols, gives an aldehyde and then an acid of the same carbon content as the alcohol, in this case EtCHO, propionaldehyde, and $EtCO_2H$, propionic acid.

n-Propyl alcohol heated with an equimolecular amount of Na n-propylate under pressure at 250° undergoes a change characteristic of primary and secondary alcohols, RCH_2CH_2OH and $RMeCHOH$,[27] in which the net result is the removal of the ONa group with an alpha hydrogen (on carbon next the carbinol group) from another molecule to form NaOH and 2-Me-1-pentanol. By more vigorous treatment it is possible to obtain "tripropyl alcohol," probably 2,4-Me_2-1-heptanol.[28]

$$PrONa + HCH(Me)CH_2OH \rightarrow PrCH(Me)CH_2OH$$
$$\rightarrow PrCH(Me)CH_2CH(Me)CH_2OH$$

1-Propanol with fused alkalies at 500° gives methane and H_2 in the ratio 4:1 and carbonates (Methanol, p. 105). Some carbonization takes place.

2. **Isopropyl alcohol,** 2-propanol, "isopropanol," Petrohol, Ikohol, dimethylcarbinol, $(CH_3)_2CHOH$, b. 82.4°.

This alcohol is available in practically unlimited amounts from the hydration of propylene from cracked gases. The propylene is dissolved in sulfuric acid, the isopropyl hydrogen sulfate is diluted and hydrolyzed by heating. The isopropyl alcohol is distilled out and the dilute sulfuric acid is concentrated for re-use. The product finds many uses formerly filled by ethyl alcohol. About 0.2% of methylisobutylcarbinol is obtained as a by-product. No n-PrOH is formed even under favorable conditions for peroxide formation.[29] Diisopropyl ether is obtained as a by-product at a cost less than that of ordinary ethyl ether.

[26] Morris. *Chem. Revs.* 1932, 465.
[27] Guerbet. *Compt. rend.* **128,** 511 (1899).
[28] Weizmann, Bergmann, Haskelberg. *Chemistry & Industry* 1937, 587.
[29] Brooks. *J. Am. Chem. Soc.* **56,** 1998 (1934).

Isopropyl alcohol cannot be used in beverages and is not subject to the legal restrictions of ethyl alcohol.

Formerly isopropyl alcohol was obtained by the reduction of acetone. Now acetone is made by the catalytic dehydrogenation of isopropyl alcohol.

Isopropyl alcohol forms an azeotrope with 12% water, b. 80.4°. It forms many other azeotropes. The following give the per cent isopropyl alcohol and the boiling point of the azeotrope with the substance indicated: chloroform, 4.2, 60.8°; CCl_4, 18, 67°; ethylene dichloride, 45, 74°; trichloroethylene, 28, 74°; tetrachloroethylene, 80.6, 81.7°; n-hexane, 22, 61°; cyclohexane, 33, 68.6°; benzene, 33.3, 71.9°; toluene, 69, 80.6°.

The reactions of isopropyl alcohol, a typical secondary alcohol, RR'CHOH, differ from those of its normal isomer in several respects.

1. Ether formation is more difficult.

2. Dehydration to olefins is easier.

3. Reaction with metals is more difficult.

4. Esterification with acids is much slower.

5. Oxidation gives the three carbon ketone, acetone, CH_3COCH_3, instead of an aldehyde as obtained from a primary alcohol. The ketone is rather stable to oxidation whereas the aldehyde is readily oxidized to an acid of the same number of carbon atoms. More vigorous oxidation converts the acetone to acids but only by breaking the carbon chain thus giving acetic and formic or carbonic acids. This difference between 1-propanol and 2-propanol on oxidation is general for all primary and secondary alcohols, RCH_2OH and RR'CHOH.

6. It reacts in a complex way with bromine giving mainly brominated acetones such as $BrCH_2COCBr_3$ and isopropyl bromide.

7. The Guerbet condensation proceeds much as with 1-propanol. Alkalies or the alcoholate and the alcohol give condensation products between two or more molecules. As usual an alpha H atom is involved. The chief products are "diisopropyl alcohol," 4-Me-2-pentanol, methylisobutylcarbinol and "triisopropyl alcohol," 4,6-Me$_2$-2-heptanol. Fused alkalies at high temperature (500°) give carbonates, methane and H_2 quantitatively without any carbonization (Methanol, p. 105).

Butyl Alcohols

Unlike the lower alcohols these are not completely soluble in water, except t-butyl alcohol.

1. **Normal butyl alcohol,** 1-butanol, "Butanol," n-propylcarbinol, $CH_3CH_2CH_2CH_2OH$, b. 117.7°.

Since World War I this has been the most largely used of the butyl alcohols. Larger amounts of acetone were needed for smokeless powder manufacture than could be made from pyroligneous acid even by converting the calcium acetate into acetone. A fermentation process was developed for producing acetone[30] (C.S.C.). This yielded twice as much normal butyl alcohol as acetone. The alcohol had no adequate outlet until the development of the modern lacquer industry. Then acetone became the by-product. Other by-products are ethanol, hydrogen and carbon dioxide. The latter two were formerly converted into methanol catalytically.

The mechanism of the butylic fermentation may involve the aldol condensation of acetaldehyde formed from pyruvic acid (p. 107).[31] The following disproportionation could take place

$$2 \ MeCHOHCH_2CHO \rightarrow MeCH_2CH_2CH_2OH + MeCOCH_2CO_2H$$

The acetoacetic acid on decarboxylation would yield CO_2 and acetone.

The supply of 1-butanol is limited only by the demand since it is a primary product and the process starts with corn or molasses.[32]

n-Butyl acetate is made by the reduction of crotonic aldehyde by metals and acetic acid.

n-Butyl alcohol is also made by the catalytic hydrogenation of crotonic aldehyde (C. and C.).

1-Butanol is found in minute amounts in fusel oil.[33] This is probably formed from the small amount of norvaline in the proteins of the molasses.

Along with other higher alcohols, n-butanol is also produced in important quantities from CO and H_2 by the Fischer-Tropsch synthesis (p. 420).

The reactions of 1-butanol are essentially like those of ethyl alcohol.

2. Isobutyl alcohol, 2-methyl-1-propanol, isopropylcarbinol,

$$(CH_3)_2CHCH_2OH, \ b. \ 108°.$$

This is the longest known of the butyl alcohols, being obtained from fusel oil formed in ordinary alcoholic fermentation. The amount available from this source is relatively small. Isobutyl alcohol is not formed from glucose but from valine, obtained by the hydrolysis of the protein impurities in the fermenting mixture. The zymase of the yeast removes CO_2 and hydrolyzes off ammonia.

$$Me_2CHCH(NH_2)CO_2H + H_2O \rightarrow Me_2CHCH_2OH + NH_3 + CO_2$$
$$\text{Valine}$$

The valine is dextro rotatory but no asymmetric carbon is left in isobutyl alcohol. Fermentation isobutyl alcohol contains a minute impurity of a nitrogen compound which gives a scarlet color with sodium pentacyanosulfitoferroate.[34] It is possible to distinguish the fermentation and synthetic isobutyl alcohols by this test in 1:1000 dilution in water.

[30] Weizmann. C. A. **13**, 1595 (1919). Fernbach, Strange. C. A. **7**, 206 (1913).
[31] Johnson, Peterson, Fred. J. Biol. Chem. **101**, 145 (1933).
[32] Killeffer. Ind. Eng. Chem. **19**, 46, (1927).
[33] Emmerling. Ber. **35**, 694 (1902).
[34] Baudisch. Biochem Z. **232**, 35 (1931).

Isobutyl alcohol is a by-product of synthetic methanol manufacture.[35] In Germany, it was prepared by a modification of the methanol synthesis using 240 atmospheres and 430° C. and a catalyst consisting of zinc and chromium oxides. The starting materials are CO and H_2 from the usual synthesis gas sources. One plant had a crude isobutanol output of 40 T/D. Actually, methanol was formed also in ratio of 5:1 to isobutanol, but was recycled until converted to isobutanol.

Isobutyl alcohol gives the reactions typical of a primary alcohol except that the presence of the branched chain next to the carbinol group makes rearrangements unusually easy. Thus if the directions for converting n-butyl alcohol to its bromide by sodium bromide and sulfuric acid are applied to isobutyl alcohol a much poorer yield is obtained. Considerable amounts of tertiary butyl bromide and isobutylene and its polymers are obtained. A good example of the tendency for rearrangement is the fact that isobutyl alcohol with HBr gives 11% tertiary butyl bromide.[36] The best preparation of isobutyl bromide is however by means of gaseous HBr. Its preparation with PBr_3 is less satisfactory. This type of reaction is usually represented as follows:

$$3 \text{ ROH} + PBr_3 \rightarrow 3 \text{ RBr} + H_3PO_3$$

As a matter of fact the reaction is much more complex. It is probable that the primary reaction involves the formation of an alkyl phosphite, $(RO)_3P$ and HX. These react in steps giving $(RO)_2POH$, $ROP(OH)_2$, $P(OH)_3$ and RX.[36a] If this is correct it would seem desirable to pass a stream of HX gas through the mixture during the reaction.

The extreme case of rearrangement of isobutyl alcohol is obtained in dehydration at high temperatures which gives normal butylenes as well as isobutylene. In this case a methyl group rearranges. It has been shown that this rearrangement depends on the presence of acid (Ipatieff). Dehydration of isobutyl alcohol with Al_2O_3 completely free from acid gives pure isobutylene.

Isobutyl alcohol with fused alkalies at 500° gives much carbonization. This is apparently characteristic of alcohols having at least the chain CCCO, and is in sharp contrast to the behavior of isopropyl and tertiary butyl alcohols.[37]

3. **Secondary butyl alcohol,** 2-butanol, methylethylcarbinol, $CH_3CH_2CHOHCH_3$, b. 100°.

This alcohol is available from the hydration of n-butylenes from cracked gases by sulfuric acid. The isobutylene is first removed by more dilute acid than that necessary to dissolve the n-butylenes.

In addition to the preparation from the butylenes, 2-butanol can be obtained as follows:

[35] Frolich, Cryder. *Ind. Eng. Chem.* 22, 1051 (1930).
[36] Michael, Zeldler. *Ann.* 393, 81 (1912).
[36a] Ann. Rep. Chem. Soc. (London) 1918, 51.
[37] Fry, Otto. *J. Am. Chem. Soc.* 50, 1122 (1928).

1. By reduction of methyl ethyl ketone obtained with acetone in wood distillation. Now the pure ketone is obtained from the secondary alcohol by dehydrogenation (Shell). The reduction of a ketone is a general method for making secondary alcohols. Some bimolecular reduction product (a pinacol) is practically always formed as a by-product (p. 209).

2. The action of ethylmagnesium halide with acetaldehyde.

$$\text{EtMgBr} + \text{MeCHO} \rightarrow \text{EtCHOHMe}$$

The same product is obtained from MeMgX and propionaldehyde but the starting materials are more expensive. In general, secondary alcohols, RR′CHOH, can be made in this way by the proper selection of the aldehyde and Grignard reagent provided the R groups are not both tertiary.[38]

Since 2-butanol has an asymmetric molecule it can exist in two enantiomorphic optically active forms, dextro-rotatory and levo-rotatory. The synethetic material contains these in equal amounts (racemic mixture) and is optically inactive. Treatment with phthalic anhydride gives the acid phthalic esters, $C_6H_4(CO_2H)CO_2R$. The acid esters formed by the d- and l-alcohols are still enantiomorphic and cannot be separated because their properties are identical except their behavior to plane polarized light. The racemic mixture of the two acid esters is then treated with an optically active base, such as d-brucine. This gives two salts which may be represented as d-acid·d-brucine and l-acid·d-brucine. These salts are no longer enantiomorphs, parts of the two molecules are asymmetric and enantiomorphic and parts are asymmetric and identical. In such pairs of substances the properties are not identical. In this case there is enough difference in solubility between the two salts to allow their separation by fractional crystallization. The progress of the separation can be followed by the optical rotation of each of the fractions. When material is obtained which has the same specific rotation after repeated crystallizations the separation is assumed to be complete. The product is then treated, first, with acid to remove the brucine, and then with alkali to liberate the alcohol.

This process for resolving racemic mixtures of asymmetric secondary alcohols is due to Pickard and Kenyon and has been widely used.[39]

The specific rotation: $[\alpha_2] = \dfrac{100\alpha}{gdl}$ in which α is the angle through which the plane of polarization of the light is rotated by a solution of g grams in 100 g. of solution of density d in a tube l decimeters long. Values for the dextro-rotatory form for the D line of the sodium spectrum are

$$[\alpha]^{20}\text{D}, \ +13.87°; \ [\alpha]^{50}\text{D}, \ +12.48°.$$

The *reactions* of 2-butanol are those of a secondary alcohol like isopropyl alcohol. Thus it esterifies more slowly than its primary isomers. With

[38] Conant, Blatt. *J. Am. Chem. Soc.* 51, 1227 (1929).
[39] "Org. Reactions," II, p. 376.

oxidizing agents it forms methyl ethyl ketone which gives acetic acid on more vigorous oxidation. With fused alkalies it gives "di-*sec*-butyl alcohol," 5-Me-3-heptanol. A hydrogen of the methyl group reacts rather than one from the alpha methylene group.

$$\text{EtMeCHONa} + \text{HCH}_2\text{CH(OH)Et} \rightarrow \text{EtCH(Me)} - \text{CH}_2\text{CH(OH)Et}$$

4. Tertiary butyl alcohol, trimethylcarbinol, $(\text{CH}_3)_3\text{COH}$, b. 82.8°, m. 25.5°. This alcohol is readily available from the hydration of isobutylene from cracked gases. Its supply is dependent on the demand for secondary butyl alcohol. Whether there are adequate uses for tertiary butyl alcohol or not, the isobutylene must be taken out to avoid contamination of the secondary butyl alcohol by the tertiary. This is done by a more dilute sulfuric acid than will absorb the *n*-butenes.

The *reactions* of tertiary butyl alcohol are typical of tertiary alcohols and are radically different from those of primary and secondary alcohols.

1. Acids.

(*a*) Even dilute inorganic acids dehydrate it on heating to give isobutylene.

(*b*) Concentrated halide acids or the gaseous hydrogen halides react rapidly and completely giving tertiary butyl halides and water. The reaction is almost like that between an acid and a base.

(*c*) Organic acids form esters very slowly and incompletely. This is the reason an impurity of a tertiary alcohol in a secondary alcohol is harmful since the chief use of the latter is in making esters for solvents.

2. Acid chlorides present an interesting series of reactions with *t*-butyl alcohol as with other tertiary alcohols. Acetyl chloride in the cold gives *t*-butyl chloride and acetic acid quantitatively. When the two reagents are heated, a nearly 1:1 mixture of chloride and acetate is obtained. The acetate can be obtained in quantitative yield by adding dimethylaniline to combine with the HCl as fast as liberated in the normal reaction.

$$\text{Me}_3\text{COH} + \text{MeCOCl} \rightarrow \text{HCl} + \text{MeCO}_2\text{CMe}_3$$

If the HCl is not removed it reacts more or less completely with the tertiary ester to give the chloride and acetic acid.

3. Another evidence that tertiary butyl alcohol is more "basic" than "acidic" is the slowness with which it reacts with the alkali metals.

4. With bromine, there is no chance for oxidation as with the primary and secondary alcohols. Instead, a dibromide is formed corresponding to the olefin which would be obtained by dehydrating the alcohol, namely, isobutylene.

$$\text{Me}_2\text{C(OH)CH}_3 + \text{Br}_2 \rightarrow \text{Me}_2\text{CBrCH}_2\text{Br} + \text{H}_2\text{O}$$

5. Fused NaOH at 500° gives sodium carbonate and methane quantitatively.[40]

[40] Fry, Otto. *J. Am. Chem. Soc.* **50**, 1122 (1928).

6. Mild oxidizing agents have no effect on tertiary butyl alcohol. More vigorous ones split it into acetone, acetic acid and CO_2. Acid oxidizing agents give a peculiar by-product, isobutyric acid, Me_2CHCO_2H.[41] Probably the first step is the dehydration to isobutylene followed by oxidation with re-arrangement. The activated form of the olefin may add an oxygen atom from the oxidizer to form a carbonium ion in which an electron pair with its attached proton migrates to form isobutyraldehyde which is then oxidized to the acid.

Similar processes take place in the oxidation of diisobutylene and triiso-butylene.[42]

t-Butyl alcohol and 30% hydrogen peroxide react to give *t*-butyl peroxide which is unusually stable.[43]

Amyl Alcohols

All eight of the theoretically possible structural isomers have been prepared.

1. **Normal amyl alcohol**, 1-pentanol, *n*-butylcarbinol, $CH_3(CH_2)_3CH_2OH$, b. 137°.

This alcohol has recently become available by the hydrolysis of 1-chloro-pentane formed by the direct chlorination of pentane from natural gasoline (Sharples). Before that, it could be prepared from a *n*-butyl Grignard reagent and formaldehyde. Before *n*-butyl alcohol was available it could be made from *n*-propyl Grignard reagent and ethylene oxide. An addition compound is first formed which on heating gives a derivative of the desired alcohol. Decomposition of the first product with acid gives propane and ethylene glycol.

$$C_2H_4O \cdot RMgX \xrightarrow{heat} C_3H_7CH_2CH_2OMgX \rightarrow C_5H_{11}OH$$

This is a general method for building up higher primary alcohols from primary and secondary halides. It fails with tertiary Grignard reagents.

Another method of preparation is the reduction of ethyl *n*-valerate with sodium and alcohol.[44] The ester is made by converting *n*-butyl halide to the cyanide and then treating with ethyl alcohol and sulfuric acid. Reduction of the ester with hydrogen and copper chromite catalyst also gives *n*-amyl alcohol.[45]

[41] Butlerow. *Z. Chem.* (2) VII, 484 (1871).
[42] Whitmore et al. *J. Am. Chem. Soc.* 56, 1128, 1397 (1934).
[43] Milas, Harris. *J. Am. Chem. Soc.* 60, 2434 (1938).
[44] Bouveault, Blanc. *Chem. Zentr.* 1905, II, 1700.
[45] Adkins. The University of Wisconsin Press, 1937.

Minute traces of *n*-amyl alcohol have been identified in fusel oil. It is probably formed from norleucine, in the same way that isobutyl alcohol is formed from valine. Older attempts failed to find this alcohol in fusel oil. Thus Tissier[46] fractionally distilled 1600 l. of fusel oil in the search for this alcohol which was believed to be present because the dehydration of fusel oil amyl alcohol gave some normal amylenes. It is now known that these come from the rearrangement of *sec*-butylcarbinol.

n-Amyl alcohol gives the reactions of primary alcohols.

2. Secondary butylcarbinol, "active amyl alcohol," 2-methyl-1-butanol, $CH_3CH_2CH(CH_3)CH_2OH$, b. 128°.

This alcohol has long been available in fusel oil, mixed with about seven times as much isoamyl alcohol. The separation has been difficult. Active amyl alcohol is so called because of its optical activity. It is formed during fermentation from isoleucine derived from the proteins in the material fermented.[47] The net change is

$$MeEtCHCHNH_2CO_2H + H_2O \rightarrow MeEtCHCH_2OH + CO_2 + NH_3$$

The isoleucine has two asymmetric carbons and is optically active as are practically all natural products which have asymmetric molecules. In the change to the alcohol only one of the asymmetric atoms is rendered symmetrical. Hence the product is still optically active. Although it is levorotatory, $[\alpha]_D^{20} = -5.9°$, it is called D-amyl alcohol because on oxidation it gives D-valerianic acid (D-methylethylacetic acid).[48] This apparent inconsistency in the naming of optically active compounds is general. The symbols D- and L- simply show the chemical relationship of the substance named to some standard reference substance. Thus D-fructose is levorotatory. It is called D- because its number 5 carbon atom has the same configuration as D-glyceraldehyde, the standard reference substance for naming all optically active compounds.

The separation of pure optically active *sec*-butylcarbinol was extremely difficult until distilling columns having the equivalent of one hundred theoretical plates were used to distill fusel oil.[49, 50]

The proportion of "active" amyl alcohol in amyl alcohol from fusel oils varies very decidedly according to the material fermented. Apparently the largest amount is obtained in the fermentation of beet-molasses.

At present the best source for racemic secondary-butylcarbinol is the hydrolysis of 1-chloro-2-methylbutane obtained by chlorinating isopentane (Sharples) and as a by-product in methanol synthesis. Obtained thus, the

[46] Tissier. *Bull. soc. chim.* (3) 9, 100 (1893).
[47] Ehrlich. *Ber.* 40, 1027 (1907).
[48] Marckwald. *Ber.* 35, 1595 (1902).
[49] Whitmore, Olewine. *J. Am. Chem. Soc.* 60, 2569 (1938).
[50] Fenske et al. *Ind. Eng. Chem.* 28, 644 (1936).

alcohol is optically inactive, being a racemic mixture of the d- and l-forms in equal amounts. It was produced in Germany by the following steps:

1. Condensation of acetaldehyde with propionaldehyde in the presence of alkali.

2. Dehydration of the resulting aldol.

3. Hydrogenation of the resulting tiglic aldehyde.

$$CH_3-CHO + CH_3CH_2CHO \xrightarrow{alkali} \begin{matrix} H \\ | \\ CH_3-COH \\ | \\ CH_3-C-CHO \\ | \\ H \end{matrix} \xrightarrow{-H_2O} \begin{matrix} CH_3-CH \\ \| \\ CH_3-C-CHO \end{matrix} \xrightarrow{H_2}$$

$$\begin{matrix} CH_3-CH_2 \\ | \\ CH_3-C-CH_2OH \end{matrix}$$

sec-Butylcarbinol has also been made synthetically by use of the secondary butyl Grignard reagent with formaldehyde.

The reactions of this alcohol are like those of isobutyl alcohol, that is, the reactions of the primary alcohol group are complicated by the adjacent tertiary H and the fork in the chain. Thus it gives some tertiary halide and on dehydration gives some normal amylenes.[51]

3. **Isoamyl alcohol,** 3-methyl-1-butanol, isobutylcarbinol,

$$(CH_3)_2CHCH_2CH_2OH, \text{ b. } 131°.$$

This is the principal amyl alcohol in fusel oil. It is formed during fermentation from leucine. The optical activity of the latter is lost during the decarboxylation and deamination. The chief use of fusel oil is for making crude amyl acetate (banana oil) as a solvent. For this purpose, it is usually not worth while to remove the n-propyl alcohol (5%) and the isobutyl alcohol (20%) before esterification. Before the introduction of nitrocellulose lacquers, this was adequate as an ester-type solvent. Then the production of amyl alcohol from fusel oil could not be expanded to meet the new need because of its by-product nature. The need has been met by n-butyl alcohol from corn or molasses (C.S.C.), by amyl alcohols from the pentanes (Sharples), by sec-butyl and sec-amyl alcohols from cracked gases (Stanco, Shell) and by n-propyl, isobutyl and higher alcohols obtained as by-products in the action of hydrogen and carbon monoxide (DuPont).

The *reactions* of isoamyl alcohol are those of a primary alcohol.

Ordinary dehydration of isoamyl alcohol gives rearranged products in addition to isopropylethylene, since that substance readily rearranges to Me$_3$-ethylene in presence of the acid dehydrating agent.

4. **Neopentyl alcohol,** dimethylpropanol, t-butylcarbinol, $(CH_3)_3CCH_2OH$, b. 113°, m. 50°.

[51] Tissier. *Bull. Soc. Chem.* (3) 9, 100 (1893).

This alcohol is prepared from t-butylmagnesium chloride and formaldehyde. Another preparation is by the reduction of an ester of trimethylacetic acid with sodium and absolute alcohol.

Neopentyl alcohol is formed in nearly quantitative yield in an unusual reaction between Me_3-acetyl chloride and an excess of t-Bu Grignard reagent.[52]

$$Me_3CCOCl + 2 Me_3CMgCl \rightarrow Me_3CCH_2OMgCl + 2 Me_2C = CH_2 + MgCl_2$$

The acid chloride is first reduced to the aldehyde which is then further reduced to the alcoholate.[53] The absence of copper salts is necessary for such reductions.[53a]

The reactions of neopentyl alcohol give the clue to the rearrangements of alcohols and related compounds. Reagents which do not disturb the electronic link C:O act normally. Thus alkali metals, oxidizing agents, organic acids and acyl halides give good yields of products containing the neopentyl system of one carbon with four carbons attached to it. On the other hand, reagents which ordinarily replace or remove the OH group, such as halide acids, inorganic acid chlorides, and dehydrating agents, produce mainly t-amyl derivatives or the related olefins such as Me_3-ethylene.[54] The alcohol dissolves in *cold* concentrated sulfuric acid. On dilution and neutralization, the alcohol can be recovered. The cold alcohol will slowly absorb slightly more than 1 mol of dry HBr gas. The latter can be driven out quantitatively by a stream of inert gas. It can also be washed out with water, leaving the unchanged alcohol. Each of these combinations probably consists of an oxonium salt which is stable at ordinary temperature. Heat gives t-amyl derivatives and amylenes. Presumably, the decomposition of the oxonium compound removes the oxygen as water with its complete octet of electrons, thus leaving the carbon with only six electrons (an "open sextet") with the formation of a neopentyl carbonium ion which immediately rearranges to the more stable t-amyl carbonium ion, which then gives the observed products (p. 298). The rearrangement is not complete since about 5% of neopentyl bromide can be obtained by allowing the alcohol, saturated with dry HBr, to stand in a sealed tube for several months. The older conception that neopentyl alcohol first gives a neopentyl chloride which is unstable and rearranges to a t-amyl chloride is exploded by the fact of the stability of neopentyl chloride (p. 83).

The fact that neopentyl alcohol reacts with organic acids to give neopentyl esters which can be hydrolyzed or saponified back to neopentyl alcohol without any formation of t-amyl derivatives shows that these processes do not involve a breaking of the C:O linkage in the alcohol or the ester. This agrees with the accumulated evidence that esterification of an alcohol and a carboxylic acid involve their H and OH groups respectively.[55]

[52] Greenwood, Whitmore, Crooks. *J. Am. Chem. Soc.* 60, 2028 (1938).
[53] Whitmore et al. *J. Am. Chem. Soc.* 63, 643 (1941).
[53a] Cook, Percival. *J. Am. Chem. Soc.* 71, 4141 (1949).
[54] Whitmore et al. 54, 3274, 3431, 3435 (1932).
[55] Norris, Cortese. *J. Am. Chem. Soc.* 49, 2640 (1925).

This mechanism for esterification and hydrolysis of esters is confirmed by the results obtained using water containing O^{18} in the acid or base catalyzed hydrolysis which results in the retention of all of the heavy oxygen by the acid.[56-58]

Thus added weight is given to the conception that both esterification and hydrolysis take place through *addition* to the carbonyl group.[59, 60]

$$MeC{=}O \ + \ ROH \rightleftharpoons Me{-}\overset{\displaystyle OH}{\underset{\displaystyle OR}{C}}{-}OH \rightleftharpoons MeC{=}O \ + \ HOH$$

The initial step in each process is the addition of H ion to the oxygen of the carbonyl group of the acid or ester.

A more modern conception can be obtained by a consideration of oxonium ions and carbonium ions and their relation to hydrogen ion (hydronium ion). These will be considered first electronically and then in abbreviated form. Water and hydrogen halide gas give halide ions and hydronium ions.

$$: \overset{..}{\underset{..}{X}} : H + : \overset{..}{\underset{H}{O}} : H \rightarrow \left[: \overset{..}{\underset{..}{X}} : H : \overset{..}{\underset{H}{O}} : H \right] \rightarrow \left[: \overset{..}{\underset{..}{X}} : \right]^{-} + \left[H : \overset{..}{\underset{H}{O}} : H \right]^{+}$$

The formation of the intermediate complex is due to *hydrogen-bonding* (p. 237) of the X and O. Similarly, alcohols can form oxonium ions. When water is removed, usually by heat, the C formerly attached to O is left with only 6 electrons instead of the stable complement of eight. This leaves the C with a unit positive charge. It should be emphasized that the resulting carbonium ion is unlike stable ions having completed octets of electrons.[61, 62]

$$\left[R : \overset{..}{\underset{H}{O}} : H \right]^{+} \quad \rightarrow \quad : \overset{..}{\underset{H}{O}} : H \quad + \quad R^{+}$$

Oxonium ion   Carbonium ion

Carbonium ions can be formed not only by such a process of "subtraction" but also by *addition* of a proton or other electrophilic agent (BF_3, $AlCl_3$ etc.) to a multiple linkage.[63] Thus

$$R{-}CO_2H \xrightarrow{\ H^+ \ } R{-}\overset{+}{C}(OH)_2$$

[56] Mumm. *Ber.* 72, 1874 (1939).
[57] Datta, Day, Ingold. *J. Chem. Soc.* 1939, 838.
[58] Polanyi, Szabo. *Trans. Faraday Soc.* 30, 508 (1934).
[59] Wegscheider. *Monatsh.* 16, 75 (1895); 18, 629 (1897).
[60] Henry. *Ann. Rep. Chem. Soc.* (London) 1910, 66.
[61] Whitmore. *J. Am. Chem. Soc.* 54, 3274 (1932).
[62] *Chem. Eng. News.* 26, 668 (1948).
[63] Treffers, Hammett. *J. Am. Chem. Soc.* 59, 1708 (1937).

Carbonium ions can readily satisfy their electronic deficiency by coordination with substances like water and alcohols to produce oxonium ions which lose protons to give stable products.

Returning to the mechanism of esterification, the initial step is the addition of a proton to the carboxyl group to give a carbonium ion. This coordinates with ROH. The resulting complex loses a proton to give the intermediate assumed by Henry.

$$H-CO_2H + H^+ \rightarrow H-\overset{+}{C}(OH)_2 \xrightarrow{\text{ROH}} H-\overset{\overset{..}{\text{ROH}}}{C}(OH)_2 \rightarrow H^+ + H-C(OH)_2(OR)$$

A proton from the catalyst can attack this complex at a free electron pair of either an OH or OR group. If the former is attacked, the resulting oxonium ion can lose water leaving the carbonium ion, $H-C^+(OH)(OR)$ which expels a proton leaving the ester $H-CO_2R$. If the attack is on the OR group, the resulting oxonium ion can lose ROH and the final result is the acid. The above mechanism also gives a rationalization of the experimentally observed instability of the groupings $C(OH)_2$ and $C(OH)(OR)$. Similar principles hold for $C(OH)Cl$ and $C(OH)(NH_2)$.

5. **2-Pentanol,** methylpropylcarbinol, $CH_3CH_2CH_2CHOHCH_3$, b. 119°, is another of the amyl alcohols from chlorinated pentane (Sharples). The hydrolysis of 2-chloropentane gives considerable olefin and consequently a poorer yield of alcohol than the hydrolysis of the primary halides.

2-Pentanol is also available from the hydration of 1-pentene obtained in the cracking process (cf. isopropyl and secondary butyl alcohols) (Shell). Commercial secondary amyl alcohol from cracked gases contains the 2- and 3-isomers in the ratio 4:1.[64]

2-Pentanol made by either of these methods always contains 3-pentanol. To obtain the pure alcohol it is best to use the general method for making a secondary alcohol, namely the action of a Grignard reagent on an aldehyde. The desired result can be obtained from acetaldehyde and *n*-propyl Grignard reagent or from *n*-butyraldehyde and methyl Grignard reagent. The *d*-form of the alcohol has been obtained, $[\alpha]_D^{20} = +13.7°$.

Another general method for making secondary alcohols is the reduction of ketones. In this case methyl propyl ketone would be used. This could be made through the acetoacetic ester synthesis using ethyl bromide as the halide and splitting the final product with *dilute* alkali and acid (p. 366). The reduction of ketones practically always gives as by-products pinacols, in this case 4,5-dimethyl-4,5-octandiol.

2-Pentanol shows the usual reactions of secondary alcohols. If halides are made from it by the ordinary methods (acid and heat) a mixture of the 2- and 3- compounds is obtained. Thionyl chloride and pyridine give the pure 2-chloropentane.

[64] Brooks. *Ind. Eng. Chem.* **27,** 278 (1935).

6. **3-Pentanol,** diethylcarbinol, $(C_2H_5)_2CHOH$, b. 117°, is obtained from 3-chloropentane in the usual way (Sharples).

Before this method was available it could be made by the Grignard reaction in two ways: (a) Propionaldehyde and EtMgX (b) Ethyl formate and EtMgX.

The action of a Grignard reagent with a formate is a general method for making symmetrical secondary alcohols of the type R_2CHOH.

Another preparation is by the reduction of diethyl ketone formed by the dry distillation of calcium propionate or by passing the vapors of propionic acid over manganous carbonate. The pinacol formed as a by-product of the reduction would be 3,4-Et₂-3,4-hexandiol.

The reactions of 3-pentanol are like those of 2-pentanol including the tendency to rearrangement.

7. **Methylisopropylcarbinol,** secondary isoamyl alcohol, 3-methyl-2-butanol, $(CH_3)_2CHCHOHCH_3$, b. 112°, is not obtained from the chlorination products of isopentane. The corresponding chloride probably loses HCl very rapidly to give trimethylethylene, etc. It is prepared by MeMgX and isobutyraldehyde. This gives a better over-all yield than the use of an isopropyl Grignard reagent with acetaldehyde.

Another method of preparation is by the reduction of methyl isopropyl ketone. This is available by the hydrolysis and rearrangement of trimethylethylene dibromide obtained from the direct bromination of t-AmOH.[65] This alcohol cannot be converted to the corresponding secondary halides. All processes used give t-amyl halides instead (pp. 83, 129).

8. **t-Amyl alcohol,** 2-methyl-2-butanol, dimethylethylcarbinol, "Amylene Hydrate," $(CH_3)_2C(OH)CH_2CH_3$, b. 102°, is available in large amounts because the chlorination of isopentane and the hydrolysis of the resulting products give large amounts of trimethylethylene. When this is hydrated by means of 50% sulfuric acid, tertiary amyl alcohol is obtained.

t-Amyl alcohol may also be made by means of suitable Grignard reagents with acetone, methyl ethyl ketone and ethyl propionate.

The reactions of tertiary amyl alcohol are practically identical with those of tertiary butyl alcohol. It is even more easily dehydrated. t-Amyl alcohol with dilute H_2SO_4 in a sealed tube gives two layers (olefin and aqueous acid) on warming. On cooling one layer is formed again. Refluxing with 15% sulfuric acid gives a 7:1 mixture of Me₃-ethylene and 1,1-MeEt-ethylene.[66]

With bromine it gives trimethylethylene dibromide directly. This, on hydrolysis, rearranges to form methyl isopropyl ketone and a trace of trimethylacetaldehyde.

t-Amyl alcohol is valuable in the dry cleaning industry for removing spots before treatment with the usual solvents and for the dry cleaning of Celanese (cellulose acetate fiber).

[65] "Org. Syntheses."
[66] Whitmore et al. *J. Am. Chem. Soc.* **64,** 2970 (1942).

A crúde mixture of the amyl alcohols obtained by the hydrolysis of chloro-pentanes by NaOH and an emulsifying agent such as soap is marketed as Pentasol for use as a solvent alone and in making Pentacetate (Sharples). Pentasol contains approximately the following percentages of amyl alcohols: sec-butylcarbinol, 32; n-amyl alcohol, 26; 3-pentanol, 18; isoamyl alcohol, 16; and 2-pentanol, 8. The method of hydrolysis converts any t-amyl chloride and 2-chloro-3-Me-butane in the chlorinated pentanes to amylenes and traces of the corresponding alcohols.

Hexyl Alcohols

All seventeen of the theoretically possible hexyl alcohols are known. Most of them have been prepared in a variety of ways. Only the most typical methods will be considered. The reactions of these alcohols are like those of the lower homologs of the corresponding classes. The alcohols are grouped as primary, secondary and tertiary.

1. **n-Hexyl alcohol,** 1-hexanol, $CH_3(CH_2)_4CH_2OH$, b. 157°. This alcohol is available commercially (C and C). It can be formed through the action of acetaldehyde with crotonic aldehyde (methyl reactive by vinyl-ogy), followed by catalytic hydrogenation.

$$MeCHO + CH_3CH = CHCHO \rightarrow MeCHOHCH_2CH = CHCHO$$

$$MeCH = CHCH = CHCHO \rightarrow Me(CH_2)_4CH_2OH$$

It can be synthesized in a variety of other ways.

a. Reduction of ethyl caproate or caproamide by sodium and absolute alcohol.[67] A modification of these reactions is the reduction of ethyl caproate by sodium in liquid ammonia. It can also be reduced by H_2 and a copper chromium oxide catalyst under pressure.

b. n-Butyl magnesium bromide with ethylene oxide (p. 311), obtained from ethylene chlorohydrin and alkali.[68]

c. A similar reaction is that of n-PrMgX with trimethylene oxide obtained from 3-chloro-1-propanol (trimethylene chlorohydrin) and alkali.

d. Since n-amyl alcohol is available it could be made from n-amyl Grignard reagent and formaldehyde.

2. **2-Methyl-1-pentanol,** $(CH_3CH_2CH_2)(CH_3)CHCH_2OH$, b. 148°. This alcohol is obtained in fair amounts as a by-product of synthetic methanol. It has also been made as follows:

a. By heating propyl alcohol with sodium propylate at 220° (Guerbet).

b. By converting 2-chloropentane to the Grignard reagent and treating with trioxymethylene (polymerized formaldehyde).

[67] Bouveault, Blanc. *Compt. rend.* 138, 148 (1904).
[68] "Org. Syntheses."

c. By the sodium and absolute alcohol reduction of methylpropylacetoacetic ester.

$$MeCOCMePrCO_2Et + Na + EtOH \rightarrow MeCO_2Na + MePrCHCH_2OH$$

The splitting of the grouping $MeCO-C-CO_2Et$ by vigorous treatment with NaOEt is characteristic.

3. 3-Methyl-1-pentanol, $(CH_3CH_2)(CH_3)CHCH_2CH_2OH$, b. 154°, is best made synthetically from secondary butylcarbinol through the Grignard reagent and formaldehyde.[69] It occurs as an ester in Roman Camomile oil. The natural alcohol is optically active $[\alpha]_D^{20.5} = +8.8$. The active material was made by reducing the active ester by $Na-EtOH$.[70] Its structure is proved by oxidation to β-Me-valeric acid.

4. Isohexyl alcohol, 4-methyl-1-pentanol, $(CH_3)_2CH(CH_2)_2CH_2OH$, b. 148°.

a. From isoamyl magnesium bromide and formaldehyde.

b. By heating isobutyl acetoacetic ester with sodium and absolute alcohol.

c. Both *n*- and *iso*-hexyl alcohols can be produced by treating a mixture of acetaldehyde and butyraldehyde with dilute sodium hydroxide under carefully controlled conditions and subsequently hydrogenating the distilled unsaturated aldehydes. The hexyl alcohols are then separated from higher boiling alcohols found in the reaction. Ratio of *n*- to *iso*- is 1:5.

5. 2-Ethyl-1-butanol, diethylcarbincarbinol, 3-methylolpentane, 3-hydroxymethylpentane, $(C_2H_5)_2CHCH_2OH$, b. 146°, is the most readily available hexyl alcohol (CC). It could be formed by the reaction of 2 mols of acetaldehyde with 2 of the alpha H atoms of a third, followed by dehydration and catalytic hydrogenation.

$$2\ MeCHO + H_2CHCHO \rightarrow (MeCHOH)_2CHCHO \rightarrow$$

$$\begin{bmatrix} MeCH=CCHO \\ | \\ CH=CH_3 \end{bmatrix} \rightarrow Et_2CHCH_2OH$$

Before it was available commercially, the best preparation was by copper chromite hydrogenation of ethyl diethylacetate formed from ethyl diethylmalonate, an intermediate for veronal.

Dehydration of diethylcarbincarbinol gives 3-hexene, 2-hexene, 3-Me-2-pentene and very little of the normal dehydration product, 2-Et-1-butene.[71] With HCl and $ZnCl_2$, the alcohol gives an even more complicated set of rearranged products (p. 85–6).

6. *t*-Amylcarbinol, 2,2-dimethylbutan-1-ol $(CH_3CH_2)(CH_3)_2CCH_2OH$, b. 135°.

a. By reduction of the ethyl ester of dimethylethylacetic acid with sodium and absolute alcohol.

[69] Norris, Cortese. *J. Am. Chem. Soc.* **49,** 2640 (1925).
[70] Levene, Marker. *J. Biol. Chem.* **91,** 77 (1931).
[71] Karnatz, Whitmore. *J. Am. Chem. Soc.* **54,** 3461 (1932).

b. From *t*-amylmagnesium chloride and ethyl formate. This involves an addition and a reduction by the Grignard reagent.

c. From the *t*-amyl Grignard reagent and formaldehyde.

This alcohol reacts abnormally with acids, undergoing the usual rearrangement of the "neopentyl alcohol" (p. 120). Dehydration gives the olefins $MeEtC = CHMe$, $Et_2C = CH_2$, $Me_2C = CHEt$ and $MePrC = CH_2$.

7. **Neopentylcarbinol,** 3,3-dimethyl-1-butanol, $(CH_3)_3CCH_2CH_2OH$, b. 143°.

The preparation of this alcohol from acetone illustrates a wide range of synthetic processes.

             reduction
1. Acetone ⟶ pinacol, $Me_2C(OH)C(OH)Me_2$

          acid
2. pinacol ⟶ pinacolone, Me_3CCOMe

            reduction
3. pinacolone ⟶ pinacolyl alcohol, $Me_3CCHOHMe$

           CS_2 MeI
4. pinacolyl alcohol ⟶ Na pinacolyl xanthate ⟶ Me pinacolyl xanthate

          NaOH

       heat
5. xanthate ⟶ *t*-butylethylene

           HBr
6. *t*-butylethylene ⟶ $Me_3CCH_2CH_2Br$

         CH_3CO_2Na
7. bromide ⟶ acetate

        KOH
8. acetate ⟶ $Me_3CCH_2CH_2OH$.

Step 6 represents an addition contrary to Markownikoff's Rule due to the presence of peroxides in the olefin.

Neopentylcarbinol forms halides with the same ease as does 1-butanol and also without rearrangement. This is in marked contrast to its lower homolog, neopentyl alcohol (p. 120).

8. **2,3-Dimethyl-1-butanol,** $(CH_3)_2CH(CH_3)CHCH_2OH$, b. 145°.

This alcohol has been made only by the reduction of the ethyl ester of methyl isopropyl acetic acid with sodium and alcohol. The ester is made through the acetoacetic ester or malonic ester synthesis for acids or by the Reformatski reaction from acetone and ethyl α-Br-propionate.

9. **2-Hexanol,** methyl-*n*-butylcarbinol, $CH_3CHOH(CH_2)_3CH_3$, b. 141°.

a. By reduction of methyl butyl ketone by sodium in a mixture of ether and water, by catalytic hydrogenation (Sabatier), and by sodium amalgam and dilute acid. The methyl butyl ketone is obtained from sodium acetoacetic ester and *n*-propyl bromide.

b. By treating 1-hexene from *n*-PrMgBr and allyl bromide with 85% sulfuric acid, diluting and distilling.

c. From *n*-butylmagnesium bromide and acetaldehyde.

10. **3-Hexanol,** ethylpropylcarbinol, $CH_3CH_2CH_2CHOHCH_2CH_3$, b. 135°.

 a. By reduction of ethyl propyl ketone by sodium amalgam and water.

The ketone can be obtained from a mixture of propionic and *n*-butyric acids either through the calcium salts or by hot manganous oxide.

 b. A better preparation is from EtMgX and *n*-butyraldehyde.

11. **3-Methyl-2-pentanol,** methyl-*sec*-butylcarbinol,

$$(CH_3CH_2)(CH_3)CHCHOHCH_3, \text{ b. } 134°.$$

By the reduction of methyl *sec*-butyl ketone by sodium and wet ether. The ketone is made by introducing a methyl and an ethyl group into acetoacetic ester in the usual way and splitting by dilute alkali.

12. **Methylisobutylcarbinol,** 4-Me-2-pentanol, $(CH_3)_2CHCH_2CHOHCH_3$, b. 132°.

This alcohol is available commercially from the catalytic dehydration and hydrogenation of diacetone alcohol.

The dextrorotatory form has been made through the brucine salt of the acid phthalic ester. $[\alpha]_D^{21.3} = +20.4°$. The levorotatory form has been obtained similarly from the brucine salt of the acid succinic ester, $[\alpha]_D^{14} = -20.8°$.

The *dl*-alcohol has been made as follows:

 a. From isovaleraldehyde and zinc dimethyl.

 b. By the reduction of mesityl oxide, $Me_2C = CHCOMe$, with sodium and alcohol or catalytically.

 c. By the reduction of methyl isobutyl ketone with sodium and a mixture of ether and water. The ketone is obtained from sodium acetoacetic ester and isopropyl bromide.

 d. From acetaldehyde and isobutylmagnesium bromide.

 e. From the action of zinc or magnesium on a mixture of isobutyl iodide and acetic anhydride. This is undoubtedly a Grignard reaction followed by reduction.

 f. From isopropyl alcohol, either by heating it with KOH or by heating it with sodium isopropylate to 200°. As usual, the ONa combines with an alpha hydrogen.

13. **Ethylisopropylcarbinol,** 2-methyl-3-pentanol,

$$(CH_3)_2CHCHOHCH_2CH_3, \text{ b. } 128°.$$

The dextrorotatory form has been obtained by means of the strychnine salt of the half phthalic ester. $[\alpha]_D^{21.4} = +12.4°$. The *dl*-alcohol has been made as follows:

 a. From isobutyraldehyde and EtMgBr.

 b. By hydrogenating ethyl isopropyl ketone.

14. **Pinacolyl alcohol,** methyl-*t*-butylcarbinol, 2,2-dimethyl-3-butanol, $(CH_3)_3CCHOHCH_3$, m. 5.5°, b. 121°.

The *d*-form has been made in the usual way. $[\alpha]_D^{20} = +7.7°$. The *dl*-material has been prepared as follows:

a. Reduction of pinacolone by sodium amalgam.

b. Acetaldehyde and tertiary butylmagnesium chloride.

This alcohol contains the neopentyl alcohol grouping (p. 120). Since there are three alpha H atoms in the molecule, it might be expected that dehydration would involve one of them to give *t*-butylethylene. In ordinary acid dehydration this normal product is obtained only to the extent of about 3%. Me₄-ethylene and *unsym*-methylisopropylethylene are obtained in yields of about 60 and 30%.[72] The normal dehydration product can be obtained by the xanthate reaction or by dehydration over alumina free from acid (p. 45).

Conversion of pinacolyl alcohol to halides gives those related to dimethylisopropylcarbinol by rearrangement of the neopentyl alcohol grouping. The pinacolyl halides can be made from *t*-butylethylene.

15. **2-Methyl-2-pentanol,** dimethyl-*n*-propylcarbinol,

$$(CH_3CH_2CH_2)(CH_3)_2COH, \text{ b. } 123°.$$

a. *n*-Butyryl chloride and zinc dimethyl. This method served as the equivalent of the Grignard reaction until about 1900.

b. Ethyl *n*-butyrate and excess methyl Grignard reagent.

c. Acetone and *n*-propylmagnesium bromide. This is a general method for tertiary alcohols. Selection of the proper ketone and Grignard reagent will give any alcohol RR′R″COH unless there is too much branching in the R groups.[73]

This tertiary alcohol, in common with all tertiary alcohols of C₆ and higher, is dehydrated by refluxing with a trace of iodine.[74] *t*-Butyl and *t*-amyl alcohols boil at too low a temperature to give this reaction. The latter is dehydrated slowly by refluxing with iodine at two atm. pressure.

16. **Methyldiethylcarbinol,** 3-methyl-3-pentanol, MeEt₂COH, b. 123°.

a. From acetyl chloride and zinc diethyl.

b. From ethyl acetate and excess ethylmagnesium bromide.

This method is general for making tertiary alcohols of the type RR₂′COH. A ketone is probably formed as an intermediate in the reaction mixture (p. 129). With more complex esters and Grignard reagents the corresponding ketone is obtained instead of the tertiary alcohol.

17. **Dimethylisopropylcarbinol,** 2,3-dimethyl-2-butanol,

$$(CH_3)_2CHC(OH)(CH_3)_2, \text{ b. } 122°.$$

This alcohol has been made by a great variety of methods.

a. By the ordinary Grignard reactions using isopropylmagnesium bromide with acetone, and methyl isobutyrate with excess methylmagnesium bromide, and trimethylethylene oxide with methylmagnesium bromide. The last two

[72] Whitmore, Meunier. *J. Am. Chem. Soc.* 55, 372 (1933).
[73] Conant, Blatt. *J. Am. Chem. Soc.* 51, 1227 (1929).
[74] Hibbert *J. Am. Chem. Soc.* 37, 1784 (1915).

methods are preferable. The forked Grignard reagent, isopropylmagnesium bromide, gives side reactions with the ketone including its reduction to isopropyl alcohol and its condensation to mesityl oxide and more complex products. Thus the yield of the desired alcohol is only 35–40%.

b. By more complex actions of MeMgX with alpha chloroisobutyraldehyde and with the chlorohydrin of trimethylethylene. The intermediate product in both cases is trimethylethylene oxide.

c. By similar reactions of zinc dimethyl with isobutyryl chloride, chloral, dichloroacetyl chloride, and alpha bromopropionyl bromide. The first of these is an ordinary "Grignard" type of reaction. The other three are somewhat more complex although the individual steps are entirely normal.

d. At the present time the way to make this alcohol is from MeMgX and methyl isopropyl ketone, since the latter is readily available from tertiary amyl alcohol.

Heptyl Alcohols

Of the 39 theoretically possible heptyl alcohols, all but two have been made, 2,2-dimethyl-1-pentanol, and 3-ethyl-1-pentanol. The reactions used are like those for the hexyl alcohols except in the case of triethylcarbinol which is prepared by adding an excess of ethylmagnesium halide to ethyl propionate or ethyl carbonate. The latter represents a general method for symmetrical tertiary alcohols, R_3COH.[75,76]

Even if less than 3 mols of Grignard reagent are used, considerable tertiary alcohol is formed because the ketone and ester formed as intermediates are considerably more reactive than ethyl carbonate.

n-**Heptyl alcohol**, $CH_3(CH_2)_5CH_2OH$, b. 176°, has long been available as a reduction product of heptaldehyde obtained by the pyrolysis of castor oil.

Methyl-*n*-amylcarbinol is made by the reduction of the corresponding ketone.

Diisopropylcarbinol, 2,4-Me$_2$-3-pentanol, $(CH_3)_2CHCHOHCH(CH_3)_2$, is the chief secondary alcohol obtained as a by-product of methanol synthesis.[77] This alcohol gives a crystalline oxonium salt $(ROH)_2 \cdot HBr$, m. 69°.[78]

While the oxonium salts of most alcohols are difficult to isolate, many highly branched alcohols give stable crystalline compounds. The formation of the simplest possible oxonium salt, $ROH \cdot HX$, such as $MeOH \cdot HBr$ which is stable only at low temperatures, may involve the coordination of a single hydrogen ion by the oxygen

$$\left[\begin{array}{c} H \\ \cdot\cdot \\ R \cdot O : H \\ \cdot\cdot \end{array} \right]^+ [Br]^-$$

[75] Tschitschibabin. *Ber.* **38**, 561 (1905).
[76] "Org. Syntheses."
[77] Graves. *Ind. Eng. Chem.* **23**, 1318 (1931).
[78] Favorsky. *J. Russ. Phys. Chem. Soc.* **45**, 1557 (1913).

The more complex oxonium salt would then be

$$\left[\begin{matrix} H \\ \cdot\cdot \\ R : O : H \\ \cdot\cdot \\ H \end{matrix} \right]^{++} \quad [Br]^- [OR]^-$$

It is not possible to convert diisopropylcarbinol to the corresponding secondary halide. Instead, the product is the tertiary halide,

$$Me_2CXCH_2CHMe_2$$

This illustrates the ease of rearrangement of a tertiary H (pp. 83, 123).

2,4-Me₂-1-pentanol and *4-Me-1-hexanol*[79] are also obtained in the methanol synthesis but in smaller amounts than the other by-product alcohols.

Pentamethylethanol, 2,3,3-Me₃-2-butanol, *Triptanol,*[80] m. 17°, b. 131°, forms a crystalline hydrate, (ROH)₂.H₂O, m. 83°, which sublimes even at room temperature. Of many preparations of this alcohol, the best is probably by the action of CO₂ with *t*-BuMgCl followed by treatment of the reaction mixture with excess MeMgX.[81]

$$Me_3CMgCl \rightarrow Me_3CCO_2MgCl \rightarrow Me_3CC(Me_2)OMgCl$$

Octyl Alcohols

Of the 89 possible isomers two-thirds are known.

1-Octanol, $CH_3(CH_2)_6CH_2OH$, b. 195°, occurs as esters in various plants. It is available commercially by hydrogenation of the lower boiling cocoanut oil acids and from the hydrogenation of the condensation product of *n*-butyraldehyde with crotonic aldehyde or of 2 mols of the latter. This preparation is like that of *n*-hexanol (p. 132). If the alpha H takes part in the aldol condensation instead of an H of the methyl group (vinylogy), the final product is **2-Et-1-hexanol** (CC).

Capryl alcohol, 2-octanol, methyl-*n*-hexylcarbinol, $CH_3CHOH(CH_2)_5CH_3$, b. 178°, has long been available from the action of bases with castor oil.[82] Heating with alkali gives di- and tri-capryl alcohols and even higher products (Guerbet). The condensation involves the methyl rather than the alpha methylene group. Thus the products are 9-Me-7-pentadecanol and 11-Me-9-*n*-Hex-7-heptadecanol.

An interesting series of 22 of the octyl alcohols related to *n*-octane and the three methylheptanes has been made and studied.[83]

The dehydration of some of the octyl alcohols throws further light on the peculiar reactions of the neopentyl alcohol grouping (p. 120). *Isopropyl-t-*

[79] Chu, Marvel. *J. Am. Chem. Soc.* **53,** 4449 (1931).
[80] Butlerow. *Ann.* **177,** 176 (1875).
[81] Calingaert, Hnizda, Shapiro. *J. Am. Chem. Soc.* **66,** 1389 (1944).
[82] "Org. Syntheses."
[83] Reid et al. *J. Am. Chem. Soc.* **63,** 3100 (1941).

butylcarbinol, $(CH_3)_2CHCHOHC(CH_3)_3$, containing a tertiary hydrogen next to the carbinol group would be expected to give the olefin, $Me_2C=CHCMe_3$. Only 2% of this is obtained using acid catalysts. Even with acid-free alumina, only 24% of this expected product is obtained. The rearrangement products with their percentages are as follows: 2,4,4-Me$_3$-1-pentene 24; 2,3,4-Me$_3$-1-pentene 29; 2,3,4-Me$_3$-2-pentene 18; 3-Me-2-*iso*Pr-1-butene 3; 3,3,4-Me$_3$-1-pentene 2.[84] It is to be noted that the first named products are the diisobutylenes but that they are formed in 1:1 ratio instead of the 1:4 equilibrium ratio (p. 46). The failure of acid-free alumina to give equilibrium mixtures of olefins is characteristic (preparation of *t*-Bu-ethylene from pinacolyl alcohol, p. 44). The dehydration of dimethylneopentylcarbinol by 15% sulfuric acid gives the diisobutylenes in 1:4.5 ratio (placing the 2-olefin first). Under similar conditions the analogous tertiary alcohol, dimethylethylcarbinol (*t*-AmOH) gave a 7:1 ratio of Me$_3$-ethylene and 1,1-MeEt-ethylene. Thus, under these mild conditions the removal of a proton from ethyl is 30 times as easy as from neopentyl.[85]

The two simplest *tertiary* alcohols containing the neopentyl alcohol system which could lose H_2O *either* with or without rearrangement are *methylethyl-t-butylcarbinol* and *dimethyl-t-amylcarbinol*. These dehydrate giving not over 20% of rearranged products.[86]

Higher Alcohols

Many higher alcohols have been made by the methods outlined.

An unusual preparation of complex tertiary alcohols was developed by Grignard himself[87] by passing CO_2 into one Grignard solution, RMgX, boiling out the excess CO_2 and then adding 2 mols of R'MgX thus obtaining RR'$_2$COH. In this way he made *diethylisoamylcarbinol* and *isobutyldiisoamylcarbinol*.

Highly branched carbinols can be made from esters or ketones, alkyl halides, and sodium. Tri-*t*-butyl carbinol, m. 94°, has been made this way.[88] Extensive use has been made of this reaction in preparing highly branched hydrocarbons for engine testing.[89]

n-Alcohols having even numbers of carbons 8 to 18 are available from the acids of cocoanut oil through the reduction of their esters by Na and higher alcohols (Me-cyclohexanols) or with Cu chromite catalyst and H_2.

Stenol (DuPont) is technical stearyl alcohol. It is obtained by the first method.

Lauryl alcohol, 1-dodecanol, is the best known of these alcohols. Mixed with the other alcohols from cocoanut oil it is sold as *Lorol*. When treated with sulfuric acid and converted to the sodium salt this mixture forms an

[84] Dixon, Cook, Whitmore. *J. Am. Chem. Soc.* **70**, 3361 (1948).
[85] Whitmore et al. *J. Am. Chem. Soc.* **64**, 2970 (1942).
[86] Whitmore, Laughlin. *J. Am. Chem. Soc.* **54**, 4011 (1932).
[87] Grignard. *Compt. rend.* **138**, 152 (1904).
[88] Bartlett, Schneider. *J. Am. Chem. Soc.* **67**, 141 (1941).
[89] Howard. *J. Research Nat. Bur. Standards* **41**, 111 (1948).

important detergent with many properties superior to those of ordinary soap (*Gardinol, S.L.S., Dreft* etc.). These salts can be used in slightly acid solutions which would precipitate the fatty acids from ordinary soaps and in salt water and hard waters. The relation to ordinary soaps is shown by the formulas, $CH_3(CH_2)_nCH_2OSO_3Na$ and $CH_3(CH_2)_mCO_2Na$, in which n and m are 10 or more. Pure sodium lauryl sulfate is not as useful a detergent as a mixture with its higher homologs. These are obtainable from oleic and stearic esters as well as from the mixture of esters obtainable from higher acids produced by oxidation of petroleum fractions. Cetyl sodium sulfate is *Avirol*.

The sodium salts of the acid sulfate of secondary alcohols of 10 to 20 carbons are used as wetting and penetrating agents[90] (*Tergitols*).

A notable preparation of an alcohol from the next higher acid is illustrated by the formation of *1-pentadecanol* from silver palmitate and I_2.[91]

$$2 C_{15}H_{31}CO_2Ag + I_2 \rightarrow 2 AgI + CO_2 + C_{15}H_{31}CO_2C_{15}H_{31}$$

Many complex alcohols will be available as by-products of synthetic methanol, of the Fischer-Tropsch synthesis[92] from carbon monoxide and hydrogen, and from the Oxo Process (diisobutylene, p. 46) involving the catalytic action of carbon monoxide and olefins. Those from the first source have been most thoroughly studied. They include normal and isopropyl alcohols, isobutyl alcohol, 2-Me-1-butanol, 2-Me-1-pentanol, 4-Me-1-hexanol, 3-Me-2-butanol, 2,3-Me$_2$-3-pentanol, 2-Me-1-hexanol, 3-Me-1-pentanol, isoamyl alcohol, 2,3-Me$_2$-1-butanol, 2-Me-3-pentanol, 2,4-Me$_2$-1-pentanol, 2,4-Me$_2$-3-pentanol, 2,4-Me$_2$-1-hexanol, 2-Me-3-hexanol, and tertiary butyl and amyl alcohols.

Waxes contain higher alcohols in the form of esters. *Cetyl alcohol* (ethal), 1-hexadecanol, $C_{16}H_{33}OH$, m. 49°, was obtained by Chevreul from spermaceti (cetyl palmitate) over a century ago.[93] It is now more cheaply prepared from ethyl palmitate by hydrogenation. *Carnaubyl alcohol*, $C_{24}H_{49}OH$, m. 69°, occurs as esters in wool grease. *Ceryl alcohol*, $C_{26}H_{53}OH$, m. 79°, occurs as ceryl cerotate in Chinese wax. *Montanyl alcohol*, $C_{28}H_{57}OH$, in montan wax and in cotton. *Ginnol*, 10-nonacosanol (C_{29}). *Melissyl* or *myricyl alcohol*, $C_{30}H_{61}OH$ or $C_{31}H_{63}OH$, m. 85°, occurs as esters in beeswax and Carnauba wax. An isomer, *gossypyl alcohol*, and its C_{32} and C_{34} homologs are found in the wax of American cotton.[94] *Lacceryl alcohol*, $C_{32}H_{65}OH$. *Psyllastearyl* or *psyllicyl alcohol*, $C_{33}H_{67}OH$, m. 79°.

Identification of Alcohols

Several series of crystalline derivatives of the alcohols have been studied. The commonest consists of the *3,5-dinitrobenzoates* obtained from the acid

[90] Wilkes, Wickert. *Ind. Eng. Chem.* **29**, 1234 (1937).

[91] Simonini. *Monatsh.* **14**, 81 (1893).

[92] *Chem. Ind.* **61**, 965 (1947).

[93] Youtz. *J. Am. Chem. Soc.* **47**, 2252 (1925).

[94] *Ann. Rep. Chem. Soc.* (London) **1925**, 69.

chloride and the alcohol alone or in the presence of pyridine. Solid aryl urethans are obtained by the action of phenyl isocyanate, α-naphthyl isocyanate, p-nitrophenyl isocyanate, or 4-biphenyl isocyanate. The alcohol used must be anhydrous since water changes the isocyanate to the corresponding diarylurea. Similarly most tertiary alcohols cannot be identified by this method because they lose a molecule of water so easily.

$$\text{ArNCO} + \text{ROH} \rightarrow \text{ArNHCO}_2\text{R}$$
$$2\ \text{ArNCO} + \text{H}_2\text{O} \rightarrow (\text{ArNH})_2\text{CO}$$

Some higher tertiary alcohols form phenylurethans. For example 3-Et-3-octadecanol, from Et palmitate and EtMgBr, reacts with phenyl isocyanate to give both a phenylurethan, m. 55°, and diphenylurea.

Alcohols which can be converted to alkyl halides without rearrangement can be identified by means of the latter (p. 86).

A special method for identifying primary and secondary alcohols is to make their esters with pyruvic acid, $\text{CH}_3\text{COCO}_2\text{H}$, and convert these to semicarbazones.[95] Tertiary alcohols are dehydrated by this procedure.

B. Olefinic Alcohols

1. Vinyl alcohol, $\text{CH}_2=\text{CHOH}$, is known mainly in the form of derivatives such as its ether, b. 35.5°, which may be made from $\beta\beta'$-dichlorodiethyl ether and alkalies.

$$(\text{ClCH}_2\text{CH}_2)_2\text{O} \rightarrow (\text{CH}_2=\text{CH})_2\text{O}$$

This is used as an anesthetic (*Vinethene*). A solution of vinyl alcohol can be obtained by the reaction of vinyl acetate with cold N NaOH and extraction with ether.

Polyvinyl alcohol is an important plastic of unusual toughness (DuPont).

Vinyl sulfide, b. 101°, can be made from mustard gas.

Reactions which would be expected to give vinyl alcohol result in the formation of acetaldehyde

$$\text{CH}_2=\text{CHOH} \rightarrow \text{CH}_3-\overset{\displaystyle H}{\underset{\displaystyle}{C}}=O$$

This type of change takes place in most cases in which a hydroxyl is attached to an ethylenic or acetylenic carbon. The change is seldom complete, however, a condition of equilibrium being reached

$$\text{C}=\text{C}-\text{OH} \rightleftharpoons \text{CH}-\text{C}=\text{O}$$
$$\text{enol form} \qquad \text{keto form}$$

[95] Bouveault. *Compt. rend.* **138**, 984 (1904).

This equilibrium system is very general. It may be expressed in a generalized form

$$M = Q - Z - H \rightleftharpoons MH - Q = Z$$

in which M, Q and Z represent various elements, especially combinations of C, N, O and S atoms. The commonest of these systems are

$$C = C - OH \rightleftharpoons CH - C = O \qquad C = C - SH \rightleftharpoons CH - C = S$$
Aldehydes, ketones, etc.   Thio analogs of aldehydes, etc.

$$C = NOH \rightleftharpoons CH - N = O$$
Oximes. In this case the "enol" is the stable form.

$$N = C - OH \rightleftharpoons NH - C = O \qquad N = C - SH \rightleftharpoons NH - C = S$$
Cyanic acid derivatives   Isothiocyanic acid derivatives

Such equilibria are called tautomeric equilibria and the substances are referred to as tautomers.

Tautomerism is not to be confused with mesomerism which is associated with the concept of resonance. Isomers, often separable as with the keto-enol forms of acetoacetic ester and which result from a rapid equilibrium in which atoms take different position in the molecule, are tautomers. When however only one substance exists, corresponding to two or more structural formulas different only in the position of electrons, it is known as a mesomer. This coalescence of the electron structures is called resonance and is accompanied by a decrease in the energy of the molecule below the lowest energy level that would be expected from any of the structural formulas.[96]

Many of the reactions of aldehydes, ketones and acids and their derivatives can best be explained on the basis that the keto system, $CH - C = O$, which they contain is in equilibrium with the enol system, $C = C - OH$.

It was formerly assumed that the change between keto and enol forms involved an actual transfer of H from one part of a given molecule to another part of that molecule. It is now generally recognized that the change is *intermolecular* rather than *intramolecular*. The change involves the addition to one part of the molecule of a hydrogen ion from the medium (the solvent or the substance itself) and the loss of a hydrogen ion from another part of the molecule. The changes may take place in the reverse order, hydrogen ion being removed from one part of the molecule and added at another point. In either case the net result is the same as in an actual intramolecular shift of H. The process may be illustrated by acetaldehyde and vinyl alcohol.

$$H^+ + CH_3 - CHO \rightleftharpoons [CH_3 - CH(OH)]^+ \rightleftharpoons H^+ + CH_2 = CH(OH)$$

Regarded electronically, the conversion of a keto to an enol form involves the addition of H to one of the electron pairs of the carbonyl oxygen to form a complex ion which can lose H either from the oxygen or from the alpha carbon.

[96] Ingold. *Nature* 141, 314 (1928); Branch and Calvin. "Theory of Organic Chemistry." Prentiss Hall, 1941. p. 70.

In the latter case, the enol results. The changes are reversible.

$$: \overset{..}{\underset{H}{C}} : \overset{..}{C} :: \overset{..}{\underset{..}{O}} + H^+ \rightleftharpoons \left[: \overset{..}{\underset{H}{C}} : \overset{..}{\underset{+}{C}} : \overset{..}{\underset{..}{O}} : H \right]^+ \rightleftharpoons H^+ + : \overset{..}{C} :: \overset{..}{C} : \overset{..}{\underset{..}{O}} : H$$

Neutral	Positive	Neutral
keto	complex	enol
form.	ion.	form.

1-Propenol, $CH_3CH = CHOH$, and 2-propenol, $CH_3COH = CH_2$, are merely the enol forms of propionaldehyde and acetone respectively, and are not capable of independent existence. There is good evidence, however, that these forms exist in equilibria with the ordinary $CH = CO$ forms.

Allyl alcohol, 3-propenol, vinyl carbinol, $CH_2 = CH - CH_2OH$, b. 97°, is prepared by heating glycerol with formic or oxalic acid and a trace of ammonium chloride to 220°.

Allyl alcohol is best prepared by hydrolysis of allyl chloride from the high temperature action of propylene and chlorine (Shell) (p. 95).

Allyl alcohol acts as a primary alcohol except that there is no tendency for dehydration. Its esters, especially the carbonate, can be polymerized to give useful resins (*Allymers*). It gives the ordinary addition reactions of an olefin and also adds $KHSO_3$ when refluxed with a concentrated solution of that substance.

$$CH_2 = CHCH_2OH + KHSO_3 \rightarrow CH_3 - CH(SO_3K)CH_2OH$$

Allyl alcohol gives the Diels-Alder reaction with conjugated dienes to form H_4-benzyl alcohol and its substitution products.

Oxidation by dilute $KMnO_4$ gives glycerol. In order to oxidize the alcohol group alone, the ethylenic linkage must be "protected." This is done by adding bromine, which can later be removed by means of zinc.

$$\begin{array}{ccccccc}
CH_2 & & CH_2Br & & CH_2Br & & CH_2Br \\
\| & & | & & | & & | \\
CH & \rightarrow & CHBr & \rightarrow & CHBr & \rightarrow & CHBr \\
| & & | & & | & & | \\
CH_2OH & & CH_2OH & & CHO & & CO_2H
\end{array}$$

By removing bromine from the last two products, acrolein, $CH_2 = CHCHO$, and acrylic acid, $CH_2 = CHCO_2H$, would be obtained.

Hypohalous acids add to allyl alcohol contrary to Markownikoff's Rule to give beta glycerol halohydrins,

$$HOCH_2CHXCH_2OH$$

Addition takes place more rapidly from aqueous solutions of Cl_2 or Br_2 than of HOCl or HOBr.[97]

[97] Taylor, Maass, Hibbert. *Can. J. Research* **4**, 119 (1931).

Allyl esters are important intermediates for plastics both alone and in a great variety of copolymers.

Methylvinylcarbinol, 1-buten-3-ol, $CH_2=CHCHOHCH_3$, b. 97°, is readily obtained from acrolein and MeMgX. Its behavior with HBr illustrates the allylic type of rearrangement[98] (p. 97).

Crotyl alcohol, crotonyl alcohol, propenylcarbinol, 2-buten-1-ol, $CH_3CH=CHCH_2OH$, b. 121°, is difficultly obtainable from crotonaldehyde by ordinary methods. Reduction by Al isopropoxide gives a 60% yield.[99]

$$3 \text{ RCHO} + \text{Al(OCHMe}_2)_3 \rightarrow (\text{RCH}_2\text{O})_3\text{Al} + 3 \text{ Me}_2\text{CO}$$

Crotyl alcohol with HBr gives crotyl bromide and the rearranged methyl-vinylcarbinyl bromide.

A similar rearrangement takes place when linalool is converted to the stereoisomers, geraniol and nerol, by heating with acetic anhydride

$$(CH_3)_2C=CHCH_2CH_2\underset{\underset{CH_3}{|}}{\overset{\overset{OH}{|}}{C}}CH=CH_2 \rightarrow (CH_3)_2C=CHCH_2CH_2\underset{\underset{CH_3}{|}}{C}=CH-CH_2OH$$

The rearrangement of the allyl group occurs in a similar manner in systems of the general type

where X and Y = CN or CO_2Et.[100]

Much work has been done on the glutaconic acids and similar compounds containing a three carbon system like that in allyl alcohol.

Allylcarbinol, 1-buten-4-ol, $CH_2=CHCH_2CH_2OH$, b. 114° has been made by heating $NH_2(CH_2)_4NH_2$ with $AgNO_2$ and HCl.

Unsaturated esters other than $\alpha\beta$ can be reduced by zinc chromite or Na and EtOH to unsaturated primary alcohols without reducing the double bond. Thus *oleyl alcohol,* 9-octadecen-1-ol, is available from oleic esters.

4-Penten-1-ol, $CH_2=CH(CH_2)_3OH$, results from the action of Mg with tetrahydrofurfuryl bromide.[101]

$$\underset{CH_2OCH-CH_2Br + Mg \rightarrow}{\overset{CH_2-CH_2}{|\qquad\quad|}}$$

$$BrMgOCH_2CH_2CH_2CH=CH_2 \rightarrow HO(CH_2)_3CH=CH_2$$

[98] Prevost. *Compt. rend.* **185**, 1283 (1927).
[99] Young, Hartung, Crossley. *J. Am. Chem. Soc.* **58**, 100 (1936).
[100] Cope et al. *J. Am. Chem. Soc.* **63**, 1843 (1941).
[101] Robinson, Smith. *J. Chem. Soc.* **1936**, 195.

This is a peculiar case of Boord's formation of an olefin from a beta-bromo ether (p. 45).

A few *higher unsaturated alcohols* are found in nature: *citronellol*, 2,6-dimethyl-1-octen-8-ol, in rose oil as the levorotatory form, in oil of citronella as the dextrorotatory form, and in geranium oil in both forms; its isomer rhodinol, 2,6-dimethyl-2-octen-8-ol; phytol, 3,7,11,15-tetramethyl-2-hexadecen-1-ol, a constituent of chlorophyll.

C. ACETYLENIC ALCOHOLS

The simplest possible member would be *ethynol* (ethinol), $HC \equiv C-OH$. This exists only in the tautomeric form, $H_2C=C=O$, ketene.

Propargyl alcohol, propiolic alcohol, propynol, $HC \equiv CCH_2OH$, b. 115°, can be made from glycerol tribromohydrin by treatment with alcoholic KOH, hydrolyzing with water and further treatment with alcoholic KOH.

$$BrCH_2CHBrCH_2Br \rightarrow CH_2=CBrCH_2Br$$
$$\rightarrow CH_2=CBrCH_2OH \rightarrow CH \equiv CCH_2OH$$

The product from the first step is easily isolated. One Br is inactive and the other unusually active being an "allyl" halogen. The removal of the last Br to give the triple bond is the most difficult step. Propargyl alcohol can be made from acetylene and formaldehyde under suitable conditions.[102]

The effect of unsaturation on physical properties is seen in the series propyl, allyl and propargyl alcohols, with m. $-127°$, $-129°$, $-17°$; b. 97°, 97°, 115°, and d. about 0.80, 0.85, 0.97. The effect of the triple bond is much more marked than that of the double bond. Propargyl alcohol shows the group reactions typical of a primary alcohol group and of a terminal acetylene linkage. Thus HBr gives Br-allyl alcohol, PBr₃ gives propargyl bromide, and ammoniacal Ag and cuprous solutions give characteristic precipitates.

Higher acetylenic alcohols may be made by the action of acetylenic Grignard reagents, $HC \equiv CMgX$ and $RC \equiv CMgX$, or the corresponding sodium derivatives with aldehydes and ketones. Another general method involves the use of sodium, potassium or copper derivatives. $NaNH_2$, KOH sand (made in butyl alcohol and xylene at 140°), and KOH acetal complexes are effective reagents. A great variety of syntheses is possible.[103],[104] The general reaction is represented thus

$$
\begin{array}{ccc}
\mathrm{CH_3} & \mathrm{CH_3} & \mathrm{CH_3} \\
| & | & \diagdown \\
\mathrm{CO} \rightarrow & \mathrm{C-ONa} \rightarrow & \mathrm{HC \equiv C-C-OH} \\
| & \| & \diagup \\
\mathrm{CH_3} & \mathrm{CH_2} & \mathrm{CH_3}
\end{array}
$$

[102] Reppe. *Modern Plastics* **23**, No. 6, 169 (1946).
[103] Allen. *Syn. Org. Chem.*, E. K. Co. 19, 3 (1947).
[104] Weizmann. *C. A.* **42**, 3772 (1948).

Traces of water greatly lower the yield of the ethynylcarbinols in these reactions.[105]

An example of the practical utilization of this process is the conversion of 2-Me-2-hepten-6-one ("methylheptenone" from citral) to 3,7-Me$_2$-1-octyn-6-ene-3-ol which is then partially hydrogenated to the perfume material linalool, 3,7-Me$_2$-1,6-octadien-3-ol.

Vinylethynylcarbinols give viscous polymers of high adhesive power.

IV. ETHERS, R—O—R, R—O—R′

The absence of the H attached to oxygen makes ethers much less reactive than the corresponding alcohols. They are, however, much more reactive than the corresponding hydrocarbons. They range from b. −25° to high boiling liquids and finally solids. The ethers differ from the alcohols and water in not having associated molecules. Thus the boiling points are much lower than those of the isomeric alcohols and near to those of the hydrocarbons of the same number of atoms in a chain. Thus: propane, −45°; CH$_3$OCH$_3$, −24.5°; CH$_3$OH, 66°; n-pentane, 36°, EtOEt, 35°.

1. **Methyl ether,** dimethyl ether, CH$_3$OCH$_3$, b. −24.5°, can be prepared from methanol and concentrated sulfuric acid. The reaction takes place in definite steps:

(a) cold $MeOH + H_2SO_4 \rightarrow MeOSO_3H + H_2O$

(b) hot $MeOH + MeOSO_3H \rightarrow Me_2O + H_2SO_4$

An alternative or supplementary mechanism for the formation of ethers from alcohols and sulfuric acid involves the addition of alcohol to sulfuric acid in the same way water adds to it.[1] Just as H_2SO_4 can add 2 H_2O to give the unstable ortho acid $S(OH)_6$ it can add 2 ROH to give $(RO)_2S(OH)_4$ which could lose R_2O and H_2O leaving sulfuric acid to start the cycle again. The action of sulfuric acid is a general method for making ethers. With higher alcohols, especially those with branched chains, the tendency to form olefins even at moderate temperatures increases greatly. This can be diminished by operating at 140–145° under pressure (Oddo).

If one alcohol is treated with sulfuric acid to make the acid ester and then a different alcohol is added during the heating, a *mixed ether*, ROR′, is obtained.

Methyl ether as now commercially available is obtained from methanol catalytically (DuPont). It is also a by-product of the preparation of dimethylaniline from aniline, MeOH and HCl.

Another method of preparation is from sodium methylate and methyl halide, preferably the iodide.[2] In the case of the methyl compounds this reaction gives an excellent yield. With higher halides there is increasing

[105] Hurd, McPhee. *J. Am. Chem. Soc.* **69**, 239 (1947).

[1] Oddo. *Gazz. chim. ital.* **31**, i, 285 (1901).

[2] Williamson. *J. Chem. Soc.* **4**, 106 (1852).

tendency to form olefins. *The Williamson preparation of simple and mixed ethers* is the best proof of their structure.

Methyl ether is soluble 600:1 (by volume) in conc. sulfuric acid. Its vapor pressure at 20° is 5 atm.

Because of its remarkable properties as a solvent and because it can be handled as readily as NH_3 and SO_2 in the liquid state, methyl ether finds many uses.

Dimethyl ether, like its homologs, forms complex compounds with many types of substances. $Me_2O \cdot C_2H_2$, $Me_2O \cdot NH_3$, $Me_2O \cdot AlCl_3$, $Me_2O \cdot 2\ AlCl_3$, $Me_2O \cdot CaCl_2$, $Me_2O \cdot HCl$,[3] $Me_2O \cdot 4\ HCl$, etc. Many of these are known to be oxonium compounds.

Of the same nature as the oxonium compounds but characterized by greater stability is the coordination compound $Me_2O \cdot BF_3$, b. 128°, now commercially available.

$$\left[\begin{array}{c} F \\ \cdot\cdot\ \ \cdot\cdot \\ Me : O : B : F \\ \cdot\cdot\ \ \cdot\cdot \\ Me\ \ F \end{array} \right]^\circ \qquad \left[\begin{array}{c} \cdot\cdot \\ Me : O : H \\ \cdot\cdot \\ Me \end{array} \right]^+ Br^-$$

In many ways these resemble ammonium salts which were formerly represented as $NH_3 \cdot HX$. Just as these are related to the unstable radical NH_4, the oxonium salts are related to an even less stable oxonium ion $[H_3O]^+$. The oxonium salts are completely ionized.[4]

In case of an over-production of by-product methyl ether, it can be converted smoothly to methanol by water and catalysts at 400°.

Methyl ethyl ether, b. 10.8°, is readily made from MeI and NaOEt or from EtI and NaOMe. If needed commercially it could readily be made from the two alcohols.

2. Ethyl ether, "ether," "sulfuric ether," $C_2H_5OC_2H_5$, b. 34.5°, is made from ethanol and sulfuric acid at 130–140°. The process is not perfectly continuous because (1) most of the water formed remains in the acid and dilutes it and (2) some of the ethanol is dehydrated to give ethylene which partly polymerizes to give materials capable of reacting with the sulfuric acid and reducing it to sulfur dioxide. Other by-products such as isethionic acid, $HOCH_2CH_2SO_3H$, are also formed. Ethyl ether is now available as a by-product of the preparation of ethanol by hydrating ethylene with sulfuric acid (CC).

A modification of the sulfuric acid preparation is to heat ethanol with an aromatic sulfonic acid, always keeping an excess of alcohol present. The process is much like that using sulfuric acid except that a higher temperature can be used without reducing the acid. The cheapest sulfonic acid is para-

[3] Friedel. *Compt. rend.* **81**, 152 (1875).
[4] Oddo, Scandola. *Gazz. chim. ital.* **40**, ii, 163 (1910).

toluene sulfonic acid (from saccharin manufacture).

$$ArSO_3H + EtOH \rightarrow ArSO_3Et + H_2O$$
$$ArSO_3Et + EtOH \rightarrow Et_2O + ArSO_3H$$

Ethyl ether is obtained in good yield when ethanol is passed over alumina at 250–300°.[5] A trace of a heavy metal salt like ferric chloride catalyzes ether formation.[6]

Ethyl ether has been made by Williamson's method (80% yield from EtI + EtONa). The formation of ethyl ether from Ag_2O and EtI is additional evidence of its structure.

Ethyl ether is a typical ether in its reactions.

1. It dissolves in concentrated H_2SO_4. At higher temperatures it reacts.

$$Et_2O + H_2SO_4 \rightarrow EtOSO_3H + EtOH$$

2. Concentrated HI solution splits it even at room temperature.

$$Et_2O + HI \rightarrow EtI + EtOH$$

Hydrobromic acid produces a similar change but less readily.

3. With sodium it does not react at its boiling point. This gives the best way of removing the last traces of water and alcohol from ether for use in making Grignard reagents. The assumption that *higher ethers* will not react with Na is erroneous. On heating they react to give alcoholates, olefins and H_2.

4. It forms "oxonium salts" with halide acids, $R_2O.HX$.

5. Ether reacts slowly with oxygen to give complex "peroxide" compounds and aldehydes. Ether thus contaminated is dangerous in anesthetic use. Evaporation of "old" ether sometimes leaves dangerous amounts of this highly explosive material. Ether has been successfully preserved by storage with asbestos impregnated with alkali and either pyrogallol or permanganate.[7]

6. Acid iodides split ethers.[8]

$$Et_2O + MeCOI \rightarrow EtI + MeCO_2Et$$

Acid chlorides and acid anhydrides in the presence of anhydrous $ZnCl_2$ give a similar splitting (Meerwein).

7. Phosphorus pentachloride reacts with ethers but, contrary to the ordinary conception, gives complex chlorinated products instead of simple alkyl chlorides.[9]

8. Chlorine and bromine react with ethers much as with paraffin hydrocarbons. The first H atom replaced is one on the C attached to O. Then

[5] Senderens. *Bull. soc. chim.* (4) 5, 480 (1909).
[6] Oddo. *Gazz. chim. ital.* 31, i, 285 (1901).
[7] Palkin, Watkins. *Ind. Eng. Chem.* 21, 863 (1929).
[8] Gustus, Stevens. *J. Am. Chem. Soc.* 55, 378 (1933).
[9] Whitmore, Langlois. *J. Am. Chem. Soc.* 55, 1518 (1933).

those of the next carbon are attacked (Oddo). Finally perchloroether, $(C_2Cl_5)_2O$, is obtained as a solid with an odor like camphor. At low temperatures ether combines with bromine to give the compound, $Et_2O.Br_2$, m. 24°. At higher temperatures this loses bromine with bromination of the ether.

The most important use of ether is as the commonest general anesthetic. This property was probably first observed by Faraday. The actual use as an anesthetic was made independently by Drs. Long, Morton, and Simpson of Georgia, Boston, and Edinburgh in 1842, 1846 and 1848 respectively.

Ether is a very important general solvent in organic chemistry because it is only slightly soluble in water (7%) and dissolves a great variety of materials which are also soluble in water. It is thus used for extracting many materials from their dilute solutions in water (Soxhlet, etc.). It is used for extracting acetic acid from pyroligneous acid (Suida Process).

Since ether is largely used because of its chemical inertness, it is useful to remember that it does not react under ordinary conditions with metals, Grignard reagents, organic acid chlorides, organic acids, phosphorus trihalides, permanganates and bases.

U.S.P. Ether contains about 4% ethanol and traces of water. Absolute, anhydrous, or "dry" ether can be made from it by removing the alcohol with solid $CaCl_2$, decanting and distilling from P_2O_5 or Na. The product is satisfactory for making Grignard reagents. Final traces of water and alcohol can be removed by refluxing with a small amount of a cheap Grignard reagent such as MeMgCl or EtMgBr and distilling the ether.

Methyl n-propyl ether, b. 39°; ethyl n-propyl ether, b. 64°; and ethyl isopropyl ether, b. 54°, have been made in the usual way.

3. (a) **n-Propyl Ether,** $C_3H_7OC_3H_7$, b. 90.4°, is made by the same reactions as ethyl ether. The tendency for olefin formation is greater both by the acid and the Williamson methods. Its reactions are like those of ethyl ether.

(b) **Isopropyl Ether,** $(CH_3)_2CHOCH(CH_3)_2$, b. 69°, is available as a by-product of the hydration of propylene to isopropyl alcohol (C and C). Its higher boiling point and its even lower mutual solubility with water gives it advantages over ethyl ether as an extracting agent. It is specially useful in extracting acetic and lactic acids from their dilute water solutions. Since it is actually cheaper than ethyl ether it will undoubtedly find wide uses as a solvent and extractant. It is apparently less useful in the Grignard reaction than ethyl ether and the higher normal ethers.

Isopropyl ether is used in high anti-knock fuels for airplanes.

4. Higher ethers can be obtained by the regular methods, the amount of olefin increasing with the size and branching of the alcohols or halides used. n-Butyl ether, b. 141°, is valuable as a high boiling ether for the Grignard reagent. It can be made by the action of 50% sulfuric acid with 1-butanol. Mixed amyl ethers, b. 170–190°, are available as a by-product of the preparation of amyl alcohols from pentanes (Sharples). Mixed higher ethers containing the t-butyl group and a Me, Et, n-Pr, n-Bu or isopropyl have been

made by heating the mixtures of alcohols with a 15% solution of sulfuric acid or NaHSO$_4$. In this process there is little or no tendency to form the simple ethers. To prepare ethers containing primary and secondary alkyl groups it is necessary to use 50% acid. All three possible ethers were formed in each case, the mixed ethers predominating.[10] This preparation of ethers is all the more important because of the good yields obtained. Thus ethyl t-butyl ether, b. 73°, was made in 95% yield and isopropyl t-butyl ether, b. 87°, in 82%.

Mixed primary-tertiary and secondary-tertiary ethers are obtained by adding primary or secondary alcohols to olefins related to tertiary alcohols, in presence of catalysts, usually acidic in nature.[11] Thus methanol and isopropyl alcohol with isobutylene give respectively MeOCMe$_3$ and Me$_2$CHOCMe$_3$.

As would be expected, ethers containing tertiary groups are readily split by acids like HCl to give a tertiary chloride and the alcohol corresponding to the other radical.

Higher ethers can be made by the action of the Grignard reagent with an alpha chloro or bromomethyl ether[12] obtained from formaldehyde, an alcohol and HX.[13]

$$ROH + CH_2O + HX \rightarrow ROCH_2X + H_2O$$
$$ROCH_2X + R'MgX \rightarrow ROCH_2R'$$

Beta halogenated ethers are less active but can be caused to react with Grignard reagents by using a higher temperature made possible by the use of a higher ether such as n-butyl ether. Thus $\beta\beta'$-dichloroethyl ether, (ClCH$_2$CH$_2$)$_2$O, gives (RCH$_2$CH$_2$)$_2$O. The former compound is *Chlorex*, first made by dehydrating ethylene chlorohydrin,[14] long a useless by-product of ethylene glycol manufacture, but now important as a solvent in the petroleum industry, an intermediate in chemical synthesis, a soil fumigant and an insecticide. Chlorex with Grignard reagents provides a convenient synthesis of ethers.

Trimethylethylene with methanol and sulfuric acid gives a methyl amyl ether.[15]

The higher ethers, because of their higher boiling points, react more than the lower members. Thus, refluxing amyl ether with phosphoric anhydride gives amylene.

Higher ethers also react with Na on heating.

$$2 RCH_2CH_2OR' + 2 Na \rightarrow 2 R'ONa + 2 RCH=CH_2 + H_2$$

Alkyl derivatives of hydrogen peroxide such as EtOOH, liquid, and EtOOEt, b. 65°, are known. They are strong oxidizing agents and are ex-

[10] Norris, Rigby. *J. Am. Chem. Soc.* 54, 2088 (1932).
[11] Edlund, Evans. *C. A.* 28, 5831 (1934).
[12] Hamonet. *Compt. rend.* 138, 813 (1904).
[13] Henry. *Compt. rend.* 113, 368 (1891).
[14] Kamm, Waldo. *J. Am. Chem. Soc.* 43, 2223 (1921).
[15] Reychler. *Bull. soc. chim. Belg.* 21, 71 (1907).

plosive. Unsaturated ethers, particularly the vinyl type, are becoming increasingly important, especially in plastics and chemical synthesis; methyl vinyl ether b. 6°, vinyl n-butyl ether b. 94°. Vinyl ether, b. 39° (Vinethene), is used as a general anesthetic and has been found particularly useful in dentistry.[16,17]

V. ALIPHATIC SULFUR COMPOUNDS

Almost all organic compounds on heating with sulfur give H_2S and substitution products containing sulfur. Usually these are of unknown complexity.

While some of the simpler sulfur compounds resemble the corresponding oxygen compounds in their formulas they differ in many ways:

1. The sulfur atom can be oxidized whereas the oxygen atom cannot. This ability makes possible compounds such as the sulfoxides, sulfones, sulfonic acids and sulfinic acids, R_2SO, R_2SO_2, RSO_3H and RSO_2H respectively.

2. The $-S-S-$ linkage is stable whereas the $-O-O-$ linkage is very unstable.

3. The -onium compounds of sulfur are very stable whereas those of oxygen are unstable.

Considerable amounts of aliphatic sulfur compounds occur in many crude petroleums. The removal of such compounds from petroleum products is troublesome and expensive. It is stated that the danger of corrosion from small amounts of sulfur compounds is greatly exaggerated.[1]

Organic sulfur compounds, under hydrogenation conditions, give up all their sulfur as H_2S which can readily be separated. The older catalytic processes were stopped by sulfur compounds but the modern sulfide catalysts used in the hydrogenation of petroleum and coal are not poisoned in this way.[2] This conversion of complex organic sulfur compounds to H_2S represents one of the chief advantages of hydrogenation as a method of refining.

Mercaptans, thioalcohols, alkanethiols, RSH, resemble alcohols only as to formulas. The lower members have remarkably disagreeable odors. Ethyl mercaptan can be detected by smell in as small an amount as 2×10^{-12} gram.[3] This is less than 1/200 of the smallest amount of Na detectable with the spectroscope. The offensive odor decreases with the homologs until the C_9 member has a pleasant odor. The lower mercaptans boil lower than the corresponding alcohols just as H_2S boils lower than H_2O. This difference steadily decreases. For the first six normal members the *differences* (°C) between the boiling points of the alcohols and mercaptans are 58, 41, 30, 20, 12, and 7. The higher mercaptans (C_7 etc.) boil at almost the same point as

[16] Carr, Kibler, Krantz. *Anesthesiology* 5, 495 (1944).
[17] *C. A.* 39, 2330 (1945).
[1] Egloff, Lowry, Truesdell. *Natl. Petroleum News* 22, No. 24, 41 (1930).
[2] Haslam, Russel. *Ind. Eng. Chem.* 22, 1030 (1930).
[3] Fischer, Penzoldt. *Ann.* 239, 131 (1887).

the alcohols. The mercaptans are insoluble in water but soluble in alcohols and ethers. The reactions of the mercaptans and alcohols differ profoundly. The reactions of the mercaptans can be generalized more safely than those of the alcohols largely because the C–S linkage is not broken as is the C–O linkage in most reactions. Even the preparation of mercaptans from alkyl halides apparently involves less rearrangement than does that of the alcohols. This may well be due to the ability of sulfur to *add* alkyl halides as in the formation of sulfonium halides.

Preparation of Mercaptans.

1. By passage of an alcohol and H_2S over hot thorium oxide.

2. By addition of H_2S to olefins.

3. By heating NaSH solution (NaOH saturated with H_2S) with a solution of a salt of an alkyl sulfuric acid (a solution of an alcohol in sulfuric acid, diluted and exactly neutralized with a base).

$$NaSH + RNaSO_4 \rightarrow RSH + Na_2SO_4$$

4. By treatment of an alkyl halide with concentrated KSH solution. This has the disadvantages of the insolubility of the alkyl halide and the basic nature of the reagent which tends to remove HX from halides. This reaction has not been studied with all types of alkyl halides (p. 73–6).

5. In special cases, by treatment of a Grignard reagent with sulfur.[4]

$$RMgX \rightarrow RSMgX \rightarrow RSH$$

In this way neopentyl mercaptan, Me_3CCH_2SH, and related compounds have been made available.

Reactions of Mercaptans. The stability of the C–S linkage is much like that of the C–N linkage thus differing very much from the C–O linkage.

1. Mercaptans form metal derivatives, mercaptides or thiolates, much more readily than alcohols form alcoholates. Thus they are soluble in base solutions and react with HgO to give RSNa, $(RS)_2Hg$, etc. They also form mixed compounds like RSHgCl, valuable for identification purposes because of their crystalline character and high melting points, and RSHgR' of value therapeutically when one of the R groups contains a solubilizing group (p. 698).

2. Gentle oxidation merely removes the hydrogen from the –SH group giving a disulfide

$$2 \ RSH + [O] \rightarrow H_2O + RS–SR$$

This process takes place even more easily with the soluble mercaptides. Thus a mercaptan in presence of ammonia solution is oxidized by air to a disulfide. The process is easily reversed by reducing agents. The oxidation-reduction system represented by the mercaptans and their disulfides is of great biological importance in the action of substances like cystine and glutathione.

The use of compounds of the isotope S^{34} is valuable in biological studies.

[4] *Ann. Rep. Chem. Soc.* (London) 1905, 105.

3. Vigorous oxidation converts the $-SH$ group to the *sulfonic acid* group $-SO_3H$. The intermediate *sulfinic acid*, RSO_2H, cannot be isolated because of its ease of oxidation.

4. Just as alcohols react with carbonyl compounds to give hemiacetals, acetals and esters, mercaptans give similar reactions but with *greater ease*. Whereas alcohols give acetals with aldehydes but do not react with unsubstituted ketones, mercaptans give *mercaptals* with aldehydes and *mercaptols* with ketones in presence of HCl.

$$=C=O \rightarrow \ =C(SR)_2$$

The RS— groups are readily removed by treatment with mercuric compounds and water. $=C(SR)_2 + HgO \rightarrow \ =C=O + (RS)_2Hg$. This gives a method of "protecting" the aldehyde group while other groups in the molecule are being studied.

With carboxylic acids, the mercaptans give monothioesters and water. The mechanism is probably like that of esterification with an alcohol. The $R-S$ linkage is not broken in the process. This agrees with the conclusion that the $R-O$ *linkage in alcohols is not broken during esterification with carboxylic acids* (p. 120).

Mercaptans do not react with halide acids as do alcohols. Thus the most prolific source of rearrangements in alcohols is lacking with the mercaptans just as it is with the amines.

5. Mercaptans add to $\alpha\beta$-unsaturated compounds.[5]

$$RCH=CHCOR' \rightarrow RCH(SR'')CH_2COR'$$

Mercaptans can be identified by the mixed sulfides formed by the action of their sodium compounds with the active chlorine atom in 2,4-dinitrochlorobenzene. The melting points for the compounds $R-S-C_6H_3(NO_2)_2$ for the first nine normal alkyl groups (°C.) are 128, 115, 81, 66, 80, 74, 82, 78, 86 and cetyl, C_{16}, 91. Potassium permanganate in acetic acid oxidizes these sulfides to sulfones with the following melting points for C_1 to C_9, 189, 160, 127, 92, 83, 97, 101, 98, 92.

Mercaptans can be estimated by titration with Cu alkyl phthalates.[6]

Methyl mercaptan, methanthiol, CH_3SH, b. 7.6°, is formed by the decay and hydrolysis of various proteins. Presumably its source is methionine or related compounds (p. 556-7).

Ethyl mercaptan, ethanthiol, CH_3CH_2SH, b. 37°, is made commercially from $NaEtSO_4$ for the preparation of the soporifics *Sulfonal* and *Trional*. t-BuSH, b. 64°, is readily available from isobutylene and H_2S. t-Alkyl sulfides are available with 4-8 and 12, 14, and 16 carbon atoms (Phillips, Socony).

n-BuSH is used as an alarm agent in mines since its penetrating odor reaches all points more surely than any ordinary alarm system.

[5] *Ann. Rep. Chem. Soc.* (London) 1905, 115.
[6] Turk, Reid. *Ind. Eng. Chem., Anal. Ed.* 17, 713 (1945).

EtSH·18H$_2$O, crystals below 8°; (EtS)$_2$Hg, m. 76°; (EtS)$_2$Pb, m. 150°; BuSH, b. 97°.

1- and 2-mercaptans through C$_9$ have been made and their properties studied in detail.[7]

Higher mercaptans to C$_{18}$ have been prepared.[8]

Monothioglycol, mercaptoethanol, HOCH$_2$CH$_2$SH and the corresponding sulfide, (HOCH$_2$CH$_2$)$_2$S, *Kromax*, are readily available from ethylene chlorohydrin. The latter is the intermediate for *mustard gas*.

Thioglycolic acid, mercaptoacetic acid, HSCH$_2$CO$_2$H, is readily prepared from chloroacetic acid. It is used as a depilatory and in the permanent waving of hair.

The thioglycollates presumably react with cystine and break the sulfide linkages, weakening the fiber; a mild oxidizing agent such as hydrogen peroxide acts as a neutralizer restoring the sulfide linkage.[9]

Mercaptans.[10]

Alkyl sulfides, thioethers, R$_2$S, are prepared by treating alkali sulfides with alkyl halides or alkyl sodium sulfates. Mixed alkyl sulfides can be made from alkyl iodides and mercaptides, RSNa + R'I → NaI + RSR'. Grignard reagents treated with one atomic equivalent of sulfur and then with R'X give alkyl sulfides. RMgX → RSMgX → RSR'.[11] The alkyl sulfides are pleasant smelling (when pure) and are insoluble in water. They boil higher than the corresponding mercaptans and ethers. The effect of −S− is about equal to that of −CH$_2$OCH$_2$−. Thus the boiling points of the following pairs (°C.) are: Me$_2$S, Et$_2$O, 38, 35, and Et$_2$S, Pr$_2$O, 92, 91.

The alkyl sulfides form addition products with a great variety of substances including mercuric salts, halogens, and alkyl halides to form R$_2$S.HgX$_2$, R$_2$S.X$_2$, and R$_3$SX. These are sulfonium compounds analogous to ammonium compounds.

$$\left[\begin{array}{c} R \\ \cdot\cdot \\ R : S : Hg \\ \cdot\cdot \end{array} \right]^{++} X_2^{--} \quad \left[\begin{array}{c} R \\ \cdot\cdot \quad \cdot\cdot \\ R : S : X : \\ \cdot\cdot \quad \cdot\cdot \end{array} \right]^{+} X^- \quad \left[\begin{array}{c} R \\ \cdot\cdot \\ R : S : R \\ \cdot\cdot \end{array} \right]^{+} X^-$$

The second compound is interesting in that one halogen atom is part of the cation and the other forms the anion.

Me$_2$S, b. 38°; Et$_2$S, b. 92°; Bu$_2$S, b. 182°.

Mild oxidation converts the sulfides to sulfoxides, R$_2$SO, and more vigorous oxidation to sulfones, R$_2$SO$_2$.

[7] Ellis, Reid. *J. Am. Chem. Soc.* **54**, 1674 (1932).
[8] Collin et al. *J. Soc. Chem. Ind.* **52**, 272T (1933).
[9] Addleston. *Chem. Ind.* **58**, 414 (1946).
[10] Malisoff, Marks, Hess. *Chem. Rev.* **7**, 493 (1930).
[11] *Ann. Rep. Chem. Soc.* (London) 1905, 105.

The splitting of a $C-S$ linkage is difficult. The sulfides can be split by treatment with cyanogen bromide.

$$RSR' + BrCN \rightarrow RSCN + R'Br$$

In most cases $R > R'$.[12]

Polymethylene dimercaptans, $HS(CH_2)_nSH$, react with polymethylene bromides, $Br(CH_2)_mBr$, and bases to give polymeric cyclic sulfides in great variety.[13]

$\beta\beta'$-Dichloroethyl sulfide, Mustard Gas, Yperite, $(ClCH_2CH_2)_2S$, b. 218°, was the most important vesicant used during World War I.[14] The two methods for its preparation start from ethylene, sodium chloride and sulfur. Thus the folly of any attempted legislation against materials for war "gases" is obvious. In the German method,[15] ethylene chlorohydrin is treated with Na_2S and the resulting thiodiglycol gives the chloride with HCl. The other method (Guthrie 1860)[16] adds sulfur monochloride, S_2Cl_2, to ethylene to give the sulfide and free S which remains in colloidal form.

$$C_2H_4 \rightarrow ClCH_2CH_2OH \rightarrow S(CH_2CH_2OH)_2 \rightarrow S(CH_2CH_2Cl)_2$$
$$C_2H_4 + S_2Cl_2 \rightarrow S + S(CH_2CH_2Cl)_2$$

This compound had been obtained even earlier by the action of chlorine on Et_2S.[17] The vesicant properties depend on the β-position of the Cl. $ClCH_2CH_2SMe$ has similar properties but $ClCH_2SMe$ does not.[18] The best method for rendering mustard gas harmless is by chlorination.[19] The potential production of mustard gas in all countries at the time of the Armistice (1918) was of the order of 300 tons a day. The amount available but not used in World War II was of an entirely higher order of magnitude. The predominance of the U.S.A. in the ethylene industry probably accounted for the failure of the Axis nations to start using Mustard.

Thiokol is a complex elastomer obtained from sodium sulfide or polysulfide with dichlorides such as ethylene chloride or 1,4-Cl_2-2-butene obtained by 1,4-addition of Cl_2 to butadiene. The resulting products contain groupings such as $-(CH_2CH_2S)_n-$, $-(SCH_2CH_2)_n-$, and $-(CH_2CH=CHCH_2)_n$. The latter can be vulcanized like the polymers of isoprene and of butadiene.

The action of sulfur with olefins has not been adequately studied in spite of the great practical importance of *vulcanization* and the large mass of applied information in this field.[20]

[12] *Ann. Rep. Chem. Soc.* (London) 1923, 87.
[13] Meadow, Reid. *J. Am. Chem. Soc.* 56, 2177 (1934).
[14] *Ann. Rep. Chem. Soc.* (London) 1920, 58; 1922, 23.
[15] *Ber.* 19, 3259 (1886); 20, 1729 (1887).
[16] Prentiss. "Chemicals in War." p. 625.
[17] Richie. *Ann.* 92, 353 (1854).
[18] Kirner. *J. Am. Chem. Soc.* 50, 2446 (1928).
[19] Lawson, Dawson. *J. Am. Chem. Soc.* 49, 3119 (1927).
[20] Armstrong, Little, Doak. *Ind. Eng. Chem.* 36, 628 (1944).

Mixed tetra-thioethers, $C(CH_2SR)_4$, have been made from the tetra bromide of pentaerythritol and mercaptides from C_5 to C_{12}.[21]

Allyl sulfide, $(CH_2=CHCH_2)_2S$, b. 140°, occurs in onions and garlic. It is readily prepared from the allyl halides.

Sulfur compounds are best removed from petroleum products by hydrogenation with various molybdenum catalysts.[22] The products are hydrocarbons and H_2S.

The sulfonium halides, RR'R''SX, are obtained from alkyl halides and dialkyl sulfides. They resemble ammonium and other -onium salts. Heating alkyl iodides with sulfur gives a similar result to that obtained with phosphorus (p. 847–8), forming periodides, R_3SI_3, which can easily be reduced to R_3SI. Treatment of sulfonium halides with AgOH gives the sulfonium hydroxides, strong soluble bases. Asymmetric sulfonium compounds have been obtained as optically active enantiomers. The sulfonium compounds form stable products with mercuric salts, halogens, etc. The electronic nature of the substances may be closely related to that of the sulfonium salts

$$\begin{bmatrix} & R & \\ & \ddots & \\ R & : S & : \\ & \ddots & \\ & R & \end{bmatrix}^{+} X^{-} \qquad \begin{bmatrix} & R & \\ & \ddots & \\ R & : S & : Hg \\ & \ddots & \\ & R & \end{bmatrix}^{+++} 3\,X^{-} \qquad \begin{bmatrix} & R & \\ & \ddots & \\ R & : S & : X : \\ & \ddots & \ddots \\ & R & \end{bmatrix}^{++} 2\,X^{-}$$

Alkyldisulfides, RS—SR, may be made by oxidizing the mercaptans cautiously or from alkali disulfides and alkyl sodium sulfates. Reduction gives mercaptans. Mild oxidation converts one sulfur to a sulfone grouping thus giving a thiosulfonic acid ester, RSO_2SR, sometimes incorrectly called a *disulfoxide*. After oxidation starts, the monosulfoxide next goes to a monosulfone before the second sulfur is attacked. Vigorous oxidation gives a disulfone, RSO_2SO_2R, and finally two molecules of a sulfonic acid, RSO_3H. The —S—S— is broken not only by reduction and oxidation but by the action of alkyl iodides, which give sulfonium compounds. Me_2S_2, b. 112°, Et_2S_2, b. 153°.

Alkyl sulfoxides, R_2SO, are obtained by the action of nitric acid with the sulfides. The first product is a "nitrate," $R_2SO.HNO_3$, which is decomposed by water or mild bases (p. 150). A better preparation is from the sulfides and an organic peroxide such as perbenzoic acid, $PhCO_3H$. The lower sulfoxides are liquids which solidify on cooling. They are soluble in water, alcohol and ether. The higher ones are solids insoluble in water. Pr_2SO, m. 15°; Bu_2SO, m. 32°; $Hept_2SO$, m. 70°. Reduction changes them back to sulfides.

Any superficial resemblance between the SO and the CO groupings must not be accepted as real. This will be clearer from a consideration of the mode

[21] Backer, Dykstra. *Rec. trav. chim.* **51,** 289 (1932).
[22] Byrns, Bradley, Lee. *Ind. Eng. Chem.* **35,** 1160 (1943).

of formation of the two types of groups. Using electronic formulas:

$$
\begin{array}{c}
: \ddot{O} : H \\
R : \ddot{C} : R' + \left[: \ddot{O} : \right] \rightarrow H : \ddot{O} : H + R : \ddot{C} : R' \\
H
\end{array}
\qquad
\begin{array}{c}
: O : \\
R : \ddot{C} : R'
\end{array}
$$

$$
R : \ddot{S} : R' + \left[: \ddot{O} : \right] \rightarrow R : \ddot{S} : R'
$$
Sulfoxide

In the case of the ketone there is a true double bond involving a sharing of four electrons by the C and O to *complete both octets*. In the case of the sulfoxide there is no double bond although this type of linkage is sometimes referred to as a *semipolar double bond*. It is more properly classed as a *coordinate link*.[23] It is also represented as $R_2S \rightarrow O$, $R_2S \rightleftharpoons O$ and $R_2S^+ - O^-$. The difference between the CO and SO groups can be shown better with space models, using tetrahedra to represent the C and S atoms.

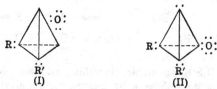

(I)   (II)

It will be seen that the ketone (I) has a plane of symmetry through R, R' and the O atom while the sulfoxide (II) has no such plane. This agrees with the experimental fact that sulfoxides can be obtained as optically active isomers.[24] This would be obviously impossible with a structure $RR'S = O$ which would have a plane of symmetry and could not exist in enantiomeric forms.

The differences between sulfoxides and ketones extends to nearly all their properties.[25] However some of the reactions and properties of sulfoxides, sulfones and sulfonic acids may be interpreted on the basis that sulfur can accommodate more than eight electrons.[26]

The difference between the SO and CO groupings persists in all compounds containing them. Another experimental evidence of this difference is given by the value of the parachors for the two types of compounds, one indicating a double bond and the other no double bond.[27,28]

[23] Sidgwick. "The Electronic Theory of Valency." Oxford, 1927. p. 60.
[24] Kenyon. *Ann. Rep. Chem. Soc.* (London) 1926, 106, 111, 126.
[25] Shriner, Struck, Jorison. *J. Am. Chem. Soc.* 52, 2060 (1930).
[26] Luder, Zuffanti. "Electronic Theory of Acids and Bases." John Wiley & Sons, 1946.
[27] Hückel. "Theoretische Grundlagen der Organischen Chemie," Vol. II. Leipzig, 1931. p. 169.
[28] Sidgwick. "The Electronic Theory of Valency." Oxford, 1927. p. 60.

The existence of the "nitrates" of the sulfoxides offers an interesting possibility in electronic formulas:

$$H^+ \begin{bmatrix} :\overset{..}{O}: \\ \overset{..}{N}:\overset{..}{O}: \\ :\overset{..}{O}: \end{bmatrix}^- + R:\overset{..}{\underset{..}{S}}:\overset{..}{O}: \rightarrow H^+ \begin{bmatrix} :\overset{..}{O}: & :\overset{..}{O}: \\ R:\overset{..}{\underset{R}{S}} & : & \overset{..}{N}:\overset{..}{O}: \\ & & :\overset{..}{O}: \end{bmatrix}^-$$

Such complex acids (not true nitrates) could not be produced by ordinary acids whose negative ions have complete octets and so could not form a coordinate link with the "free" electron pair of the sulfoxide.

Further evidence for the coordinate linking in sulfoxides is given by the existence of *cis* and *trans* forms of certain disulfoxides such as those of 1,4-dithian. The double bond formula is inadequate to express this isomerism. The coordinate link formulas are shown with the edge of the plane of the ring represented by the heavy line.

Alkyl sulfones, R_2SO_2, are stable, crystalline, odorless solids obtained by vigorous oxidation of the sulfides or by treating sodium alkyl sulfinates with alkyl halides. They are not attacked by reducing agents or by PCl_5. Their properties are reflections of the effect of the two coordinate links and the resulting strong electrical fields within the materials.

Me_2SO_2, m. 109°, b. 238°; Et_2SO_2, m. 70°, b. 248°.

Soporifics of the *sulfonal* type are disulfones obtained by oxidizing mercaptols of ketones.

Sulfonal, *Sulfonmethane* (NNR), $(CH_3)_2C(SO_2C_2H_5)_2$, m. 125°, is made by oxidizing the mercaptol of acetone with $KMnO_4$.

$$Me_2CO + 2\ EtSH \rightarrow Me_2C(SEt)_2 \rightarrow Me_2C(SO_2Et)_2$$

Many modifications are known. The best known are *trional* and *tetronal* which contain three and four Et groups. They can be made from MeEt-ketone and Et_2-ketone but are more readily prepared from acetaldehyde and propionaldehyde.

$$RCHO \rightarrow RCH(SEt)_2 \rightarrow RCH(SO_2Et)_2 \xrightarrow[\text{EtBr}]{\text{Na}} REtC(SO_2Et)_2$$

The H on the carbon between two sulfone groups is replaceable like that between two carbonyl groups.

Alkyl sulfonic acids, RSO_3H, are soluble, strong acids. Their formation by oxidation of mercaptans gives strong evidence of the radical difference between RSH and ROH. Omitting the intermediate steps, the oxidation may be represented as follows:

$$R : \overset{..}{\underset{..}{S}} : H + 3\left[\overset{..}{\underset{..}{O}} :\right] \rightarrow \left[\begin{array}{c} : \overset{..}{O} : \\ R : \overset{..}{\underset{..}{S}} : \overset{..}{O} : \\ : \overset{..}{O} : \end{array}\right]^{-} H^{+}$$

Any representaton of double bonds between sulfur and oxygen as ordinarily given in the "structure" formulas of sulfuric acid and sulfonic acids is contrary to fact.[29,30] The difference between this type of oxidation and that of the oxidation of an alcohol may be emphasized as follows:

$$\begin{array}{c} H \\ .. \quad .. \\ R : \overset{..}{\underset{..}{C}} : \overset{..}{\underset{..}{O}} : H \\ H \end{array} + 2\left[\overset{..}{\underset{..}{O}} :\right] \rightarrow \left[\begin{array}{c} : \overset{..}{O} : \\ R : \overset{..}{\underset{..}{C}} : \overset{..}{O} : \\ : \overset{..}{O} : \end{array}\right]^{---} 3H^{+} \rightarrow H_2O + \left[\begin{array}{c} : \overset{..}{O} : \\ R : \overset{..}{\underset{..}{C}} : \overset{..}{O} : \\ \end{array}\right]^{-} H^{+}$$

This illustrates important differences between sulfur and oxygen and sulfur and carbon:

a. The difference between the small kernel of oxygen which can only take electrons and the larger one of sulfur which can also give electrons.

b. The difference between sulfur with six valence electrons and carbon with four.

c. The difference between the small kernel of carbon and the large one of sulfur in that the former cannot have a stable "orthocarboxylate" *ion,* $RCO_3{}^{---}$ analogous to RSO_3^{-}.

Contrary to the usual conception, sulfonic acids can be distilled: n-$BuSO_3H$, b. 145° (0.2 mm.); isoAmSO_3H, b. 178° (0.2 mm).[31] They are obtainable in certain cases by direct sulfonation of paraffin hydrocarbons with fuming sulfuric acid.[32] Secondary and, especially, tertiary hydrogen atoms are

[29] Lewis. *J. Am. Chem. Soc.* **38**, 762 (1916).
[30] Sidgwick. "The Electronic Theory of Valency." Oxford, 1927. p. 286.
[31] v. Braun, Weissbach. *Ber.* **63**, 2836 (1930).
[32] Worstall. *Am. Chem. J.* **20**, 664 (1898).

attacked more readily. Thus normal paraffin hydrocarbons can be freed from isomers by long treatment with oleum or chlorosulfonic acid.[33]

Better preparations for the sulfonic acids are: 1. Oxidation of mercaptans, which proves the $C-S$ linkage in the sulfonic acids; 2. Action of an alkyl halide (Class 1, p. 74) with an alkali sulfite

$$K_2SO_3 + RX \rightarrow KX + RSO_3K$$

The formation of a $C-S$ compound, presumably by replacement of K in a $K-O$ linkage, formerly presented considerable theoretical difficulties. Recognition of the ionic nature of such a salt and the fact that there is really no linkage between K and O in the compound but merely a polar attachment of the ions K^+ and SO_3^{--}, has cleared up the process. It involves the following changes:

$$2K^+ \begin{bmatrix} :\overset{..}{O}: \\ :\overset{..}{S}:\overset{..}{O}: \\ :\overset{..}{O}: \end{bmatrix}^{--} + R:\overset{..}{X}: \rightarrow \begin{bmatrix} :\overset{..}{X}: \end{bmatrix}^- + \begin{bmatrix} :\overset{..}{O}: \\ R:\overset{..}{S}:\overset{..}{O}: \\ :\overset{..}{O}: \end{bmatrix}^- + 2K^+$$

The alkyl carbonium ion formed by the minimal ionization of the alkyl halide unites with the free electron pair of the sulfite ion to form the more stable alkylsulfonate ion. A similar change occurs between the alkyl sodium sulfates and an alkali sulfite.

Additional evidence on this process is given by the fact that Ag_2SO_3 and alkyl halides give alkyl esters of the alkyl sulfonic acids, RSO_2OR. In this case after one R has attached itself to the lone pair of electrons, the other has to attach itself to one of the oxygen atoms thus forming an ester.

3. Grignard reagents with SO_2 give sulfinic acids which are readily oxidized to sulfonic acids.

The monosulfonic acids of methane and of ethane are available commercially. They are said to be useful additives in metal plating baths.

Alkyl sulfonyl chlorides may be prepared directly by the chlorination of S-alkyl isothioureas. The reaction is liable to become dangerously *explosive*.[34]

Alkyl sulfonic acids, C_9 to C_{14}, have been made.[35]

Mixtures of sulfonic acids are obtained in the sulfuric acid refining of petroleum fractions and in the refining of medicinal *white oils* with oleum. Part of these stay in the oil layer and part in the water-acid layer. These are called respectively *mahogany acids* and *green* acids. Neutralization of the oil layer and extraction gives *mahogany soaps*.[36] (Petronates, Petronix, Avitone).

[33] Shepard, Henne. *Ind. Eng. Chem.* 22, 356 (1930).

[34] Folkers, Russell, Bost. *J. Am. Chem. Soc.* 63, 3530 (1941).

[35] Noller, Gordon. *J. Am. Chem. Soc.* 55, 1090 (1933).

[36] Brooks, Peters, Lykens. *Ind. Eng. Chem., Anal. Ed.* 18, 544 (1946).

The alkyl sulfonic acids are stable to hot acids and alkalies. With PCl_5, they give sulfonyl chlorides, RSO_2Cl. These react with water, bases, alcohols, ammonia and amines to give RSO_3H, RSO_3Na, RSO_3R', RSO_2NH_2, RSO_2NHR' and $RSO_2NR'R''$. With zinc and acid they give mercaptans while with zinc in neutral or slightly alkaline solution they give zinc sulfinates, $(RSO_2)_2Zn$.

The alkyl esters of the alkyl sulfonic acids, RSO_2OR', are isomeric with the dialkyl sulfites, $(RO)_2SO$. The former on hydrolysis either in acid or base lose only one R group. Vigorous hydrolysis of the latter by excess of acid or base removes both R groups forming SO_2 or alkali sulfites.

The presence of a coordinate link (p. 149) in a compound greatly increases its dielectric constant. The value of this constant for a hydrocarbon is about 2, for a carboxylic ester about 7, for diethyl sulfite 19, and for the isomeric ethyl ester of ethyl sulfonic acid 42. The last two compounds contain one and two coordinate links respectively. The presence of coordinate links tends to decrease volatility by giving a strong electrical field in the liquid. Thus $(EtO)_2SO$, b. 161° and $EtSO_3Et$, b. 207°.[37]

Heating with concentrated base solutions removes the sulfonic acid group as a sulfite in the same way that fusion with alkalis converts aromatic sulfonic acids to phenols.

$$RSO_3Na + NaOH \rightarrow ROH + Na_2SO_3$$

The stability of the $C-S$ linkage in alkyl sulfonic acids has been studied for C_1 to C_5 normal alkyls and isopropyl and sec-butyl.[38]

Chloroiodomethane sulfonic acid has been obtained in optically active form.[39]

Abrodil, (Skiodan), ICH_2SO_3Na, is used in X-ray diagnosis in medicine because of the opaqueness of the iodine atom.

Di- and tri-sulfonic acids of methane have been extensively studied.[40] Methionic acid, $CH_2(SO_3H)_2$, can be made by hydrolyzing the product from the action of acetylene with fuming sulfuric acid.

$$(HO)_2CHCH(SO_3H)_2 \rightarrow HCO_2H + CH_2(SO_3H)_2$$

Carbyl sulfate is the inner anhydride of ethionic acid, formed by the action of SO_3 with ethylene.

Vinyl sulfonic acid, $CH_2=CHSO_3H$, can be prepared as follows: ethylene chlorohydrin is converted to the mercaptan and oxidized to *isethionic acid*, $HOCH_2CH_2SO_3H$. When the acetyl derivative of a salt of this acid is boiled, a salt of vinyl sulfonic acid results. The double bond is not reduced by sodium amalgam as would be the case in $\alpha\beta$-unsaturated carboxylic acids. Thus the double bond in vinyl sulfonic acid is not part of a conjugated system.

[37] Sidgwick. "The Electronic Theory of Valency." Oxford, 1927. p. 123.
[38] Wagner, Reid. *J. Am. Chem. Soc.* 53, 3407 (1931).
[39] Read, McMath. *J. Chem. Soc.* 1932, 2723.
[40] Backer, Klaassens. *Rec. trav. chim.* 49, 1107 (1931).

It is worth while to contrast the apparently similar C and S systems.

$$CH_2 :: CH : C :: \overset{..}{\underset{..}{O}} \quad \overset{H^+}{\longrightarrow} \quad CH_2 :: CH : \overset{+}{C} : \overset{..}{\underset{..}{O}} : H \quad \rightarrow \quad \overset{+}{C}H_2 : CH :: C(OH)_2$$
$$\underset{..}{\overset{..}{O}H} \qquad\qquad\qquad \underset{..}{\overset{..}{O}H}$$

$$CH_2 :: CH : \overset{\overset{..}{:O:}}{\underset{\underset{..}{OH}}{\overset{..}{\underset{..}{S}}}} : \overset{..}{\underset{..}{O}} : \quad \overset{H^+}{\rightleftharpoons} \quad \left[CH_2 :: CH : \overset{\overset{..}{:O:}}{\underset{\underset{..}{OH}}{\overset{..}{\underset{..}{S}}}} : \overset{..}{\underset{..}{O}} : H \right]^+$$

In the first case, the beta carbon becomes activated whereas in the second, no change takes place except in the sulfonic acid group.

Alkyl thiosulfonic acids, RSO_2SH, and **alkyl thiosulfuric acids,** $RSSO_2OH$, exist as their salts.

Alkyl sulfinic acids, RSO_2H, are syrupy, water-soluble compounds of strong reducing power. Their salts are available in a variety of ways: 1. From alkyl sulfonyl chlorides with zinc dust and water or with sodium mercaptides; 2. From zinc dialkyls or Grignard reagents with SO_2 or with sulfuryl chloride, SO_2Cl_2. The old discussion as to the formula of the sulfinic acids as

$$RS\overset{\overset{O}{\diagup\diagup}}{-}OH \quad \text{or} \quad RS\overset{\overset{O}{\diagup\diagup}}{\underset{\diagdown H}{=}}O$$

and the action of the sodium salt to give sulfones with alkyl halides and the existence of esters $RSO(OR')$ isomeric with the sulfones, all become clear in electronic notation.

$$\left[\overset{\overset{..}{:O:}}{\underset{\underset{..}{:O:}}{R : \overset{..}{S} :}} \right]^- Na^+ + R'X \rightarrow Na^+ + \left[\overset{\overset{..}{:O:}}{\underset{\underset{..}{:O:}}{R : \overset{..}{S} : R'}} \right]^\circ + X^-$$

Sulfinic esters of the type, RSO_2R', have been obtained in optically active forms in the aromatic series.[41] The older formula containing a double bond does not represent an asymmetric molecule. Moreover the parachors of these esters indicate the absence of any double bond.[42] Thus the electronic formula is indicated on two counts.

$$R\text{—}\overset{\overset{O}{\|}}{S}\text{—}OR' \qquad\qquad R : \overset{\overset{..}{:O:}}{\underset{..}{S}} : \overset{..}{\underset{..}{O}} : R'$$

[41] Phillips. *Ann. Rep. Chem. Soc.* (London) 1925, 111.
[42] Sidgwick. "The Electronic Theory of Valency." Oxford, 1927. p. 127.

Sodium sulfinates, with halogens, give sulfonyl halides.

$$\left[\begin{array}{c} :\overset{\cdot\cdot}{O}: \\ R:\overset{\cdot\cdot}{S}: \\ :\overset{\cdot\cdot}{O}: \end{array} \right]^{-} Na^{+} + :\overset{\cdot\cdot}{X}:\overset{\cdot\cdot}{X}: \rightarrow Na^{+} + \left[:\overset{\cdot\cdot}{X}: \right]^{-} + R:\overset{\cdot\cdot}{S}:\overset{\cdot\cdot}{X}: \begin{array}{c} :\overset{\cdot\cdot}{O}: \\ \\ :\overset{\cdot\cdot}{O}: \end{array}$$

The sulfonyl chlorides are becoming increasingly important as tannins and in the manufacture of detergents and resins. They are readily available through the Reed reaction in which paraffins react with SO_2 and Cl_2 in the presence of actinic light to give alkyl sulfonyl chlorides.[43]

Organic compounds of selenium and tellurium analogous to those of sulfur are known. The bivalent compounds are even less pleasant. These of tellurium are actually dangerous in that they persist in the body for long periods and give an unbearable odor.

A type of tellurium compound which is sufficiently different from the sulfur compounds to deserve mention is dimethyl telluronium diiodide, Me_2TeI_2, which was reported as existing in two stereoisomeric forms.[44] This has been shown to be incorrect.[45] The β-form is really $Me_3TeI \cdot MeTeI_3$. The change postulated to explain the β-form involves a pinacolic type of rearrangement.

VI. ESTERS OF INORGANIC ACIDS

These are obtained either by direct esterification of the acid with an alcohol (reversible) or by the action of the acid chloride with an alcohol.

Since structure formulas are often written for inorganic acids merely by analogy to organic compounds and without any direct evidence for the assigned structures and since the esters of the inorganic acids give means for judging these formulas it is well to consider the experimental evidence regarding these structures. A purely physical demonstration of the existence of a double bond was given by Sugden in 1924 in his introduction of the conception of the *parachor* of a compound (essentially, the product of its molecular volume by the fourth root of its surface tension) as a truly additive property dependent only on the kinds and numbers of atoms and linkages in a molecule.[1] Measurements on large numbers of non-associated liquids give the following values for individual atoms and linkages: H, 17.1; C, 4.8; N, 12.5; O, 20.0; F, 25.7; Cl, 54.3; Br, 68.0; I, 91.0; S, 48.2; P, 37.7; double linkage, 23.2; triple linkage, 46.4; 3-ring, 15.5; 4-ring, 11.6; 5-ring, 9.3; 6-ring, 7.7. The most valuable use of the parachor has been as a physical means of determining the presence of multiple linkages or their absence in given compounds. The value

[43] Lockwood. *Chem. Ind.* **62**, 760 (1948).
[44] Sidgwick. "The Electronic Theory of Valency." Oxford, 1927. p. 286.
[45] *Ann. Rep. Chem. Soc.* (London) **1929**, 80.
[1] Sidgwick. "The Electronic Theory of Valency." Oxford, 1927. p. 125.

23.2 holds for all double linkages, $C=C$, $C=O$, $C=N$, $C=S$, etc., and the value 46.4 holds for $C\equiv C$, $C\equiv N$, etc. In many cases in which double bonds were written under the older conceptions, determinations of the parachor show that there are not double bonds but coordinate linkages. Thus olefins, ketones, esters, alkyl nitrites, CS_2, and isothiocyanates are found to contain the double bonds usually indicated. Nitro compounds and alkyl nitrates are found to have only *one* double linkage instead of the two usually indicated. $POCl_3$, alkyl phosphates, $SOCl_2$, SO_2Cl_2, alkyl sulfites and alkyl sulfonates are found to contain *no* double linkages. These conceptions may be illustrated by bond and electronic formulas:

$$H_2C=CH_2 \qquad H : \overset{H}{\underset{\cdot\cdot}{C}} :: \overset{H}{\underset{\cdot\cdot}{C}} : H$$

$$S=C=S \qquad \overset{\cdot\cdot}{\underset{\cdot\cdot}{S}} :: C :: \overset{\cdot\cdot}{\underset{\cdot\cdot}{S}}$$

$$\begin{matrix} RO \\ \diagdown \\ \diagup \\ RO \end{matrix} SO \qquad \begin{matrix} R : \overset{\cdot\cdot}{O} : \\ : S : \overset{\cdot\cdot}{O} : \\ R : \overset{\cdot\cdot}{\underset{\cdot\cdot}{O}} : \end{matrix} \quad \text{or} \quad \begin{matrix} RO \\ \diagdown \\ \diagup \\ RO \end{matrix} S \rightarrow O$$

$$\begin{matrix} OR \\ | \\ OSO \\ | \\ OR \end{matrix} \qquad \begin{matrix} R \\ : \overset{\cdot\cdot}{O} : \\ : \overset{\cdot\cdot}{O} : S : \overset{\cdot\cdot}{O} : \\ : \overset{\cdot\cdot}{\underset{\cdot\cdot}{O}} : \\ R \end{matrix} \quad \text{or} \quad \begin{matrix} OR \\ | \\ O \leftarrow S \rightarrow O \\ | \\ OR \end{matrix}$$

Similarly the older "structural" formulas of acids like perchloric acid, sulfuric acid and orthophosphoric acid must be replaced by formulas which agree with the fact that these acids do not contain double linkages.[2] In each case the ion of the acid is more correctly represented by

$$\left[\begin{matrix} : \overset{\cdot\cdot}{O} : \\ : \overset{\cdot\cdot}{O} : A : \overset{\cdot\cdot}{O} : \\ : \overset{\cdot\cdot}{\underset{\cdot\cdot}{O}} : \end{matrix} \right]^{-n}$$

$n = 32 - (24 + v) = 8 - v$, in which v is the number of valence electrons in an atom of A.

Esters of nitric acid can be made from nitric acid and an alcohol in the presence of urea to remove nitrous acid.

$MeONO_2$, b. 66°; $EtONO_2$, b. 87°; $PrONO_2$, b. 110°. Nitric acid esters are explosive.

[2] Lewis. *J. Am. Chem. Soc.* **38**, 762 (1916).

Esters of nitrous acid are obtained from an alcohol with a nitrite and sulfuric acid. MeONO, b. $-12°$; EtONO, b. $17°$; PrONO, b. $57°$. Ethyl nitrite and amyl nitrite are used in medicine, the former as "sweet spirits of nitre." Amyl nitrite is probably the most effective antidote for cyanide poisoning. Alkyl nitrites in alcohol solution can be used as a source of nitrous acid for diazotizations and for making isonitroso compounds from ketones.

The difference in the boiling points of the nitrites and nitrates, for the first three members, $78°$, $70°$ and $53°$, is greater than would be expected for a difference of one oxygen. The explanation is that the nitrites contain only ordinary covalences while the nitrates contain a coordinate link as evidenced by the values of their parachors (Sidgwick).

$$Me : \overset{..}{\underset{..}{O}} : \overset{..}{N} :: \overset{..}{\underset{..}{O}} \qquad Me—O—N=O$$

$$
\begin{array}{c}
\overset{..}{\underset{..}{O}} : \\[2pt]
Me : \overset{..}{\underset{..}{O}} : \overset{..}{N} :: \overset{..}{\underset{..}{O}}
\end{array}
\qquad
\begin{array}{c}
O \\
\uparrow \\
Me—O—N=O
\end{array}
$$

Esters of hypochlorous acid, ROCl, are violently explosive. They are readily obtained from hypochlorous acid and alcohols either in aqueous solution or in mixed water and carbon tetrachloride solution. They decompose in the cold and explode on exposure to sunlight or heat. Tertiary hypochlorites are less unstable than those of primary and secondary radicals.

Ethyl hypochlorite finds use in the preparation of pure calcium hypochlorite, $Ca(OCl)_2$, (HTH), as contrasted with bleaching powder, $CaCl(OCl)$.

$$EtOH \rightarrow EtOCl \xrightarrow{Ca(OH_2)} Ca(OCl)_2 + 2\ EtOH$$

$MeOCl$,[3] b. $12°$; EtOCl, b. $36°$.

Esters of acids of phosphorus.

Alkyl phosphites, $P(OR)_3$, are obtained from PCl_3 and NaOR. The use of alcohols instead of the alcoholates give unstable acid esters and alkyl chlorides. $(EtO)_3P$, b. $156°$, heated with MeI at $220°$ gives methyl phosphinic acid, $MePO_3H_2$, ethyl iodide and ethylene. The alkyl phosphites with sulfur trioxide yield phosphates.

Alkyl phosphates, $(RO)_3PO$, are made by the action of $POCl_3$ with alcohols. $(EtO)_3PO$, b. $216°$. The increase of $50°$ in the b.p. as compared with the phosphite is related to the coordinately linked oxygen in the phosphate.[4] Tributyl phosphate is an important plasticizer. Tetraethyl pyrophosphate is unusually effective against aphids, mites, and certain crop pests difficult or impossible to control otherwise.[5]

[3] *Ann. Rep. Chem. Soc.* (London) 1923, 66; 1925, 69.
[4] Sidgwick. "The Electronic Theory of Valency." Oxford, 1927. p. 123.
[5] *Chem. Ind.* 62, 78 (1948).

Esters of arsenious and arsenic acids are known. $(EtO)_3As$, b. 166°; $(EtO)_3AsO$, b. 149° (60 mm.).

Boric acid esters, $(RO)_3B$, are obtained from BCl_3 and alcohols and from boric anhydride, B_2O_3, and alcohols.[6] BF_3 with alcohols gives strongly acid $H(BF_3OR)$. 1,2-Glycols form complex highly ionized acids with H_3BO_3. Thus an alkaline solution of borax, $Na_2B_4O_7$, becomes acid on addition of glycerol or glycol and can be titrated with a base. The boric acid formed by hydrolysis unites with the 1,2-hydroxyl groups. This property is also used in determining configurations in sugars.

The ions formed resemble the stable 32 electron systems found in ions such as ClO_4^- (p. 157).

$B(OMe)_3$, b. 68.7°, m. −29.3°; $B(OEt)_3$, b. 111.8°, m. −84.8°.

Esters of silicic acid are known.

$(EtO)_4Si$, ethyl orthosilicate, b. 165°; $(EtO)_3SiOSi(OEt)_3$, hexaethyl disilicate, b. 236°.

Esters of sulfuric acid.

1. The acid esters or alkyl sulfuric acids, $ROSO_3H$, are made by simply dissolving the alcohol in sulfuric acid. If the compound is desired pure the mixture can be neutralized with barium hydroxide and filtered hot from the $BaSO_4$ formed. Cooling or evaporation of the filtrate gives $Ba(RSO_4)_2$. Treatment of this with the exact amount of dilute H_2SO_4 to precipitate the barium leaves a solution of $RHSO_4$. The salts such as $KEtSO_4$ are well crystallized. They are used in many reactions in place of alkyl halides because of their ready solubility in water.

2. Dialkyl sulfates. Methyl sulfate, Me_2SO_4, b. 188°, can be obtained by heating methyl hydrogen sulfate or by treating dimethyl ether with SO_3. It is *odorless and extremely poisonous.* It is a more vigorous alkylating agent than $NaMeSO_4$.

[6] Webster, Dennis. *J. Am. Chem. Soc.* **55,** 3233 (1933).

Ethyl sulfate, Et_2SO_4, b. 208°, is made from ethylene and sulfuric acid under pressure[7] or by the vacuum distillation of $NaEtSO_4$. It is an important ethylating agent.[8]

Di-isopropyl sulfate is stable if free from acid. Methylene sulfate, CH_2SO_4, and ethylene sulfate, $C_2H_4SO_4$, are known. Mixed alkyl sulfates have been made.[9]

Esters of sulfurous acid are best obtained from thionyl chloride, $SOCl_2$, and alcohols. The alkali alkyl sulfites, $NaRSO_3$, are obtained from alcoholates and SO_2. The peculiar relations of the sulfites and the sulfonic acids are discussed under the latter (p. 152).

VII. ALIPHATIC NITRO AND NITROSO COMPOUNDS

Aliphatic nitro compounds, RNO_2, are distinguished from the isomeric alkyl nitrites in several ways.

1. They boil about 100° higher. This is because of the semipolar double bond or coordinate link which they contain. The parachor for nitromethane shows that it contains only *one* double bond. Thus the ordinary formula which is written with two double bonds is unsound.[1] The formulas of the nitrates, nitrites and nitro compounds may be expressed as follows:

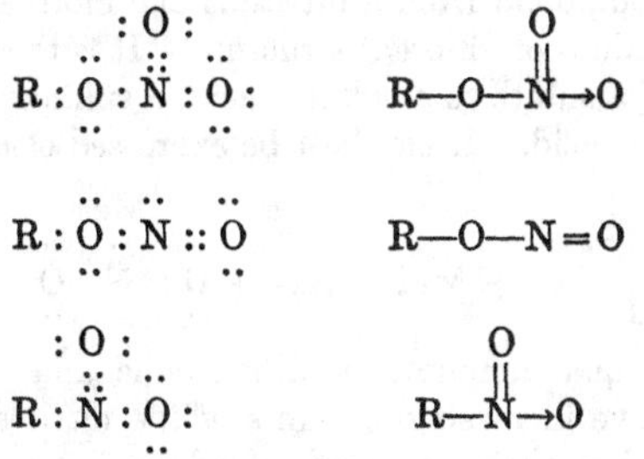

The apparent lack of symmetry in the electronic structure of the nitro group does not agree with the facts as determined by dipole moment measurements. Thus the two oxygen atoms are probably equivalent because of a condition of resonance

2. Reduction gives a primary amine, RNH_2, proving the $C-N$ linkage.
3. They cannot be hydrolyzed by dilute acids and bases as are the nitrites.

[7] Curme. *C. A.* **14**, 2001 (1920).
[8] Cade. *Chem. Met. Eng.* **29**, 319 (1923); *C. A.* **20**, 768 (1926).
[9] *Ann. Rep. Chem. Soc.* (London) 1924, 65.
[1] Sidgwick. "The Electronic Theory of Valency." Oxford, 1927. pp. 123, 127.

4. If the alkyl radical is primary or secondary, sodium salts can be formed. Such nitro compounds are pseudo acids.[2,3]   An alpha H undergoes a tautomeric change to give an acinitro form.

$$\underset{RCH_2N=O}{\overset{O}{\uparrow}} \rightleftharpoons \underset{RCH=N—OH}{\overset{O}{\uparrow}}$$

When R is phenyl, the true nitro and the acinitro forms have been isolated. The latter is formed by acidifying the sodium derivative at low temperatures. It is instantly soluble in NaOH while the true nitro compound dissolves only slowly.

Preparation of Alkyl Nitro Compounds.

1. Direct nitration of paraffins.   This was formerly limited to tertiary hydrocarbons[4] but is now extended to propane and even to methane.[5,6]

2. **Nitromethane**, CH_3NO_2, b. 101°, is available as a by-product of the commercial nitration of propane.   When needed in sufficient amounts it can be made from cheap natural gas by nitration at about 100 psi.[7]

It can be made from MeI and silver or mercurous nitrite but better from sodium chloroacetate and sodium nitrite.

$$ClCH_2CO_2Na \rightarrow O_2NCH_2CO_2Na \rightarrow CH_3NO_2 + NaHCO_3$$

The formation of a nitro compound from a nitrite is one more example of an ion giving a non-ionized product of different structure.   It is the same type of change as the conversion of a sulfite to a sulfonic acid, a sulfinate to a sulfone and an arsenite to an arsonic acid.   It can best be expressed electronically.

$$\left[\overset{..}{\underset{..}{O}} :: \overset{..}{N} : \overset{..}{\underset{..}{O}} : \right]^{-} Ag^+ + MeI \rightarrow AgI + \overset{Me}{\overset{..}{O} :: \overset{..}{N} : \overset{..}{\underset{..}{O}} :}$$

Ethyl iodide gives about equal amounts of nitroethane and ethyl nitrite. Higher iodides in general give increasing amounts of the nitrites.   Bromides give more nitro compound than the corresponding iodides.[8]

It is apparently possible for the nitrite ion to attach itself to the alkyl in two ways:

$$R : \overset{..}{\underset{..}{I}} : + Ag \left[: \overset{..}{\underset{..}{O}} : \overset{..}{N} :: \overset{..}{\underset{..}{O}} \right] \rightarrow Ag : \overset{..}{\underset{..}{I}} : + \begin{matrix} \overset{R}{} \\ : \overset{..}{O} : \overset{..}{N} :: \overset{..}{O} \\ + \\ R : \overset{..}{\underset{..}{O}} : \overset{..}{N} :: \overset{..}{\underset{..}{O}} \end{matrix}$$

[2] Hantzsch.   *Ber.* **32**, 575 (1899).
[3] Branch.   *Ann. Rep. Chem. Soc.* (London) 1927, 107.
[4] Ellis.   "Chemistry of Petroleum Derivatives."   Reinhold, 1934.   p. 1040.
[5] Hass et al.   *Ind. Eng. Chem.* **28**, 393 (1936); **35**, 1146 (1943).
[6] Hass, Riley.   *Chem. Revs.* **32**, 373-430 (1943).
[7] Hass et al.   *Ind. Eng. Chem.* **39**, 919 (1947).
[8] Reynolds, Adkins.   *J. Am. Chem. Soc.* **51**, 279 (1929).

The larger group in general tends to attach itself to one of the oxygen electron pairs at the ends of the ion rather than to the lone electron pair of the N in the middle.

3. The nitration of propane gives small amounts of nitromethane and nitroethane, b. 114°, and good yields of 1-nitropropane, b. 132° and 2-nitropropane, b. 120°.[9-11]

4. Tertiary nitro compounds can be made from the amine of a tertiary radical by oxidation. A better method is the action of a zinc alkyl on the bromination product of a secondary nitro compound.[12]

$$Me_2BrCNO_2 + Me_2Zn \rightarrow Me_3CNO_2 + MeZnBr$$

Reactions of Nitro Compounds.

1. All are reduced to primary amines.

2. Primary and secondary nitro compounds, RCH_2NO_2 and $RR'CHNO_2$, are characterized by reactions due to *alpha* hydrogen.

(*a*) Formation of salts of the acinitro form (*nitronic acid*), $-C = NO(ONa)$ (p. 160).

(*b*) Very ready bromination and chlorination. This is like that of the $\alpha - H$ in an aldehyde, ketone, or a carboxylic acid. As in such cases, an "enolization" process is involved. In this case a base is used to form the salt of the acinitro form of the nitro compound which is treated with halogen.

$$R_2CNO_2Na + X_2 \rightarrow NaX + R_2CXNO_2$$

R may be H or an organic group. The process can be continued as long as alpha H atoms are available. Other H atoms can be replaced by using intense illumination and the exclusion of base and moisture.[13]

$$CH_3CH_2NO_2 + Cl_2 \rightarrow ClCH_2CH_2NO_2$$

(*c*) Action with nitrous acid. The primary compounds up to C_8 give *nitrolic* acids, $RC(NO_2) = NOH$, crystalline acids which give deep *red* salts. The secondary compounds up to C_5 give *pseudo nitroles*, $[R_2C(NO_2)NO]_2$, white crystalline materials which are *blue* when dissolved or fused, due to dissociation into the monomolecular nitroso compounds.

These reactions are the basis for the "red, white and blue test" for the nature of alcohols. The alcohol to be tested is converted to the iodide, treated with silver nitrite, made alkaline, treated with $NaNO_2$, acidified and then made alkaline again. A red color indicates a primary alcohol, a blue color secondary and no color tertiary because in the last there is no alpha H and no action. This test is largely of historical interest since it is successful only with lower alcohols which are now well-known.

[9] Gabriel. *Chem. Inds.* **45**, 664 (1939); *Ind. Eng. Chem.* **32**, 887 (1940).
[10] Hass. *Ind. Eng. Chem.* **35**, 1146 (1943).
[11] Hass, Riley. *Chem. Revs.* 32, 373-430 (1943).
[12] Bewad. *Ber.* 26, 129 (1893).
[13] Hass. *Ind. Eng. Chem.* **35**, 1146 (1943).

(d) The alpha H atoms give the aldol condensation with formaldehyde giving such compounds as $RCH(NO_2)CH_2OH$, $RC(NO_2)(CH_2OH)_2$ and $RR'C(NO_2)CH_2OH$. The nitropropanes give 2-nitro-1-butanol, 2-nitro-2-Et-1,3-propanediol, and 2-nitro-2-Me-1-propanol. These alcohols can be converted to acetates and higher esters. Reduction of the nitro group gives amino alcohols and glycols. Higher aldehydes with nitro compounds give nitro secondary alcohols and the corresponding nitro olefins. Thus benzaldehyde and nitroethane can be used to make benzedrine.[14]

(e) Reduction with stannous chloride, $SnCl_2$, gives oximes, $RCH=NOH$ and $RR'C=NOH$, which can be hydrolyzed to carbonyl compounds.[15]

3. Primary nitro compounds, boiled with sulfuric acid, give carboxylic acids and hydroxylamine sulfate. This involves a complicated series of changes probably going through the hydroxamic acid.

$$RCH_2NO_2 \rightarrow R\overset{H}{\underset{|}{C}}=\overset{O}{\underset{\uparrow}{N}}-OH \rightarrow R\overset{OH}{\underset{|}{C}}=NOH \rightarrow R\overset{OH}{\underset{|}{C}}=O + H_2NOH$$
$$\text{Hydroxamic acid}$$

The formation of the hydroxamic acid would involve a peculiar intramolecular oxidation of the alpha H by the coordinately linked oxygen. With aliphatic nitro compounds readily available by the vapor-phase nitration of paraffin hydrocarbons this reaction gives a practical method for making acids like propionic and butyric acids from the corresponding hydrocarbons.

4. Conc. KOH gives the following changes:

$$2\ CH_3NO_2 \rightarrow HON=CH-CH=NO_2K \rightarrow$$
$$\text{Salt of } methazonic \ acid$$

$$N\equiv C-CH=NO_2K \overset{H_2O}{\rightarrow} O_2NCH_2CO_2H$$
$$\textit{Nitroacetic acid}, \text{ m. } 87°$$

$EtNO_2$, b. 115°; n-$PrNO_2$, b. 131°; iso-$PrNO_2$, b. 118°; n-$BuNO_2$, b. 151°; n-$OctNO_2$, b. 208°.

Dinitromethane, $CH_2(NO_2)_2$, is obtained by the reducing action of sodium arsenite with $Br_2C(NO_2)_2$ which can be made by the vigorous action of nitric acid on various dibromo compounds. It is strongly acid in solution, probably existing as $O_2NCH=NO_2H$. It is unstable and gives a potassium salt which explodes at 205°. This behavior is reminiscent of the strong tendency of 1:3 dicarbonyl compounds to enolize to $-COCH=C(OH)-$.

$Cl_2C(NO_2)_2$ and $Br_2C(NO_2)_2$ are obtained by the action of nitric acid on the trihalogen anilines. These are good halogenating agents for cases in which the formation of HX is undesirable.

$$Br_2C(NO_2)_2 + 2\ HQ \rightarrow 2\ QBr + CH_2(NO_2)_2$$

[14] Hass. *Ind. Eng. Chem.* **35**, 1146 (1943).
[15] v. Braun, Sobeiki. *Ber.* **44**, 2526 (1911).

The Ag compound of *dinitroethane* gives two different reactions with MeI, one the expected alkylation to give dinitropropane (I) and the other the production of formaldehyde and ethyl nitrolic acid (II).

$$[MeC(NO_2)_2]Ag \xrightarrow{MeI} \begin{cases} Me_2C(NO_2)_2 \quad (I) \\ MeC=NOH + H_2CO \\ \quad | \\ \quad NO_2 \\ \quad (II) \end{cases}$$

The latter reaction can be regarded as an internal oxidation-reduction process taking place in an O—Me compound formed simultaneously with the C—Me compound (I).

$$MeC=NO_2Me$$
$$\quad |$$
$$\quad NO_2$$

$$\overset{..}{\underset{..}{:O:}} \qquad\qquad \overset{..}{:O:H}$$
$$Me:C::N:O:CH_3 \rightarrow Me:C::N: \quad + H_2CO$$
$$\underset{..}{\overset{..}{}} \qquad\qquad \underset{..}{}$$
$$NO_2 \qquad\qquad\qquad NO_2$$

Trinitromethane, nitroform, $CH(NO_2)_3$, m. 15°, is obtained by the action of bases on $C(NO_2)_4$, a reaction like that of CI_4 with bases. It is explosive on heating. In water it forms a stable strongly acid solution. It can be obtained in the aciform, $(O_2N)_2C=NO_2H$, m. 50°.

Tetranitromethane, $C(NO_2)_4$, m. 13°, b. 126°, is best made from acetic anhydride and fuming nitric acid.[16] It reacts as though one of the four groups were different from the other three. This is much like the action of carbon tetrahalides. Thus KOMe gives K nitroform and KNO_2. Potassium ferrocyanide gives the K salt of acitrinitromethane, $(O_2N)_2C=NO_2K$.[17]

Tetranitromethane in petroleum ether gives intense colorations with $C=C$ compounds of all types including enols.[18] In methyl alcohol it adds to double bonds:[19]

$$C=C + MeOH + C(NO_2)_4 \rightarrow \underset{OMeNO_2}{C\!\!-\!\!-\!\!-\!\!C} + CH(NO_2)_3$$

The temptation to assign an unsymmetrical formula to tetranitromethane is probably unsound since the Raman spectrum indicates a symmetrical molecule, as does the low value of 2.3 for its dielectric constant.[20]

[16] Chattaway. *J. Chem. Soc.* **97,** 2099 (1910).
[17] *Ann. Rep. Chem. Soc.* (London) 1916, 94.
[18] Ostromisslenskii. *J. prakt. Chem.* **84,** 489 (1911).
[19] *Ann. Rep. Chem. Soc.* (London) 1912, 106.
[20] Sidgwick. "The Electronic Theory of Valency." Oxford, 1927. p. 124.

Tetranitromethane can also be made from nitric acid and acetylene.[21]

Dinitroparaffins are not obtained by direct nitration. They are formed by the addition of nitrogen tetroxide to olefinic linkages and from nitroparaffins and ketones.[22] Thus acetone and nitromethane give dinitroneopentane, $O_2NCH_2CMe_2CH_2NO_2$.

Nitroethylene, $CH_2=CHNO_2$, b. 98°, is made from β-nitroethanol and P_2O_5. It is a lachrymator. Like $CH_2=CHCHO$ it polymerizes readily. Its addition reactions have been studied.[23] The reactions of nitro olefins have been extensively studied.[24]

1-Nitroisobutylene, $(CH_3)_2C=CHNO_2$, b. 155°, is readily obtained by the action of nitric acid with t-butyl alcohol. On hydrolysis it gives acetone and CH_3NO_2.

Other $\alpha\beta$-unsaturated nitro compounds can be obtained by condensing aldehydes with aliphatic nitro compounds (p. 162).

The products from formaldehyde are best dehydrated by pyrolysis of the acetates. A typical case is

$$CH_3CH_2CH_2NO_2 \rightarrow CH_3CH_2CH(NO_2)CH_2OH$$
$$\rightarrow CH_3CH_2CH(NO_2)CH_2OAc \rightarrow CH_3CH_2C(NO_2)=CH_2$$

The resulting 2-nitro-1-butene, like all alpha-beta unsaturated nitro compounds give typical 1,4-addition reactions by which the double bond is saturated as in $\alpha\beta$-unsaturated carbonyl compounds (p. 224). Thus ammonium bisulfite adds to give ammonium 2-nitrobutane-1-sulfonate (Visking Corp., Chicago).

The nitro-olefins are also formed by the action of bases on the esters. Thus $RCHNO_2CH_2OAc + NaHCO_3 \rightarrow RC(NO_2)=CH_2 + NaOAc$ etc.

Chloropicrin, Cl_3CNO_2, b. 112°, is obtained from bleaching powder and picric acid or, better, from $MeNO_2$ and $NaOH$ and Cl_2. It is a powerful lachrymator and nauseant agent (PK, "puke stuff"). It is much used as a fumigant and soil-sterilizer. In contrast to the action of carbon tetrachloride, chloropicrin reacts readily with sodium ethylate to give ethyl orthocarbonate, $C(OEt)_4$, and with ammonia to give guanidine, $HN=C(NH_2)_2$.

Nitroethane readily yields $MeCCl_2NO_2$, *Ethide*, a more pleasant fumigant and insecticide than chloropicrin. Many chloro and dichloro compounds have been made from the available nitroparaffins, NPs.

Hyponitrous esters, $RON=NOR$, can be made from alkyl halides and silver hyponitrite. Hydrolysis involves oxidation-reduction,

$$MeON=NOMe \rightarrow N_2 + MeOH + CH_2O$$

[21] Orton, McKie. *J. Chem. Soc.* 117, 283 (1920).
[22] Fraser, Kon. *J. Chem. Soc.* 1934, 604.
[23] Flürscheim, Holmes. *J. Chem. Soc.* 1932, 1453.
[24] Buckley et al. *J. Chem. Soc.* 1947, 1471 ff.

Aliphatic nitroso compounds, RNO, are known only when R is a tertiary radical. t-Nitrosoisobutane, Me_3CNO, can be made from the amine by oxidation with Caro's acid (monopersulfuric acid, H_2SO_5). As solids, the nitroso compounds are colorless dimers. In solution or fused, they are blue monomers. The relation is like that between brown NO_2 and colorless N_2O_4.

The relation of the nitroso compounds and their dimers may be expressed electronically as follows:

$$R : \overset{\displaystyle ..}{\underset{\displaystyle ..}{N}} : \qquad \qquad R : N :: N : R$$

Reactions which would be expected to give primary or secondary nitroso compounds, give the *oximes* instead. Hence these are sometimes called *isonitroso* compounds. The grouping $=CHNO$ becomes $=C=NOH$, a complete "enolization." An exception to this generalization is the compound, 1-chloro-1-nitrosoethane, $CH_3CHCl(NO)$, obtained from chlorine and acetaldoxime. It exists as a colorless dimer which melts to the blue monomer at 85°. On long heating it becomes colorless again by conversion to the oxime of acetyl chloride, $MeCCl=NOH$.

Nitroso chloroform, Cl_3CNO, is a blue oil obtained in good yield by the action of nitric acid on trichloromethyl sulfinic acid, Cl_3CSO_2H, obtained from the sulfone chloride by reaction with H_2S. Oxidation by chromic acid gives chloropicrin.

VIII. AMINES, ALKYL DERIVATIVES OF AMMONIA

Depending on the number of H atoms replaced in NH_3, the amines are primary, secondary and tertiary, RNH_2 (amino group, $-NH_2$), $RR'NH$ (imino group, $=NH$), and $RR'R''N$ respectively. They have the basic properties of ammonia but form more strongly basic solutions probably because of the greater stability of their "hydrates" as compared with that of ammonia. The primary and secondary amines are "alcohols of the ammonia system"[1] while the tertiary amines are "ethers" on the same basis. The lower amines have characteristic odors somewhat resembling that of NH_3. The three methylamines and primary ethylamine boil below 20°. The others are liquids with boiling points increasing with their molecular weights to heptadecylamine at 340° and tri-n-octylamine at 366°. Branching lowers the boiling points as in the alcohols. This is shown by the following isomeric (metameric) amines: Et_3N, 90°; Pr_2NH, 110°; $HexNH_2$, 129°. The densities of the amines range from 0.66 to 0.77.

The preparations and reactions of the amines can be more safely generalized than can those of the alcohols. There are few reactions of amines in which the

[1] Franklin. "The Nitrogen System of Compounds." Reinhold Publishing Corp., 1935.

C$-$N linkage is broken whereas most reactions of alcohols involve the breaking of the C$-$O linkage. The outstanding example of the breaking of a C$-$N linkage, the action of nitrous acid with a primary amine, presents even more irregularities than any alcohol reaction (p. 172).

The amines unite with HX in the same way that NH_3 does, forming salts such as NH_4Cl, RNH_3Cl, R_2NH_2Cl, R_3NHCl, etc. The substituted ammonium salts (amine hydrochlorides etc.) have much the same properties as the parent substance. They are very soluble crystalline compounds. They are more soluble in organic liquids than the simple ammonium salts. Ammonia and the amines add certain alkyl halides to give ammonium salts having one more alkyl group than the starting material. The last possible stage is a *quaternary ammonium* compound R_4NX. The corresponding hydroxides RNH_3OH, R_2NH_2OH, R_3NHOH, R_4NOH, act as stronger bases than the parent ammonium hydroxide probably because of their greater stability. Me_4NOH can be obtained as a solid resembling KOH in physical and chemical properties. On heating, however, it decomposes giving Me_3N and MeOH. Secondary amines give stronger bases than either the corresponding primary or tertiary amines.[2]

The valence of nitrogen and the "shape" of its atom have long been a source of discussion.[3] This is a case where the electronic conceptions of valence are helpful in organic chemistry. This is illustrated by NH_3 and an ammonium salt.

$$\begin{array}{cc} & H \\ & \overset{..}{} \\ H:\overset{..}{N}: & \\ & \overset{..}{} \\ & H \end{array} \qquad \left[\begin{array}{c} H \\ \overset{..}{} \\ H:\overset{..}{N}:H \\ \overset{..}{} \\ H \end{array} \right]^+ \; [:\overset{..}{\underset{..}{X}}:]^-$$

The related organic compounds simply have R groups in place of H atoms. Thus the peculiar additive power of ammonia and the amines is due to their active electron pair which can combine with a positively charged group to give a positive ammonium ion, either simple or substituted. The ammonium compounds give good examples of *polar* and *non-polar* linkages. The former exist between the two charged ions each of which has a complete octet while the latter exist between the nitrogen and the organic groups in which a pair of electrons is shared by the two atoms, each having a complete octet only by means of the shared pair.

The tetrasubstituted ammonium ion is undoubtedly tetrahedral rather than pyramidal as was formerly supposed.[4] This is confirmed by the occurrence of various complex ammonium compounds in enantiomorphic optically active forms. It is easy to see that the four electron pairs and their attached groups must assume a tetrahedral arrangement in space. The older conception of a

[2] Hall, Sprinkle. *J. Am. Chem. Soc.* **54**, 3469 (1932).

[3] Gilman. "Organic Chemistry," Vol. I. John Wiley & Sons, New York, 1943.

[4] *Ann. Rep. Chem. Soc.* (London) 1927, 102.

pyramidal nitrogen atom with the "fifth" group attached at the apex becomes impossible. The fifth group is attached only ionically as is chlorine to sodium. Werner in 1902 in his coordination theory hit upon the true nature of the ammonium compounds long before the electronic theory.

While nitrogen was believed to be quinquevalent in the sense that carbon is quadrivalent, many attempts were made to prepare compounds such as Me_5N but these failed until more complex groups were used and $Me_4NC(C_6H_5)_3$ and $Me_4NCH_2C_6H_5$ were obtained.[5] These proved not to be true quinquevalent nitrogen compounds but merely a new type of ammonium "salt," the properties resembling those of an organic sodium compound. Water hydrolyzes them giving Me_4NOH with triphenylmethane and toluene respectively.

Preparation of Amines. A mixture of the three classes can be obtained by treating certain alkyl halides with ammonia, usually in alcoholic solution. The halides which work best are those of Class 1 and isopropyl and isobutyl halides. Even with Class 1 halides, the higher members become increasingly sluggish and give lesser amounts of tertiary and secondary amines. The other classes of halides either do not react or merely lose HX to give olefins and NH_4X.

The action of alkyl chlorides with NH_3 is accelerated by a trace of an inorganic iodide.

The addition of alkyl halides to NH_3 and amines is greatly influenced by the solvent used.[6] Thus isopropyl bromide and Et_2NH react very slowly without a solvent but readily in glycerol or ethylene glycol.[7]

The reaction of NH_3 with RX may be pictured as:

$$H : \overset{\cdot\cdot}{\underset{\cdot\cdot}{N}} : + R : \overset{\cdot\cdot}{\underset{\cdot\cdot}{X}} : \rightarrow \left[H : \overset{\cdot\cdot}{\underset{\cdot\cdot}{N}} : R \right]^+ + [: \overset{\cdot\cdot}{\underset{\cdot\cdot}{X}} :]^-$$

The mixing of a suitable alkyl halide with a solution of NH_3 can give all four possible ammonium halides (p. 77). Treatment with a base converts the first three to amines leaving the quaternary ammonium salt unchanged in the solution. These can be distilled out and, except in the case of the methylamines, separated by distillation. The boiling points (°C.) of the three ethyl amines and the three n-butyl amines are 19, 56, 90 and 78, 160, 216 respectively. Thus the Hinsberg method (p. 629) of separating different classes of amines becomes important only historically.

Primary amines, free from other amines, can be prepared in a variety of ways.

1. The hydrolysis of alkyl isocyanates was the method used by Wurtz in his discovery of the first amine.

$$RN = C = O \rightarrow RNHCO_2H \rightarrow CO_2 + RNH_2$$

[5] Schlenk. *Ann. Rep. Chem. Soc.* (London) 1916, 111; 1917, 117.
[6] Menschutkin. *Z. Physik. Chem.* 6, 41 (1890).
[7] Caspe. *J. Am. Chem. Soc.* 54, 4457 (1932).

The difficulty is that isocyanates can be prepared readily only from alkyl halides of Class 1.

2. Hofmann's degradation of an acid amide by a hypochlorite or hypobromite or by halogen and a base. This gives the net change

$$RCONH_2 \rightarrow RNH_2.$$

The intermediate steps involve the formation of an N-bromoamide and its action with the basic solution to give an isocyanate which then gives the primary amine.

$$RCONH_2 \rightarrow RCONHBr \rightarrow RNCO \rightarrow RNHCOOH \rightarrow RNH_2$$

With anhydrous bases it is possible to remove HBr from the N-bromoamide with formation of the isocyanate. The mechanism of this rearrangement is probably like that in which neopentyl alcohol gives t-amyl derivatives (p. 120). The isocyanate adds water (p. 450), and then decomposes to the amine.

If the Hofmann degradation is not carefully controlled some of the amine may unite with unchanged isocyanate to give a disubstituted urea,

$$RNHCONHR.$$

Also the amine may react with any excess of hypohalite to give a cyanide, $RCH_2NH_2 \rightarrow RCN$. These side reactions are easily avoided and yields approaching 100% of primary amine are obtainable from an acid amide in spite of the many steps involved. The process is successful for *all types of primary amines*. Thus although tertiary butyl amine, Me_3CNH_2, and neopentyl amine, $Me_3CCH_2NH_2$ cannot be made from the corresponding halides they are readily obtained from the amides of trimethylacetic and t-butylacetic acids. The advantage of this method is obvious from the fact that t-butyl chloride and NH_3 give only a 3% yield of amine.[8]

3. The Curtius conversion of an acid azide, $RCON_3$, to RNH_2. As in the Hofmann reaction the isocyanate is an intermediate product. This method is best carried out by heating an acid chloride with sodium azide, NaN_3, in toluene, filtering off the NaCl and heating with an excess of HCl to give the amine hydrochloride. In some cases the isocyanate can be isolated (p. 576) or the primary addition product of the isocyanate with the solvent may be isolated (p. 694).

The Curtius Reaction and the related reactions of Hofmann, of Lossen, and of Schmidt have been studied in vast detail.[9]

4. Reduction of an oxime, $C=NOH$, or a hydrazone, either unsubstituted or substituted, $C=NNH_2$, $C=NNHQ$, gives $C-NH_2$. Thus pinacolylamine (2,2-Me$_2$-3-aminobutane), $Me_3CCH(NH_2)Me$, cannot be made by any of the ordinary ways because the necessary intermediates are lacking or react in other ways. Pinacolone, Me_3CCOMe, however, can readily be converted to its

---

[8] Brander. *Rec. trav. chim.* **37**, 67 (1917).
[9] "Org. Reactions," III.

oxime which is easily reduced to the primary amine. A similar result can be obtained with the hydrazone, or the phenylhydrazone.

5. The reduction of a cyanide (nitrile), $RCN \rightarrow RCH_2NH_2$. Acid reducing conditions are avoided as they favor the formation of secondary amines probably by the hydrolysis of the intermediate product $RCH=NH$ to give the aldehyde, $RCHO$, which combines with the amine to form $RCH=NCH_2R$, a Schiff's base. This is reduced to $(RCH_2)_2NH$.

6. Reduction of an acid amide with sodium and absolute alcohol,

$$RCONH_2 \rightarrow RCH_2NH_2.$$

7. Gabriel's phthalimide synthesis[10] is useful in preparing special primary amines. It consists in the action of an alkyl iodide (usually of Class 1) with potassium phthalimide. The resulting N-alkyl phthalimide is hydrolyzed by HCl under pressure.

In cases in which hydrolysis is difficult, hydrazine can be used to give the amine.[11]

8. The best preparation of a primary amine of a tertiary alkyl is that from the Grignard reagent and chloroamine, $ClNH_2$.[12] This method gives over 60% yields of t-amyl amine, Me_2EtCNH_2.

9. The reduction of nitroso and nitro compounds gives primary amines. The latter is becoming increasingly important in making amino alcohols from nitro alcohols obtained by condensing formaldehyde with aliphatic nitro compounds (CSC).

The best preparation of t-butylamine is by the reduction of the nitroso compound from the action of nitrosyl chloride on the Grignard reagent.

10. Alkyl halides heated with hexamethylenetetramine, $(CH_2)_6N_4$, give only primary amines (Delepine). Presumably this process would be most successful with Class 1 halides.

Secondary amines, free from other types, may be prepared:

1. From the N-disubstituted anilines through the para nitroso compounds which are easily prepared by the action of nitrous acid and are readily hydrolyzed by bases to give p-nitrosophenol and the secondary amine.

$$R_2NC_6H_5 \rightarrow R_2NC_6H_4NO \rightarrow HOC_6H_4NO + R_2NH$$

2. From disodium cyanamide, Na_2NCN, and an alkyl halide. The sodium compound is obtained by the action of sodium carbonate on commercial

[10] Gabriel. *Ber.* 20, 2224 (1887).
[11] Ing, Manske. *J. Chem. Soc.* 1926, 2348.
[12] Coleman, Yager. *J. Am. Chem. Soc.* 51, 567 (1929).

calcium cyanamide (Kalkstickstoff). The dialkyl cyanamide readily undergoes acid hydrolysis $R_2NCN + 2\ HCl + 2\ H_2O \rightarrow R_2NH_2Cl + NH_4Cl + CO_2$.

These two methods are useful for the preparation of substances like diethylamine and di-n-butylamine for Novocaine and Butyn manufacture.

3. Secondary amines, especially mixed amines, RR'NH, can be made by the reduction of the condensation products of aldehydes with primary amines (Schiff's bases), $RCH = NR'$.

Secondary amines of secondary and tertiary alkyl groups are unknown except diisopropylamine, b. 84°.

Tertiary amines are best prepared by the action of an excess of alkyl halide with ammonia and a base. Distillation of the alkaline solution gives the tertiary amine which can be distilled with steam. The residual quaternary salt may be evaporated and distilled with solid KOH to give more of the tertiary amine. The fourth group is removed as an olefin except in the case of methyl which appears as MeOH.

A special preparation of tertiary amines is the action of a Grignard reagent with an α-dialkylamino-ether obtainable from a secondary amine and an alpha chloro ether which can be made from HCl, an aldehyde and an alcohol.[13]

$$RMgX + R'R''NCH_2OEt \rightarrow RCH_2NR'R'' + EtOMgX$$

Tertiary amines containing three secondary or tertiary groups are unknown except that tri-isopropyl amine is reported as one of the reduction products of acetone oxime, $Me_2C = NOH$.

Trialkyl amines such as n-Bu and n-Am are good inhibitors of corrosion.[14]

The dialkylaminomethylphenols are important tertiary amines, used as, or for the preparation of corrosion inhibitors, detergents, emulsifiers, germicides etc. (DMP, Rohm and Haas).

Tri-n-butylamine is used as an anti-oxidant for gasoline (DuPont).

The separation of the three amines of a given alkyl radical has been the subject of much study.

1. For ethyl and higher, the best method is by distillation through effective columns.

2. The Hinsberg Method (1890) (p. 773), formerly used for separation in special cases, is chiefly valuable as part of the identification of amines.

3. The three classes of amines can also be separated by heating with diethyl oxalate, with which the tertiary amines do not react, the primary amines give crystalline diamides, $(CONHR)_2$, and the secondary amines give oily monoamides, EtO_2CCONR_2. The latter compounds are easily separated and can be hydrolyzed to the amines. With higher primary amines there is difficulty due to the stopping of the reaction at $EtO_2CCONHR$ which would appear with the secondary amine compound.

---

[13] *Ann. Rep. Chem. Soc.* (London) 1923, 87.
[14] Mann. *Trans. Electrochem. Soc.* 69, 14 pp. (1936).

4. Primary amines form insoluble Schiff's bases $C_6H_5CH = NR$ with benzaldehyde, while secondary and tertiary amines do not react. Acid hydrolysis regenerates $RNH_2.HX$.

Tests for the three classes of amines have been devised in great numbers.

Reactions Given by All Three Classes. These are due to the basic properties of the nitrogen atom, ability to combine with hydrogen ions (protons). They all combine with acids to form substituted ammonium salts, $[(Amine)H]_nQ$, in which n is the valence of the acid radical Q. This includes complex acids such as chloro-auric and chloroplatinic acids, $HAuCl_4$ and H_2PtCl_6. They combine with alkyl halides, especially those of Class 1, to form substituted ammonium halides containing one more alkyl group than the amine used.

Reactions Given by Both Primary and Secondary Amines. These are due to the N—H grouping.

1. Acid chlorides and anhydrides give N-alkylated acid amides. The active H of the amine adds to the oxygen of the carbonyl and the N adds to the C.

$$RC{=}O \rightarrow \begin{bmatrix} Cl \\ RC{-}OH \\ NHR \end{bmatrix} \rightarrow HCl + RC{=}O \\ NHR$$

The HCl unites with another molecule of the amine to form its hydrochloride. Tertiary amines form addition compounds with acid chlorides which are decomposed by water whereas the amides are not. The action of sulfonyl chlorides has already been considered.

2. They can be further alkylated to give more highly substituted ammonias while tertiary amines are converted directly to quaternary ammonium salts. Amines may be methylated by MeI, by Me_2SO_4 or by heating with formalin solution (H_2CO).

3. With alkali metals, some form RNHNa, R_2NNa etc. substances analogous in the ammonia system to NaOEt in the water system. At ordinary temperatures most amines can be stored over metallic Na.

4. They react with chlorine and with bromine to give N-halogen derivatives, $RNHX$, RNX_2 and R_2NX and complex compounds with an excess of halogen. The halogen attached to nitrogen is "positive" and resembles that attached to oxygen as in hypohalous acids. Indeed they act as the hypohalous "acids" of the ammonia system. Thus they are powerful oxidizing agents. One halogen attached to N liberates two iodine atoms from an iodide solution. This is the *test for positive halogen*.

5. With CS_2 they give amine salts of N-alkylated dithiocarbamic acids, $RNHCS_2NH_3R$ and $R_2NCS_2NH_2R_2$.

6. Grignard reagents form RNHMgX and R_2NMgX from which the amines can be regenerated by acid.[15]

[15] Hibbert. *J. Chem. Soc.* **101**, 344 (1912).

7. Nitroamines, $RNHNO_2$ and $RR'NNO_2$, are obtainable only by indirect means because nitric acid oxidizes the amines.

Nitroamines.[16]

Reactions Given Only by Primary Amines

These are due to the amino group $-NH_2$.

1. Primary amines with chloroform and a base give the characteristically vile odor of isocyanides (carbylamines), RNC (p. 419).

2. Nitrous acid with primary amines gives a quantitative yield of N_2 gas (Van Slyke), the other products consisting of alcohols (related to the alkyl or rearranged), olefins and other products. Thus neopentylamine with nitrous acid gives t-AmOH and isoamylenes by rearrangement. The decomposition of the unstable intermediate diazonium compound leaves the neopentyl group without an electron pair thus forming an unstable carbonium ion which immediately rearranges (pp. 120, 298). n-Butylamine gives a mixture of the alcohols, chlorides and traces of the nitrites of n-Bu and sec-Bu together with 1- and 2-butenes when treated with HCl and $NaNO_2$.[17] It is important to observe that the action of a primary amine with nitrous acid is *never a good preparative method for an alcohol* (p. 79). Even with methylamine, the chief product is methyl ether instead of methanol.

Secondary amines give oily nitroso compounds, R_2NNO.

3. Primary amines can be changed to the corresponding halides by benzoylation and treatment with PBr_5 or PCl_5 (p. 79).[18]

4. A solution of a primary amine containing copper powder absorbs air on warming and forms an aldehyde which can be distilled out and tested.[19]

5. Primary amines, RCH_2NH_2, with monopersulfuric acid (Caro's acid, H_2SO_5), give hydroxylamines, RCH_2NHOH, oximes, $RCH=NOH$, and finally hydroxamic acids, $RC(OH)=NOH$ which form violet colors with ferric chloride. Amines of secondary groups, $RR'CHNH_2$, are converted to ketoximes, $RR'C=NOH$, while those of tertiary groups give nitroso and nitro compounds.

6. Thionyl chloride, $SOCl_2$, gives thionylamines, $RNSO$. These are liquids of sharp odor which unite with water to form thionaminic acids, $RNHSO_2H$. Excess water gives hydrolysis to the amine sulfite.

7. Reaction with CS_2, followed by treatment with a heavy metal salt such as $HgCl_2$, gives mustard oils, $RNCS$, of characteristic odor.

Reactions Given Only by Secondary Amines

These are due to the imino group, $=NH$.

1. Nitrous acid gives insoluble oily nitrosamines, R_2NNO, from which the

[16] Ahrens. *Sammlung Chemische-Technischen Vortrage* 1912, 119.

[17] Whitmore, Langlois. *J. Am. Chem. Soc.* 54, 3441 (1932).

[18] v. Braun, Irmisch. *Ber.* 65B, 880 (1932).

[19] *Ann. Rep. Chem. Soc.* (London) 1906, 106.

amines can be regenerated by concentrated HCl. Primary amines give N_2 gas, and tertiary amines do not act in the cold.

The nitrosamines, R_2NNO, when heated with phenol and sulfuric acid, diluted and made alkaline, give a blue-violet color (Liebermann reaction).

2. Oxidation with hydrogen peroxide forms hydroxylamines, RR'NOH.

Reactions of Tertiary Amines

1. Nitrous acid, in the cold, forms a soluble nitrite. On heating, this gives a nitrosamine, RR'NNO and ROH or related products.

2. Acid chlorides form compounds, $RCOCl.R_3N$.[20] Heating under pressure gives $RCONR_2$ and RCl. Treatment with water gives RCO_2H and R_3NHCl. The best diagnosis of a tertiary amine as distinguished from a primary or secondary amine is its failure to give substituted amides with *cold* acetyl chloride.

3. Hypochlorous acid gives secondary amines (or their N-chloro derivatives) and aldehydes or related compounds.

4. Hydrogen chloride under pressure gives RCl and NH_4Cl.

5. Cyanogen bromide gives R_2NCN and RBr.[21]

6. Oxidation by hydrogen peroxide forms *amine oxides*, R_3NO. A formula, $R_3N=O$, indicating pentavalent nitrogen does not agree with the properties of these substances. The value of the parachor for the amine oxides shows no increment such as would correspond to the *double bond*.[22] Thus further experimental evidence is given for the electronic structure.

Thus the $N=O$ linkage in the amine oxides is impossible as is the $S=O$ linkage in the sulfoxides (p. 149) and similar compounds.

Amine oxides have been obtained as optically active enantiomers.[23]

Identification of Amines

Primary and secondary are best converted to crystalline derivatives.

1. Sulfonamides, by treatment with sulfonyl chlorides such as *p*-toluene-, benzyl-, and methyl-.[24]

2. Ureas, by isocyanates such as phenyl-, *p*-nitrophenyl- and α-naphthyl.

$$ArNCO + RNH_2 \rightarrow ArNHCO(NHR)$$

[20] Dehn. *J. Am. Chem. Soc.* **34**, 1399 (1912).
[21] v. Braun. *Ber.* **33**, 1438, 2734 (1900).
[2] Hückel. "Theoretische Grundlagen der Organischen Chemie," Vol. II. 1931. p. 169.
[23] Meisenheimer. *Ann. Rep. Chem. Soc.* (London) **1911**, 75.
[24] Marvel, Helfrick, Belsley. *J. Am. Chem. Soc.* **51**, 1272 (1929).

Potassium cyanate and acid give substituted ureas, $RNHCONH_2$ and $RR'NCONH_2$.

3. Thioureas, by means of phenyl-, p-tolyl-, o-tolyl- and α-naphthyl-mustard oils.

$$ArNCS + RNH_2 \rightarrow ArNHCS(NHR)$$

4. Acetyl and benzoyl derivatives, CH_3CONHR, and $-NR_2$, C_6H_5CONHR, and $-NR_2$, of many primary and secondary amines have characteristic melting points.

5. All amines form chloroplatinates and chloroaurates, just as NH_3 forms $(NH_4)_2PtCl_6$ and NH_4AuCl_4.

6. They also form characteristic picrates (with trinitrophenol) and picrolonates (with 1-p-nitrophenyl-3-Me-4-nitropyrazolone-5).

Quaternary Ammonium Compounds and Exhaustive Methylation

As has been seen, the treatment of an alkyl halide with ammonia gives some quaternary ammonium halide, R_4NX. This can be produced more effectively from ammonia or amines by alkylating agents such as alkyl halides, alkyl sulfates, etc., and a base. A quaternary ammonium chloride with alcoholic KOH gives a precipitate of KCl and a solution of R_4NOH. The stability of tetraalkyl ammonium hydroxides has been widely studied.[25-27] The following results of heating these substances are typical:

1. $Me_4NOH \rightarrow Me_3N + MeOH$

2. $EtMe_3NOH \rightarrow Me_3N + H_2C=CH_2 + H_2O$

3. $(Me_2CHCH_2)Me_3NOH \rightarrow Me_3N + Me_2C=CH_2 + H_2O$

4. $(Me_3CCH_2)Me_3NOH \rightarrow MeOH + Me_3CCH_2NMe_2$

The formation of olefins cannot be regarded as due to dehydration of alcohols first formed because the temperatures used are far below those required for such dehydrations. The process may be regarded as analogous to the decomposition of ordinary NH_4OH.

$$\begin{bmatrix} H \\ H : \overset{\cdot\cdot}{N} : H \\ H \end{bmatrix} \begin{bmatrix} \overset{\cdot\cdot}{:O} : H \end{bmatrix} \rightarrow H : \overset{\cdot\cdot}{N} : + H : \overset{\cdot\cdot}{O} : H$$

The new combination is more stable.

$$Me : \overset{\cdot\cdot}{N} : R \rightarrow Me : \overset{\cdot\cdot}{N} : + [R^+]$$

with Me above and below each N.

[25] Ahrens. *Sammlung Chemische-Technischen Vortrage* 1899, 54.
[26] v. Braun, Buchman. *Ber.* **64B**, 2610 (1931); v. Braun, Anton. *Ber.* **64B**, 2865 (1931).
[27] Ingold et al. *J. Chem. Soc.* **1933**, 66 ff.

If the hypothetical R^+ contains a hydrogen on the carbon alpha to the carbon which was left with only six electrons by the decomposition of the complex ammonium ion, then that hydrogen is expelled as a proton (H^+) and an olefin is formed $R^+ \rightarrow H^+ +$ olefin.   If, however, such a loss of a proton is not possible as with the methyl (1) or the neopentyl group (4) the R^+ combines with the OH^- to give ROH.

Whereas the decomposition of a quaternary ammonium hydroxide results in the detachment of the largest group, the more difficult decomposition of the corresponding halide gives a methyl halide if that is possible.

The useful process of *exhaustive methylation* is based on the ready decomposition of quaternary ammonium hydroxides.   The process will be illustrated by several examples.

1. An amine, $C_6H_{15}N$, with nitrous acid, gives no N_2 but gives an oily nitroso compound from which the original amine can be regenerated.   There are some 14 secondary amines of this formula.   Treatment with an excess of MeI and NaOH gives a quaternary ammonium salt, $C_8H_{20}NI$.   The ammonium hydroxide, obtained by Ag_2O and water, decomposes on heating to isobutylene and a tertiary amine, $C_4H_{11}N$, which must be ethyldimethylamine.   Thus the original amine was either ethylisobutylamine or ethyl-*t*-butylamine.   The identity as between these two would have to be decided by synthesis and the preparation and comparison of derivatives.   If the tertiary amine obtained cannot be identified directly it is exhaustively methylated again to split off another group.

2. Pyrrolidine, when treated with excess of methyl iodide and base and then converted to the quaternary ammonium salt, etc. undergoes the following processes:

$$\text{CH}_2\text{—CH}_2$$
$$\begin{array}{c} | \\ | \\ | \end{array} \quad \text{NH} \rightarrow \quad \text{NMe} \rightarrow \quad \text{NMe}_2\text{I} \rightarrow \quad \text{NMe}_2(\text{OH}) \rightarrow$$
$$\text{CH}_2\text{—CH}_2$$

$$\begin{array}{c} \text{CH}=\text{CH}_2 \\ | \\ \text{CH}_2\text{—CH}_2\text{NMe}_2 \end{array} \rightarrow \text{—NMe}_3\text{I} \rightarrow \text{—NMe}_3(\text{OH}) \rightarrow \begin{array}{c} \text{CH}=\text{CH}_2 \\ | \\ \text{CH}=\text{CH}_2 \end{array}$$

3. Many naturally occurring products contain cyclic structures having a nitrogen bridge as in tropine and related compounds.   The first formation of a quaternary base and its decomposition removes the N from one of its attachments across the ring and the second exhaustive methylation detaches it entirely.   Thus tropidine would go through the following steps to 1,3,5-cycloheptatriene.

$$\begin{array}{c} \text{CH}_2\text{—CH—CH} \\ | \quad | \quad \| \\ | \quad \text{NMe CH} \rightarrow \\ | \quad | \quad | \\ \text{CH}_2\text{—CH—CH}_2 \end{array} \quad \begin{array}{c} \text{CH}=\text{CH—CH} \\ | \quad \| \\ | \quad \text{NMe}_2\text{CH} \rightarrow \\ | \quad | \quad | \\ \text{CH}_2\text{—CH—CH}_2 \end{array} \quad \begin{array}{c} \text{CH}=\text{CH—CH} \\ | \quad \| \\ | \quad \text{CH} + \text{Me}_3\text{N} \\ | \quad | \\ \text{CH}_2\text{—CH}=\text{CH} \end{array}$$

Free tetra-alkyl ammoniums are very unstable. They are obtained by electrolysis of the salts at low temperatures. They decompose to give tertiary amines and products corresponding to the free radical involved. With a mercury cathode tetramethylammonium amalgam has been prepared.[28] It resembles ammonium amalgam.

A solution of free tetramethylammonium has been formed in liquid ammonia by treating a solution of tetramethylammonium chloride with metallic lithium.

Quaternary ammonium halides form complex polyhalides of the types R_4NX_n in which n is 3, 5, 7, 9, and 13.[29]

The compound $Et_3N \cdot BF_3$, m. 29°, is undoubtedly formed by a *coordinate link* made possible by the free electron pair of the amine and the ability of the boron atom to share such a pair.

$$
\begin{array}{cc}
\text{Et} & \text{F} \\
\vdots & \vdots \\
\text{Et}:\text{N}: + \text{B}:\text{F} \\
\vdots & \vdots \\
\text{Et} & \text{F}
\end{array}
\rightarrow
\begin{array}{cc}
\text{Et} & \text{F} \\
\vdots & \vdots \\
\text{Et}:\text{N}:\text{B}:\text{F} \\
\vdots & \vdots \\
\text{Et} & \text{F}
\end{array}
$$

The non-equivalence of the valences of quinquevalent nitrogen has been shown by making isomers, $Me_3N(OH)(OR)$.[30] Me_3NO treated with MeI and then NaOH gives a substance which on heating forms Me_3N, HCHO and H_2O while the isomer obtained by Me_3N treated with HCl and then with NaOMe gives Me_3NO and MeOH. These are easily understood in the ammonium ion notation

$$
\left[
\begin{array}{c}
\text{Me} \\
\vdots \ \vdots \\
\text{Me}:\text{N}:\text{O}:\text{Me} \\
\vdots \ \vdots \\
\text{Me}
\end{array}
\right]^{+} \text{OH}^{-}
\qquad
\left[
\begin{array}{c}
\text{Me} \\
\vdots \ \vdots \\
\text{Me}:\text{N}:\text{O}:\text{H} \\
\vdots \ \vdots \\
\text{Me}
\end{array}
\right]^{+} \text{OMe}^{-}
$$

Individual Amines

Methylamines are best obtained by the catalytic dehydration of a mixture of methanol and ammonia (Sabatier). The relative amounts of the three amines can be modified somewhat by controlling the concentrations of the reactants and by changing the conditions. The resulting mixture can be distilled under increased pressure to avoid low temperature distillation although that is possible according to the technique developed for separating propane and the butanes from natural gas. The boiling points of the three methylamines are −7°, +7° and +3.5°, the tertiary being exceptional. All other tertiary amines boil considerably higher than the corresponding secondary amines.

[28] McCoy, Moore. *J. Am. Chem. Soc.* **33**, 273 (1911).
[29] *Ann. Rep. Chem. Soc.* (London) 1923, 87.
[30] Meisenheimer. *Ann.* **397**, 273 (1913).

The three methylamines are now available commercially. Many uses are being found for them.[31]

In the laboratory, methylamine and trimethylamine can be made from formalin solution and paraformaldehyde with ammonium chloride.[32]

$$2 \ H_2CO + NH_4Cl \rightarrow CH_3NH_2 . HCl + HCO_2H$$
$$3 \ (H_2CO)_3 + 2 \ NH_4Cl \rightarrow 2 \ (CH_3)_3NHCl + 3 \ CO_2 + 3 \ H_2O$$

Di- and trimethylamines are formed in decaying fish and similar organic materials. The latter is also formed in dry distillation of beet residues after the sugar has been extracted.

Methylamine has an odor much like ammonia and is very soluble in water. It is inflammable. Its hydrochloride, m. 225°, resembles ammonium chloride. Picrate, m. 215°.

Several individual compounds related to methylamine may be considered here:

1. **N,N-Dichloromethylamine,** CH_3NCl_2, b. 59°, is made from methylamine with chlorine and alkali. It reacts with water and alcohols to give hypochlorous acid and its esters. With iodides it liberates iodine. Thus the halogen atoms are "positive."

2. **Thionylmethylamine,** $CH_3N=S=O$, b. 58°, is obtained from methylamine and $SOCl_2$. It reacts with active H compounds giving CH_3NH_2 and SO_2 or a sulfite derivative.

3. **Methyl thionamic acid,** CH_3NHSO_2H, a white unstable solid, is obtained directly from the amine and SO_2.

4. **Methylnitroamine,** CH_3NHNO_2, m. 38°, is prepared by nitrating N-methylurethan and treating the product with ammonia. It gives explosive salts.

$$\underset{}{MeNHCO_2R} \quad \overset{HNO_3}{\rightarrow} \quad \underset{\underset{NO_2}{|}}{MeNCO_2R} \quad \overset{NH_3}{\rightarrow} \quad MeNHNO_2 + NH_2CO_2R$$

The imino H is replaceable by alkali metals and thus by organic radicals.

$$Me-NH-NO_2 \rightarrow \underset{\underset{OH}{|}}{MeN=NO} \rightarrow \underset{\underset{ONa}{|}}{Me-N=NO} \overset{RX}{\rightarrow} \underset{\underset{R}{|}}{Me-N-NO_2}$$

The resulting disubstituted nitroamines may be reduced to unsymmetrical *disubstituted hydrazines*, CH_3RNNH_2.

Dimethylamine forms a hydrochloride which is soluble in chloroform. The nitroso compound, Me_2NNO, b. 149°, forms a hydrochloride which is hydrolyzed by water. The nitroamine, Me_2NNO_2, is a solid, m. 57°, b. 187°.

[31] Riddle. *Chem. Ind.* 1944, 209.
[32] "Org. Syntheses."

Trimethylamine when obtained as a by-product can be utilized by conversion to methyl chloride by heating with HCl under pressure.

$$Me_3N + 4\ HCl \rightarrow NH_4Cl + 3\ MeCl$$

Ethylamines are best prepared from ethyl chloride and ammonia under pressure. Their separation offers no difficulty. The preparation from ethanol and NH_3 is difficult because of the large amount of olefin formed. This applies even more to higher alcohols. Ethyl, propyl and butyl amines are commercially available.

Ethylamine on pyrolysis gives ethylene and acetonitrile as the chief products. At higher temperatures methane is the chief hydrocarbon.[33]

C-Halogenated amines are obtainable by the standard methods for preparing ordinary amines. Ethylene dibromide through the phthalimide synthesis, gives beta-bromoethylamine, $BrCH_2CH_2NH_2$. Such a substance, containing two groups which can react with each other but are too close to form a stable ring within the molecule, tends to form a ring between two molecules. In this case the product is the dihydrobromide of piperazine,

$$HBr.NH(CH_2CH_2)_2NH.HBr.$$

Gamma-halogenatedpropylamines give similar 8-membered rings. Delta-halogenatedbutylamines readily form pyrrolidines. 3-Bromopropyl- and 4-bromobutyl-diethylamines have been prepared by a series of reactions illustrative of the possibilities in making compounds containing two very active groupings.[34]

$$Br(CH_2)_3Br \rightarrow PhO(CH_2)_3Br \rightarrow PhO(CH_2)_3NEt_2 \rightarrow Br(CH_2)_3NEt_2$$
$$PhO(CH_2)_3Br \rightarrow PhO(CH_2)_3CN \rightarrow PhO(CH_2)_4NH_2 \rightarrow Br(CH_2)_4NEt_2$$

The splitting of the phenyl ether takes place with HBr without difficulty. The amines are obtained as their salts. The free amines form cyclic quaternary compounds. The bromine gives the reactions of Class 1 halides including that with sodium malonic ester.

A general method for making omega (ω) halogenated tertiary amines depends on the following reactions:

$$R_2NH + RMgX \rightarrow R_2NMgX + RH$$
$$R_2NMgX + Tol-SO_3(CH_2)_nCl \rightarrow R_2N(CH_2)_nCl + Tol-SO_3MgX$$

Unsaturated amines are obtained by different reactions depending on the relation of the double bond to the amino group.

1. **Δ-1,2-Amines, Vinylamine,** $CH_2=CHNH_2$, would be expected from the removal of HBr from bromoethyl amine. The product obtained is prob-

[33] Hurd. "The Pyrolysis of Carbon Compounds." Chemical Catalog Co., New York, 1929.
[34] Marvel, Zartman, Bluthardt. *J Am. Chem. Soc.* 49, 2299 (1927).

ably the cyclic ethylene imine, $(CH_2)_2NH$ corresponding to ethylene oxide.[35] Neurine is a related quaternary base, $CH_2 = CH - NMe_3OH$. The formula, $CH_2 = CHNH_2$, corresponds to an enol $CH_2 = CHOH$. Nitrous acid would be expected to give this enol which would tautomerize to acetaldehyde. As a matter of fact the product is ethylene glycol showing the relation of the substance to ethylene diamine.

Alpha beta unsaturated amines, $-CH = CHNH_2$, are in equilibrium with the form, $-CH_2 - CH = NH$, which is readily hydrolyzed to $-CH_2 - CHO$. Thus the Hofmann degradation of $\alpha\beta$-unsaturated acid amides gives the next lower aldehyde instead of the unsaturated amine,

$$RCH = CH - CONH_2 \rightarrow RCH_2CHO^{[36]}$$

Conjugated unsaturated amines, $RCH = CHCH = CHNH_2$, have been obtained.[37] The amino group is only weakly basic.

2. Δ-2,3-Amines, Allylamine, $CH_2 = CH - CH_2NH_2$, b. 57°, can be made from the halides but is most readily obtained by hydrolyzing allyl isothiocyanate (mustard oil). It is only about one tenth as strong a base as n-propylamine. Its quaternary compounds, $CH_2 = CHCH_2NMe_3X$ are much less poisonous than its lower homolog, the naturally occurring neurine.

Δ-3,4- and Δ-4,5-Amines are obtainable from the corresponding C-bromo amines. They show a decided tendency to ring formation. Thus 5-dimethyl-amino-1-pentene, with bromine gives the quaternary bromide of 2-bromo-methyl-NN-dimethylpyrrolidine.

The Br with only six electrons probably adds to the extra electron pair of the olefinic linkage, thus leaving the second carbon with only six electrons. The second carbon can satisfy this deficiency by union with the unshared electron pair of the nitrogen atom in the 1,5-position to it. The resulting quaternary ammonium ion unites with the Br⁻.

IX. ALKYLHYDRAZINES AND RELATED COMPOUNDS

Alkylhydrazines, derivatives of hydrazine, H_2NNH_2 (diamide), are less important than the corresponding aromatic compounds. The alkylhydrazines resemble the amines in most of their properties but are much less volatile and partake of the properties of hydrazine itself in being powerful reducing agents.

---

[35] Euler. *Chem. Zentr.* **1904** I, 999.
[36] *Ann. Rep. Chem. Soc.* (London) **1920**, 80.
[37] Muskat, Grimsley. *J. Am. Chem. Soc.* **55**, 3762 (1933).

Primary hydrazines, $RNHNH_2$, are preparard:

1. From alkyl sulfates and excess of hydrazine and a base.

$$R_2SO_4 + N_2H_4 + NaOH \rightarrow RNHNH_2 + NaRSO_4 + H_2O$$
$$RNaSO_4 + N_2H_4 + NaOH \rightarrow RNHNH_2 + Na_2SO_4 + H_2O.$$

An excess of alkylating agent adds more alkyl groups to the same N, the final result being an amino quaternary ammonium salt, $[R_3NNH_2]X$.

2. From primary amines through substituted ureas by treatment with nitrous acid, reduction and hydrolysis by conc. HCl.

$$RNH_2 \xrightarrow{COCl_2} RNHCONHR \xrightarrow{HNO_2} RN(NO)CONHR \rightarrow$$

$$RN(NH_2)CONHR \xrightarrow{HCl} CO_2 + RNH_2 \cdot HCl + RNHNH_2 \cdot HCl$$

N-alkyl urethans can be used similarly.

$$RNHCO_2R \xrightarrow{HNO_2} \underset{\overset{|}{NO}}{RNCO_2R} \xrightarrow{4[H]} \underset{\overset{|}{NH_2}}{RNCO_2R} \xrightarrow{H_2O}$$

$$RNHNH_2 + CO_2 + ROH$$

In the same way that the amines are more basic than NH_3, the alkylated N in $RNHNH_2$ is more strongly basic than the other N. Thus MeI adds to the first N. Probably HCl does the same

$$\begin{matrix} H : \overset{..}{\underset{..}{N}} : R \\ H : \overset{..}{\underset{..}{N}} : H \end{matrix} \rightarrow \begin{bmatrix} Me \\ H : \overset{..}{\underset{..}{N}} : R \\ H : \overset{..}{\underset{..}{N}} : H \end{bmatrix} I \rightarrow \begin{bmatrix} Me \\ Me : \overset{..}{\underset{..}{N}} : R \\ H : \overset{..}{\underset{..}{N}} : H \end{bmatrix} I$$

The quaternary iodide will add no more MeI under ordinary conditions. Heated at 125° with excess MeI it reacts with a splitting of the $N-N$ linkage to give Me_4NI.

The inertness of the free NH_2 in the hydrazines is further evidenced by their behavior with isocyanates and mustard oils which give $RN(CONHR')NH_2$ etc. Thus the H attached to the substituted N is more reactive than the other two. A reaction of the free NH_2 group is that with $SOCl_2$ giving thionyl derivatives, $RNHNSO$.

The alkylhydrazones, formed with aldehydes and ketones, are usually liquids and consequently are not valuable for identification purposes as are the aromatic compounds.

Methylhydrazine, CH_3NHNH_2, b. 87°, may be obtained by reducing nitro or nitroso N-methylurethan and hydrolyzing the product. Methylhydrazine resembles methylamine in many respects but is a powerful reducing agent, reacting with Fehling's solution even in the cold.   The oxidation of methyl-hydrazine by HgO gives considerable amounts of mercury dimethyl, Me_2Hg (very poisonous).   Methylhydrazine reacts with potassium pyrosulfate to give potassium methylhydrazine sulfonate.

$$MeNHNH_2 + K_2S_2O_7 \rightarrow MeNHNHSO_3K + KHSO_4$$

EtNHNH$_2$, b. 102°.

Unsymmetrical dialkylhydrazines, RR'NNH$_2$, are readily available by the reduction of the nitroso compounds of secondary amines.   A surprising reaction is that with the calculated amount of nitrous acid which gives N$_2$O instead of N$_2$, and a secondary amine.

$$R_2NNH_2 + HNO_2 \rightarrow R_2NH + N_2O + H_2O$$

Mercuric oxide gives a tetrazone, $R_2NN = NNR_2$.
Alkylation gives the same unsymmetrical products as the monoalkyl hydrazines.

$$Me_2NNH_2, \text{ b. } 63° \qquad Et_2NNH_2, \text{ b. } 98°$$

Symmetrical dialkylhydrazines, hydrazoalkanes, RNHNHR, are difficult to prepare.   They are not available by the alkylation of hydrazine nor a monoalkylhydrazine because this takes place asymmetrically.   The diformyl derivative of hydrazine can, however, be alkylated on both nitrogen atoms.[1] Treatment of the product with conc. HCl gives a poor yield of the desired product.

$$HCONHNHCHO \rightarrow HCONRNRCHO \rightarrow 2 HCO_2H + RNHNHR$$

More vigorous treatment with HCl gives RCl+NH$_4$Cl.   Mercuric oxide with the ethyl compound gives diethyl mercury and N$_2$ instead of the expected azo-compound.   Nitrous acid splits the compound to give ethyl nitrite.   Thus the symmetrical molecule has a much less stable N−N linkage than the unsymmetrical ones.

Careful oxidation of sym-Me$_2$-hydrazine dihydrochloride with potassium dichromate solution gives azomethane, MeN = NME, b. 2°.   Its decomposition (thermal, photochemical and catalytic), have been carefully studied.[2]   The thermal decomposition gives mainly CH$_4$ and N$_2$.

Azomethane with HCl gives formaldehyde and methylhydrazine hydro-chloride.   This is due to a tautomeric change in which the N = N linkage plays the part usually played by C=O.   H$^+$ adds to one N and is expelled from the carbon in the 3-position.   Hydrolysis then gives the observed products.

$$CH_3-N=N-CH_3 \xrightarrow{H^+} [CH_3\overset{+}{N}-NHCH_3]^+ \rightarrow$$

$$H^+ + CH_2=N-NHCH_3 \rightarrow CH_2O + CH_3NHNH_2 \cdot HCl$$

[1] Harries, Klamt. *Ber.* **28**, 503 (1895).
[2] Emmett, Harkness. *J. Am. Chem. Soc.* **54**, 538 (1932).

sym-Diisopropylhydrazine and its oxidation product "azoisopropane" have been prepared.[3] The decomposition of the latter is homogeneous and unimolecular, giving N_2 and diisopropyl as the chief products.[4] EtNHNHEt, b. 85°.

Alkylhydroxylamines, derivatives of H_2NOH, occur as O-derivatives (*alpha*) and N-derivatives (*beta*).

Direct alkylation of hydroxylamine with methyl iodide gives the quaternary ammonium salt, $Me_3N(OH)I$. It is to be noted that the hydroxyl group is not ionized but forms part of the inner complex. The electronic linkage $N:O$ is like the $O:O$ linkage in hydrogen peroxide.

$$\left[\begin{array}{c} Me \\ \cdot\cdot \quad \cdot\cdot \\ Me : N : O : H \\ \cdot\cdot \quad \cdot\cdot \\ Me \end{array} \right] I \qquad\qquad H : \overset{\cdot\cdot}{\underset{\cdot\cdot}{O}} : \overset{\cdot\cdot}{\underset{\cdot\cdot}{O}} : H$$

With ethyl and higher halides, it is possible to interrupt the alkylation at R_2NOH, the N,N- or $\beta\beta$-dialkylhydroxylamine.

Alkyl nitrites with Grignard reagents give N,N-dialkylhydroxylamines.[5]

$$2 \, RMgX + R'ONO \rightarrow R_2NOMgX + R'OMgX$$

When heated with acetic acid N,N-diethylhydroxylamine gives ethylamine acetate and acetaldehyde. The formation of the latter is due to the oxidizing action of the peroxide-like oxygen.

O-Alkylhydroxylamines, alkoxylamines, H_2NOR, cannot be obtained by direct alkylation because of the additive power of the nitrogen for all alkylating agents. On the other hand oximes, $R_2C=NOH$, can be alkylated and then hydrolyzed to give the desired products. The alkoxylamines can be split by HCl. They can be alkylated to N,O-dialkyl derivatives, R'NHOR, and finally to quaternary salts $[R_3NOR]X$.

N-Alkylhydroxylamines, RNHOH, can be obtained by the electrolytic reduction of nitroparaffins and by the splitting of N,O-dialkylhydroxylamines.

The N-alkyl derivatives are more powerful reducing agents than the O-alkyl compounds.

β-Methylhydroxylamine, CH_3NHOH, m. 41°, is obtained by the reduction of nitromethane by zinc dust and water. It reduces Fehling's solution. **α-Methylhydroxylamine,** NH_2OCH_3 does not reduce Fehling's solution. When heated with HCl it gives formaldehyde.

$EtONH_2$, b. 68°; hydrochloride, m. 128°.

EtNHOH, m. 60° dec; Et_2NOH, b. 134° dec; EtNHOEt, b. 83°.

[3] Lochte, Noyes, Bailey. *J. Am. Chem. Soc.* **44**, 2556 (1922).
[4] Ramsperger. *J. Am. Chem. Soc.* **50**, 715 (1928).
[5] Bewad. *Ber.* **40**, 3065 (1907).

Diazoparaffins, $RCHN_2$, are of little importance compared with the corresponding aromatic compounds. The simplest member, diazomethane, azomethylene, CH_2N_2[6,7] is important as a special methylating agent. It is a yellow poisonous gas which liquefies at $-24°$. It is used in ether solution. It is prepared from nitrosomethylurea and alkali. The steps are as follows, starting with methylamine.[8]

$$MeNH_2 + H_2NCONH_2 \rightarrow MeNHCONH_2 \rightarrow MeN(NO)CONH_2 \rightarrow CH_2N_2$$

An alcoholic solution of nitrosomethylurethan with NaOEt can be used as nascent diazomethane.[9] It can also be formed from hydrazine, $CHCl_3$ and KOH and from hydroxylamine and N-N-dichloromethylamine. The structure of diazomethane is written as (I) and (II).[10] The second or its electronic form III is more probable.

(I) $CH_2 \begin{smallmatrix} N \\ \| \\ N \end{smallmatrix}$ (II) $CH_2 = N \equiv N$ (III) $H : \overset{..}{C} :: \overset{H}{N} : \overset{..}{\underset{..}{N}} :$

This indicates the ease with which a molecule of N_2 can be liberated leaving the active methylene residue. Thus, with $HgCl_2$ and organomercuric chlorides it gives N_2 and $ClCH_2HgCl$ and $ClCH_2HgR$.[11] Iodine gives N_2 and CH_2I_2, while HCl gives methyl chloride.

Diazomethane converts acids and phenols to methyl esters and methyl ethers. This action is valuable in the treatment of costly or sensitive substances.

It reacts with acid chlorides and with aldehydes to give diazo ketones and methyl ketones respectively.[12]

The Arndt-Eistert synthesis involves the reaction of an acid chloride with diazomethane to yield a diazoketone, which on acid hydrolysis undergoes the Wolff rearrangement to give an acid with one more carbon than the initial acid chloride.[13]

Acetone, in presence of a little water, reacts with diazomethane to give a 40% yield of methyl ethyl ketone.[14]

It should be remembered that diazomethane fails to methylate in many special cases which are not as yet thoroughly understood.

[6] v. Pechmann. *Ber.* **27**, 1888 (1894).
[7] "Org. Reactions," I. p. 50.
[8] "Org. Syntheses."
[9] *Ann. Rep. Chem. Soc.* (London) 1919, 86.
[10] *ibid.* 1917, 87.
[11] Hellerman, Newman. *J. Am. Chem. Soc.* **54**, 2859 (1932).
[12] *Ann. Rep. Chem. Soc.* (London) 1928, 86.
[13] "Org. Reactions," I.
[14] *Ann. Rep. Chem. Soc.* (London) 1928, 89.

Sometimes diazomethane reacts without losing its nitrogen. Thus, with acetylene and ethylene it gives pyrazole and pyrazoline.

$$
\begin{array}{ll}
\begin{array}{l} CH{-}NH \\ \| \qquad\quad N \\ CH{-}CH \end{array} &
\begin{array}{l} CH_2{-}NH \\ | \qquad\quad N \\ CH_2{-}CH \end{array}
\end{array}
$$

Similar reactions occur with substituted acetylenes and olefins especially those with the unsaturation $\alpha\beta$- to a $C{=}O$ grouping.

Polymerization of diazomethane gives C,C-dihydrotetrazine,

$$CH_2(N{=}N)_2CH_2$$

Diazoethane has been made from nitroso-N-ethylurethan.

Higher homologs are obtained by HgO oxidation of hydrazones of aldehydes and ketones. Me_2CN_2 is a red oil.

Alkyl azoxy compounds have been prepared by condensing the β-hydroxylamine and nitroso compound in the presence of alkali. Hydrolysis of these compounds with acid gives nitrogen, olefin and an alcohol.[15]

Alkyl azides, alkyl azimides, azido alkanes, RN_3, are obtainable from alkyl iodides or sulfates and sodium azide, NaN_3. They explode at high temperatures. MeN_3, b. about 20°. As with the aliphatic diazo compounds, these triazo compounds probably have a linear rather than a cyclic structure. The resonance hybrid which includes the two electronic formulas is probably nearer the truth than the older formulas.

$$
R{-}N\begin{array}{c} N \\ \| \\ N \end{array} \qquad
R{-}N{=}N{\equiv}N \qquad
R:\overset{..}{N}:\overset{-}{\underset{..}{\overset{..}{N}}}:\overset{+}{N}: \qquad
R:N::\overset{+}{\underset{..}{N}}::\overset{-}{\underset{..}{N}}:
$$

X. ALDEHYDES AND KETONES

Their relations as dehydrogenation products of the primary and secondary alcohols have been considered. Both contain the reactive grouping $C{=}O$, carbonyl, a grouping characterized by extreme unsaturation or ability to add a great variety of reagents. This resembles the unsaturation of the olefinic linkage but is greater because of the more polar character of the O atom.

$$
\overset{..}{\underset{..}{C}}::\overset{..}{\underset{..}{O}} \quad \leftrightarrow \quad \overset{..}{\underset{..}{C}}:\overset{..}{\underset{..}{O}}:
$$

The activated form will add H or a metal atom from a great variety of reagents. The rest of the addend may add to the electronically deficient carbon or elsewhere in the molecule if a shift first takes place (1:4-addition).

[15] Aston et al. *J. Am. Chem. Soc.* **54**, 1530 (1932); **56**, 1387 (1934); **60**, 1930 (1938).

Water is often lost giving a product C = Q which gave the basis for the older conception that it was the result of a mysterious condensation in which the O of the carbonyl united with the H atoms of H_2Q, thus leaving the remaining fragments to unite. Although it is still convenient to indicate such a change by the familiar "lasso sign," it should be remembered that the reaction undoubtedly goes through an initial addition step.

The additive power of the carbonyl group is modified by the attached groups. Thus the aldehydes which have H attached to the carbonyl group are considerably more reactive than the ketones which have two alkyl groups attached to it. In turn, the methyl ketones containing the grouping CH_3CO are more reactive than those which contain two larger groups. Branching of the alkyl groups, especially near the carbonyl has a profound effect both on the activity of that group and on the production of such aldehydes and ketones.[1]

Aldehydes and ketones are anhydrides of *gem*-dihydroxyl compounds. Hydrates of carbonyl compounds can be isolated only in special cases such as chloral hydrate, $Cl_3CCH(OH)_2$. In water solution, the lower aldehydes and ketones probably exist in equilibrium with their hydrates.[2]

$$= C = O + H_2O \rightleftharpoons = C(OH)_2$$

The aliphatic aldehydes range in volatility from gaseous formaldehyde through liquids to solids at the C_{12} member and in odor from the very sharp smelling lower aldehydes through perfume materials to odorless solids. The ketones range from liquids to solids. They have sweet odors. The intermediate ones are valuable in perfumes.

A. SATURATED ALDEHYDES

These have the formula, $R - C \underset{\diagup}{\overset{H}{=}} O$, and are named from the acids which they give on oxidation.

Formaldehyde, methanal, HCHO, b. $-21°$, is prepared from methanol by oxidation or dehydrogenation. Because of its volatility, the first reaction is readily carried out although formaldehyde is much more readily oxidized than methanol. There is excellent evidence that the first step in the oxidation of a substance like methanol is the conversion of the H of the group CHOH to hydroxyl, giving $C(OH)_2$. Water is lost leaving a carbonyl group.

$$CH_3OH + [O] \rightarrow H_2C(OH)_2 \rightarrow H_2C = O + H_2O$$

Dehydrogenation is carried out in the presence of heated platinum, silver or copper. By admitting air with the methanol vapors, part of the hydrogen is oxidized to give the heat necessary for the reaction. Formaldehyde is available

[1] Conant, Blatt. *J. Am. Chem. Soc.* 51, 1227 (1929).
[2] Walker. "Formaldehyde." Reinhold, N.Y.C., 1944. p. 29.

as *formalin*, a 35–40% solution in water and methanol. This also contains varying amounts of impurities such as *formal*, formaldehyde dimethyl acetal, $H_2C(OMe)_2$. Formaldehyde has a characteristic penetrating odor. Formaldehyde is also obtained in small yields but in large amounts by the partial oxidation of natural gas. It is also an important product of the non-catalytic high temperature oxidation of butane (p. 18) in which it is obtained in dilute water solution admixed with volatile organic compounds. By distillation under reduced pressure, water and the other materials are removed at a temperature at which $CH_2(OH)_2$ is stable. After this separation the pressure is raised and CH_2O distilled out.[3]

Formaldehyde is said to occur in air in minute amounts as a result of its formation during combustion processes.[4]

Formaldehyde can also be produced by heating certain formates,

$$(HCO_2)_2Ca \rightarrow CaCO_3 + H_2CO$$

This again is dismutation, half of the formate is oxidized to carbonate and the other half is reduced to formaldehyde.

There is excellent evidence that the first product formed from carbon dioxide by the chlorophyll of green plants is formaldehyde,

$$CO_2 + H_2O \rightarrow H_2CO + O_2{}^5.$$

The ratio of oxygen given off by the plant to the carbon dioxide absorbed has been found experimentally to be 1:1 (Willstätter). The formaldehyde, being extraordinarily reactive, undergoes a variety of changes. If the living plant needs an oxidizer or a reducer, the formaldehyde can serve in either capacity, changing to methanol and to formic acid or carbonic acid respectively. The first of these changes is indicated by the presence of methyl derivatives in many plant products. Most of the formaldehyde probably undergoes *condensation*, the H from one molecule adding to the carbonyl group of another, to form successively glycolic aldehyde, glyceric aldehyde, and hexoses (pp. 460–1, 486).

The reactions of formaldehyde are partly typical aldehyde reactions and partly peculiar to itself.

1. Oxidation is very easy.

$$H_2C{=}O \rightarrow HC\overset{OH}{=}O \rightarrow HO{-}C\overset{OH}{=}O \rightarrow CO_2 + H_2O$$

In this case the mechanism of oxidation through the change of $C-H$ to $C-OH$ is apparent. In common with other aliphatic aldehydes and in contrast to ketones, formaldehyde reduces ammoniacal silver solution and Fehling's

[3] Levey. *Chem. Inds.* **50**, 204 (1942).
[4] Goldman, Yagoda. *Ind. Eng. Chem., Anal. Ed.* **15**, 378 (1943).
[5] *Ann. Rep. Chem. Soc.* (London) 1906, 83.

solution (alkaline cupric tartrate solution). In alkaline solution hydrogen peroxide oxidizes it quantitatively to a formate.

2. Reduction is also easy, $CH_2O \rightarrow CH_3OH$.

3. Polymerization. Formaldehyde readily changes reversibly into solid polymers. *Paraformaldehyde*, paraform, $(CH_2O)_x$, is obtained as an amorphous white solid by evaporation of an aqueous solution of CH_2O. It is probably $HOCH_2(OCH_2)_nCHO$.[6,7] Rapid condensation of CH_2O vapors can give α-trioxymethylene, *trioxane*, m. 61°, b. 115°, which is a definite 6-membered ring compound. It is crystalline and readily soluble in water, alcohol and ether[8] (DuPont).

$$
\begin{array}{c}
CH_2 \\
O \diagup \quad \diagdown O \\
| \qquad\qquad | \\
CH_2 \qquad CH_2 \\
\diagdown \quad \diagup \\
O
\end{array}
$$

Treatment of formaldehyde solution with sulfuric acid in various ways gives several different polyoxymethylenes of unknown molecular weight and structure. All these polymers are readily converted to formaldehyde by heat. "Paraform" candles are used for fumigating, part of the paraformaldehyde burning and part being depolymerized to formaldehyde.

4. With bases.

a. Concentrated strong bases cause dismutation[9] (Cannizzaro).

$$2\ H_2CO + KOH \rightarrow HCO_2K + CH_3OH$$

This reaction is characteristic of aldehydes which have no *alpha* hydrogen, that is, no hydrogen on the carbon *next* to the carbonyl group.

b. Dilute and weak bases cause aldol type condensations as discussed under the behavior of formaldehyde in plants.

Formaldehyde supplies a very reactive carbonyl group for aldol condensations with α−H compounds such as acetaldehyde and isobutyraldehyde. With ketones and secondary amines it replaces the α−H atoms with the group $-CH_2NR_2$ (Mannich). Thus acetone with formalin and dimethylamine gives condensation products which range from $Me_2NCH_2CH_2COCH_3$ to $(Me_2NCH_2)_3CCOC(CH_2NMe_2)_3$.

5. Alcohols in the presence of a trace of acid give *formals*, ethers of the hypothetical hydrate of formaldehyde $H_2C(OH)_2$.

$$H_2CO + 2\ ROH \rightarrow H_2C(OR)_2 + H_2O$$

[6] Staudinger et al. *Ann.* **474**, 145 (1929).
[7] Sauter. *Z. physik. Chem.* **B21**, 186 (1933).
[8] Walker, Carlisle. *Chem. Eng. News* **21**, 1250 (1943).
[9] Fry, Uber, Price. *Rec. trav. chim.* **50**, 1061 (1931).

Methylal is methylene dimethyl ether, $CH_2(OMe)_2$, b. 42°. It is obtained by the partial oxidation of methanol at low temperatures with acid oxidizing agents. It may also be made from a methylene halide and sodium methylate. The formals differ from ordinary ethers in being readily hydrolyzed by strong acids. Hemiformals, $CH_2(OH)OR$, are less stable. Glycols give cyclic formals, *dioxolanes*, the parent substance (I) being obtained from formaldehyde and ethylene glycol. The first product may be a hemiacetal with reactive groups in the 1,5-position, thus making ring closure easy.

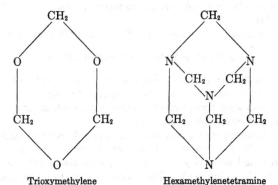

6. Ammonia forms hexamethylenetetramine, $(CH_2)_6N_4$. The first product undoubtedly results from the addition of NH_3 to the carbonyl group in the usual manner. This product loses water to form the imino analog of formaldehyde, $H_2C=NH$. This polymerizes much as formaldehyde does except that the higher valence of nitrogen makes possible a more complex polymer.

Trioxymethylene   Hexamethylenetetramine

The probability of the latter formula becomes very obvious when it is made from atomic models. It forms a symmetrical structure containing four 6-membered rings. Hexamethylene tetramine is used in medicine as a urinary antiseptic (Urotropine, *Methanamine*, NNR). In acid solution it slowly gives ammonium salts and formaldehyde which acts as a disinfectant. It is also used with phenols to form resins. In the rubber industry it finds use as an accelerator (Hexa).

With anhydrous nitric acid, hexamethylene tetramine forms the explosive trimethylenetrinitramine, "hexogen" or "cyclonite."[10]

$$O_2N-N\overset{\overset{\displaystyle CH_2}{\diagup\quad\diagdown}}{\underset{\underset{\displaystyle CH_2}{\diagdown\quad\diagup}}{\quad}}N-NO_2$$

$$\underset{\underset{\displaystyle NO_2}{\displaystyle |}}{N}$$

Any one of the four similar six-membered rings in the starting material can supply the ring in the product, the other nitrogen and three of the methylene groups being oxidized.

In World War II it found special uses in "block-buster" bombs, in torpedos and the like.

As has been seen, formaldehyde reacts with ammonium chloride to give the three methylamines.

The extreme activity of the carbonyl group of formaldehyde in aldol type condensations is shown by its action with phosphine and HCl to give

$$(HOCH_2)_4PCl.$$

7. With NH_4Cl and KCN it gives two trimers of methyleneaminoace-tonitrile.[11]

8. Formaldehyde reacts with amines to give more or less complex products. Thus, with isobutylamine, it forms triisobutyltrimethylenetriamine, a substance having RN in place of each O of trioxymethylene. The intermediate $RN=CH_2$ polymerizes readily. The reaction of formaldehyde with amino acids is of considerable practical importance in determining these substances. Since they contain both acid and basic groups they are amphoteric and cannot be titrated either with bases or acids. Glycine may be taken as an example. It exists mainly as an inner salt (like betaine).

$$\underset{\displaystyle CO-OH}{\overset{\displaystyle CH_2-NH_2}{|}} \qquad \underset{\displaystyle CO-O}{\overset{\displaystyle CH_2-NH_3}{|\quad|}} \quad \text{or} \quad {}^+NH_3CH_2CO_2{}^-$$

Treatment with formaldehyde eliminates the amino group so that the acid can be titrated with a base and indicator (*Formol titration*)

$$\underset{\displaystyle CO_2H}{\overset{\displaystyle CH_2NH_2}{|}} \rightarrow \underset{\displaystyle CO_2H}{\overset{\displaystyle CH_2NH-CH_2OH}{|}} \rightarrow \underset{\displaystyle CO_2H}{\overset{\displaystyle CH_2N=CH_2}{|}}$$

[10] Hale. *J. Am. Chem. Soc.* **47**, 2754 (1925).

[11] Johnson, Rinehart. *J. Am. Chem. Soc.* **46**, 768 (1924).

The action of formaldehyde with amino groups undoubtedly explains its ability to harden and preserve biological specimens.

Dimethylamine and H_2CO give $HOCH_2NMe_2$ and $Me_2NCH_2NMe_2$[12]. The former is probably an intermediate in the Mannich Reaction of active H compounds, HQ, to form $Q-CH_2NMe_2$.

9. With primary and secondary amines and alcohols it gives ethers of the types $MeN(CH_2OR)_2$ and Me_2NCH_2OR.[13] Mercaptans give similar thio ethers.

10. **Angeli-Rimini aldehyde reactions.** When heated with the sodium salt of nitrohydroxylamine, aldehydes give hydroxamic acids which react with ferric chloride to form deep purplish colors.

$$RCHO + HON{=}NO_2Na \rightarrow RC\overset{OH}{{=}}NOH + NaNO_2$$

Treatment of an aldehyde with N-benzenesulfonylhydroxylamine, $PhSO_2NHOH$, and a base gives a hydroxamic acid.[14] This reaction fails, as does the Angeli reaction, with aldehydes having OH, NH_2, or CO groups in the gamma or delta position. These evidently exist in cyclic structures and have no free aldehyde group.

11. Grignard reagents form primary alcohols. For example,

$$Me_3CMgCl + H_2CO \rightarrow Me_3C-CH_2OMgCl \rightarrow Me_3CCH_2OH$$

This is another case of addition to the carbonyl group, the $-MgCl$ adding to the O and the tertiary butyl group adding to the C.

12. Sodium bisulfite forms an addition compound as with many carbonyl compounds.

$$H_2C{=}O + NaHSO_3 \rightleftharpoons H_2C\overset{OH}{-}SO_2ONa$$

Formerly this was believed to have the structure of a sulfite because of the ease with which acids reverse the reaction. The present tendency is to regard

[12] Friedlander. "Fortschritte der Teerfarbenfohrikation," Vol. II. Springer, Berlin. p. 783.

[13] *Ann. Rep. Chem. Soc.* (London) 1921, 76; 1923, 87.

[14] *ibid.* 1909, 79.

it as a sulfonic acid derivative with the S attached to C.[15,16] This agrees with the modern conception of the "structure" of sulfites (p. 152).

$$\left[\begin{array}{c} :\overset{\cdot\cdot}{O}: \\ :\overset{\cdot\cdot}{O}:\overset{\cdot\cdot}{S}: \\ :\overset{\cdot\cdot}{O}: \end{array}\right]^{--} \overset{Na^{\oplus}}{\underset{H^{\oplus}}{}} + \begin{array}{c} H \\ C:H \\ :\overset{\cdot\cdot}{O}: \end{array} \rightleftharpoons \left[\begin{array}{ccc} :\overset{\cdot\cdot}{O}: & & H \\ :\overset{\cdot\cdot}{O}:\overset{\cdot\cdot}{S}: & : & C:H \\ :\overset{\cdot\cdot}{O}: & & :\overset{\cdot\cdot}{O}: \\ & & H \end{array}\right]^{-} Na^{+}$$

The hydroxy sulfonic acid structure of the bisulfite products of carbonyl compounds involves the conception that the C—S linkage is weakened by an OH group on the same carbon; thus a *hydroxyl* and a *sulfonic acid group* on the *same carbon* form an unstable arrangement like the combination of two hydroxyl groups with one carbon. The structure of the bisulfite addition product as a *sulfonate*, $HOCH_2SO_3Na$, rather than as a sulfurous ester is shown by many of its reactions to give known sulfonic acids. Treatment with NH_3 and acidification gives $NH_2CH_2SO_3H$. Condensation of the bisulfite addition product with acetoacetic ester followed by splitting with alkali or acid gives $HO_3SCH_2CH_2CO_2H$ or $CH_3COCH_2CH_2SO_3H$[17]. The formation of a C—S linkage from a carbonyl compound and a bisulfite is entirely analogous to that from an alkyl halide and a sulfite (p. 152). The bisulfite compound reverts to the aldehyde when treated with reagents which react with $NaHSO_3$ such as acids, bases, carbonates, etc.

Aldehydes can also unite with sulfurous acid itself. This is the basis for Schiff's test to distinguish aldehydes from ketones. A solution of the dye fuchsin (magenta) is just decolorized with sulfurous acid. Addition of an aldehyde gives a pink color.

Aldehyde bisulfite compounds.[18,19]

When formaldehyde-sodium bisulfite is treated with zinc dust it is converted to "sodium formaldehyde sulfoxylate," $HOCH_2SO_2Na$, or, less probably, $HOCH_2OSONa$. This is used as a bleaching agent (Rongalite C, Sulfoxite C, Formopon) with vat dyes and in converting Salvarsan to Neosalvarsan. Its most probable structure is that of a salt of a sulfinic acid.

$$\left[\begin{array}{ccc} & H & \\ H:\overset{\cdot\cdot}{O}:\overset{\cdot\cdot}{C} & : & \overset{\cdot\cdot}{S}:\overset{\cdot\cdot}{O}: \\ & H & :\overset{\cdot\cdot}{O}: \end{array}\right]^{-} Na^{+}$$

[15] Lauer, Langkammerer. *J. Am. Chem. Soc.* **57,** 2369 (1935).
[16] Shriner, Land. *J. Org. Chem.* **6,** 888 (1941).
[17] *Ann. Rep. Chem. Soc.* (London) **1927,** 65.
[18] *ibid.* **1928,** 74.
[19] Gibson. *Chem. Rev.,* 435 (1934).

A better method than that starting with sodium bisulfite is to treat SO_2 solution with zinc dust to get zinc hydrosulfite, ZnS_2O_4, which reacts with formaldehyde to give zinc formaldehyde sulfoxylate. Treatment with sodium carbonate gives the sodium derivative.

The formaldehyde sulfoxylates are important in discharge printing with azo dyes and in the general use of vat or indigoid dyes. Both classes of dyes are reduced to colorless materials, the first irreversibly and the latter reversibly.

13. Halide acids form unstable halogenated methanols, XCH_2OH, and more stable ethers, XCH_2OCH_2X. In presence of alcohols, mixed ethers, XCH_2OR, are obtained. Higher aldehydes give similar α-halogen ethers.[20] The halogens in these ethers react metathetically with Grignard reagents. (p. 44).

14. Hydrogen sulfide gives the cyclic trimer of thioformaldehyde, $(CH_2S)_3$, *trithiane*. Hydrogen selenide gives an analogous compound which is odorless and has high anti-knock value but, unfortunately, gives odorous combustion products. Ammonium sulfide gives the 6-membered ring, $CH_2(SCH_2)_2NH$.

15. Acid chlorides form chloromethyl esters, for example $CH_3CO_2CH_2Cl$, b. $115°$. The reactions of this type of substance illustrate the greater ease of addition than of metathetical reactions. In most cases it is not possible to *replace* the chlorine in these esters because *addition* takes place with the carbonyl group of the ester instead. This may be illustrated by the action with water and with ammonia.

$$CH_3C\!\!\!\overset{O}{\underset{}{\diagup\!\!\!\!/}}\!\!\!-OCH_2Cl \quad \xrightarrow{H_2O} \quad CH_3-\!\!\overset{\overset{\displaystyle OH}{|}}{\underset{}{C}}\!\!\overset{OH}{\diagup}\!\!-OCH_2Cl \rightarrow$$

$$CH_3CO_2H + HOCH_2Cl \rightarrow HCl + CH_2O$$

$$\xrightarrow{NH_3} CH_3C\!\!\overset{\overset{\displaystyle NH_2}{|}}{\underset{}{}}\!\!\overset{OH}{\diagup}\!\!-OCH_2Cl \rightarrow$$

$$CH_3CONH_2 + HOCH_2Cl$$

A metathetical replacement of the chlorine in chloromethyl esters has been accomplished with NaSCN.

16. With urea, it forms both simple and complex condensation products, $HOCH_2NHCONHCH_2OH$, dimethylolurea, etc., until the molecules become so complex that resins are formed. Thiourea gives resins even more readily, perhaps because of the polymerizing tendency of sulfur. Melamine gives plastics which are specially resistent to heat and moisture.

17. With casein, important plastic resins are obtained.

18. The formation of resins (Bakelite, Redmanol, Indur, etc.) with phenols and related compounds starts with the usual addition to the carbonyl group

[20] Gebauer, Fuelnegg, Moffett. *J. Am. Chem. Soc.* 56, 2009 (1934).

of formaldehyde. The H atoms in the ortho and para positions in phenols are reactive. One of these adds to the O and the rest of the phenol molecule adds to the C.

$$C_6H_5OH + H_2CO \rightarrow HOCH_2C_6H_4OH$$

These molecules can condense with more phenol and more formaldehyde forming molecules with several reactive groups which continue to react until the final molecules are too large and complex to crystallize and consequently form resins.[21] In general any organic molecule which contains at least two reactive groups can produce a resin.

19. With other aromatic compounds, products are formed in which two aromatic nuclei are linked by a methylene group. These reactions take place with surprising ease and are of great value in making certain classes of dyes and syntans, as in the union of two dimethylaniline or two naphthalene sulfonic acid residues. Under suitable conditions, the linkage of CH_2 to aromatic nuclei can be continued to the formation of hydrocarbon-formaldehyde resins.

20. Formaldehyde, as shown above and in other connections, forms the most important single intermediate in the enormous and rapidly expanding plastics industry. Treatment of aromatic compounds like benzene and naphthalene with formaldehyde, and HCl in presence of $ZnCl_2$ introduces the chloromethyl group, $-CH_2Cl$.[22] Iron compounds even in traces must be excluded. Otherwise the molecules act with each other to give unmanageable resins.

21. Formaldehyde reacts with olefins in presence of strongly acid catalysts to give 1,3-glycols or their secondary products. Thus propylene and HCl react with it to give 1,3-butandiol.[23] Isobutylene with $SnCl_4$ in $CHCl_3$ gives a 90% yield of 3-Me-3-butenol.[24]

Formaldehyde is produced in very large amounts commercially for the manufacture of resins and other organic products and as a disinfectant.[25] It is an important war chemical. Because of the reactivity of formaldehyde with bases, it was formerly difficult to remove traces of acid and heavy metal compounds from formalin. This is now readily achieved by treatment with ion-exchange resins.

Formalin was one of the first *pickling inhibitors*. It was added to muriatic acid by plumbers for cleaning rusty pipes. The aldehyde inhibits the action of the acid on metal but not on the oxides. It is finding increasing use against sulfide corrosion in oil wells. The addition of 0.2% formalin to water gives "artificial ice," especially useful for icing fish.

Because of its occasional use as a harmful preservative for foods, very delicate tests have been evolved for traces of formaldehyde. In the presence

[21] Wansheidt, Itenberg, Andreeva. *Ber.* **69B**, 1900 (1936).
[22] Rueggeberg et al. *Ind. Eng. Chem.* **38**, 478 (1946).
[23] Fitzky. *C. A.* **33**, 2914 (1939).
[24] Mikeska, Arundale. *C. A.* **37**, 3450 (1943).
[25] Ellis. "Chemistry of Petroleum Derivatives." Reinhold, 1934. p. 860.

of milk, addition of ferric chloride and underlaying with conc. sulfuric acid give a violet ring if formaldehyde is present in as much as 5 p.p.m. (parts per million). A test for formaldehyde in larger amounts depends on mixing with a dilute solution of resorcinol and pouring it over the surface of conc. sulfuric acid to form a red ring. A more delicate test is that of Schryver with phenylhydrazine hydrochloride, potassium ferricyanide and conc. hydrochloric acid to give a red color.

Typical derivatives of formaldehyde are:

a. Oxime, $CH_2 = NOH$, b. 84°; b. Phenylhydrazone, alpha and beta forms, m. 168° and 211°; c. p-Nitrophenylhydrazone, m. 182°.

Formaldehyde can be determined by spectrophotometer.[26, 27]

Acetaldehyde, "Aldehyde," ethanal, CH_3CHO, b. 21°, is readily obtained by the oxidation of ethyl alcohol. Because of its volatility, the acetaldehyde escapes from the oxidizing mixture before it can be oxidized to acetic acid. It is also contained in the "first runnings" in the distillation of fermented liquids. Acetaldehyde can be made by the controlled catalytic air oxidation of ethanol.

Acetaldehyde is made from acetylene by catalytic hydration by means of an acid solution of mercuric sulfate.

$$HC \equiv CH \rightarrow H_2C = CHOH \rightarrow CH_3CHO$$

An intermediate organic mercury compound is formed which reacts with the acid to give acetaldehyde and regenerate the mercuric salt. The process is carried out by passing acetylene up through a packed tower down which the catalyst solution is flowing. Some of the mercuric salt is reduced to metallic mercury which collects at the bottom of the tower. This is periodically converted to mercuric sulfate by nitric and sulfuric acids. A modification of this process uses a suspension of HgO in dilute sulfuric acid. The mercury gradually accumulates in an organic sludge. This is electrolyzed in alkaline solution to regenerate the HgO.

In Germany the shortage of mercury during World War II forced the successful development of a catalyst consisting of 120 g. of H_2SO_4 and 90 g. of Fe as sulfates per liter. Used at 94–98°, this gives a 96% yield of acetaldehyde.[28]

Another preparation developed in Germany (Reppe) consisted in bubbling acetylene under pressure at about 200° through an alcohol to form the vinyl ether, $CH_2 = CHOR$, which is readily hydrolyzed by dilute acid to give acetaldehyde and ROH which can be reused.

Acetaldehyde is obtained along with formic acid by warming lactic acid, $MeCHOHCO_2H$, with dilute sulfuric acid. This is a general method for converting an *alpha hydroxy acid to the next lower aldehyde.* If the former has a branch at the alpha carbon, the product is a *ketone.*

[26] Bricker, Johnson. *Ind. Eng. Chem., Anal. Ed.* 17, 400 (1945).
[27] Walker. "Formaldehyde," ACS Monograph. Reinhold, N.Y.C., 1944. p. 400.
[28] Office of Publication Board. No. 515, 1945.

The *reactions of acetaldehyde* are typical of those of most aliphatic aldehydes.

1. Oxidation is very easy. In air, peroxides are first formed.[29, 30] Commercially, pure acetaldehyde is oxidized to glacial acetic acid by air and a manganous oxide catalyst. Ammoniacal silver solution is reduced to a silver mirror.

$$CH_3CHO + 2\ Ag(NH_3)_2OH + 2\ H_2O \rightarrow CH_3CO_2NH_4 + 2\ Ag + 3\ NH_4OH$$

Oxidation in water solution probably involves the dehydrogenation of the aldehyde hydrate. Thus finely divided palladium in the absence of oxygen converts a small amount of acetaldehyde in water solution into acetic acid. When the palladium is saturated with hydrogen the reaction stops until O_2 is admitted to oxidize the hydrogen in the palladium.

A most interesting oxidation of acetaldehyde is that by means of selenium dioxide, SeO_2, which gives a good yield of glyoxal,[31] $(CHO)_2$. The attack on the alpha H atoms and the failure to attack the easily oxidized aldehyde H indicate that the SeO_2 attacks the enol form. This valuable reagent has been studied with a great variety of aldehydes and ketones. Its specificity for the conversion of an alpha CH_2 to carbonyl is unusual. It suggests the specificity of enzyme reactions. Just as an enzyme adds to a specific configuration, the SeO_2 may add to the enolic double bond

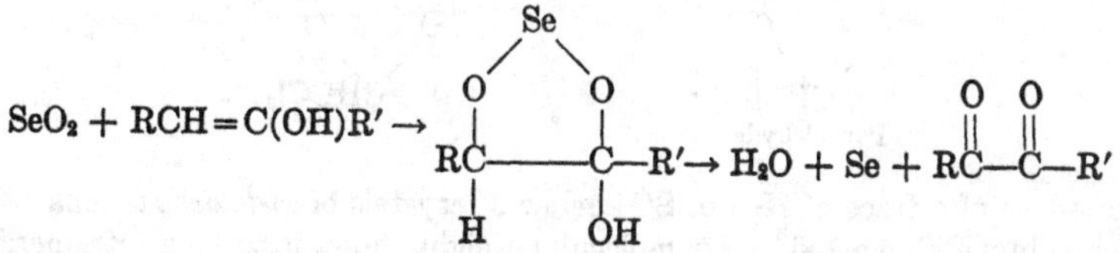

in which R and R' may be H, alkyl or aryl. Propionaldehyde and butyraldehyde give 30 and 45% yields of MeCOCHO and EtCOCHO by this process.

2. Reduction is also easy. There is always a possibility that ethyl alcohol may be produced commercially this way. It is largely a matter of cheap acetylene. In addition to reduction to a primary alcohol, there is some product of a bimolecular reduction to give a substituted glycol.

$$2\ CH_3CHO \rightarrow CH_3CHOHCHOHCH_3$$

This may be the source of the β-butylene glycol formed in some fermentation processes. The reductions which give a maximum of the bimolecular product are usually by means of metals. The processes involved may be as follows:

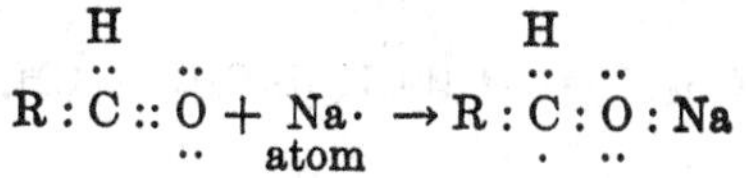

[29] Hatcher, Steacie, Howland. *Can. J. Research* **7**, 149 (1932).
[30] Wieland, Richter. *Ann.* **495**, 284 (1932).
[31] Riley, Morley, Friend. *J. Chem. Soc.* **1932**, 1875.

The resulting intermediate product is a free radical (with an odd electron). With some aromatic compounds another Na atom can unite to give a metal ketyl. Ordinarily an H atom adds to the free radical giving the primary alcohol. It is also possible for two of the free radicals to combine to give

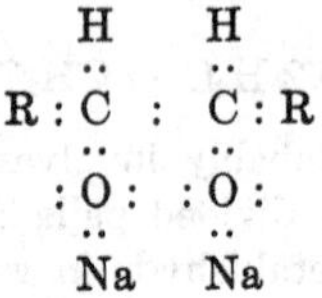

$$
\begin{array}{ccc}
& H & H \\
& \cdot\cdot & \cdot\cdot \\
R:&C&:\quad C:R \\
& \cdot\cdot & \cdot\cdot \\
& :O: & :O: \\
& \cdot\cdot & \cdot\cdot \\
& Na & Na
\end{array}
$$

3. It polymerizes reversibly. A trace of sulfuric acid converts it to *paraldehyde*, b. 124°, a trimer, Me₃-trioxymethylene. This is used as a soporific in cases where all others fail, as in delirium tremens. It reverts to acetaldehyde on heating with dilute acid. It is used in many reactions in place of acetaldehyde itself. The polymerization of acetaldehyde is increased by very high pressures. *Mixed paraldehydes* are obtainable from mixtures of aldehydes.[32]

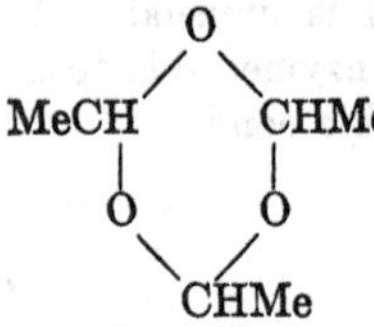

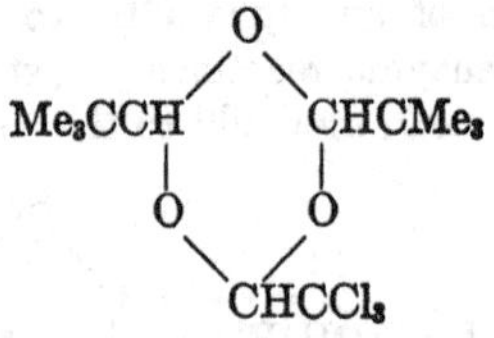

In presence of a trace of HCl or SO₂ below 0° crystals of *metaldehyde* separate which sublime above 150°. Its molecular weight shows it to be a tetramer.[33] Heating with sulfuric acid depolymerizes it. Metaldehyde is used as an ant poison.

4. Bases. *a.* Dilute or weak bases produce the true *aldol condensation*[34] from which all other similar processes are named. A trace of amino acids accelerates many aldol condensations. The most generally accepted mechanism for this type reaction proposes that the base removes an alpha hydrogen and that the carbanion thus formed adds to the carbonyl carbon of another aldehyde molecule.

$$CH_3CHO + B \rightarrow :CH_2CHO + BH$$

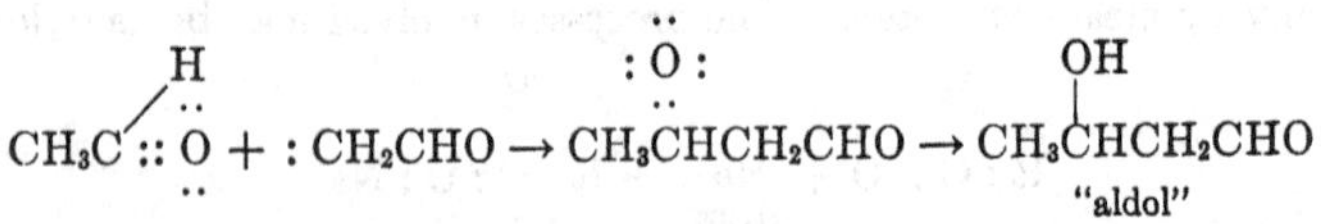

[32] Hibbert, Gillespie, Montanno. *J. Am. Chem. Soc.* **50**, 1950 (1928).
[33] Hantzsch. *Ann. Rep. Chem. Soc.* (London) **1907**, 80.
[34] Wurtz. *Bull. soc. chim.* (2) **17**, 436 (1872).

The addition intermediate takes a proton from the medium and becomes "aldol." This reaction takes place with great vigor and, because of the volatility of acetaldehyde (b. 21°) is likely to become dangerous. In commercial practice the product contains not only aldol but also its dehydration product, crotonic aldehyde, and more complex aldol condensation products obtained from it.

$$CH_3CH = CHCHO + HCH_2CHO \rightarrow CH_3CH = CHCHOHCH_2CHO$$
Crotonic aldehyde

This would readily lose water to give $CH_3CH = CHCH = CHCHO$.

The aldol condensation can be controlled by rapid circulation at 20° through a cooled horizontal tubular heat exchanger using 0.1% of KOH as catalyst. Side reactions are prevented by keeping the conversion below 50% per pass.[35]

Acetaldehyde may be used to supply the *carbonyl group* for an aldol condensation with another substance which supplies the *alpha hydrogen*. Thus $CH_3CHO + HCH_2COCH_3 \rightarrow CH_3CHOHCH_2COCH_3$. Each of the initial materials would also condense with itself. Such aldol combinations are of the utmost value in synthetic chemistry. The β-hydroxy carbonyl compounds obtained in this way are readily dehydrated to give $\alpha\beta$-unsaturated carbonyl compounds. Thus the two compounds listed above give $CH_3CH = CHCHO$, crotonaldehyde, and $CH_3CH = CHCOCH_3$, 3-penten-2-one.

Acetaldehyde may also supply the *alpha hydrogen* for condensation with the *carbonyl group* of another compound. Thus

$$H_2C = O + HCH_2CHO \rightarrow HOCH_2CH_2CHO$$

The other alpha hydrogens can react with more formaldehyde finally giving $(HOCH_2)_3CCHO$. If an excess of formaldehyde is used, it reduces the $-CHO$ group to $-CH_2OH$, forming pentaerythritol, $C(CH_2OH)_4$. The actual process is far simpler than its description, acetaldehyde and an excess of formalin merely being stirred with lime. A modification of this reaction using formaldehyde and dimethylamine with acetaldehyde gives replacement of the three alpha H atoms to form $(Me_2NCH_2)_2C(CH_2OH) - CHO$.

This is a special case of the *Mannich Reaction* in which reactive H atoms in aldehydes, ketones, acids, esters, phenols, acetylenes and alpha-picolines and quinaldines are replaced by condensation with formaldehyde and ammonia, a primary or a secondary amine.[36]

When a mixture of two different aldehydes, RCH_2CHO, is subjected to the aldol condensation four products are possible but the chief reaction generally consists in the union of the carbonyl of the smaller one with an alpha H from the larger one. In general, the ease of supplying the α-H for the condensation is in the order $R_2CHCHO > RCH_2CHO > CH_3CHO$.

[35] Monrad. *Chem. Met. Eng.* **53**, 120 (1946).
[36] "Org. Reactions," I. p. 303.

The aldol condensation is given by certain α-H compounds other than aldehydes and ketones. In general any $C-H$ compound which will give a sodium derivative, such as malonic ester, acetoacetic ester, cyanoacetic ester and nitromethane will condense with aldehydes in the presence of a trace of base or acid. The "aldols" formed lose water readily if they contain α-H. This gives important preparations for $\alpha\beta$-unsaturated compounds

$$RCHO + H_2C(CO_2Et)_2 \rightarrow RCHOHCH(CO_2Et)_2 \rightarrow$$
$$RCH = C(CO_2Et)_2 \rightarrow RCH = CHCO_2H$$

$$RCHO + CH_3NO_2 \rightarrow RCHOHCH_2NO_2 \rightarrow RCH = CHNO_2$$
$$\underline{} (RCHOH)_2CHNO_2 \text{ etc.}$$

Compounds, ROMgX, cause the aldol condensation and at the same time give ester formation by dismutation[37] (see 5 below).

The aldol condensation can also be obtained with the H which is "alpha" to a conjugatedly unsaturated aldehyde system. $CH_3(CH = CH)_nCHO$ (p. 225).

b. Concentrated strong bases give *aldehyde resins*.[38] As in the case of all resins these are simply molecules too large and complex to be true crystalline solids. Resins result from combinations of organic compounds which have a number of points of reactivity. Acetaldehyde answers this description admirably. It has three alpha hydrogen atoms which can react with the carbonyl groups of three other molecules of acetaldehyde. The resulting molecule will still have a carbonyl group which can react with an alpha hydrogen of another acetaldehyde molecule, etc., etc. Moreover, the resulting products are secondary alcohols and secondary alcohols may condense with each other in the presence of strong bases to give higher secondary alcohols. Finally some of the products contain no H atoms alpha to the $-CHO$ group and, consequently, can give the Cannizzaro reaction with the strong base used.

Apparently an aldehyde must have at least two α-H atoms to give resins as isobutyraldehyde, Me_2CHCHO, gives none but undergoes the aldol condensation and Cannizzaro reaction.

5. Aluminum alcoholates (alkoxides). Although the reactivity of its α-H atoms prevents acetaldehyde from giving the Cannizzaro reaction, a similar dismutation may be achieved by means of aluminum ethylate which converts it to ethyl acetate catalytically[39] $2 MeCHO \rightarrow MeCO_2CH_2Me$. This reaction takes place with higher aldehydes as well. Addition compounds such as $AlCl_3 \cdot 3 Al(OEt)_3$ and $HgCl_2 \cdot 2 Al(OEt)_3$ are even more effective.[40]

[37] *Ann. Rep. Chem. Soc.* (London) 1928, 96.
[38] *ibid.* 1912, 81.
[39] Tischtschenko. *J. Russ. Phys. Chem. Soc.* 38, 398 (1906).
[40] *Ann. Rep. Chem. Soc.* (London) 1925, 72.

Aluminum ethylate in absolute ethyl alcohol reduces higher aldehydes smoothly to primary alcohols.[41]

$$3 \ RCHO + Al(OEt)_3 \rightarrow (RCH_2O)_3Al + 3 \ CH_3CHO$$

The reaction apparently goes through the stage $RCH(OEt)OAl(OEt)_2$ which loses acetaldehyde and forms $RCH_2OAl(OEt)_2$, etc. $EtOMgCl$ acts like $Al(OEt)_3$. Aluminum isopropoxide is an even better special reducing agent for aldehydes[42] $3 \ RCHO + Al(OCHMe_2)_3 \rightarrow (RCH_2O)_3Al + 3 \ Me_2CO$. This type of reduction is of wide application. Since it involves the mutual interconversion of carbonyl compounds and the related alcohols it can be run in either direction using the aluminum alkoxide as a catalyst with an excess of the alcohol or ketone to be used as the reducer or oxidizer.[43]

$$=C=O + Me_2CHOH \xrightleftharpoons{Al(OCHMe_2)_3} =CHOH + Me_2C=O$$

6. With alcohols and a trace of acid, *acetals* are formed.

$$CH_3CHO + 2 \ RCH_2OH \rightarrow CH_3CH(OCH_2R)_2 + H_2O$$

"Acetal," $CH_3CH(OC_2H_5)_2$, b. 104°, is also obtained by the partial oxidation of ethyl alcohol with acid oxidizers. A study of the rate of acetal formation with a variety of alcohols in presence of HCl gives surprising results.[44] Methyl alcohol acts most slowly and tertiary alcohols more rapidly. Heptaldehyde reacts several hundred times as rapidly as butyraldehyde. Catalysts for acetal formation include the chlorides of zinc, iron, calcium and boron. One of the best catalysts is NH_4Cl. Before isolating the product the mixture is exactly neutralized with NaOEt.[45]

The term *acetal* is applied not only to the compounds $CH_3CH(OR)_2$ but also to all similar compounds formed from aldehydes, $RCH(OR')_2$. The aldehydes are regenerated by boiling with an excess of dilute acid. Acetals are stable to bases. H^+ ions apparently add to an O atom giving an oxonium salt whereas OH^- ion has no point of attachment as it has in esters.

Acetals are readily hydrogenated to alcohols with copper chromite and equal molar amounts of water at 200°. This affords a convenient synthesis of primary alcohols since acetals are readily available from Grignard reagents and triethyl orthoformate.

Cyclic acetals, *dioxolanes*, are given by glycols (p. 304). Similar products are formed with glycerol (p. 318). Acetals of polyvinyl alcohol are important commercial plastics (Monsanto).

The union of one mol of alcohol with a carbonyl group gives a *hemiacetal*, $RCH(OH)OR'$. These are usually very unstable.[46]

[41] Meerwein. *Ann. Rep. Chem. Soc.* (London) 1925, 71.
[42] Young, Hartung, Crossley. *J. Am. Chem. Soc.* 58, 100 (1936).
[43] "Org. Reactions," II. p. 178.
[44] Adkins. *Ann. Rep. Chem. Soc.* (London) 1925, 69.
[45] *Ann. Rep. Chem. Soc.* (London) 1922, 68.
[46] Adkins, Broderick. *J. Am. Chem. Soc.* 50, 499 (1928).

Mercaptans form mercaptals, $CH_3CH(SR)_2$. The aldehyde can be regenerated by treatment with HgO or $HgCl_2$ and water.

7. Aldehydes with alcohols and an excess of HCl or HBr give α-halogen ethers of great value in synthetic work because they react nearly quantitatively with the Grignard reagent (p. 44).

$$RCHO + R'OH + HX \rightarrow RCHX-OR'$$

8. Acid anhydrides react with aldehydes to give esters of the hydrated aldehydes, such as

$$CH_3CH(OCOMe)_2,$$

acetaldehyde diacetate, ethylidene diacetate. These are hydrolyzed by dilute acids. Ethylidene diacetate is obtained commercially from acetylene and acetic acid with a Hg catalyst. On heating, it gives acetic anhydride and MeCHO.[47] The esters of the aldehydes are readily decomposed by acids and bases.

Aldehydes boiled with acetic anhydride give acetates of their enol forms, $RCH=CHOAc$.[48]

9. Hydrogen sulfide forms thioacetaldehyde which polymerizes readily.

$$CH_3CHO + H_2S \rightarrow CH_3CH(OH)SH \rightarrow H_2O + CH_3CHS \rightarrow (CH_3CHS)_3$$

The formation of the thioaldehyde depends on the fact that the $C-S$ linkage is stronger than the $C-O$ linkage. Consequently the first addition product loses H_2O rather than H_2S.

Ammonium sulfide gives the cyclic compound $MeCH(S-CHMe-)_2NH$, m. 40°, "thialdine" (CC).

10. Ammonia adds to form "aldehyde ammonia." The reaction is usually carried out in dry ether in which the product is insoluble. Treatment with acid regenerates the aldehyde. Thus the process is sometimes used for isolating acetaldehyde from a mixture. The first step is the usual addition to the carbonyl group to give $MeCH(OH)NH_2$. This grouping is unstable and loses NH_3 to give back acetaldehyde or changes to a trimer. The cyclic formula analogous to that of paraldehyde has been questioned because of the easy change of the trimer to a dimer in solution.[49] Either of these substances reacts with acids to give ammonium salts and aldehyde.

11. Various nitrogen bases related to ammonia give simpler condensation products than ammonia. Typical examples are hydroxylamine, NH_2OH, phenylhydrazine, $C_6H_5NHNH_2$, and semicarbazide, $NH_2NHCONH_2$. The reactions consist of an addition to the carbonyl group followed by the loss of

[47] Coffin. Can. J. Research 5, 636 (1931).
[48] Ann. Rep. Chem. Soc. (London) 1909, 79.
[49] ibid. 1915, 83.

water to leave a $C=N$ linkage.   Thus

$CH_3CHO + NH_2OH \rightarrow CH_3CH=NOH$, acetaldehyde oxime, acetaldoxime.

$\qquad C_6H_5NHNH_2 \rightarrow CH_3CH=NNHC_6H_5$, acetaldehyde phenylhydrazone.

$\qquad NH_2NHCONH_2 \rightarrow CH_3CH=NNHCONH_2$, acetaldehyde semicarbazone.

The last two substances are useful for identifying acetaldehyde because they are easily purified and give definite melting points.   Acetaldehyde phenylhydrazone exists in α- and β-forms, m. 100° and 57° respectively.[50]

12. Primary amines form Schiff's bases, $MeCH=NR$.   Treatment with acids gives back the original substances.   Reduction in alkaline solution gives secondary amines, $EtNHR$.   Secondary amines give resinous products with aldehydes.

13. Sodium bisulfite forms the addition compound, $MeCH(OH)SO_3Na$. The rate of addition of $NaHSO_3$ to higher aldehydes decreases with length and branching of the alkyl group.

14. Hydrocyanic acid, HCN, forms $MeCH(OH)CN$, a cyanohydrin, the nitrile of an alpha hydroxy acid.[51]   Traces of NH_3 or amines catalyze the addition.   Hydrolysis gives the α-OH-acid.

$$MeCHOHCN + 2 H_2O + H_2SO_4 \rightarrow NH_4HSO_4 + MeCHOHCO_2H$$

Treatment with ethyl alcohol and H_2SO_4 gives ethyl lactate directly.   This is used in lacquer solvents since it has the solvent properties of an alcohol and an ester in the same substance.

A better *preparation* of the cyanohydrin is to make the bisulfite addition compound and treat it with a soluble cyanide,

$$MeCH(OH)SO_3Na + NaCN \rightarrow Na_2SO_3 + MeCHOHCN.$$

This might either throw doubt on the sulfonic acid structure of the bisufitel compound or be a further indication of the lability of the $C-S$ linkage in the presence of OH on the same carbon.   The formation of a cyanide from a sulfonic acid is common in the aromatic series but requires fusion of the sodium salt with NaCN.

15. An ammonium salt and an alkali cyanide give an alpha amino nitrile from which an alpha amino acid (alanine) can be prepared.   This is the *Strecker amino acid synthesis*.

$$RCHO + NH_4Cl + KCN \rightarrow RCH(NH_2)CN + KCl$$

16. Grignard reagents give secondary alcohols, $MeCHOHR$.

17. Reactive aromatic compounds such as phenols and naphthalene compounds condense to give substances $CH_3CHOHAr$ and CH_3CHAr_2.

[50] *Ann. Rep. Chem. Soc.* (London) 1911, 98.
[51] Lapworth, Peters.   *J. Chem. Soc.* 1931, 1382.

18. Halogens replace the alpha hydrogens much more easily than in hydrocarbons. This is presumably because the *enol* form reacts.

$$CH_3CHO \rightleftharpoons CH_2=CHOH \rightarrow BrCH_2CH(OH)Br \rightarrow BrCH_2CHO$$

Since halogen and hydroxyl cannot exist on the same carbon, HX is lost leaving bromoacetaldehyde, $BrCH_2CHO$. It is not necessary to assume that the bromine atom adds to the aldehyde carbon and then splits off. The positive Br may add to the activated double bond in the enol, thus leaving the other C electronically deficient. This is overcome by the attraction of the electron pair holding the H thus forming a carbonyl group and expelling a proton which forms HBr. The change is thus like the formation of allyl chloride by the chlorination of propene.

The best preparation for α-bromoaldehydes is the action of the trimers such as paraldehyde with bromine in the cold.[52] Chlorine reacts with paraldehyde in the cold to give $CH_3CHClCCl_2CHO$, *butyl chloral*, 2,2,3-trichlorobutanal.

19. Phosphorus pentachloride replaces the O by Cl_2 giving 1,1-dichloroethane, ethylidene chloride, CH_3CHCl_2. This reaction is important with higher aldehydes as a source of intermediates for making 1-acetylenes, $RC \equiv CH$. PBr_5 tends to replace the alpha H atoms by bromine because of its easy dissociation to give Br_2 and PBr_3.

20. The thermal decomposition of acetaldehyde to CH_4 and CO is of theoretical interest. It is catalyzed by such diverse substances as I_2 and H_2S.[53]

Important Derivatives of Acetaldehyde

a. Acetaldoxime, $CH_3CH=NOH$, m. 47°, b. 115°;

b. Phenylhydrazone, alpha and beta forms, m. 100° and 57°;

c. p-NO_2-phenylhydrazone, m. 128°.

Propionaldehyde, propanal, CH_3CH_2CHO, is prepared in the usual way by the oxidation of n-propyl alcohol. It can also be prepared by dehydrogenating the alcohol by passing the vapors over a heated copper or brass catalyst. This avoids the danger of further oxidation to propionic acid.

The reactions of propionaldehyde are practically like those of acetaldehyde. It must be remembered that only the two alpha hydrogen atoms are active in replacement and condensation reactions. Thus:

$$CH_3CH_2CHO + HCHCHO \rightarrow CH_3CH_2CHOHCHCHO$$
$$\qquad\qquad\qquad |\qquad\qquad\qquad\qquad\qquad\qquad |$$
$$\qquad\qquad\qquad CH_3\qquad\qquad\qquad\qquad\qquad\qquad CH_3$$

In this case, the remaining alpha H is also tertiary. The product is, therefore, very easily dehydrated to 2-methyl-2-penten-al.

[52] *Ann. Rep. Chem. Soc.* (London) 1927, 63.
[53] Roth, Rollefson.   *J. Am. Chem. Soc.* **64,** 1707 (1942).

Butyraldehyde, butanal, "Butalyde," $CH_3CH_2CH_2CHO$, is commercially available as an oxidation product of n-butyl alcohol or by the partial reduction of crotonaldehyde obtained from acetaldehyde.

$$2 \text{ MeCHO} \rightarrow \text{MeCHOHCH}_2\text{CHO} \rightarrow \text{MeCH}\!=\!\text{CHCHO} \rightarrow \text{MeCH}_2\text{CH}_2\text{CHO}$$

Isobutyraldehyde, methylpropanal, dimethylacetaldehyde, $(CH_3)_2CHCHO$, is obtained by oxidizing isobutyl alcohol or by dehydrogenating it. The presence of the fork in the alpha position presents slight complications in both cases, cutting down the yields as compared with the normal compound. It can also be obtained by hydrolysis (with rearrangement) of isobutylene bromide, Me_2CBrCH_2Br, obtained by direct bromination of t-BuOH. By a similar rearrangement, methallyl alcohol is hydrated by dilute sulfuric acid to give isobutyraldehyde (Shell).

$$CH_2=\underset{\underset{CH_3}{|}}{C}-CH_2OH \xrightarrow{H^+} (CH_3)_2\overset{+}{C}-CH_2OH \xrightarrow{H\sim}$$

$$Me_2CH-\overset{+}{C}HOH \rightarrow Me_2CH-CHO + H^+$$

The reactions of isobutyraldehyde are like those of its lower homologs except that the forked chain introduces variations. Thus it is possible to oxidize the tertiary H rather than the aldehyde group, by ceric sulfate, $Ce(SO_4)_2$, to give $Me_2C(OH)CHO$.[54] It gives an aldol condensation with formaldehyde.

$$Me_2CHCHO + H_2CO \rightarrow HOCH_2CMe_2CHO$$

An excess of formaldehyde gives 2,2-dimethyl-1,3-propanediol. Isobutyraldehyde with concentrations of alkali which would give aldehyde resins with most aldehydes undergoes an aldol condensation to give $Me_2CHCHOHC(Me)_2CHO$. This can undergo a Cannizzaro reaction with excess of base to form the corresponding primary alcohol and a salt of isobutyric acid.[55] Reactions of isobutyraldehyde with primary Grignard reagents give the expected products but more complex Grignard reagents give a side reaction which is not noticeable with simpler ones. This consists in the *reduction* of a portion of the aldehyde to the primary alcohol. Thus

$$2 \text{ Me}_2\text{CHCHO} + 2 \text{ Me}_3\text{CMgCl}$$
$$\rightarrow \text{Me}_2\text{CHCHOHCMe}_3 + \text{Me}_2\text{CHCH}_2\text{OH} + \text{Me}_2\text{C}=\text{CH}_2$$

The yields of the normal and abnormal products are about 40 and 20%.[56] With n-butyraldehyde the yield of normal product becomes 60%.

[54] Conant, Aston. *J. Am. Chem. Soc.* 50, 2783 (1928).
[55] *Ann. Rep. Chem. Soc.* (London) 1927, 92.
[56] Conant, Blatt. *J. Am. Chem. Soc.* 51, 1227 (1929).

Aldehydes RR'CHCHO can be made by treating the ketone RR'CO with ClCH$_2$CO$_2$Et and solid NaOEt. The intermediate glycidic acid loses CO$_2$ and gives the aldehyde.[57]

Methylethylacetaldehyde, 2-Me-butanal, MeEtCHCHO, reacts with alcoholic KOH to give MeEtCHCHOHC(Me)(Et)CH$_2$OH and MeEtCHCO$_2$K. The first is the result of an aldol condensation followed by reduction of the aldehyde group at the expense of a mole of the initial aldehyde. This type of reduction is like that in fermentation processes (p. 107).

n-Valeraldehyde, pentanal, CH$_3$(CH$_2$)$_3$CHO, and isovaleraldehyde, 3-methylbutanal, (CH$_3$)$_2$CHCH$_2$CHO, are made from the corresponding primary alcohols in the usual way. The oxidation of the corresponding primary alcohol offers complications because of the ease of oxidation of the aldehyde which stays in the reaction medium because of its high boiling point. Use of insufficient oxidizing agent merely leads to ester formation, 2 RCH$_2$OH → RCO$_2$CH$_2$R. Dehydrogenation with copper catalysts gives better results but complications are caused by the temperatures required. Before n-amyl alcohol was available, n-valeraldehyde could be made by the destructive distillation of a mixture of calcium n-valerate, made from n-butyl bromide through the cyanide, and calcium formate.

$$(BuCO_2)_2Ca + (HCO_2)_2Ca \rightarrow 2\ BuCHO + 2\ CaCO_3$$

Other products would be dibutyl ketone and formaldehyde. A similar result is obtained by passing the vapors of n-valeric acid and formic acid over heated manganous oxide as a catalyst. This changes to the salts which decompose as do the calcium salts. At the temperature used the manganous carbonate decomposes as fast as formed to give MnO again. TiO$_2$ has also been used.

Trimethylacetaldehyde, pivalic aldehyde, neovaleraldehyde, 2,2-Me$_2$-propanal, (CH$_3$)$_3$CCHO, can be made by oxidizing or dehydrogenating the corresponding alcohol Me$_3$CCH$_2$OH. Since this does not involve breaking the C−O linkage in neopentyl alcohol, no rearrangement takes place. A better method of preparation is from a tertiary butyl Grignard reagent and an excess of ethyl or methyl formate at a low temperature.

$$Me_3CMgCl + HCO_2R \rightarrow Me_3C-CH(OMgCl)OR \rightarrow Me_3CCHO + ROMgCl$$

Although an aldehyde normally reacts with a Grignard reagent, this is prevented in this case by the following factors:

(1) The low temperature retards the decomposition of the first addition product. (2) The excess of formate is in competition with the less active aldehyde. (3) The highly branched Grignard reagent reacts more slowly with the highly branched aldehyde. Any of the former which does react, reduces the aldehyde to alcohol instead of adding to it.

Trimethylacetaldehyde is also formed in small amounts from the hydrolysis of trimethylethylene dibromide (by ~ of Me).

[57] *Ann. Rep. Chem. Soc.* (London) **1904,** 62.

Since neovaleraldehyde has no alpha hydrogen atoms it fails to give many of the ordinary aldehyde reactions such as resin formation and aldol condensation with carbonyl groups. It gives the Cannizzaro reaction with alkalies. It gives the ordinary carbonyl reactions except that it reacts smoothly only with normal Grignard reagents. With branched Grignard reagents it is reduced to the alcohol. With light of about 3000 A it gives CO and isobutane.[58]

Me_3CCHO and isovaleraldehyde have been isolated from the "ketone oil" from wood distillation.[59]

Higher aldehydes have similar preparations and reactions to the ones described with like groupings around the alpha carbon. Many unusual aldehydes can be prepared in excellent yields by the Oxo reaction of an olefin and CO[60] (cf. diisobutylene p. 46).

Heptaldehyde, oenanthol, $CH_3(CH_2)_5CHO$, is formed in about 60% yield by the destructive distillation of castor oil under reduced pressure.[61] The catalytic hydrogenation of heptaldehyde gives heptyl alcohol, n-heptane, n-hexane and other products depending on the catalyst and the conditions.

Caprylic aldehyde (C_8) and **pelargonic aldehyde** (C_9) are found in the essential oils of many plants.

Many *higher aldehydes*, including the C_6, C_9, C_{12}, C_{14}, C_{16} and C_{17} members, have been made by heating the lactides from α-hydroxy acids.[62]

$$2 \ RCHOHCO_2H \rightarrow \underset{\displaystyle O-CO-CHR}{RCH \overset{\displaystyle CO-O}{\diagdown \diagup}} \rightarrow 2 \ CO + 2 \ RCHO$$

Grignard reagents with ortho formates give excellent yields of acetals which are readily hydrolyzed to the aldehydes.[63]

α-Hydroxy acids are also converted to aldehydes by treatment with PbO_2 and sulfuric acid.

α-Amino acids with NaOCl give N-chloro amino acids which lose HCl to form imino acids. Hydrolysis of the latter gives aldehydes.

α-Bromo acid amides in the Hofmann reaction (Br_2 and KOH) give aldehydes. Thus stearic acid is readily converted to margaric aldehyde, $C_{16}H_{33}CHO$.[64] This resembles the degradation of gluconic amide to arabinose. In these cases the rearrangement step gives RCHBrNCO and RCH(OH)NCO which on hydrolysis give compounds having a Br or an OH on the same carbon as the NH_2 group. These form aldehydes readily.

More difficultly obtainable aldehydes may be made by the action of an

[58] Conant, Webb, Mendum. *J. Am. Chem. Soc.* **51,** 1250 (1929).
[59] *Ann. Rep. Chem. Soc.* (London) **1923,** 67.
[60] Adkins, Kroek. *J. Am. Chem. Soc.* **70,** 383 (1948).
[61] Jones, Pyman. *J. Chem. Soc.* **127,** 2597 (1925).
[62] Blaise. *Ann. Rep. Chem. Soc.* (London) **1904,** 62.
[63] Bachmann. "Org. Syntheses," Coll. Vol. II, 1943. p. 323.
[64] *Ann. Rep. Chem. Soc.* (London) **1922,** 69.

acid chloride with hydrogen and a palladium catalyst poisoned by sulfurized quinoline.[65-67]

Aldehydes and ketones react with $ClCH_2CO_2Et$ and NaOEt to give ethylene oxidic esters (glycidic esters)

$$RCH\text{---}CHCO_2Et \quad \text{and} \quad RR'C\text{------}CHCO_2Et$$
$$\diagdown O \diagup \qquad\qquad \diagdown O \diagup$$

The free acids lose CO_2 and rearrange to give RCH_2COR and $RR'CH_2COR$.[68, 69]

n-Octylic, myristic, and stearic aldehydes have been obtained in nearly quantitative yields from the corresponding nitriles.[70]

$$RCN \xrightarrow{HCl} \xrightarrow{SnCl_2} RCH{=}NH.HCl \rightarrow RCHO$$

Ether containing a minute trace of water is used.

The normal 8–10 carbon aldehydes have been obtained from the Jeffrey pine.[71]

Some aldehydes are obtainable by decomposition of ozonides of olefinic compounds.[72] Thus oleic ester gives pelargonic aldehyde and the half aldehyde of azelaic acid.

Higher aldehydes give the reactions of the lower ones according to their structures as RCH_2CHO, $RR'CHCHO$, and $RR'R''CCHO$. They all give the Angeli-Rimini hydroxamic acid reaction (p. 190).

Aldehydes and ketones can be converted to the corresponding hydrocarbons by heating their hydrazones, phenylhydrazones or semicarbazones with dry NaOEt.[73] The same result can be achieved by heating the carbonyl compound with hydrazine sulfate and excess NaOEt.[74]

Chloral, trichloroethanal, Cl_3CCHO, b. 97°, is obtained by the thermal chlorination of ethyl alcohol or acetaldehyde with chlorine. It differs from most carbonyl compounds in adding water and alcohols to its carbonyl group to form stable compounds, chloral hydrate, $Cl_3CCH(OH)_2$, m. 57°, and chloral alcoholate, $Cl_3CCH(OH)OC_2H_5$, a hemi-acetal. It also forms acetals, $Cl_3CCH(OR)_2$. With glycols it gives cyclic acetals readily.

Chloral gives the usual aldehyde reactions with oxidizing agents, ammonia, bisulfites, hydrocyanic acid, acetic anhydride, etc.

[65] Rosenmund. *Ann. Rep. Chem. Soc.* (London) 1918, 52; 1922, 68; 1928, 74.
[66] Glattfeld, Straitff. *J. Am. Chem. Soc.* 60, 1384 (1938).
[67] Mosettig. "Org. Reactions," Vol. IV, 1948. p. 362.
[68] Erlenmeyer. *Ber.* 13, 307 (1880).
[69] Yarnall, Wallis. *J. Org. Chem.* 4, 270 (1939).
[70] Stephen. *Ann. Rep. Chem. Soc.* (London) 1925, 70.
[71] Foote. *J. Am. Pharm. Assoc.* 18, 350 (1929).
[72] Harries. *Ber.* 31, 1933 (1903).
[73] Wolff. *Ann. Rep. Chem. Soc.* (London) 1927, 121.
[74] "Org. Reactions," IV. p. 378.

With Al(OEt)₃ it gives Cl₃CCH₂OH.⁷⁵ Tribromoethanol, Br₃CCH₂OH (Avertin, an anesthetic (NNR)), is made similarly from *bromal* and aluminum isopropoxide.

Chloral hydrate reacts with hydroxy acids in presence of sulfuric acid to give chloralides.⁷⁶ Lactic acid gives MeCH−O−CHCCl₃. *Chloralide* itself

$$\text{MeCH} - \text{O} - \text{CHCCl}_3$$
$$\mspace \underset{\text{CO} \longrightarrow \text{O}}{}$$

has Cl₃C in place of the Me group, m. 115°, b. 273°. It is formed by an unusual reaction with oleum, involving 3 mols of chloral one of which contributes only the CO to hold the other two together. Chloralide, with KOH, gives 2 CHCl₃ and HCO₂K, and with EtOH at 150° gives chloral hemiacetal and ethyl Cl₃-lactate.

Chloral hydrate reacts with salicylic acid even without a condensing agent to give a 6-ring analog of the chloralides.

Chloral gives the *haloform reaction* (p. 90).

$$\text{Cl}_3\text{CCHO} + \text{NaOH} \rightarrow \text{CHCl}_3 + \text{HCO}_2\text{Na}$$

The most important reaction of chloral is with chlorobenzene in presence of sulfuric acid or other suitable condensing agent to give 1,1,1-Cl₃-2,2-bis-*p*-Cl-phenylethane, the important insecticide *DDT* which made World War II different by blocking the pandemic of louse-borne typhus fever which followed all earlier wars. It also did its share in cutting the incidence of tropical malaria by controlling *Anopheles* larvae.

Butylchloral, 2,2,3-Cl₃-butanal, CH₃CHClCCl₂CHO, is obtained as the hydrate from the reaction mixture from dry paraldehyde and chlorine.

$$\text{CH}_3\text{CHO} \rightarrow \text{ClCH}_2\text{CHO} \rightarrow [\text{CH}_3\text{CHOHCHClCHO}]$$
$$\rightarrow [\text{CH}_3\text{CH}=\text{CClCHO}] \rightarrow \text{CH}_3\text{CHClCCl}_2\text{CHO}$$

The hydrate is used as a sedative (NNR) and as a readily available chemical intermediate. It is better prepared by adding dry HCl to crotonaldehyde at 10°, followed by treatment with chlorine at 50° to give the 2,3-Cl₂ compound. Steam distillation gives alpha-chlorocrotonaldehyde which readily adds chlorine to give the desired product.

$$\text{MeCH}=\text{CHCHO} \rightarrow \text{MeCHClCH}_2\text{CHO} \rightarrow \text{MeCHClCHClCHO}$$
$$\rightarrow \text{MeCH}=\text{CClCHO} \rightarrow \text{MeCHClCHCl}_2\text{CHO}$$

B. Saturated Ketones

Aliphatic ketones, R₂CO and RR′CO, are named according to the two organic groups. The simple ketones are also named from the acids which can be used to prepare them by the thermal decomposition of the calcium salts.

---

⁷⁵ *Ann. Rep. Chem. Soc.* (London) **1925,** 72.
⁷⁶ *ibid.* **1927,** 84.

Thus:

Acid	Ketone	
Acetic	acetone	Me_2CO
Propionic	propionone	Et_2CO
Butyric	butyrone	Pr_2CO
Isobutyric	isobutyrone	$(Me_2CH)_2CO$
Pivalic	pivalone	$(Me_3C)_2CO$
Stearic	stearone	$(C_{17}H_{35})_2CO$

They have much the same reactions as the aldehydes, namely, those of the carbonyl group and of alpha hydrogen atoms. The carbonyl group is less reactive than in the aldehydes. This loss of activity increases with the size and, especially, with the branching of the alkyl groups. The absence of the H in direct attachment to the carbonyl group causes marked differences in properties. Thus ketones do not polymerize, are not readily oxidized and do not give the Angeli-Rimini reaction (p. 190).

Acetone, propanone, $(CH_3)_2C=O$, b. 56.3°, is contained in pyroligneous acid in small amounts. This supply soon proved inadequate and more was obtained by heating the crude calcium acetate obtained from the acid. $(MeCO_2)_2Ca \rightarrow CaCO_3 + Me_2CO$. Heating acetic anhydride with CaC_2 gives acetone more readily than heating calcium acetate.[77] During World War I the demand for acetone for the manufacture of explosives and for "dopes" for cloth airplane wings increased so that new sources were necessary. Acetaldehyde made from acetylene was oxidized catalytically to glacial acetic acid which was then converted to acetone by passing over hot lime, or better, $MnO.2\ MeCO_2H \rightarrow Me_2CO + CO_2 + H_2O$. Thorium oxide, ThO_2, at 400° converts acids[78, 79] and acid anhydrides to ketones. Thoria aerogel only requires 310°. Iron filings at 600° give a quantitative yield of acetone.[80] Fe_2O_3 at 480° gives similar results.[81]

Another source developed to meet the war demand for acetone was the special fermentation of starch from corn to give acetone and n-butyl alcohol in the ratio 1:2.[82] This process is more used than ever but the acetone is now a by-product and molasses is used in place of corn. More recently still acetone has been produced commercially starting with propylene from cracked gases.[83] It can also be prepared from ethanol in about 80% yield at 400° over a tin oxide catalyst promoted by a rare earth oxide or an alkaline earth carbonate.[84, 85]

[77] *Ann. Rep. Chem. Soc.* (London) 1906, 81.
[78] *ibid.* 1912, 80; 1928, 79.
[79] Kistler, Swann, Appel. *Ind. Eng. Chem.* **26**, 388 (1934).
[80] Curme. *C. A.* **13**, 2882 (1919).
[81] *Ann. Rep. Chem. Soc.* (London) 1914, 113.
[82] Peterson, Fred. *Ind. Eng. Chem.* **24**, 237 (1932).
[83] Ellis. "Chemistry of Petroleum Derivatives." Reinhold, 1934. p. 404.
[84] Meerwein, Morschel. *C. A.* **27**, 1896 (1933).
[85] Office of Publication Board, no. 170, 4115. 1945.

The *Roka Process* (1928) involves a similar reaction produced by passing a preheated 2:1 mixture of steam and ethanol vapor at 400° over a sponge iron-iron oxide catalyst.

$$2\ CH_3CH_2OH + H_2O \rightarrow CH_3COCH_3 + 4\ H_2 + CO_2$$

The exit gases contain only small amounts of CH_4 and unsaturates. The process can be used for converting higher alcohols RCH_2OH to R_2CO.

Acetone is formed in small amounts in animal metabolism by the decomposition of acetoacetic acid, $CH_3COCH_2CO_2H$, formed by the "beta oxidation" of fats. In diabetics the amounts of this acid and acetone increase greatly.

The reactions of acetone are largely those due to the activity of alpha H atoms and the carbonyl group. In these respects its reactions resemble closely those of the aldehydes.

1. Oxidizing agents do not act readily because the carbon attached to oxygen does not also have an H attached to it. That is, it does not contain

the $-\overset{\displaystyle H}{\underset{\displaystyle |}{C}}-O-$ combination. On vigorous oxidation, the carbon chain is broken

with the formation of smaller acids. The oxidation probably consists in the addition of 2 $-OH$ groups to the enol form of the ketone, followed by ordinary oxidation processes. The splitting of a ketone into simpler acids is valuable in determining the structures of higher ketones. Thus a ketone which gives C_2 and C_9 acids on oxidation is a methyl nonyl ketone. The structure of the nonyl group is determined by identification of the C_9 acid.

Ketones do not reduce Fehling's solution nor ammoniacal silver solution.

Oxidation with selenium dioxide converts an alpha CH_2 to CO forming pyruvic aldehyde, CH_3COCHO. This process works with any compound containing the grouping $COCH_2$ (p. 195).

2. Reducing agents change acetone to isopropyl alcohol. This process was formerly of commercial importance. Most reducing agents give as a by-product a bimolecular reduction product, pinacol ("pinacone"), b. 174° (p. 308). Amalgamated magnesium gives pinacol as the chief product.[86] This reaction was important in Germany during World War I because pinacol was the intermediate for "methyl rubber" (p. 62). A perhaps fanciful picture of the reaction with Mg involves the action of the latter with two molecules of acetone followed by a closing of a 5-membered ring through the free valences (single electrons) of the carbonyl carbon atoms.[87]

$$2\ Me_2C{=}O + Mg \rightarrow$$

[86] "Org. Syntheses."
[87] Couturier. *Ann. Rep. Chem. Soc.* (London) **1906**, 81.

The name *pinacol* is applied to any tetrasubstituted ethylene glycol. Under the influence of acids, pinacols undergo the *pinacolone rearrangement* so-called from the name of the substance obtained from pinacol itself.

$$Me_2C(OH)C(OH)Me_2 \rightarrow Me_3CCOMe + H_2O$$

Pinacolone (pinacoline)

Vigorous reduction of ketones by amalgamated zinc and conc. hydrochloric acid gives the corresponding hydrocarbon (Clemmensen).[88, 89] This process is best done by powdering pure zinc, treating it with Hg and then washing three times with hot HCl before adding the keto compound with more HCl. It is most successful with higher ketones. The reaction is hastened by the addition of one of the many available commercial wetting agents.

3. Pyrolysis of acetone vapor at 700° gives ketene, $H_2C = C = O$, and CH_4.[90]

4. Acetone does not polymerize like aldehydes. This is apparently because of the lesser activity of the carbonyl group in acetone. *Thioketones*, however, polymerize to such an extent that the monomeric forms are unknown. The conversion of acetone to its sulfur analog is accomplished by the addition of hydrogen sulfide to its carbonyl group, in presence of a trace of HCl.

$$Me_2C = O \rightarrow Me_2C(SH)OH \rightarrow H_2O + Me_2C = S \rightarrow$$

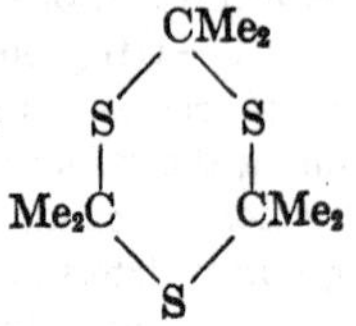

This complete polymerization of the thioacetone is further evidence of the failure to form a double bond between C and S (p. 151).

5. Acetone gives the *aldol condensation* readily. Contact with solid barium hydroxide converts it to diacetone alcohol, $MeCOCH_2C(OH)Me_2$, 2-Me-2-pentanol-4-one.

CaC_2 and CaH_2 also cause the aldol condensation with acetone and with methyl ketones in general but not with other ketones such as diethyl ketone.[91]

Treatment with stronger alkaline reagents which would resinify aldehydes containing alpha H atoms converts acetone to *isophorone*, 3,5,5-Me_3-2-cyclo-hexenone (p. 561).

With vigorous condensing agents such as HCl, one molecule of acetone condenses with two others and the product is dehydrated to give *phorone*, $Me_2C = CHCOCH = CMe_2$. With concentrated sulfuric acid or longer treatment with HCl, acetone forms mesitylene, 1,3,5-trimethylbenzene. This change probably involves the following steps:

[88] Clemmensen. *Ann. Rep. Chem. Soc.* (London) 1914, 123.
[89] "Org. Reactions," Vol. I.
[90] Hurd. "Org. Syntheses," Coll. Vol. I.
[91] *Ann. Rep. Chem. Soc* (London) ·1925, 73.

a. The diacetone alcohol is dehydrated to *mesityl oxide,* $Me_2C = CHCOMe$. If the COMe group acts with acetone, phorone is formed.

b. One of the beta Me groups in mesityl oxide can react with acetone to give $Me_2C = CHC(Me) = CHCOMe$.

c. The carbonyl group and the Me group at the other end of the conjugated system in this molecule are in the 1,6-position and can approach each other in space. They take part in an intermolecular aldol condensation. Loss of water gives mesitylene

In view of the varied possibilities at each step, it is not surprising that the yield of mesitylene is small (30%).[92]

Acetone also gives aldol condensations with other ketones and with aldehydes. In the latter case, the aldehyde supplies the more reactive carbonyl group and the acetone supplies the $\alpha - H$ atom. Thus a mixture of acetaldehyde and acetone treated with dilute bases gives aldol and $MeCHOHCH_2COMe$ but no $Me_2C(OH)CH_2CHO$. Formaldehyde and acetone give 1-butanol-4-one and more complex products due to the reaction of more $\alpha - H$ atoms with CH_2O. Under ordinary conditions 90% of the product consists of resins. Under carefully controlled conditions di-hydroxymethyl-tetrahydropyrone is a product. This is formed by inner ether formation between OH groups in the 1,5-position in sym-tetra-hydroxymethyl-acetone.

In contrast to acetone, Me Et ketone and Et_2 ketone, with formaldehyde, give 5% and less than 1% resin respectively.

6. One of the best-known "aldol" condensations of acetone is that with chloroform in presence of a solid base to give *Chloretone* (p. 91) (*Chlorobutanol, Chlorobutol* NNR).

[92] Ipatieff, Dolgov, Volnov. *Ber* **63B, 3072** (1930).

7. The Claisen condensation of ketones with esters in presence of NaOEt is another form of the aldol condensation.

$$Me_2CO + CH_3CO_2Et + EtONa \rightarrow EtOH +$$

$$\left[\begin{array}{c} ONa \\ | \\ CH_3C \!\!-\!\!\!-\!\!\!-\!\! CH_2COMe \\ | \\ OEt \end{array} \right] \rightarrow EtONa + CH_3COCH_2COMe$$
$$\text{acetylacetone}$$

8. A reaction related to the aldol condensation is that of a ketone with chloroacetic ester in presence of NaOEt to give a glycidic ester. The latter decomposes readily to give a disubstituted acetaldehyde.

$$Me_2CO + ClCH_2CO_2Et \rightarrow Me_2C\!\!-\!\!CHCO_2Et \rightarrow$$
$$\qquad\qquad\qquad\qquad\quad | \quad |$$
$$\qquad\qquad\qquad\qquad OH \ Cl$$

$$Me_2C\!\!-\!\!\!-\!\!\!-\!\!CHCO_2Et \rightarrow Me_2CHCHO$$
$$\qquad\quad \backslash \ O \ /$$

The last step is a common rearrangement of ethylene oxides to related carbonyl compounds.

9. A peculiar condensation involving $\alpha-$H atoms is that with R_2NCH_2OH, from formaldehyde and a secondary amine. From one to six of the H atoms in acetone can be replaced by the group $-CH_2NR_2$ (Mannich).[93]

10. Although water and alcohols do not add to the carbonyl group of acetone to form a hydrate or *hemiacetals* it is possible to obtain *acetals* by heating with orthoformates.

$$Me_2CO + HC(OR)_3 \rightarrow Me_2C(OR)_2 + HCO_2R$$

This reaction is deeply influenced by the size and branching of the alkyl groups. While acetone gives a 90% yield, Me *t*-Bu ketone gives only a 12% yield.[94]

Cyclic acetals, dioxolanes, are readily formed with compounds having hydroxyls on two adjacent carbons with unrestricted rotation (pp. 312, 318, 481).

11. Mercaptans add readily to ketones in the presence of dilute acids, to give *hemimercaptols*, $Me_2C(SR)OH$. These react readily with more RSH to form *mercaptols*, $Me_2C(SR)_2$. Whereas acetals of ketones are readily split by dilute acid, the mercaptols are stable to acid but can be split by Hg salts. Oxidation changes the mercaptols to disulfones, soporifics of the Sulfonal type (p. 150).

12. While acetone does not form acetals with ordinary alcohols, it forms *cyclic acetals* with 1,2-glycols such as ethylene glycol, glycerol, and the various

[93] "Org. Reactions," Vol. I.
[94] Adkins. *Ann. Rep. Chem. Soc.* (London) 1928, 75.

carbohydrates. This is an example of the stabilizing effect of ring formation of 5 or 6 atoms. Thus with an ordinary alcohol, the equilibrium is so far on the side of the acetone that there is no opportunity to react with another molecule of alcohol to give the stable acetal.

$$Me_2C=O + ROH \rightleftharpoons Me_2C(OH)OR$$

On the other hand, when a glycol is used, the second OH may assume a favorable position in respect to the OH of the hemiacetal. Water is lost and the stable cyclic acetal results.

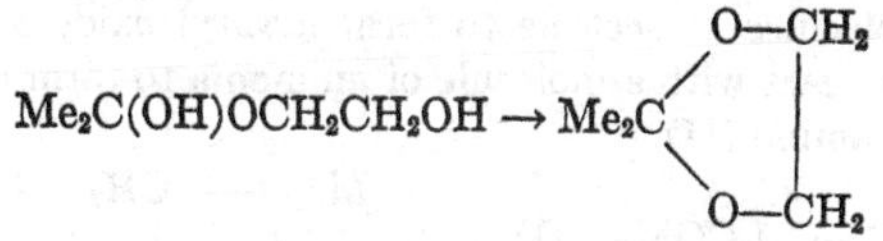

$$Me_2C(OH)OCH_2CH_2OH \rightarrow Me_2C \underset{O-CH_2}{\overset{O-CH_2}{\big\langle}}$$

A related reaction is that with α-hydroxy acid amides to give keto-tetrahydro-oxazoles.

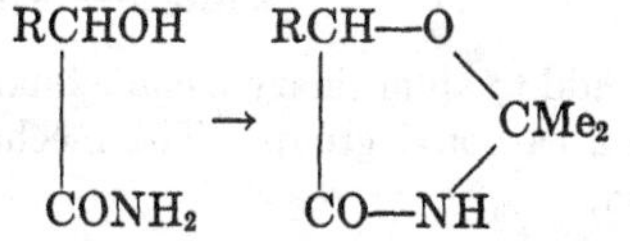

$$\underset{CONH_2}{\overset{RCHOH}{|}} \rightarrow \underset{CO-NH}{\overset{RCH-O}{|}} \diagdown CMe_2$$

13. Sodium bisulfite forms a crystalline addition compound with acetone. This can be recrystallized and then treated with sodium carbonate to regenerate the acetone. Methyl ketones, MeCOR, form such compounds with decreasing ease as the other alkyl group increases in length, especially if the alkyl group is branched. If both groups in the ketone are ethyl or higher the reaction with sodium bisulfite is too slow to be of any practical value.[95] Acetone slowly restores the color to Schiff's reagent (p. 191). Higher ketones act more slowly or not at all.

14. Hydrocyanic acid gives a *cyanohydrin*, which on hydrolysis gives alpha hydroxyisobutyric acid, and on alcoholysis gives the corresponding ethyl ester, which is used as a solvent. $Me_2C(OH)CN \rightarrow Me_2C(OH)CO_2Et$. Dehydration gives ethyl methacrylate, the intermediate for important transparent plastics (Plexiglas, Lucite). Ketones react with NH_4 salts and cyanides to give $RR'C(NH_2)CN$ which on hydrolysis form the corresponding amino acids.

15. Grignard reagents react with acetone to give tertiary alcohols of the type $(CH_3)_2C(OH)R$. Complications appear here because of the basic nature of the Grignard reagent, and of the primary product, RMe_2COMgX. Both cause the acetone to undergo the aldol condensation. Thus the best way to make an alcohol, RMe_2COH, is not from acetone but from RCO_2Et and an excess of MeMgCl. A good example is the preparation of dimethylisobutyl-carbinol.[96] Ethyl isovalerate with excess MeMgBr gives an 86% yield whereas

[95] Stewart. *J. Chem. Soc.* **87**, 185 (1905).
[96] de Graef. *Bull. soc. chim. Belg.* **40**, 315 (1931).

acetone and i-BuMgBr gives only a 40% gield with the following by-products: mesityl oxide, phorone, isophorone, di-isobutyl, diacetone alcohol and the compounds $Me_2C=CHC(Me)=CHCMe_2$, $Me_2C(OH)CH_2COCH=CMe_2$, and $Me_2C(OH)CH_2C(OH)(Me)CH_2CMe_2$.

Ketones with zinc and α-bromo esters give beta hydroxy esters (Reformatzky) (p. 344). The simplest case is that of acetone and bromoacetic ester giving $Me_2C(OH)CH_2CO_2Et$.

16. With ammonia, acetone gives even more complicated products than do aldehydes. The ammonia first acts as a base inducing aldol condensations of two and three molecules of acetone to form mesityl oxide and phorone, respectively. These react with a molecule of ammonia to form diacetone amine (I) and triacetone amine (II).

$$
\begin{array}{cc}
Me_2C\text{---}CH_2COMe \quad (I) & \begin{array}{cc} Me_2C\text{------}CH_2 \\ | \quad\quad | \\ NH \quad CO \quad (II) \\ | \quad\quad | \\ Me_2C\text{------}CH_2 \end{array} \\
| & \\
NH_2 &
\end{array}
$$

While ammonia does not add to an ordinary olefinic linkage, it adds when that linkage is alpha beta to a carbonyl group. The mechanism is 1:4 addition followed by "ketonization,"

$$C=C\text{---}C=O \xrightarrow{NH_3} H_2N\text{---}C\text{---}C=C\text{---}OH \rightarrow H_2N\text{---}C\text{---}CH\text{---}C=O.$$

In the case of phorone, the addition of NH_3 would give

$$Me_2C(NH_2)CH_2COCH=CMe_2$$

in which the NH_2 and the farther olefinic carbon are in the 1:6 position. Intramolecular addition takes place to form triacetone amine. Triacetonamine is an intermediate for various drugs. Thus the benzoate of the corresponding secondary alcohol is Beta Eucaine, a cocaine substitute.

17. Hydroxylamine, phenylhydrazine and semicarbazide act with acetone as with aldehydes, giving oximes, phenylhydrazones and semicarbazones.

Acetone oxime, acetoxime, $Me_2C=NOH$, m. 69°.
Acetone 2,4-dinitrophenylhydrazone, m. 128°.

18. Nitrous acid from RONO + HCl reacts with acetone to give isonitrosoacetone. The reaction probably resembles an aldol condensation with the $N=O$ group serving the part of the carbonyl group.

$$Me_2CO + HON=O \rightarrow MeCOCH_2N(OH)_2 \rightarrow$$
$$[MeCOCH_2N=O] \rightarrow MeCOCH=NOH$$

As has been seen, the grouping $=CHNO$ changes to $=C=NOH$. It is interesting that formyl acetone, $HCOCH_2COMe$, exists only as the Na derivative of the enol, $NaOCH=CHCOMe$. Acidification gives complex products including triacetylbenzene. Isonitrosoacetone is the mono oxime of pyruvic aldehyde, MeCOCHO. It gives that substance and hydroxylamine on hydrolysis with dilute acids.

19. Sodium and sodamide form sodium derivatives of the enol form of acetone, $CH_3C(ONa)=CH_2$.[97] When sodium is used, the hydrogen evolved reduces some of the acetone to isopropyl alcohol. This can be prevented by adding CS_2.[98] Moreover there is a tendency to form the disodium derivative of pinacol.

$$2\ Me_2C{=}O \rightarrow 2\ Me_2C\!\!\!\begin{array}{c}\diagup\\ \diagdown\end{array}\!\!\!\begin{array}{c}\\ ONa\end{array} \rightarrow \overset{\overset{\displaystyle ONa}{|}}{Me_2C}\!\!-\!\!\!-\!\!\overset{\overset{\displaystyle ONa}{|}}{CMe_2}$$

Sodamide is the better reagent for making the sodium derivative of acetone. This sodium derivative reacts with alkyl halides or alkyl sulfates to give ketones of the form CH_3COCH_2R.[99] If the treatment with $NaNH_2$ and alkyl halides is repeated, the final product will be R_3CCOCR_3, all the alpha H atoms having been replaced. With larger or forked groups the reaction fails before this point is reached. When a ketone containing no alpha H is heated with sodamide and the products are treated with water, the carbon chain is split.

$$R_3C-CO-CR_3 \rightarrow R_3CCONH_2 + R_3CH$$

The reaction of sodium acetone compound with RX might be expected to give $CH_3C(OR)=CH_2$, an unsaturated ether. The fact that CH_3COCH_2R is obtained has been used as an argument that the sodium compound has the structure CH_3COCH_2Na, the metal being attached to carbon. Both these conceptions are at fault because they are based on the assumption that the reaction is a simple metathesis like the action of solutions of sodium chloride and silver nitrate. The negative ion of the sodium-acetone may be represented by the following

$$\left[\begin{array}{c} Me \\ \overset{..}{C}:\overset{..}{O}: \\ \overset{::}{}\ \overset{..}{} \\ H:\overset{..}{C}:H \end{array}\right]^{-} \longleftrightarrow \left[\begin{array}{c} Me \\ \overset{..}{C}::\overset{..}{O} \\ \ \ \overset{..}{} \\ H:\overset{..}{C}:H \\ \overset{..}{} \end{array}\right]^{-}$$

A.   B.

In the course of the reaction the alkyl halide $R:\overset{..}{\underset{..}{X}}:$ parts with the halogen, the latter taking the electron octet. The R left with two electrons less than an octet is in a position to unite with a free electron pair on the O in A or on the

[97] *Ann. Rep. Chem. Soc.* (London) **1907**, 83.
[98] Wertheim. *J. Am. Chem. Soc.* **53**, 4037 (1931).
[99] *Ann. Rep. Chem. Soc.* (London) **1913**, 61.

C in B. The energy relations are such that a new $C-C$ linkage is formed instead of a $C-O$ linkage. This is the same type of process as that involved in the action of alkyl halides with such compounds as the sodium derivatives of malonic ester, acetoacetic ester, nitromethane, sulfurous acid, sulfinic acids, arsenious acid and arsinoxides (p. 152).

Acetone in contact with heavy water exchanges H atoms for deuterium atoms.[100]

$$\overset{\text{D}_2\text{O}}{CH_3COCH_3 \;\rightleftharpoons\; CH_3C(OH)=CH_2 \;\rightleftharpoons\; CH_3C(OD)=CH_2 \;\rightleftharpoons\; CH_3COCH_2D}$$

20. Halogens replace the alpha hydrogens in acetone much as in aldehydes.[101] This process presumably takes place through the enol form. The conception that enolization is a bimolecular rather than a monomolecular process (p. 202) is strengthened by the fact that the rate of halogenation of ketones in solution is independent of the concentration of halogen but is proportional to the hydrogen ion concentration. The electronic system in the case of acetone and its enol would be as follows:

$$\text{H}^+ + \;\overset{\text{H}}{\underset{\text{H}:\ddot{\text{O}}:}{\text{H}:\ddot{\text{C}}:\text{C}:\text{Me}}}\;\rightleftharpoons\; \left[\overset{\text{H}}{\underset{\underset{\text{H}}{\text{H}:\ddot{\text{O}}:}}{\text{H}:\ddot{\text{C}}:\text{C}:\text{Me}}}\right]\;\rightleftharpoons\; \text{H}^+ + \;\underset{\underset{\text{H}}{\ddot{\text{H}}\;\;:\ddot{\text{O}}:}}{\text{H}:\ddot{\text{C}}::\text{C}:\text{Me}}$$

Stabilized monochloroacetone, $MeCOCH_2Cl$, b. 119°, is available commercially (CC). The monohalogen acetones are severe lachrymators. Treatment with hypohalites or with halogens and a base gives a haloform, CHX_3. Thus acetone gives the iodoform test and is used in manufacturing iodoform.

$$\overset{\text{NaOH}}{Me_2CO \rightarrow MeCOCX_3 \longrightarrow MeCO_2Na + CHX_3}$$

21. Phosphorus pentachloride reacts with acetone to give a poor yield of "acetone dichloride," $CH_3CCl_2CH_3$. A large amount of $CH_3CCl=CH_2$ is obtained. Similar complications appear with higher ketones. PBr_5 replaces alpha H by Br probably because of its dissociation into Br_2 and PBr_3. This effect has been observed with the following ketones: Me_2, Me Et, Me *iso*Pr, Et *t*-Bu, di-*iso*Pr, *iso*Pr *t*-Bu and *iso*Pr *iso*Bu. Pinacolone, Me *t*-butyl ketone, gives some of the ketone dibromide but, in addition, the surprising products $Me_2CBrCOCH_2Me$ and $Me_2CBrCOCHBrMe$.[102, 103]

22. Sunlight causes several peculiar reactions with acetone (Ciamician and Silber).

[100] Halford, Anderson, Bates. *J. Am. Chem. Soc.* **56**, 491 (1934).
[101] Watson. *Chem. Revs.* **7**, 173 (1930).
[102] Favorsky et al. *J. Russ. Phys. Chem. Soc.* **44**, 1339 (1912).
[103] Favorsky. *J. prakt. Chem.* (2) **88**, 641 (1913).

a. Water gives methane and acetic acid.

b. Methanol gives isobutylene glycol, isopropyl alcohol and ethylene glycol.

c. Ethanol gives Me_3-ethylene glycol, 2,3-butandiol, and isopropyl alcohol.

d. Isopropyl alcohol gives pinacol (Me_4-ethylene glycol).

e. The photolysis of acetone to CO and ethane has been much studied because of its low quantum yield.[104]

23. Acetone condenses with reactive aromatic compounds in the presence of acids, bridging together two aromatic radicals by the group $(CH_3)_2C$. Thus:

$$2\ HO\langle\ \rangle + MeCOMe \rightarrow HO\langle\ \rangle\underset{Me}{\overset{Me}{\underset{|}{\overset{|}{C}}}}\langle\ \rangle OH$$

24. Acetone can be purified by conversion to the crystalline compound $NaI \cdot 3\ Me_2CO$. The purified solid is treated with water and the acetone is separated by fractional distillation and drying with anhydrous sodium sulfate. Acetone forms compounds with $CaCl_2$.[105]

25. Acetone can be readily identified by conversion to dibenzalacetone, $(C_6H_5CH=CH)_2CO$, m. 112°. It can be determined by ultraviolet spectrophotometry.[106]

A similar type reaction produces ethyl sodium acetone-oxalate,

$$CH_3COCH=C(ONa)COOC_2H_5,$$

which readily condenses to 1,3,5-hydroxy toluic acid (U.S.I.).

Methyl ethyl ketone, butanone, $CH_3COCH_2CH_3$, b. 79.6°, is obtained from pyroligneous acid in even smaller amounts than acetone. It is manufactured by catalytic dehydrogenation of 2-butanol from cracked gases.[107]

Its reactions are practically identical with those of acetone.

Vigorous oxidation gives mainly acetic acid. Very little propionic acid and CO_2 are obtained. This is probably because the enolization takes place mainly with the methylene group rather than with the methyl group.

$$\underset{\text{Very little}}{CH_2=C(OH)CH_2Me} \rightleftharpoons CH_3COCH_2Me \rightleftharpoons \underset{\text{Mainly}}{CH_3C(OH)=CHMe}$$

The last product formed before the breaking of the carbon chain is probably diacetyl, $CH_3COCOCH_3$. This is readily oxidized to acetic acid.

Of the five alpha hydrogen atoms, the two in the methylene group are generally more reactive with nitrous acid, in enolization as with $NaNH_2$, and

[104] Herr, Matheson, Walters. *J. Am. Chem. Soc.* **63**, 1464 (1941).
[105] *Ann. Rep. Chem. Soc.* (London) 1917, 68.
[106] Barthauer, Jones, Metler. *Ind. Eng. Chem., Anal. Ed.* **18**, 354 (1946).
[107] Ellis. "Chemistry of Petroleum Derivatives." Reinhold, 1934. p. 415.

with halogens. Thus

$$CH_3COCH_2CH_3 + HONO \rightarrow CH_3COC(=NOH)CH_3$$
$$CH_3COCH_2CH_3 + NaNH_2 + CH_3I \rightarrow CH_3COCH(CH_3)_2$$
$$CH_3COCH_2CH_3 + Br_2 \rightarrow CH_3COCHBrCH_3 \text{ (mainly)}$$

The product with nitrous acid is the half oxime of diacetyl. Treatment with hydroxylamine gives the di-oxime, $CH_3C(=NOH)C(=NOH)CH_3$, *dimethylglyoxime*, an important analytical reagent for nickel.

In the aldol condensation with HCl, the methylene group is reactive giving mainly $MeEtC=C(Me)COMe$ while with NaOMe, the alpha methyl group reacts to give $MeEtC=CHCOEt$.[108,109] Butyl Grignard reagent acts like NaOMe giving a good yield of the product from the action of the methyl group. CaH_2 and CaC_2 cause condensation of the CH_3 group. This is in keeping with the fact that they cause aldol condensations of methyl ketones but are without effect on diethyl ketone.[110]

Sodium bisulfite adds, but less rapidly than to acetone.[110a] Hydrocyanic acid adds normally. The reactions with ammonia and related compounds are like those of acetone. Phosphorus pentachloride gives an even poorer yield of the dichloride than with acetone. This is probably because of the greater tendency to split HX out with a secondary H atom than with a primary one, $CH_3CCl_2CH_2CH_3 \rightarrow CH_3CCl=CHCH_3$.

Bromination in water in presence of $CaCO_3$ gives two monobromides, MeCOCHBrMe, b. 134°, and $BrCH_2COEt$, b. 146°.

Methyl ethyl ketone condenses with normal aldehydes up to C_7 to give RCHOHCHMeCOMe, the CH_2 group entering the condensation. With a fork at the alpha carbon of the aldehyde as with isobutyraldehyde, MeEt-acetaldehyde, Et_2-acetaldehyde and EtBu-acetaldehyde the CH_3 group reacts giving $RRCHCHOHCH_2COEt$.[111]

Methyl *n*-propyl ketone, 2-pentanone, $CH_3COCH_2CH_2CH_3$, can be obtained in a variety of ways typical for the preparation of higher ketones.

1. By the oxidation of 2-pentanol with chromic acid mixture. Even though the alcohol is pure there is danger of some rearrangement under the influence of the acid to give some 3-pentanone.

2. The destructive distillation of a mixture of the calcium salts of acetic and butyric acids gives methyl propyl ketone along with acetone and dipropyl ketone. A similar result is obtained by passing the vapors of the acids over heated MnO.

[108] Colonge. *Bull. soc. chim.* (4) **49**, 426 (1931).
[109] Powell, Seymour. *J. Am. Chem. Soc.* **53**, 1049 (1931).
[110] *Ann. Rep. Chem. Soc.* (London) **1925**, 73.
[110a] Stewart. *J. Chem. Soc.* **87**, 185 (1905).
[111] Powell, Baldwin. *J. Am. Chem. Soc.* **58**, 1871 (1936).

3. From a *n*-propyl Grignard reagent and acetonitrile or acetamide:

$$MeCN + RMgX \rightarrow RMeC=NMgX \rightarrow RMeC=NH \rightarrow RMeC=O$$
$$MeCO(NH_2) + 2\ RMgX \rightarrow$$
$$RMeC(OMgX)(NHMgX) \rightarrow RMeC(OH)NH_2 \rightarrow RMeCO$$

4. From sodium acetoacetic ester and ethyl bromide, followed by treatment with *dilute* alkali and acid. The first steps of this process duplicate those of alkylating acetone by sodamide and alkyl halides. The last step depends on the instability of acids containing the grouping $-COCH_2CO_2H$, beta-keto acids, which lose CO_2 readily to give $-COCH_3$.

$$
\begin{array}{ccccccc}
MeC=O & MeCONa & MeCO & MeCO & & & MeCO \\
| & \| & | & | & & & | \\
CH_2 & \rightarrow & CH & \rightarrow & CHEt \rightarrow & CHEt \rightarrow CO_2 & + & CH_2 \\
| & & | & | & | & & | \\
CO_2R & & CO_2R & CO_2R & CO_2H & & Et
\end{array}
$$

By similar reactions, monosubstituted acetones, CH_3COCH_2R, may be prepared from alkyl halides of Class 1 (p. 74) and a few others such as isopropyl and isobutyl. After one alkyl group has been introduced, another can be put in by treatment with NaOEt and the proper alkyl halide. The second alkyl group to enter can usually belong only to Class 1.

(5) From sodium oxalylacetone formed from acetone and diethyl oxalate and NaOEt.[112]

$$Me_2CO + EtOCOCO_2Et \rightarrow MeCOCH_2COCO_2Et$$

This is converted to the sodium derivative, treated with ethyl halide and then split by bases and acid. This method is not limited to methyl ketones.

Methyl isopropyl ketone, 3-methyl-2-butanone, $CH_3COCH(CH_3)_2$, can be made as follows:

(1) By oxidation of methylisopropylcarbinol prepared by the Grignard reaction.

(2) From acetoacetic ester by converting it to the sodium derivative and treating with methyl iodide; converting again to the sodium derivative and again treating with methyl iodide; finally saponifying with dilute alkali and acidifying. The last step is

$$CH_3COC(CH_3)_2CO_2H \rightarrow CO_2 + CH_3COCH(CH_3)_2$$

(3) Best, by hydrolyzing and rearranging trimethylethylene dibromide or dichloride obtained directly from tertiary amyl alcohol and bromine or chlorine or from Me_3-ethylene (Sharples).

$$Me_2CBrCHBrMe + H_2O \rightarrow Me_2CHCOMe + 2\ HBr$$

The reactions of methyl isopropyl ketone are like those of the other methyl ketones except that they are complicated by the presence of the forked chain

[112] *Ann. Rep. Chem. Soc.* (London) 1913, 61.

and the tertiary H atom in the alpha position. The replacement of this H by Br or NO is very easy indeed. Thus slightly less than 1 mole of Br_2 at $0°$ gives a 90% yield of $MeCOCBrMe_2$.[113] PBr_5 gives a similar result. Phosphorus pentachloride would probably give none of the dichloride. Because of the activity of the tertiary H atom, the product would probably be

$$MeCCl = CMe_2$$

Ethyl nitrite and HCl give the bimolecular form of the tertiary nitroso compound, $[MeCOC(NO)Me_2]_2$. This is colorless but on heating turns blue, due to the formation of the monomolecular true nitroso compound. Reduction gives first an azoxy compound, $R : N :: N : R$, and then hydrazine, H_2NNH_2,

$$: \overset{\cdot\cdot}{\underset{\cdot\cdot}{O}} :$$

and Me isoPr ketone.[114] Nitrous acid reacts to give only $[MeCOC(NO)Me_2]_2$.

The tertiary H does not take part in aldol condensations, apparently because of the difficulty for a relatively large group to add such a "crowded" atom (*steric hindrance*). It gives an aldol condensation involving an alpha H atom in the methyl group

$$2\ Me_2CHCOCH_3 \rightarrow Me_2CHC(OH)(CH_3)CH_2COCHMe_2 \rightarrow$$
$$Me_2CHC(Me) = CHCOCHMe_2$$
$$\text{2,3,6-Me}_3\text{-3-hepten-5-one}$$

This change is also brought about by treatment with a tertiary butyl Grignard reagent. The action of methyl isopropyl ketone with *t*-butyl magnesium chloride is a good example of the influence of branching on an addition reaction. Depending on the branched structures of the Grignard reagent and the ketone, their action may give: (*a*) A tertiary alcohol by their union; (*b*) A secondary alcohol by reduction of the ketone; (*c*) A mesityl oxide homolog by aldol condensation and dehydration; (*d*) An MgX derivative of the enol form of the ketone with evolution of RH.[115]

Diethyl ketone, propionone, 3-pentanone, $CH_3CH_2COCH_2CH_3$, may be made as follows:

(1) By the destructive distillation of calcium propionate or by passing the vapors of propionic acid over hot MnO.

(2) By oxidation of 3-pentanol obtained by the action of an excess of ethyl Grignard reagent with an ester of formic acid.

(3) By the Roka reaction on 1-propanol and steam (p. 209).

Its reactions are typical ketone reactions except that its carbonyl group is less reactive than in simple methyl ketones. It is one of the few ketones, other than methyl ketones, which react even slowly with sodium bisulfite.[116]

[113] Favorski. *J. prakt. Chem.* (2) **88**, 641 (1913).
[114] Aston, Menard, Mayberry. *J. Am. Chem. Soc.* **54**, 1530 (1932)
[115] Conant, Blatt. *J. Am. Chem. Soc.* **51**, 1227 (1929).
[116] Stewart. *J. Chem. Soc.* **87**, 185 (1905).

A good example of the effect of structure on the addition reactions of the carbonyl group is given by the rate constants for semicarbazone formation which are 0.086 and 0.0086 for acetone and Et_2 ketone respectively. Me t-Bu ketone gives an even smaller value of 0.0010. With ethyl acetate and NaOEt at room temperature diethyl ketone gives MeCOCHMeCOEt, the expected product, but at higher temperatures MeCOCH$_2$COEt is also formed.[117] The latter indicates the formation of Me Et ketone in the mixture. This is due to the ordinary alkaline splitting of a 1,3-diketone, especially one substituted in the 2 position. This is familiar in substituted acetoacetic esters. The split can occur in the 1:2 or 2:3 position.

$$\underset{\substack{| \\ O}}{MeC}-\underset{\substack{| \\ Me}}{CH}-\underset{\substack{\| \\ O}}{CEt} \xrightarrow{\;NaOEt\;} MeCO_2Et + MeCH{=}C(ONa)Et$$

$$+ EtCO_2Et + MeCH{=}C(ONa)Me$$

The last product is the Na derivative of Me Et ketone.

Diethyl ketone with NaNH$_2$ and MeI gives Et isoPr ketone and di-isoPr ketone.[118]

Diethyl ketone, condensed with ethyl oxalate and EtONa, treated with RX and then split with bases gives ketones EtCOCHMeR.[119]

Methyl n-butyl ketone, 2-hexanone, $CH_3CO(CH_2)_3CH_3$, can be made (1) from sodium acetoacetic ester and n-propyl halide, (2) by oxidation of 2-hexanol, (3) from acetic and n-valeric acids, and (4) from n-butyl Grignard reagent and acetyl chloride. Its reactions are those of methyl ketones retarded by the increased size of the molecule.

Ethyl n-propyl ketone, 3-hexanone, $CH_3CH_2COCH_2CH_2CH_3$, can be made by the usual methods except the acetoacetic ester synthesis. Its carbonyl group is even more sluggish than that in its methyl isomer.

Methyl isobutyl ketone, 4-methyl-2-pentanone, $CH_3COCH_2CH(CH_3)_2$, can be made by the usual reactions including the sodium acetoacetic ester synthesis although the action of isopropyl halides with that substance gives poorer yields than do the halides of Class 1. A better preparation is by a special method, the reduction of mesityl oxide by an acid reducing agent or by a catalytic method which does not act on the carbonyl group.

$$MeCOCH{=}CMe_2 \rightarrow MeCOCH_2CHMe_2$$

Another preparation would be the oxidation of the corresponding alcohol obtained by heating isopropyl alcohol with bases (Guerbet).

Ethyl isopropyl ketone, 2-methyl-3-pentanone, $CH_3CH_2COCH(CH_3)_2$, can be made from isobutyronitrile and ethyl Grignard reagent or by the oxidation

[117] *Ann. Rep. Chem. Soc.* (London) 1925, 73.
[118] Haller, Bauer. *Ann. Rep. Chem. Soc.* (London) 1913, 61.
[119] *Ann. Rep. Chem. Soc.* (London) 1913, 61.

of the corresponding secondary alcohol. The preparation from two acids, always unsatisfactory because of the three products, would be especially so because in this case both of the acids are rather expensive.

Treatment with nitrous acid gives the surprising result that the replacement of s- and t- H takes place in the ratio 4:1 giving $MeC(=NOH)COCHMe_2$ and $[MeCH_2COC(NO)Me_2]_2$. Reduction of the latter compound gives hydrazine, H_2NNH_2, indicating the linkage of $N-N$ in the bimolecular form of the nitroso ketone.[120]

A peculiar formation of ethyl isopropyl ketone is from n- or iso-pentane and CO in presence of $AlCl_3$.

Methyl _sec_-butyl ketone, 3-methyl-2-pentanone, $CH_3COCH(CH_3)CH_2CH_3$, can be made by the acetoacetic ester synthesis by use of methyl and ethyl halides in succession. The ethyl group should be introduced first because it is easier to purify $MeCOCHEtCO_2R$ than $MeCOCHMeCO_2R$.

The best preparation would probably be from a sec-butyl Grignard reagent and acetyl chloride in the presence of copper salts.

Methyl _t_-butyl ketone, pinacolone, "pinacoline," 3,3-Me_2-butanone, $CH_3COC(CH_3)_3$, b. 106°, is obtained by the rearrangement of pinacol by acids. A related preparation is by the warming of Me_4-ethylene dibromide with water. It can readily be made by the action of t-BuMgCl on acetyl chloride in the presence of cuprous chloride. If a smaller group is to be introduced in place of the Cl of an acid chloride, excess acid chloride, cuprous chloride and temperatures around 0° are necessary for good yields.

Pinacolone gives the expected methyl ketone reactions such as the haloform reaction to give Me_3-acetic acid and condensation with benzaldehyde to give benzalpinacolone, $C_6H_5CH=CHCOCMe_3$.

Methyl _n_-amyl ketone is obtained by the dehydration and partial reduction of the condensation product of acetone with n-butyraldehyde (CC).

$$PrCHO + CH_3COMe \rightarrow PrCHOHCH_2COMe \rightarrow$$
$$PrCH=CHCOMe \rightarrow PrCH_2CH_2COMe$$

Dipropyl ketone, butyrone, $(CH_3CH_2CH_2)_2CO$, is obtained in the usual ways and gives the reactions of a higher ketone. It can also be made from n-BuOH by the Roka Process (p. 209).

Diisopropyl ketone, isobutyrone, $(CH_3)_2CHCOCH(CH_3)_2$, is obtained by oxidation of the corresponding secondary carbinol which is an important by-product of synthetic methanol (p. 103). It can also be made from isobutyric acid and by the action of $NaNH_2$ and MeI with acetone, methyl ethyl ketone, diethyl ketone and methyl isopropyl ketone.

Diisopropyl ketone was used in a novel synthesis of triptane during World War II. The corresponding chloro-ketone was converted with KOH under

[120] Aston, Menard, Mayberry. _J. Am. Chem. Soc._ **54**, 1530 (1932).

anhydrous conditions to dimethylisopropylacetic acid which was reduced with a nickel-tungsten sulfide catalyst to triptane in an overall yield of 40%.[121]

Treatment with PCl_5 replaces one of the α-tertiary H atoms instead of giving the expected dichloride.[122]

Di-isobutyl ketone is available by the partial hydrogenation of phorone made from acetone (CC).

Hexamethylacetone, pivalone, $Me_3CCOCMe_3$, b. 151°, is obtained by exhaustive methylation of methyl isopropyl ketone or diisopropyl ketone with $NaNH_2$ and MeI or Me_2SO_4.[123] It can be made in excellent yield from pivalyl chloride, Me_3CCOCl and t-BuMgCl in the presence of cuprous chloride.[124,124a] It gives no derivatives with hydroxylamine, hydrazines, etc. It is best identified by treatment with MeMgX to give Me_7-isopropyl alcohol, m. 42°.

Methyl neopentyl ketone, $MeCOCH_2CMe_3$, b. 124°, is available from the oxidation of the lower boiling diisobutylene (p. 46). Its reactions are those of an ordinary methyl ketone.

Higher ketones have been made in large numbers by the reactions outlined above and by other special reactions.

Higher methyl ketones may be obtained by the destructive distillation of the barium salt of a higher acid with Ba acetate.[125] In this way Me C_{18} and Me C_{19} ketones have been made. It will be recalled that margaric acid (C_{17}) for intarvin manufacture is made from the Me C_{17} ketone prepared in a similar way (Krafft) (p. 255). Thus methyl heneicosyl ketone, $MeCOC_{21}H_{43}$, has been made from behenic acid obtained by the hydrogenation of erucic acid.[126]

Ketones of the type $RCOCH_2R$ can be made from *acyloins*, RCOCHOHR, obtained by the action of Na on esters in ether solution. Reduction of the acyloin by Na and absolute alcohol gives RCHOHCHOHR and RCH_2CHOHR. Treatment with dilute sulfuric acid rearranges the glycol to the desired ketone and oxidation gives the same product from the alcohol.[127]

The ketones are readily reduced to alcohols by heating with sodium isopropylate. Thus Me n-nonyl ketone gives a quantitative yield of 2-undecanol and acetone.[128] Heating a ketone with hydrazine sulfate and an excess of NaOEt reduces it to the hydrocarbon by converting CO to CH_2 (Wolff, Kishner).[129]

[121] TOM Microfilm Reel no. 138.
[122] Favorsky. *J. prakt. Chem.* (2) 88, 641 (1913).
[123] Haller, Bauer. *Ann. Rep. Chem. Soc.* (London) 1913, 61.
[124] Whitmore, Stahly. *J. Am. Chem. Soc.* 55, 4155 (1933).
[124a] Cook, Percival, *ibid.* 71 (1949).
[125] *Ann. Rep. Chem. Soc.* (London) 1925, 81.
[126] *ibid.*
[127] *ibid.* 1906, 82.
[128] *Ann. Rep. Chem. Soc.* (London) 1925, 72.
[129] "Org. Reactions," IV, 1948.

C. Unsaturated Aldehydes and Ketones

These give the reactions of the saturated carbonyl compounds and of olefins and, in addition, give other reactions if the unsaturation is conjugated with the carbonyl giving groupings, $C=C-C=O$, $C=C-C=C-C=O$, etc.

1. Certain groups which ordinarily add to the carbonyl carbon add to the carbon at the end of the conjugated system by a process of 1,4-addition. The net result is the addition to the $\alpha\beta$-double bond of groups which do not ordinarily add to an olefinic linkage.

$$C=O + NH_3 \rightarrow \underset{\underset{NH_2}{|}}{C}-OH \qquad C=C-C=O \rightarrow \underset{\underset{NH_2\ H}{|\ \ \ |}}{C-\!\!-\!\!-C-C}=O$$

The electronic mechanism for addition to a conjugated system involving a carbonyl group may include the following steps:

$$\underset{A}{C::C:C::O} \ \rightleftharpoons \ \underset{B}{\underset{-}{C::C:C:O:}} \ \rightleftharpoons \ \underset{C}{\underset{+\ \ \ -}{C:C::C:O:}}$$

Phase A is the stable arrangement. B and C are formed by combination with H^+ or a group like MgX. In B the carbon of the carbonyl group is in a position to add an atom or group having a complete octet such as Br^- or NH_2. In C the carbon at the end of the conjugated system can add such a group. In the addition of HBr or NH_3 to the unsaturated carbonyl compound, reaction with B would give products containing the unstable groupings $C(OH)Br$ and $C(OH)NH_2$ which can lose HBr and NH_3. On the other hand, addition to C is irreversible. Hence the 1:4 addition product is obtained. With the Grignard reagent, the 1:2 addition product on the carbonyl group is stable and is the only product unless the steric relations are such as to favor the addition of the organic radical in the 4 rather than the 2-position.[130] The electronic shifts in such a system are often indicated as $C=C-C=O$.[131]

A similar process is the addition of beta keto esters and related compounds under the influence of NaOEt and similar catalysts (Michael). Thus sodio-malonic ester is represented as adding to methylene acetone as follows:

$$CH_2=CHCOCH_3 \rightarrow (EtO_2C)_2CHCH_2CH=C(ONa)CH_3 \rightarrow$$
$$(EtO_2C)_2CHCH_2CH_2COCH_3$$

Only a small amount of NaOEt is needed since the malonic ester is more "acidic" than the enol. The net result is the addition of malonic ester to the olefinic double bond. Acetoacetic ester, cyanoacetic ester, acetylacetone and such compounds give this reaction.

[130] Kohler. *Am. Chem. J.* **38**, 511 (1907).
[131] *Ann. Rep. Chem. Soc.* (London) **1927**.

Another important reagent which does not add to ordinary double bonds but does react with $\alpha\beta$-unsaturated compounds is sodium bisulfite.

$$-CO-CH=CH- \rightarrow -CO-CH_2CH(SO_3Na)-$$

Because of the solubilizing effect of the new group, this reaction makes available many types of detergents and other surface-active agents (p. 391).

2. The carbon "alpha" to the end of the conjugated system has the properties of an ordinary alpha carbon in that its H atom can take part in an aldol condensation.

$$MeCHO + CH_3CH=CHCHO \rightarrow MeCHOHCH_2CH=CHCHO$$

This reaction is general for the H which is "alpha" to a conjugatedly unsaturated aldehyde system. $CH_3(CH=CH)_nCHO$;[132] *Vinylogy*.[133, 134] This gives an important method for making aldehydes with several ethylenic linkages in conjugated relation. This may be illustrated:

$$2\ Me_3C=CHCHO \rightarrow Me_3C=CHCH=CHC(Me)=CHCHO$$

This transmittal of the "alpha effect" through a conjugated system is responsible for the reactivity of the methyl group in piperitone.[135] It also explains the observation that acetaldehyde in presence of K_2CO_3 gives a product from which n-octoic acid can be obtained. It gives a preparation for n-hexyl alcohol and n-octyl alcohol either from acetaldehyde directly or through crotonic aldehyde.

3. Another peculiarity of the conjugated carbonyl system is the Diels-Alder addition of conjugated dienes to the $\alpha\beta$-double bond.

Cumulative double bonds, as found in the *ketene* grouping, $C=C=O$, give even more unusual mutual effects between the olefinic and carbonyl linkages than those shown by conjugated compounds.

Acrolein, acrylic aldehyde, propenal, $CH_2=CHCHO$, b. 52°, is formed in the thermal decomposition of fats. Its odor is very irritating. It is best prepared by heating glycerol with a mild dehydrating agent such as $KHSO_4$ or $MgSO_4$.[136] The process involves dehydration and rearrangement, perhaps

[132] *Ann. Rep. Chem. Soc.* (London) 1907, 80; 1932, 115.
[133] Fuson, Christ. *Science* 84, 294 (1936).
[134] Blatt. *J. Org. Chem.* 1, 154 (1936).
[135] Simonsen. "The Terpenes." University Press. Cambridge, Eng., 1947. p. 368.
[136] *Ann. Rep. Chem. Soc.* (London) 1912, 81.

including the following steps:

$$HOCH_2CHOHCH_2OH \rightarrow HOCH_2CH=CHOH \rightarrow$$
$$HOCH_2CH_2CHO \rightarrow CH_2=CHCHO$$

It is available commercially (Shell).

Acrolein polymerizes to resins unless stabilized by various "antioxidants" such as hydroquinone. It gives resins with phenols.[137]

Acrolein gives both the reactions of olefins and of aldehydes. Its conjugated system of double bonds modifies its addition reactions somewhat. The ethylenic bond is more easily reduced due to 1:4 addition of hydrogen. Reduction also gives a 25% yield of the dimolecular product, divinyl glycol, $CH_2=CHCHOHCHOHCH=CH_2$. Halide acids add entirely contrary to Markownikoff's Rule which would predict that the halogen would take the alpha position. This also may be explained on the basis of 1,4-addition, the H adding first to the O.

Acrolein with H_2O at 100° gives hydracrylic aldehyde, $HOCH_2CH_2CHO$, b. 90° (18 mm.).[138]

Acrolein reacts with Grignard reagents normally to give unsaturated secondary alcohols, $CH_2=CHCHOHR$ (Bouis). Ethyl malonate gives a 1,4-condensation product, $(EtOOC)_2HC-C=C-CHO$.[139]

Both the olefinic and aldehyde groups are sensitive to oxidation. Thus mild oxidation gives glyceric acid, $CH_2OHCHOHCO_2H$, and more vigorous oxidation breaks the chain. Silver oxide does not attack olefinic linkages and thus oxidizes acrolein to acrylic acid. If ordinary oxidizing agents are to be used to oxidize only one of the reactive groups the other must be "protected." The $C=C$ can be protected by adding bromine.

$$\overset{\text{Zn}}{CH_2BrCHBrCHO \rightarrow CH_2BrCHBrCO_2H \rightarrow CH_2=CHCO_2H}$$

The CO can be protected by making the acetal with EtOH and HCl gas. Thus

$$\overset{\text{KOH}}{CH_2=CHCHO \rightarrow ClCH_2CH_2CH(OEt)_2 \longrightarrow}$$
$$CH_2=CHCH(OEt)_2 \rightarrow CH_2OHCHOHCH(OEt)_2$$

The acetal groups can be removed by cold dilute acid, thus giving glyceric aldehyde.[140]

Crotonaldehyde, crotonic aldehyde, 2-butenal, $CH_3CH=CHCHO$, b. 104°, is obtained by allowing aldol to stand at room temperature with a trace of a dehydrating agent. The yield is usually not over 50% but can be increased to

[137] Maksorov, Andrianov. *Ind. Eng. Chem.* **24,** 827 (1932).
[138] Nef. *Ann.* **335,** 219 (1904).
[139] Warner, Mol. *J. Am. Chem. Soc.* **70,** 3470 (1948).
[140] "Org. Syntheses."

80% by using iodine as the catalyst.[141]   The product consists entirely of the *trans* form.[142]

Its reactions are practically identical with those of acrolein except that the CH_3 next to the conjugated system can take part in aldol condensations. Thus crotonaldehyde can be condensed to an aldol which on dehydration gives octatrienal, $Me(CH=CH)_3CHO$.[143]   With acetaldehyde, crotonaldehyde gives hexadienal, $Me(CH=CH)_2CHO$.[144]

The autoxidation of crotonaldehyde is stopped by 10 p.p.m. of hydroquinone.[145]

The reduction of crotonaldehyde by the Zn−Cu couple gives *n*-BuOH, crotonyl alcohol, $MeCH=CHCH_2OH$, but mainly the bimolecular reduction product, dipropenyl glycol, $MeCH=CHCHOHCHOHCH=CHMe$.

Vinylacetaldehyde, 3-butenal, $CH_2=CHCH_2CHO$, is unknown probably because of its easy rearrangement to the conjugated compound crotonic aldehyde.   The tendency to form a conjugated system resembles the tendency for an enol form to change to a keto form.

$$-CH=CH-CH_2-C=O \rightarrow -CH_2-CH=CH-C=O$$
$$C=C-OH \rightarrow CH-C=O$$

In both cases this is probably the net result of a bimolecular reaction with hydrogen ion (p. 216). The $C=C-C=O$ and $C=O$ groups resemble each other in stability as compared with the non-conjugated and enolic forms. They confer a similar reactivity on H atoms attached to a carbon *alpha* to them.

Tiglic aldehyde, α-methylcrotonic aldehyde, αβ-dimethylacrolein, 2-methyl-2-butenal, $CH_3CH=C(CH_3)CHO$, b. 118°, is obtained from a mixture of acetaldehyde and propionaldehyde through the aldol condensation. The other product of the "mixed" condensation, $EtCH=CHCHO$, is formed in only small amounts. Thus the lower aldehyde tends to supply the carbonyl group rather than the alpha H for the aldol condensation.

β-Methylcrotonaldehyde, ββ-Me₂-acrolein, $(CH_3)_2C=CHCHO$, b. 133°, is made by brominating the acetal of isovaleraldehyde, removing HBr by alkali and hydrolyzing the unsaturated acetal. It gives the reactions of crotonaldehyde including aldol condensations involving the beta methyl groups (vinylogy).   Thus it gives an aldol which can be dehydrated to 3,7-Me₂-2,4,6-octatrienal.[146]   In dilute acid, Me₂-acrolein is in equilibrium with its hydration product beta-hydroxyisovaleraldehyde.[147]

αγ-Dimethylcrotonic aldehyde, α-methyl-β-ethylacrolein, 2-methyl-2-pentenal, $CH_3CH_2CH=C(CH_3)CHO$, is obtained by aldol condensation of pro-

[141] Hibbert. *J. Am. Chem. Soc.* **37**, 1748 (1915).
[142] Young. *J. Am. Chem. Soc.* **54**, 2498 (1932).
[143] *Ann. Rep. Chem. Soc.* (London) 1932, 115.
[144] Kuhn, Hoffer. *Ber.* **64**, 1977 (1931).
[145] *Ann. Rep. Chem. Soc.* (London) 1922, 69.
[146] *ibid.* 1932, 115.
[147] Lucas, Stewart, Pressman. *J. Am. Chem. Soc.* 1818 (1944).

pionaldehyde. Similarly, *α-ethyl-β-propylacrolein* is obtained from *n*-butyraldehyde. It is related to 2-Et-1-hexanol (C and C).

Higher alpha-beta unsaturated aldehydes and **ketones** are readily obtained from the corresponding aldol condensation products.

Citronellal, (Rhodinal), 3,7-Me$_2$-7 or 6-octenal, b. 202°, 82° (2 mm.), is found in *d*- and *l*-forms in various essential oils (p. 563). Ethyl citronellol has been produced for commercial use by the Grignard reaction.

The essential oil of *Eryngium foetidum* is mainly 2-dodecenal.

Citral (Geranial, Neral), 3,7-Me$_2$-2,6-octadienal (*cis* and *trans* forms), b. 228°, occurs in many essential oils. Heating with mild alkali gives acetaldehyde and methylheptenone. Treatment with dilute sulfuric acid closes a 6-membered ring and dehydrates it to give cymene, 1,4-methylisopropylbenzene. Citral is used in making α- and β-ionones for perfumery.

Violet leaves contain a perfume oil which consists largely of 2,6-nonadienal.[148]

Propargylic aldehyde, propynal, $CH \equiv CCHO$, can be made by adding bromine to acrolein acetal, treating with excess of alkali and hydrolyzing the propargylic acetal with excess of dilute acid.

Higher αβ-acetylenic aldehydes can be made from acetylenic Grignard reagents and orthoformic ester.

$$RC \equiv CH \xrightarrow{\text{MeMgCl}} RC \equiv CMgCl \xrightarrow{\text{HC(OEt)}_3} RC \equiv CCH(OEt)_2$$
$$\rightarrow RC \equiv CCHO$$

They can also be made from $RC \equiv CNa$ and formic esters.[149]

Vinyl methyl ketone, b. 81°, methylene acetone, $CH_2 = CHCOCH_3$, is made by hydrating vinyl acetylene, $CH_2 = CHC \equiv CH$. It polymerizes much like acrolein. The dimer formed on heating methyl vinyl ketone is believed to be 6-acetyl-5,6-dihydro-2-methyl-1,4-pyran.[150]

Because of the conjugated system it adds NH_3, primary and secondary amines, HCN and sodium bisulfite.

$$CH_3COCH = CH_2 \rightarrow CH_3COCH_2CH_2NH_2 \text{[151]}$$
$$\rightarrow CH_3COCH_2CH_2CN \text{[152]}$$
$$\rightarrow CH_3COCH_2CH_2SO_3Na$$

[148] Späth, Kesztler. *Ber.* **67B**, 1496 (1934).
[149] Moureau, Delange. *Compt. rend.* **133**, 105 (1901).
[150] Alder, Offermanns, Rüden. *Ber.* **74B**, 905 (1941).
[151] Friedländer. "Fortschritte der Teerfarbenfohrikalion," Vol II. Springer, Berlin. p. 785.
[152] Lapworth. *J. Chem. Soc.* **85**, 1214 (1904).

These are examples of 1,4-addition (p. 224). Grignard reagents add normally (1:2) to the carbonyl group, whereas the related compound PhCOCH = CHPh adds PhMgBr to give PhCOCH$_2$CHPh$_2$ (p. 224).

Methyl vinyl ketone is toxic and strongly lachrymatory. It is commercially available as an azeotrope (b. 76°) containing 15% water (DuPont).

Vinyl alkyl ketones can be made from β-chloropropionyl chloride with RZnI (Blaise).

$$ClCH_2CH_2COCl \rightarrow ClCH_2CH_2COR \rightarrow CH_2 = CHCOR$$

3-Hexen-2-one has been found in acetone oil from wood distillation.[153]

Methyl unsaturated ketones of the type CH$_3$COCH = CHR can be made by dehydrating the aldol condensation product of acetone and an aldehyde. Many aldehydes, however, condense with themselves much more readily than with acetone.

Mesityl oxide, 2-Me-2-penten-4-one, b. 130°, is readily available by the dehydration of diacetone alcohol. It is a typical $\alpha\beta$-unsaturated ketone. It gives the reactions of the conjugated system as given by methyl vinyl ketone. With HOCl it gives Me$_2$C(OH)CHClCOMe, which with alkali gives the $\alpha\beta$-ethylene oxide.[154] ROH in presence of KOH adds 1:4 to give an ether of diacetone alcohol.[155] With monobutyl oxalate it gives 1,4-addition followed by ring closure to form 2,2-Me$_2$-6-carbobutoxy-2,3-H$_2$-1,4-pyrone (p. 778).

As usual with conjugated systems, reduction is easy. Various conditions give methylisobutylcarbinol and methyl isobutyl ketone which are available commercially (CC). Mild reduction gives a bimolecular product which undergoes internal aldol condensation.

Phorone, 2,6-Me$_2$-2,5-heptadien-4-one, b. 197°, is also obtained by aldol condensation of acetone. It gives the same reactions as mesityl oxide with the added possibility of internal condensation to form a 6-ring as in triacetonamine. Mild reduction gives a dimolecular condensation product analogous to that from mesityl oxide. Vigorous reduction gives 2,6-Me$_2$-4-heptanol (diisobutylcarbinol). Reduction by H$_2$ and Pt gives diisobutyl ketone, diisobutylcarbinol or methylisobutylcarbinol depending on the pressure used.[156]

[153] Ann. Rep. Chem. Soc. (London) 1923, 67.
[154] ibid. 1918, 54.
[155] Simms. C.A. 26, 152 (1932).
[156] Ann. Rep. Chem. Soc. (London) 1911, 92.

$\alpha\beta$-Unsaturated ketones give saturated ketones with H_2 and Pd or with H_2 and Ni.[157, 158]

Beta-gamma unsaturated ketones have been studied with special reference to their relation to the alpha-beta isomers into which they are readily changed by acid catalysts.[159]

The addition of propionyl chloride to Me_3-ethylene in presence of $SnCl_4$ gives about equal amounts of the $\alpha\beta$- and $\beta\gamma$-unsaturated ketones, 2,3-Me_2-2-hexen-4-one and its 1-isomer.[160]

Methylheptenone, 2-Me-2-hepten-6-one, $Me_2C = CHCH_2CH_2COCH_3$, b. 171°, is obtained from citral and dilute bases. This unusual reaction probably consists of the rare 1:4-hydration of a conjugated system, followed by a reversal of the aldol condensation as occurs with diacetone alcohol (p. 338). The latter apparently depends on the beta-branching.

$$O = CHCH = \overset{|}{\underset{Me}{C}} - R \rightarrow HOCH = CH\overset{OH}{\underset{Me}{\overset{|}{C}}} - R \rightarrow O = CHCH_2\overset{OH}{\underset{Me}{\overset{|}{C}}} - R$$

$$O = CHCH_3 + MeCO - R$$

in which R is $Me_2C = CHCH_2CH_2 -$.

Methylheptenone gives reactions of carbonyl and olefin groupings. It gives a peculiar reaction with sulfuric acid in which it condenses internally to form 5,6-H_2-m-xylene. The acid may cause the rearrangement of the double bond to the 1-position as occurs in the diisobutylenes (p. 46). The addition of a proton to the carbonyl O leaves its C electronically deficient and thus in a position to act with the free electron pair of the olefinic linkage which can approach it because of the 1:6 relation.

α-*Methylheptenone* with the grouping $CH_2 = C(Me) -$ instead of $Me_2C =$ has been prepared, b. 168°. Its oxidation does not give acetone as does that of ordinary methylheptenone. In the terpene and related series there is much

[157] Covert, Connor, Adkins.   *J. Am. Chem. Soc.* **54**, 1658 (1932).
[158] *Ann. Rep. Chem. Soc.* (London) 1909, 77.
[159] Kon, Linstead.   *J. Chem. Soc.* **127**, 815 (1925).
[160] Colonge, Mostafavi.   *Bull. soc. chim.* **6**, 335 (1939).

confusion between these groupings probably because one can be converted into the other in the presence of acid catalysts.

2,6-Me$_2$-7-octen-4-one, dihydrotagatone,

$$(CH_3)_2CHCH_2COCH_2CH(CH_3)CH=CH_2,$$

b. 186°, does not react with 20% sulfuric acid or with NaHSO$_3$ (no conjugated carbonyl system). With 70% sulfuric acid it gives MeEtCO and Me-isoBuCO. Apparently the acid rearranges the double bond to the 5-position conjugated with the carbonyl group. A reversed aldol condensation would then give the two ketones as with diacetone alcohol (p. 338).

Tagatone, 2-Me-6-methylene-7-octen-4-one,

$$(CH_3)_2CHCH_2COCH_2C(=CH_2)CH=CH_2,$$

b. 210°, is related to the above ketone. With Na it gives red salts.

Artemesia ketone, $Me_2C=CHCOC(Me_2)CH=CH_2$, is one of the few naturally occurring substances containing a *neo* carbon.

Pseudoionone, 2,6-Me$_2$-2,6,8-undecatrien-10-one, is obtained by aldol condensation of citral and acetone. By ring closure it gives α- and β-ionone. Other ketones have been used instead of acetone.[161]

Special unsaturated ketones can be obtained from acid chlorides or anhydrides with olefins in presence of ZnCl$_2$. Thus acetic anhydride with Me$_3$- and Me$_4$-ethylenes give respectively 3,4-Me$_2$-3 and 4-penten-2-one and 3,3,4-Me$_3$-4-penten-2-one. This reaction gives good yields with most tertiary olefins. Triptene gives poor results.

Chelation of unsaturated ketones. The fact that acetoacetic ester in presence of NaOEt will add to dibenzalacetone but not to dicinnamalacetone can be explained on the basis of residual affinities at the ends of conjugated systems (*chelate ring* formation).[162]

αβ-Acetylenic ketones, $RC\equiv CCOR'$, are obtained from acyl halides, R'COX, and $RC\equiv CNa$ or $RC\equiv CMgX$.[163]

Because of their manifold relationships to other organic compounds such as olefins, acetylenes, alcohols, and acids, the identification of specific aldehydes

[161] Hibbert, Cannon. *J. Am. Chem. Soc.* 46, 119 (1924).

[162] *Ann. Rep. Chem. Soc.* (London) 1910, 70.

[163] Hurd, Cohen. *J. Am. Chem. Soc.* 53, 1071 (1931).

and ketones has been the subject of much study. Reagents which give derivatives of suitable properties include substituted hydrazines and semi-carbazides. Among the best of these are 2,4-dinitrophenylhydrazine[164-166] and 4,4-diphenylsemicarbazide.[167]

D. KETENES

Ketenes are obtained by the pyrolysis of acids, anhydrides, ketones, and esters; by removing X_2 from alpha halogen acid halides; and by removing HX from acid halides.[168]

Ketenes of two types are known, aldoketenes, $RCH = C = O$, and keto-ketenes, $R_2C = C = O$. The carbonyl group is even more reactive than in formaldehyde. The reactions with themselves (polymerization), with oxygen, with active H compounds (water, ammonia, acids, alcohols, etc.), and with Grignard reagents are unusually violent.

Ketene, carbomethylene, methylene ketone, $CH_2 = C = O$, b. $-41°$, is made by the pyrolysis of acetic anhydride or, preferably, acetone.[169] A preparation which is used for higher ketenes is the action of zinc with a bromoacyl bromide. This preparation also throws light on the structure of the ketenes.

$$\underset{\underset{Br}{\overset{\displaystyle |}{}}\ \underset{Br}{\overset{\displaystyle |}{}}}{Me-CH-C=O} + Zn \rightarrow ZnBr_2 + MeCH=C=O$$

Ketene polymerizes very rapidly. Its dimer, b. $127°$, is most probably an equilibrium mixture of vinylaceto-β-lactone and β crotonolactone,

$$\underset{\underset{O-C=O}{\overset{\displaystyle |\qquad |}{}}}{CH_2=C-CH_2} \rightleftarrows \underset{\underset{O-C=O}{\overset{\displaystyle |\quad |}{}}}{CH_3-C=CH}$$

Ketene dimer treated with N-bromosuccinimide followed by alcoholysis gives $CH_3COCHBrCOOEt$.[170, 171] Further polymerization gives dehydroacetic acid. This may involve a Diels-Alder reaction.

[164] Campbell. *Analyst* 61, 391 (1936).
[165] Roberts, Green. *J. Am. Chem. Soc.* 68, 214 (1946).
[166] Clark, Kaye, Parks. *Ind. Eng. Chem., Anal. Ed.* 18, 310 (1946).
[167] Toschi. *Gazz. chim. ital.* 45, I, 205 (1915).
[168] "Org. Reactions," III.
[169] "Org. Syntheses."
[170] Blomquist, Baldwin. *J. Am. Chem. Soc.* 70, 29 (1948).
[171] Miller, Koch *J. Am. Chem. Soc.* 1890. (1948).

Ketene dimer is an important intermediate for the preparation of methyl and ethyl acetoacetates by the action of MeOH and EtOH and of acetoacetanilide by the action of aniline (C and C).

The addition reactions of ketenes are of two types, those which involve the olefinic linkage directly and those which involve it only indirectly, addition taking place first at the carbonyl linkage and being followed by rearrangement of the resulting enol to a keto form. To the first type belongs the action of halogens and of oxygen. The former give the alpha halogen acyl halides from which the ketenes can be made by the action of zinc. Oxygen gives very explosive peroxides which decompose to form CO_2 and aldehydes or ketones. Grignard reagents and active H compounds add to the carbonyl group of ketene homologs in the usual way. Ketene itself usually gives resins but with some Grignard reagents gives fair yields of ketones.[172] With zinc alkyls it gives addition products.[173] The primary addition product is a metal derivative of an enol form.

$$RCH=C=O + RMgX \rightarrow RCH=\overset{R}{\underset{|}{C}}-OMgX \rightarrow$$

$$RCH=\overset{R}{\underset{|}{C}}-OH \rightarrow RCH_2-\overset{R}{\underset{|}{C}}=O$$

The net result is the addition of RH to the olefinic bond much as in the 1,4-addition of a Grignard reagent to benzalacetophenone. Similarly H_2O, ROH, NH_3, RNH_2, R_2NH, HCl, etc., add to the carbonyl group with great readiness. Ketonization then gives acids, esters, acid amides, acid chlorides, etc. Again, the net result is the addition of HQ to the olefinic bond.

$$CH_2=C=O + H-Q \rightarrow CH_2=\overset{Q}{\underset{|}{C}}-O-H \rightarrow CH_3-\overset{Q}{\underset{|}{C}}=O$$

The fact that the initial and final products both contain the carbonyl group might be used as an argument that the latter group is inactive in ketenes. The same conclusion might be drawn erroneously from the fact that carbonyl reagents like phenylhydrazine do not react in the ordinary way with ketenes to give derivatives of the phenylhydrazone type. The reactions of acetone and of ketene with phenylhydrazine may be compared.

$$Me_2CO + PhNHNH_2 \rightarrow Me_2C(OH)NHNHPh \rightarrow H_2O + Me_2C=NNHPh$$
$$CH_2=CO + PhNHNH_2 \rightarrow CH_2=C(OH)NHNHPh \rightarrow$$
$$CH_3CO-NHNHPh \leftrightharpoons CH_3C(OH)=NNHPh$$

The product from ketene is simply acetyl phenylhydrazine.

[172] Dashkevick. *C. A.* 1948, 7244.
[173] Hurd, Williams. *J. Am. Chem. Soc.* 58, 962 (1936).

Practically all reactions of ketene give resinous by-products due to the great reactivity of the substance with itself and products formed in the reactions.

Glacial acetic acid and ketene form acetic anhydride. This reaction is responsible for the chief commercial preparation of acetic anhydride by "cracking" acetic acid at high temperature.

As would be expected of such a reactive substance, ketene is very poisonous. Attempts to find adequate commercial uses for ketene are being pushed vigorously.

Higher ketenes are made by the action of zinc with alpha bromo acid bromides. The *aldoketenes* are colorless, are not auto-oxidizable, and are polymerized by pyridine probably to give homologs of acetylketene, $RCH_2COC(R)=C=O$. The *ketoketenes* are colored, are easily auto-oxidized to give peroxides, and form addition compounds with cyclic tertiary bases like pyridine, quinoline, and acridine. These are definite compounds formed by a peculiar co-polymerization involving two mols of the ketoketene and the $N=C$ linkage of the cyclic base. The dimers of the ketoketenes are tetra-substituted 1,3-cyclo-butandiones.

Carbon suboxide, C_3O_2, $O=C=C=C=O$, m. $-107°$, b. $7°$, a diketene, can be obtained by the action of phosphoric anhydride with malonic acid or of zinc with dibromomalonyl bromide. Probably the best preparation is by the pyrolysis of diacetyltartaric anhydride.[174] It gives the typical ketene reactions. Thus water, alcohol, and ammonia give malonic acid, its ester and amide. A less expected reaction is that with formic acid to give a substance which may be $CH_2(COCO_2H)_2$.[175] It has been reported that C_3O_2 at $200°$ gives CO_2 and C_2, a red gas which rapidly polymerizes.[175a] This and other suboxides of carbon have been widely studied.[176, 177]

A cyclic structure containing a triple bond was proposed for C_3O_2 but was almost immediately abandoned.[178]

E. OXIMES

Aldoximes, $RCH=NOH$, and ketoximes, $RR'C=NOH$, are obtained from aldehydes and ketones with hydroxylamine, H_2NOH.[179] Reactions which should give primary and secondary nitroso compounds give oximes instead (p. 165).

$$=C=O + H_2NOH \rightarrow [=C(OH)NHOH] \rightarrow =C=NOH$$

[174] Hurd, Pilgrim. *J. Am. Chem. Soc.* **55**, 757 (1933).
[175] *Ann. Rep. Chem. Soc.* (London) 1908, 85.
[175a] Klemenc, Wechsberg, Wagner. *Z. Electrochem.* **40**, 488 (1934).
[176] Reyerson. *Kobe. Chem. Revs.* **7**, 479 (1930).
[177] Ahrens. *Sammlung Chemische-Technischen Vorträge* 1924, 17.
[178] Michael. *Ber.* **41**, 925 (1908).
[179] Meyer. *Ber.* **15**, 1324 (1882).

The velocity of formation of oximes depends on the length and branching of the groups attached to the carbonyl groups.[180, 181] In many cases the oximes are crystalline compounds of value in *identifying* the *carbonyl compounds*. By the *Beckmann rearrangement* the ketoximes give a method of degrading ketones to simpler products. The oximes can exist in several *isomeric forms*.

$$RCH_2CH = NOH \rightleftharpoons RCH = CH—NHOH \rightleftharpoons RCH_2—CH\diagdown_{O}\diagup NH$$

$$\qquad A \qquad\qquad\qquad B \qquad\qquad\qquad C$$

Forms *A* and *B* can each exist in geometrically isomeric forms, *cis* and *trans*, or *syn* and *anti*.

$$\begin{array}{ccc} RCH_2—C—H & RCH_2—C—H & R—C—H \\ \| & \| & \| \\ N—OH & HO—N & H—C—NHOH \end{array}$$

$$\begin{array}{c} R—C—H \\ \| \\ HONH—C—H \end{array}$$

In the aliphatic series, with primary and secondary groups, it is not possible to isolate the individual compounds because of the lability of $\alpha - H$ atoms. In the aromatic series, however, it is possible to isolate the isomers (p. 678).[182]

Type *C* represents the *isoximes*. See nitroso compounds (p. 165).

Reactions.

1. Hydrolysis regenerates the carbonyl compound and H_2NOH.
2. Bases give soluble compounds such as $RCH = NONa$.
3. HCl gives hydrochlorides.
4. Reduction gives primary amines, RNH_2. This is valuable in cases where the corresponding bromide is either unobtainable or reacts abnormally. Examples are $Me_3CCH_2NH_2$ and $Me_3CCH(NH_2)Me$.
5. Ketoximes give stable acetates $RR'C = NOCOMe$. Alpha-aldoximes give acetates which are less stable, being converted by heat to nitriles.

$$RCH = COCOMe \rightarrow RC \equiv N + MeCO_2H$$

Beta-aldoximes give even less stable acetates (p. 678).

6. Ketoximes, with acetyl chloride, PCl_5, HCl, conc. sulfuric acid in glacial acetic acid, and other acid reagents, give the Beckman rearrangement, resulting in a monosubstituted acid amide. BF_3 causes the same change.[183]

$$RR'C = NOH \rightarrow RCONHR' + R'CONHR$$

[180] Petrenko-Kritschenko. *J. Russ. Phys. Chem. Soc.* **36**, 1505 (1907).
[181] Stewart. *J. Chem. Soc.* **87**, 185 (1905).
[182] Freudenberg. "Stereochemie," p. 982.
[183] Meerwein. *Ber.* **66B**, 411 (1933).

In general the larger group migrates to the greater extent. This is like the behavior in 1,2-carbon rearrangements such as the pinacolone rearrangement. Thus the Beckmann rearrangement of the oximes of methyl *t*-amyl ketone and methyl pinacolyl ketone (2,2,3-Me₃-4-pentanone) gives as the chief products N-*t*-amylacetamide, MeCONHCMe₂Et, and N-pinacolylacetamide, MeCONHCH(Me)CMe₃.[184] Similarly the oximes of methyl cyclohexyl ketone and methyl undecyl ketone give only N-substituted acetamides. Hydrolysis of the substituted amides gives acids and primary amines which can be identified in the usual ways.

The Beckmann rearrangement involves essentially the same type of change as the Hofmann rearrangement of N-bromoamides, RCONHBr, the Lossen rearrangement of hydroxamic acids, RCONHOH, and the Curtius rearrangement of acid azides, RCON₃. These processes have been exhaustively studied by Julius Stieglitz and Lauder W. Jones and their students.

7. Oxidation of aldoximes with monopersulfuric acid forms *hydroxamic*

$$acids, \; RC \overset{\displaystyle NOH}{\underset{\displaystyle OH}{\Big\backslash}}$$

, which give color reactions with ferric chloride.

8. With bromine water, the ketoximes give bromonitroso compounds, RR′C(NO)Br.

9. Alkylation may take place on the N or on the O giving derivatives which, on hydrolysis, form RNHOH or H₂NOR. Thus alkylation of acetoxime, Me₂C=NOH, with MeI gives exclusively the N−Me compound while MeI in the presence of NaOMe gives the O−Me compound as well.

10. An exceptional reaction is that of acetaldoxime with chlorine to give MeCHCl(NO) which exists as a blue liquid above 85°. Below that temperature, it forms a colorless dimer (p. 165). Long heating changes the blue liquid colorless by conversion to the oxime of acetyl chloride.

$$MeCH = NOH \rightarrow MeCHCl - N = O \rightarrow MeCCl = N - OH$$

The first change is like that of isobutylene with chlorine to give a "substitution" product (p. 40).

Formaldoxime, CH₂=NOH, b. 85°, tends to polymerize to a trimer.

Acetoxime, m. 59°, b. 135°.

The melting points of the aldoximes, RCH=NOH, for ethyl to hexadecyl (°C.) are 21, liq., 51, 52, 58, 59, 63, 69, 72, 73, 80, 82, 86, 88, 89. Those for the ketoximes, RR′C=NOH for combinations of alkyl groups are di-isopropyl, 7, di-*n*-heptyl, 20, di-*n*-octyl, 12,[185] di-*n*-undecyl, 40, di-*n*-tridecyl, 51, di-*n*-pentadecyl, 59, and di-*n*-heptadecyl, 63.

[184] Drake, Kline, Rose. *J. Am. Chem. Soc.* **56**, 2076 (1934).
[185] Kipping. *J. Chem. Soc.* **63**, 452 (1893).

XI. ALIPHATIC MONOBASIC ACIDS

A. Saturated Acids, $C_nH_{2n+1}CO_2H$

These acids are oxidation products of the primary alcohols, RCH_2OH, and the aldehydes, $RCHO$; hydrolysis products of the alkyl cyanides, RCN; and reaction products of the Grignard reagents, $RMgX$ with CO_2. The name fatty acids is due to the occurrence of higher homologs such as palmitic, C_{16}, and stearic, C_{18}, acids in natural fats in the form of glycerides.

The carboxyl group, $-C\overset{OH}{=}O$, is represented as a combination of a carbonyl group and a hydroxyl group. Many properties of acids indicate that this is not an accurate picture. The absorption spectra of carboxylic acids and their esters differ radically. The acids are *associated* while the esters are not. The parachor is less than would be expected. Alpha ketonic acids and esters are colorless instead of being yellow as would be expected from compounds having adjacent carbonyl groups as in diacetyl, MeCOCOMe, and benzil, PhCOCOPh. The peculiar nature of the carbonyl group in the carboxylic acids has been indicated in various ways.[1, 2]

Mesohydric form (Oddo 1906, 1932)[3]   (Hantzsch, Rep. 1918, 55)[4]   (Lewis 1916)[5]

In more modern terms this might be thought of as involving a coordinate link between the H and the carbonyl oxygen; $R-C\underset{O}{\overset{O}{\diamond}}H$. Many facts about the carboxylic acids and their esters make this improbable.[6] Coordinate

[1] *Ann. Rep. Chem. Soc.* (London) 1912, 191.
[2] Bülow, Ahrens. *Sammlung Chemische-Technischen Vorträge*, 1919, 189.
[3] Oddo. *Gazz. chim. Ital.* 36, i, 83 (1906); 61, 699 (1931).
[4] Hantzsch. *Ann. Rep. Chem. Soc.* (London) 1918, 55.
[5] Lewis. *J. Am. Chem. Soc.* 38, 762 (1916).
[6] Sidgwick. "The Electronic Theory of Valency.' Oxford, 1927. pp. 252–5.

links are possibly formed between two molecules, thus causing the well-known association of acetic acid and its homologs.[7, 8]

Accurate studies on the vapor densities of acetic acid indicate the presence of a small amount of a higher polymer which is suggested to be a tetramer involving a higher degree of hydrogen bonding.[9]

In view of the uncertainty as to the exact nature of linkages within the carboxy group, the formula, $R-CO_2H$, is perhaps better than the older formula, $R-CO(OH)$ or $R-COOH$. In aqueous solution the carboxyl group is at least partially hydrated.

$$RCO_2H + H_2O \rightleftharpoons RC(OH)_3$$

The hydrogen atoms on the alpha carbon atom (next the carboxyl group) are reactive somewhat as are the alpha H atoms of aldehydes and ketones. The degree of activity is less. As in other alpha H compounds the activity is explained partly on the basis of enolization as in the ready replacement by halogens and in the racemization of compounds having an asymmetric alpha carbon and partly as the effect of the adjacent group as in the condensation of acid anhydrides with aldehydes in presence of sodium salts of carboxylic acids (Perkin reaction) (p. 703).

A related reaction is that of the esters of carboxylic acids with Na to give sodioketoesters (Claisen) (p. 367).

The carboxylic acids range from liquids to solids varying in boiling point from 101° at 760 mm. to about 300° at 100 mm. for the C_{19} member. The

[7] Huggins. *J. Org. Chem.* 1, 407 (1936).
[8] Smith, Hitchcock. *J. Am. Chem. Soc.* 54, 4631 (1932).
[9] Ritter, Simons. *J. Am. Chem. Soc.* 67, 757 (1945).

lower members have a sharp odor, the middle members odors of the goat-like type and the higher members practically no odor. The melting points fall into two series for the odd and even members, the former being slightly lower than the latter. This alternation in melting points is found in various homologous series and is an indication of the slightly greater symmetry of the even numbered molecules.[10]

NORMAL FATTY ACIDS

C atoms	Name	m.p. °C.	b.p. °C.
1	Formic	8.3	101.
2	Acetic	16.67	119.
3	Propionic	− 36.	141.
4	Butyric	− 4.7	163.
5	Valeric	− 34.5	186.
6	Caproic	− 1.5	205.
7	Oenanthic	− 10.5	223.
8	Caprylic	16.5	237.
9	Pelargonic	12.5	254.
10	Capric	31.4	269.
11	Undecanoic	28.	212. (100 mm.)
12	Lauric	43.6	225. "
13	Tridecanoic	40.5	236. "
14	Myristic	58.	248. "
15	Pentadecanoic	52.1	257. "
16	Palmitic	63.1	268. "
17	Margaric	62.	277. "
18	Stearic	70.1	287. "
19	Nonadecanoic	69.4	298. "
20	Arachidic	76.2	328.
21	Heneicosanoic	75.2	
22	Behenic	80.5	
23	Tricosanoic	79.6	
24	Lignoceric (Carnaubic)	84.7	
25	Pentacosanoic	83.2	
26	Hexacosanoic	87.8	
27	Carboceric	81.5	
28	Octacosanoic	90.9	
29	Montanic	90.3	
30	Triacontanoic	93.6	
32	Dotriacontanoic	96.0	
33	Psyllastearic	95.	
34	Tetratriacontanoic	98.2	
36	Hexatriacontanoic	99.9	
46	Hexatetracontanoic	106.8	

Cluytinic acid, m. 68.5°, is a mixture of the C_{21} acid with higher and lower members of the series. Cerotic acid, m. 78.5°, is a mixture of C_{26} and C_{28} acids. Melissic acid, m. 94°, is a mixture of C_{30} and C_{31} acids. Lacceric acid, m. 94°, is a mixture of acids about C_{32}.

[10] *Ann. Rep. Soc. Chem.* (London) **1927**, 95.

Many of the higher normal fatty acids have been synthesized.[11, 12]

The first three members of the fatty acid series are completely soluble in water. The solubilities from n-butyric to n-heptoic acid drop from 6 to 0.2 g. per 100 cc. of water. The acids to C_6 are readily volatile with steam, those from C_{16} up are not volatile at all with steam. The volatility with steam of the acids from various fats is useful in detecting adulteration. Thus butter fat gives a larger amount of volatile acids than does oleomargarine.

Formic acid appears about 5% ionized in 0.1 N solution while its homologs are only about 1% ionized.

Formic acid, methanoic acid, HCO_2H, b. 100.5, m. 8.3°, has been made in ways which are of interest as bearing on its structure and its relation to its homologs.

1. Hydrolysis of HCN by dilute acid

$$HCN + 2 H_2O + HCl \rightarrow HCO_2H + NH_4Cl$$

2. Saponification of methine halides.

$$CHCl_3 + 4 NaOH \rightarrow HCO_2Na + 3 NaCl + 2 H_2O$$

3. Hydrogen and carbon dioxide in the corona discharge (silent electric discharge).

$$H_2 + CO_2 \rightarrow HCO_2H$$

Formic acid is prepared *commercially* by special processes.

1. By the action of CO under pressure at about 200° on NaOH solution or soda lime.

$$CO + NaOH \rightarrow HCO_2Na$$

2. Methyl formate is similarly obtained from CO and methanol. Treatment with NH_3 gives formamide which is hydrolyzed to formic acid by water and sulfuric acid.

3. From oxalic acid, by heating with glycerol. Heated alone, oxalic acid gives very little formic acid but mainly CO and CO_2. The glycerol presumably forms an oxalic ester (I) which loses CO_2 leaving a formic ester which then reacts with more oxalic acid to give formic acid and regenerate the oxalic ester.

$$Q-OH \rightarrow Q-OCOCO_2H \rightarrow CO_2 + Q-OCHO \rightarrow HCO_2H + Q-OCOCO_2H$$
$$\text{(I)} \qquad\qquad\qquad\qquad\qquad\qquad\qquad\qquad \text{(I)}$$

A more volatile alcohol would give esters which would distill instead of decomposing. The formic acid distills over with the water formed. Since the glycerol oxalate (I) is constantly regenerated, only a small amount of glycerol is needed unless too high a temperature is reached and allyl alcohol is formed.

---

[11] Francis. *J. Am Chem. Soc.* **61**, 578 (1939).

[12] *Ann. Rep. Chem. Soc.* (London) 1926, 100.

Obviously this process is available only when oxalic acid is much cheaper than formic acid. Since one of the commonest ways of making oxalic acid is from formates, this is not usually the case. In the laboratory, cellulose (cotton) has been used in place of glycerol.[13]

In many of its *reactions*, formic acid differs from its homologs.

1. It is a stronger acid. Its ionization constant is 21×10^{-5} whereas those for its homologs are of the order of 1×10^{-5}.

It esterifies alcohols readily, serving as its own catalyst.

2. Containing an aldehyde group, it is a reducing agent.

$$HCO_2H + [O] \rightarrow H_2O + CO_2$$

It reduces a solution of mercuric chloride to a precipitate of calomel.

$$2 HgCl_2 + HCO_2H \rightarrow Hg_2Cl_2 + CO_2 + 2 HCl$$

Its reducing power is utilized in a process for separating Cu and Cd analytically.[14] The solution is treated with potassium formate, evaporated and heated at 160°. The copper salt is reduced to the metal while the Cd salt is not.

3. It reacts with dehydrating agents.

$$HCO_2H \rightarrow H_2O + CO$$

This is the best way to make pure carbon monoxide on a laboratory scale.[15]

4. It decomposes on heating at a relatively low temperature (160°).

$$HCO_2H \rightarrow H_2 + CO_2$$

This is a case of oxidation and reduction within the molecule, the carbon being oxidized from the valence in CO to that in CO_2 and the hydrogen being reduced to the elementary state.

Alkali formates, heated under reduced pressure, give oxalates and H_2. This is an important preparation of oxalic acid.

Lead formate is used in heavy duty brake linings. The heat generated decomposes the formate giving metallic lead, the coefficient of friction of which increases at high temperature. Most materials become less effective at higher temperatures.

The vapor of formic acid is a mixture of single and double molecules.[16]

The azeotrope with 22.5% water is unusual in that it boils higher than either component, namely at 107°.

The photochemical decomposition of formic acid gives CO and CO_2.[17]

Anhydrous formic acid has been used as a solvent for the study of the action

[13] Montgomery. *J. Am. Chem. Soc.* **53**, 2700 (1931).

[14] Fulmer. *Ind. Eng. Chem., Anal. Ed.* **3**, 257 (1931).

[15] Schierz, Ward. *J. Am. Chem. Soc.* **50**, 3240 (1928).

[16] Coolidge. *J. Am. Chem. Soc.* **50**, 2166 (1928).

[17] Herr, Noyes. *J. Am. Chem. Soc.* **50**, 2345 (1928).

of acids.   In it, sulfuric acid acts as a strong monobasic acid and Na formate and aniline act as strong bases.[18]

Formic acid containing small amounts of water is a good solvent for kinetic studies on inert reagents such as neopentyl bromide.[19]

Because it is the strongest of the common organic acids, formic acid finds many industrial uses in connection with rubber, textiles, leather, laundry operations and the like.   It is also used in silage.

Acetic acid, ethanoic acid, CH_3CO_2H, b. 119°, m. 16.67°, is obtained in a great variety of ways.

1. By oxidation of many organic compounds since it is more resistant to oxidation than most such compounds and their primary oxidation products.

2. By the hydrolysis of methyl cyanide obtained from methyl halides.

3. By the action of carbon dioxide on sodium methyl obtained from mercury dimethyl.

$$CH_3Na + CO_2 \rightarrow CH_3CO_2Na$$

A more practical form of this method utilizes the Grignard reaction.

$$CH_3MgCl + CO_2 \rightarrow CH_3CO_2MgCl$$

The best results are obtained at low temperatures.

Another modification is the use of ethyl carbonate with the Grignard reagent (Tschitschibabin).

$$MeMgCl + (EtO)_2C=O \rightarrow (EtO)_2C(Me)OMgCl \rightarrow MeCO_2Et + EtOMgCl$$

The reaction is carried out rapidly at a low temperature to prevent the further action of the Grignard reagent to form acetone and tertiary butyl alcohol.

4. By the action of carbon monoxide on sodium methylate at 160°.

$$CH_3ONa + CO \rightarrow CH_3CO_2Na$$

A modification of this reaction is the catalytic action of carbon monoxide under high pressure with methanol to give acetic acid directly.[20]

5. By the action of alkalies on trihalogenated acetone.

$$CH_3COCX_3 + NaOH \rightarrow CH_3CO_2Na + CHX_3$$

6. By the splitting of acetoacetic ester by *concentrated* alcoholic potash.

$$CH_3COCH_2CO_2Et + 2\ KOH \rightarrow 2\ CH_3CO_2K + EtOH$$

With α-substituted acetoacetic esters this gives mono- and di-substituted acetic acids.

7. By the thermal decomposition of malonic acid.

$$CH_2(CO_2H)_2 \rightarrow CO_2 + CH_3CO_2H$$

[18] Hammett.   *J. Am. Chem. Soc.* **54**, 4239 (1932).
[19] Dostrovsky, Hughes.   *J. Chem. Soc.* **1946**, 171.
[20] Krase, Singh.   *Ind. Eng. Chem.* **27**, 909 (1935).

Substituted malonic acids give RCH_2CO_2H and $RR'CHCO_2H$.

These methods of formation combine to settle definitely the structure of acetic acid and, consequently, throw light on the structure of the many organic substances related to it.   They are also applicable to the preparation of certain of its *higher homologs*.

Three methods are used for the *commercial preparation of acetic acid*.

1. Until recently the chief source was pyroligneous acid from wood distillation.   This liquid contains about 10 per cent acetic acid.

(*a*) The older method consists in neutralizing the liquid with lime and distilling off the volatile materials including methyl alcohol, acetone, higher ketones, etc.   The residue is "brown" or "gray" acetate of lime depending on the amount of tarry impurities.   It is distilled with dilute sulfuric acid to give dilute acetic acid (28 per cent).

(*b*) Because of the competition of newer processes, modifications have been made in the separation of acetic acid from pyroligneous acid.   This is now accomplished by a continuous extraction process using solvents such as ethyl ether or, more recently, isopropyl ether.   The latter is less soluble in the aqueous layer and is more readily recovered by distillation.   The Suida process uses an oil distilled from the pyroligneous acid itself as an extractant for the acetic acid.[21]

The wood carbonization industry is not only surviving but is prospering in the face of competition from synthetic processes for its best known products, acetic acid, methanol and acetone.   Improved methods have lowered the cost of production of these and have added a multitude of byproduct acids, acetals, amines, alcohols, aldehydes, ethers, esters, furans, hydrocarbons, ketones, phenols and pyrones of various types totalling nearly two hundred compounds.[22]   About a score of these are already commercially available.

2. The oldest method of producing acetic acid is the acetic fermentation of dilute ethyl alcohol.   This process has been known since early times in the souring of alcoholic liquors and in the making of vinegar.   This method is not suitable for making acetic acid itself because of the large amount of water involved.   By recent developments, the dilute acetic acid (about 10%) so obtained can be converted directly into alkyl acetates, the chief form in which acetic acid is used.

3. The best method for pure acetic acid (*glacial acetic acid*) is the catalytic oxidation of acetaldehyde obtained from acetylene.   Air and manganese acetate wet with acetic acid are used at 70° to absorb the acetaldehyde.   Yields of better than 80% from acetylene to acetic acid are obtained.[23]

The *reactions of acetic acid* are typical of the saturated aliphatic acids.

1. It forms soluble salts of all the metals.   With many of the heavy metals it forms "basic salts."   Some of these are very complex such as the blood red

[21] Poste. *Ind. Eng. Chem.* **24**, 722 (1932).
[22] Goos, Retter. *Ind. Eng. Chem.* **38**, 131 (1946).
[23] *Ann. Rep. Chem. Soc.* (London) **1920**, 53.

basic ferric acetate, $[Fe_3(OH)_2(OAc)_6]OAc$. These compounds probably contain chelate rings as in basic beryllium acetate.[24]

Sugar of lead is $Pb(OAc)_2 . 3 H_2O$. *Verdigris* is basic copper acetate. *Paris Green* contains copper acetate and arsenite. Acetates of various heavy metals such as Al, Cr and Fe are used as mordants with dyes like alizarin.

Lead tetraacetate is a valuable oxidizing agent for converting reactive CH to COAc, and C=C to the diacetate of the corresponding glycol.[25]

2. With alcohols it forms esters. This reaction results in an equilibrium. The first step probably consists in the usual addition to a carbonyl group (p. 120). The attainment of the equilibrium is catalyzed by hydrogen ions.

Secondary alcohols react much less rapidly to form esters.

When tertiary alcohols are heated with acetic acid, the chief result is catalytic dehydration of the alcohol.

The *acetylation* of a hydroxyl group by means of acetic acid or one of its derivatives is an important means of *protecting* the hydroxyl. The protecting group can be removed by basic or acid hydrolysis.

3. Sodium acetate heated with soda lime gives methane.

$$CH_3CO_2Na + NaOH \rightarrow CH_4 + Na_2CO_3$$

This is *not* a general reaction. Na salts of higher acids give complex mixtures with rapidly decreasing amounts of the paraffin, RH, related to RCO_2H.

The photochemical decarboxylation of gaseous acetic acid gives CH_4 and CO_2 with a quantum yield of 1.

4. When calcium acetate is heated, acetone is obtained. This process has had considerable commercial value.

$$(CH_3CO_2)_2Ca \rightarrow CH_3COCH_3 + CaCO_3$$

When a mixture of calcium acetate and another calcium salt is heated, a mixed ketone, CH_3COR, is obtained along with acetone and the ketone RCOR.

The same result is achieved by passing the vaporized acid or a mixture of acids over heated manganous oxide and other catalysts.

$$2 CH_3CO_2H \rightarrow CH_3COCH_3 + CO_2 + H_2O$$

Another method is to pass acetic acid or, better, acetic anhydride over hot calcium carbide.[26]

[24] Sidgwick. "The Electronic Theory of Valency." Oxford, 1927. p. 255.
[25] *Ann. Rep. Chem. Soc.* (London) 1923, 90.
[26] *Ann. Rep. Chem. Soc.* (London) 1906, 81.

5. When it is heated alone to 500°, ketene is obtained (Hurd).

$$CH_3CO_2H \rightarrow H_2O + CH_2{=}C{=}O$$

6. When a solution of its sodium salt is electrolyzed, ethane is obtained.[27] The acetate ion is discharged at the anode and the resulting "free radicals" undergo a series of changes ending with ethane in 80% yield.[28]

$$CH_3CO_2^- \rightarrow [CH_3CO_2] \rightarrow CO_2 + [CH_3] \rightarrow CH_3{-}CH_3$$

Salts of the C_3 and C_4 acids on electrolysis give poor yields of the saturated hydrocarbons but the normal acids from C_5 to C_{18} give 75–90% yields. A branch at the α-position causes the production of olefin.[29]

7. Distillation of a concentrated solution of ammonium acetate gives acetamide.

$$CH_3C\overset{ONH_4}{=}O \rightarrow H_2O + CH_3C\overset{NH_2}{=}O$$

The mechanism of this dehydration may resemble the mechanism suggested for esterification, involving addition to the carbonyl group.

$$CH_3C\overset{ONH_4}{=}O \rightleftharpoons NH_3 + CH_3C\overset{OH}{=}O \rightleftharpoons CH_3\underset{NH_2}{\overset{OH}{C{-}OH}} \rightleftharpoons H_2O + CH_3C\overset{NH_2}{=}O$$

8. With acid chlorides, acetyl chloride is formed.

$$
\begin{aligned}
CH_3CO_2H + PCl_5 &\rightarrow CH_3C\overset{Cl}{=}O + HCl + POCl_3 \\
3\,CH_3CO_2H + PCl_3 &\rightarrow 3\,CH_3COCl + H_3PO_3 \\
CH_3CO_2H + SOCl_2 &\rightarrow CH_3COCl + SO_2 + HCl \\
CH_3CO_2H + COCl_2 &\rightarrow CH_3COCl + CO_2 + HCl
\end{aligned}
$$

9. When sodium acetate is treated with half the amount of acid chloride needed to form the acyl chloride, acetic anhydride is obtained.

$$6\,CH_3CO_2Na + POCl_3 \rightarrow 3\,(CH_3CO)_2O + 3\,NaCl + Na_3PO_4$$

Sulfur monochloride, S_2Cl_2, is used in place of the more expensive acid chlorides.

10. Chlorine and bromine replace alpha hydrogen atoms in acetic acid especially in the presence of catalysts like the chlorides or bromides of sulfur and phosphorus. This probably takes place through the enol form.[30]

[27] Kolbe. *Ann.* **69**, 257 (1849).
[28] *Ann. Rep. Chem. Soc.* (London) **1926**, 102.
[29] Petersen. *Z. physik. chem.* **33**, 99, 295 (1900).
[30] *Ann. Rep. Chem. Soc.* (London) **1923**, 67.

Treatment of anhydrous acetic acid with dry bromine and phosphorus gives bromoacetyl bromide, $BrCH_2COBr$ (Hell-Volhard-Zelinsky). Di- and tri-halogen acetic acids are best prepared by indirect means.

11. When its sodium salt and its anhydride are heated with an aldehyde, an alpha-beta unsaturated acid results (Perkin).

Glacial acetic acid is finding wide use as a non-aqueous solvent for studying problems of solutions.[31] Solutions of acids in it show "super-acid properties." For instance they can be used to titrate extremely weak "bases" such as acetamide.[32] The purest acid has a specific conductance of 0.7×10^{-8} mho. A pH scale has been worked out for solutions in glacial acetic acid.[33]

Addition of water to glacial acetic acid at first increases its density. A maximum is reached corresponding to the ortho acid $CH_3C(OH)_3$.

While pure acetic acid freezes at 16.67° small amounts of water depress the freezing point greatly. Thus 99% and 87% acids freeze at 14.8° and −0.2°.

Much study has been devoted to the dissociation constant of acetic acid, 1.75×10^{-5} at 25°.[34, 35]

Acetic acid is sold in various concentrations. The British official acetic acid is 36%, glacial acetic acid contains at least 99%. Commercial glacial acid contains about 80%. A technical grade containing 28% acid is also known as No. 8 acetic acid because it can be diluted with water to give eight volumes of official 4% vinegar (British Excise system).

Peracetic acid, $\begin{matrix} O-OH \\ \diagup \\ CH_3C=O, \end{matrix}$ now commercially available, is best obtained by the action of conc. hydrogen peroxide with acetic anhydride. It is a powerful oxidizing agent, with unusual properties being capable of converting manganese salts to permanganates.[36]

Propionic acid, propanoic acid, methylacetic acid, $EtCO_2H$, is prepared from ethanol and CO under pressure using a BF_3 catalyst. It can also be prepared by oxidation of n-propyl alcohol. During World War I it was available as a by-product in the extraction of potash from kelp. The mixture of aliphatic acids obtained by the fermentation of the kelp was separated by the distillation of the ethyl esters. It can be salted out of its water solution as an oil. Hence its name as the "first" fatty acid.

The reactions of propionic acid are like those of acetic acid. It is a weaker acid than the latter. Only its two alpha H atoms are readily replaced by halogens and take part in the Perkin reaction.

$$CH_3CH_2CO_2H \rightarrow CH_3CHBrCO_2H \rightarrow CH_3CBr_2CO_2H$$
$$RCHO + CH_3CH_2CO_2Na + (CH_3CH_2CO)_2O \rightarrow RCH=C(Me)CO_2H$$

[31] Hall, Werner. *J. Am. Chem. Soc.* 50, 2367 (1928).
[32] Hall. *J. Am. Chem. Soc.* 52, 5115 (1930).
[33] Conant, Hall. *J. Am. Chem. Soc.* 49, 3062 (1928).
[34] MacInnes, Shedlovsky. *J. Am. Chem. Soc.* 54, 1429 (1932).
[35] Harned, Ehlers. *J. Am. Chem. Soc.* 54, 1350 (1932).
[36] *Ann. Rep. Chem. Soc.* (London) 1912, 91.

Sodium and calcium propionates are used to inhibit mold and fungus in baked goods (*Mycoban*). They find similar uses with other vegetable materials including tobacco.

n-**Butyric acid,** butanoic acid, ethylacetic acid, $CH_3CH_2CH_2CO_2H$, occurs in butter (2–3%). It is made by the butyric fermentation of carbohydrates and by the oxidation of *n*-butyl alcohol. The latter process was carried out in Germany during World War II by passing the alcohol vapor through a mixture of the alcohol and 17% excess of NaOH at 230–70° (TOM). The same process was applied to isobutyl alcohol and to the normal 5–12 carbon alcohols.

$$RCH_2OH + NaOH \rightarrow RCO_2Na + 2 H_2$$

In the fermentation process, the intermediate acetaldehyde probably goes to aldol which undergoes a dismutation to the acid. Certain microorganisms convert lactic acid to *n*-butyric acid by similar changes. Its reactions are like those of propionic acid. In addition it shows the phenomenon of "beta oxidation." Thus, with hydrogen peroxide it goes through the following steps.[37]

$$CH_3CH_2CH_2CO_2H \rightarrow CH_3CHOHCH_2CO_2H \rightarrow$$
$$CH_3COCH_2CO_2H \rightarrow CO_2 + CH_3COCH_3$$

A similar oxidation of fatty acids occurs in living organisms. Its mechanism has never been satisfactorily explained.[38] Alpha oxidation is also possible.[39]

1-Nitrobutane reacts with sulfuric acid to give *n*-butyric acid and hydroxylamine sulfate (CSC).

Electrolysis of potassium butyrate solution gives mainly propylene together with isopropyl butyrate. Presumably the *n*-propyl radical forms propylene which combines with butyric acid to form isopropyl butyrate.

Isobutyric acid, methylpropanoic acid, dimethylacetic acid,

$$(CH_3)_2CHCO_2H,$$

b. 155°, occurs in carob beans (St. John's Bread). It is cheaply available from isobutylene (Shell). Treatment with chlorine gives mainly methallyl chloride which reacts with lime and water to form isobutyraldehyde, readily convertible to the acid by catalytic oxidation.

Its reactions are complicated by the fork in the chain in the alpha position. Thus it is much more sensitive to oxidation than any of its lower homologs except formic acid.

$$Me_2CHCO_2H \rightarrow Me_2C(OH)CO_2H \rightarrow Me_2CO + CO_2 + H_2O$$

Sodium isobutyrate, on electrolysis, gives large amounts of propylene and isopropyl isobutyrate. This is in contrast to the *n*-butyrate which gives mainly propylene and very little ester.

[37] Dakin. *Am. Chem. J.* **44**, 46 (1910).
[38] Harmon, Marvel. *J. Am. Chem. Soc* **54**, 2515 (1932).
[39] Witzemann. *J. Am. Chem. Soc.* **48**, 202, 208, 211 (1926).

Ethyl isobutyrate does not condense with itself under the influence of NaOEt (Hurd). Sodium triphenylmethyl gives a 35% yield of 2,2,4,4-tetramethylacetoacetic ester.[40]

While Ca isobutyrate is more soluble in hot water than in cold, Ca n-butyrate is more soluble in cold than in hot water. One salt crystallizes with 5 H_2O and the other with 1 H_2O.

n-Valeric acid, n-valerianic acid, pentanoic acid, n-propylacetic acid, $CH_3CH_2CH_2CH_2CO_2H$, is made by the oxidation of n-amyl alcohol or from n-butyl alcohol through the bromide and the cyanide. An 80% yield has been obtained from BuMgBr and CO_2 at $-20°$. At higher temperatures considerable Bu_2CO results.[41] It is also obtained by the vigorous reduction of levulinic acid obtained from sucrose and hydrochloric acid,

$$MeCOCH_2CH_2CO_2H \rightarrow MeCH_2CH_2CH_2CO_2H$$

Isovaleric acid, isovalerianic acid, isopropylacetic acid, 3-Me-butanoic acid, $(CH_3)_2CHCH_2CO_2H$, b. 174°, is obtained by oxidation and hydrolysis of the corresponding alcohol and cyanide respectively. It is available in reasonable amounts from pyroligneous acid.

Electrolysis of its sodium salt gives 60% of diisobutyl and 29% of isobutylene.

Heating calcium isovalerate gives mainly isovaleraldehyde instead of isovalerone.[42]

Active valeric acid, methylethylacetic acid, 2-methylbutanoic acid, $C_2H_5(CH_3)CHCO_2H$, b. 177°, is obtained by the oxidation of "active" amyl alcohol. If d-amyl alcohol (levorotatory) from fusel alcohol is used, the product is dextrorotatory. If secondary-butylcarbinol from the chlorination of isopentane is used, the product is a "racemic mixture" of the two optically active isomers. A similar optically inactive mixture is obtained if the acid is made by the malonic ester synthesis or the acetoacetic ester synthesis.

Such a mixture of enantiomorphic forms cannot be separated by any ordinary means because the physical properties of the d- and l-isomers are identical except for the action on plane polarized light. Such a mixture of optically active acids can be resolved into the optical isomers by forming salts with an optically active base such as d-brucine. Two salts will be formed which are no longer enantiomorphic and which differ enough in solubilities to make their separation by fractional crystallization possible. These two salts may be represented as

d-Acid.d-Brucine   and   l-Acid.d-Brucine

After they have been crystallized into two fractions the optical rotations of which do not change on further crystallization, the optically active acids can

[40] Hauser, Renfrow. *J. Am. Chem. Soc.* 59, 1823 (1937).
[41] Gilman, Parker. *J. Am. Chem. Soc.* 46, 2816 (1924).
[42] *Ann. Rep. Chem. Soc.* (London) 1906, 81.

be liberated by treatment with hydrochloric acid.

$$d\text{-Acid}.\,d\text{-Brucine} + HCl \rightarrow d\text{-Brucine}.\,HCl + d\text{-Acid}$$
$$l\text{-Acid}.\,d\text{-Brucine} + HCl \rightarrow d\text{-Brucine}.\,HCl + l\text{-Acid}$$

The wide occurrence of isovaleric and active valeric acids in nature is due to their formation by enzymes from the amino acids, leucine and isoleucine. Many higher optically active acids have been obtained.[43]

Trimethylacetic acid, pivalic acid, 2,2-dimethylpropanoic acid,

$$(CH_3)_3CCO_2H,$$

b. 163°, m. 35°, has been made:

(a) From tertiary butyl cyanide made from the chloride and the double salt $KCN.\,Hg(CN)_2$; (b) From the Grignard reagent and CO_2; (c) From acetone through pinacol and pinacolone.[44] It gives the usual acid reactions except that the accumulated groups on the alpha carbon and the absence of $\alpha - H$ atoms modify them. Thus its velocity of esterification is much less than that of isobutyric acid which, in turn, esterifies much more slowly than n-butyric acid (Menschutkin). These facts have sometimes been improperly generalized in a statement that primary, secondary and tertiary acids can be distinguished by their speeds of esterification. This is a safe method only if acids of the *same molecular weight* are compared.

Over thoria catalyst at 490° trimethylacetic acid is decarboxylated to t-butyl isobutyl ketone. In the absence of *alpha* hydrogens the *beta* hydrogens apparently become activated at the higher temperatures producing the unusual results.[45]

With phosphorus pentoxide, Me_3-acetic acid gives a 60% yield of CO and no CO_2.[46]

$$Me_3CCO_2H + P_2O_5 \rightarrow 2\ HPO_3 + CO + (Me_2C{=}CH_2)_n$$

Under similar conditions isobutyric and n-butyric acids give small yields of mixtures of CO and CO_2 in the ratios 5:1 and 1:5 respectively.

Trisubstituted acetic acids are difficultly obtainable. Special methods have been developed for their preparation. This problem is of unusual interest because it had been suspected that *phthioic acid* from tubercle bacilli[47] was ethyldecyldodecylacetic acid.[48] The most general method is that of Haller and Bauer,[49] starting with a phenyl alkyl ketone, $C_6H_5COCH_2R$ and replacing the alpha H atoms by successive treatments with $NaNH_2$ and RX. It is best to use the largest alkyl in the starting ketone, then to put in the smallest alkyl

[43] Levene, Marker. *J. Biol. Chem.* **93**, 749 (1931).
[44] "Org. Syntheses."
[45] Whitmore, Miller. *J. Am. Chem. Soc.* **72**, 2732 (1950).
[46] Whitmore, Laughlin. *J. Am. Chem. Soc.* **54**, 4462 (1932).
[47] Anderson. *J. Biol. Chem.* **83**, 169 (1929).
[48] Birch, Robinson. *J. Chem. Soc.* 1942, 488.
[49] Carter, Slater. *J. Chem. Soc.* 1946, 130.

and, finally, the third alkyl using the iodide. The trisubstituted acetophenone is then split by vigorous treatment with sodamide, followed by hydrolysis to the acid.

$$C_6H_5COCRR'R'' + NaNH_2 \rightarrow RR'R''CCONH_2 + C_6H_6$$

This process works well for dimethylacetic acids and fairly well if there is one ethyl group and the two other groups are not too large. In the latter case, the trisubstituted acetophenone is difficult to obtain and, on splitting, gives benzamide and the aliphatic hydrocarbon instead of the desired products. Thus it fails in the preparation of ethyldecyldodecylacetic acid.

Another method is to introduce a tertiary group into a furan nucleus and then oxidize the latter to leave only a carboxyl group attached to the tertiary group.[50]

Isobutyric and related esters can be alkylated in the alpha position by means of tritylsodium and alkyl halides.[51]

MeEt$_2$-acetic acid is obtained by a special method from ethylene, carbon monoxide, BF$_3$.H$_2$O, and benzoyl peroxide.[52]

Caproic acid, n-hexylic acid, n-butylacetic acid, hexanoic acid,

$$CH_3(CH_2)_4CO_2H.$$

At the time caproic acid was first needed for the production of hexyl-resorcinol, n-butyl alcohol was readily available. The acid was made by the acetoacetic acid synthesis. The entire process starting with ethyl acetate is as follows:

$$2\ CH_3CO_2Et + Na + trace\ NaOEt \rightarrow CH_3C(ONa)=CHCO_2Et + EtOH$$

The trace of sodium ethylate causes an aldol condensation between an alpha H of one ester molecule with the carbonyl group of another. The resulting product loses alcohol, enolizes and forms the sodio derivative. This is treated with absolute alcohol and n-butyl bromide forming n-butylacetoacetic ester or, more properly, ethyl n-butylacetoacetate. As usual the new C—C linkage is not formed where the Na was attached. A consideration of only the reacting portions of the molecules gives

[50] Reichstein, Rosenberg, Eberhardt. *Helv. Chim. Acta,* 18, 721 (1935).
[51] Hudson, Hauser. *J. Am. Chem. Soc.* 62, 2457 (1940).
[52] U. S. Patent No. 2,378,009 (1945).

The butylacetoacetic ester is then split by heating with alcoholic potash to give the potassium salts of acetic acid and of n-butylacetic acid.

$$MeCOCH(Bu)CO_2Et + 2\ KOH \rightarrow MeCO_2K + BuCH_2CO_2K + EtOH$$

The mixture is acidified and the caproic acid floats as an oily layer. A similar treatment gives any substituted acetic acid RCH_2CO_2H in which R corresponds to a halide of Class 1 (p. 74) and a few members of the other classes (*not* tertiary). Similarly the even-numbered normal acids $C_{20}-C_{30}$ have been prepared.

When n-amyl alcohol became available, the process for caproic acid was changed to go through the corresponding bromide or chloride and cyanide.

Probably the best present method for caproic acid is from the aldol condensation products of acetaldehyde. The unsaturated aldehyde can be hydrogenated to the saturated aldehyde or alcohol which can then be oxidized to the acid.

$$3\ CH_3CHO \rightarrow CH_3(CH=CH)_2CHO \rightarrow Acid.$$

The making of caproic acid for hexyl resorcinol is typical of the development of organic chemical processes. If it had been needed before World War I, it would have been very difficult to obtain. Since that time C_4, C_5 and C_6 compounds have been developed which can be used in its preparation.

Another preparation of caproic acid is by the oxidation of capryl alcohol, methylhexylcarbinol obtained from castor oil.

Isocaproic acid, isobutylacetic acid, 4-methylpentanoic acid,

$$(CH_3)_2CHCH_2CH_2CO_2H,$$

can be obtained by the acetoacetic ester synthesis using isobutyl bromide. The yield is cut because of the alkaline nature of the sodium acetoacetic ester, much of the bromide being changed to olefin. Better methods of preparation are from the corresponding cyanide and from an isoamyl Grignard reagent and carbon dioxide.

Methyl-n-propylacetic acid, 2-methylpentanoic acid, $PrMeCHCO_2H$, is best made by the acetoacetic ester synthesis. Acetoacetic ester in absolute alcohol is treated with one equivalent each of NaOEt and MeI and refluxed until a test sample mixed with water is neutral. The solution then contains methyl acetoacetic ester, $MeCOCH(Me)CO_2Et$. Then another equivalent of NaOEt and one of n-PrBr are added. The final product is methyl-n-propylacetoacetic ester. A better yield of the ester is obtained this way than if the n-propyl group is introduced first. The product is split by alcoholic KOH as usual.

$$MeCOC(Me)(Pr)CO_2Et + 2\ KOH \rightarrow MeCO_2K + MePrCHCO_2K$$

The splitting effect of concentrated alcoholic KOH on the groupings

$$-CO-CH_2-CO-\quad -CO-CHR-CO-\quad -CO-CRR'-CO-$$

is analogous to the splitting of the grouping $Cl_3C - CO -$ even by dilute alkali.

The *malonic ester synthesis of this acid* is entirely similar to this process. Treatment with NaOEt, MeI and again with NaOEt and finally with PrBr gives methyl-*n*-propylmalonic ester, $MePrC(CO_2Et)_2$. When this is saponified with alcoholic KOH and acidified, the corresponding disubstituted malonic acid is obtained. On heating, this loses CO_2 giving the desired disubstituted acetic acid.

Another preparation of MePr-acetic acid is by the oxidation of the alcohol obtained by heating *n*-propyl alcohol with its sodium derivative (Guerbet).

$$PrONa + MeCH_2CH_2OH \rightarrow PrMeCHCH_2OH \rightarrow PrMeCHCO_2H$$

This acid has been obtained in optically active form $[\alpha]_D = +8.9°$.

Diethylacetic acid, 2-ethylbutanoic acid, $(CH_3CH_2)_2CHCO_2H$, was formerly prepared by introducing two ethyl groups into acetoacetic ester or malonic ester and splitting in the usual manner. It is now produced commercially from the corresponding aldehyde obtained by the partial hydrogenation of the unsaturated aldehyde obtained by the condensation of two molecules of *n*-butyraldehyde (C and C).

sec-**Butylacetic acid,** 3-methylpentanoic acid, $CH_3CH_2(CH_3)CHCH_2CO_2H$, would be obtained in poor yield by either the malonic ester or the acetoacetic ester synthesis. It is best made through the cyanide from *sec*-butyl carbinol.

Dimethylethylacetic acid, 2,2-dimethylbutanoic acid,

$$CH_3CH_2(CH_3)_2CCO_2H,$$

cannot be made by the acetoacetic ester or malonic ester synthesis. It is made by the action of carbon dioxide on a *t*-amyl Grignard reagent.

t-**Butylacetic acid,** 3,3-dimethylbutanoic acid, $(CH_3)_3CCH_2CO_2H$, cannot be made by the acetoacetic ester or malonic ester synthesis. It cannot be made by the cyanide method because the necessary halide is inactive. It is best made by the action of Br_2 and NaOH on methyl neopentyl ketone, $MeCOCH_2CMe_3$, from the oxidation of diisobutylene (p. 46).

Methylisopropylacetic acid, 2,3-dimethylbutanoic acid,

$$(CH_3)_2CHCH(CH_3)CO_2H,$$

can be made in poor yields by the malonic ester or acetoacetic ester synthesis. The isopropyl group must be introduced first and followed by the methyl group.

n-**Heptanoic acid,** *n*-heptylic acid, oenanthic acid, heptanoic acid, $CH_3(CH_2)_5CO_2H$, is readily available by the oxidation of heptaldehyde obtained by the destructive distillation of castor oil.

n-**Octanoic acid,** caprylic acid, $CH_3(CH_2)_6CO_2H$, has been obtained from natural fats and by the oxidation and reduction of the aldol condensation product of two molecules of aldol, $Me(CH=CH)_3CHO$.[53]

2-Ethylhexanoic acid, $CH_3(CH_2)_3CH(C_2H_5)CO_2H$, b. 230, is commercially available from the aldol from two molecules of butanal (C and C).

Methylneopentylacetic acid, $(CH_3)_3CCH_2CH(CH_3)CO_2H$, b. 218°, is a by-product of the chromic acid oxidation of isobutylene. This involves the same rearrangement as the acid oxidation of isobutylene and of the tri-iso-butylenes to give acids of the same number of carbon atoms as the olefins (p. 255).

n-Nonanoic acid, pelargonic acid, $CH_3(CH_2)_7CO_2H$, is obtained by the oxidation of oleic acid. A preparation useful in making many complex acids is by the reduction of the corresponding $\alpha\beta$-unsaturated acids with sodium amalgam. The unsaturated acids can be made from malonic acid or ester and aldehydes by a modification of the aldol condensation (Knoevenagel 1894, 1904).

$$C_6H_{13}CHO + CH_2(CO_2H)_2 \xrightarrow{Et_2NH} C_6H_{13}CH=C(CO_2H)_2 \rightarrow$$
$$C_6H_{13}CH=CHCO_2H \rightarrow C_6H_{13}CH_2CH_2CO_2H$$

It can also be obtained by fusion of the potassium salt of undecylenic acid with KOH. This represents a shift of the double bond from one end of the chain to the $\alpha\beta$-position at the other end. The $\alpha\beta$-unsaturated acid is split by the fused alkali.

$$CH_2=CH(CH_2)_8CO_2K + KOH + H_2O \rightarrow$$
$$CH_3(CH_2)_7CO_2K + CH_3CO_2K + H_2$$

A similar process changes oleic acid to palmitic acid. Before the possibilities in the shift of a double bond were understood, this reaction was believed to indicate oleic acid as an $\alpha\beta$-unsaturated acid.

n-Decanoic acid, capric acid, $CH_3(CH_2)_8CO_2H$, is obtained from lauric acid by oxidation of the $\alpha\beta$-unsaturated acid obtained from the α-bromo acid. This is a general method of removing two carbons from an acid.

$$RCH_2CH_2CO_2H \rightarrow RCH_2CHBrCO_2H \rightarrow RCH=CHCO_2H \rightarrow RCO_2H$$

Capric acid is readily available from the seed fat of the California bay tree.[54]

n-Undecanoic acid, $CH_3(CH_2)_9CO_2H$, is best prepared by hydrogenation of undecylenic acid, $CH_2=CH(CH_2)_8CO_2H$, obtained by the destructive distillation of castor oil.

Lauric acid, n-dodecanoic acid, $CH_3(CH_2)_{10}CO_2H$, is readily obtained from cocoanut oil which contains glycerides of lower acids than most oils and fats. Conversion of the glycerides to methyl esters and the fractionation of 130

[53] *Ann. Rep. Chem. Soc.* (London) 1907, 80.
[54] Noller, Millner, Gordon. *J. Am. Chem. Soc.* **55**, 1227 (1933).

kg. of these esters gave the following acids (%): caproic, C_6, 0.4; caprylic, C_8, 8.7; capric, C_{10}, 5.6; *lauric*, C_{12}, 45.0; myristic, C_{14}, 17; C_{16} and higher 23. The catalytic reduction of lauric acid or its ester gives lauryl alcohol which is valuable in detergents containing sodium lauryl sulfate (S.L.S.).

Electrolysis of sodium laurate gives a 90% yield of n-$C_{22}H_{46}$.

Butlerow's Acids, $C_{11}H_{23}CO_2H$, **dineopentylacetic acid,**

$$(Me_3CCH_2)_2CHCO_2H$$

and **methyl-t-butylneopentylacetic acid,** $(Me_3CCH_2)(Me_3C)MeCCO_2H$, are obtained by the acid oxidation of triisobutylene ($C_{12}H_{24}$).[55–57] Such a formation of an acid from an olefin *without loss of carbon* involves a rearrangement.

Myristic acid, $CH_3(CH_2)_{12}CO_2H$, is best prepared from its glyceride found in nutmegs.[58]

Palmitic acid, hexadecanoic acid, $C_{15}H_{31}CO_2H$, and **stearic acid,** octadecanoic acid, $C_{17}H_{35}CO_2H$, occur as glycerides in animal and vegetable fats and oils. In general the higher the melting point of the fat the larger the proportion of the glycerides of these saturated acids. Similarly the "hardening" of liquid fats and oils by catalytic hydrogenation increases the proportion of these acids in the fats.

They are prepared from the glycerides by hydrolysis in presence of a base (lime) or with an acid catalyst (Twitchell's reagent, a complex product from the action of naphthalene, oleic acid and sulfuric acid). The free acids are allowed to crystallize and the liquid acids such as oleic acid are pressed out. The resulting mixture of solid acids is *stearin*, formerly used for candle making. It is also converted to the sodium salts (soap). If the pure acids are desired, the mixture can be fractionated at about 5 mm. pressure.[59] Separation is more readily obtained in the laboratory by distillation of the Me or Et esters.[60]

Greases are made from higher petroleum fractions and sodium, calcium, or lithium soaps of higher fatty acids. Aluminum stearate is used in waterproofing and in thickening petroleum fractions for use in incendiary bombs.

Hydrogenation with a Cu−Cr oxide catalyst gives the C_{16} and C_{18} alcohols whose sulfates are valuable as detergents (Dreft, etc.). Until the development of this catalytic method, acids could be reduced only by conversion to the esters and treatment with Na and absolute EtOH (Bouveault and Blanc).

Silver salts of the fatty acids undergo a peculiar reaction to form the ester of the acid with the next lower alcohol.

$$2\ RCH_2CO_2Ag + I_2 \rightarrow 2\ AgI + CO_2 + RCH_2CO_2CH_2R$$

[55] Conant, Wheland. *J. Am. Chem. Soc.* 55, 2499 (1933).
[56] Whitmore, Laughlin. *J. Am. Chem. Soc.* 56, 1128 (1934).
[57] Whitmore, Wilson. *J. Am. Chem. Soc.* 56, 1397 (1934).
[58] "Org. Syntheses."
[59] Stingley. *Chemistry & Industry* 49 (1941).
[60] *Ann. Rep. Chem. Soc.* (London) 1925, 78.

Hydrolysis of the ester gives the lower alcohol RCH_2OH. This is a special way of going down the series.

$$RCH_2CO_2H \rightarrow RCH_2OH$$

An isomer of stearic acid, 3,5,7-Me$_3$-pentadecanoic acid, has been made starting with methyl n-octyl ketone and applying the Reformatzky reaction, dehydration, hydrogenation, conversion to the acid chloride and treatment with dimethylzinc to give a methyl ketone on which the cycle of changes is *twice* repeated.[61]

$$RCOMe + BrCH_2CO_2Et + Zn \rightarrow RMeC(OH)CH_2CO_2Et \rightarrow$$
$$RMeC = CHCO_2Et \rightarrow RMeCHCH_2CO_2Et \rightarrow$$
$$RMeCHCH_2COCl \rightarrow RMeCHCH_2COMe \rightarrow Etc.$$

A very important series of isomers of C_{16} and C_{18} acids has been made by means of the malonic ester synthesis. Hexadecanoic and octadecanoic acids, $RR'CHCO_2H$, in which R is one of the first eight n-alkyls or isopropyl, isobutyl, sec-butyl or sec-amyl have been studied as to their bactericidal action.[62]

Three stereoisomeric forms of 10-methyl octadecanoic acid have been synthesized from the three corresponding isomers of 3-methylhendecanoic acid. The $(-)$-10-methyloctadecanoic acid is identical with tuberculostearic acid.[63]

Other high molecular weight fatty acids have been made by the malonic ester synthesis.[64]

Alpha methyl derivatives of the normal C_{15} to C_{18} acids have been made by treating the α-bromo acids in ether solution with an excess of MeMgX.[65]

Margaric acid, daturic acid, $C_{16}H_{33}CO_2H$, differs from the naturally occurring fatty acids which contain an even number of carbon atoms. The latter are probably built up in the organism by some such process as the aldol condensation which would give products containing $2n$ carbons. The natural fats on digestion undergo beta oxidation and lose two carbon atoms at a time thus eventually giving $CH_3COCH_2CO_2H$ and CH_3COCH_3. In diabetic conditions these products accumulate. Hence margaric acid, an odd-numbered acid which cannot give these products, has been made as a source of artificial fats for the use of diabetics (Intarvin). Its preparation is by the method devised by Krafft (1882) for making higher acids. The Ba or Ca salts of stearic acid and acetic acid are destructively distilled to give acetone and methyl heptadecyl ketone, $MeCOC_{17}H_{35}$, which on oxidation gives acetic acid and margaric acid.

The method of degrading the naturally occurring fatty acids by conversion to methyl ketones and oxidation has been performed from stearic acid to

[61] Rupe, Willi. *Helv. Chim. Acta* 15, 842 (1932).
[62] Stanley, Jay, Adams. *J. Am. Chem. Soc.* 51, 1261 (1929).
[63] Prout, Cason, Ingersoll. *J. Am. Chem. Soc.* 70, 298 (1948).
[64] Bleyberg, Ulrich. *Ber.* 64B, 2504 (1931).
[65] Morgan, Holmes. *J. Soc. Chem. Ind.* 46, 152T (1927).

capric acid, C_{10}, which was proved by synthesis to have a normal chain. This is the evidence for the normal chain in the natural acids. A fork in the chain would produce a ketone instead of an acid.

$$RR'CHCOMe \rightarrow MeCO_2H + RR'CO$$

Various *higher fatty acids* are obtained by the oxidation of paraffin wax and other petroleum products by air or oxygen.[66, 67] The production of *synthetic fatty acids* was put on a plant scale in Germany during World War II. Paraffins, from the hydrogenation of lignite, were oxidized by air at the rate of about 100,000 tons per year. The C_1-C_4 acids were scrubbed from the effluent gases, the formic acid being used in silage, the acetic and butyric acids for making cellulose esters, and the propionic acid being used as the calcium salt as a mold inhibitor in bread. The C_5-C_7 fraction was reduced to the alcohols for use in modified glyptal resins, the C_7-C_9 found similar uses and also in foamite type fire extinguishers, and the C_9-C_{11} fraction was used in mineral flotation. Distillation at 3 mm. separated the higher acids, a $C_{10}-C_{18}$ fraction being used for soaps, a C_9-C_{16} fraction for edible fats, and a $C_{18}-C_{24}$ fraction for Na, Ca, and Li salts for greases. The residual pitch can be hydrogenated directly to high molecular weight plasticizers, or can be converted to ketones which give still higher hydrocarbons on hydrogenation. Thus a substitute for vaseline is obtained.

Dicetylacetic acid, $(C_{16}H_{33})_2CHCO_2H$, m. 70°, has been made by the malonic ester synthesis.

Tricosanoic (C_{23}) and tetracosanoic (C_{24}) acids have been synthesized by the Clemmensen reduction of the 13-keto acids from the action of the chlorides of undecanoic and lauric acids on the sodium derivative of an ester obtained from sodium acetoacetic ester and ethyl 11-Br-undecoate. While the melting points of the acids are 79° and 84° respectively, mixtures containing up to 40% of the second melt 78.5–79°.

n-Triacontanoic acid (C_{30}) has been synthesized by a method which involves ω-bromo esters, sodio acetoacetate and acyl chlorides, in this case ω-bromo-undecylenic acid and stearyl chloride

$$Br(CH_2)_nCO_2Et \xrightarrow[\text{NaOEt}]{\text{MeCOCH}_2\text{CO}_2\text{Et}} MeCOCH(CO_2Et)(CH_2)_nCO_2Et$$

$$\xrightarrow[\text{Me(CH}_2)_m\text{COCl}]{\text{NaOEt}} Me(CH_2)_mCOC(COMe)(CO_2Et)(CH_2)_nCO_2Et \rightarrow$$

$$Me(CH)_mCO(CH_2)_{n+1}CO_2H \xrightarrow[\text{HCl}]{\text{Zn} \times \text{Hg}} Me(CH_2)_mCH_2(CH_2)_{n+1}CO_2H.$$

The last step is a Clemmensen reduction (p. 252).

[66] Ellis. "Chemistry of Petroleum Derivatives." Reinhold, 1934. pp. 930–960.
[67] Penniman, James, Burwell. *Ann. Rep. Chem. Soc.* (London) 1920, 52.

Identification of Acids

This may be achieved by conversion to a variety of solid derivatives of definite melting points such as: *a*. amides, $RCONH_2$; *b*. anilides, $RCONHC_6H_5$; *c*. para halogen phenacyl esters, $RCO_2CH_2COC_6H_4X$, obtained from the sodium salts and $BrCH_2COC_6H_4X$; *d*. *p*-phenylphenacyl esters,

$$RCO_2CH_2COC_6H_4C_6H_5.$$

Even the formic ester of the latter melts at 74°; *e*. *p*-nitrobenzyl esters, $RCO_2CH_2C_6H_4NO_2$; *f*. a method valuable in special cases is the reaction of the sodium salts with chloroacetone and formation of the semicarbazone of the resulting ester, $RCO_2CH_2C(Me)=NNHCONH_2$ (Locquin 1904, Freylon 1910); *g*. the diamides from p,p'-$(NH_2)_2$-diphenylmethane.[68]

The partition coefficients of acids between immiscible solvents such as ether and water are helpful in separation and identification.[69]

B. OLEFINIC ACIDS, $C_nH_{2n-1}CO_2H$

1. **Acrylic acid,** propenoic acid, $CH_2=CHCO_2H$, m. 12°, b. 142°.
Preparation. 1. Oxidation of acrolein by silver oxide.

$$CH_2=CHCHO + Ag_2O \rightarrow CH_2=CHCO_2H + 2\ Ag$$

More powerful oxidizers attack the double bond as well as the aldehyde hydrogen.

2. By oxidation of allyl alcohol after protecting the double bond.

$$CH_2=CHCH_2OH \rightarrow CH_2BrCHBrCH_2OH \rightarrow$$
$$CH_2BrCHBrCO_2H \rightarrow CH_2=CHCO_2H$$

3. By dehydration of hydracrylic acid, beta hydroxypropionic acid. It is not obtained from lactic acid or its esters which give a lactide instead. Since its main uses are based on the polymerization or copolymerization of alkyl acrylates, these are generally prepared without making the acid. Beta-hydroxypropionitrile is made from ethylene chlorohydrin and NaCN or from ethylene oxide and HCN. This is alcoholized in the presence of sulfuric acid to give the ester, $HOCH_2CH_2CO_2R$. This is dehydrated with P_2O_5 or by cracking the acetate.

4. From α- or β-halogen propionic acids and alcoholic KOH.
5. It is reported that acrylic acid can be made from ethylene and CO_2.[70]
6. Acrylic acid can be made from lactic acid.

Reactions. 1. Acrylic acid gives the reactions of carboxylic acids, sometimes complicated by the alpha beta unsaturation (conjugated with the unsaturation of the carbonyl group).

[68] Ralston, McCorkle. *J. Am. Chem. Soc.* 61, 1604 (1939).
[69] Dermer. *J. Am. Chem. Soc.* 65, 1633 (1943).
[70] Ellis. "Chemistry of Petroleum Derivatives." Reinhold, 1934. p. 586.

2. It gives olefin reactions with unusual ease. It behaves as a typical alpha-beta unsaturated keto compound.[70a] It is readily reduced by Na $\times$ Hg, whereas olefinic hydrocarbons are not. It adds HX contrary to Markownikoff's Rule. Under suitable conditions, the acid or its esters will add NH_3, HCN, ROH, and the Na derivatives of 1,3-diketo compounds such as malonic and acetoacetic esters. All these processes involve 1,4-addition with the H or Na adding to the carbonyl O and the other group adding to the beta C(p. 224). With a substance like hypochlorous acid which adds to olefins and contains no group which can add to oxygen, a mixture of the two possible isomers is obtained.

$$CH_2 = CHCO_2H \rightarrow HOCH_2CHClCO_2H + ClCH_2CHOHCO_2H$$

A study of the addition of HOCl to the hexenoic acids shows the influence of the position of the double bond on the direction of the addition.[70b] With the $\alpha\beta$-compound the Cl adds 100% α, with the $\beta\gamma$-compound 90% β and with the $\gamma\delta$-compound 95% δ.

3. It polymerizes readily. This tendency can be overcome by adding a trace of an inhibitor such as hydroquinone.

4. When heated with very concentrated alkali it is split at the double bond.

$$CH_2 = CHCO_2K + KOH + H_2O \rightarrow HCO_2K + CH_3CO_2K + H_2$$

This is characteristic of all alpha beta unsaturated acids.

5. Acrylic esters, alone or with methacrylic esters are polymerized to valuable colorless transparent plastics by catalysts such as benzoyl peroxide[71, 72] (*Plexiglas*, *Lucite*).

6. Polymers from higher esters of acrylic acid are valuable additives for lubricants and hydraulic fluids (*Acryloids*, RH).

7. Copolymers of acrylates with small amounts of allyl maleate or chloroethyl vinyl ether can be vulcanized.[73]

8. Acrylic esters react with Na and alcohols to give β-alkoxy propionic esters, and with diazomethane and diazoacetic ester to give pyrazoline-3-carboxylic ester and pyrazoline-3,5-dicarboxylic ester respectively. With pure NaOMe, methyl acrylate gives α-methyleneglutaric ester. With conjugated dienes, substituted tetrahydrobenzoic acids are obtained (Diels, Alder) (p. 57). This is general for $\alpha\beta$-unsaturated compounds.

9. Acrylic acid is about twice as much ionized as propionic acid.

β-**Methylacrylic acids**, 2-butenoic acids, Δ^2-butenoic acids,

$$CH_3CH = CHCO_2H,$$

exist in *cis* and *trans* forms, isocrotonic acid and crotonic acid respectively.

[70a] *Ann. Rep. Chem. Soc.* (London) **1932**, 107.
[70b] *Ann. Rep. Chem. Soc.* (London), **1932**, 106.
[71] Neher.   *Ind. Eng. Chem.* **28**, 267 (1936).
[72] Price, Kell, Krebs.   *J. Am. Chem. Soc.* **64**, 1103 (1942).
[73] Mast, Smith, Fisher.   *Ind. Eng. Chem.* **36**, 1027 (1944).

Some of their physical properties follow:

cis acid		trans acid
14.6°	M. Pt.	72°
171.9°	B. Pt.	185°
1.027 at 15°	Density	0.964 at 79.7°
	Solubility	
40.	g./100 cc. H_2O	8.3
	at 25°	
	Heat of combustion	
486	Cal./mole	478

The boiling point relationship of the isomers differs from that of pairs of olefinic isomers in which the *cis* form has the higher value. In the latter the olefinic linkage is more "exposed" and gives a higher boiling point as in the case of 1-olefins. In the *trans* acid, the greater "exposure" of the carboxyl group with the added tendency for hydrogen bonding (p. 238) overcomes the effect of the exposure of the olefinic linkage in the *cis* acid.

Crotonic acid, *cis*-$CH_3CH=CHCO_2H$, is available commercially as a product from acetaldehyde obtained from acetylene (Niacet, SCL).

Preparation. 1. By catalytic oxidation of crotonaldehyde from aldol.[74]

2. From alpha bromo-*n*-butyric ester and alcoholic KOH or a tertiary amine.

3. From malonic ester with paraldehyde in acetic anhydride or with acetaldehyde and secondary amines (Knoevenagel).

$$MeCHO + CH_2(CO_2R)_2 \rightarrow MeCHOHCH(CO_2R)_2 \rightarrow MeCH=C(CO_2R)_2 \rightarrow$$
$$MeCH=C(CO_2H)_2 \rightarrow MeCH=CHCO_2H$$

4. From allyl cyanide by hydrolysis and rearrangement. The cyanide is made from allyl bromide and cuprous cyanide.[74a]

$$CH_2=CHCH_2CN \rightarrow [CH_2=CHCH_2CO_2H] \rightarrow CH_3CH=CHCO_2H$$

This shift of a double bond to the alpha beta position in which it is conjugated with the carbonyl group is general. The reverse shift is favored only in presence of a branch in the beta or gamma position.[75]

Reactions. The reactions of crotonic acid closely resemble those of acrylic acid but are more readily available because it is less easily polymerized. Moreover, the methyl group at the end of the conjugated system gives special reactions.

Crotonic acid is readily hydrogenated to *n*-butyric acid by Na × Hg and water, Zn and acid, Ni–Al alloy (Raney) and 10% NaOH, or H_2 and Ni or Pt at 190°. These easy reductions probably go by 1,4-addition.

The conjugation of the olefinic and carbonyl linkages decreases the un-

[74] Young. *J. Am. Chem. Soc.* **54**, 2498 (1932).
[74a] Org. Synthesis.
[75] *Ann. Rep. Chem. Soc.* (London) 1927, 111; 1932, 136–144.

saturation of each. The effect on the former is shown by the slower addition of bromine. Thus the velocity constant at 15° for the addition of bromine to 3-pentenoic acid is about 2×10^6 times that of crotonic acid (2-butenoic acid). This effect can be largely overcome by bromination in water solution in which the carboxyl group is largely hydrated to $-C(OH)_3$. Fused crotonic acid and HBr give beta-bromobutyric acid quantitatively. This has been interpreted as 1,4-addition initiated by attack of H on the carbonyl oxygen or by attack of H on the alpha carbon. The latter seems unlikely because of the lessened unsaturation of the olefinic linkage and the fact that even 20% hydrochloric acid adds to crotonic acid giving beta-chlorobutyric acid and the beta-hydroxy acid. The process may involve the following steps:

$$MeCH=CH-CO_2H \xrightarrow{H^+} MeCH=CH-\overset{+}{C}(OH)_2 \xrightarrow{\text{1,3-shift}} Me\overset{+}{C}H-CH=C(OH)_2$$

The latter carbonium ion combines with Cl^-, or coordinates with H_2O and then loses a proton to the water or another molecule of acid. The resulting products are the "enol" forms which revert to the acids.

The reactivity of the conjugated system to reagents which can attack the carbonyl group is shown by the ready conversion of sodium crotonate by water or ammonia to beta-hydroxy- and -amino-butyric acids.

Reagents which can only attack the olefinic linkage, do so with the positive part of the addend combining mainly with the alpha carbon giving substituted butyric acids as follows:

	Substituents:	
Addend.	*alpha*	*beta*
Hypohalous acids	halogen	hydroxy
Iodine monochloride	iodo	chloro
NCl_3	chloro	$-NCl_2$
N_2O_4	nitro	hydroxy
I_2 in MeOH	iodo	methoxy
$Hg(OAc)_2$ in MeOH	$-HgOAc$	methoxy

The effect of the conjugated olefinic linkage in lowering the reactivity of the carbonyl group is shown by the slower esterification of the acid and the slower hydrolysis of the esters, as compared with the corresponding saturated compounds.

The gamma H atoms in such an alpha-beta unsaturated acid have some of the properties of alpha H[76,77] (Vinylogy). Thus the esters of α-Me- and β-Me-crotonic acids react with oxalic ester and KOEt the same as ethyl acetate does.[77a]

$$EtO_2CCO_2Et + CH_3CO_2Et \rightarrow EtO_2CCOCH_2CO_2Et$$
$$\text{Oxalacetic ester}$$
$$EtO_2CCO_2Et + CH_3CH=CMeCO_2Et \rightarrow EtO_2CCOCH_2CH=CMeCO_2Et$$

[76] Fuson. *Chem. Rev.* 16, 1 (1935).
[77] Blatt. *J. Org. Chem.* 1, 154 (1936).
[77a] *Ann. Rep. Chem. Soc.* (London) 1923, 73.

The product is stable as the enol containing an 8-atom conjugated system, $O = C(OEt) - C(OH) = CH - CH = C(Me) - C = O$. Treatment with HCl gives the α-pyrone,
$$\underset{OEt}{}$$

$$\begin{array}{c} CH-CH=CMe \\ \parallel \qquad\qquad | \\ EtO_2CC \underline{\qquad} O-CO \end{array}$$

On the other hand, the $\gamma - H$ atoms will not take part in the Perkin reaction with benzaldehyde. Benzaldehyde and crotonic anhydride react in the presence of tertiary bases but the $\alpha - H$ is involved instead of the $\gamma - H$ to give $C_6H_5CH = C(CO_2H)CH = CH_2$.[78]

Crotonic acid fused with isobutylamine gives a good sizing or sealing material for wood.

Isocrotonic acid, *cis*-2-butenoic acid, *cis*-Δ^2-butenoic acid,

$$CH_3CH = CHCO_2H,$$

is obtained by the reduction of chloroisocrotonic ester with sodium amalgam.

$$CH_3COCH_2CO_2Et \rightarrow CH_3CCl_2CH_2CO_2Et \rightarrow$$
$$CH_3CCl = CHCO_2Et \rightarrow CH_3CH = CHCO_2Et$$

This replacement of a halogen by hydrogen under conditions which would be expected to hydrogenate the double bond is common with acids containing the grouping $C = CX$.

Isocrotonic and crotonic acids are identical *structurally* since both give *n*-butyric acid on hydrogenation, potassium acetate on heating with conc. KOH, and acetic and oxalic acids on oxidation. Isocrotonic acid melts at 15° and boils at 172° whereas crotonic acid melts at 72° and boils at 189°. Of these the labile form is isocrotonic acid which is converted to the stable form, crotonic acid, by action of various catalysts such as nitrous acid and also by traces of I_2 or Br_2 in sunlight. This type of change is general for geometric isomers.

α-Methylacrylic acid, methacrylic acid, methylpropenoic acid,

$$CH_2 = C(CH_3)CO_2H,$$

is prepared by removing halogen acid from α-bromoisobutyric acid. It is chiefly valuable as its esters which, alone or with acrylates, are polymerized to transparent plastics.[79-80] The esters are made by converting acetone to its cyanohydrin and changing that to the esters of α-hydroxyisobutyric acid by alcoholysis in presence of sulfuric acid. The esters are dehydrated by phosphorus pentoxide or by pyrolysis.

[78] *Ann. Rep. Chem. Soc.* (London) **1932**, 115.
[79] Neher. *Ind. Eng. Chem.* **28**, 267 (1936).
[80] Strain, Kennelly, Dittmar. *Ind. Eng. Chem.* **31**, 382 (1939).

The reactions of methacrylic acid resemble those of acrylic acid except that HX adds according to Markownikoff's rule to give an alpha halogen compound, thus illustrating the unusual additive power of olefinic linkages of the type $CH_2 = CR_2$. Other reagents give 1,4-addition reactions as with acrylic acid.

Vinylacetic acid, 3-buten-oic acid, Δ^3-butenoic acid, $CH_2 = CH - CH_2CO_2H$.
Preparation. 1. From allyl magnesium bromide and CO_2.
2. From allyl cyanide after protecting the double bond.

$$CH_2 = CHCH_2CN \rightarrow BrCH_2CHBrCH_2CN \rightarrow$$
$$BrCH_2CHBrCH_2CO_2H \rightarrow CH_2 = CHCH_2CO_2H$$

Reactions. Vinylacetic acid gives the reactions of a carboxyl compound and of an olefin practically independent of each other because there is no conjugation of the unsaturation in the two groups. It adds Br_2 more rapidly than does the $\alpha\beta$-unsaturated acid.[81] Thus the greater activity of the $\alpha\beta$-acids with substances of the type HQ is due to the initiation of the reaction by addition of the H to the O of the conjugated system. Bromine cannot add to O in this way. It must add to C directly. The $\beta\gamma$-double bond is thus more unsaturated than the $\alpha\beta$-double bond when the latter acts independently of the carboxyl.

On boiling with dilute acids or alkalies, it changes to crotonic acid. Gaseous HBr even at 0° causes this change. This is another indication of the activity of its double bond.

$$H^+ + CH_2 = CH - CH_2CO_2H \rightarrow$$
$$CH_3 - \overset{+}{C}H - CH_2CO_2H \rightleftharpoons H^+ + CH_3CH = CHCO_2H$$

Concentrated alkalies give two molecules of acetate by shifting the double bond and splitting the alpha-beta unsaturated compound as it is formed. The change from beta-gamma to alpha-beta unsaturation in a straight chain acid is easy because of the greater lability of alpha H as compared with gamma H. This is important in biological "beta oxidation" of fats and in the action of higher olefinic acids with strong alkalies which always results in the removal of two C atoms. The presence of a fork in the gamma position makes possible an equilibrium between the $\beta\gamma$- and $\alpha\beta$-acids.[82] The older conception that the shift of the double bond took place by the addition and removal of water is probably incorrect.[83]

In one reaction, vinylacetic acid shows a complication between its two reactive groups. It readily changes to butyrolactone because the carboxyl

[81] *Ann. Rep. Chem. Soc.* (London) 1910, 70.
[82] *Ann. Rep. Chem. Soc.* (London) 1927, 111.
[83] *ibid.* 1927, 108.

and ethylene linkages can approach each other closely in space.[84]

$$\begin{array}{ccc}
\begin{array}{c} CH_2-CO-OH \\ | \\ | \\ CH=CH_2 \end{array} & \rightarrow & \begin{array}{c} CH_2CO \\ | \qquad\diagdown \\ | \qquad\quad O \\ CH_2CH_2 \diagup \end{array}
\end{array}$$

Lactone formation is a general property of $\beta\gamma$-unsaturated acids.

β-Ethylacrylic acid, 2-penten-oic acid, Δ^2-pentenoic acid,

$$CH_3CH_2CH=CHCO_2H.$$

Like other alpha beta unsaturated acids, it can be made (1) by removing HX from the alpha halogen acid, (2) by the condensation of an aldehyde with sodium acetate and acetic anhydride, and (3) through the condensation of an aldehyde with malonic ester in the presence of secondary amines (Knoevenagel). Because of the availability of n-amyl alcohol the first method would be the best. Its reactions are like those of crotonic acid.

$\beta\beta$-Dimethylacrylic acid, 3-methyl-2-butenoic acid, senecioic acid,

$$Me_2C=CHCO_2H,$$

is found in roots of the genus *Senecio*. It is made (1) from acetone and malonic acid, (2) from the corresponding α-bromoisovaleric acid and (3) by dehydrating β-hydroxyisovaleric acid obtained from acetone, bromoacetic ester and zinc (Reformatzky). This is a general method for $\alpha\beta$-unsaturated acids of the types $RR'C=CHCO_2H$ and $RR'C=CR''CO_2H$.

The *reactions* of dimethylacrylic acid are like those of its lower homologs. The H atoms of the methyl groups show alpha H reactions due to vinylogy.[84a] Thus ethyl oxalate condenses as it would with ethyl acetate but then leads to more complex products.

$$(CO_2Et)_2 + Me_2C=CHCO_2Et \rightarrow EtO_2CCOCH_2C(Me)=CHCO_2Et \rightarrow$$

$$\begin{array}{c}
\qquad\qquad\qquad\qquad\qquad\qquad\qquad CH-CMe=CH \\
EtO_2CC(OH)=CHC(Me)=CHCO_2Et \rightarrow EtO_2C\overset{\|}{C}\text{------}O\text{------}\overset{\|}{C}O
\end{array}$$

The formation of the unsaturated lactone (α-pyrone) takes place spontaneously. This represents an internal alcoholysis like an ester exchange involving the conversion of a methyl to an ethyl ester.

$$R-CO_2Me + EtOH \rightleftharpoons R-\overset{\overset{\displaystyle OEt}{|}}{\underset{\underset{\displaystyle OMe}{|}}{C}}-OH \rightleftharpoons MeOH + R-CO_2Et$$

[84] *ibid.* 1932, 108.
[84a] *Ann. Rep. Chem. Soc.* (London) 1923, 73.

β-Pentenoic acid, $CH_3CH=CHCH_2CO_2H$, is not available by any of the ordinary methods. A beta halogen acid loses HX entirely with an alpha H atom. A gamma halogen acid when treated with alkali gives a gamma lactone instead of the desired unsaturated acid. This is another example of the ease of formation of a five membered ring. The removal of the halogen atom leaves a carbonium fragment which coordinates with the spatially near carboxyl group and produces the lactone. In common with other *beta gamma unsaturated acids*, it can be made by distilling the proper alkyl paraconic acid obtained by a modification of the Perkin reaction from an aldehyde, sodium succinate and acetic anhydride (p. 382).

Another method of making the Δ^3-pentenoic acid is by cracking the lactone obtained by reducing levulinic acid, 4-pentanonic acid.

The unsaturated acid gives separate reactions of an olefin and of a carboxylic acid. It also gives reactions involving both groups. Thus with sulfuric acid it gives γ-valerolactone and on fusion with KOH it gives potassium acetate and propionate, showing the shift of the double bond to the alpha beta position.

Allylacetic acid, Δ^4-pentenoic acid, $CH_2=CHCH_2CH_2CO_2H$, may be prepared readily by the malonic ester or acetoacetic ester synthesis for acids using allyl bromide. Another preparation is from tribromopropane. (Perkin.)

$$BrCH_2CHBrCH_2Br + Na \text{ Malonic Ester} \rightarrow CH_2=CBrCH_2CH(CO_2R)_2$$

In this process one of the Br atoms reacts "metathetically," one is removed as HBr by the alkaline action of the sodium malonic ester and the third remains. By hydrolysis and heating, the substituted malonic ester is converted to 4-bromo-4-pentenoic acid. This is vigorously reduced by alcohol and sodium.

$$CH_2=CBr-CH_2CH_2CO_2Na \rightarrow CH_2=CH-CH_2CH_2CO_2Na$$

The reactions of allylacetic acid are like those of its lower homologs which have no conjugated system. With sulfuric acid, it forms methylbutyrolactone (γ-valerolactone). This reaction is in accord, however, both with Markownikoff's Rule and Bayer's Strain Theory. The tendency to form a six-membered ring is not enough to make the addition go contrary to Markownikoff's Rule with the formation of δ-valerolactone.

Angelic and tiglic acids, *cis-trans* isomers of 2-methyl-Δ^2-butenoic acid, $CH_3CH=C(CH_3)CO_2H$, are found in the roots of *Angelica archangelica* and in *Croton tiglium* respectively. Their melting points are 45° and 64.5° and their boiling points 185° and 198° respectively.

$$\begin{array}{ccc} CH_3-C-H & & H-C-CH_3 \\ \| & and & \| \\ CH_3-C-CO_2H & & CH_3-C-CO_2H \end{array}$$

The two acids are obtained in 25% and 17% yields by the dehydration of α-hydroxy-α-methylbutyric acid obtained from methyl ethyl ketone by the cyanohydrin synthesis. They add HI giving the β-iodo-α-methylbutyric acids, m. 58° and 86° respectively which decompose on boiling with sodium carbonate solution to form the two stereomers of 2-butene.

Hydrosorbic acid, Δ^3-hexenoic acid, $CH_3CH_2CH = CHCH_2CO_2H$, is obtained by the distillation of ethyl paraconic acid (p. 383) and by the partial reduction (1:4) of sorbic acid obtained by the condensation of crotonaldehyde with malonic acid. This reduction takes place more easily than with a similar 4-carbon system not adjacent to a carbonyl group. Thus it is highly probable that the process here is really 1,6-addition followed by the usual change to the keto form.

$$CH_3CH = CH\!-\!CH = CH\!-\!\underset{\displaystyle}{C} \overset{\displaystyle OH}{=} O \;\rightarrow$$

$$CH_3CH_2\!-\!CH = CH\!-\!CH = \underset{\displaystyle}{C}\overset{\displaystyle OH}{-}OH \rightarrow$$

$$CH_3CH_2CH = CH\!-\!CH_2\!-\!\underset{\displaystyle}{C}\overset{\displaystyle OH}{=} O$$

The partial reduction of similar diolefinic acids is a general method for making *$\beta\gamma$-unsaturated acids.*

Hydrosorbic acid gives the usual reactions of a $\beta\gamma$-unsaturated acid, including rearrangement and splitting by alkalies and lactone formation with sulfuric acid.

Isohydrosorbic acid, Δ^2-hexenoic acid, $CH_3CH_2CH_2 - CH = CHCO_2H$, is obtained from *n*-caproic acid through the alpha bromo compound and by heating hydrosorbic acid with dilute alkali.

Pyroterebic acid, $(CH_3)_2C = CHCH_2CO_2H$, b. 207°, is obtained from terebic acid by heating. Reduction by HI gives isocaproic acid.

Teracrylic acid, $(CH_3)_2C = C(CH_3)CH_2CO_2H$, b. 218°, is obtained by heating terpenylic acid.

A **decenoic acid,** having a branched chain has been obtained from Capsaicin.[85]

Undecylenic acid, Δ^{10}-undecenoic acid, $CH_2 = CH(CH_2)_8CO_2H$, is obtained by the destructive distillation of castor oil under reduced pressure (15% yield). The yield can be increased by heating the castor oil with rosin. Its constitution is proved by its conversion to sebacic acid, $HO_2C(CH_2)_8CO_2H$, by oxidation. When undecylenic acid is fused with KOH it gives potassium acetate and nonylate in the usual way. Thus the action of alkali shifts the double bond from one end of the chain to the other. The change

$$-CH = CH - CH_2 - \;\rightarrow\; -CH_2 - CH = CH -$$

is reversible until the beta position is reached. Then an alpha beta double

[85] *Ann. Rep. Chem. Soc.* 1920, 85.

bond is established and cannot change except by splitting in presence of the fused alkali. This is another example of a change which takes place as a result of a complex series of equilibrium reactions connected with one process which is not reversible. Thus the double bond may be regarded as wandering up and down the long carbon chain purely at random except that each molecule in which the double bond reaches the alpha beta position can not longer change except by splitting. A similar rearrangement of the double bond occurs with sulfuric acid which produces undecalactone, a gamma lactone,

$$Me(CH_2)_6CHCH_2CH_2CO.$$
$$\underline{\qquad O \qquad}$$

The addition of HBr to undecylenic acid in ether follows Markownikoff's Rule while in toluene the ω-bromo acid is obtained.[86]

Undecylenic acid and its zinc salt (*Desenex*) are used in treatment of "Athlete's Foot."

Undecylene amide and its N-butyl derivative are insecticides.

ω-Olefinic acids of C_{12} and C_{13} have been synthesized.[87]

Myristolenic (C_{14}) and **palmitolenic (zoomaric)** (C_{16}) **acids** with the double bond mainly in the 9-position have been obtained from certain whale oils.[88]

Oleic acid, rapic acid, Δ^9-octadecenoic acid, $CH_3(CH_2)_7CH = (CH_2)_7CO_2H$, m. 13–14°, occurs as glyceride in most fats and oils. Its constitution is proved by its hydrogenation to stearic acid and its ozonation to pelargonic aldehyde, $CH_3(CH_2)_7CHO$ and the half aldehyde of azelaic acid, $OCH(CH_2)_7CO_2H$. At 100° it reacts with oxygen to give a very complex mixture of acids including 15% 9,10-dihydroxystearic acid, 35% nonylic acid, 1% acetic acid and 0.5% formic acid.

Fusion with KOH converts oleic acid to the $\alpha\beta$-unsaturated salt and splits that into the acetate and palmitate (C_{16}). Oxidation with nitric acid gives all the lower fatty acids from C_{10} to C_2 with the corresponding dibasic acids. This indicates shifts of the double bond under the influence of the acid. It is noteworthy that the shift takes place in either direction from the position in the middle of the chain.

Very dilute $KMnO_4$ solution adds two hydroxyl groups to produce 9,10-dihydroxystearic acid. More vigorous oxidation with permanganate gives pelargonic acid and azelaic acid. The oxidation of ethyl oleate in acetone or glacial acetic acid solution gives the two acids in 59 and 95% yields.[89] Evidently there is no shift of the double bond as in the presence of acids or fused alkalies.

The formation of oleic acid from stearolic acid by careful hydrogenation with nickel indicates its *cis* configuration.[90]

[86] *Ann. Rep. Chem. Soc.* (London) 1932, 107.
[87] *ibid.* 1927, 87.
[88] *ibid.* 1925, 79; 1928, 84.
[89] *ibid.* 1925, 79.
[90] *ibid.* 1926, 100.

Addition of sulfuric acid followed by hydrolysis gives mainly 10-hydroxy-stearic acid.[91] The addition product before hydrolysis can be converted to a disodium salt, "sodium sulfo oleate," *Thigenol* (NNR). Benzene adds to oleic acid under the influence of sulfuric acid to give equal amounts of 9- and 10-phenylstearic acids.[92]

Treatment with a trace of nitrous acid converts the liquid oleic acid, m. 14°, into its solid *trans* stereoisomer, *elaidic acid*, m. 51°.[93] The structural identity of the two acids is proved by the products obtained by hydrogenation and by vigorous oxidation. Mild oxidation by hydrogen peroxide converts methyl oleate and methyl elaidate into the two dihydroxystearic acids of m.p. 95° and 132° respectively.[94]

Oleic acid with zinc chloride changes into a β-γ-unsaturated acid which then undergoes lactone formation to give stearolactone.[95]

Oleic acid polymerizes to dimers which react with $(CH_2)_2(NH_2)_2$ to give *norelac*.

Lead oleate differs from the lead salts of stearic and palmitic acids in being soluble in ether. Pure oleic acid can be prepared from this salt or from the lithium salt or from the product of addition of mercuric acetate.[96] Crude oleic acid is Red Oil.

Reduction of oleic esters with Na and EtOH gives oleyl alcohol without changing the double bond.

Oleic acid has been synthesized.[97]

Iso-oleic acid, m. 45°, with Δ^{10}, is obtained by hydrating oleic acid and dehydrating the hydroxystearic acid formed.

Petroselic acid, m. 34°, is an isomer of oleic acid having the double bond in the 6,7-position. It is thus related to lactarinic acid (6-keto-) and tariric acid (6,7-acetylenic). Petroselic acid is changed by a trace of nitrous acid to its *trans* isomer, m. 54°[98]. Oxidation gives lauric and adipic acids. The Δ^{11}, liquid, and Δ^{12}, m. 35°, isomers of oleic acid are obtained from the bromide of 12-hydroxystearic acid formed by the catalytic hydrogenation of ricinoleic acid.[99]

Erucic and **brassidic acids,** are *cis-trans* isomers of Δ^{13}-docosenoic acid, $CH_3(CH_2)_7CH=CH(CH_2)_{11}CO_2H$.[100] Erucic acid is present as a glyceride in the oils from rape-seed and grape-seed. It melts at 34°. Oxidation gives nonylic and brassylic acids. Nitrous acid converts erucic acid to brassidic

[91] DeGroote et al. *Ind. Eng. Chem., Anal. Ed.* 3, 243 (1931).
[92] Harmon, Marvel. *J. Am. Chem. Soc.* 54, 2515 (1932).
[93] Smith. *J. Chem. Soc.* 1939, 974.
[94] *Ann. Rep. Chem. Soc.* (London) 1928, 83.
[95] *ibid.* 1926, 100.
[96] *ibid.* 1927, 86.
[97] *ibid.* 1925, 80.
[98] *ibid.* 1927, 89.
[99] *ibid.* 1912, 84.
[100] *ibid.* 1922, 72.

acid, m. 65°. The *cis* form, erucic acid, is hydrogenated more rapidly than the *trans* form.[101] Isoerucic acid is a mixture of the Δ^{12} and Δ^{14} isomers.[102]

Cetoleic acid, an isomer of erucic acid having Δ^{11}, is found in certain animal oils.[103]

Nervonic acid, erucylacetic acid, $Me(CH_2)_7CH = CH(CH_2)_{13}CO_2H$, has been isolated from a cerebroside from human brains. It has been synthesized from erucyl alcohol through the bromide and the malonic ester synthesis.[104] It exists in *cis* and *trans* forms, m. 39° and 61°.

C. DIOLEFINIC ACIDS, $C_nH_{2n-3}CO_2H$

β-Vinylacrylic acid, 2,4-pentadienoic acid, $CH_2 = CHCH = CHCO_2H$, m. 80°, is made from acrolein and malonic acid in presence of pyridine. This is a general method for making 2,4-diolefinic acids.

Vinylacrylic acid with H_2 and PtO_2[105] gives ethylacrylic acid while reduction by Na_xHg gives 1,4-addition to form 3-pentenoic acid with a lesser amount of 1,2-addition to give vinylpropionic acid.

Thus catalytic hydrogenation apparently initiates at the terminal carbon while reduction starts at the carbonyl oxygen as part of the 1,6-conjugated system.

Vinylacrylic acid adds Br_2 at the terminal double bond instead of at the ends of the conjugated system. The reaction is evidently initiated at the end C.

Oxidation by $KMnO_4$ produces a split in the 4-position. Thus vinylacrylic acid gives formic and tartaric acids. The influence of the 6-atom conjugated system is seen in the addition of two mols of NH_3 to give diaminovaleric acid when vinylacrylic acid is heated with ammonium hydroxide solution. Since ammonia does not add to ordinary olefinic linkage, to butadiene, or to βγ-unsaturated acids, the process probably includes the following steps, involving a 1:6- and then a 1:4-addition:

$$CH_2 = CHCH = CHC = O(OH) \rightarrow NH_2CH_2CH = CHCH = C(OH)_2$$
$$\rightarrow NH_2CH_2CH = CHCH_2C = O(OH)$$
$$\rightarrow NH_2CH_2CH_2CH = CHC = O(OH)$$
$$\rightarrow NH_2CH_2CH_2CH(NH_2)CH = C(OH)_2$$
$$\rightarrow NH_2CH_2CH_2CH(NH_2)CH_2CO_2H$$

The change from the βγ- to the αβ-unsaturated acid takes place as usual in the alkaline medium.

The addition of hypochlorous acid takes place exclusively at the terminal olefinic linkage forming $ClCH_2CHOHCH = CHCO_2H$, thus following Mark-

[101] *ibid.* 1930, 84.
[102] *ibid.* 1927, 87.
[103] *ibid.* 1928, 84.
[104] *ibid.* 1926, 100.
[105] "Org. Syntheses."

ownikoff's Rule. The $\alpha\beta$-unsaturation remains as part of the conjugated system with the carbonyl group.[106]

Vinylacrylic acid and other 2,4-diolefinic acids polymerize on heating, probably by the Diels-Alder reaction, the butadiene system of one molecule reacting with the $\alpha\beta$-double bond of another. Vinylacrylic acid gives ethylbenzene on vigorous heating.[107]

Vinylacrylic ester with sodiomalonic ester gives less than 2% addition at the $\alpha\beta$-bond but mainly $(MeO_2C)_2CHCH_2CH = CHCH_2CO_2Me$ due to either $\alpha\delta$- or 1:6-addition.[108]

A related substance, cinnamalmalonic ester, $PhCH = CHCH = C(CO_2Et)_2$, adds HBr, HCN and cyanoacetic ester in the $\alpha\beta$-position but malonic ester in the $\alpha\delta$-position.

Sorbic acid, 2,4-hexadienoic acid, $CH_3CH = CHCH = CHCO_2H$, m. 134°, occurs in unripe sorb apples. It is best prepared from crotonic aldehyde and malonic acid in pyridine. It gives reactions like those of vinylacrylic acid. Partial hydrogenation gives hydrosorbic acid (3-hexenoic acid) and, to a lesser extent, the $\alpha\beta$- and $\gamma\delta$-addition products[108a]. Thus the chief addition is 1:6, initiated at the carbonyl oxygen. Hypochlorous acid, which cannot initiate such a reaction, adds to the 4-ethylenic linkage, the Cl adding to the $5 - C$.[109]

Cyanoacetic ester adds to sorbic ester mainly in the $\alpha\delta$-position and only about 10% in the $\alpha\beta$-position. This may be assumed to be due to 1:6- and 1:4-addition initiated at the carbonyl oxygen or to 1:4- and 1:2-addition to the conjugated four carbon system adjacent to the carboxyl group.

The H atoms of the methyl at the end of the 6-atom conjugated system show some properties of $\alpha - H$ atoms. Thus sorbic ester condenses with oxalic ester to give the remarkably conjugated compound,[110]

$$O = C(OEt)C(OH) = CHCH = CHCH = CHC = O(OR)$$

Sorbic acid heated with barium hydroxide gives o-propyltoluene, the first step being undoubtedly a Diels-Alder addition.[111]

Alcoholic KOH at 170° converts sorbic acid to 3-hexynoic acid which on oxidation gives malonic, propionic and oxalic acids.

Geranic acid, 3,7-Me$_2$-2,6-octadienoic acid,

$$(CH_3)_2C = CH(CH_2)_2C(CH_3) = CHCO_2H,$$

b. 153° (13 mm.), is obtained by oxidation of the corresponding aldehyde (citral) by silver oxide or by the hydrolysis of the nitrile obtained from the oxime of citral and acetic anhydride. Under the action of 70% sulfuric acid

[106] *Ann. Rep. Chem. Soc.* (London) 1932, 109.
[107] Kuhn, Deutsch. *Ber.* 65, 43 (1932).
[108] Kohler, Butler. *J. Am. Chem. Soc.* 48, 1036 (1926).
[108a] *Ann. Rep. Chem. Soc.* (London) 1932, 109.
[109] *ibid.* 1932, 110.
[110] *ibid.* 1932, 114.
[111] *ibid.* 1932, 112.

a 6-membered ring is closed to form α-cyclogeranic acid, 2,2,6-Me$_3$-1,2,3,4-tetrahydrobenzoic acid.

Linoleic acid, linolic acid, 9,12-octadecadienoic acid,

$$CH_3(CH_2)_4CH = CHCH_2CH = CH(CH_2)_7CO_2H,$$

occurs as glycerides in almost all oils and fats, especially in corn oil, cottonseed oil and tall oil. The latter is produced from the kraft pulping of wood and consists mainly of about equal amounts of fatty and resin acids. Linseed oil does not contain as much linoleic as was formerly believed. Linoleic acid exists in the four possible stereoisomeric forms, *trans-trans, trans-cis, cis-cis* and *cis-trans.* Oxidation with alkaline permanganate gives *sativic acid,* tetrahydroxy-stearic acid, which can be reduced to stearic acid by HI. More vigorous oxidation gives caproic acid, oxalic acid and azelaic acid. With benzoyl peroxide, it gives a di-ethylene oxide, m. 32°.

The fatty acids from dehydrated castor oil (*Isoline*) contain mainly 9,12-linoleic acid with about 25% of the 9,11-acid. Treatment of this mixture with aqueous NaOH gives a 10,12-linoleic acid, m. 57°; Me ester, m. 25°.[112]

Fats from linoleic acid and more highly unsaturated acids are necessary in metabolism. A diet consisting entirely of glycerides of saturated acids and oleic acid is insufficient.

The *8,10-isomer* of linoleic acid has been made from oleic acid by adding Cl$_2$ and removing 2 HCl by alcoholic KOH.

D. TRIOLEFINIC ACIDS, C$_n$H$_{2n-5}$CO$_2$H

Dehydrogeranic acid, 3,7-Me$_2$-2,4,6-octatrienoic acid, (A),

$$(CH_3)_2C = CHCH = CHC(CH_3) = CHCO_2H,$$

m. 186°, occurs in nature. The high melting point probably indicates a *cis-cis* form (H,H and CH$_3$,H on same side of double bond) which would allow the carboxyl group and the 6-olefinic linkage to approach each other in space. The *trans-trans* form would be a linear molecule, the ends of which could not approach each other while the *cis-trans* form would allow only the methyls or a methyl and H to approach each other. It has been found in essential oils as an ester with geraniol. It has been synthesized.

$$Me_2C = CHCHO + Me_2CO \rightarrow Me_2C = CHCH = CHCOMe$$

$$\begin{array}{l} BrCH_2CO_2R + Zn \\ \xrightarrow{\hspace{3cm}} Me_2C = CHCH = CHC(OH)(Me)CH_2CO_2R \rightarrow (A) \end{array}$$

Linolenic acid, 9,12,15-octadecatrienoic acid,

$$CH_3(CH_2CH = CH)_3(CH_2)_7CO_2H,$$

is present in large amounts in the glycerides of linseed oil. It has also been

[112] Mikusch. *Ind. Eng. Chem.* **32,** 1314 (1940).

found in Cassava starch. It is usually isolated as the hexabromide. Treatment with alkali gives the 10,12,14-acid, *pseudoeleostearic acid*.[113]

Isolinolenic and *pecorinic acids* are also isomers of linolenic acid.

Elaeostearic acid ("eleomargaric acid"), m. 44°, formerly regarded as a stereoisomer of linolenic acid, is the 9,11,13-isomer; α-form, m. 47°; β-form, m. 67°. As glycerides, it constitutes over 90% of Chinese wood oil.[114] It exists in several stereo forms.

Licanic acid, 4-keto-9,11,13-octadecatrienoic acid, constitutes about 50% of the acids in Brazilian *oiticica oil*.

Clupanodonic acid, $C_{18}H_{28}O_2$, contains four double bonds. It is found in sardine oil. A similar acid, $C_{22}H_{36}O_2$, is found in various algae.[115]

Drying oils are usually mixtures of glycerol esters of fatty acids. They derive their drying properties from the presence in the mixture of a large proportion of the more highly unsaturated linoleic, linolenic, and elaeostearic acids. It has long been known that drying oils change from liquids to solids upon exposure to air. Initial absorption of oxygen by drying oils has been shown to be in the form of peroxides, which are often represented by the structure:

$$\begin{array}{c} | \\ -\mathrm{C}-\mathrm{O} \\ | \quad | \\ -\mathrm{C}-\mathrm{O} \\ | \end{array}$$

These peroxides then react further to produce a complex structure, resulting in solidification of the oil. The mechanism of chemical oxidation and polymerization responsible for the hardening process is very complex and poorly understood.

Commercial drying oils usually are given a preliminary treatment in which they receive a partial oxidation to speed up the drying process. These products are classified as blown oils, boiled oils, etc., according to the conditions of treatment. Boiling linseed oil for nine hours at above 300° C. increases the molecular weight from 700 to about 1600 and the absorption of oxygen is accompanied by a decrease in iodine number.

Drying Oils.[116]

E. ACETYLENIC ACIDS, $C_nH_{2n-3}CO_2H$

These are made by standard reactions. Some members differ physically from the olefinic acids in having higher melting points. Their reactions are those of their groups. When the acetylenic group is αβ- it shows the usual increased additive power of such conjugated compounds. The acetylenic acids differ from the olefinic acids and especially from the diolefinic acids in not being oxidized by air.

[113] Kass, Burr. *J. Am. Chem. Soc.* 61, 3292 (1939).
[114] *Ann. Rep. Chem. Soc.* (London) 1925, 79.
[115] *ibid.* 1925, 80.
[116] Sutheim. *Chem. Ind.* 62, 65 (1948).

Propiolic acid, propargylic acid, acetylene carboxylic acid, propynoic acid, $HC{\equiv}CCO_2H$, m. 9°, b. 144° (dec.), can be prepared as follows:

1. From excess *alcoholic* KOH with dibromopropionic acid formed from acrolein, acrylic acid or allyl alcohol.

$$BrCH_2CHBrCO_2K \rightarrow CH_2{=}CBrCO_2K \rightarrow CH{\equiv}CCO_2K$$

2. From the monosodium derivative of acetylene,[117] or the mono-Grignard derivative with CO_2.

3. Probably best, from maleic acid by addition of Br_2, removal of 2 HBr and loss of CO_2 on heating the acid potassium salt of acetylene dicarboxylic acid.

$$HO_2CC{\equiv}CCO_2K \rightarrow CO_2 + HC{\equiv}CCO_2K$$

The effect on physical properties of the triple bond as compared with a double or single bond is interesting. For the three acids, propionic, acrylic and propiolic, the melting points are −19°, +12°, +9°; the boiling points are 141°, 142° and 144° (dec.); the densities are 0.987, 1.06 and 1.33; while the boiling points of the ethyl esters are 99°, 100° and 120°.

Propiolic acid shows the properties of acetylene and of a carboxylic acid. The two groups influence each other as is shown by the easy reduction of the triple bond to form acrylic acid and then propionic acid. The esters and salts of $\alpha\beta$-acetylenic acids also add one or two mols of sodium bisulfite or ammonia or primary or secondary amines, the H going to the α-carbon and the rest of the addend to the β-carbon. Thus the mechanism probably involves 1,4-addition initiated at the carbonyl group.

Sodium propiolate reacts with ammoniacal cuprous solution to give $(NaO_2CC{\equiv}CCu)_2$. Oxidation of this compound with potassium ferricyanide gives the sodium salt of diacetylenedicarboxylic acid.

Polymerization of propiolic acid in sunlight gives a small yield of trimesic acid, benzene-1,3,5-tricarboxylic acid.

Tetrolic acid, methylpropiloic acid, 2-butynoic acid, $CH_3C{\equiv}CCO_2H$, m. 76°, b. 203°, is prepared from crotonic acid through the dibromide or from allylene through its Na or Grignard derivative. It gives the expected reactions. Hydrogenation with Pd at 20° gives isocrotonic acid (the *cis* form).[118] Many homologs of the type $RC{\equiv}CCO_2H$ have been prepared. The best known one is made from heptaldehyde from castor oil and is used in the form of its methyl ester as methyl heptyne carbonate, an artificial violet perfume. The steps are as follows:

$$C_6H_{13}CHO \xrightarrow[CO_2]{PCl_5} \xrightarrow[acid]{NaOH} \xrightarrow[MeOH]{NaNH_2} \xrightarrow{Na} \longrightarrow \longrightarrow \longrightarrow C_5H_{11}C{\equiv}CCO_2Me$$

[117] *Ann. Rep. Chem. Soc.* (London) 1926, 98.
[118] *ibid.* 1925, 75.

Similarly, methyl octyne carbonate can be made from capryl alcohol (MeHexcarbinol) from castor oil by first oxidizing it to the ketone and then using a similar set of changes.

4-Pentynoic acid, $CH{\equiv}CCH_2CH_2CO_2H$, is made from 1,2,3-tribromopropane (tribromohydrin) and sodiomalonic ester.[119] A more efficient way would be to convert the tribromide to 2-bromoallyl bromide by alkali and use that with the malonic ester to give $CH_2{=}CBrCH_2CH(CO_2Et)_2$ which gives the desired product on treatment with alcoholic KOH and decarboxylation.

Dehydro-undecylenic acid, 10-undecynoic acid, $CH{\equiv}C(CH_2)_8CO_2H$, b. 175° (15 mm.), m. 43°, can be made from the dibromide of undecylenic acid by careful treatment with alkali.[120] Its structure is proved by oxidation to sebacic acid and by the formation of typical acetylenic silver compounds by its esters. When heated with bases to 180° it is converted to *undecolic acid,* m. 59°, 9-undecynoic acid. This is converted by oxidation to azelaic acid. Its esters form no Ag or Cu compounds.

Stearolic acid, 9-octadecynoic acid, $CH_3(CH_2)_7C{\equiv}C(CH_2)_7CO_2H$, m. 48°, is made from oleic or elaidic acid through their dibromides. It reacts with conc. sulfuric acid to give 9- and 10-ketostearic acids.[121] Reduction by H_2 and Ni gives oleic acid (*cis*) while zinc and acetic acid give elaidic acid (*trans*). Oxidation by HNO_3 or $KMnO_4$ gives first stearoxylic acid, a 9,10-diketo acid, and then pelargonic and azelaic acids.

Tariric acid, m. 50°, the 6-isomer of stearolic acid, is found in nature. Its reactions are entirely analogous to those of stearolic acid. Its final oxidation products are lauric and adipic acids.

Behenolic acid, 13-docosynoic acid, $CH_3(CH_2)_7C{\equiv}C(CH_2)_{11}CO_2H$, m. 57°, is made from the dibromides of erucic and brassidic acids. It resembles stearolic acid in its reactions. Mild reduction gives erucic acid (*cis*) and more vigorous reduction gives brassidic acid (*trans*).[122]

F. Halogenated Aliphatic Acids

In general, the preparations and reactions of the halogenated carboxylic acids resemble those of alkyl halides of analogous structure. An exception is that the alpha hydrogen atoms in an acid are more easily replaced by chlorine and bromine than are those of hydrocarbons. The properties of the halogen acids vary with the position of the halogen in relation to the carboxyl group. Halogen in the alpha position increases the dissociation of an acid greatly. Thus trichloroacetic acid is nearly as strong as the inorganic acids. Halogen atoms introduced farther away from the carboxyl group have little effect on its strength.

[119] *Ann. Rep. Chem. Soc.* (London) 1907, 92.
[120] *ibid.* 1927, 88.
[121] *ibid.* 1926, 99.
[122] *Ann. Rep. Chem. Soc.* (London) 1926, 99.

The action of bases on halogenated acids varies greatly with the position of the halogen.

1. Alpha halogen acids give alpha hydroxy acids and alpha beta unsaturated acids.

$$RCH_2CHXCO_2H \rightarrow RCH_2CHOHCO_2H + RCH = CHCO_2H$$

2. Beta halogen acids, if there is no substitution at the alpha carbon, give mainly $\alpha\beta$-unsaturated acids with lesser amounts of $\beta - OH$-acids.

$$RCHXCH_2CO_2H \rightarrow RCH = CHCO_2H + RCHOHCH_2CO_2H$$

With an α-substituted acid, carbon dioxide is lost and an olefin formed.

$$RCHXCHQCO_2H + NaOH \rightarrow NaBr + CO_2 + RCH = CHQ$$

This reaction occurs whether Q is an alkyl group or a halogen. Thus it is given by α-Me-β-Br-butyric acid, $\alpha\alpha$-Me$_2$-β-Br-butyric acid and by $\alpha\beta$-Br$_2$-butyric acid.

3. Gamma and delta halogen acids form the corresponding lactones.

4. When the halogen is farther away than the delta position, it acts like that in an alkyl halide. It can react intermolecularly with the carboxyl of another molecule. Thus highly polymeric esters can be formed.

Halogenated Fatty Acids

Chloroformic acid, chlorocarbonic acid, chloromethanoic acid, $ClCO_2H$, and the other halogen analogs are unknown in the free state since they decompose to HX and CO_2 (p. 428).

Monochloroacetic acid, chloroethanoic acid, $ClCH_2CO_2H$, m. 63°.

Preparation. 1. Chlorination of acetic acid in sunlight or with halogen carriers like sulfur and phosphorus. 2. Hydrolysis of trichloroethylene obtained from acetylene.[123]

$$ClCH = CCl_2 \rightarrow ClCH_2 - CCl_2(OH) \rightarrow ClCH_2COCl \rightarrow ClCH_2CO_2H$$

Reactions. 1. As an acid, it is stronger than acetic acid but gives similar reactions. 2. As an alpha halogen compound it gives the reactions of alkyl halides but is more reactive. With the free acid or one of its salts, the Cl can be replaced by (1) iodine by NaI in alcohol, (2) CN by NaCN, (3) OH by water or alkali, (4) sulfonic acid group by sodium sulfite, (5) amino by ammonia or Gabriel's phthalimide reaction, and (6) nitro by $AgNO_2$. If $NaNO_2$ is used, the resulting sodium nitroacetate decomposes to nitromethane.

Chloroacetic acid is used in making indigo from aniline. It is used in large quantities in the preparation of carboxymethyl cellulose and 2,4-D.[124]

Bromoacetic acid can be prepared like chloroacetic acid. **Iodoacetic acid** is best prepared from chloroacetic acid with sodium iodide in alcohol or acetone.

[123] *Ann. Rep. Chem. Soc.* (London) 1922, 73.
[124] *Chem. Ind.* 62, 733 (1948).

The resulting sodium chloride is insoluble in the organic solvent. Esters of bromoacetic acid, in common with other alpha bromo esters, give the *Reformatzky reaction* with ketones and zinc to give beta-hydroxy esters.

Iodoacetic acid apparently depletes the sulfur-containing amino acids in the organism.[125, 126] Sodium iodoacetate is readily obtained from chloroacetic acid with NaI in dry acetone. Filtration from the NaCl and adjustment of the pH at 6.8 with NaOH gives a solution from which the pure salt is readily obtained.[127]

Sodium fluoroacetate is the rat killer "1080." It is also poisonous for larger animals.

Dichloroacetic acid, Cl_2CHCO_2H, b. 194°, can be prepared by chlorination but is more readily obtained from chloral hydrate and NaCN.[128, 129] A still better method is from trichloroacetic acid and copper in water solution.[130]

The reactions of dichloroacetic acid are those of an acid and methylene chloride. Boiling with alkalies gives oxalates and acetates. This reaction represents an extreme case of dismutation. The process is surprising since the related glyoxylic acid gives the normal dismutation with alkalies, forming oxalic acid and glycollic acid.

Trifluoroacetic acid, F_3CCO_2H, is prepared by a method which illustrates the stability and inactivity of the $C-F$ linkage. Benzotrichloride is treated with Ag or Sb fluoride and the resulting benzotrifluoride is vigorously oxidized, the benzene ring being destroyed.

$$F_3CC_6H_5 \rightarrow F_3CCO_2H$$

The benzene ring is more easily destroyed if it is first nitrated and the resulting nitro compound reduced. This is because of the greater ease of oxidation of an aromatic amine as compared with the hydrocarbon.

Trichloroacetic acid, Cl_3CCO_2H, b. 195°, m. 55°, is best prepared by the oxidation of chloral hydrate with nitric acid. It is a much stronger acid than most carboxylic acids. It is dangerously corrosive on the skin. Hot alkalies split it to give chloroform and carbonates. The effect of halogenation on acid strength is interesting:

acid (.03M)	% ionization
Acetic	2
Chloroacetic	20
Dichloroacetic	70
Trichloroacetic	90
Trifluoroacetic	—
2,2,3-Trichlorobutyric	>90

[125] White. *Science* **86**, 588 (1937).
[126] White. *J. Biol. Chem.* 112, 503 (1936).
[127] Goldberg. *Science* **98**, 385 (1943).
[128] Pucher. *J. Am. Chem. Soc.* 42, 2251 (1920).
[129] Doughty. *J. Am. Chem. Soc.* **47**, 1091 (1925).
[130] *ibid.* **53**, 1594 (1931).

Esters of trichloroacetic acid heated with alcohols give $CHCl_3$ and alkyl carbonates.

Trichloroacetic acid gives peculiar types of esters. Thus with ethylene glycol the product is related to the ortho acid. The benzyl ester decomposes on heating to give an apparent reversal of the Tischenko reaction.[131]

$$Cl_3CCO_2H + HOCH_2CH_2OH \rightarrow Cl_3C-C-O-CH_2$$
$$HO \quad O-CH_2$$

$$Cl_3CCO_2CH_2Ph \rightarrow Cl_3CCHO + PhCHO$$

Alpha-chloropropionic acid, 2-chloropropanoic acid, $CH_3CHClCO_2H$.

Preparation. 1. By chlorination with catalysts. Only the alpha H atoms are replaced readily. The process of replacing these atoms may involve enolization. The chlorination of acid halides and esters is more rapid than that of the free acids, probably because of the greater enolization of these substances. This, in turn, depends on the fact that the carbonyl group in these substances is more reactive than the corresponding group in the acids.

One of the best methods of alpha chlorination is the use of anhydrous acid with chlorine (dried with H_2SO_4) and phosphorus to give the alpha chloro acid chloride (Hell-Volhard-Zelinsky).[132] This may be purified by distillation and treated with water or alcohol to give the pure alpha chloro acid or ester.

Alpha halogen acids undergo a substitution of the halogen by other groups such as hydroxyl or amino but they do not lose HX simply to give alpha beta unsaturated acids on treatment with alkali. On the other hand, pyrolysis at a very high temperature will remove HX leaving unsaturated acids. The reactions of the α-hydroxy acids are similar (p. 341).

α-Bromopropionic acid, b. 205°, m. 25°, is prepared in a similar way. For preparation of the acid bromide, all materials must be dried with sulfuric acid or phosphorus pentoxide (a given amount of water will convert eleven times its weight of the bromo acid bromide to the bromo acid).

The *iodo acid* is made from the chloro acid and NaI.

2. From lactic acid by treatment with inorganic acid chlorides.

$$CH_3CHOHCO_2H + SOCl_2 \rightarrow CH_3CHClCO_2H$$

An excess gives the α-chloro acid chloride.

Reactions. 1. With suitable reagents the alpha halogen is readily replaced as in chloroacetic acid. Hydrogen and palladium remove α-chlorine but not β-chlorine.

2. With aqueous alkalies the main product is lactic acid with some acrylic acid, $CH_2=CHCO_2H$. With alcoholic KOH the latter is the chief product.

Beta-chloropropionic acid, 3-chloropropanoic acid, $ClCH_2CH_2CO_2H$.

[131] *Ann. Rep. Chem. Soc.* (London) 1931, 96.
[132] *ibid.* 1912, 83.

Preparation. 1. From ethylene by the following steps:

$$\text{Ethylene} \rightarrow \text{chlorohydrin} \rightarrow \text{cyanohydrin} \rightarrow \beta\text{-hydroxy acid} \xrightarrow{\text{HCl}} \beta\text{-chloro acid}$$

2. From trimethylene glycol by replacing one hydroxyl by means of HCl and oxidizing with nitric acid.[133]

3. From acrolein by addition of HCl (1,4-addition) followed by oxidation.[134]

4. From acrylic acid and HCl.

5. Propionyl chloride in CCl$_4$ solution reacts with chlorine in sunlight to give beta substitution, ClCH$_2$CH$_2$COCl (Michael).

Reactions. 1. Replacement of the beta halogen by other groups is not as easy as with the alpha compound.

2. Bases give the alpha beta unsaturated acid with smaller amounts of the beta hydroxy acid. Other alkaline reagents have a similar tendency.

αα-Dichloropropionic acid, 2,2-dichloropropanoic acid, CH$_3$CCl$_2$CO$_2$H, can be prepared by direct chlorination with catalysts. Its silver salt, on boiling with water, gives pyruvic acid, CH$_3$COCO$_2$H.

αα-Br$_2$-propionic acid with bases gives pyruvic acid, α-Br-acrylic acid and a polymer (C$_3$H$_4$O$_3$)$_x$.

αβ-Dichloropropionic acid, 2,3-dichloropropanoic acid, ClCH$_2$CHClCO$_2$H, is made by (1) adding chlorine to acrylic acid and (2) chlorinating β-chloro-propionic acid.

Boiling *aqueous* alkalies give carbonates and vinyl chloride. This is a general reaction of αβ-dihalogen acids. The beta halogen reacts rather than the alpha.

$$\text{RCHXCHXCO}_2\text{Na} \rightarrow \text{NaX} + \text{RCH}{=}\text{CHX} + \text{CO}_2$$

Milder treatment of αβ-Br$_2$-propionic acid with aqueous bases gives glyceric acid, α-Br-β-OH-propionic acid and a small amount of pyruvic acid. Silver carbonate gives α-Br-β-OH-propionic acid. It is to be noted that the beta halogen reacts more readily. Moreover the αβ-dibromo compound reacts more readily than the αα-.[134a] Excess *alcoholic* KOH gives propiolic acid, HC≡CCO$_2$H. Zinc or inorganic iodides remove both halogens giving acrylic acid.

ββ-Dichloropropionic acid, 3,3-dichloropropanoic acid, is obtained from the αβ-dichloro-compound by removing one HCl by *alcoholic* KOH and then adding HCl to the resulting β-chloroacrylic acid, ClCH=CHCO$_2$H. Boiling with water gives the half aldehyde of malonic acid which is stable as the enol form.

$$\begin{array}{ccc} \text{CHCl}_2 & \text{CHO} & \text{CHOH} \\ | & \rightarrow \quad | & \rightarrow \quad || \\ \text{CH}_2\text{CO}_2\text{H} & \text{CH}_2\text{CO}_2\text{H} & \text{CHCO}_2\text{H} \end{array}$$

Alcoholic KOH gives propiolic acid.

[133] "Org. Syntheses."

[134] *ibid.*

[134a] Lassen, *Ann. Rep. Chem. Soc.* (London) **1906**, 98.

Halogenated Butyric Acids.

a. The alpha halogenated *n*-butyric acids resemble the corresponding compounds of propionic acid in preparation and properties.

b. The beta acids are best prepared by adding HX to the alpha beta unsaturated acid. Their reactions are like those of the β-halogenated propionic acids. Loss of HX always takes place with an alpha H rather than with a gamma H.

c. Gamma halogen *n*-butyric acids are best obtained by treating butyrolactone with HX

$$\begin{matrix} CH_2CH_2 \\ | \qquad \diagdown \\ \qquad \qquad O + HX \rightarrow XCH_2CH_2CH_2CO_2H \\ | \qquad \diagup \\ CH_2CO \end{matrix}$$

The lactone is made by hydrolyzing the cyanide from 3-chloro-1-propanol. It is not possible to replace the gamma halogen by any other group because of the tendency for lactone formation. Either O of the carboxyl group is in the 1,5-position to the carbon bearing the halogen and can approach it in space, thus facilitating the removal of the halogen and closing of the lactone ring. This internal reaction is easier than an intermolecular change.

d. Alpha halogen isobutyric acids are formed with unusual ease because the alpha H in isobutyric acid is also tertiary. The reactions are like those of the other alpha halogen acids except for an increased tendency to lose HX instead of replacement of X by other groups. All reagents of alkaline nature give mainly α-methylacrylic acid or its reaction products. Thus sodiomalonic ester with α-Br-isobutyric ester gives $(EtO_2C)_2CHCHCMeCO_2Et$, the product which would be obtained from methacrylic ester.

2,2,3-Cl₃-butyric acid, m. 60°, b. 236°, is a slightly stronger acid than Cl_3-acetic acid. It gives esters of possible value in perfumes and plasticizers.

Halogenated valeric acids resemble the corresponding butyric compounds. δ-Halogen valeric acids, on heating alone or with alkalies, give delta valerolactone (6-membered ring). In higher acids, having the halogen farther away from the carboxyl group than the delta position, the halogen acts independently of the carboxyl.

Halogenated caproic acids follow the usual reactions. An exception is 3-Br-2,2-Me₂-butyric acid which gives Me₃-ethylene on conversion to its sodium salt. The reaction resembles that of αβ-dihalogen acids with bases.

Higher alpha halogen acids are made from the corresponding fatty acids. If the acid is a monosubstituted acetic acid and has to be synthesized, it is best to use the malonic ester synthesis and brominate the resulting malonic acid before decomposing it.[135]

$$RCH(CO_2H)_2 \xrightarrow{\text{Br}_2} RCBr(CO_2H)_2 \xrightarrow{\text{heat}} RCHBrCO_2H$$

[135] "Org. Syntheses."

The fact that $RCH(CO_2H)_2$ is brominated more easily than RCH_2CO_2H is due to the increased activity of a tertiary H atom and to the greater enolization in the malonic acid derivative.

Methyl Cl_5-stearate is a plasticizer for polyvinyl plastics and films.

Higher omega (ω) halogen acids are made from the polymethylene glycols which are available by the reduction of esters of dibasic acids with sodium and absolute alcohol (Bouveault and Blanc). One carbinol group is converted to a halide and the other is then oxidized to a carboxyl. When the ester salts of the dibasic acids are available they can be reduced to the ω-hydroxy acids. The fact that the ester group is reduced while the salt group remains unchanged is due to the presence of a carbonyl group in the former and, perhaps, none at all in the latter (p. 238).

Many higher halogenated fatty acids are obtained by adding Br_2, HBr, or HI to olefinic acids or I_2 to acetylenic acids. *Sabromin* and *Sajodin* are thus obtained by adding Br_2 and HI respectively to behenolic acid.

Halogen-substituted Unsaturated Acids

α-Chloroacrylic acid, $CH_2 = CClCO_2H$, m. 65°, is obtained from dichloropropionic acid (2,3 or 2,2) with alcoholic KOH. HCl adds to give the 2,3-Cl_2-compound by 1,4-addition.

β-Chloroacrylic acid, $ClCH = CHCO_2H$, m. 85°, is obtained by reducing chloralide with Zn and aqueous-alcoholic HCl. It adds HCl to form the 3,3-Cl_2-compound.

α-Chlorocrotonic acid, $CH_3CH = CClCO_2H$, m. 99°, b. 212°, is made from butyrchloral and potassium ferrocyanide.

$$CH_3CHClCCl_2CHO \rightarrow CH_3CH = CClCO_2H$$

α-Chloroisocrotonic acid, m. 66°, is obtained from 2,3-Cl_2-butyric acid by cautious treatment with bases. Vigorous treatment gives 1-Cl-1-propene. Heat converts the acid partly to its stereoisomer. Reduction with Na amalgam gives isocrotonic acid.

β-Chlorocrotonic acid, $CH_3CCl = CHCO_2H$, m. 94°, b. 211°, is obtained from its chloride formed by PCl_5 with acetoacetic ester. It is less volatile with steam than its stereomer.

β-Chloroisocrotonic acid, m. 60°, b. 195°, is obtained like its stereomer and is separated by its ready volatility with steam.

The solubilities of the chloro-2-butenoic acids in parts of cold water are about as follows:

	crotonic	isocrotonic
Alpha Cl	47.	15
Beta Cl	35.	52

Iodostarin is the product from 1 mol of I_2 and tariric acid.

XII. DERIVATIVES OF ALIPHATIC ACIDS

A. Carboxylic Esters, RCO_2R'

While the formulas of the esters resemble those of the salts, RCO_2M, their properties are radically different. In general, they are volatile non-ionized sweet-smelling liquids which are insoluble in water but soluble in organic liquids. They are good solvents for a variety of materials including nitrocellulose.

Boiling Points of Typical Esters, °C.

n-alkyls	Formates	Acetates	Propionates	n-Butyrates	n-Valerates	n-Caproates	n-Heptoates	n-Octoates
Methyl..	32	57	80	102	127	149	172	193
Ethyl ...	54	77	99	121	145	167	187	206
Propyl ..	81	102	123	143	167	185	206	225
Butyl ...	107	126	145	164	186	204	225	240
Amyl....	130	148	164	185	204	222	123/19	100/5
Hexyl...	154	169	109/56	205	224	247	137/19	145/15
Heptyl..	177	191	208	225	244	259	274	289
Octyl....	198	210	226	244	260	268	290	306

Formates are readily prepared by refluxing formic acid with primary or secondary alcohols.

$$HCO_2H + CH_3OH \rightleftharpoons HCO_2CH_3 + H_2O$$

The reaction is reversible and consequently reaches a condition of equilibrium if none of the products is allowed to escape. Because of its low boiling point (32°), methyl formate is easily distilled out of the mixture as rapidly as formed. Hydrogen ions catalyze both the esterification and the hydrolysis reactions. Since formic acid is a relatively strong acid no other catalyst need be added.

The reaction takes place through an addition to the carbonyl group[1] (p. 120).

Much study has been devoted to the mechanism of the formation and hydrolysis of esters under the influence of catalysts.[2]

When the hydrolysis of the ester is assisted by the presence of a base and the ester is completely converted to a salt, the process is called *saponification* because it is the process used for soap making. The term saponification is incorrectly used as synonymous with hydrolysis.

Ethyl formate, b. 54°, can be made similarly by refluxing a mixture of the acid and alcohol under a column equivalent to about 10 theoretical plates to separate the ester from the ethanol, b. 78°.

Reactions. 1. Hydrolysis and saponification. In addition to the usual methods involving acid and basic reagents, esters can be split in neutral ether solution by MgI_2.[3]

$$RCO_2R' + MgI_2 \rightarrow RCO_2MgI + R'I$$

[1] Henry. *Ann. Rep. Chem. Soc.* (London) 1910, 66.
[2] Ingold, Ingold. *J. Chem. Soc.* 1932, 756.
[3] *Ann. Rep. Chem. Soc.* (London) 1916, 82.

2. Ammonolysis resembles hydrolysis

$$\begin{array}{ccc} \overset{\displaystyle OR}{\underset{}{HC}}=O + NH_3 \rightleftharpoons & \overset{\displaystyle OR}{\underset{NH_2}{HC{-}OH}} \rightleftharpoons ROH + & \overset{\displaystyle}{\underset{NH_2}{HC}}=O \end{array}$$

With higher esters this process may become very slow.   Thus ethyl isobutyrate reacts at room temperature only after about eight weeks.   Similar reactions occur with primary and secondary amines, hydroxylamine, hydrazine, and substituted hydrazines.   In each case ROH rather than H_2O is eliminated.[4]

3. Alcoholysis.[5]   The ester of one alcohol can be transformed into that of another alcohol by refluxing with an excess of the second alcohol in the presence of a trace of a catalyst such as an alcoholate or an acid.

$$\begin{array}{ccc} \overset{\displaystyle OEt}{\underset{}{HC}}=O + MeONa \rightleftharpoons & \overset{\displaystyle OEt}{\underset{OMe}{HC{-}ONa}} \rightleftharpoons EtONa + & \overset{\displaystyle}{\underset{OMe}{HC}}=O \end{array}$$

$$MeONa + EtOH \rightleftharpoons EtONa + MeOH$$

This change can be produced even with a trace of NaOH as catalyst.   In general it is easier to replace a higher alcohol by a lower one.[6]   This reaction is valuable in cases where ordinary hydrolysis or saponification would give undesirable side reactions.   Thus $MeCHClCH_2OAc$ on refluxing with an excess of MeOH gives MeOAc and $MeCHClCH_2OH$ whereas treatment with water or bases would have involved the chlorine atom as well as the acetate group.

4. Other carbonyl reactions.

(a) With Grignard reagents, formic esters give aldehydes and secondary alcohols depending on the conditions used.

$$\begin{array}{ccc} \overset{\displaystyle OCH_3}{\underset{}{HC}}=O + RMgX \rightarrow & \overset{\displaystyle OCH_3}{\underset{R}{HC{-}OMgX}} \rightarrow & \overset{\displaystyle H}{\underset{}{RC}}=O \end{array}$$

$$RCHO + RMgX \rightarrow R_2CHOMgX$$

Orthoformates give aldehyde acetals.

$$HC(OEt)_3 + RMgX \rightarrow EtOMgX + RCH(OEt)_2$$

[4] Gilman.   "Org. Chemistry," 2nd Ed.   p. 653.
[5] *Ann. Rep. Chem. Soc.* (London) 1917, 73; 1921, 63.
[6] Bellett.   *Compt. rend.* 193, 1020 (1931).

(b) With substances having alpha hydrogen atoms. The presence of traces of sodium alcoholates induces an "aldol" condensation

$$HC{\overset{OC_2H_5}{\nwarrow}}=O + HCH_2COCH_3 \rightarrow \left[HC{\overset{OC_2H_5}{\nwarrow}}{\underset{OH}{\searrow}}CH_2COCH_3 \right] \rightarrow [HCOCH_2COCH_3]$$

The resulting product, which would be expected to be an aldehyde, exists only as the Na derivative of the *enol* form, hydroxymethyleneacetone, 2-buten-3-on-1-ol.

$$CH_3COCH_2C{\overset{H}{\nwarrow}}=O \rightarrow CH_3COCH=CHONa$$

On liberation from the Na compound, the product undergoes condensation to triacetylbenzene.

Similarly,

$$HCO_2Et + CH_3CO_2Et \rightarrow NaOCH=CHCO_2Et$$
$$HCO_2Et + CH_2(CO_2Et)_2 \rightarrow NaOCH=C(CO_2Et)_2$$

The resulting products behave as salts of fairly strong acids.

Methyl formate is made from CO and methanol under pressure (DuPont).

Orthoformates, $HC(OR)_3$, are made from $CHCl_3$, ROH and Na.

Ethyl orthoformate, like other orthoformates, $RC(OR')_3$, is stable to bases but is sensitive to acid. Its hydrolysis has been studied in relation to catalysts.[7] The effective catalyst is the hydronium ion, $(H_3O)^+$. The first step of the acid hydrolysis is practically instantaneous to give the ordinary ester and alcohol.

Orthoformates react with carbonyl compounds to give acetals.

$$R_2CO + HC(OEt)_3 \rightarrow R_2C(OEt)_2 + HCO_2Et$$

With Grignard reagents they give acetals of aldehydes.

$$RMgX + HC(OEt)_3 \rightarrow RCH(OEt)_2 + EtOMgX$$

They condense with active methylene compounds to give $EtOCH=CQ_2$ in which the Q groups are combinations of CH_3CO, CN, CO_2R, etc.

$HC(OMe)_3$, b. 102°; $HC(OEt)_3$, b. 145°. Orthoformic esters of Pr, isoPr, Bu, isoBu and isoAm alcohols have been made.[8]

Ortho esters of higher acids can be made from the imido ethers obtained from the nitriles, HCl and EtOH.

$$RC(OEt)=NH \cdot HCl + 2\ EtOH \rightarrow RC(OEt)_3 + NH_4Cl$$

[7] Harned, Samaras. *J. Am. Chem. Soc.* 54, 1 (1932).
[8] Sah, Ma. *J. Am. Chem. Soc.* 54, 2964 (1932).

Acetates.

Preparation.[9] 1. From acetic acid and primary and secondary alcohols saturated with HCl gas or with a small amount of conc. H_2SO_4 as catalyst. When equivalent amounts of acetic acid and ethanol are used an equilibrium is reached when two-thirds of the material is converted to ester and water. Thus the intermediate $MeC(OH)_2OEt$ tends to lose H_2O rather than EtOH (p. 121).

The per cent esterification of acetic acid by various alcohols in one hour at 155° is as follows: Me, 56; Et, 47; Bu, 47; Octyl, 47; isoPr, 26; MeHexCHOH, 21; Et_2CHOH, 17; t-Bu, 1.4; t-Am, 0.8. It is not safe to assume from these results that all alcohols can be separated into their three classes by the difference in their speeds of esterification. In the higher alcohols, branching in an alkyl group has a profound effect. Thus Et_2CHCH_2OH does not esterify like a primary but rather like a tertiary alcohol. Studies on the speed of esterification of acids of the types RCH_2CO_2H, R_2CHCO_2H and R_3CCO_2H, show that the velocity decreases with the complexity of the acid.[10] The per cent esterification of isobutyl alcohol by various acids in one hour at 155° follows: formic, 62; acetic, 44; caprylic, 31; isobutyric, 29; Bu_2-acetic, 3; Me_3-acetic, 8; Me_2Et-acetic, 3; Et_3-acetic, 1.4; dineopentylacetic, 1.4; Me-t-Bu-neopentylacetic, 0.8. It is thus not correct to assume that the primary, secondary or tertiary nature of a complex acid can be determined with certainty by its speed of esterification. This is possible only in comparing acids of the same molecular weight.

The structure and molecular weight of the acid and alcohol portions of esters have similar profound effects on the rates of hydrolysis of the esters. Again, partial comparisons may lead to erroneous results. Thus the rate of hydrolysis of esters of cyclohexane carboxylic acid is much greater than that of the same esters of diethylacetic acid.[11]

2. Commercially, ethyl acetate is made from 10% acetic acid ("quick vinegar" process) and 50% alcohol ("high wines") by a distillation process. A ternary mixture of ethyl acetate, water and ethyl alcohol distills first. This is diluted with water to give two layers, nearly pure ethyl acetate and dilute alcohol. The latter is returned to the process. The "bottoms" are practically pure water and are discarded.

3. Acetic acid and ethanol passed over a silica gel catalyst at 150° give an equilibrium mixture containing 85% ethyl acetate.[12]

4. Still another commercial process for ethyl acetate is the dismutation of acetaldehyde by catalysts such as Al ethylate and its compounds with $AlCl_3$

[9] Keyes. *Ind. Eng. Chem.* **24**, 1096 (1932).

[10] Menschutkin. *Bull. soc. chim.* (3), **34**, 87 (1880); *Ann. chim.* (5), **23**, 14 (1881).

[11] Smith, Steele. *J. Amer. Chem Soc.* **63**, 3466 (1941).

[12] Reid. *J. Am. Chem. Soc.* **53**, 4353 (1931).

and $HgCl_2$.[13]

$$2\ MeCHO \rightarrow MeCO-OCH_2Me$$

Similarly ethanol with a dehydrogenating catalyst gives ethyl acetate.[14]

5. Acetates of some secondary alcohols are made by adding olefins from cracked gases to acetic acid in the presence of sulfuric acid.

$$CH_3CO_2H + CH_3CH=CHCH_3 \rightarrow CH_3CO_2CH(CH_3)CH_2CH_3$$

6. Higher acetates may be made from acetyl chloride or acetic anhydride. Acetic anhydride is used for making cellulose acetate for nonexplosive photographic film, and for artificial silk (Celanese).

Ketene, $CH_2=C=O$, acts as an anhydride of acetic acid and reacts with alcohols to form acetates.

The preparation of acetates of tertiary alcohols is best from acetyl chloride with a solution of the alcohol in pyridine. The base combines with the liberated HCl. Otherwise the latter produces mainly the t-alkyl chloride.

7. In extreme cases, acetates are sometimes prepared from silver acetate and alkyl halides, preferably the iodides. This is a poor preparative method even with halides of Class 1 (p. 74). The use of thallous salts has been recommended.[15]

$$MeCO_2Ag + RI \rightarrow AgI + MeCO_2R$$

8. Ethyl acetate can be made catalytically directly from ethanol.

$$2\ CH_3CH_2OH \rightarrow CH_3CO_2CH_2CH_3 + 2\ H_2$$

Reactions. The esters of acetic acid give the same reactions as the corresponding formates. In addition, an acetate can act as both components in an aldol condensation (Claisen).

$$\overset{OR}{MeC}=O + HCH_2CO_2R \xrightarrow[\text{NaOR}]{\text{Trace}} \overset{OR}{MeC}\underset{OH}{-}CH_2CO_2R \rightarrow$$

$$ROH + MeCOCH_2CO_2R \xrightarrow{\text{Na}} MeC(ONa)=CHCO_2R$$
$$\text{Keto form}$$

Much work has been done on the mechanism of this condensation but the above conception seems as good as any.[16-20] The reaction is carried out by

[13] *Ann. Rep. Chem. Soc.* (London) 1925, 72.
[14] Zeisberg. *C. A.* **23**, 2449 (1929).
[15] *Ann. Rep. Chem. Soc.* (London) 1926, 99.
[16] *ibid.* 1927, 90.
[17] Scheibler, Tutundzitsch. *Ber.* **64B**, 2916 (1931).
[18] Adickes. *Ber.* **65B**, 522 (1932).
[19] Cox, Kroeker, McElvain. *J. Am. Chem. Soc.* **56**, 1173 (1934).
[20] Hauser. *J. Am. Chem. Soc.* **69**, 295 (1947).

treating purified ethyl acetate with clean sodium. If the ethyl acetate has been purified to the point at which it contains no ethyl alcohol, a trace has to be added to start the reaction. Ethyl acetate will react similarly with other alpha hydrogen compounds. For instance, acetone gives acetylacetone, $MeCOCH_2COMe$.

With higher esters, only the $\alpha - H$ atoms take part in the Claisen reaction. Thus propionic ester gives $EtCOCHMeCO_2Et$.

Butyl and amyl acetates are of the utmost importance as lacquer solvents. "Pentacetate" is a mixture of amyl acetates (Sharples). Esters of secondary as well as primary alcohols are used.[21]

Esters of higher acids may be made like those of acetic acid. With less reactive acids such as stearic, good yields are obtained by heating with an excess of alcohol and a little sulfuric acid and aluminum sulfate at 100°.

In case the acid is to be synthesized through the *cyanide*, the latter can be *changed directly to the ester by alcoholysis* instead of being hydrolyzed to the acid and then esterified.

$$RCN + EtOH + H_2O + H_2SO_4 \rightarrow RCO_2Et + NH_4HSO_4$$

Esters made by this method are often cheaper than the free acids. Examples are those of malonic acid, lactic acid, and α-hydroxyisobutyric acid.

Methyl esters are available by the action of diazomethane (poison).

$$RCO_2H + CH_2N_2 \rightarrow N_2 + RCO_2CH_3$$

A cheaper but less effective reagent is Me_2SO_4 (poison).[22]

Esters of higher acids are reduced to the corresponding higher alcohols by sodium and absolute alcohol.[23] Thus ethyl laurate gives lauryl alcohol.

$$RCO_2Et + 4 Na + 2 EtOH \rightarrow RCH_2ONa + 3 EtONa$$

A modification of this process involves the addition of sodium and acetic acid to an ether solution of the ester in contact with sodium acetate solution.[24]

Reduction of esters by sodium has become less important since the development of $Cu - Cr$ oxide catalysts for direct hydrogenation.[25] A special form of Raney nickel will also reduce esters.[26] Lithium aluminum hydride converts both esters and acids quantitatively to alcohols.[27]

Sodium acts with esters in ether solution to give *acyloins*. The first step is like that in the formation of a pinacol from a ketone and a metal. A free

[21] Park, Hopkins. *Ind. Eng. Chem.* 22, 826 (1930).
[22] *Ann. Rep. Chem. Soc.* (London) 1904, 85.
[23] Bouveault, Blanc. *Compt. rend.* 136, 1676 (1903).
[24] *Ann. Rep. Chem. Soc.* (London) 1924, 63.
[25] Folkers, Adkins. *J. Am. Chem. Soc.* 54, 1145 (1932).
[26] Adkins, Billica. *J. Am. Chem. Soc.* 70, 3121 (1948).
[27] Nystrom, Brown. *J. Am. Chem. Soc.* 69, 1197 (1947).

radical (having the odd electron due to the Na atom) is formed and then gives a dimolecular product.

$$2 \, RCO(OEt) + 2 \, Na \rightarrow 2 \, RC(ONa)(OEt) \rightarrow \begin{array}{c} RC(ONa)(OEt) \\ | \\ RC(ONa)(OEt) \end{array} \rightarrow$$

$$2 \, NaOEt + \begin{array}{c} RCO \\ \| \\ RCO \end{array} \xrightarrow{2 \, Na} \begin{array}{c} RCONa \\ \| \\ RCONa \end{array} \xrightarrow{acid} \begin{array}{c} RCOH \\ \| \\ RCOH \end{array} \rightarrow \begin{array}{c} RCO \\ | \\ RCHOH \end{array}$$

The disodium derivative formed by the 1,4-addition of Na to the diketone can be obtained and changed to a diacetate. The free enediol is unstable and ketonizes to the acyloin at once.

The reaction of esters with excess Grignard reagent is a step in the Barbier-Wieland degradation of an acid to the next lower acid.[28] Using a phenyl Grignard reagent, the following steps are involved:

$$RCH_2CO_2Me \xrightarrow{2 \, PhMgBr} RCH_2C(OH)Ph_2 \xrightarrow{acid} RCH=CPh_2 \xrightarrow{CrO_3 \, mixture} RCO_2H$$

Esters of higher acids and higher alcohols are *waxes*. These occur widely distributed in nature both in animal and vegetable products. Spermaceti contains the palmitates of cetyl (C_{16}) and octadecyl alcohols. Wool fat contains esters of carnaubyl alcohol (C_{24}). Carnauba wax and beeswax contain myricyl (C_{30}) cerotate (C_{26}) and palmitate. Chinese wax is largely ceryl cerotate. Artificial waxes have been made from a variety of higher alcohols and acids.[29] The oil of the "castor oil fish" consists mainly of liquid waxes, esters of cetyl and oleyl alcohols with oleic and hydroxyoleic acid.[30]

Vapor pressure curves have been determined for methyl esters of caproic, caprylic, capric, lauric, myristic, palmitic, stearic, oleic, and linoleic acids.[31]

Vinyl esters, $RCO_2CH=CH_2$, are obtained by heating an ethylene dihalide with the corresponding sodium salt under pressure. A better method is the addition of an acid to acetylene in presence of a Hg catalyst.[32] The ethylidene diester is also formed. Vinyl acetate is an important intermediate for Vinylite resins.

Many esters are used in perfumes (Bogert).[33]

Hydroxyacids give esters of various types. The product of two molecules of an α-hydroxy acid is a *lactide*. The internal esters from γ- and δ-hydroxy acids are lactones. When the hydroxyl and carboxyl are farther separated the product consists of cyclic or linear esters of high molecular weight. The

[28] Gilman. "Org. Chemistry," 2nd Ed. p. 1357.
[29] *Ann. Rep. Chem. Soc.* (London) 1926, 102.
[30] Cox, Reid. *J. Am. Chem. Soc.* 54, 220 (1932).
[31] Althouse, Triebold. *Ind. Eng. Chem., Anal. Ed.* 16, 605 (1944).
[32] Ellis. "Chemistry of Petroleum Derivatives." Reinhold, 1934. p. 679.
[33] "Synthetic Organic Chemistry in the Study of Odorous Compounds," Columbia Univ. Press.

smaller molecules can be removed by means of a molecular still. The polymerization of these esters is reversible.[34]

Esters of unsaturated acids are valuable in reactions in which it is desirable to mask the carboxyl group. Thus $\alpha\beta$-unsaturated esters react with esters having active methylene groups in the presence of sodium ethylate (Michael).[35] The net result is the addition of the active ester to the double bond with H in the alpha position (1,4-addition).

$\alpha\beta$-Unsaturated esters add NH_3 and HCN similarly.

Esters of halogenated acids.

Preparation. 1. Chlorination or bromination of the esters in sunlight to give alpha substitution.

2. Esterification of the halogen acids. This is easier than with the unsubstituted acids because the halogen increases the strength of the acids.

3. Treatment with absolute alcohol of the alpha halogen acyl halides prepared from the dry acids, bromine and phosphorus (Hell-Volhard-Zelinsky).

4. Iodo esters from the chloro esters and sodium iodide in alcohol.

5. Peroxide catalyzed addition of α-bromo esters to olefins.[36]

$$R-C=C + Br-C-COOEt \rightarrow R-C-\overset{\overset{\displaystyle Br}{|}}{C}-C-C-COOEt$$

Reactions. 1. The usual reactions of the corresponding acids except that the ester grouping prevents lactone formation except under conditions which would hydrolyze the ester.

2. Alpha halogen esters react with ketones and esters in the presence of zinc by a reaction resembling the Grignard reaction except that the intermediate organic zinc compound has never been prepared (Reformatzky).[37]

$$RR'CO + BrCH_2CO_2C_2H_5 + Zn \rightarrow RR'C(OH)CH_2CO_2C_2H_5$$

Esters such as ethyl oxalate and ethyl acetate will supply the reactive carbonyl group as well as the ketones. Oxalic ester and bromoacetic ester give oxaloacetic ester, $EtO_2CCOCH_2CO_2Et$. The esters may also be of the types

$$RCHBrCO_2Et \text{ and } RR'CBrCO_2Et.$$

Thus α-bromoisobutyric ester and acetoacetic ester react to give hydroxy-trimethylglutaric acid.

$$Me_2CBrCO_2Et + MeCOCH_2CO_2Et \rightarrow EtO_2CC(Me_2)C(OH)(Me)CH_2CO_2Et$$

Esters of unsaturated alcohols are important as monomers for Vinylite Resins which find wide uses in electrical insulation, safety glass, phonograph records, combs, belts, beer can linings, floor tiles, and Vinyon yarn and felt. Typical intermediates of this type are allyl chloroformate and allyl carbonate.

[34] Carothers, Dorough, Van Natta. *J. Am. Chem. Soc.* **54**, 761 (1932).
[35] *Ann. Rep. Chem. Soc.* (London) 1922, 71.
[36] Kharasch. *J. Am. Chem. Soc.* **70**, 1055 (1948).
[37] "Org. Reactions," I.

B. Acid Halides

An acid halide, either organic or inorganic, is derived from an acid by replacing the "OH groups" in the acid by halogens. Acyl halides do not ionize to yield halide ions but are readily hydrolyzed, probably by initial addition to the very active carbonyl group (p. 289).

Acid	Acid Chloride
H_3PO_3	PCl_3
H_3PO_4	$POCl_3$
$[H_5PO_5]$	PCl_5
H_2SO_4	SO_2Cl_2
H_2SO_3	$SOCl_2$
H_2CO_3	$COCl_2$
HNO_2	$NOCl$

The hydrolysis of an acid chloride differs from that of a chloride salt of a weak base in that it is not reversed by HCl. Thus

$$POCl_3 + 3\ H_2O \rightarrow H_3PO_4 + 3\ HCl$$
$$AlCl_3 + 3\ H_2O \rightleftharpoons Al(OH)_3 + 3\ HCl$$

The inorganic acid halides are valuable in organic chemistry for converting a hydroxyl group to a halogen and for forming inorganic esters. Thionyl chloride, $SOCl_2$, is especially valuable for replacing an alcoholic hydroxyl by chlorine with a minimum of rearrangement. Thionyl bromide is even more effective for this purpose.[38] Sulfuryl chloride, SO_2Cl_2, with peroxides, is an important chlorinating agent for paraffin hydrocarbons, for aliphatic acids, and for side-chains in aromatic compounds.[39] Organic acid halides of the type

$$RC\!\!\underset{\diagup}{\overset{X}{=}}\!O$$

are *acyl halides.*

Preparation. 1. An acid or its salt is treated with a suitable inorganic acid halide.

Inorganic acid halide + organic acid → Organic acid halide
$\qquad\qquad\qquad\qquad\qquad\qquad\qquad$ + inorganic acid (or its anhydride)

Thionyl chloride is the best reagent.[40]

$$RCO_2H + SOCl_2 \rightarrow HCl + SO_2 + RCOCl$$

PCl_3, the most commonly used reagent, does not react quantitatively according to the equation.

$$3\ CH_3CO_2H + PCl_3 \rightarrow H_3PO_3 + 3\ CH_3COCl$$

[38] Elderfield et al. *J. Am. Chem. Soc.* **68**, 1579 (1946).
[39] Kharasch et al. *J. Am. Chem. Soc.* **62**, 2393 (1940).
[40] Ackley, Tesoro. *Ind. Eng. Chem., Anal. Ed.* **18**, 444 (1946).

Much HCl is evolved due to side reactions.   A better yield is obtained from PCl_3 and a mixture of the acid and its anhydride. PCl_5 may also be used for converting acids to acid chlorides.[41]

2. The chlorination of aldehydes in the absence of water, catalysts and sunlight gives acyl chlorides.

$$CH_3\overset{H}{\overset{/}{C}}{=}O + Cl_2 \rightarrow CH_3\overset{Cl}{\overset{/}{C}}{=}O + HCl$$

Reactions.   1. Acyl halides react readily and completely with compounds containing H attached to an element other than carbon, "active H compounds."

a. Water reacts rapidly to give HX and the organic acid.   This reaction between water and a non-ionized halide is surprising and in sharp contrast to the general inactivity of alkyl halides with water even on heating.   The most reasonable explanation of this reactivity is that it does not involve metathesis but a primary addition to the reactive carbonyl group.   Thus reactions which cannot take place in this way are not rapid.   An example is the action with AgCN to give RCOCN

$$RCOCl + HOH \rightarrow HCl + RCO_2H \quad \text{rapid}$$
$$RCOCl + AgCN \rightarrow AgCl + RCOCN \text{ slow}$$

The rapid reaction is often pictured as involving the steps:

$$R\overset{Cl}{\overset{/}{C}}{=}O + HOH \rightarrow R\overset{Cl}{\overset{/}{\underset{\backslash}{C}}}{-}OH \rightarrow HCl + R\overset{}{\underset{\backslash OH}{C}}{=}O$$

The addition may be initiated by a hydrogen-bonding of a molecule of water to the carbonyl oxygen.

$$R\overset{Cl}{\overset{|}{C}}::\overset{..}{\underset{..}{O}} \rightarrow R\overset{Cl}{\overset{|}{C}}\underset{+}{:}\overset{..}{\underset{..}{O}}:H \rightarrow R\overset{Cl}{\overset{|}{C}}:\overset{..}{\underset{..}{O}}:H$$

b. Ammonia gives acid amides, $RCONH_2$, by a similar mechanism.

Higher acid chlorides give more complex results.   Thus diethylacetyl chloride gives a 50% yield of the amide and smaller amounts of the ammonium salt and the nitrile Et_2CHCN, and a trace of the diacyl amide.   In this case the amide is dehydrated to the nitrile by an excess of the acid chloride.

[41] "Org. Reactions," II.

c. Primary and secondary alcohols give esters, RCO_2R'. When a tertiary alcohol is treated with acid chlorides without any precaution, the chief product is a tertiary chloride. That this is due to the ready action of HCl with tertiary alcohols and their esters is shown by the formation of the tertiary ester from the alcohol and acyl halide in presence of a basic substance like pyridine or dimethylaniline which can combine with the HX as rapidly as it is formed.

2. With sodium salts of organic acids they yield anhydrides.

$$CH_3COCl + CH_3CO_2Na \rightarrow NaCl + (CH_3CO)_2O$$

3. With silver cyanide, mercurous cyanide or cuprous cyanide, they give acyl cyanides, RCOCN, which can be hydrolyzed to alpha keto acids, $RCOCO_2H$.

4. With Grignard reagents they give ketones and then tertiary alcohols. Earlier, a similar result was achieved with organic zinc compounds. In the latter case the reaction does not go beyond the ketone, if propyl or higher compounds of zinc are used. These reactions have been ordinarily regarded as purely metathetical in the step which gives the ketone, but in reality the reaction is probably an ordinary carbonyl addition.

$$RC\!\!\!\overset{X}{=}\!\!\!O + R'MgX \rightarrow RC\!\!\!\underset{R'}{\overset{X}{—}}\!\!\!OMgX \rightarrow MgX_2 + RCOR'$$

5. Acyl halides may be reduced to aldehydes with H_2 and a palladium catalyst.[42] Palladium is considerably cheaper than platinum. Again, the reaction probably proceeds through the carbonyl group.

6. Acyl halides react with other halide acids. For instance one acid halide can be converted to a different one by an excess of the proper hydrogen halide.[43]

$$RC\!\!\!\overset{Cl}{=}\!\!\!O + HI \rightleftharpoons RC\!\!\!\underset{I}{\overset{Cl}{—}}\!\!\!OH \rightleftharpoons RC\!\!\!\underset{I}{=}\!\!\!O + HCl$$

The reaction can be run to completion in either direction.

7. Acyl halides react with aromatic hydrocarbons and some of their derivatives in the presence of equimolar amounts of anhydrous aluminum chloride to give mixed aliphatic-aromatic ketones (Friedel and Crafts). For instance:

$$CH_3COCl + C_6H_6 \xrightarrow{\text{AlCl}_3} CH_3COC_6H_5 + HCl$$

Methyl phenyl ketone

[42] Rosenmund, Zetzsche. *Ber.* **54B**, 425 (1921).
[43] *Ann. Rep. Chem. Soc.* (London) 1913, 62.

The mechanism of such processes has never been worked out. Definite compounds are known between all of the materials involved in the reaction, $RCOCl(AlCl_3)_n$, $RCOAr(AlCl_3)_m$, $ArH(AlCl_3)_p$.

8. Surprisingly, acyl halides do not react as readily as do alkyl halides with metals like Na and MgI (MgI_2 and Mg, Gomberg)(p. 337). Higher members of the series form diesters of enediols formed from two molecules.[44]

$$4 \text{ RCOCl} + 4 \text{ Na} \rightarrow \begin{array}{c} RCOCOR \\ \| \\ RCOCOR \end{array}$$

9. Acid chlorides react with diazomethane to give diazomethyl ketones, the intermediates for the Arndt-Eistert method of going up the series from a given acid to its next homolog.[45]

$$RCOCl + 2 CH_2N_2 \rightarrow RCOCHN_2 + N_2 + CH_3Cl$$

Treatment with catalysts such as colloidal silver and H_2O, ROH, NH_3, or RNH_2 give respectively the next higher acid, its ester, its amide or a substitution product of the latter.

$$RCOCH_2N_2 + HOH \xrightarrow{Ag} RCH_2CO_2H + N_2$$

Formyl chloride, HCOCl, is unstable." A mixture of HCl and CO acts with aromatic hydrocarbons as formyl chloride.

$$C_6H_6 + CO + HCl \xrightarrow{AlCl_3} C_6H_5CHO + HCl$$

Acetyl chloride, CH_3COCl, b. 50.9°, is a very important acetylating agent for converting $-OH$, $-NH_2$, and $=NH$ compounds to their acetyl derivatives. Commercially it is prepared from sodium acetate and a mixture of SO_2 and Cl_2. If ketene ever becomes cheap enough it can be used with HCl gas to give acetyl chloride

$$CH_2 = C = O + HCl \rightarrow CH_2 = C(OH)Cl \rightarrow CH_3COCl$$

Acetyl fluoride, b. 20°, is made from the chloride and zinc fluoride. The *bromide*, b. 77°, is made from the acid and PBr$_3$. It reacts with Mg to give RCOMgBr.[47] The *iodide*, b. 108°, is made from the chloride and HI. Acetyl iodide splits ethers.[48]

$$ROR + CH_3COI \rightarrow RI + CH_3CO_2R$$

The boiling points of the *n*-acid chlorides from C_3 to C_7 are (°C.), 80, 102, 128, 153, 175.

[44] Ralston, Selby. *J. Am. Chem. Soc.* **61**, 1019 (1939).
[45] "Org. Reactions," I. p. 38.
[46] Krauslopp, Rollefson. *J. Am. Chem. Soc.* **56**, 2542 (1934).
[47] Tishchenko. *Bull. soc. chim.* **37**, 623 (1925).
[48] Gustus, Stevens. *J. Am. Chem. Soc.* **55**, 378 (1933).

C. Acid Anhydrides

Formic anhydride, $(HCO)_2O$, is unknown.
Acetic anhydride, $(CH_3CO)_2O$, b. 139°.
Preparation.   1. Ketene and acetic acid

$$CH_2=C=O + CH_3CO_2H \rightarrow (CH_3CO)_2O$$

This reaction, without isolating the ketene, is probably the best method of preparation. Acetic acid is cracked at high temperature using a liquid catalyst containing phosphoric acid or one of its esters to give ketene (b. $-56°$) which readily separates from the water and unchanged acid and is passed up through a descending column of glacial acetic acid.[49]

Ketene reacts with other acids to give mixed anhydrides,

$$RCO-O-COMe.^{50}$$

When these are distilled the simple anhydrides result.

2. Sodium acetate with enough of an inorganic acid chloride, such as $POCl_3$, $SOCl_2$, SO_2Cl_2, or $COCl_2$, to convert half of the salt to acetyl chloride which acts with sodium acetate as rapidly as formed. Sulfur monochloride, S_2Cl_2, is used commercially.[51]

3. Heating acetaldehyde diacetate obtained from acetylene and acetic acid.

$$HC\equiv CH \rightarrow \underset{\text{vinyl acetate}}{CH_3\overset{O}{\overset{\|}{C}}-OCH=CH_2} \rightarrow \underset{\text{acetaldehyde diacetate}}{(CH_3CO_2)_2CHCH_3} \rightarrow$$

$$(CH_3CO)_2O + CH_3CHO$$

4. Acetyl chloride heated with sodium acetate. Because of the volatility of acetyl chloride this method must be carried out under pressure.

Reactions. Acetic anhydride resembles acetyl chloride in its reactions. With any active hydrogen compound, HQ, it forms a molecule of acid and an acid derivative, $CH_3\overset{Q}{\overset{\diagup}{C}}=O$. The initial step involves the usual addition compound.

$$\begin{array}{c} Me-CO \\ \diagdown \\ O + HQ \rightarrow \\ \diagup \\ Me-CO \end{array} \left[\begin{array}{c} Q \\ Me-\overset{\diagup}{C}-OH \\ | \\ O \\ | \\ Me-CO \end{array} \right] \rightarrow MeC\overset{Q}{\overset{\diagup}{=}}O + MeCO_2H$$

[49] Sixt, Magdan. *C. A.* **32,** 2961 (1938).
[50] Hurd, Dull. *J. Am. Chem. Soc.* **54,** 3427 (1932).
[51] *Ann. Rep. Chem. Soc.* (London) **1909,** 79.

In this way it acts with (1) water, (2) halide acids, (3) ammonia, (4) amino compounds, (5) imino compounds, (6) hydroxyl compounds. Acetic anhydride is the most important *acetylating agent* both for laboratory and commercial use. The largest amount of it is used in acetylating cotton for use in photographic film and artificial silk. A trace of sulfuric acid catalyzes the acetylating action of acetic anhydride.

Sometimes it reacts violently with hydroxyl compounds. Thus dangerous explosions have been obtained with it and ethylene glycol.

Chlorine reacts with acetic anhydride to give chloroacetyl chloride.[52]

Acetic anhydride reacts with aldehydes to give their diacetates.

$$RC\!\!\stackrel{H}{=}\!\!O + (MeCO)_2O \rightarrow RCH(OCOMe)_2$$

Although sodium amalgam will not reduce acids, it reduces anhydrides to aldehydes and then to alcohols.

$$Me\!-\!\underset{\underset{\text{OCOMe}}{|}}{\overset{H}{C}}\!\!=\!\!O + 2[H] \rightarrow Me\!-\!\underset{\underset{\text{OCOMe}}{|}}{\overset{H}{C}}\!-\!OH \rightarrow MeCHO + MeCO_2H$$

hemiacetate of
acetaldehyde

Because of its ability to combine with water, acetic anhydride is much used as a dehydrating reagent, for example in the Perkin reaction.

Higher anhydrides may be obtained by the same methods and also by treating the acid or its salt with acetyl chloride or acetic anhydride in the presence of a catalytic trace of H_2SO_4. The reactions of the higher anhydrides are like those of acetic anhydride but are progressively slower as the molecular weight increases. Propionic anhydride is used in making *cellulose propionate* which is rapidly becoming commercially important.

Mixed organic and inorganic anhydrides are known.

Acetyl nitrate, $CH_3CO(ONO_2)$, is a powerful nitrating agent. It is dangerously explosive.

Acetyl sulfuric acid, $CH_3CO-OSO_3H$, is both an acetylating and sulfonating agent.[53]

$$C_6H_2Br_3OH + CH_3CO_2SO_3H \rightarrow C_6H_2Br_3OCOCH_3 + H_2SO_4$$
$$C_6H_6 + CH_3CO_2SO_3H \rightarrow C_6H_5SO_3H + CH_3CO_2H$$

Acetyl peroxide, $(CH_3CO)_2O_2$, is obtained from BaO_2 and acetic anhydride. It is a syrupy liquid. It is a powerful oxidizer and is dangerously explosive.[54] Its chief use is as a polymerization catalyst for acrylic esters and the like.

[52] *Ann. Rep. Chem. Soc.* (London) 1923, 68.
[53] *Ann. Rep. Chem. Soc.* (London) 1921, 63.
[54] Kuhn. *Chem. Eng. News* **26**, 3197 (1948).

The peroxides of tertiary radicals such as *t*-butyl and triptyl are more stable.[55] Their chemistry may be illustrated as follows:

Triethylmethyl hydroperoxide, Et_3COOH, m. 3°, b. 28°/2 mm., is made from triethylcarbinol and 70% sulfuric acid and 27% hydrogen peroxide.

Di-triethylmethyl peroxide is obtained from the above compound and a solution of Et_3COH in cold 70% sulfuric acid.

$$Et_3COOH + Et_3COSO_3H \rightarrow Et_3COOCEt_3 + H_2SO_4$$

Pyrolysis of the peroxide at 250° gives 60% of diethyl ketone together with *n*-butane, ethane, and ethylene. Evidently the first product is the free triethylmethyl radical which loses an ethyl radical leaving the ketone. The ethyl radicals combine or disproportionate to give the other products.

D. Acid Amides, $RCONH_2$

The replacement of the OH of a carboxylic acid by NH_2 produces a surprising change in properties. The melting and boiling points are increased greatly.

	m. °C.		b. °C.	
	Acid	Amide	Acid	Amide
Formic.................	8.	2.5	101.	200. dec.
Acetic.................	17.	82.	118.	222.
Propionic.............	−36.	79.	141.	213.
n-Butyric.............	− 8.	115.	162.	216.
n-Valeric.............	−34.5	115.	186.	
n-Caproic.............	− 9.5	98.	205.	255.

The melting points of the members C_7 to C_{12} are (°C.) 95, 97, 99, 108, 103 and 110; those of the even members C_{14} to C_{20} are 102, 104, 109 and 108.

The NH_2 group has entirely different effects in acyl and alkyl compounds, $RCONH_2$ and RNH_2. This can be seen from the boiling points of the C_2 compounds, ethanol 78°, ethyl amine 19°, acetic acid 118°, and acetamide 222°. In the first pair the boiling point is *lowered* and in the second *raised* by the substitution of NH_2 for OH. The replacement of the amino H atoms by methyl has the same effect as the replacement of the hydroxyl H in acetic acid. The boiling points of NN-dimethylacetamide and methyl acetate are 167° and 57°.

The basic properties of the nitrogen atom are practically neutralized by the "acid" properties of the acyl group, RCO. In glacial acetic acid, however, acid amides can be titrated with perchloric acid.[56] They are also very weak acids.[56a] In liquid ammonia they act as acids of the ammonia system, neu-

[55] Milas et al. *J. Am. Chem. Soc.* 68, 205, 1938 (1946).
[56] Hall. *J. Am. Chem. Soc.* 52, 5115 (1930).
[56a] Branch, Clayton. *J. Am. Chem. Soc.* 50, 1689 (1928).

tralizing the bases of that system, the metal amides (Franklin).[57]   The amides react in tautomeric forms.

$$RC\underset{\diagup}{\overset{NH_2}{=}}O \rightleftharpoons RC\underset{\diagdown}{\overset{NH}{-}}OH$$

N-alkyl and O-alkyl derivatives of the acid amides are known.

The lower acid amides are electrolytic conductors when fused.   They are also ionizing solvents for electrolytes.

The most important reaction of acid amides is the *Hofmann Rearrangement*[58, 59] with hypohalites or bases and halogens to give primary amines.   The net result is the removal of CO from the molecule.   In addition to its practical importance this reaction has been the incentive for many theoretical studies.[60–64]   It is known that the steps involved are amide to N-halogen-amide to isocyanate to primary amine.   The organic group moves from C to the adjacent N with its electron pair.   Thus *t*-Bu-acetamide gives neopentyl-amine quantitatively without any rearrangement of the carbon skeleton.[65]

$$Me_3CCH_2CONH_2 \rightarrow Me_3CCH_2NH_2$$

This failure of a neopentyl system to form the usual *t*-amyl derivatives is diagnostic of the conditions necessary for this type of rearrangement.   The neopentyl group must not be deprived of its electron pair as in the action of neopentyl alcohol with HX and of neopentyl halides with bases[66] (pp. 120, 172). In the Hofmann rearrangement the neopentyl group is *not* so deprived. Further evidence along the same line is given by the migration of an optically active group in the Hoffmann rearrangement without loss of optical activity.[67, 68]

With higher amides, an excess of hypohalite must be avoided or the primary amine first formed may be oxidized to the corresponding cyanide.

$$RCH_2NH_2 \rightarrow [RCH_2NX_2] \rightarrow RC\equiv N$$

The Hofmann rearrangement offers an important method of "going down the series" provided an alcohol is not involved in one of the steps (p. 172). Thus stearic acid may be converted to its next lower homolog, margaric acid, by the following steps:

$$C_{17}H_{35}CO_2H \rightarrow RCOCl \rightarrow RCONH_2 \rightarrow RNH_2 \rightarrow C_{16}H_{33}CN \rightarrow C_{16}H_{33}CO_2H$$

[57] Franklin.   "The Nitrogen System of Compounds."   Reinhold Publishing Corp., 1935.
[58] "Org. Reactions," III.
[59] Hofmann.   *Ber.* 14, 2725 (1881); 15, 407 (1882).
[60] Tiemann.   *Ber.* 24, 4162 (1891).
[61] Stieglitz.   *Am. Chem. J.* 18, 751 (1896).
[62] Jones.   *Am. Chem. J.* 50, 414 (1913).
[63] Wallis, Nagel.   *J. Am. Chem. Soc.* 53, 2787 (1931).
[64] Mauguin.   *Ann. chim.* (8), 22, 297 (1911).
[65] Whitmore, Homeyer.   *J. Am. Chem. Soc.* 54, 3435 (1932).
[66] Whitmore et al.   *J. Am. Chem. Soc.* 54, 3274, 3431, 3435 (1932); 61, 1586 (1939).
[67] Wallis, Nagel.   *J. Am. Chem. Soc.* 53, 2787 (1931).
[68] Wallis, Moyer.   *J. Am. Chem. Soc.* 55, 2598 (1933).

The Hofmann reaction with α-halogen acid amides gives aldehydes or ketones

$$Me_2CBrCONH_2 + Br_2 + KOH \rightarrow Me_2C=O + KBr + K_2CO_3$$

A curious method of going down the series is the action of α-bromo acid amides with bases to give aldehydes.

$$RCHBrCONH_2 + 2\ NaOH \rightarrow RCHO + NaBr + NaCN + 2\ H_2O$$

The crystalline character and definite melting points of the amides make them useful in *identifying* the related *acids*.

Formamide, HC=O, m. 2°, b. 193°. (with NH$_2$ attached)

Preparation. 1. Heating ammonium formate.

$$HC\underset{=O}{\overset{ONH_4}{}} \rightarrow H_2O + HC\underset{=O}{\overset{NH_2}{}}$$

The equilibria involved are

$$HC\underset{=O}{\overset{ONH_4}{}} \rightleftarrows NH_3 + HC\underset{=O}{\overset{OH}{}} \rightleftarrows \left[HC\underset{NH_2}{\overset{OR}{}}OH \right] \rightleftarrows H_2O + HC\underset{NH_2}{\overset{}{}}=O$$

At the temperature used, the water is removed as steam and the process goes to completion.

2. Formic esters with ammonia.

$$HC\underset{=O}{\overset{OR}{}} + NH_3 \rightarrow \left[HC\underset{NH_2}{\overset{OR}{}}OH \right] \rightarrow ROH + H-C\underset{NH_2}{\overset{}{}}=O$$

The methyl and ethyl esters react most rapidly.

3. Commercially, formamide is made by the high pressure catalytic reaction of CO and NH_3, a reaction analogous to the action of CO and NaOH to form sodium formate.

4. N-Substituted formamides are obtained from isonitriles (carbylamines) and organic acids.

$$R-N=C + CH_3CO_2H \rightarrow R-N=C\overset{H}{\underset{OCOCH_3}{}}$$

This is an addition entirely on the bivalent carbon. The product adds another molecule of acetic acid at the $N=C$ linkage to give acetic anhydride and the N-substituted formamide.

$$RN = CHOCOCH_3 \rightarrow RNH - CH(OCOCH_3)_2 \rightarrow (CH_3CO)_2O + RNH - CHO$$

Reactions. 1. Hydrolysis is the reverse of the first preparation.

2. Vigorous dehydrating agents like phosphorus pentoxide give hydrocyanic acid.

$$\underset{HC=O}{\overset{NH_2}{\diagup}} \rightleftharpoons \underset{HC-OH}{\overset{NH}{\diagup\!\!\!\!\diagup}} \overset{P_2O_5}{\longrightarrow} HC \equiv N + 2\,HPO_3$$

Commercially this important change is produced by pyrolysis.

3. It dissolves mercuric oxide.

$$2\,HCONH_2 + HgO \rightarrow H_2O + (HCONH)_2Hg$$

There is good evidence that the Hg is attached to N in this compound.

4. It reacts with sodium giving a derivative in which the attachment of the sodium is not clear. It may be ionic with a mesohydric ion (p. 285).

5. With nitrous acid, the NH_2 is first replaced by OH and the formic acid is then oxidized. The net result is:

$$3\,HCONH_2 + 5\,HNO_2 \rightarrow 3\,CO_2 + 4\,N_2 + 7\,H_2O$$

6. Hypohalites or the calculated amount of halogen and base give an N-halogen amide, HCONHX. These react with water to give the amide and HOX. With excess base they give ammonia and a carbonate by the Hofmann rearrangement.[69]

7. Unstable "salts" are formed with strong acids like HCl or HNO_3.

The Raman spectrum indicates that formamide is largely associated.[70]

Formamide finds wide industrial uses as solvent, hygroscopic agent, preservative and antioxidant, plasticizer, and in the production of formic acid and hydrocyanic acid (DuPont).

Acetamide, CH_3CONH_2, m. 82°, b. 222°.

Preparation. 1. Similarly to formamide, (a) from ammonium acetate, (b) from methyl or ethyl acetate.

2. From the action of ammonia with acetyl chloride or acetic anhydride. The use of primary or secondary amines in place of NH_3 gives N-substituted acetamides.

3. By the partial hydration of methyl cyanide by means of conc. H_2SO_4 or of alkaline hydrogen peroxide.

$$CH_3C \equiv N + H_2O \rightarrow \underset{CH_3C=NH}{\overset{OH}{\diagup}} \rightleftharpoons \underset{CH_3C=O}{\overset{NH_2}{\diagup}}$$

[69] "Org. Reactions," III.
[70] Saksena. *Chem. Zentr.* 1940, I, 3244.

Reactions. 1. Similar to those of formamide except that (a) acetonitrile, CH_3CN, (b) acetic acid, and (c) methylamine are obtained by dehydration by nitrous acid and by bromine and KOH (Hofmann).

2. With PCl_5 to give acetonitrile.

$$CH_3CONH_2 \rightarrow CH_3CCl_2NH_2 \rightarrow CH_3CCl = NH \rightarrow CH_3CN$$
$$\text{Acetamido chloride} \quad \text{Acetimido chloride}$$

The intermediate chlorides are very unstable.

3. Nitrous acid forms N_2 and acetic acid. This reaction proceeds quantitatively even with highly branched compounds like Me_3-acetamide.[71]

Diacetamide and **triacetamide** are made from acetonitrile

$$CH_3CN + CH_3CO_2H \rightarrow (CH_3CO)_2NH, \text{ m. } 78°, \text{ b. } 223°$$
$$CH_3CN + (CH_3CO)_2O \rightarrow (CH_3CO)_3N, \text{ m. } 79°$$

Acetamide is finding wide uses because of its unusual solvent properties[72] (Niacet).

N-Bromoacetamide (acetbromoamide), is a brominating agent.[73]

$$C_6H_5NMe_2 + MeCONHBr \rightarrow BrC_6H_4NMe_2 + MeCONH_2$$

With silver carbonate or mercury alkyls or aryls, it gives methyl isocyanate, MeNCO, an intermediate in the Hofmann conversion of $RCONH_2$ to RNH_2.[74]

Higher acid amides resemble acetamide in preparation and properties. Their preparation from esters becomes more difficult with increase in complexity. Thus ethyl isobutyrate and ammonium hydroxide react very slowly and methyl trimethylacetate does not react at all.[75]

Acids above acetic when heated in a stream of dry NH_3 give good yields of amides.[76] Longer heating with ammonia dehydrates the amide to the nitrile.

Higher amides are available commercially including those of the saturated normal acids C_6 to C_{18} and 9-octadecenoic (oleic) and 9,12-octadecadienoic (linoleic) acids.

Large amounts of stearamide are used in the important *Zelan* waterproofing process for textiles.[77] The amide is used in place of the earlier employed stearyl alcohol. Treatment of it with pyridine hydrochloride and formaldehyde gives the chloride of the quaternary pyridonium compound, $C_{17}H_{35}CONHCH_2NC_5H_5$, by the Mannich reaction. The fabric is treated with a dilute solution of the Zelan compound, sodium acetate, and alcohol. After drying, it is baked at 150° for less than four minutes. This process

[71] Whitmore, Langlois. *J. Am. Chem. Soc.* **54**, 3438 (1932).
[72] Stafford. *J. Am. Chem. Soc.* **55**, 3987 (1933).
[73] *Ann. Rep. Chem. Soc.* (London) 1922, 101.
[74] Kharasch. *J. Am. Chem. Soc.* **43**, 1888-94 (1922).
[75] *Ann. Rep. Chem. Soc.* (London) 1906, 107.
[76] Mitchell, Reid. *J. Am. Chem. Soc.* **53**, 1879 (1931).
[77] Mullin. *Chem. Inds.* **46**, 557 (1940).

drives out pyridine, leaving the stearamide held to the molecules of the fabric by $-CH_2-O-$ or $-CH_2-N-$ links, the last atom in each case being supplied by the cellulose or protein respectively. N-isobutyl undecyleneamide is a valuable insecticide.

The *identification* of primary amides $RCONH_2$ is possible by heating in acetic acid with xanthydrol to form N-xanthylamides of definite melting points.[78]

RCONH

Other methods of identification of amides include reaction with phthalyl chloride[79] and formation of the mercury derivatives $(RCONH)_2Hg$ which have characteristic high melting points.[80]

E. HYDRAZIDES, $RCONHNH_2$, AND AZIDES, $RCON_3$

Hydrazine, H_2N-NH_2, reacts with esters and acid chlorides just as NH_3 does.

$$CH_3C \overset{\diagup OCH_3}{=}O + NH_2NH_2 \rightarrow CH_3C \overset{\diagup NHNH_2}{=}O + MeOH$$
Acethydrazide

The reactions of hydrazides are similar to those of the amides except that they (1) are more easily hydrolyzed, (2) have reducing properties, (3) have basic properties and form stable salts with strong acids, (4) react with aldehydes and ketones to give crystalline derivatives such as $MeCONHN=CHMe$, (5) react with nitrous acid to give acid azides.

$$CH_3CONHNH_2 + HNO_2 \rightarrow 2 H_2O + CH_3CON_3$$

The acid azides are acyl derivatives of hydrazoic acid, HN_3. This is now known to have the linear structure $H-N=N\equiv N$ instead of the earlier accepted cyclic structure. Acid halides with sodium azide, NaN_3, give acid azides. On heating with water or alcohols, the acid azides undergo a rearrangement (Curtius) like the Hofmann rearrangement of acid amides with bromine and bases.

$$\begin{array}{l} \overset{H_2O}{\longrightarrow} RNH_2 + CO_2 + N_2 \\ R-\overset{O}{\overset{\diagdown\diagdown}{C}}-N_3 + \overset{EtOH}{\longrightarrow} RNH-\overset{O}{\overset{\diagdown\diagdown}{C}}-OEt + N_2 \end{array}$$
An alkyl urethan

[78] Phillips, Pitt. *J. Am. Chem. Soc.* 65, 1355 (1943).
[79] Evans, Dehn. *J. Am. Chem. Soc.* 51, 3651 (1929).
[80] Williams, Rainey, Leopold. *J. Am. Chem. Soc.* 64, 1738 (1942).

The series of changes, ester to hydrazide to azide to amine is valuable in going down the series with compounds which are sensitive to more violent reactions as in the polypeptides and proteins.[81]

$$RCO_2H \xrightarrow{CH_2N_2} RCO_2Mc \xrightarrow{N_2H_4} RCONHNH_2 \xrightarrow{HNO_2} RCON_3 \xrightarrow{MeOH}$$

$$RNHCO_2Me \xrightarrow{HCl} CO_2 + MeOH + RNH_2 . HCl \xrightarrow{NaOMe} RNH_2$$

Hydrazoic acid is used in a great variety of ways with organic acids and other carbonyl compounds (Schmidt Reaction).[82] In its simplest form it goes from $-CO_2H$ to $-NH_2$ without isolation of any intermediates.

F. Hydroxamic Acids

Hydroxylamine, NH_2OH, reacts with esters, acid chlorides, and acid amides.

A solution of sodium hydroxylamine sulfonate obtained from sodium sulfite and nitrite solutions, converts aldehydes to hydroxamic acids[83]

$$RCHO + NaO_3SNHOH \rightarrow RC(OH)=NOH + NaHSO_3$$

The hydroxamic acids give a distinctive red or purplish coloration with ferric chloride (p. 190).

With strong acids they undergo a rearrangement.[84]

G. Imino Ethers and Acid Amidines

The unstable acid imido chlorides, $RCCl=NH$, obtained from acid amides and PCl_5 or from nitriles and HCl, react readily with alcohols to give imino ethers, $RC(OEt)=NH . HCl$. The best preparation is by passing dry HCl gas into a solution of nitrile in the anhydrous alcohol. The imino ethers are readily hydrolyzed to esters. With excess alcohol they give ortho esters.

[81] Curtius. *Ann. Rep. Chem. Soc.* (London) 1915, 83.
[82] "Org. Reactions," III.
[83] Angeli. See bibliography of *Yale Chem. Rev.* **33**, 209 (1943).
[84] Lossen. *Ann.* **161**, 347 (1872); *Yale Chem. Rev.* **33**, 209 (1943).

With NH_3 they give acid amidines[85]

$$CH_3C=NH.HCl + 2NH_3 \rightarrow CH_3{-}C=NH + C_2H_5OH + NH_4Cl$$
$$OC_2H_5 \qquad\qquad NH_2$$

Acetamidine

The free amidines are readily hydrolyzed to acid amides and decompose to nitriles. They are strong bases and their salts with strong acids are stable. They are not decomposed by nitrous acid.

H. THIOACIDS

1. The introduction of one sulfur into a carboxyl group might be expected to give two different products.

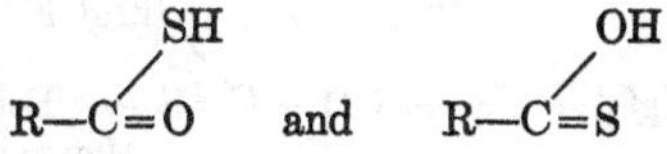

$$\begin{array}{ccc} SH & & OH \\ / & & / \\ R{-}C=O & \text{and} & R{-}C=S \end{array}$$

One of these would be called a *thiolic* acid and the other a *thionic* acid. As a matter of fact there is no evidence for the existence of such isomers. Several explanations might be given for this fact:

(1) The system $HS{-}C=O \rightleftharpoons S=C{-}OH$ would act as a typical tautomeric system.

(2) If it is true that sulfur does not readily form a "double bond" with carbon, the thiolic form, $HS{-}C=O$ would be the only one to exist.

(3) Most probably there is no "double bond" in most of the molecules and the sulfur and oxygen occupy equivalent places in the ion. This may be a "tetragonal mesohydric nucleus."[86]

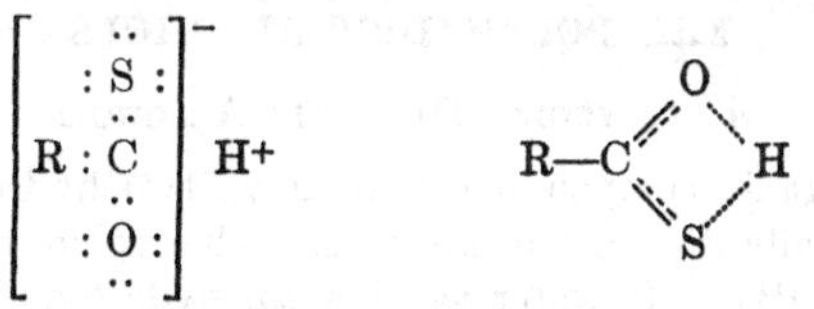

$$\begin{bmatrix} : \overset{..}{S} : \\ R : \overset{..}{C} \\ : \overset{..}{O} : \end{bmatrix}^{-} H^{+} \qquad\qquad R{-}C \overset{\displaystyle O}{\underset{\displaystyle S}{\diagup\kern-0.3em\diagdown}} H$$

Thio acids are prepared (1) from the acid with a sulfide of phosphorus and (2) from an acyl halide with NaSH.

Thioacetic acid is an unstable colorless, evil-smelling liquid, b. 93°. It may be regarded as a "mixed anhydride" of acetic acid and H_2S. It gives the typical reactions of acetic anhydride with water, alcohols and ammonia,

---

[85] *Ann. Rep. Chem. Soc.* (London) 1918, 71.
[86] Oddo. *Gaz. chim. ital.* 61, 699 (1931).

producing an acyl derivative and hydrogen sulfide.

$$CH_3\overset{SH}{\underset{}{C}}=O + HQ \rightarrow \left[CH_3\overset{SH}{\underset{Q}{C}}-OH \right] \rightarrow CH_3\overset{SH}{\underset{Q}{C}}=O + H_2S$$

This reaction is not appreciably reversible. Sodium thioacetate has been suggested for use in qualitative analysis. Acidification liberates H_2S directly in the solution.

Thiol esters, $RCO(SR')$, are readily obtained from mercaptans, acid anhydrides and the corresponding salt. CH_3COSEt, b. $116°$.

2. Carboxyl derivatives having two sulfur atoms in place of the two oxygens are obtained from carbon disulfide and Grignard reagents.[87]

$$CH_3MgCl + CS_2 \rightarrow CH_3\overset{SMgCl}{\underset{}{-C}}=S \rightarrow CH_3\overset{SH}{\underset{}{C}}=S$$

Methyl carbithionic acid
Dithioacetic acid

The $-CS_2H$ group is more strongly acidic than the $-CO_2H$ group. It may involve a structure in which the two S atoms are negative and the crowded C atom positive giving a stable anion:

$$\left[\begin{array}{c} :\overset{..}{\underset{..}{S}}: \\ R:C \\ :\overset{..}{\underset{..}{S}}: \end{array} \right]^{-} H^+$$

The dithio acids are very sensitive to hydrolysis and oxidation.

XIII. POLYHYDRIC ALCOHOLS

A. GLYCOLS, DIHYDRIC ALCOHOLS

Usually an organic compound cannot have two hydroxyls on the same carbon. Occasionally this arrangement is stable as in the case of chloral hydrate, $Cl_3CCH(OH)_2$. In water solution, the hydrated form of the carbonyl group probably exists in varying amounts. The hydrated form of formaldehyde, $CH_2(OH)_2$, is stable enough under a few atmospheres pressure to allow substances like acetone and methanol to be distilled away leaving it behind. Then, heating at ordinary pressure drives out CH_2O leaving the water behind. This process is valuable in separating formaldehyde from the complex mixture obtained by oxidizing n-butane (Celanese).[1]

[87] Ann. Rep. Chem. Soc. (London) 1907, 99.
[1] Bludworth. C. A. 36, 497 (1942).

Aldehydes and ketones give derivatives of the dihydroxyl form, namely, hemiacetals, acetals, and diacetates.

$$=C=O + H_2O \rightleftharpoons =C(OH)_2, \quad =C(OH)OR, \quad =C(OR)_2, \quad =C(OAc)_2.$$

The commonest glycols have the two hydroxyl groups on adjacent carbon atoms because of their easy preparation from olefins and by bimolecular reduction of carbonyl compounds.

1,4- and 1,5-glycols readily form inner ethers with 5- and 6-rings.

$\alpha\omega$-Glycols are obtained by reduction of esters of dibasic acids with Na and absolute alcohol.

The glycols are high boiling liquids. The simplest member has a higher boiling point than any of the higher members having the two hydroxyls on adjacent carbons. This is because it is a di-primary alcohol whereas they have one or more secondary or tertiary hydroxyls. As with the simple alcohols the latter have lower boiling points. Most of the glycols are completely soluble in water and in organic solvents. They are heavier than water.

Ethylene glycol, "Glycol," 1,2-ethandiol, $HOCH_2CH_2OH$, b. 197°.[2]

Preparation. 1. Aqueous ethylene chlorohydrin solution from chlorine water and ethylene may be heated with sodium bicarbonate under slight pressure to produce the glycol.

$$ClCH_2CH_2OH + NaHCO_3 \rightarrow NaCl + HOCH_2CH_2OH + CO_2$$

The carbon dioxide is passed into NaOH to make bicarbonate for another batch. The aqueous glycol solution is concentrated by distillation.

2. The use of stronger bases such as lime and slightly different conditions will give ethylene oxide which distills out of the mixture (b. 12°). If it is kept in under pressure it reacts with water to give ethylene glycol which reacts with more ethylene oxide to give diethylene glycol and higher polyethylene glycols, $HOCH_2CH_2(OCH_2CH_2)_nOCH_2CH_2OH$. If the ethylene oxide is distilled from the mixture and heated with water and a trace of acid under pressure, ethylene glycol is formed smoothly. Ethylene glycol is also made from formaldehyde and carbon monoxide[3] (DuPont).

$$HCHO + CO + H_2O \rightarrow HOCH_2COOH \rightarrow HOCH_2COOMe \rightarrow$$
$$CH_2OH - CH_2OH$$

Reactions. The reactions of ethylene glycol are essentially those of the primary alcohols except that they are complicated by the presence of two such groups and by the higher boiling points of glycol and its derivatives.

1. Oxidation probably yields all the theoretically possible products but these are difficult to isolate in quantity because they are more readily oxidized than the glycol itself and are too high boiling to be removed readily from the sphere of oxidation.

[2] Ellis. "Chemistry of Petroleum Derivatives." Reinhold, 1934. p. 506.
[3] *Chem. Ind.* 62, 381 (1948).

The theoretical oxidation series from glycol would be:

$$\begin{array}{ccccc} CH_2OH & CH_2OH & CH_2OH & CHO & CO_2H \\ | & \to & | & \to & | & \to & | & \to & | & \to 2\ CO_2 \\ CH_2OH & CHO & CO_2H & CO_2H & CO_2H \end{array}$$

Ethylene   Glycolic   Glycolic   Glyoxylic   Oxalic
glycol   aldehyde   acid   acid   acid

$$\begin{array}{c} CHO \\ | \\ CHO \end{array} \qquad \nearrow$$

Glyoxal

Ethylene glycol has been oxidized directly to glyoxal.[4]

Cold nitric acid oxidizes one of the alcohol groups giving glycolic acid. Heating the mixture gives oxalic acid.

A special oxidation with periodic acid, HIO_4, breaks the $C-C$ link between the adjacent hydroxyls.

$$(CH_2OH)_2 + HIO_4 \to 2\ H_2CO + H_2O + HIO_3.$$

This Periodic Acid Oxidation is general for compounds having two hydroxyl groups or a hydroxyl and an amino group on adjacent carbons.[5] It is specially valuable in the carbohydrate field (p. 478).[6]

2. With dehydrating agents to give acetaldehyde. This may be regarded as an ordinary dehydration to give the enol of acetaldehyde or as a rearrangement which is typical of the higher glycols.

$$HOCH_2CH_2OH \to H_2O + CH_2=CHOH \to CH_3CHO$$

Heating alone at 500° gives the same change. The ethers which would be expected,

$$\begin{array}{cc} \begin{array}{c} CH_2 \\ | \qquad \diagdown \\ \qquad\qquad O \\ | \qquad \diagup \\ CH_2 \end{array} & \text{and} & \begin{array}{c} CH_2{-}O{-}CH_2 \\ | \qquad\qquad | \\ | \qquad\qquad | \\ CH_2{-}O{-}CH_2 \end{array} \end{array}$$

Ethylene oxide   Diethylene oxide
                       (Dioxan)

are both known but only the latter is obtained by dehydration of glycol. They are made from ethylene chlorohydrin.

3. Halide acids react readily with one hydroxyl giving a halohydrin.

$$HOCH_2CH_2OH + HX \to XCH_2CH_2OH$$

The resulting substance does not react readily with more HX.

[4] *ibid.*
[5] "Org. Reactions," II. pp. 341–74.
[6] Jackson, Hudson. *J. Am. Chem. Soc.* **58**, 378 (1936).

4. Halides of phosphorus replace both hydroxyls. In the case of iodine, the resulting product tends to lose I_2 and give ethylene.

5. The alkali metals replace the alcohol H atoms. The mono sodium compound is formed readily but much less vigorously than with a monohydric alcohol. The disodium compound is only obtained by vigorous heating.

6. Aldehydes and ketones form dioxolanes (p. 312).

$$\begin{array}{l}\mathrm{CH_2OH} \\ | \\ \mathrm{CH_2OH}\end{array} + \mathrm{CH_3CHO} \rightarrow \mathrm{CH_3CH}\begin{array}{c} \diagup\mathrm{O\!-\!CH_2} \\ | \\ \diagdown\mathrm{O\!-\!CH_2}\end{array}$$

4-Me-dioxolane

Acetal formation between a variety of aldehydes and dipropenylglycol, 4,5-dihydroxy-2,6-octadiene, has shown that n-aldehydes to C_7 show little difference (85–90% yield).[7] Strangely chloral, benzaldehyde, cinnamaldehyde and crotonaldehyde give no acetal although acetone gives 70% yield. Dioxolanes have been obtained from aldehydes and aromatic substituted glycols.[8]

7. With organic acids to form mono- and di-esters. The former combine in the same molecule the solvent properties of a primary alcohol and an ester, $HOCH_2CH_2OCOR$. Acid anhydrides are used to advantage in making these esters. The reaction of ethylene glycol with acetic anhydride may be dangerously explosive.

With dibasic acids highly polymeric esters (M.W. about 3000) are obtained.[9] Similar polymeric esters from dibasic acids and higher $\alpha\omega$-glycols (to C_{10}) have been made.

Terylene,[10] a British developed fiber, is made from terephthalic acid and ethylene glycol. It is highly resistant to microorganisms and bacteria, and possesses very low moisture adsorption.

8. Nitric acid (with sulfuric acid) reacts with both hydroxyls, giving *glycol dinitrate*, $O_2NOCH_2CH_2ONO_2$, ("nitroglycol,") which is valuable in dynamite to lower the otherwise relatively high freezing point of the nitroglycerin).

9. Fused alkalies give hydrogen and oxalates.

10. With boric acid it forms a complex acid which is a good electrolyte and can be titrated with indicators[11]

$$\mathrm{H_3BO_3} + 2\,\mathrm{HOCH_2CH_2OH} \rightarrow 3\,\mathrm{H_2O} + \begin{bmatrix}\mathrm{CH_2O\cdot\ \ \cdot OCH_2} \\ | \quad\ B\quad | \\ \mathrm{CH_2O\cdot\ \ \cdot OCH_2}\end{bmatrix}^{-} \mathrm{H^+}$$

The complex has the tetrahedral grouping of five atoms and 32 electrons which characterizes the ions of H_2SO_4, H_3PO_4 and $HClO_4$. Similar products are

[7] Burt, Howland. *J. Am. Chem. Soc.* 52, 217 (1930).
[8] Read, Lathrop, Chandler. *J. Am. Chem. Soc.* 49, 3116 (1927).
[9] Carothers, Arvin. *J. Am. Chem. Soc.* 51, 2560 (1929).
[10] *Chem. Inds.* 62, 380 (1948).
[11] Boeseken. *Ann. Rep. Chem. Soc.* (London) 1917, 64; 1930, 97.

formed with arsenious acid. Whenever two hydroxyls on adjacent carbons can form this tetrahedral arrangement with a boron or an arsenic atom the conductivity is increased. The only thing which can prevent this union is the location of the two adjacent hydroxyls on opposite sides of the plane of a ring compound such as the furanose and pyranose forms of sugars.

11. With ketones and aldehydes and a trace of acid it forms cyclic acetals, dioxolanes. This is the reaction involved in the formation of acetone derivatives of sugars.

Enormous amounts of ethylene glycol are used as anti-freeze in automotive radiators (Prestone). It has a lower molecular weight and lower viscosity than glycerol and is much less volatile than ethanol or methanol. When taken internally it is as dangerous as methanol.[12]

Propylene glycol, 1,2-propandiol $CH_3CHOHCH_2OH$, b. 187°.[13]

Preparation. Propylene is converted to the chlorohydrin by chlorine water and that is changed to the glycol by sodium carbonate solution.

$$CH_3CH=CH_2 \rightarrow CH_3CHOHCH_2Cl \rightarrow CH_3CHOHCH_2OH$$

It can also be obtained by heating glycerol with NaOH.

Reactions. These are like those of its lower homolog except that a secondary alcohol group is involved.

1. Again oxidation is not a practical means for obtaining the substances theoretically related to it as oxidation products.

2. Dehydration and rearrangement, in the presence of acids or on heating, gives propionaldehyde. This is because of the greater reactivity of the secondary hydroxyl. Otherwise, acetone would be formed.

3. Its molecule is asymmetric but its preparation gives an optically inactive mixture (racemic). When fermented it becomes levorotatory because the organism attacks the dextro form preferentially. Levo propylene glycol

[12] Hunt. *Ind. Eng. Chem.* **24**, 361, 836 (1932).
[13] Ellis. "Chemistry of Petroleum Derivatives." Reinhold, 1934. p. 529.

can also be obtained from hydroxyacetone (acetol) by reduction by yeast.[14] This method has been applied to the synthesis of other optically active 1,2-glycols. The series of reactions is worth giving in detail. Chloroacetaldehyde (from the chlorination of paraldehyde) is treated with RMgX to give the 1-Cl-2-OH-compound. This is oxidized to the 1-Cl-ketone by dichromate and 5% sulfuric acid. Treatment with potassium formate in absolute methanol gives the 1-OH-ketone which on reduction by fermentation gives the optically active 1,2-glycol.

Taken internally, propylene glycol is harmless. This probably is because its oxidation gives pyruvic and acetic acids.

Propylene glycol is used in aerosols for sterilizing air.[15]

Trimethylene glycol, 1,3-propandiol, $HOCH_2CH_2CH_2OH$, b. 215°.

Preparation. By the fermentation of glycerol by *Schizomycetes*. It is also a by-product of the fermentation of glucose to glycerol.

Reactions are those of a di-primary alcohol with the hydroxyls far enough apart not to influence each other. Careful oxidation gives beta hydroxypropionic acid.

An interesting homolog of trimethylene glycol is its 2-*gem*-dimethyl derivative, *pentaglycol*, 2,2-dimethyl-1,3-propandiol $HOCH_2C(CH_3)_2CH_2OH$, m. 129°, obtained by the action of isobutyraldehyde with an excess of formaldehyde in presence of lime. The HCHO supplies a carbonyl group for an aldol condensation and also acts as a reducing agent for the −CHO group.

$$Me_2CHCHO + 2\ HCHO + NaOH \rightarrow HOCH_2CMe_2CH_2OH + HCO_2Na$$

One or both of the OH groups can be replaced by treatment with HBr *without rearrangement*. This is radically different from the behaviour of the corresponding alcohol, Me_3CCH_2OH, which gives mainly rearranged products whenever the OH is removed (p. 128).

Alpha butylene glycol, 1,2-butandiol, $CH_3CH_2CHOHCH_2OH$, b. 192°.

Preparation. 1. Pure 1-butene may be oxidized to the glycol by dilute $KMnO_4$.

2. Pure 1,2-dibromobutane may be converted to the diacetate of the glycol by heating with anhydrous sodium acetate. The acetate may then be saponified by alcoholic KOH. If the dibromide is to be converted directly to the glycol only very mild alkaline reagents can be used. Concentrated alkalies, either aqueous or alcoholic, tend to remove HBr to give $CH_3CH_2CBr{=}CH_2$, 2-bromo-1-butene, and then ethyl acetylene.

Reactions. Like those of propylene glycol.

Beta butylene glycol, 2,3-butandiol, $CH_3CHOHCHOHCH_3$, b. 180°, m. 27°, is prepared (1) from the pure olefin, (2) from the pure dibromide, or (3) best by a special fermentation which also produces acetoin, $MeCOCHOHMe$, and diacetyl, $MeCOCOMe$. This has also been called the *psi* butylene glycol.

---

[14] "Org. Syntheses."
[15] Robertson et al. *Science* **93,** 213 (1941).

Its reactions are those of a di-secondary alcohol. Oxidation gives two molecules of acetic acid, probably through the formation of diacetyl. It can be determined by means of mild oxidation with bromine water to diacetyl, treatment with hydroxylamine and conversion of the resulting dimethylglyoxime to the insoluble nickel compound.[16]

This glycol has been much studied as a possible intermediate for butadiene.[17, 18]

1,3-Butanediol (formerly *beta* butylene glycol), $CH_3CHOHCH_2CH_2OH$, b. 207°, is prepared by the careful hydrogenation of acetaldol. It is also formed directly from aqueous acetaldehyde and Mg_xHg. Its reactions are those which would be expected from a primary and a secondary alcohol. Its oxidation gives acetone through formation of acetoacetic acid.

Tetramethylene glycol, 1,4-butanediol, $HOCH_2CH_2CH_2CH_2OH$, b. 108° (4 mm.) is obtained by the reduction of ethyl succinate with sodium and alcohol. 1,4-Butanediol is available in any desired amount from butadiene, through 1,4-addition of chlorine, hydrolysis and hydrogenation (DuPont). Its reactions are typical of its primary alcohol groups. With dehydrating agents, it readily gives tetrahydrofuran (tetramethylene oxide). Similarly, pentamethylene oxide is formed from the corresponding glycol.

Isobutylene glycol, methylpropanediol, $(CH_3)_2C(OH)CH_2OH$, b. 177°, is obtained from isobutylene by the usual reactions. Its reactions are complicated by the tertiary hydroxyl and the fork in the chain. In the presence of even dilute acids it rearranges readily to give isobutyraldehyde.

1,5-Pentanediol is obtained by the catalytic hydrogenolysis of H_4-furfuryl alcohol.

3-Methyl-1,3-butanediol is obtained from isobutylene and formaldehyde with an acid catalyst (DuPont).

Pinacol, tetramethylglycol, 2,3-Me_2-2,3-butanediol, "pinacone,"

$$(CH_3)_2C(OH)C(OH)(CH_3)_2,$$

m. 42°, b. 172°, is obtained from the bimolecular reduction of acetone by amalgamated magnesium. It forms a hydrate with 6 H_2O, m. 46°. Its most important reaction is its rearrangement to pinacolone, Me_3CCOMe, on dehydration with acid reagents (p. 209). Treatment with HBr gives the dibromide of tetramethylethylene and some pinacolone. By catalytic dehydration it is possible to obtain 2,3-Me_2-1,3-butadiene ("methyl isoprene") from which "methyl rubber" can be made.

2-Methyl-2,4-pentanediol ("2-2-4") is made by the low temperature high pressure catalytic hydrogenation of diacetone alcohol (CSC). It is used in brake fluids, cutting oils, plasticizers and as a chemical intermediate.

[16] Matignon, Moureu, Dode. *Bull. soc. chim.* (5) 1, 411 (1934).
[17] Strohmaier, Lovell. *Ind. Eng. Chem.* 38, 721 (1946).
[18] Lees, Fulmer, Underkofler. *Iowa State Coll. J. Sci.* 18, 359 (1944).

Many higher glycols have been prepared by the methods given above. The most readily available are the substituted ethylene glycols having the hydroxyls on adjacent carbons and the glycols having the hydroxyls at the two ends of a carbon chain ($\alpha\omega$). The former are made from the corresponding olefins either through the dibromides or by direct oxidation with permanganate or barium chlorate[19] and also by the bimolecular reduction of ketones, certain aldehydes or esters.

1. The best known process is that of the reduction of ketones to pinacols. Hundreds of ketones have been thus reduced to pinacols. All pinacols on treatment with acids undergo the *pinacolone* (*"pinacoline"*) *rearrangement*, $R_2C(OH)C(OH)R_2 \rightarrow R_3CCOR$. If the pinacol contains two different groups the general tendency is for the larger group to migrate. Many theories have been evolved to explain this and similar rearrangements.[20] The first attack is undoubtedly by a proton on one of the free pairs of one of the hydroxyls to form an oxonium salt. When this decomposes with the loss of water the resulting carbonium ion $Me_2C(OH)C^+Me_2$ undergoes the usual rearrangement (p. 120) involving the transfer of an electron pair with its attached Me group to form $MeC^+(OH)CMe_3$ which loses a proton to form $MeCOCMe_3$.

2. Aldehydes can be reduced to glycols but less readily than ketones.

$$RC{=}O \rightarrow R{-}\overset{\overset{\displaystyle H}{|}}{\underset{|}{C}}{-}O{-}M \rightarrow R{-}CHOH{-}CHOH{-}R$$

3. Esters in dry ether react with metallic sodium to give acyloins, $R-CHOHCO-R$, which can be reduced to glycols. Reduction by yeast gives optically active materials.

Glycols RCHOHCHOHR can exist in stereoisomeric forms (enantiomers). Usually the meso form can be separated from the DL-form by patient application of fractionation by distillation and crystallization. Thus *sym*-dipropylglycol has been obtained in *meso-* and DL-forms, m. 124° and 24°, di-3,5-dinitrobenzoates, m. 200° and 125°.

Glycols RCHOHCHOHR on acid treatment, undergo rearrangements to give $RCOCH_2R$ and R_2CHCHO depending on the conditions and the groups involved. These products are due to the rearrangement of an H and an alkyl group respectively.

Oxidation of ethylene glycols splits the molecule between carbinol groups. This is probably the mechanism by which oxidation of an olefin splits it at the double bond. Lead tetra-acetate produces this splitting.

$$R_2C(OH)C(OH)R_2 + Pb(OAc)_4 \rightarrow 2\ R_2CO + 2\ HOAc + Pb(OAc)_2$$

R may be any group or H. An even better reagent is periodic acid.[21]

[19] Braun. *J. Am. Chem. Soc.* **54**, 1133 (1932).

[20] Whitmore. *J. Am. Chem. Soc.* **54**, 3274 (1932).

[21] Hann, Richtmyer. "The Collected Papers of C. S. Hudson." Academic Press Inc., 1946.

1,12-Octadecandiol (*Diolin*) is obtained by the catalytic hydrogenation of castor oil.

$\alpha\omega$-Glycols are obtained from the reduction of dibasic esters with Na and absolute EtOH. They are polymethylene glycols. The bps. for the members C_6 to C_{10} are (°C.) 239, 250, 259, 172 (20 mm.), 179 (10 mm.). In most reactions the primary carbinol groups act independently. An exception is the action of dehydrating agents. Thus tetradecamethylene glycol with 50% sulfuric acid gives an inner ether which on oxidation with permanganate in acetic acid gives *n*-decanoic acid. Thus one of the hydroxyls has rearranged to the position to give an inner ether with a 6-membered ring[22]

$$HO(CH_2)_{14}OH \rightarrow CH_3(CH_2)_8CH \begin{array}{c} CH_2 - CH_2 \\ \diagup \qquad \qquad \diagdown \\ \qquad \qquad \qquad CH_2 \\ \diagdown \qquad \qquad \diagup \\ O - CH_2 \end{array}$$

Glycols $R_2C(OH)(CH_2)_nC(OH)R_2$ are made from dibasic esters with an excess of RMgX.[23]

B. Ethylene Oxide and Related Compounds

Ethylene oxide, $\overset{\displaystyle \lceil O \rceil}{CH_2CH_2}$, b. 12°, is obtained by heating a solution of ethylene chlorohydrin with a base and distilling the oxide as formed.[23a] A by-product is diethylene oxide, dioxan, $O(CH_2CH_2)_2O$ (C. and C.). A preparation without the use of chlorine consists in the air oxidation of ethylene with a silver on alundum catalyst at about 225°.[23b] A yield of about 50% per pass is obtained. The product is condensed out and the unchanged ethylene is recycled.[24]

Heat converts ethylene oxide to acetaldehyde much as cyclopropane is changed to propylene. Ethylene oxide is more stable than cyclopropane because the valence angle of oxygen involves less strain in a three membered ring than does that of carbon.[25] It is soluble in water without change in the absence of catalysts. The three-membered ring in ethylene oxide opens on proper treatment with active H compounds, HQ.

$$\begin{array}{c} CH_2 \\ | \qquad \diagdown \\ \qquad \quad O + HQ \rightarrow \\ | \qquad \diagup \\ CH_2 \end{array} \qquad \begin{array}{c} CH_2OH \\ | \\ | \\ CH_2Q \end{array}$$

[22] Franke, Kroupa, Panzer. *Monatsh.* 60, 106 (1932).
[23] *Ann. Rep. Chem. Soc.* (London) 1904, 73.
[23a] Ellis, "Chemistry of Petroleum Derivatives," Reinhold, 1934, P. 533–551.
[23b] Technical Oil Mission reports.
[24] McBee, Hass, Wiseman. *Ind. Eng. Chem.* 37, 432 (1945).
[25] Hibbert, Allen. *J. Am. Chem. Soc.* 54, 4115 (1932).

Such compounds include water (heat, pressure, and acid catalyst) acids, alcohols, ammonia, and amines. Many of the products obtained are important commercially (C. and C.).[23a]

Reagent	Product
Water	Ethylene glycol
HBr	Ethylene bromohydrin
Acetic acid	Ethylene glycol monoacetate
HCN	$HOCH_2CH_2CN$ Ethylene cyanohydrin
$NaHSO_3$	$HOCH_2CH_2SO_3Na$, Na isethionate
Acyl halides	β-halogen-ethyl esters, $RCO_2CH_2CH_2X$
Ethanol	Ethylene glycol monoethyl ether, Cellosolve
Methanol and butanol	Me and Bu Cellosolves
Glycol	Diethylene glycol $HOCH_2CH_2OCH_2CH_2OH$
Diethylene glycol	Polyethylene glycols $HOCH_2CH_2(OCH_2CH_2)_nOH$, $n = 1 - 9$
Cellosolve	Monoethyl ether of diethylene glycol, Carbitol, b. 198°
Me and Bu cellosolves	Me and Bu Carbitol (b. 222°)
Ammonia	Mono-, di- and tri-ethanolamines
RNH_2, R_2NH	Substituted ethanolamines

Lengthening chains through use of ethylene oxide to form groupings $-(CH_2CH_2O)_2-$ has much the same effect in detergents and surface-active agents as the long carbon chains in the higher fatty acids and the related alcohols. Thus *Triton 720* has the structure

$$Me_3CCH_2C(Me_2)C_6H_4O(CH_2CH_2O)_2CH_2CH_2SO_3Na.$$

Many of the glycols have been found to have germicidal properties, particularly for air-borne bacteria. The most promising of these is triethylene glycol.

Diethylamine and ethylene oxide do not unite dry but react readily in methanol at 40–60°.[26] β-Diethylaminoethanol, an intermediate for novocaine, is obtained in good yield. Some of it reacts with more ethylene oxide to give $Et_2NCH_2CH_2(OCH_2CH_2)_nOH$ in which n is 1 to 4. These are all soluble in water.

Ethylene oxide reacts with Grignard reagents of primary and secondary alkyls. The first product is an oxonium compound which reacts with water or acids like the original components. The ordinary practice of heating the mixture with benzene or toluene and driving off the ether may give a dangerously explosive change. It is safer to use two mols of the ethylene oxide.[27]

$$RMgX + 2 C_2H_4O \rightarrow RCH_2CH_2OMgOCH_2CH_2X$$

[26] Horne, Shriner. *J. Am. Chem. Soc.* 54, 2925 (1932).
[27] Huston, Ogett. *J. Org. Chem.* 6, 123 (1941).

This may disproportionate to $(RCH_2CH_2O)_2Mg$ and $(XCH_2CH_2O)_2Mg$. Treatment with water and acid gives RCH_2CH_2OH and the halohydrin in yields of about 80 and 70% respectively. If only one mol of ethylene oxide is used and the mixture is heated to 60°, the yields are about 50 and 20%.

Ethylene oxide reacts with aldehydes and ketones in presence of stannic chloride to give cyclic acetals, dioxolanes, containing a 5-membered ring. Thus methyl hexyl ketone with ethylene oxide and heptaldehyde with propylene oxide give the following:[28]

2-Me-2-Hex-1,3-dioxolane
b. 97° (23 mm).

4-Me-2-Hex-1,3-dioxolane
b. 103° (23 mm).

With bromine, ethylene oxide forms a red crystalline compound, m. 65°, b. 95°. This gives dioxan with HgO.

With solutions of hydrolyzable salts, it acts as a base, uniting with the acid formed.

$$2 C_2H_4O + 2 H_2O + MgCl_2 \rightarrow 2 Mg(OH)_2 + C_2H_4(OH)Cl$$

Ethylene oxide with 90% liquid CO_2 is an important fumigant, especially for food stuffs (Carboxide, CC.).

Polyethylene oxides obtained by heating ethylene oxide with a base have been studied extensively.[29-31]

Propylene oxide, CH_3CHCH_2, with epoxide O bridge, is made from propylene chlorohydrin. Its reactions are like those of ethylene oxide. Addition usually gives a 2-hydroxyl, the other group adding to the 1-carbon.

Ethylene Oxides.[32]

Trimethylene oxide, $CH_2CH_2CH_2$, with epoxide O bridge, b. 50°, is not obtainable from the glycol but is made from trimethylene chlorohydrin with bases. It gives addition reactions like those of ethylene oxide but with slightly less ease.

Tetramethylene oxide, tetrahydrofuran, b. 67° and **pentamethylene oxide,** b. 81°, are obtained from the glycols by heating with 60% sulfuric acid. They are both soluble in water but differ from the lower oxides by being stable and not changing back to the glycols. H_4-furan is available commercially (Du-Pont). In Germany, it is made from the glycol formed by hydrogenation of

[28] Bogert, Roblin. *J. Am. Chem. Soc.* **55**, 3741 (1933).
[29] Staudinger. *Ber.* **65B**, 267 (1932).
[30] Sauter. *Z. physik. Chem.* **B21**, 161 (1933).
[31] Hibbert, Perry. *Can. J. Research* **8**, 102 (1933).
[32] Ahrens. *Sammlung Chemische-Technischen Vorträge* **1920**, 83 pp.

butynediol from the aldol condensation of 2 mols of formaldehyde with acetylene. It is an excellent solvent. It reacts with ammonia, hydrogen sulfide, hydrogen halides, chlorine, copper at 200°, water and carbon monoxide, and hexamethylene diisocyanate to give respectively pyrrolidine, H_4-thiophene, 1,4-halohydrins, 2,3-Cl_2-H_4-furan, butyrolactone, adipic acid, and a plastic containing the chain $-CONH(CH_2)_6NHCO_2(CH_2)_4O-$. Thus it is a polyurethan in contrast to the polyamides used in Nylon and related plastics.

3,4-Epoxy-1-butene is available from butadiene (CCD).

1,4-Dioxan, diethylene oxide, $O(CH_2CH_2)_2O$, b. 101°, is obtained by heating ethylene glycol with a little sulfuric acid and as a by-product of glycol manufacture. It gives oxonium compounds with acids, halogens, mercuric salts, iodoform, etc. Its *sulfate,* m. 101°, is *insoluble.* Dioxan is completely miscible with water. It forms an azeotrope with 20% water, b. 87°. Dioxan is toxic. It has valuable solvent properties (C. and C.).

Chlorination of dioxan gives di- and tetra-chloro derivatives.[33-35] The dichloro derivative is the vicinal isomer. Hydrolysis gives glycol and glyoxal. This is a practical source of derivatives of the latter substance. 2,3-Cl_2-dioxan with glycol forms *cis* and *trans* 1,4,5,8-naphthodioxan.

C. HALOHYDRINS

Halohydrins contain both hydroxyl and halogen. The commonest type is the *alkylene halohydrins* with the OH and X on adjacent carbon atoms due to their formation from the olefin and a hypohalous acid. The other types are made from di- or poly-hydroxyl compounds by replacement of one hydroxyl by halogen. The result is a mixture of the desired product with polyhalogen compounds and unchanged material.

Ethylene chlorohydrin, glycol chlorohydrin, β-chloroethyl alcohol, 2-chloroethanol, $ClCH_2CH_2OH$, b. 132°, is readily obtainable in aqueous solution.[36] Chlorine and ethylene are passed into water under carefully controlled conditions. The water and chlorine react as usual.

$$Cl_2 + H_2O \rightleftharpoons HCl + HOCl$$

The products correspond in decreasing amounts to the addition of HOCl, Cl_2, and HCl to the ethylene. The proportions seem strange until the probable mechanism of the processes is considered. The most reactive material in the solution is HOCl because of the possibility of the action of its positive Cl with the extra electron pair of the ethylene (p. 27). The most plentiful chemical species in the mixture is H_2O. The first step thus gives a carbonium ion:

$$H : \overset{..}{O} : \overset{..}{Cl} : + \; : CH_2 : \overset{+}{C}H_2 \rightarrow H : \overset{..}{O} : + \; : \overset{..}{Cl} : CH_2\overset{+}{C}H_2$$

[33] Böeseken, Tellegen, Henriques. *Rec. trav. chim.* 50, 909 (1931).
[34] Butler, Cretcher. *J. Am. Chem. Soc.* 54, 2987 (1932).
[35] Summerbell, Christ. *J. Am. Chem. Soc.* 54, 3777 (1932).
[36] Ellis. "Chemistry of Petroleum Derivatives." Reinhold, 1934. p. 487.

The best way to stabilize this ion is for it to coordinate with a free pair of a water molecule. This intermediate loses a proton to give ethylene chlorohydrin, the chief product. If the carbonium ion reacts with a chloride ion, the result is ethylene chloride. Of course this is also formed from the direct union of Cl_2 and ethylene but there is much greater probability of reaction of the former with water. The chance of a proton from a hydronium ion being able to add to ethylene under these conditions is very small. This corresponds to the negligible amount of EtCl formed.

When the concentration of chlorohydrin reaches about 10% the process is interrupted and the product is distilled to obtain the constant boiling mixture of ethylene chlorohydrin and water containing 42 per cent of the former. Further concentration of the chlorohydrin is usually unnecessary since the solutions can be used to make ethylene oxide, ethylene glycol and related products (C. and C.).

Reactions. The halogen atom is unusually reactive.

1. Hydrolysis proceeds readily in the presence of weak bases.

2. Stronger bases remove HCl to give ethylene oxide.

3. Metathetical reactions go smoothly. Thus cyanides give *ethylene cyanohydrin*, $HOCH_2CH_2CN$, an intermediate for the preparation of hydracrylic acid, acrylic acid and its esters.

Sodium sulfhydrate gives *monothioglycol*, $HOCH_2CH_2SH$, b. 70°/28 mm., an important non-nitrogenous sulfhydryl reagent for protein studies.[37] Sodium sulfide gives *thiodiglycol*, $(HOCH_2CH_2)_2S$, the intermediate for the original preparation of mustard gas.[38] Ethylene chlorohydrin is also an intermediate for the manufacture of malonic acid, phenylethyl alcohol, novocaine and indigo. For these purposes it is available in anhydrous form (C. and C.).

Ethylene chlorohydrin is a valuable plant stimulant.[39]

Ethylene bromohydrin and iodohydrin are prepared similarly to the chlorine compound.

Alpha propylene chlorohydrin, 1-chloro-2-propanol, $ClCH_2CHOHCH_3$, b. 127°, is the chief product of the addition of hypochlorous acid to propylene and of the action of HCl with propylene oxide. It is readily formed by the hydration of allyl chloride with sulfuric acid. Heating with water at 150° gives acetone and propionaldehyde, products of rearrangements. Heat alone converts it to acetone and propylene chloride.

Beta propylene chlorohydrin, 2-Cl-1-propanol, $CH_3CHClCH_2OH$, b. 134°, is formed in small amounts from propylene and HOCl. In spite of the difficulty in separating it from the alpha compound, this is probably the best method to obtain it. The method of Henry[40] for converting the alpha com-

[37] Olcott. *Science* 96, 454 (1942).

[38] Meyer. *Ber.* 19, 3259 (1886).

[39] Denny, Miller. *Contrib. Boyce Thompson Inst.* 4, 513 (1932); 6, 31 (1934); Denny. *Contrib. Boyce Thompson Inst.* 5, 435 (1933); *Ind. Eng. Chem.* 20, 578 (1928).

[40] Henry. *Bull. acad. roy. Belg.* 6, 397 (1903).

pound by successive treatment with potassium acetate, thionyl chloride and alcoholysis with absolute MeOH gives mainly the alpha instead of the claimed beta compound.[41]

Trimethylene chlorohydrin, 3-Cl-1-propanol, $ClCH_2CH_2CH_2OH$, b. 161°, is obtained from an excess of the glycol and a suitable reagent for replacing hydroxyl by Cl. The desired product is easily separated from the trimethylene dichloride and the unchanged glycol which boil at 120° and 214°.

Higher chlorohydrins are made from the olefins and hypochlorous acid. As the olefins decrease in solubility in water an alternative method becomes valuable. This is the action of N-chlorourea in acetic acid.[42]

Halohydrins, especially iodohydrins, react with silver nitrate to give a rearrangement much like the pinacolone rearrangement.[43]

D. TRIHYDRIC ALCOHOLS

Organic ortho acids, $HC(OH)_3$ and $RC(OH)_3$, are known only as alkyl derivatives, the orthoformates, orthoacetates, etc., obtained by the action of NaOR with the grouping $-CCl_3$. The orthoformates of the normal alcohols C_1 to C_4 boil 104° to 246°.[44]

Glycerol, glycerin, 1,2,3-propantriol, $HOCH_2CHOHCH_2OH$, m. 17°, b. 290°, 170° (12 mm.), is obtained as a by-product of soap manufacture and by fermentation of glucose by carefully controlling the pH of the solution by adding Na_2CO_3 or $NaHSO_3$ to cut down the formation of ethanol and CO_2 and increase that of glycerol and acetaldehyde (p. 107).[45]

Glycerol can also be made by a special hydrogenolysis of glucose. This process was used in Germany in World War II.[46] Inexpensive sugar was first inverted and then hydrogenated at 200° and 300 atm. over Ni on pumice to give 40% glycerol, 40% propylene glycol, and 20% hexahydric alcohols.

Glycerine is made from allyl chloride (see propylene) by either of two processes: (1) hydrolysis to allyl alcohol followed by chlorination and hydrolysis or (2) direct hydrochlorination followed by hydrolysis. The latter is probably the commercial synthesis.[47–49] In Germany, propane is chlorinated to dichlorides which are converted to allyl chloride.[49a]

Glycerol is completely miscible with water and alcohol but practically insoluble in ether. It is a sweet viscous liquid, $d = 1.265$ at 15°.

[41] Smith. *Z. physik. Chem.* **93**, 59 (1919).
[42] Detoeuf. *Bull. soc. chim.* **31**, 169, 176 (1922).
[43] Tiffeneau et al. *Bull. soc. chim.* **49**, 1595 ff. (1931).
[44] Sah, Ma. *J. Am. Chem. Soc.* **54**, 2964 (1932).
[45] *Ann. Rep. Chem. Soc.* (London) 1925, 68.
[46] Office Publication Board Report No. 691.
[47] Williams. *Ind. Eng. Chem.*, *News Ed.* **16**, 630 (1938).
[48] *Chem. Ind.* **63**, 374 (1948).
[49] Stockman. *Chem. & Met. Eng.* **52**, No. 4, 100 (1945).
[49a] Technical Oil Mission report.

It is non-toxic and is used in medicinal, cosmetic, and food preparations, as an anti-freeze, in making nitroglycerine and for many uses depending on its strong hygroscopic properties such as in non-drying inks, tobacco, cellophane and the like.

Reactions. These include the normal reactions corresponding to its primary and secondary alcohol groups.

1. Replacement of hydroxyl H by alkali metals is accomplished by heating with the metal or its hydroxide. The first H is fairly readily replaced, the second one with some difficulty and the third not at all. Even heavy metal hydroxides such as $Cu(OH)_2$ dissolve when heated with glycerol. Thus a copper solution containing glycerol gives no precipitate with bases.

2. Esterification of the hydroxyl groups by acids.

(*a*) Inorganic acids. Hydrogen chloride gives two glycerol monochloro-hydrins and two dichlorohydrins. Treatment of any of these with PCl_5 gives *glycerol trichlorhydrin*, 1,2,3-trichloropropane, b. 158°. While isomeric glycerol chlorohydrins cannot be separated, the mixture of mono- (b. above 200°) can be separated from the mixture of di-chlorohydrins (b. about 180°). The dinitrates of the monochlorohydrins have a very low freezing point and have been used as antifreeze in dynamite.

Treatment of the monochlorohydrin mixture with bases removes HCl to give *glycidol*, glycide alcohol, 2,3-epoxy-1-propanol, $\overset{O}{\overset{\diagup\diagdown}{CH_2CHCH_2OH}}$, b. 162°, which resembles glycerol in being soluble in alcohol and water but differs from it in being soluble in ether. Similarly, removal of HCl from the dichlorohydrin mixture gives *epichlorhydrin*, 1-chloro-2,3-epoxypropane, $ClCH_2CH\overset{O}{\overset{\diagup\diagdown}{-}}CH_2$, b. 117° (Shell). The reactions of this substance are mainly those of the ethylene oxide ring rather than of the chlorine, illustrating the greater ease of *addition* as compared to metathesis in organic compounds. Thus epichlorhydrin with RMgX gives $RCH_2CH(OMgX)CH_2Cl$.[50] It has been used in an interesting synthesis of cyclopropyl alcohol (p. 530).

Epiethylin, glycidyl ethyl ether, $\overset{O}{\overset{\diagup\diagdown}{CH_2CHCH_2OEt}}$, b. 126°, is made from the dichlorohydrin mixture and alcoholic NaOH. It has solvent properties for shellac, resins and gums.

The pure *alpha monochlorohydrin* of glycerol, 1-Cl-2,3-propandiol, $ClCH_2CHOHCH_2OH$, b. 213°, 139° (18 mm.), is obtained by adding water to epichlorhydrin (Shell); the *beta monochlorohydrin*, $HOCH_2CHClCH_2OH$, b. 146° (18 mm.), by addition of HOCl to allyl alcohol; the *alpha dichlorohydrin*, $ClCH_2CHOHCH_2Cl$, 1,3-dichlorohydrin, from epichlorhydrin and HCl; and

⁵⁰ Koelsch, McElvain. *J. Am. Chem. Soc.* **52,** 1164 (1930).

the *beta dichlorohydrin*, $ClCH_2CH_2ClCH_2OH$, b. 182°, from allyl chloride and HOCl (Shell). It is to be noted that the additions to the allyl compounds proceed contrary to Markownikoff's Rule in that the negative hydroxyl adds to the CH_2 instead of mainly to the CH as in propylene.[51]

Nitric acid gives *nitroglycerine*,[52] glycerol trinitrate, $C_3H_5(ONO_2)_3$, m. 12°, an outstanding explosive both alone and with other materials; absorbed in porous materials as *dynamite*;[53] with nitrocellulose in smokeless powders (double-base powders, cordites, ballistite, rocket propellants). It was used in many forms in World War II for demolition work against heavy concrete and armor. For such uses it had to be desensitized against explosion by rifle bullets by addition of suitable organic liquids. A spectacular use was in clearing paths through land mine fields. A flexible hose was carried over by a rocket, then filled with nitroglycerine which was finally detonated. It is also used in blasting gelatine (Nobel Explosive 808).

The esters of glycerol and phosphoric acid[54] are found widely in animal and vegetable tissues combined with fatty acids and alkamines in phosphatides (lipoids, lipins) such as lecithins and cephalins.[55] The formula of a lecithin follows:

$$C_{17}H_{35}CO_2CH_2CH(OCOC_{17}H_{33})CH_2O\overset{\displaystyle O}{\underset{\displaystyle OH}{\overset{\|}{P}}}OCH_2CH_2NMe_3(OH)$$

Hydrolysis gives stearic, oleic and phosphoric acids, glycerol and choline. Many other fatty acids are found in lecithins. Other possible modifications depend on the attachment of the phosphoric acid to glycerol in the beta position and on optical isomerism. The cephalins contain aminoethyl alcohol in place of choline (p. 326–7).

Glycerol forms ester acids with weak acids like boric acid. Peculiarly these ester acids are much stronger than the original ones. Thus borax, $Na_2B_4O_7$, which is strongly alkaline, combines with glycerol to give a strong acid. This is another example of the formation of the complex ion of a strong acid by hydroxy groups on adjacent carbons (p. 305).

(b) Organic acids react with one, two or three of the hydroxyls to give esters. The preparation and identification of pure partially esterified glycerols are very difficult.[56] The tri-esters with the higher aliphatic acids occur as *fats* such as glyceryl tristearate (tristearin) etc. Many artificial glyceryl esters are known. *Monoformin* and *diformin* are the esters of glycerol with one and two molecules of formic acid. The three *acetins* are obtained by heating

[51] Kharasch, Mayo. *J. Am. Chem. Soc.* 55, 2468 (1933).
[52] Sobrero. *L'Institut* 15, 53 (1847).
[53] Nobel. Eng. Pat. No. 1345 (1867).
[54] *Ann. Rep. Chem. Soc.* (London) 1916, 74.
[55] Pryde. "Recent Advances in Biochemistry," 3d Ed. Blakiston, 1931. p. 225–40.
[56] *Ann. Rep. Chem. Soc.* (London) 1929, 75.

glycerol with acetic acid at suitable temperatures. They have value as high boiling solvents.

3. Ethers of glycerol can be obtained in the usual ways.

Alpha monoethers $ROCH_2CHOHCH_2OH$ of methyl, n-butyl, isoamyl, and phenyl are important intermediates for modified alkyd resins and special plasticizers. The alpha gamma ethers $ROCH_2CHOHCH_2OR$ are high boiling solvents and plasticizers.

Triethylin, glyceryl triethyl ether, $C_3H_5(OEt)_3$, b. 185°, illustrates the effect of ether linkages on volatility. Heating glycerol with aqueous HCl gives diglycerol, $(HOCH_2CHOHCH_2)_2O$. The tetranitrate of this substance was formerly used to lower the freezing point of nitroglycerin for use alone or in dynamite because of the greater sensitivity of the crystalline material. This purpose is now served by glycol dinitrate.

4. Aldehydes and ketones, with acid condensing agents, give 5- and 6-ring acetals with glycerol.[57] The latter is the more stable form. Thus glycerol and acetone with anhydrous $CuSO_4$ give a product which on methylation and hydrolysis gives glycerol alpha methyl ether showing that the acetal has formed in the 1,2-position. Treatment of the 1,2-acetal with a trace of HCl converts it to the 1,3-compound.[58]

5. Heating with formic or oxalic acid gives allyl alcohol, the net result being a dehydration and reduction

$$C_3H_5(OH)_3 + HCO_2H \rightarrow 2\ H_2O + CH_2\text{=}CHCH_2OH + CO_2$$

6. Distillation with dehydrating agents like anhydrous $KHSO_4$ or $MgSO_4$ gives acrolein.[59]

$$C_3H_5(OH)_3 \rightarrow 2\ H_2O + CH_2\text{=}CHCHO$$

7. Oxidation. Of the ten theoretically possible oxidation products containing three carbon atoms five can be made by suitable choice of reagents and conditions.

(a) Mild oxidation by HNO_3 or bromine gives glycerose,[60] a mixture of glyceraldehyde, $CH_2OHCHOHCHO$, and dihydroxyacetone, $CH_2OHCOCH_2OH$.

[57] *Ann. Rep. Chem. Soc.* (London) 1915, 68; 1929, 77.
[58] Hibbert, Morazain. *Can. J. Research* **2**, 35 (1930).
[59] "Org. Syntheses."
[60] Fischer, Tafel. *Ber.* **21**, 2634 (1889).

These are the simplest possible isomeric aldose and ketose. The sorbose bacterium gives dihydroxyacetone alone.

(b) More vigorous oxidation gives glyceric acid, $CH_2OHCHOHCO_2H$ and tartronic acid, $HO_2CCHOHCO_2H$.

(c) Oxidation with bismuth nitrate yields mesoxalic acid, HO_2CCOCO_2H.

8. Fusion with KOH gives formates and acetates.

9. Iodine and phosphorus give allyl iodide if all reagents are anhydrous and isopropyl iodide if water is present. Some propylene is formed in both cases. PBr_3 and PCl_3 replace all three hydroxyls by halogen.

10. Various fermentations of glycerol give trimethylene glycol, dihydroxyacetone, lactic acid, propionic acid, 1-butanol, n-butyric acid, succinic acid and n-caproic acid.

11. Glycerol reacts with phthalic anhydride and with polybasic acids to give important plastics of the *alkyd* or *glyptal* type. The three reactive alcohol groups and the carboxyl groups offer possibilities for the formation of very large molecules, the condition necessary for the formation of resins or plastics.

Glycerol and the Glycols, J. W. Lawrie, 337 pp., New York, 1928.

1,2,3-Butantriol, $CH_3CHOHCHOHCH_2OH$, is obtained from crotonaldehyde, $CH_3CH=CHCHO$, through the corresponding alcohol and its dibromide. *Pentaglycerol*, 1,1,1-trihydroxymethylethane, $CH_3C(CH_2OH)_3$, m. 199°, is made by condensing propionaldehyde and an excess of formaldehyde with lime. Two hydroxymethyl groups are introduced by aldol condensations and the third is formed by the reduction of the aldehyde group by formaldehyde in the alkaline medium. This is really a Cannizzaro reaction and resembles the biological processes by which aldehydes are converted to acids and alcohols. The reactions of pentaglycerol are the normal reactions of primary alcohols. It does not undergo the rearrangements which might be expected from the neopentyl grouping which it contains (p. 120).

E. HIGHER POLYHYDRIC ALCOHOLS

Orthocarbonic acid, $C(OH)_4$, is known only in its esters, $C(OR)_4$. The hydrated form of glyoxal, $(HO)_2CHCH(OH)_2$, exists in the solution of the monomeric substance and explains the failure of polymerization in solution.

Since the existence of more than one hydroxyl on a carbon atom is unusual, the first true members of this series contain four carbon atoms with a hydroxyl group on each. They are known collectively as the tetritols and exist in three stereoisomeric forms: the meso form erythritol and the enantiomorphous threitols.

Erythritol, erythrite, *i*-tetritol, *i*-1,2,3,4-butanetetrol,

$$\underset{\underset{OH\ OH}{|\ \ \ \ |}}{HOH_2C-\overset{\overset{H\ \ H}{|\ \ \ \ |}}{C}-C-CH_2OH,}$$

m. 126°, is found free and as esters in certain lichens and algae. It gives reactions analogous to those of glycerol. It is optically inactive. Moreover it cannot be resolved into dextrorotatory and levorotatory forms. It is thus the internally compensated or meso form, indicated as *meso-* or *i-*.

Erythritol and D,L-threitol have been synthesized starting with 1,3-butadiene and going through its 1,4-dibromide, $BrCH_2CH=CHCH_2Br$, which is obtained in two isomeric forms, one solid and the other liquid at ordinary temperatures. Oxidation with $KMnO_4$ converts the solid to a glycol,

$$BrCH_2CHOHCHOHCH_2Br,$$

m. 87°, and the liquid to an isomeric glycol, m. 135°. Treatment of the di-

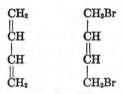

bromohydrins with KOH and ether gives di-epoxy compounds, $CH_2CHCHCH_2$, m. 14° and −15°. Treatment of these with dilute acid gives tetritols, the first giving the racemic material, m. 72°, and the other the common erythritol, m. 126° (cf. 322). The space relations of these compounds can best be shown by tetrahedral models for the carbon atoms. The changes will be discussed at length because the principles involved are of the utmost importance in the study of the carbohydrates and other optically active substances.

$$\begin{array}{cc} CH_2 & CH_2Br \\ \| & | \\ CH & CH \\ | & \| \\ CH & CH \\ \| & | \\ CH_2 & CH_2Br \end{array}$$

The olefinic linkage formed by the 1,4-addition can exist in *cis* and *trans* forms (A) and (B)

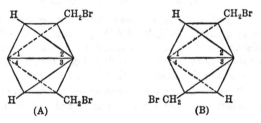

(A)   (B)

In one, the like groups can approach each other by a folding of the molecule along the double bond while in the other the like groups cannot approach each other. Oxidation by $KMnO_4$ adds two−OH groups to the double bond.

1. Addition of −OH groups to (A), the *cis* form.

(*a*) Opening the double bond at the 2,3-end would give (C).

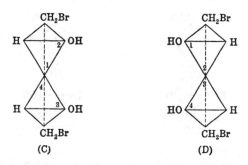

(C)   (D)

(*b*) Opening the double bond at the 1,4-end would give (D). Inspection of those models shows them to be identical rather than enantiomorphous since the rotation of (D) 180° in the plane of the paper will give (C). An observer at 1,4 of (C) looking toward the top tetrahedron would see that the groups H,OH,CH₂Br are arranged counterclockwise; if the observer would then turn and look at the bottom tetrahedron he would see the same groups in clockwise arrangement. Thus one carbon is the mirror image of the other and (C) is the internally compensated *meso-* form. Since removal of HBr from the adjacent OH and CH₂Br groups and the addition of H₂O to the epoxy grouping does not change the spatial arrangement around the two asymmetric carbon atoms, (C) would give the meso tetritol or erythritol. Since the liquid 1,4-dibromide gives erythritol it must be the *cis* form of configuration (A).

2. Addition of −OH groups to (B), the *trans* form

(*a*) At 2,3 to give (E)

(*b*) At 1,4 to give (F).

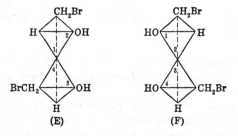

(E)   (F)

Models (E) and (F) are enantiomers. There is free rotation possible about the joined apices 1,4 and 2,3 and it is the accepted convention to rotate the lower tetrahedra so that the groups are in alignment and projecting upward from the depicting plane, as in (E') and (F').

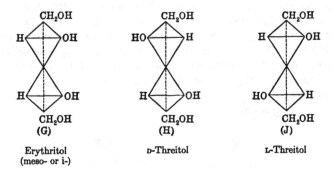

Models (E') and (F') are enantiomers. Rotation of (F') in the plane of the paper through 180° gives (F'') which is not identical with (E'). Conversion of the equimolar mixture of (E) and (F) to the tetritol would give the racemic or D,L-threitol which could be separated into the optically active isomers. Thus the solid dibromide of butadiene which gives D,L-threitol must be the *trans* form (B).

The models of the possible stereoisomeric tetritols are as follows:

The configurational symbols D and L are employed for substances that have been related to the standard of reference glyceraldehyde by unambiguous chemical methods.

Space formulas are seldom used in printed discussion of stereoisomers because of the expense and trouble involved. Instead, formalized "projections" of the space formulas are employed, following certain rules laid down by Emil Fischer. To obtain the Fischer projection the tetrahedra are aligned vertically on their back edges with the secondary hydroxyl groups in alignment and projecting upward. Thus formula (G) gives the projection (K) which may be further conventionalized as shown in (L) and (M); likewise with (H) and (J).

(G)

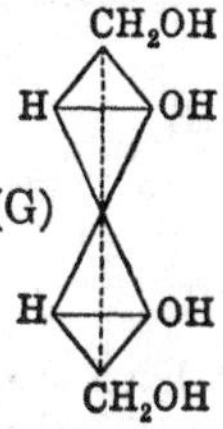

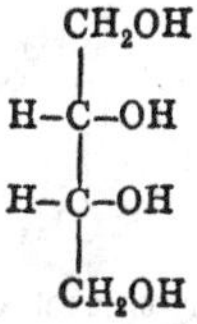

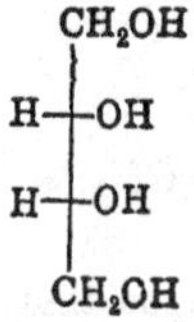

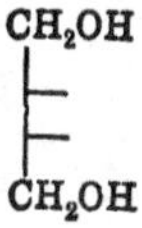

CH_2OH
H—C—OH
H—C—OH
CH_2OH
(K)

CH_2OH
H—OH
H—OH
CH_2OH
(L)

CH_2OH

CH_2OH
(M)

(H)

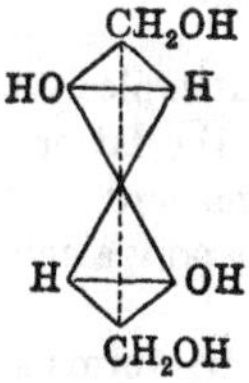

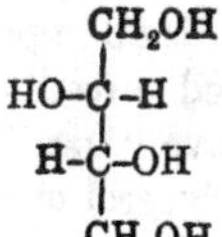

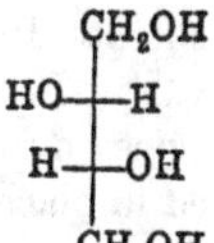

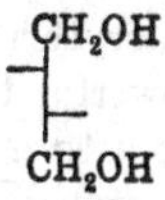

CH_2OH
HO—C—H
H—C—OH
CH_2OH

CH_2OH
HO—H
H—OH
CH_2OH

CH_2OH

CH_2OH

(J)

CH_2OH
H—C—OH
HO—C—H
CH_2OH

CH_2OH
H—OH
HO—H
CH_2OH

CH_2OH

CH_2OH

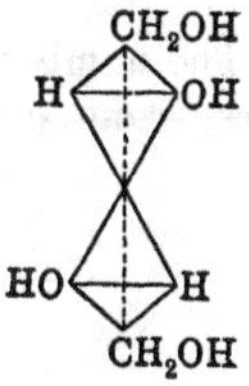

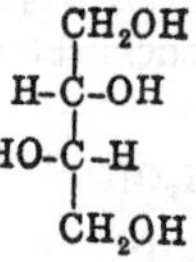

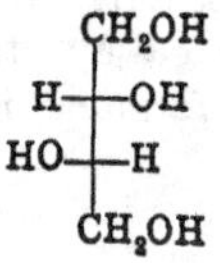

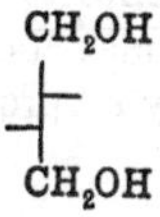

Erythritol tetranitrate, *erythrol nitrate*, is used in place of nitroglycerine for the acute spasms of *angina pectoris*.

Pentaerythritol, tetrahydroxymethylmethane, tetrahydroxyneopentane, $C(CH_2OH)_4$, m. 263°, is made from acetaldehyde, formaldehyde and lime.[61] The three $\alpha - H$ atoms are converted to methylol groups by aldol condensation and the $-CHO$ is reduced by another mol of formaldehyde. As ordinarily prepared, there is an impurity of about 15% of di-pentaerythritol,

$$(HOCH_2)_3CCH_2OCH_2C(CH_2OH)_3.$$

Similarly, tri-pentaerythritol contains eight primary alcohol groups and two ether oxygens.

Pentek is crude PE.

All four OH groups in pentaerythritol can be replaced by bromine by means of HBr. These can be replaced by acetate groups which can be hydrolyzed to give pentaerythritol, showing that rearrangement does not take place as in the case of neopentyl alcohol (p. 120). The tetrabromide reacts

[61] "Org. Syntheses."

with zinc to give mainly methylenecyclobutane. After years of study of this peculiar reaction, some of the originally assumed and expected product, spiropentane, has been isolated and identified (p. 538).

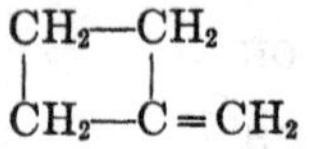

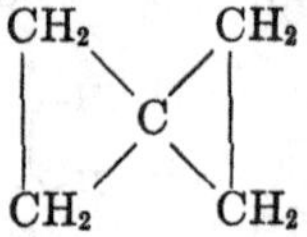

$$CH_2—CH_2$$
$$CH_2—C=CH_2$$

$$
\begin{array}{ccc}
CH_2 & & CH_2 \\
| & C & | \\
CH_2 & & CH_2
\end{array}
$$

At one time it was thought that pentaerythritol did not have the typical tetrahedral structure but consisted of a pyramidal structure with the central C at the apex and the four methylol groups in *one plane*. This has been shown to be incorrect.[62]

Pentaerythritol tetranitrate, PETN, is an important explosive considerably more powerful than TNT.[63] It is used mixed with the latter as Pentalite. It is also used in a high speed destructive "fuse," Primacord.

Pentaerythritol is used in place of glycerol or other poly alcohols in special alkyd type resins (*Pentalyn*).

Esters of unsaturated oils with PE and its di- and tri- forms are important synthetic drying oils.[64]

Pentahydric alcohols, $CH_2OH(CHOH)_3CH_2OH$. The number of possible isomers may be predicted by stereochemical studies. One possibility, as shown by a "projection formula" is

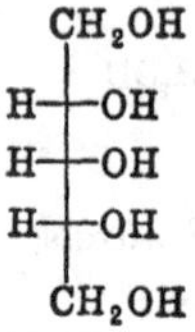

$$
\begin{array}{c}
CH_2OH \\
H——OH \\
H——OH \\
H——OH \\
CH_2OH
\end{array}
$$

(a)

This formula is symmetrical and is identical with its mirror image. It is optically inactive by internal compensation. It represents *adonitol* which can be obtained by the reduction of either D- or L-ribose. Another possibility is obtained by interchanging the H and OH on the upper asymmetric C.

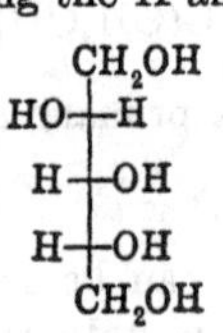

$$
\begin{array}{cc}
CH_2OH & CH_2OH \\
HO——H & H——OH \\
H——OH & HO——H \\
H——OH & HO——H \\
CH_2OH & CH_2OH \\
(b) & (c)
\end{array}
$$

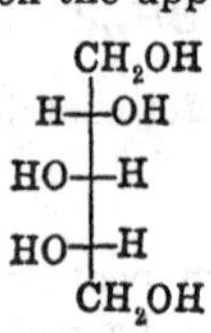

[62] *Ann. Rep. Chem. Soc.* (London) 1929, 75. Orthner, Freyss. *Ann.* **484**, 131 (1930).
[63] Office Publication Board Report No. 320.
[64] Burrell. *Ind. Eng. Chem.* **37**, 86 (1945).

Formula (b) is not symmetrical and is enantiomorphous with formula (c). Thus (b) and (c) represent a pair of optical isomers. Formula (b) represents D-*arabitol* which can be obtained by the reduction of D-arabinose or D-lyxose. Formula (c) is L-*arabitol*.

If in formula (a) the H and OH on the lower asymmetric carbon are interchanged formula (c) is obtained. If, however, the interchange is on the middle carbon, formula (d) is obtained.

(d)

This formula is symmetrical and is identical with its mirror image. It represents another optically inactive form, *xylitol*, m. 61°, which is obtained by the reduction of either D- or L-xylose.[65]

Thus the pentan-pentols exist in four stereoisomeric forms or one pair of optical isomers and two meso forms.

Hexahydric Alcohols, $CH_2OH(CHOH)_4CH_2OH$. A study of the stereochemical possibilities shows that four pairs of optical isomers and two meso forms correspond to this structural formula. D-Mannitol, sorbitol, D-iditol and dulcitol are formed in various plant products. The first three are feebly levorotatory and the last is optically inactive and cannot be resolved. It is thus a *meso* form. Sorbitol is converted to sorbose by the sorbose-bacterium whereas D-iditol is not, thus indicating that the former has the 5- and 6-OH groups on the same side of the chain whereas the latter does not. L-Talitol was the last of the isomer hexitols to be obtained in crystalline form.[66] It was made by hydrogenation of *beta*-L-altrose with Ni catalyst. Most of the possible hexan-hexols have been made by reduction of the corresponding hexoses. These relations establish the configuration of the asymmetric carbons.

Mannitol gives interesting 1,6-benzoylated anhydrides, a mono 2,5-oxide and a di-2,4:3,5-oxide. The latter contains two stable 4-member rings.[67]

[65] Wolfrom, Kohn. *J. Am. Chem. Soc.* **64**, 1739 (1942).
[66] Humoller, Wolfrom, Lew, Goepp. *J. Am. Chem. Soc.* **67**, 1226 (1945).
[67] Brigl, Gruner. *Ber.* **66B**, 1945 (1933). See, however, Hockett et al. *J. Am. Chem. Soc.* **68**, 935, 937 (1946).

Mannitol and sorbitol are available commercially by electrolytic and catalytic reduction.[68-70] Mannitol hexanitrate causes a lowering of the blood pressure (NNR). Large amounts of sorbitol are used in the manufacture of surface active agents, alkyl resins, glues, etc.

Styracitol is 1,5-anhydro-D-mannitol.[71]

Heptahydric alcohols, perseitol and *volemitol* are known.

The catalytic hydrogenation of bioses gives the corresponding polyhydric compounds such as *cellobiitol, lactitol, melibiitol,* and *maltitol.*[72]

Desoxy-polyhydric alcohols such as 1-desoxy-D-glucitol (1-desoxysorbitol, L-gulo-methylitol), the 2-isomer, and 1-desoxy-D-galactitol (L-fucitol) can be made by Raney nickel hydrogenation of the corresponding mercaptals (thioacetals) of acetylated aldoses and ketoses.[73]

XIV. ALKAMINES AND DIAMINES

A. ALKAMINES

These contain hydroxyl and amino groups. The primary product of the action of ammonia with a carbonyl compound having these groups on the same carbon is unstable.

$$=C=O \rightleftharpoons =C(OH)NH_2 \rightarrow complex\ products$$

Sometimes the latter reaction is also reversible. If the functional groups are on different carbons the amino alcohols are stable compounds and show the reactions of both groups. The stronger basic properties of the nitrogen atom cause the reactions of the hydroxyl group with acid reagents to be largely masked. Thus an amino alcohol is less readily dehydrated or esterified than is an ordinary alcohol. The methods of introducing the OH and NH$_2$ groups follow the ordinary reactions.

Mono-ethanolamine, β-aminoethyl alcohol, colamine, β-hydroxyethylamine, $HOCH_2CH_2NH_2$, is available mixed with di- and tri-ethanolamines $(HOCH_2CH_2)_2NH$, $(HOCH_2CH_2)_3N$, formed from ethylene oxide and ammonia (C. and C.). The last substance is strongly basic and is a valuable emulsifying agent.

Colamine is obtained by the hydrolysis of cephalin, a *phosphatide* obtained from brains. *Cephalin* is a mixed ester of higher fatty acids, glycerol, phos-

[68] Creighton. *C. A.* 29, 1725 (1935); *Trans. Electrochem. Soc.* 75, 18 (1939)
[69] Killeffer. *Ind. Eng. Chem., News Ed.* 15, 489 (1937).
[70] Johnson. *Chem. Ind.* 61, 618 (1947).
[71] Zervas, Papadimitriou. *Ber* 73B, 174 (1940).
[72] Wolfrom, Gardner. *J. Am. Chem. Soc.* 62, 2553 (1940).
[73] Wolfrom, Karabinos. *J. Am. Chem. Soc.* 66, 909 (1944).

phoric acid and colamine. It corresponds to the general formula.

$$\begin{array}{ccc}
\text{CH}_2\text{OCOR} & & \text{CH}_2\text{OCOR} \\
| & & | \\
\text{CHOCOR} & \text{or} & \text{CHOCOR} \\
| & & | \\
\text{CH}_2\text{OPO}_2\text{CH}_2\text{CH}_2\text{NH}_2 & & \text{CH}_2\text{OPO}_2\!-\!\text{CH}_2 \\
| & & | \\
\text{OH} & & \text{ONH}_3\text{CH}_2
\end{array}$$

The R groups may be alike or different and are usually the radicals of palmitic, stearic or oleic acids. Isomers are also known in which the phosphoric acid is attached to the beta position of the glycerol.

N-Methylcolamine is obtained by the degradation of morphine and other alkaloids.

NN-Disubstituted alkamines, $R_2NCH_2CH_2OH$, are intermediates in the preparation of novocaine and related drugs. They are readily obtained from secondary amines and ethylene chlorohydrin or more economically, ethylene oxide.

By means of the acetoacetic ester ketone synthesis, beta diethylaminoethyl bromide, $Et_2NCH_2CH_2Br$, can be converted to *Noval ketone,* 5-Et_2N-2-pentanone, which can be converted by the usual reactions to *Noval alcohol, Noval bromide* (by thionyl bromide on the alcohol[1] and *Noval diamine,* important intermediates for antimalarials of the atabrine type.

A special preparation of substituted aminoethanols is as follows :[2]

$$HOCH_2CH_2Cl + COCl_2 \rightarrow ClCO_2CH_2CH_2Cl \rightarrow$$
$$RNHCO_2CH_2CH_2Cl \rightarrow RNHCH_2CH_2OH$$

The last step takes place with excess of KOH.[3]

The conversion of alkylene dibromides to alkamines can be done by condensation with urea followed by the hydrolysis of the resulting imino-oxazolidines.

$$\begin{array}{ccccc}
\text{RCHBr} & & \text{RCHNH} & & \text{RCHNH}_2 \\
| & \rightarrow & | & & | \\
& & & \!\!\!\!\!\!\!\!\!\! C\!=\!NH \rightarrow & \\
| & & | & & | \\
\text{R'CHBr} & & \text{R'CH}\!-\!\text{O} & & \text{R'CHOH}
\end{array}$$

Choline, ethanoltrimethylammonium hydroxide, bilineurine, sincaline, $HOCH_2CH_2NMe_3(OH)$, is found combined with fatty acids and glycerophosphoric acid in *lecithins,* distributed almost universally through the animal and vegetable kingdoms (p. 317).

Choline can be synthesized from ethylene oxide, trimethyl amine and water.

[1] Elderfield et al.   *J. Am. Chem. Soc.* **68,** 1516 (1946).
[2] *Ann. Rep. Chem. Soc.* (London) **1928,** 89.
[3] Pierce.   *J. Am. Chem. Soc.* **50,** 241 (1928).

Although choline has little physiological action, it is increasingly recognized as an important dietary factor.[4][5] It may be a member of the Vitamin B Complex.

Acetylcholine is produced in nerve activity and has a stimulating effect on ganglia and muscles.[6-8] *Mecholyl chloride* is the beta methyl derivative $MeCO_2CHMeCH_2NMe_3Cl$ (NNR).

Neurine, vinyltrimethylammonium hydroxide, $CH_2=CHNMe_3(OH)$, is poisonous whereas choline is not. It is presumably one of the ptomaines formed by putrefaction.

Muscarine was formerly believed to be choline aldehyde hydrate, $(HO)_2CHCH_2NMe_3(OH)$, but has more recently been assigned the formula $EtCHOHCH(CHO)NMe_3(OH)$. It is the active principle of poisonous mushrooms (toad stools), genus *Amanita*.

A great variety of substituted beta aminoethanols is available by the reduction of the condensation products from aldehydes and ketones with nitroparaffins (CSC).

Morpholine, $HN(CH_2CH_2)_2O$, combines the properties of a cyclic ether and a secondary amine. The six-membered ring is easily closed starting either with diethanolamine or $\beta\beta'$-dichlorodiethyl ether and ammonia.

Morpholine is an important dispersing agent for waxes and the like (CC).

More complex *aminoethyl alcohols* can be obtained by

(*a*) The reduction of alpha amino esters by sodium and alcohol to give substances, $HOCH_2CH(R)NH_2$.

(*b*) The action of Grignard reagents on alpha amino aldehydes, ketones and esters to give substances, $R-CHOHCHNH_2-R'$, $RR'C(OH)CHNH_2-R''$, and $R_2C(OH)CHNH_2-R$. *Stavaine*, a cocaine substitute, is a benzoyl ester of such a compound. Its preparation is as follows:

$$MeCOCH_2Cl \rightarrow MeCOCH_2NMe_2 \rightarrow$$
$$MeEtC(OH)CH_2NMe_2 \rightarrow MeEt(Me_2NCH_2)CCO_2C_6H_5$$

Taurine, 2-aminoethyl sulfonic acid, $H_2NCH_2CH_2SO_3H$ or more properly $[\oplus H_3NCH_2CH_2SO_3\ominus]$ is found combined as taurocholic acid in bile. Its structure follows from its preparation starting with ethylene chlorohydrin

$$HOCH_2CH_2Cl \rightarrow HOCH_2CH_2SH \rightarrow HOCH_2CH_2SO_3H \rightarrow$$
$$ClCH_2CH_2SO_2Cl \rightarrow ClCH_2CH_2SO_3H \rightarrow H_2NCH_2CH_2SO_3H$$

[4] Welch. *Science* 88, 333 (1938).
[5] Best. *Science* 94, 523 (1941).
[6] Beutner, Barnes. *Science* 94, 211 (1941).
[7] Fulton, Nachmansohn. *Science* 97, 569 (1943).
[8] Nachmansohn, John. *Science* 102, 250, 359 (1945).

B. Diamines

Those having the amino groups on the same carbon are unstable.

$$=C(NH_2)_2 \rightarrow \; =C=NH \xrightarrow{H_2O} \; =C(OH)NH_2 \rightarrow \; =C=O$$

While methylene diamine, $CH_2(NH_2)_2$, is unstable, its dihydrochloride is stable and can be made from formamide.[8a]

$$2\ HCONH_2 + CH_2O \rightarrow CH_2(NHCHO)_2 \rightarrow CH_2(NH_3Cl)_2$$

Those with the amino groups on different carbons are stable and show ordinary amine reactions.

Ethylene diamine, 1,2-diaminoethane, $H_2NCH_2CH_2NH_2$, b. 116°, is readily available because of cheap ethylene dichloride obtained as a by-product in making ethylene chlorohydrin. Diethylene diamine (piperazine),

$$HN(CH_2CH_2)_2NH,$$

and triethylene diamine, $N(CH_2CH_2)_3N$, are also known. Ethylene diamine forms a monohydrate which can be distilled unchanged. When its hydrochloride is heated, two molecules form the 6-ring, piperazine

$$2\ HCl.NH_2CH_2CH_2NH_2.HCl \rightarrow HCl.NH(CH_2CH_2)_2NH.HCl + 2\ NH_4Cl$$

The first step is the formation of a molecule $NH(CH_2CH_2NH_2)_2$, in which the terminal NH_2 groups can approach closely in space and so can react more readily with each other than with NH_2 groups in other molecules.

Ethylene diamine and CO_2 react under heat and pressure to give ethylene urea, useful in the manufacture of condensation polymers with polyhydroxy compounds.[9]

Amberlite is a resin of ethylene diamine, urea and formaldehyde.

In general, diamines can be made by catalytic reduction of amino nitriles. Thus aldehydes readily give $RCH(NH_2)CH_2NH_2$ while ketones less readily give $RR'C(NH_2)CH_2NH_2$. In the latter case the yield of amino nitrile is poorer and the final product also tends to form the 6-ring tetra-alkyl piperazine with elimination of the tertiary NH_2 groups from two molecules of the first product.

Trimethylene diamine, 1,3-diaminopropane, $H_2N(CH_2)_3NH_2$, b. 137°, prepared from the dibromide and ammonia gives the usual amine reactions. Its best preparation is by Gabriel's phthalimide synthesis (p. 169).

Heat has little effect on trimethylene diamine. It acts with diethyl-carbonate to give the 6-ring trimethylene urea,

$$\begin{array}{c} CH_2\!-\!CH_2\!-\!CH_2 \\ |\qquad\qquad\quad| \\ NH\!-\!CO\!-\!NH \end{array}$$

[8a] *Ann. Rep. Chem. Soc.* (London) 1915, 82.
[9] *Chem. Ind.* **62,** 78 (1948).

Tetramethylene diamine, 1,4-diaminobutane, putrescine, $H_2N(CH_2)_4NH_2$, b. 158°, is a product of decaying flesh produced by the bacterial decomposition following protein hydrolysis.

$$\underset{\text{Arginine}}{\overset{\overset{\displaystyle NH_2}{|}}{NH=C-NH(CH_2)_3CHNH_2CO_2H}} \rightarrow$$

$$\underset{\text{Ornithine}}{NH_2(CH_2)_3CHNH_2CO_2H} \rightarrow CO_2 + NH_2(CH_2)_4NH_2$$

It can be prepared from the corresponding dibromide or, better, by reduction of *ethylene dicyanide* with Na and absolute ethanol. It can be prepared by the Hofmann degradation of the diamide of adipic acid which is readily available as an intermediate for Nylon. Besides giving the ordinary primary amine reactions, putrescine forms a five-membered ring on heating.

$$\begin{array}{ccc}
CH_2CH_2NH_2 . HCl & CH_2-CH_2 \\
| & | & \\
& \rightarrow & | \qquad NH . HCl + NH_4Cl \\
| & | & \diagup \\
CH_2CH_2NH_2 . HCl & CH_2-CH_2
\end{array}$$

Pyrrolidine, H₄-pyrrole

Pentamethylene diamine, cadaverine, $H_2N(CH_2)_5NH_2$, b. 179°, is formed in the decay of lysine,

$$NH_2(CH_2)_4CHNH_2CO_2H \rightarrow CO_2 + NH_2(CH_2)_5NH_2$$

Heat converts it, by loss of NH_3, to pentamethylene imine, $(CH_2)_5NH$, piperidine (hexahydropyridine), another example of the ease of closure of a 6-ring.

Hexamethylene diamine, $H_2N(CH_2)_6NH_2$, b. 205°, is important as a component of Nylon. It is made by catalytic hydrogenation of the diamide of adipic acid.

Two *higher amines* have been isolated from sperm namely spermidin, $NH_2(CH_2)_3NH(CH_2)_4NH_2$, and spermin,

$$NH_2(CH_2)_3NH(CH_2)_4NH(CH_2)_3NH_2.$$

The latter has been synthesized by a simple series of reactions.

$$\underset{\text{Phenol}}{C_6H_5OH} + \underset{\text{In excess}}{Br(CH_2)_3Br} \rightarrow C_6H_5O(CH_2)_3Br$$

$$BrCH_2CH_2Br \rightarrow CNCH_2CH_2CN \rightarrow NH_2(CH_2)_4NH_2$$

$$2 C_6H_5O(CH_2)_3Br + NH_2(CH_2)_4NH_2 \rightarrow$$

$$C_6H_5O(CH_2)_3NH(CH_2)_4NH(CH_2)_3OC_6H_5 \overset{HBr}{\longrightarrow}$$

$$Br(CH_2)_3NH(CH_2)_4NH(CH_2)_3Br \overset{NH_3}{\longrightarrow} spermin.$$

Ethylene imine, $\overline{CH_2CH_2NH}$, b. 56°, is made by treating beta aminoethanol with conc. sulfuric acid, heating the resulting sulfate to form aminoethyl sulfuric acid, and distilling the latter with lime.[10, 11] It is toxic.

XV. HYDROXY ALDEHYDES AND KETONES

In addition to giving the typical reactions of alcohols and aldehydes these compounds show in a remarkable way the tendency for reaction between alcohols and aldehydes. It is apparently very difficult to have these two groups in the free condition in the same molecule. There is a constant tendency for ring formation, between two molecules if the groups are close together or within the same molecule if there are at least two carbon atoms between the carbonyl and the carbinol groups. When the two groups are adjacent, an excess of phenylhydrazine gives an *osazone*, a reaction of the utmost importance in the carbohydrate field. The hydroxy aldehydes and ketones are readily soluble in water.

A. HYDROXY ALDEHYDES

Glycolic aldehyde, hydroxyacetaldehyde, ethanolal, glycolose, $HOCH_2CHO$, is obtained by the careful hydrolysis of bromoacetaldehyde. It exists only in aqueous solution. A crystalline dimer can be made by heating dihydroxymaleic acid.[1] Its freshly prepared solution shows the molecular weight of the dimer but on standing three days the solute has the single molecular weight. A very interesting series of changes is involved in these processes.

The decarboxylation leaves an *enediol* which changes to the carbonyl form. Two molecules unite as a double *hemiacetal* having a six-membered ring and a high degree of symmetry, thus giving the high melting point. The crystalline dimer can be vaporized under diminished pressure giving the monomeric vapor which condenses to the crystalline product again. On standing in aqueous solution, the stable hydrated monomer is formed.

Careful oxidation converts it to glycolic acid $HOCH_2CO_2H$. Dilute alkalies cause an aldol condensation producing trihydroxybutyraldehyde, an aldotet-

[10] Wenker. *J. Am. Chem. Soc.* **57**, 2328 (1935).
[11] Danehy, Pflaum. *Ind. Eng. Chem.* **30**, 778 (1938).
[1] Fenton, Jackson. *J. Chem. Soc.* **75**, 575 (1899).

rose. Further condensation gives a hexose. Phenylhydrazine forms the osazone of glyoxal, m. 169°. This is a general reaction in the carbohydrate field for giving crystalline derivatives of compounds containing the grouping $-CHOH-CO-$, with either H or C at the ends. The steps in the process, each involving 1 mol of $C_6H_5NHNH_2$, follow:

$$HOCH_2CHO \rightarrow HOCH_2CH=NNHC_6H_5 \rightarrow$$

Phenylhydrazone of glycolic
aldehyde

$$OCHCH=NNHC_6H_5 \rightarrow C_6H_5NHN=CHCH=NNHC_6H_5$$

Mono-phenylhydrazone   Di-phenylhydrazone or phenyl
of glyoxal   *osazone* of glyoxal

The second step involves the oxidation of the CHOH group to a carbonyl group by the phenylhydrazine, the latter being reduced to aniline and ammonia. Since phenylhydrazine does not ordinarily oxidize alcohols the process calls for a special explanation. This oxidation takes place only when the carbinol group is *alpha* to a carbonyl group. Thus $MeCOCH_2CH_2OH$ does not give an osazone. As is so often the case, when peculiar reactions are given by a carbon alpha to a carbonyl group, enolization or a similar process is involved. Once again a system of the type $HM-Q=Z \rightleftharpoons M=Q-ZH$ appears. Thus

$$HOCH_2CH=NNHAr \rightleftharpoons HOCH=CH-NHNHAr$$

The latter is an enol, a type of substance very easily oxidized.

Glycolic aldehyde reduces ammoniacal silver solution and alkaline copper solutions (Fehling's solution). With alpha naphthol and sulfuric acid it gives the violet color characteristic of sugars.[2] It gives a yellow color when heated with alkalies.

Lactic aldehyde, 2-propanolal, α-hydroxypropionaldehyde,

$$CH_3CHOHCHO,$$

exists in a crystalline dimeric form, m. 105°. To prepare it, methylglyoxal, $MeCOCHO$, is treated with 2% alcoholic HCl to give the diethyl acetal, $MeCOCH(OEt)_2$, b. 162°. This is reduced by sodium and alcohol to lactic aldehyde diethyl acetal, $MeCHOHCH(OEt)_2$, which is converted to the dimer of the aldehyde on standing with excess 0.1 N sulfuric acid. In common with other alpha hydroxyaldehydes, lactic aldehyde can be prepared from the proper polymeric aldehyde. Thus parapropionaldehyde, $(MeCH_2CHO)_3$ treated with bromine at $-10°$ and then with EtOH gives the acetal,

$$MeCHBrCH(OEt)_2.$$

Careful hydrolysis with water gives a solution of lactic aldehyde.

[2] Molisch. *Monatsh.* 7, 198 (1886).

Distillation of the dimeric form under diminished pressure gives the monomer which quickly reverts to the dimer. Distillation at higher temperatures gives an isomerization to acetol, $MeCOCH_2OH$. The hydrolysis of the acetyl derivative of lactic aldehyde prepared from Ag acetate and alpha iodopropionaldehyde gives acetol.

Lactic aldehyde under the influence of very dilute alkali gives acetylcarbinol (acetol).[3] This is the same change as the conversion of glucose to mannose and probably goes through the enediol.[4]

$$MeCHOHCHO \rightleftharpoons MeC(OH) = CHOH \rightleftharpoons MeCOCH_2OH$$

This type of change is very important in the interconversion of aldoses and ketoses.

In solution, the dimer of lactic aldehyde slowly changes to the monomer, probably in hydrated form.

$$\begin{array}{c} MeCH\!-\!O\!-\!CHOH \\ | \qquad\quad | \\ HOCH\!-\!O\!-\!CHMe \end{array} \rightleftharpoons 2\,MeCHOHCHO \overset{H_2O}{\rightleftharpoons} MeCHOHCH(OH)_2$$

With phenylhydrazine, it first gives the phenylhydrazone of the aldehyde and then the osazone of methylglyoxal, $MeC(=NNHPh)CH(=NNHPh)$, m. 145°.

β-Hydroxypropionaldehyde, hydracrylic aldehyde, 3-propanolal,

$$HOCH_2CH_2CHO,$$

b_{18}. 90°, can be made from acrolein, $CH_2 = CHCHO$, and water at 100°. It differs from the alpha hydroxyaldehydes in not forming a dimer. Such a process would involve the formation of an 8-membered ring. It reduces ammoniacal silver solution but not Fehling's solution. Dehydration gives acrolein. It forms no osazone.

α-Hydroxyisobutyraldehyde, 2-methyl-2-propanolal, $(CH_3)_2C(OH)CHO$, b. 137°, can be made by hydrolysis of the corresponding alpha bromo aldehyde with water. It forms a *hydrate*, m. 70°. The probability is that all aldehydes form such hydrates. In this case the molecule is sufficiently symmetrical to form crystals readily. Bases cause the Cannizzaro reaction. On standing, it gives a crystalline polymer, m. 65°. It can give no osazone.

Acetaldol, aldol, 3-butanolal, β-hydroxybutyraldehyde,

$$CH_3CHOHCH_2CHO,$$

b_{20}. 83°, is formed by the action of a trace of very weak acid or alkaline reagent such as dilute HCl or sodium carbonate on acetaldehyde.[5] It can be distilled under reduced pressure but above 85° it loses water because of the active alpha

[3] Nef. *Ann.* **335**, 191 (1904).
[4] Gustus, Lewis. *J. Am. Chem. Soc.* **49**, 1512 (1927).
[5] Wurtz. *Compt. rend.* **74**, 1361 (1872).

hydrogen atoms and forms crotonic aldehyde, MeCH=CHCHO. It reduces ammoniacal silver solution, being oxidized to β-hydroxybutyric acid.

Acetaldol can be reduced catalytically or by aluminum amalgam and water to 1,3-butylene glycol. Reduction by yeast gives the *dextrorotatory* glycol.

On standing, aldol forms a dimer, m. 90° (Paraldol) which reverts to the monomeric form on distillation. The high melting point of the dimer would indicate a symmetrical structure such as

$$\begin{array}{ccc} \text{MeCH} & \!\!-\!\text{O}-\!\! & \text{CHOH} \\ | & & | \\ \text{CH}_2 & & \text{CH}_2 \\ | & & | \\ \text{HOCH} & \!\!-\!\text{O}-\!\! & \text{CHMe} \end{array}$$

Depolymerization would involve a reversal of the aldol condensation which is entirely possible especially when the $\alpha-$H is also alpha to a carbinol group.

Paraldol, on heating at 170°, gives the ester,

$$\text{MeCHOHCH}_2\text{CO}_2\text{CH}_2\text{CH}_2\text{CHOHMe}$$

While such a disproportionation of aldehydes, 2 RCHO → RCO$_2$CH$_2$R, is well known, it is strange that there is no dehydration.

Paraldol, heated with a trace of iodine, gives an 80% yield of croton-aldehyde.[6]

Aldol forms an oily phenylhydrazone but *not* an osazone. The alcohol group must be alpha to the hydrazone group to undergo the oxidation by phenylhydrazine necessary for osazone formation.

p-Bromophenylhydrazone of aldol, m. 128°.

Products related to aldol are dialdan, C$_8$H$_{14}$O$_3$, m. 140°, and isodialdan (tetraldan), C$_{16}$H$_{28}$O$_6$, m. 113°. The structures usually assigned are not reasonable considering their high melting points. Dialdan can be converted to an alcohol and an acid. It also forms a diacetate. Isodialdan shows no reactions of aldehydes, alcohols or double bonds.

Formisobutyraldol, 2,2-dimethylpropanolal, pentaldol,

$$\text{HOCH}_2\text{C(CH}_3)_2\text{CHO},$$

b. 173° dec., can be made from isobutyraldehyde and 1 mol of formaldehyde at low temperature with weak alkalies.[7] More vigorous treatment reduces the aldehyde group to CH$_2$OH. It exists as a dimer, m. 90°. Under reduced pressure the vapors are also dimeric. In view of the blocking effect of the alpha methyl groups, it would appear that this dimer must contain an 8-

---

[6] Hibbert. *J. Am. Chem. Soc.* **37**, 1748 (1915).

[7] Späth. *Ber.* **76**, 949 (1943).

membered ring.

$$CH_2\text{---}O\text{---}CHOH$$
$$Me_2C \qquad\qquad CMe_2$$
$$HOCH\text{---}O\text{---}CH_2$$

The existence of such a structure is further evidence of the inability of hydroxyl and carbonyl groups to exist as such in the same molecule. This is of the utmost importance in the sugars. Another explanation of dimers which apparently involve 8-membered rings is that they are really 1,3-propylene oxides polymerized through the unsaturated affinity of the bridge oxygen.[8]

The reactions of formisobutyraldol are those to be expected from a primary alcohol and an aldehyde with no $\alpha - H$. Oxidation gives first hydroxypivalic acid, $HOCH_2CMe_2CO_2H$, and then dimethylmalonic acid. Reduction gives the 1,3-glycol. Alkali gives the Cannizzaro reaction. With acetaldehyde and K_2CO_3 it gives 4,4-Me_2-3,5-pentanediolal.

Higher hydroxy aldehydes which have the hydroxyl in the gamma or delta position exist as inner hemiacetals.[9] The properties of these substances are somewhat like those of the aldopentoses and aldohexoses. The aldehyde reactions are modified. Methyl alcoholic HCl, instead of giving an acetal $-CH(OMe)_2$, gives a methyl derivative of the cyclic hemiacetal exactly analogous to the methyl glycosides. Even when the hydroxyl group is farther away, ring formation is possible. Thus ω-hydroxynonylic aldehyde with MeOH and HCl gives a cyclic acetal, $(CH_2)_7CHOMe$.[10] Evidently such

$$CH_2\text{---}O$$

a compound cannot have a ring lying in a plane. The cyclic forms of the hydroxy aldehydes and ketones are *lactoles*. Typical examples are γ-butyrolactole, δ-valerolactole, and γ-valerolactole related to 4-butanolal, 5-pentanolal and 4-pentanolal.

The compound $C_{15}H_{31}CHOHCH_2CH_2CHO$ exists in dynamic isomerism with the aldolactole.

B. HYDROXY KETONES

Acetol, hydroxyacetone, methyl ketol, pyruvic alcohol, propanolone, acetylcarbinol, CH_3COCH_2OH, b. 147°, is prepared from chloro- or bromoacetone by heating with pure dry potassium formate in absolute methanol.[11, 12]. The halogen is first replaced by the formate group and the latter is then removed by alcoholysis. $ROCHO + MeOH \rightarrow ROH + HCO_2Me$. This procedure is useful for converting to alcohols, halides which give undesirable

[8] *Ann. Rep. Chem. Soc.* (London) **1924**, 74.
[9] *ibid.* **1922**, 78.
[10] *ibid.* **1925**, 71.
[11] Nef. *Ann.* **335**, 191 (1904).
[12] "Org. Syntheses."

reactions with alkaline reagents. Acetol reduces Fehling's solution and ammoniacal silver solution. In common with other alpha hydroxy carbonyl compounds, it owes this property to its ability to form an enediol, very sensitive to oxidation, $MeC(OH)=CHOH$. As is well known, the $C=C$ group is sensitive to oxidation. Moreover a carbon already attached to O is more readily oxidized than one not so attached. Thus an enediol should be readily oxidized. The oxidation during osazone formation is probably due to a similar process (p. 332).

The product of the oxidation of acetol with alkaline cupric solutions is lactic acid.[13]

Acetol gives a phenylhydrazone and the osazone of methylglyoxal, m. 145°. Biochemical reduction by means of a fermenting sugar solution gives levorotatory propylene glycol[14] (Levene). The reverse process takes place when the Sorbose-bacterium acts on propylene glycol. This organism has the ability to convert $-CHOHCH_2OH$ to $-COCH_2OH$. With HCl and EtOH, acetol gives the ether, $MeCOCH_2OEt$. With HCl and MeOH, however, it gives the dimethyl ether of the cyclic dimer, m. 127°. This does not reduce Fehling's solution. Heating with dilute acid converts it to acetol. The monomeric methyl ether can be obtained from methyl propargyl ether, $MeOCH_2C \equiv CH$, and mercuric salts. It reduces Fehling's solution in the cold. A substance of the structure, $MeOCH_2COMe$ would not be expected to be so strong a reducing agent since neither ketones nor ethers are readily oxidized. It is probable that this easy oxidation is due to the formation of the very reactive enediol ether

$$MeCOCH_2OMe \rightleftharpoons MeC(OH)=CHOMe.$$

Propionylcarbinol, 1-butanol-2-one, ethyl ketol, ethyl hydroxymethyl ketone, $CH_3CH_2COCH_2OH$, b. 156°, can be made from chloromethyl ethyl ketone through its formate.

The ethers of alpha hydroxy ketones can be prepared as follows:

$$ROH + H_2CO + HCl \rightarrow ROCH_2Cl \rightarrow ROCH_2CN \xrightarrow{R'MgX} ROCH_2COR'.$$

$$ROH + R'CHO + HCl \rightarrow ROCHClR' \rightarrow$$

$$ROCH(CN)R' \xrightarrow{R''Mg X} ROCH(R')COR''$$

Each step gives a good yield.

Another useful preparation for monoalkyl ketols, $RCOCH_2OH$, starts with chloroacetaldehyde and the Grignard reagent, followed by oxidation of the secondary alcohol group by chromic acid mixture and the replacement of the chlorine through the formate.

[13] Kling. *Compt. rend.* **135**, 970 (1902).
[14] "Org. Syntheses."

Acetoin, acetylmethylcarbinol, dimethyl ketol, $CH_3COCHOHCH_3$, m. 15°, b. 145°, can be made by the partial reduction of diacetyl with Zn and acid but it is available commercially as a product of fermentation. It is presumably formed in this process from acetaldehyde by a dismutation (disproportionation) resembling the benzoin condensation of aromatic aldehydes

$$2 \text{ MeCHO} \rightarrow \text{MeCOCHOHMe.}$$

It is the intermediate product for the diacetyl, MeCOCOMe, and 2,3-butyleneglycol formed in the same fermentation.

At least three crystalline dimers of acetoin have been described. These all revert to the monomer on distillation. In water solution it exists in the hydroxy ethylene oxide form. The nature of the dimers is not settled.[15]

Acetoin gives a phenylhydrazone, m. 84°, and the osazone of diacetyl, m. 242°.

Acetoin is the simplest of the acyloins, RCOCHOHR, obtainable from acid chlorides and from organic esters by the action of Na in moist ether.[16, 17] The reaction probably takes place through the formation of the disodium derivative of the enediol. When the latter is liberated, it ketonizes.[18]

$$2 \text{ RCOCl} \xrightarrow{\text{Na}} \text{NaCl} + \underset{\underset{\text{ONa ONa}}{|\quad|}}{\text{RC}=\text{CR}} \rightarrow \underset{\underset{\text{OH OH}}{|\quad|}}{\text{RC}=\text{CR}} \rightarrow \text{RCOCHOHR}$$

Acyloins, on heating, undergo a strange dismutation giving a diketone, RCOCOR, and an aldehyde.

Mixed acyloins may be made from aldehyde cyanohydrins and the Grignard reagent. One mol of the latter is wasted by action with the OH group.

$$\text{RCHO} \rightarrow \text{RCH(OH)CN} \rightarrow \text{RCHOHCOR}'$$

Mixed acyloins in which R' is larger than R undergo a rearrangement to RCOCHOHR' when heated in alcohol solution with a trace of acid. The tendency is apparently to form the smallest possible acyl group.[19]

β-Acetyl ethyl alcohol, acetonylcarbinol, 4-butanol-2-one

$$CH_3COCH_2CH_2OH,$$

b_{30}. 110°, can be made from acetone and formaldehyde with potassium carbonate solution. As usual with this type of reaction the difficulty is to stop it after the first step. The other five alpha hydrogen atoms of the acetone tend to react. The water soluble product, on standing, changes to a viscous insoluble oil.

[15] *Ann. Rep. Chem. Soc.* (London) **1930**, 95.
[16] Corson, Benson, Goodwin. *J. Am. Chem. Soc.* **52**, 3988 (1930).
[17] Hansley. *C. A.* **29**, 3354 (1935).
[18] Bouveault, Locquin. *Bull. soc. chim.* (3) **35**, 629 ff. (1906).
[19] *Ann. Rep. Chem. Soc.* (London) **1927**, 65.

Methylacetonylcarbinol, 4-pentanol-2-one, $CH_3CHOHCH_2COCH_3$, b. 177°, can be made from acetone and acetaldehyde with KCN solution. On heating at 250°, it becomes ethylidene acetone, $MeCH=CHCOMe$. The difficulty with which it loses water is in sharp contrast to the behavior of the product from benzaldehyde and acetone which loses water spontaneously. Phenylhydrazone, m. 103°. No osazone.

Methyl γ-hydroxypropyl ketone, β-hydroxyethylacetone, 5-pentanol-2-one, $CH_3COCH_2CH_2CH_2OH$, b. 145°/100 mm., can be made from sodium acetoacetic ester and excess ethylene dibromide followed by the usual ketone splitting. The second bromine atom is hydrolyzed in the process. On heating, it gives 2-Me-4, 5-dihydrofuran (III). It reduces ammoniacal silver solution *on heating.* Since neither ketones nor primary alcohols give this reaction, the formula (I) does not agree with the facts. The hydroxyl and carbonyl group are far enough apart to react through ring formation to give an inner hemiacetal (II).

This type of substance is more readily oxidizable, as in the case of gamma sugars. The primary alcohol group shows normal reactions, replacement on treatment with HBr and oxidation to carboxyl to give levulinic acid. The carbonyl group can be reduced to give 1,4-pentanediol and reacts normally with the Grignard reagent.

Diacetone alcohol, dimethylacetonylcarbinol, 4-Me-4-pentanol-2-one, $(CH_3)_2C(OH)CH_2COCH_3$, b. 166°, is obtained by the aldol condensation of acetone in the presence of a solid base.[20] The reverse change of diacetone alcohol to acetone has been studied in great detail.[21, 22] On dehydration, it gives mesityl oxide, $Me_2C=CHCOMe$, a typical alpha beta unsaturated ketone.

Rapid low temperature hydrogenation with careful exclusion of dehydrating conditions gives the glycol 2-Me-2,4-pentanediol ("2-2-4," CSC). Without such precautions, mesityl oxide is formed and eventually gives methyl isobutyl ketone or the corresponding carbinol.

5-Hexanol-2-one and **6-hexanol-2-one** (b./100 mm. 142° and 157°) differ strikingly in that the first reduces Fehling's solution and ammoniacal silver solution on heating while the latter does not. This recalls the difference in activity of the furanose (gamma) and pyranose sugar derivatives. Both these

[20] "Org. Syntheses."
[21] Akerlöf. *J. Am. Chem. Soc.* 48, 3046, 49, (1926); 49, 2955 (1927); 50, 733, 1272 (1928).
[22] Miller, Kilpatrick. *J. Am. Chem. Soc.* 53, 3217 (1931).

ketones exist as cyclic compounds:

$$\begin{array}{ccc}
\text{CH}_2\text{---CH}_2 & & \text{CH}_2 \\
& & \text{CH}_2 \quad \text{CH}_2 \\
\text{MeCH} \quad \text{C(OH)Me} & & \text{CH}_2 \quad \text{C(OH)Me} \\
\text{O} & & \text{O}
\end{array}$$

They can be regarded as tetradesoxy-fructofuranose and -fructopyranose respectively.

Both these hydroxy ketones, on long heating, give anhydrides, the first 2,5-Me$_2$-2,3-dihydrofuran, b. 70° (red violet color with pine shaving and HCl) and the second 2-Me-5,6-H$_2$-pyran,

$$\begin{array}{c}
\text{MeC}\!\!=\!\!\text{CH---CH}_2. \\
\text{O---CH}_2\text{---CH}_2.
\end{array}$$

Polyhydroxy aldehydes and ketones, even more nearly related to the sugars, have been made by Helferich and his students.[22a] A typical example is $\alpha\gamma$-dihydroxycaproic aldehyde which gives two compounds with methyl alcohol ("glycosides").

$$\begin{array}{cc}
\text{CH}_2\text{---CHOH} & \\
\text{EtCH} \quad \text{CHOMe} \quad \text{and} & \text{EtCHOHCH}_2\text{CH}\text{---CHOMe} \\
\text{O} & \text{O}
\end{array}$$

Two mols of the aldehyde form a "biose."

$$\begin{array}{cc}
\text{CH}_2\text{---CHOH} & \text{CHOH---CH}_2 \\
\text{EtCH} \quad \text{CHO---CH} & \text{CHEt} \\
\text{O} & \text{O}
\end{array}$$

or

$$\begin{array}{cc}
\text{CH}_2\text{---CHOH} & \text{CH}_2\text{---CHEt} \\
\text{EtCH} \quad \text{CH---O---CH} & \text{O} \\
\text{O} & \text{O}
\end{array}$$

These would correspond to a non-reducing and a reducing biose respectively since the second contains a potential aldehyde group, the hemiacetal group.

A diketoalcohol, 2,5-hexanedion-3-ol, $\text{CH}_3\text{COCHOHCH}_2\text{COCH}_3$, is obtained by the action of methylglyoxal with acetoacetic acid.[23] It exists as a dimer, possibly 2,5-Me$_2$-2,5-(OH)$_2$-3,6-diacetonyldioxan.

[22a] *Ann. Rep. Chem. Soc.* (London) 1923, 78 etc.
[23] Henze. *Z. physiol. Chem.* **232**, 123 (1935).

XVI. HYDROXY MONOBASIC ALIPHATIC ACIDS

A. MONO-HYDROXY ACIDS

These acids are found widely distributed in nature in fruit juices. They show the individual properties of alcohols and acids and also special properties due to the presence of both these reactive groups in the same molecule. The nature of these reactions depends on the relative position of the two groups. Thus the hydroxyacids are naturally classified as alpha, beta, gamma, delta, etc. depending on the position of the hydroxyl. The compound having only OH and CO_2H is carbonic acid. It behaves as a very weak dibasic acid rather than as a hydroxyacid.

Alpha Hydroxyacids

Glycolic acid, hydroxyacetic acid, $HOCH_2CO_2H$, occurs in dimorphous forms, m. 63° and 80°.

Preparation. 1. Commercially by the high pressure catalytic reaction of carbon monoxide and formaldehyde solution (DuPont).

2. By boiling monochloroacetic acid with water in presence of a substance like $CaCO_3$.

3. By oxidizing glycol with dil HNO_3.

4. By electrolysis of oxalic acid solution with lead electrodes. Such a reduction of a carboxyl group is unusual.

5. An even more unusual preparation is that from "potassium carbonyl."[1] CO is passed through a solution of metallic potassium in liquid NH_3 at −50° and the product is cautiously hydrolyzed.

$$2 \ KNH_3 + 2 \ CO \rightarrow KOC \equiv COK \rightarrow HOCH_2CO_2K$$

Reactions. 1. It gives the usual reactions of a primary alcohol and a carboxylic acid.

2. Several "anhydrides" of glycolic acid are known.

(*a*) Diglycolic acid, $O(CH_2CO_2H)_2$, is obtained by boiling a solution of chloroacetic acid with lime. It is a crystalline solid and gives two series of salts. As would be expected, it can form a 6-ring imide, $O(CH_2CO)_2NH$. While this is tasteless, its N-alkyl derivatives are sweet. This property is not utilized because the substances are hygroscopic and easily hydrolyzed.[2]

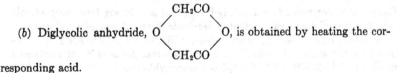

(*b*) Diglycolic anhydride, $O \underset{CH_2CO}{\overset{CH_2CO}{\diagup\diagdown}} O$, is obtained by heating the corresponding acid.

[1] Joannis. *Compt. rend.* **158**, 874 (1914).
[2] *Ann. Rep. Chem. Soc.* (London) **1921**, 77.

(c) "Glycolic anhydride" is really an ester $HOCH_2CO_2CH_2CO_2H$, formed by heating glycolic acid at 100°.

(d) Glycolide, a cyclic di-ester, $\begin{matrix} O—CH_2—CO \\ | \qquad\qquad | \\ CO—CH_2—O \end{matrix}$, m. 87°, formed by vacuum distillation of sodium bromoacetate.

(e) Polyglycolide, $(-OCH_2CO-)x$, a solid, large ring polymer.

Glycolic acid is available in 70% solution (DuPont) for use by tanners and dyers in place of lactic, formic, and acetic acids.

Lactic acid, α-hydroxypropionic acid, $CH_3CHOHCO_2H$. Because of its asymmetric molecule due to a carbon attached to four different groups, lactic acid can exist in D- and L-forms. Most lactic acid is prepared in the DL-form by various fermentations of carbohydrates in the presence of calcium carbonate. Dextrorotatory lactic acid, *sarcolactic acid*, is formed in the muscles from glycogen. If certain types of *Penicillium* are allowed to grow in a solution of ammoniun DL-lactate, the dextrorotatory form remains unchanged. Levorotatory lactic acid is formed by the fermentation of sucrose by *Bacillus acidi levolactici*. In the systematic classification of optically active substances, levorotatory lactic acid is related to D-glyceraldehyde acid and is consequently properly named D-lactic acid while the lactic acid of muscle is L-lactic acid.

$$\begin{matrix} COOH \\ | \\ H—C—OH \\ | \\ CH_3 \end{matrix}$$

D-lactic acid
(levorotatory)
(salts dextrorotatory)

Ethyl lactate for use as a lacquer solvent is made directly from the cyanohydrin of acetaldehyde.

$$MeCHOHCN + H_2SO_4 + EtOH + H_2O \rightarrow MeCHOHCO_2Et + NH_4HSO_4$$

Reactions. 1. The usual acid and secondary alcohol reactions.

2. Heat alone or with dehydrating agents changes it to *lactide*.

$$\begin{matrix} CH_3CH—O—CO \\ | \qquad\qquad | \\ CO—O—CHCH_3 \end{matrix}$$

The failure to remove the alpha hydroxyl with a beta hydrogen by heat or dehydration to give an alpha beta double bond is characteristic of alpha hydroxy acids and their esters and is in sharp contrast to the ease of dehydration of the corresponding beta hydroxy compounds. Thus to obtain an acrylate from a lactic ester, it is necessary to esterify the hydroxyl of the latter and then pyrolize or crack the product.[3]

$$CH_3CH(OAc)CO_2R \rightarrow HOAc + CH_2=CHCO_2R$$

3. With acids at about 130° it loses a molecule of formic acid.

$$CH_3CHOHCO_2H \rightarrow HCO_2H + CH_3CHO$$

These reactions are typical of alpha hydroxy acids.

The configurational relations of optically active forms of hydroxyacids, amino acids, halogen acids and simple carbohydrates have been studied in great detail.[4]

Lactic acid is widely used in industry.[5]

α-Hydroxybutyric acid, $CH_3CH_2CHOHCO_2H$, is made (a) from *n*-butyric acid through the α-halogen acid or (b) from propionaldehyde through its cyanohydrin. Its reactions are like those of lactic acid.

α-Hydroxy*iso*butyric acid, $(CH_3)_2C(OH)CO_2H$ is made (a) by hydrolyzing the cyanohydrin of acetone, (b) by mild oxidation of isobutyric acid (the tertiary H is easily oxidized by dilute $KMnO_4$), (c) by oxidation of trimethylethylene glycol by nitric acid and (d) by hydrolysis of "acetone chloroform," $(CH_3)_2C(OH)CCl_3$, obtained by an aldol condensation between acetone and chloroform.

The reactions of α-hydroxyisobutyric acid are like those of other α-hydroxy acids except that the fork in the chain makes decomposition and oxidation easier. Iodine does not cause the dehydration of the tertiary alcohol group. The esters with P_2O_5 give methacrylic esters, $CH_2=CMeCO_2R$, illustrating the greater ease of removal of a tertiary hydroxyl even from the alpha position (contrast the action of lactic acid with dehydrating agents).

α-Hydroxy-*n*- and *iso*-valeric acids are made through the α-halogen acids or the cyanohydrins. Their reactions are normal, except that the iso acid on dehydration gives $\beta\beta$-dimethylacrylic acid instead of a lactide, thus indicating the reactivity of a tertiary H.

$$Me_2CHCHOHCO_2H \rightarrow Me_2C=CHCO_2H$$

The salts of α-hydroxyisohexoic acid are sweet, the intensity of the taste varying in the optical isomers in the order D- > DL- > L-[5a].

The conductivity of α-hydroxyacids is increased by addition of boric acid.[6] This is due to a highly ionized complex formed between the boron atom and the adjacent hydroxyls. The effect on the conductivity is less in $RR'C(OH)CO_2H$ than in $RCHOHCO_2H$.

Cerebronic acid is an $\alpha-(OH)C_{25}$ acid with a branched chain.[7]

[3] Rehberg, Faucette, Fisher. *Ind. Eng. Chem.* **36**, 469 (1944).

[4] Levene. *Chem. Rev.* **2**, 179 (1926).

[5] Smith, Claborn. *Ind. Eng. Chem., News Ed.* **17**, 370 (1939).

[5a] *Ann. Rep. Chem. Soc.* (London) 1920, 59.

[6] *Ann. Rep. Chem. Soc.* (London) 1922, 71.

[7] *Ann. Rep. Chem. Soc.* (London) 1922, 72.

The amides of α-hydroxyacids react with ketones and HCl to give oxazole derivatives, $\begin{array}{c} RCH-O \\ | \quad\quad | \\ OCNHCR'R'' \end{array}$. Probably the first step involves an "aldol" condensation between an amido H and the CO group to give

$$RCHOHCONHC(OH)R'R''.$$

This has two hydroxyl groups in the 1,4 position so that dehydration gives a 5-ring.

α-Hydroxyacid amides with bromine and KOH give aldehydes.[8] The Hofmann reaction would give $RCH(OH)NH_2$ which readily loses NH_3 to form the aldehyde.

Beta Hydroxyacids

Hydracrylic acid, ethylene lactic acid, β-hydroxy propionic acid, $HOCH_2CH_2CO_2H$.

Preparation. 1. By hydrolysis of ethylene cyanohydrin.

2. By partial oxidation of trimethylene glycol.

3. By heating acrylic acid with alkalies. The addition of water takes place contrary to Markownikoff's Rule, by 1,4-addition.

As an intermediate for acrylic esters for polymers and plastics, hydracrylic esters are prepared by alcoholysis of ethylene cyanohydrin in presence of sulfuric acid.

$$HOCH_2CH_2CN + ROH + H_2O \rightarrow HOCH_2CH_2CO_2R + NH_3$$

Reactions. 1. Ordinary primary alcohol and carboxylic acid reactions.

2. Heat, dehydrating agents, and even heating with acids give acrylic acid.

$$HOCH_2CH_2CO_2H \rightarrow CH_2=CHCO_2H + H_2O$$

There is no tendency to form cyclic compounds because these would have either 4- or 8-membered rings. This ready conversion to alpha-beta unsaturated acids is characteristic of beta hydroxy acids and is dependent on the unusual reactivity of alpha H.

The lactone of hydracrylic acid is β-propiolactone, which is made from ketene and formaldehyde. It polymerizes readily with acids, bases or salts to polyester acids ranging from colorless viscous oils to white solids with average molecular weights of 800–1000, containing one carboxy group per molecule

$$\begin{array}{c} CH_2-CH'_2 \\ | \quad\quad | \\ O-C=O \end{array} \rightarrow RCOO(CH_2CH_2COO)_xCH_2CH_2COOH$$

The polymer breaks down to hydrocrylic acid in boiling aqueous alkali, to

[8] *ibid.* 1922, 77.

hydracrylate esters on treatment with alcohols, and to acrylic acid on pyrolysis.[9-11]

β-Hydroxybutyric acid, $CH_3CHOHCH_2CO_2H$ is obtained (a) from acetoacetic ester by reduction and hydrolysis, (b) by the partial oxidation of aldol, and (c) by the direct oxidation of butyric acid by hydrogen peroxide. Its reactions are like those of hydracrylic acid except that the hydroxyl is on a secondary carbon and is, consequently, more easily removed to give an unsaturated compound. This removal of the OH takes place entirely with the alpha H and not at all with the gamma H. This acid occurs with acetoacetic acid in diabetic urine. Presumably it is more difficult to oxidize in the body than the α-hydroxyacids.[12]

β-Hydroxyisobutyric acid, $HOCH_2CH(CH_3)CO_2H$, cannot be prepared by ordinary methods. These give α-methylacrylic acid instead, thus illustrating the reactivity of a hydrogen which is both tertiary and alpha. A 40% yield of the ester can be obtained by the action of paraformaldehyde and ethyl α-Br-propionate with zinc.

$$MeCHBrCO_2Et + HCHO + Zn \xrightarrow{H_2O} MeCH(CH_2OH)CO_2Et + BrZnOH$$

This is an example of the Reformatzky reaction of carbonyl compounds with alpha bromo esters and zinc to give beta hydroxy esters. It is a modified Grignard reaction of wide application.

$$RR'CO + R''CHBrCO_2Et \rightarrow RR'C(OH)CH(R'')CO_2Et$$

The substituent groups may be H or any primary alkyl group.[13] A lactone, $MeCHCH_2$, can be obtained by the action of silver nitrate on sodium β-
$O=C—O$ (with vertical bonds joining)
iodoisobutyrate.

β-Hydroxy-n-valeric acid, $CH_3CH_2CHOHCH_2CO_2H$, is best made from the chlorohydrin of 1-butene through the cyanohydrin. Treatment of the β-halogen valeric acids with alkalies gives mainly the αβ-unsaturated acid.

β-Hydroxyisovaleric acid, $(CH_3)_2C(OH)CH_2CO_2H$, is obtained by the direct oxidation of isovaleric acid. It is best made as the ester from acetone and bromoacetic ester with amalgamated zinc (Reformatzky).

Similarly, methyl ethyl ketone and α-bromopropionic ester would give αβ-Me₂-β-OH- valeric acid.

Beta lactones are possible when the alpha carbon has two alkyl groups. These prevent the ordinary dehydration to give an alpha beta unsaturated acid and also favor ring closure.

[9] *Chem. Ind.* 61, 977 (1947).
[10] Gresham, Shaver. *C. A.* 42, 209 (1948).
[11] Gresham, Jansen, Shaver. *J. Am. Chem. Soc.* 70, 998, 999, 1001 (1948).
[12] Witzemann. *J. Am. Chem. Soc.* 4B, 211 (1926).
[13] "Org. Reactions," I. pp. 1–37.

Gamma Hydroxyacids

γ-**Hydroxybutyric acid**, $HOCH_2CH_2CH_2CO_2H$, exists only in solution and as its derivatives such as salts, esters, etc. The free acid readily forms an internal ester or lactone because the carboxyl and hydroxyl groups can approach each other closely in space.

$$
\begin{matrix}
CH_2\!\!-\!\!CO_2H \\
| \\
CH_2\!\!-\!\!CH_2OH
\end{matrix}
\quad \rightleftharpoons \quad H_2O \; + \;
\begin{matrix}
CH_2\!\!-\!\!CO \\
| \qquad\qquad \diagdown \\
| \qquad\qquad\qquad O \\
CH_2\!\!-\!\!CH_2 \diagup
\end{matrix}
$$

The product is *butyrolactone* (1,4-butanolide), a gamma lactone, commercially available (Dow).

Butyrolactone is made from ethylene chlorohydrin through the acetoacetic ester or malonic ester synthesis. A modification of this method uses vinyl beta-chloroethyl ether. The starting material is the very cheap dichlorodiethyl ether obtained in the ethylene industry (CC). This is treated with alkali to remove one HCl.

Na Malonic Ester $+$ $ClCH_2CH_2OCH\!=\!CH_2 \rightarrow$

$$CH_2\!=\!CH\!-\!OCH_2CH_2CH(CO_2R)_2 \xrightarrow{\text{NaOH}}$$

$$CH_2\!=\!CH\!-\!O\!-\!CH_2CH_2CH(CO_2Na)_2 \xrightarrow{\text{Acid}}$$

$$
\left[
\begin{matrix}
CH_3\!\!-\!\!CH\!\!-\!\!OCH_2CH_2CH(CO_2H)_2 \\
| \\
OH
\end{matrix}
\right] \rightarrow
$$

(a)

$$[CH_3CHO + HOCH_2CH_2CH(CO_2H)_2] \rightarrow$$

(b)

$$
\begin{matrix}
CH_2.CH_2 \\
| \qquad\quad | \\
OCOCHCO_2H
\end{matrix}
\xrightarrow{\text{heat}} CO_2 +
\begin{matrix}
CH_2\!-\!CH_2 \\
| \qquad\quad | \\
O.CO.CH_2
\end{matrix}
$$

Compound (a) is formed by the hydration of the double bond in the presence of acid. It is a hemiacetal and consequently unstable. Thus the ether linkage $CH_2\!=\!CHOR$ differs from other ether linkages in being easily split by dilute acids. Compound (b) is a gamma hydroxy acid and immediately changes to the lactone.

Reactions. Butyrolactone reacts with (a) bases, (b) alcohols (acid catalysts), (c) halogen acids, (d) ammonia, (e) amines, etc., to give derivatives of γ-hydroxybutyric acid.

The presence of substituents changes the ease of opening of the lactone ring.

Metallic sodium and a trace of sodium ethylate cause a typical "aldol" condensation between the carbonyl group of one molecule of butyrolactone and the alpha H of another.   Treatment with acid dehydrates the first product formed.   Removal of CO_2 leaves "oxetone," a "lactone" or "inner acetal" of a $\gamma\gamma'$-dihydroxy ketone.

All the steps involve well-known reactions except the last one in which the grouping $C=C-C-C-O$ adds to itself to give a 5-membered ring. A proton from the acid probably adds to the olefinic linkage according to Markownikoff's rule forming a carbonium ion $^+C-C-C-C-O$, in which the electronically deficient carbon can approach the O in space and become coordinated with one of its free electron pairs. Loss of a proton leaves a tetrahydrofuran ring. This same tendency causes $\beta\gamma$-unsaturated acids to change to lactones.

The oxetone consists of two five-membered rings at right angles to each other in space with one carbon common to both rings. Such a combination of two rings with one atom in common is a *spiro* system. Homologous oxetones are obtained from higher gamma lactones.

The type of ring in oxetone resembles that found in many carbohydrates. Like many of the simpler sugars, oxetone reduces ammoniacal silver nitrate. This reaction is not given by ketones or alcohols alone. Oxetones react with fuming HBr to give the corresponding $\gamma\gamma'$-dibromoketones.

A reaction of hydroxylated gamma lactones of special value in the study of carbohydrates is their reduction by sodium amalgam and acid to form hydroxy aldehydes or, rather, the lactones or inner hemiacetals of these. A simple lactone without additional OH groups cannot be reduced in this way.

γ-**Hydroxy-n-valeric acid,** $CH_3CHOHCH_2CH_2CO_2H$, exists only in solution and as derivatives. The free acid changes into γ-valerolactone, which resembles butyrolactone very closely. The presence of the methyl group makes the closing of the lactone ring easier and its opening more difficult. It is commercially available by the reduction of levulinic acid (gamma ketovaleric acid) (Mon.). It is also obtained by the action of dilute acids on allylacetic acid.

Delta Hydroxyacids

δ-**Hydroxyvaleric acid,** $HO(CH_2)_4CO_2H$, is obtained as its delta lactone by means of the malonic ester or acetoacetic ester synthesis using trimethylene bromohydrin. The six-membered ring in delta valerolactone is not closed as easily as the five-membered ring in the gamma lactones. It is also opened more readily. Thus delta valerolactone on boiling with water gradually becomes acidic.

ε-**Hydroxycaproic acid,** $HO(CH_2)_5CO_2H$, does not close its lactone ring as readily as its lower homolog. Treatment of ε-bromocaproic acid with NaOEt gives mainly a polymeric lactone[14] $[O(CH_2)_5CO_2(CH_2)_5CO]_x$.

Gamma-, delta-, and epsilon-lactones behave as esters in acting with Na and EtOH to give the corresponding glycols.[15]

Lactones.[16]

Hydroxyl derivatives of many higher aliphatic acids have been made. In general, their properties follow those of the corresponding classes of simpler hydroxyacids. As the hydroxyl group is farther removed than the delta position, it and the carboxyl tend to react independently. When the hydroxyl group is at least twelve carbons away from the carboxyl, it again becomes possible for lactones to exist. Such higher cyclic compounds have their atoms so arranged in space as to give "strainless rings." An example of such a lactone is 1,15-pentadecanolide, a musk-like material occurring in certain plant products. It can be obtained from the omega-bromo-acid and silver oxide, or,

[14] Marvel, Birkhimer. *J. Am. Chem. Soc.* **51**, 260 (1929).
[15] *Ann. Rep. Chem. Soc.* (London) **1906**, 120.
[16] Ahrens. *Sammlung Chemische-Technischen Vortrage* **1903**, 64; **1920**, 36.

more readily, by treatment of cyclopentadecanone with Caro's acid, H_2SO_5,

$$
\begin{array}{ccc}
(CH_2)_{12}CH_2 & & (CH_2)_{12}\!\!-\!\!CH_2 \\
\mid \qquad \mid & \rightarrow & \mid \qquad\qquad \mid \\
CH_2\!\!-\!\!CO & & CH_2\!\!-\!\!O\!\!-\!\!CO
\end{array}
$$

Treatment with a base and then with acid would give the hydroxyacid. The preparation from the dibasic acid through the cyclic ketone thus involves the change of a CO_2H to OH, a special method for going to the next lower homolog. In this way the omega hydroxyacids, $C_{13}-C_{17}$, have been made.[17]

Ambrettolide, found in musk, is a large ring lactone of an unsaturated acid,[18]

$$
\begin{array}{l}
CH(CH_2)_5CO. \\
\| \qquad\qquad \mid \\
CH(CH_2)_8O
\end{array}
$$

By ozonolysis of suitable unsaturated esters, aldehydo-esters can be obtained, $OCH(CH_2)_nCO_2R$. By catalytic reduction of these aldehydes the omega hydroxyacids, C_9-C_{13}, have been made.[19]

The four isomeric hydroxystearic acids having the OH in positions 9, 10, 11, and 12 have been made.[20] The melting points of the 9-, and 10-acids and a 3:7 mixture of the two are 74-75°, 81-82° and 74-76° respectively.

Hydroxyacids in which neither small nor large rings can be formed give chain polymers on heating. Thus ω-hydroxydecanoic acid gives polymers of M.W. 1000-9000 and the general formula

$$HO(CH_2)_9CO_2[(CH_2)_9CO_2]_x(CH_2)_9CO_2H.$$

No cyclic compounds were found.[21]

Sabinic and **juniperic acids** are ω-hydroxy derivatives of the normal C_{12} and C_{16} acids.

Jalapinolic acid is 11-OH-hexadecanoic acid. It has been synthesized from *n*-amyl-MgBr and methyl 10-aldehydodecanoate.[22]

10-Hydroxystearic acid is obtained by hydrating oleic acid with sulfuric acid. No evidence for the simultaneous formation of the 9-OH compound exists.

B. Polyhydroxy Acids

Glyceric acid, $HOCH_2CHOHCO_2H$, is obtained by careful oxidation of glycerol. D-Glyceric acid has the same configuration as D-lactic acid.

$$
\begin{array}{c}
CO_2H \\
\mid \\
H\!\!-\!\!\!\mid\!\!-\!\!OH \\
\mid \\
CH_2OH
\end{array}
$$

[17] Ruzicka et al. *Helv. Chim. Acta* 11, 496, 670, 686 (1928).
[18] Ruzicka. *Ann. Rep. Chem. Soc.* (London) **1927**, 89.
[19] Lycan, Adams. *J. Am. Chem. Soc.* 51, 625 (1929).
[20] *Ann. Rep. Chem. Soc.* (London) **1927**, 87.
[21] Lycan, Adams. *J. Am. Chem. Soc.* 51, 3451 (1929).
[22] Davies, Adams. *J. Am. Chem. Soc.* 50, 1749 (1928).

The designations D- and L- refer to the configuration as related to D- and L-glyceraldehyde. In both cases the salts of the D-forms are dextrorotatory while the corresponding free acids are levorotatory. Besides the usual reactions expected from its primary and secondary alcohol groups and its carboxyl group, glyceric acid in either of its optically active forms undergoes a process of *racemization* on heating with water, alone or in the presence of an organic base such as quinoline. Since this process takes place readily only with an alpha asymmetric carbon, it probably goes through an enol form.

$$HOCH_2CHOHCO_2H \rightleftharpoons HOCH_2C(OH){=}C(OH)_2 \rightleftharpoons HOCH_2CHOHCO_2H$$
D- or L-　　　　　　　　　　　　　　　　　　　DL-

The result is a 50:50 or racemic mixture of the D- and L-forms. The enol is symmetrical as regards the two OH groups and the double bond.

Homologs of glyceric acid have been made from unsaturated acids by mild oxidation or through the dibromides or chlorohydrins.

Cis- and *trans*-9,10-(OH)$_2$-stearic acid, m. 95° and 132°, are obtained by oxidizing oleic and elaidic acids with H_2O_2 in acetic acid or acetone. By alkaline oxidation, oleic acid gives the 132° acid.[23]

3,12-Dihydroxypalmitic acid is obtained along with D-glucose and L-rhamnose from rhamnoconvolvulic acid.

Trihydroxy n-butyric acids, $HOCH_2CHOHCHOHCO_2H$, exist in four optically active forms.

D-erythronic acid　　　L-erythronic　　　D-threonic　　　L-threonic acid
acid　　　　　acid

The names show the relation to the aldo tetroses, erythrose and threose. These acids have been prepared from pentoses and hexoses. They may be built up from the glyceraldehydes by the cyanohydrin synthesis.

$$HOCH_2CHOHCHO \rightarrow HOCH_2CHOHCHOHCN \rightarrow$$
$$HOCH_2CHOHCHOHCO_2H$$

The addition of HCN to the carbonyl group creates a new asymmetric carbon. Since the carbonyl group may open in two ways, two new compounds are obtained. Ordinarily when a new asymmetric carbon atom is formed the chance of the two possible configurations is 50:50. When an optically active substance like D-glucose has a new asymmetric carbon added to it, the two possibilities are *not* formed in equal amounts. In some cases one form appears

[23] *Ann. Rep. Chem. Soc.* (London) **1926**, 99.

almost to the complete exclusion of the other form.   A process in which a new asymmetric carbon is created in other than 50:50 proportions because the initial material is already optically active is called an *asymmetric synthesis*. Many such processes have been carried out.   They represent the normal course of events in natural processes in which it is decidedly unusual to have a DL mixture formed.

The results from the two glyceraldehydes by the cyanohydrin synthesis are as follows:

$$
\begin{array}{ccc}
\text{CHO} & \text{CO}_2\text{H} & \text{CO}_2\text{H} \\
\text{H}-\!\!-\text{OH} \longrightarrow & \text{H}-\!\!-\text{OH} \quad \text{and} & \text{HO}-\!\!-\text{H} \\
\text{CH}_2\text{OH} & \text{H}-\!\!-\text{OH} & \text{H}-\!\!-\text{OH} \\
 & \text{CH}_2\text{OH} & \text{CH}_2\text{OH}
\end{array}
$$

| D-glyceric aldehyde | D-erythronic acid | D-threonic acid |

Compounds related to each other like these two acids are called *epimers*. Epimeric acids of this type are related (1) by being obtained by adding a new asymmetric carbon to the same substance and (2) by being interconvertible by racemization of the alpha carbon by heating with an organic base.   Thus, either D-erythronic acid or D-threonic acid when treated in this way gives an equilibrium mixture of the two.

The tetronic acids readily change to gamma lactones.   As is usual with such lactones, these can be reduced by sodium amalgam.   Enough acid is added to prevent the opening of the lactone ring by an accumulation of alkali. The resulting sodium salts are not reducible.   This is probably another example of the activity of the carbonyl group in an ester.

$$
\begin{array}{ccc}
\text{CHOH}-\text{CH}_2 & \text{CHOH}-\text{CH}_2 & \text{CHOH}-\text{CH}_2\text{OH} \\
\qquad\qquad\searrow\text{O} \quad [\text{H}] & \qquad\qquad\searrow\text{O} \rightleftharpoons & \\
\text{CHOH}-\text{C}=\text{O} \longrightarrow & \text{CHOH}-\text{C}-\text{OH} & \text{CHOH}-\text{CHO} \\
 & \qquad\qquad\quad\text{H} &
\end{array}
$$

Lactone form of a
γ-hydroxy aldehyde,
an internal hemiacetal

The situation may not be as simple as this, since unhydroxylated lactones are not thus reducible.

Trihydroxyisobutyric acid, $(\text{HOCH}_2)_2\text{C(OH)CO}_2\text{H}$, is obtained from dihydroxy acetone by the cyanohydrin synthesis.

Aleuritic acid is 9,10,16-$(\text{OH})_3$-palmitic acid.   Dilute alkali splits it to give azelaic acid and 7-OH-heptoic acid.[24]

[24] *Ann. Rep. Chem. Soc.* (London) 1927, 85.

Sativic acid is 9,10,12,13-(OH)$_4$-stearic acid.[25]

The acids HOCH$_2$(CHOH)$_n$CO$_2$H are called **aldonic acids** because they are formed by cautious oxidation of the corresponding aldoses. They can also be made from the next lower aldoses by the cynohydrin synthesis. Their lactones are readily reduced to the corresponding aldoses. Since the addition of one new asymmetric carbon takes place in two ways, the building up of the series gives the following optical isomers, 8 pentonic acids, 16 hexonic acids, 32 heptonic acids, etc. The *pentonic acids* are the D- and L- forms of *arabonic*, *xylonic*, *ribonic* and *lyxonic* acids, named from the aldopentoses from which they can be made by oxidation. In a similar way the *hexonic acids* are *gluconic*, *mannonic*, *galactonic*, *gulonic*, *talonic*, *idonic*, *altronic*, and *allonic acids*.

Gluconic acid on prolonged heating at a relatively low temperature gives the delta-lactone. More vigorous heating of the acid gives the more stable gamma lactone.

Gluconic acid is available commercially by electrolytic and by enzymatic oxidation of glucose.

Special bacterial fermentation converts glucose to 5-ketogluconic acid and to 2-keto-gluconic acid, the first in 90% yield with *acetobacter suboxylans* and the second in 82% yield with an unnamed bacteria.[26]

D-Galactonic and D-talonic acids have been obtained in crystalline form.[27] The other hexonic acids exist as gums or as crystalline lactones.

Sodium chlorite is a valuable reagent for the preparation of aldonic acids from the aldoses.[28]

Each of the aldohexoses gives an "alpha" and a "beta" heptonic acid which is named after it. The α- and β- refer to the order of isolation of the products formed and *not* to a position in the chain. Thus D-glucose gives a crystalline α-D-glucoheptonic and a syrupy β-D-glucoheptonic lactone. The α-form has the configuration H + OH for the new asymmetric carbon. The configuration of the α-form was proved by its oxidation to an optically inactive inseparable (*i*- or *meso*) pentahydroxy pimelic acid, HO$_2$C(CHOH)$_5$CO$_2$H, configuration $++-++$. The heptonic acid lactones can be reduced to aldoheptoses.

Treatment of aldonic acids with *o*-phenylene diamine gives the usual benzimidazole formation (652).

The Benzimidazole Rule[29] predicts the configuration of aldonic acids and related compounds: whenever the hydroxyl group on the second (or alpha) carbon atom of an aldonic acid is written on the right in the conventional projection formula, the rotation of the derived benzimidazole is positive, and

[25] *ibid.* 1922, 72.
[26] Stubbs et al. *Ind. Eng. Chem.* 32, 1626 (1940).
[27] Hedenburg, Cretcher. *J. Am. Chem. Soc.* 49, 478 (1927).
[28] Jeanes, Isbell. *J. Research, Nat. Bur. Standards* 27, 125 (1941).
[29] Richtmyer, Hudson. *J. Am. Chem. Soc.* 64, 1612 (1942).

conversely, when the hydroxyl group is written on the left, the rotation of the benzimidazole derivative is negative.

C. Hydroxy Unsaturated Acids

Hydroxy olefinic acids are isomeric with the aldehydic and ketonic acids and with glycidic acid, CH_2—$\overset{\displaystyle O}{\overbrace{}}$CHCO$_2$H, and its homologs. They are very rare except for ricinoleic acid in the glycerides of castor oil.

β-Hydroxyacrylic acid, HOCH = CHCO$_2$H, is stable only as the Na derivative of the ester obtained from formic and acetic esters and sodium in ether. The free ester changes to a benzene derivative, ethyl trimesate.

α-Hydroxyvinylacetic acid, ethenylglycolic acid, 3-butene-2-oloic acid, CH_2 = CHCHOHCO$_2$H, m. 40°, can be made from acrolein through its cyanohydrin. It gives the reactions of an olefin and of an α-hydroxy acid. At 190° it loses CO$_2$.

β-Hydroxyisocrotonic acid, CH$_3$C(OH) = CHCO$_2$H, exists only as the ether acids CH$_3$C(OR) = CHCO$_2$H, or the ether acid esters CH$_3$C(OR) = CHCO$_2$R, formed from β-chloroisocrotonic acid or its esters and Na alcoholates. Methyl ether acid, m. 128°, ethyl ether acid, m. 137°. Acetoacetic acid is its keto form.

γ-Hydroxy-3 or 4-pentenoic acids exist only as the lactones, the alpha and beta anhydrides of levulinic acid, obtained by heating it until a mole of water is evolved. The changes are standard but interesting.

$$
\begin{array}{ccc}
\text{CH}_2\!-\!\!-\text{CH}_2 & & \text{CH}_2\!-\!\!-\text{CH}_2 \\
| \quad\quad | & \rightarrow & | \quad\quad | \\
\text{Me—CO HOCO} & & \text{Me—C—O—CO} \\
& & | \\
& & \text{OH}
\end{array}
$$

$$
\begin{array}{ccc}
\text{CH}\!-\!\!-\!\!-\text{CH} & & \text{CH}_2\!-\!\!-\text{CH}_2 \\
\|\quad\quad\| & \text{and} & | \quad\quad | \\
\text{Me—C—O—C—OH} & & \text{CH}_2\!=\!\text{C—O—CO} \\
\text{alpha} & & \text{beta}
\end{array}
$$

The more volatile and stable alpha "anhydride" is probably 2-Me-5-OH-furan as indicated.

Ambrettolic acid, 16-hydroxy-7-hexadecenoic acid,

$$\text{HOCH}_2(\text{CH}_2)_7\text{CH} = \text{CH}(\text{CH}_2)_5\text{CO}_2\text{H},$$

is contained in small amounts in musk-kernel oil.[30] Ozonolysis gives pimelic and azelaic acids. Its lactone, *ambrettolide,* has a 17-membered ring and smells like musk. The dihydro derivative has a similar odor.

[30] *Ann. Rep. Chem. Soc.* (London) **1927,** 88.

Ricinoleic acid, 12-hydroxy-9-octadecenoic acid,

$$CH_3(CH_2)_5CHOHCH_2CH = CH(CH_2)_7CO_2H,$$

m. 5°, in the form of glycerides is the chief component of castor oil. It can be purified by crystallization of its lithium salt. The pure acid is dextrorotatory and shows neutralization and iodine numbers of 188.1 and 84.7[31] (Calc. 188.2 and 85.0). The structure of ricinoleic acid has been proved in a variety of ways. One of the most interesting uses the steps: a. Treatment with bromine, b. Conversion to an acetylenic compound, c. Hydration to a ketone and conversion to the oxime, d. Beckmann rearrangement of the oxime and hydrolysis of the products.[32] The substances identified were:

$Me(CH_2)_5CHOHCH_2CH_2CO_2H$ and its lactone; $NH_2(CH_2)_7CO_2H$;

α-hexyltrimethyleneimine, $Me(CH_2)_5\overline{CHCH_2CH_2N}H$; and azelaic acid, $HO_2C(CH_2)_7CO_2H$.

Ozonolysis of ricinoleic acid gives 3-OH-pelargonic acid and azelaic acid and its half aldehyde. Hydrogenation gives 12-OH-stearic acid. Ricinoleic acid polymerizes readily. Nitrous acid converts it to ricinelaidic acid (*trans*), m. 53°.

The pyrolysis of castor oil under reduced pressure gives undecylenic acid, $CH_2 = CH(CH_2)_8CO_2H$, and heptaldehyde in yields of 10–20% and 30–60%. The same splitting occurs when methyl ricinoleate is heated.[33] A yield of the unsaturated ester of 68% has been obtained.[34] The unmanageable polymer usually obtained in these processes can be avoided by adding rosin before heating.[35]

Alkali salts of ricinoleic acid (soaps), when heated with excess alkali give capryl alcohol (methylhexylcarbinol), sodium sebacate, and H_2.[36]

The splitting of ricinoleic acid at C_{11} by heat and at C_{10} by alkali fusion is noteworthy.

Sodium ricinoleate has valuable properties in counteracting various bacterial toxins.

Ricinoleic acid is used in modified glyptal type resins with sebacic acid and glycerol (Paraplex).

By acetylating the hydroxyl groups and esterifying the carboxyl group, high boiling liquids are obtained with plasticizing, emulsifying, lubricant and detergent properties (CSC).

[31] André, Vernier. *J. Rheology* **3**, 336 (1932).
[32] Goldsobel. *Ber.* **27**, 3121 (1894).
[33] *Ann. Rep. Chem. Soc.* (London) **1928**, 83.
[34] Grün, Wirth. *Ber.* **55**, 2206 (1922).
[35] Bruson, Robinson. *C. A.* **27**, 1365 (1933).
[36] "Org. Syntheses."

Another hydroxyoleic acid has been obtained from the liquid waxes of the castor oil fish.[37] A dihydroxyoleic acid has been found in whale oil.[38]

XVII. DICARBONYL COMPOUNDS

A. DIALDEHYDES

Glyoxal, oxalic aldehyde, ethandial, $O=CHCH=O$, is formed in various solid and liquid polymeric forms in poor yields by the oxidation of glycol, or acetaldehyde by nitric acid. Such an oxidation of a methyl group in preference to the easily oxidizable aldehyde group is remarkable. It is favored by certain catalysts such as silica gel. This same oxidation can be achieved by means of selenium dioxide which does not oxidize an aldehyde group but converts $\alpha - CH_2$ to CO. Ethanol with SeO_2 in an autoclave at 200° gives a 40% yield of glyoxal. When glyoxal is needed as an intermediate it can be obtained in the reaction mixture by the hydrolysis of Cl_2-dioxan.[1]

$$\begin{array}{c} ClCH-O-CH_2 \\ | \qquad\qquad | \\ ClCH-O-CH_2 \end{array} + 2\,H_2O \rightarrow \begin{array}{c} CHO \\ | \\ CHO \end{array} + 2\,HCl + (CH_2OH)_2$$

The vapor phase oxidation of ethylene glycol gives glyoxal.[2]

Monomeric glyoxal can be obtained as a green vapor by heating the polymer with P_2O_5. This vapor condenses to yellow crystals, m. 15°, b. 50°. The yellow crystals become colorless on strong cooling.[3] Glyoxal is the simplest colored organic compound. Its chromophore is $O=C-C=O$ (Will).

Glyoxal gives the usual aldehyde reactions except that it has no alpha hydrogen. Thus alkalies convert it to glycolic acid (Cannizzaro). It reacts surprisingly with ammonia to give imidazole (I) (glyoxaline), the parent substance of a great variety of substances obtainable from "1,2" dicarbonyl compounds, ammonia and aldehydes and from "1,2" diamines and aldehydes.

$$\begin{array}{c} CHNHCH \\ \| \qquad \| \qquad \text{I} \\ CH——N \end{array}$$

With phenylhydrazine, glyoxal gives a di-phenylhydrazone or *osazone*, m. 168°. This type of compound is very important in the sugar series.

Glyoxal unites with *o*-phenylene diamine to give quinoxaline, parent substance of many compounds formed from 1,2-diamines and 1,2-dicarbonyl

[37] Cox, Reid. *J. Am. Chem. Soc.* **54,** 220 (1932).
[38] Moore. *J. Soc. Chem. Ind.* **38,** 320T (1919).
[1] Butler, Cretcher. *J. Am. Chem. Soc.* **54,** 2987 (1932).
[2] Fields. *Chem. Inds.* **60,** 960 (1947).
[3] *Ann. Rep. Chem. Soc.* (London) **1907,** 81.

compounds (p. 806).

The conjugation of the new double bonds with the conjugated system in the benzene ring is notable.

A solution of monomeric glyoxal reduces ammoniacal silver solution but not Fehling's solution. On the other hand trimeric glyoxal reduces the latter, suggesting the structure OCHCHOHCOCHOHCOCHO or one of the completely enolized forms OCHC(OH)=C(OH)C(OH)=C(OH)CHO or HOCH=C(OH)C(OH)=C(OH)COCHO or an intermediate enediol. Any of these forms could act like a reducing sugar.

Glyoxal is available commercially in 30% aqueous solution (CC). This solution has been suggested as a less disagreeable preservative than formalin for histological material. It is used as a shrink-proofing reagent for rayon.

Malonic aldehyde, propandial, OCHCH₂CHO, is known only in water solution and there probably only in the enol form, HOCH=CHCHO. The steps by which its preparation has been attempted are as follows:

$$\text{Acrolein dibromide} \xrightarrow[\text{Acid}]{\text{EtOH}} \text{BrCH}_2\text{CHBrCH(OEt)}_2 \xrightarrow{\text{base}}$$

$$\text{CH}\equiv\text{CCH(OEt)}_2 \xrightarrow{\text{NaOEt}} \text{EtOCH}=\text{CHCH(OEt)}_2 \xrightarrow{\text{H}_2\text{O}}$$

$$\text{EtOCHOHCH}_2\text{CH(OEt)}_2 \rightarrow \text{O}=\text{CHCH}_2\text{CH}=\text{O} \rightleftharpoons \text{HOCH}=\text{CHCHO}$$

The resulting water solution is strongly acid indicating the presence of the enol form. With benzene diazonium chloride it gives a derivative of malonic aldehyde, C₆H₅NHN=C(CHO)₂.

Succindialdehyde, butandial, OCHCH₂CH₂CHO, is obtained (1) by the ozonization of diallyl, made from allyl iodide by the Wurtz reaction, and (2) from pyrrole.

$$\begin{array}{c}\text{CHNHCH} \\ \parallel \qquad \parallel \\ \text{CH}\text{---}\text{CH}\end{array} \xrightarrow{\text{NH}_2\text{OH}} \text{HON}=\text{CHCH}_2\text{CH}_2\text{CH}=\text{NOH} \xrightarrow{\text{HNO}_2} \text{HCOCH}_2\text{CH}_2\text{CHO}$$

The first step probably consists in the addition of hydroxylamine at the double bonds, followed by the loss of NH₃, the grouping $\text{C}\overset{\displaystyle \text{NH}_2}{\diagup}\text{NHOH}$ being unstable like the groupings C(NH₂)₂, C(OH)NH₂, and C(OH)₂.

In addition to the usual aldehyde reactions, succindialdehyde readily forms the 5-membered ring compounds, *furan, pyrrole,* and *thiophene,* when treated with dehydrating agents, ammonia and sulfides of phosphorus respectively.

$$
\begin{array}{c}
\text{OCH CHO} \\
|\quad\ | \\
\text{CH}_2\text{CH}_2
\end{array}
\xrightarrow{\ \text{P}_2\text{O}_5\ }
\begin{array}{c}
\text{CHOCH} \\
\|\quad\ \| \\
\text{CH—CH}
\end{array}
+ 2\,\text{HPO}_3
$$

This probably involves an enolization of one of the aldehyde groups followed by the formation of a gamma lactonic ring with the other aldehyde group.

$$
\begin{array}{c}
\text{HOCH CHO} \\
|\quad\ | \\
\text{CH—CH}_2
\end{array}
\rightarrow
\begin{array}{c}
\text{CHOCHOH} \\
\|\quad\ | \\
\text{CH—CH}_2
\end{array}
\rightarrow
\begin{array}{c}
\text{CHOCH} \\
\|\quad\ \| \\
\text{CH—CH} \\
\text{furan}
\end{array}
$$

In the formation of pyrrole and thiophene the mechanism is the same, NH and S taking the place of O.

Glutaric dialdehyde, $OCH(CH_2)_3CHO$, b. $72°/10$ mm., is obtained by decomposing the ozonide of cyclopentene. It readily changes to a glassy polymer which gives the monomer on vacuum distillation.

Higher dialdehydes can be made from the chlorides of the dibasic acids.[4]

$$
\text{ClCO(CH}_2)_n\text{COCl} \xrightarrow[\text{Pd}]{\text{H}_2} \text{HCO(CH}_2)_n\text{CHO} + 2\,\text{HCl}
$$

B. DIKETONES

Diacetyl, butandione, $CH_3COCOCH_3$, b. $88°$, is available commercially along with acetoin and 2,3-butylene glycol from a special fermentation which apparently involves the following steps:[5]

$$
\text{Glucose} \rightarrow \text{MeCOCO}_2\text{H} \rightarrow \text{MeCHO} \rightarrow \text{MeCHOHCOMe} \rightarrow
$$
$$
\text{MeCOCOMe} + \text{MeCHOHCHOHMe}
$$

It is also available from the modern wood distillation industry.

Diacetyl can be made from methyl ethyl ketone in several ways:

1. By conversion to the isonitroso compound, the monoxime of diacetyl

$$
\text{MeCOCH}_2\text{Me} + \text{HNO}_2 \rightarrow \text{MeCOC(NOH)Me}
$$

This can be converted to diacetyl by hydrolysis with HCl or by careful treatment with nitrous acid.

2. By bromination in ethyl bromide, followed by hydrolysis to methylacetylcarbinol with bicarbonate solution and oxidation by ferric chloride to diacetyl.

3. By oxidation with selenium dioxide.

[4] Rosenmund, Zetzsche. *Ber.* **54B**, 2888 (1921).
[5] Pryde. "Recent Advances in Bio-Chemistry." Blakiston's Son & Co., 1931. p. 142.

Reactions. 1. With strong oxidizing agents diacetyl gives acetic acid. The splitting of an olefin by oxidation probably goes through similar steps.

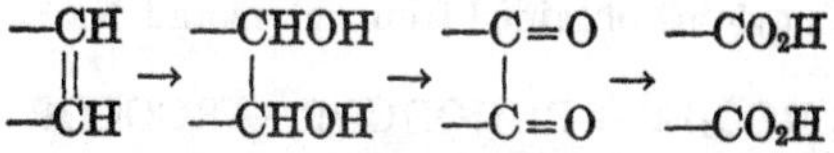

$$\begin{matrix} -CH \\ \| \\ -CH \end{matrix} \rightarrow \begin{matrix} -CHOH \\ | \\ -CHOH \end{matrix} \rightarrow \begin{matrix} -C=O \\ | \\ -C=O \end{matrix} \rightarrow \begin{matrix} -CO_2H \\ \\ -CO_2H \end{matrix}$$

2. Diacetyl gives imidazoles (glyoxalines) and quinoxalines.

3. Diacetyl reacts with alkalies to give first an ordinary aldol condensation between two molecules and then a similar condensation *within* the large molecule forming a six membered ring. In this way substituted para *quinones* are obtained from many alpha diketones.

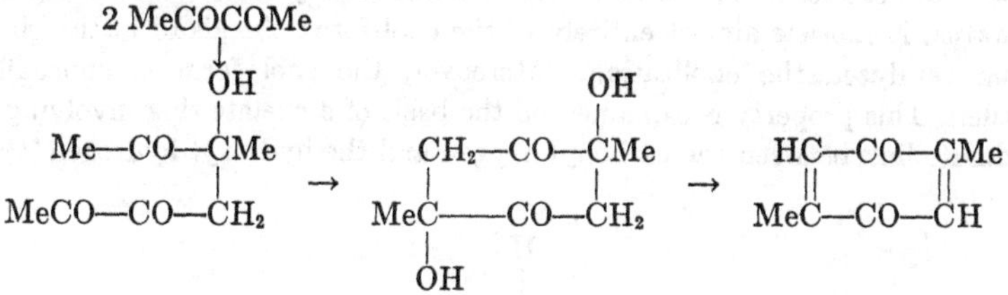

2 MeCOCOMe

$$\begin{matrix} & \overset{\overset{\displaystyle OH}{|}}{} & & & \overset{\overset{\displaystyle OH}{|}}{} \\ Me-CO-CMe & & CH_2-CO-CMe & & HC-CO-CMe \\ & & | & & \| & \| \\ MeCO-CO-CH_2 & \rightarrow & MeC-\!\!-\!\!-CO-CH_2 & \rightarrow & MeC-CO-CH \\ & & \overset{\displaystyle |}{OH} & & \end{matrix}$$

p-xyloquinone

The dioxime of diacetyl (*dimethylglyoxime*) is prepared from the isonitroso compound of methyl ethyl ketone and hydroxylamine. It is an important analytical reagent for nickel.

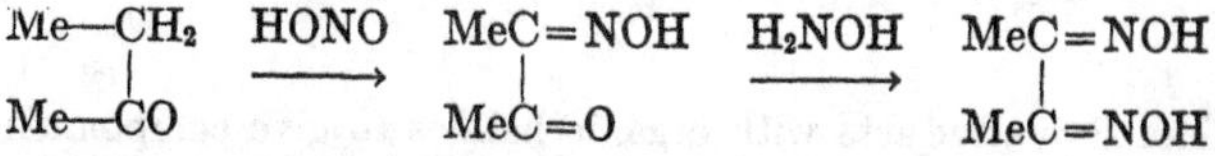

$$\begin{matrix} Me-CH_2 & & MeC=NOH & & MeC=NOH \\ | & \xrightarrow{HONO} & | & \xrightarrow{H_2NOH} & | \\ Me-CO & & MeC=O & & MeC=NOH \end{matrix}$$

The red precipitate obtained with nickel salts contains two molecules of dimethyl glyoxime for each atom of nickel. It exists in stereoisomeric forms.[6] The compound is pictured as involving two ordinary and two subsidiary or coordination valences of nickel (Werner).[7]

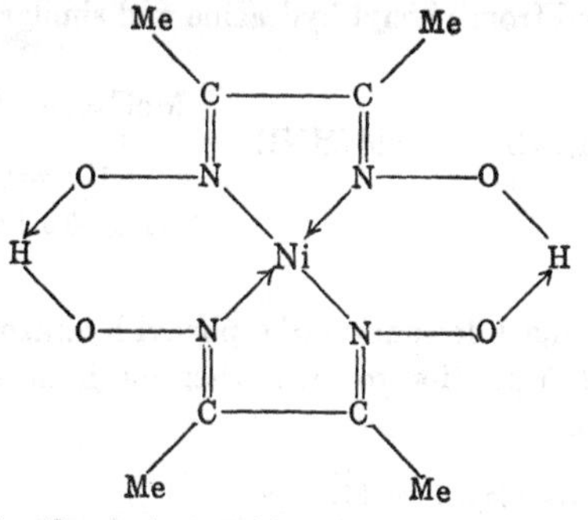

[6] *Ann. Rep. Chem. Soc.* (London) 1923, 86.
[7] Brady. *J. Chem. Soc.* 1930, 1599.

Diacetyl may be purified by means of the compound $(MeCO)_2 . 2\ H_3PO_4$.

A general preparation of alpha diketones (1,2-diketones) is by catalytic dehydrogenation of acyloins obtained from esters and Na.[8]

$$2\ RCO_2Et \rightarrow RCHOHCOR \rightarrow RCOCOR$$

Acetylacetone, 2,4-pentandione, $CH_3COCH_2COCH_3$, b. 136°, is prepared by condensing ethyl acetate with acetone (CC).

Acetylacetone is a typical *beta diketone* or *1,3-diketone*. It is a typical tautomeric or desmotropic substance. Under ordinary conditions it consists of the enol-keto forms in the ratio 3:1. When freshly distilled from a glass apparatus, it consists almost entirely of the enol form, the alkali in the glass having catalyzed the enolization. Moreover, the enol form is unusually volatile. This property is explained on the basis of a chelate ring involving a coordinate link between the carbonyl oxygen and the hydroxyl hydrogen.[9, 10]

Its sodium derivative acts with organic halides to give compounds of the types $(CH_3CO)_2CHR$ and $(CH_3CO)_2CRR'$. Its reactions are typical of the large number of beta diketones, beta ketonic esters, etc.

It reacts with a great variety of compounds to give heterocyclic rings. Among the best known of these are the pyrazoles (pyrroles with the $\alpha - CH$ replaced by N) obtained from phenyl hydrazine and similar compounds.

$$MeCOCH_2COMe + PhNHNH_2 \rightarrow$$

$$\begin{array}{c} MeC=N-NPh \\ | \qquad | \\ CH=\!\!=\!\!CMe \end{array}$$

1-phenyl-3,5-dimethylpyrazole

The first step is the ordinary formation of a phenyl hydrazone which contains NH and C=O in the 1:5 position to each other and is thus capable of intramolecular condensation.

[8] *Ann. Rep. Chem. Soc.* (London) 1906, 82.

[9] Sidgwick. "The Electronic Theory of Valency." Oxford, 1927.

[10] *Ann. Rep. Chem. Soc.* (London) 1926, 127.

With guanidine, substituted pyrimidines are formed.

$$CH_3-\underset{\underset{O}{\|}}{C}-CH_2-\underset{\underset{O}{\|}}{C}-CH_3 \ + \ \underset{NH_2}{\overset{NH_2}{}}\hspace{-0.5em}{}C{=}NH \longrightarrow$$

Pentanedione   Guanidine

2-Amino-4,6-dimethyl
Pyrimidine

In addition to the alkali salts of its enolic form, acetylacetone gives compounds with other metals including copper, iron, aluminum and beryllium. The latter are of special interest because of their volatility. These salts contain one acetylacetone residue for each valence of the metal and are pictured as involving both normal and auxiliary valences of the metals. Thus the copper salt is formulated as containing two six-membered chelate rings both including the Cu atom.[11]

The formulation with the Cu atom as the center of a spiro system of two unsymmetrical six-membered rings would indicate the possibility of enantiomorphism. This is evident when it is recognized that the planes of the two rings must be about at right angles to each other. A simple diagram will show the possibilities

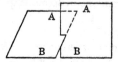 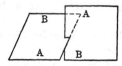

These models each having two identical asymmetric planes at right angles are enantiomorphous.

The correctness of this conception of the metal salts of beta diketones is shown by the resolution of a similar compound of beryllium into its optical isomers.[12] Beryllium forms a compound with two molecules of benzoyl pyruvic acid, $PhCOCH_2COCO_2H$, which still has free carboxyl groups. By means of optically active bases like brucine, d- and l-forms of the beryllium

[11] Sidgwick. "Electronic Theory of Valency." Oxford, 1927.
[12] Mills, Gotts. *J. Am. Chem. Soc.* **48**, 3121 (1926).

complex can be separated.

The formulation of the metal atom as being held by two ordinary "bonds" and two "coordinate links" is made because of our respect for ordinary valence conceptions. The properties of the substances themselves would not indicate such a difference in the four attachments of the metal atom.

Beta diketones having one or both of the methylene H atoms substituted are split by Grignard reagents.[13]

Acetonylacetone, 2,5-hexandione, $CH_3COCH_2CH_2COCH_3$, b. 188°, is made from acetoacetic ester by treating its sodio compound with I_2 and decarboxylating the resulting product.

Bimolecular electrolytic oxidation of acetoacetic ester gives a similar result. It can also be made from acetoacetic ester and monochloroacetone although the yield is poor. As usual with the $-COCH_2X$ grouping, an alkaline reagent tends to cause the reaction of the halogen from one molecule with the enolic H of another.

Acetonylacetone is available commercially (CC.).

Like all *gamma* dicarbonyl compounds, acetonylacetone readily forms five-membered rings giving substituted furans, pyrroles and thiophenes.

Symmetrical diketones, including the highly branched ones, can be made from ketones by diacetyl peroxide.[14]   Thus

[13] Kohler, Erickson.   *J. Am. Chem. Soc.* **53**, 2301 (1931).
[14] Kharasch, McBay, Urry.   *J. Am. Chem. Soc.* **70**, 1269 (1948).

Higher symmetrical diketones, $RCO(CH_2)_nCOR$, in which n > 2, can be made from RZnX and dibasic acyl chlorides. The zinc compounds do not react with the ketones as would RMgX.

δ-Diketones (1:5) are rare in the aliphatic series. Reactions which would be expected to give them produce cyclic products as the result of condensation within the molecule.

$$2\ MeCOCH_2CO_2R + H_2CO \xrightarrow{\text{amines}} \underset{\underset{CO_2R\ \ CO_2R}{|\quad\quad|}}{MeCOCHCH_2CHCOMe}\ \text{dil. acid heat} \xrightarrow{\quad\quad} \underset{\text{base}}{\longrightarrow} \longrightarrow$$

$$[CH_3COCH_2CH_2CH_2COCH_3] \rightarrow \underset{\underset{CH_2-CO-CH_2}{|}}{\overset{\overset{OH}{|}}{MeC-CH_2-CH_2}} \rightarrow \underset{\underset{CH-CO-CH_2}{|\qquad\qquad|}}{\overset{MeC-CH_2-CH_2}{\underset{\|}{}}}$$

3-Methyl-cyclohexen-2-one
1-Methyl-3-keto-cyclohexene

2,6-Heptanedione has been made by chromic acid oxidation of 6-heptanol-2-one.[15] It readily gives a cyclic condensation product.

ε-Diketones (1:6) can exist as such without ring closure. Thus $MeCO(CH_2)_4COMe$, obtained from Na acetoacetic ester and ethylene bromide is stable. The corresponding 4,4,5,5-Me_4-compound exists only in the cyclic form.[16] $\underset{\underset{Me_2C \!-\!-\!-\! CHCOMe}{|\qquad\quad|}}{Me_2CCH_2C(OH)Me}$ This shows the influence of methyl groups in

favoring ring closure. Many reactions of the 1:6-diketones give ring compounds. Thus reduction gives a considerable amount of pinacol formation within the molecule and alkaline condensing agents give the aldol condensation intramolecularly.

$$\underset{\underset{CH_2CH_2C(OH)Me}{|}}{CH_2CH_2C(OH)Me} \leftarrow \underset{\underset{CH_2CH_2COMe}{|}}{CH_2CH_2COMe} \rightarrow \underset{\underset{CH_2 - CHCOMe}{|\qquad\ |}}{CH_2CH_2C(OH)Me}$$

A *1:7 diketone* undergoes the internal aldol condensation to give a 6-ring while a 1:8 diketone gives no intramolecular reaction.

Aliphatic *triketones* are rare. Only two examples will be given, one with the CO groups together and the other with them separated.

Triketopentane, pentantrione, $CH_3(CO)_3CH_3$, b. 55° (12 mm), is obtained by condensing acetylacetone with p-nitrosodimethylaniline and hydrolyzing the product with acid

$$(MeCO)_2CH_2 + ONC_6H_4NMe_2 \rightarrow (MeCO)_2C=N\phi \rightarrow (MeCO)_2CO + \phi NH_2$$

[15] Fargher, Perkin. *J. Chem. Soc.* 105, 1354-5 (1914).
[16] *Ann. Rep. Chem. Soc.* (London) 1927, 114.

The free ketone is orange red. It forms a colorless hydrate. Bases give acetates and formaldehyde readily. It combines readily with acetyl acetone to give $(MeCO)_2C(OH)CH(COMe)_2$.

Diacetylacetone, 2,4,6-heptantrion, $(CH_3COCH_2)_2CO$, does not exist as the triketone but in a cyclic form related to dimethyl pyrone.[17]

C. KETOALDEHYDES

Pyruvic aldehyde, methylglyoxal, pyroracemic aldehyde, propanalone, $(CH_3COCHO)_4$, is obtained only in polymeric form or in solution:

1. From acetone through the isonitroso compound.

$$MeCOCH_3 + HON=O \rightarrow MeCOCH_2N(OH)_2 \rightarrow MeCOCH=NOH \xrightarrow{NaHSO_3}$$

$$\xrightarrow{acid} MeCOCHOHSO_3Na \longrightarrow MeCOCHO \text{ solution}$$

2. From ozonolysis of mesityl oxide, $Me_2C=CHCOMe$.[18]
3. By oxidation of acetone with SeO_2

$$CH_3COCH_3 + SeO_2 \rightarrow CH_3COCHO + H_2O + Se$$

The monomeric form of pyruvic aldehyde is difficultly obtainable.[19]

With formaldehyde and ammonia, Me-glyoxaline is obtained.

$$MeCOCHO + 2 NH_3 + HCHO \rightarrow MeCCH_2N$$
$$\qquad\qquad\qquad\qquad\qquad \underset{N - CH}{\|\qquad\|}$$

It is significant that sugars heated with NH_4OH give the same product.[20]

With alcohols, pyruvic aldehyde forms dialkyl and tetra-alkyl acetals, and with glycols high boiling polymeric acetals. It is also useful as an insolubilization reagent reacting with protein materials and forming cross linked water insoluble resins with polyhydroxy compounds.[21]

Various organisms and alkalies convert pyruvic aldehyde to lactic acid by dismutation.

In general, α-keto aldehydes, RCOCHO, can be made as follows:
1. From methyl ketones.

$$RCOCH_3 \rightarrow RCOCH_2Br \rightarrow RCOCH_2OCOCH_3 \rightarrow$$

$$RCOCH(Br)OCOCH_3 \xrightarrow{heat} CH_3COBr + RCOCHO$$

In the first step, the isomeric bromoketone is formed and has to be separated from the desired bromomethyl ketone.

[17] Collie, Reilly. *J. Chem. Soc.* 121, 1984 (1922).
[18] Harries. *Ann. Rep. Chem. Soc.* (London) 1905, 74.
[19] *ibid.* 1907, 83.
[20] *ibid.* 1917, 68.
[21] *Chem. Inds.* 61, 72 (1947).

2. From dichloroacetic ester.[22]

$$Cl_2CHCO_2Et \rightarrow (EtO)_2CHCO_2Et \xrightarrow[CH_3CO_2Et]{NaOEt + Na}$$

$$(EtO)_2CHCOCH_2CO_2Et \xrightarrow{Na} \xrightarrow{RX}$$

$$(EtO)_2CHCOCHRCO_2Et \xrightarrow{Acid} O=CHCOCH_2R$$

3. By oxidation of the α-CH$_2$ group in aldehydes by SeO$_2$

$$RCH_2CHO + SeO_2 \rightarrow RCOCHO + Se + H_2O$$

The α-ketoaldehydes show the yellow-green color characteristic of glyoxal. They show a corresponding tendency to polymerize.

3-Butanon-al, (CH$_3$COCH$_2$CHO), a *beta* keto aldehyde which would be expected from the condensation of a formic ester with acetone in the presence of sodium ethylate (Claisen) exists only as the sodium salt of the enol form, 1-butene-1-ol-3-one.

$$HC\!\!\stackrel{OR}{=}\!\!O + CH_3COCH_3 \rightarrow \left[HC\!\!\stackrel{OR}{\underset{OH}{-}}\!\!CH_2COCH_3 \right] \rightarrow$$

$$\left[\underset{O}{\overset{\parallel}{HC}}\!\!-\!\!CH_2COCH_3 \right] \rightarrow NaOCH=CHCOCH_3$$

The sodium salt is very soluble in water. When it is treated with acid to liberate the free aldehyde or enol, triacetyl benzene is obtained. Three molecules condense by the usual aldol condensation mechanism and then an alpha H and a carbonyl group in the large molecule are near enough to each other to condense to a ring.

The Na salt of the enol also reacts with hydrazine to form methylpyrazole and with hydroxylamine to give methylisoxazoles.

$$\begin{array}{ccc}
CH\text{------}C\text{---}CH_3 & CH\text{------}CH & CH\text{------}C\text{---}CH_3 \\
\parallel \qquad \parallel & \parallel \qquad \parallel & \parallel \qquad \parallel \\
CH\text{---}NH\text{---}N & CH_3\text{---}C\text{---}O\text{---}N & CH\text{---}O\text{---}N \\
 & \alpha\text{-Methyl} & \gamma\text{-Methyl} \\
 & \text{chief product} &
\end{array}$$

The first step in each case is the ordinary reaction with one of the carbonyl groups. This is followed by an internal condensation to give a five-membered ring. Of the two carbonyl groups or their equivalent, the terminal one would

[22] *Ann. Rep. Chem. Soc.* (London) **1914, 78.**

be expected to be most reactive. This is confirmed by the predominance of the α-methyl isoxazole. In the case of the pyrazole there is no chance for isomerism since

$$
\begin{array}{cc}
\text{CH} \text{------} \text{C—CH}_3 \\
\parallel \qquad\qquad \parallel \\
\text{CH—NH—N}
\end{array}
\quad \text{and} \quad
\begin{array}{cc}
\text{CH} \text{------} \text{CH} \\
\parallel \qquad\qquad \parallel \\
\text{CH}_3\text{—C—NH—N}
\end{array}
$$

are tautomeric by prototropic shift of H from one N to the other with corresponding shifts of the double bonds.

In general, *beta ketoaldehydes* are known only as the metal derivatives of hydroxymethylenemethyl ketones, $NaOCH = CHCOR$.

Levulinic aldehyde, γ-ketovaleraldehyde, $CH_3COCH_2CH_2CHO$, is obtained by ozonolysis of caoutchouc or, better, of Me-heptenone, $Me_2C = CH(CH_2)_2$-COMe. It gives the reactions expected of a dicarbonyl compound except that it does not give the Angeli-Rimini reaction for aldehydes. It reduces Fehling's solution even in the cold.

Mesoxalic dialdehyde, $OCHCOCHO$, has been obtained in aqueous solution, as the stable hydrate and as a solid polymer.[23]

1. By ozonolysis of phorone, $Me_2C = CHCOCH = CMe_2$, in chloroform followed by careful hydrolysis of the explosive diozonide.

2. By the action of nitrous acid on di-isonitrosoacetone,

$$
HON = CHCOCH = NOH.
$$

XVIII. ALDEHYDE ACIDS AND KETONE ACIDS

A. ALDEHYDE ACIDS

Glyoxalic acid, ethanal acid, $OCHCO_2H$, $(HO)_2CHCO_2H$, is made (1) by the hydrolysis of dichloroacetic acid and (2) as a by-product in the electrolysis of oxalic acid with lead electrodes.[1] In the solid state and in its salts it always contains a molecule of water which probably indicates the hydration of the carbonyl group as in chloral hydrate and in mesoxalic acid. Thus it is sometimes called dihydroxyacetic acid.

In solution, it gives typical aldehyde and acid reactions. With bases, it disproportionates to form salts of oxalic and glycolic acids. Reduction gives tartaric acid in addition to the expected glycolic acid.

The *half aldehyde of malonic acid*, propanal acid, formylacetic acid, $OCHCH_2CO_2H$, exists only as the sodium derivative of β-hydroxyacrylic ester or hydroxymethyleneacetic ester, $NaOCH = CHCO_2R$, formed by condensing esters of formic and acetic acids. The Na compound reacts readily with acetyl chloride to give the acetate, $CH_3CO_2CH = CHCO_2R$. The sodium compound, on acidification, condenses to a benzene-1,3,5-tricarboxylic ester. As usual three molecules condense giving a compound in which an internal

[23] *Ann. Rep. Chem. Soc.* (London) 1905, 79.
[1] *ibid.* 1926, 98.

condensation can give a six-membered ring. The most important use of this sodium derivative is with guanidine to make the intermediate for sulfadiazine (ACC).

The *half aldehyde of succinic acid,* butanal acid, $OCHCH_2CH_2CO_2H$, exists in the lactone form (inner hemi-acetal). It can be made by the malonic ester or acetoacetic ester synthesis using the acetal of monochloroacetaldehyde.

$$(EtO)_2CHCH_2Cl \rightarrow (EtO)_2CHCH_2—CH_2CO_2H \rightarrow$$

$$OCHCH_2CH_2CO_2H \rightleftharpoons \begin{array}{c} CH_2—CHOH \\ | \qquad\qquad O \\ CH_2—CO \end{array}$$

The lactone form of the aldehyde is in equilibrium with the free aldehyde form since it gives typical aldehyde reactions.

Higher aldehyde acids have been prepared by ozonizing unsaturated acids.[2] Thus oleic and erucic acids give $OCH(CH_2)_7CO_2H$, nonanal acid, and $OCH(CH_2)_{11}CO_2H$, tridecanal acid.

B. KETO ACIDS

1. **Pyruvic acid,** pyroracemic acid, propanone acid, keto propionic acid, CH_3COCO_2H, m. 13°, b. 165°. The enol form $CH_2=COHCO_2H$ is called *glucic acid.*

Preparation. 1. By heating tartaric acid either alone or with $NaHSO_4$.

$$\begin{array}{c} HOCH—CO_2H \\ | \\ HOCH—CO_2H \end{array} \rightarrow \begin{array}{c} CH—CO_2H \\ || \\ HOC—CO_2H \end{array} \rightarrow \begin{array}{c} CH_2—CO_2H \\ | \\ CO—CO_2H \end{array} \rightarrow CO_2 + CH_3COCO_2H$$

The last step involves the usual loss of CO_2 by a beta carbonyl acid on heating.

2. From acetyl chloride through the cyanide. The intermediate acetyl cyanide behaves peculiarly on hydrolysis. With water or bases it gives acetic acid and HCN but with concentrated HCl it gives the amide of pyruvic acid which can then be hydrolyzed to the acid itself.

$$CH_3C{\overset{\displaystyle CN}{=}}O \;+\; H_2O \rightarrow CH_3C{\overset{\displaystyle CN}{\underset{\displaystyle OH}{—}}}OH \rightarrow HCN + CH_3C{\overset{}{\underset{\displaystyle OH}{=}}}O$$

$$CH_3\overset{\displaystyle O}{\overset{||}{C}}—C{\equiv}N + HCl \rightarrow \left[CH_3\overset{\displaystyle O\;\;Cl}{\overset{||\;\;|}{C}—C}=NH \right] \rightarrow$$

$$\left[CH_3CO\overset{\displaystyle OH}{\overset{|}{C}}=NH \right] \rightleftharpoons CH_3COCONH_2$$

[2] Noller, Adams. *J. Am. Chem. Soc.* **48**, 1074 (1926).

Another interpretation is that in the lower concentration of hydrion, the attack of the latter is on the carbonyl double linkage, whereas in the higher concentration it is on the cyanide group to give the last two compounds.

Reactions. 1. Usual acid and ketone reactions.

2. It reduces ammoniacal silver solutions. Its reducing properties are important in its role as an intermediate in alcoholic fermentation.

3. Heated with dilute sulfuric acid, it gives CO_2 and acetaldehyde. Concentrated H_2SO_4 gives CO and acetic acid.

4. Heated alone or with HCl, it gives methylsuccinic acid (pyrotartaric acid). This peculiar change probably involves an "aldol" condensation followed by loss of CO_2 and an intramolecular oxidation and reduction.

5. It polymerizes to a non-volatile syrup.

6. With reducing agents it gives DL-lactic acid and DL-dimethyltartaric acid.

2. **Acetoacetic acid,** 3-butanone acid, β-ketobutyric acid, acetone carboxylic acid, $CH_3COCH_2CO_2H$, is very unstable, changing to CO_2 and acetone readily. In striking contrast to the instability of acetoacetic acid is the fact that $F_3CCOCH_2CO_2H$ can be distilled without decomposition.[3] This may be due to a stabilizing of a chelate ring by the trifluoromethyl group.

$$\begin{array}{c} O \\ F_3C-C \qquad H \\ H-C \qquad O \\ C \\ OH \end{array}$$

Acetoacetic acid occurs with acetone in diabetic urine.

Acetoacetic esters are stable and are among the most useful reagents in organic chemistry. The preparation of ethyl acetoacetate is the classical example of the *Claisen condensation.*[4] Two molecules of ethylacetate are brought into reaction by means of metallic sodium in the presence of a trace of alcohol. Sodium ethylate is the "condensing agent" (p. 197).

$$2\ CH_3CO_2Et \rightarrow CH_3COCH_2CO_2Et \rightarrow CH_3C(ONa)=CHCO_2Et$$

The percentages of the enol and keto forms under different conditions have been determined by chemical means,[5] and by physical means.[6, 7] The ordinary ester is more than 90% keto. It has been possible to prepare the practically pure forms.[8] The enol reacts instantly with ferric chloride giving

[3] Swarts. *Ann. Rep. Chem. Soc.* (London) 1927, 90.
[4] "Org. Reactions," I. p. 266.
[5] Meyer. *Ber.* 45, 2843 (1912).
[6] Bruhl, Schroder. *Ber.* 38, 1871 (1905).
[7] Perkin. *J. Chem. Soc.* 61, 808 (1892).
[8] Knorr et al. *Ber.* 44, 1138 (1911).

an intense violet-red color. The pure keto form gives this color only on standing. The keto form is solid at $-78°$ while the enol is still liquid at that temperature.

Acetoacetic Ester, ethyl acetoacetate $CH_3COCH_2CO_2Et$, b. 181° is available commercially as an intermediate for dyes, drugs and perfumes.

Esters of acetoacetic acid are perhaps best prepared from the alcohol and the dimer of ketene, acetylketene.

$$CH_3COCH=C=O + ROH \rightarrow CH_3COCH_2CO_2R$$

Methyl acetoacetate is also available commercially.

Reactions. 1. "Hydrolysis."

(a) With dilute bases, the ester is saponified in the usual way.

$$CH_3COCH_2CO_2R + NaOH \rightarrow ROH + CH_3COCH_2CO_2Na$$

On acidification, the liberated acetoacetic acid decomposes.

$$CH_3COCH_2CO_2H \rightarrow CO_2 + CH_3COCH_3$$

Alpha mono and disubstituted acetoacetic esters give *mono and disubstituted acetones*, CH_3COCH_2R and $CH_3COCHRR'$, by these reactions.

(b) With concentrated bases, the carbon chain is split between the acetyl group and the rest of the molecule. This is therefore a reversal of the original Claisen reaction by which the ester was made. This splitting is characteristic of the grouping

$$-CO-\overset{\displaystyle |}{\underset{\displaystyle |}{C}}-CO-.$$

$$CH_3COCH_2CO_2R + 2\ KOH \rightarrow 2\ CH_3CO_2K + ROH$$

This splitting is not due to the enolization of the grouping, $-COCH_2CO-$, because alpha mono and disubstituted acetoacetic esters are split in the same way by concentrated alkalies. In fact, the disubstituted esters are most easily split. From the substituted esters *mono and disubstituted acetic acids*, RCH_2CO_2H and $RR'CHCO_2H$ are obtained.

$$CH_3COCRR'CO_2Et + 2\ KOH \rightarrow CH_3CO_2K + RR'CHCO_2K + EtOH$$

The grouping $-CO-\overset{\displaystyle |}{\underset{\displaystyle |}{C}}-CO-$ is thus *formed* under the influence of traces of

alkaline reagents such as NaOEt and $NaNH_2$ and is *split* by concentrated alkaline reagents such as alcoholic KOH. This splitting is much like that of the grouping Cl_3C-CO- by KOH to give $CHCl_3$ and $-CO_2K$. A similar splitting also results from the action of ketones with an excess of sodamide. The decarboxylation of acids by heating their sodium salts with soda-lime probably follows a similar course.

2. The most striking reaction of acetoacetic ester and of all other beta dicarbonyl compounds having at least one alpha H is their ready reaction with alkali metals to give metallic derivatives of the enols.

$$\underset{\overset{\displaystyle \|}{\text{O}}}{-\text{C}-}\underset{\overset{\displaystyle |}{\text{H}}}{\text{C}-}\underset{\overset{\displaystyle \|}{\text{O}}}{\text{C}-} \rightleftharpoons \underset{\overset{\displaystyle |}{\text{OH}}}{-\text{C}=}\underset{}{\text{C}-}\underset{\overset{\displaystyle \|}{\text{O}}}{\text{C}-} \rightarrow \underset{\overset{\displaystyle |}{\text{ONa}}}{-\text{C}=}\underset{}{\text{C}-}\underset{\overset{\displaystyle \|}{\text{O}}}{\text{C}-}$$

Such a shift of the H (proton) is called *prototropy*.[9] It actually takes place by the ordinary bimolecular mechanism of enolization, a proton or its equivalent attacking one of the free electron pairs of the carbonyl oxygen and another proton being expelled from the alpha carbon. Acetoacetic ester reacts with sodium, sodium alcoholates and sodamide to give sodium acetoacetic ester. Acidification of the sodium derivatives gives the enol form which rapidly changes to the equilibrium keto-enol mixture.[10]

The enol form of acetoacetic acid differs from the keto form in being more volatile, less soluble in water and more soluble in organic solvents.[11] Such properties do not correspond with the ordinary formulas

$$CH_3COCH_2CO_2Et \quad \text{and} \quad CH_3C(OH)=CHCO_2Et.$$

Usually a hydroxyl derivative is less volatile than a corresponding carbonyl compound. Thus the boiling points of *n*-propyl, isopropyl, allyl, and propargyl alcohols are 97°, 82°, 96° and 115° while those of propionaldehyde, acetone and acrolein are 48°, 56° and 52°. The lower boiling point of the enol form of acetoacetic ester is explained by a *chelate ring* structure in which the H of the hydroxyl is *coordinated* with the carbonyl oxygen. The coordination of the H *intramolecularly* prevents its coordination with the oxygen of another molecule which would cause *association* and increase in boiling point (p. 238).

Calcium acetoacetic ester is obtained by heating the ester with CaC_2. Acetoacetic ester gives a copper compound soluble in organic liquids. This

[9] *Ann. Rep. Chem. Soc.* (London) 1927, 106.
[10] *ibid.* 1911, 100; 1912, 108; 1922, 66.
[11] *Ann. Rep. Chem. Soc.* (London) 1925, 70.

is a chelate compound.[12]

$$
\begin{array}{ccc}
\text{Me—C—O} & \text{O—C—Me} \\
\text{HC} \quad \quad \text{Cu} \quad \quad \text{CH} \\
\text{RO—C=O} \quad \quad \text{O=C—OR}
\end{array}
$$

A similar compound is formed with beryllium.

Sodium acetoacetic ester acts with reactive halides such as the alkyl and acyl halides, halogen derivatives of esters, ketones, acetals, etc., to give sodium halide and a new organic molecule. The fact that the new molecule has its parts linked through C rather than through oxygen caused much confusion and discussion as to the position of the Na in sodium acetoacetic ester. Two facts seem to emerge: (1) the Na is attached to O; (2) the entering carbon group which "replaces" the Na is attached to C. In other words the conception of simple replacement must be revised. As we have seen, the apparently conflicting facts may be reconciled by assuming that the sodium derivative gives an organic ion which is capable of rearrangement[13] (p. 152).

After one radical has been introduced a sodium derivative may again be made and another radical introduced. In using these reactions it must constantly be kept in mind that the sodium derivative is a strongly alkaline reagent and that organic halides can react with alkaline reagents to give olefins. Thus better yields of substituted acetoacetic esters are obtained with primary halides than with secondary halides. Tertiary halides give olefins almost entirely. Steric influences also limit the usefulness of these reactions. Thus isopropyl halides give a fair yield of isopropyl acetoacetic ester. The sodium derivative of isopropyl acetoacetic ester reacts with more isopropyl halide to give propylene entirely.

An important use of the acetoacetic ester ketone synthesis involves beta diethylaminoethyl chloride or bromide to produce 5-Et$_2$N-2-pentanone, *Noval ketone*, which can be converted to *Noval alcohol, halides,* and *amine* by suitable reactions. These are intermediates for important antimalarials.

Sodium acetoacetic ester reacts with halogens to give diacetosuccinic ester. This may be regarded as an oxidation of the organic ions by the halogen.

$$
2\ \underset{|}{\overset{\text{ONa}}{\text{MeC}}}=\text{CHCO}_2\text{R} \rightarrow 2\left[\underset{|}{\overset{\text{O}\ominus}{\text{MeC}}}=\text{CH—CO}_2\text{R} \right] \rightleftharpoons
$$

$$
\left[\underset{|}{\overset{\text{O}}{\underset{}{\text{MeC}}}}\overset{\ominus}{-}\text{CH—CO}_2\text{R} \right] \xrightarrow{\text{I}_2} 2\left[\text{MeCO}\underset{|}{-}\text{CH—CO}_2\text{R} \right] + 2\ \text{I}^- \rightarrow \begin{array}{c} \text{MeCOCHCO}_2\text{R} \\ | \\ \text{MeCOCHCO}_2\text{R} \end{array}
$$

[12] Sidgwick. "The Electronic Theory of Valency." Oxford, 1927. p. 262.
[13] *Ann. Rep. Chem. Soc.* (London) 1908, 81; 1927, 108.

It may also be regarded as involving the formation of alpha iodoacetoacetic ester which then reacts with another mol of sodium acetoacetic ester to form the observed product.

Diacetosuccinic ester, as a *beta keto ester* gives reactions like those of acetoacetic ester itself and as a *gamma diketone* it gives 5-ring heterocyclic compounds. Thus with acid, with ammonia, and with primary amines it gives 2,5-Me$_2$-derivatives of 3,4-dicarboxylic derivatives of furan, of pyrrole and of N-substituted pyrroles.

$$\begin{array}{cc} CH\!-\!\!-\!CH & CH\!-\!\!-\!\!-\!CH \\ \| \quad \| & \| \qquad \| \\ Me\!-\!C\!-\!O\!-\!C\!-\!Me & Me\!-\!C\!-\!NH\!-\!C\!-\!Me \end{array}$$

$$\begin{array}{c} CH\!-\!\!-\!\!-\!CH \\ \| \qquad \| \\ Me\!-\!C\!-\!NR\!-\!C\!-\!Me \end{array}$$

3. Acetoacetic ester reacts with aldehydes in the presence of secondary amines[14] to form compounds of the type, MeCOCH—CHR—CHCOMe which

$$\qquad\qquad\qquad\qquad\quad \overset{|}{CO_2Et} \qquad \overset{|}{CO_2Et}$$

undergo the ketone and acid splittings characteristic of beta ketonic compounds and also give the reactions of delta keto esters and of delta (1:5) diketones to form 6-ring compounds. Thus from acetaldehyde can be obtained 3,5-dimethyl-4-carbethoxy-5-cyclohexenone.[15]

4. The ketone group in acetoacetic ester gives the usual reactions. In many cases the resulting product has an active H near the ester group. An internal condensation results in ring closure.

(*a*) Ammonia gives β-amino-crotonic ester.

$$\begin{array}{ccc} O & NH_2 \quad OH & NH_2 \\ \| & \diagdown\!\!\diagup & | \\ Me\!-\!C\!-\!CH_2CO_2R & \rightarrow \quad Me\!-\!C\!-\!CH_2\!-\!CO_2R & \rightarrow \quad Me\!-\!C\!=\!CH\!-\!CO_2R \end{array}$$

$$\qquad\qquad NH_2$$
$$\qquad\qquad |$$
The form $-C\!=\!C-$ is the ammonia analog of an enol. The equilibria

$$\begin{array}{cccccc} O & & OH & & & \\ \| & & | & & & \\ -C\!-\!CH & \rightleftharpoons & C\!-\!CH & \rightleftharpoons & -C\!=\!C & \rightleftharpoons & -C\!-\!CH \\ & & | & & | & & \| \\ & & NH_2 & & NH_2 & & NH \end{array}$$

are of great importance in the biological interconversion of ketonic and amino acids.

[14] Knoevenagel. *Ber.* **31**, 735 (1898); *Ann.* **281**, 104 (1894).
[15] *Org. Syn.* **27**, 24 (1947).

(b) Hydroxylamine.

$$\begin{array}{c} MeCO \\ | \\ CH_2CO_2R \end{array} \rightarrow \begin{array}{c} MeC=N-OH \\ | \\ CH_2CO_2R \end{array} \rightarrow$$

Methylisoxazolone

(c) Similarly, phenylhydrazine gives phenylmethylpyrazolone,

and sym-phenylmethylhydrazine gives phenyldimethylpyrazolone.

5. PCl_5 gives β-chlorocrotonic ester, $MeCCl=CHCO_2R$. This may indicate action with the enol form or the loss of HCl from the α—H and one of the Cl atoms of a dichloride related to the keto form. It may be recalled that acetone with PCl_5 gives Me_2CCl_2 and $MeCCl=CH_2$ with the latter predominating.

6. Acid halides in presence of pyridine react with the ester or its sodium derivative to give an ester of the enol form instead of a C— compound.

$$MeC(ONa)=CHCO_2Et \rightarrow MeC(OCOR)=CHCO_2Et$$

7. Strong heating of acetoacetic ester gives *dehydroacetic* acid, (Adams 1924), $MeC=CHCO$ ─, a keto form of 6-Me-3-acetyl-2-OH-γ-pyrone.

$$\begin{array}{c} | \\ O-COCHCOMe \end{array}$$

8. H_2O_2 with the sodium derivative oxidizes the α—H to give

$$MeCOCHOHCO_2Et \quad or \quad MeC(OH)=C(OH)CO_2Et.[16]$$

9. Among miscellaneous reactions of acetoacetic ester may be included those with the following:

        (a) Aldehyde ammonias → dihydropyridines
        (b) Ureas → uracils.
        (c) Amidines → pyrimidines.
        (d) Aromatic amines → dyes.
        (e) Phenols → coumarins.
        (f) Quinones → coumarones.

[16] *Ann. Rep. Chem. Soc.* (London) **1926**, 98.

Ethyl α-Br-acetoacetate is converted to the γ-Br-compound in presence of dry HBr.[17]

Many other *beta ketonic acids and esters* have been obtained in the same way as acetoacetic ester by the Claisen condensation. It must be remembered that the condensation involves the $\alpha-H$ of one molecule and the carbonyl group of another. In some cases in which NaOR is not an effective condensing agent, tritylsodium, $(C_6H_5)_3CNa$, is useful.[18]

Another preparation of *beta ketonic acids* involves the hydration of an $\alpha\beta$-triple bond by means of sulfuric acid. The preparation from an aliphatic acid follows:

$$RCH_2CH_2CO_2H \rightarrow RCH_2CHBrCO_2H \rightarrow RCH=CHCO_2H \rightarrow$$
$$RCHBrCHBrCO_2H \rightarrow RC\equiv CCO_2H \rightarrow$$
$$RC(OH)=CHCO_2H \rightarrow RCOCH_2CO_2H$$

Levulinic acid, γ-ketovaleric acid, β-acetylpropionic acid,

$$CH_3COCH_2CH_2CO_2H,$$

m. 33°, b. 246°.

Preparation. 1. It is available commercially from the action of acid on starch.[19]

2. From sucrose, dextrose and, especially levulose, by the action of dilute HCl or H_2SO_4.[20]

$$C_6H_{12}O_6 \rightarrow CH_3COCH_2CH_2CO_2H + CO + H_2O$$

3. From acetoacetic ester by means of an ester of a monohalogen acetic acid.

Reactions. Its reactions indicate that levulinic acid exists chiefly in a lactone form.

$$\begin{array}{c} CH_2{-}CO{-}Me \\ | \\ CH_2{-}CO_2H \end{array} \rightleftharpoons \begin{array}{c} OH \\ | \\ CH_2{-}C{-}Me \\ | \quad\quad O \\ | \quad / \\ CH_2{-}CO \end{array}$$

1. It forms salts and esters normally.

2. With acid chlorides, it gives γ-chloro-γ-valerolactone instead of a true acid chloride.

3. Its esters react with ammonia to give γ-amino-γ-valerolactone instead of an acid amide.

4. Acetic anhydride gives a definite crystalline acetate.

[17] Conrad. *Ber.* **29**, 1042 (1896).
[18] Renfrow, Hauser. *J. Am. Chem. Soc.* **60**, 463 (1938).
[19] Hands, Whitt. *J. Chem. Soc. Chem. Ind.* **66**, 415 (1948).
[20] Thomas, Schuette. *J. Am. Chem. Soc.* **53**, 2324 (1931).

5. Long heating gives monomolecular "anhydrides" (p. 352).

6. It can also be dehydrated to *angelica lactone*, MeC—O—CO. This

$$\underset{\underset{CH\text{———}CH_2}{|}}{MeC}\overset{||}{\text{—}}O\text{—}CO$$

forms dimers of unknown structure. Alcohols add readily in presence of traces of mineral acid to give cyclic pseudo esters, MeC—O—CO. These

differ from ordinary levulinic esters in being more soluble in water and in being quantitatively split by 0.1 N base.

7. Benzaldehyde condenses with a beta H in acid solution and with a delta H in basic solution (both are "alpha" to the CO group).

$$C_6H_5CHO + CH_3COCH_2CH_2CO_2Na \rightarrow C_6H_5CH = CH - COCH_2CH_2CO_2Na$$

8. Sodium amalgam gives sodium γ-hydroxyvalerate.

9. Neutral or acid reduction gives gamma valerolactone.

The commercial production of levulinic acid is about half a million pounds per year. It is used for calcium levulinate in pharmaceuticals, as esters in solvents and plasticizers, and for making valerolactone. As its uses expand, it will probably be made from cellulosic wastes.

Mesitonic acid, a homolog of levulinic acid is obtained by boiling mesityl oxide with KCN. Undoubtedly HCN first adds to the conjugated system and the resulting cyanide is then hydrolyzed.

$$MeCOCH = CMe_2 \rightarrow MeC(OH) = CHC(CN)Me_2 \rightarrow MeCOCH_2CMe_2CO_2H$$

γ-**Acetobutyric acid,** δ-ketocaproic acid, $CH_3COCH_2CH_2CH_2CO_2H$, is made by oxidizing the corresponding primary alcohol obtained by the acetoacetic ester synthesis from trimethylene dibromide.

$$Br(CH_2)_3Br \rightarrow MeCO(CH_2)_4Br \rightarrow MeCO(CH_2)_4OH \rightarrow MeCO(CH_2)_3CO_2H$$

It is best prepared through the Michael reaction of methyl acrylate and acetoacetic ester.[21] Reduction of the acid gives δ-hydroxy acid which then forms δ-caprolactone.

Higher ketoacids can be obtained by a variety of reactions. Two of these involve the acid chlorides of ester acids, $ClCO(CH_2)_nCO_2Et$:

1. Alkyl zinc halides give $RCO(CH_2)_nCO_2Et$. Unlike the Grignard re-

[21] Albertson. *J. Am. Chem. Soc.* **70,** 669 (1948).

agent, the alkyl zinc compounds do not react readily with ester or ketone groups. Cadmium alkyls may also be used.[22]

2. Sodiomalonic ester and sodioalkylmalonic esters, followed by hydrolysis and decarboxylation give $CH_3CO(CH_2)_nCO_2Et$ and $RCH_2CO(CH_2)_nCO_2Et$. Thus the ester chloride of adipic acid and malonic ester give a 48% yield of ϵ-ketoheptoic acid.

A third method is the action of solutions of mercuric salts or of sulfuric acid with acetylenic acids.[23] Thus 10-ketostearic acid can be obtained in the following steps starting with oleic acid.

$$Me(CH_2)_7CH = CH(CH_2)_7CO_2H \rightarrow -CHBrCHBr \rightarrow$$
$$-C \equiv C- \rightarrow Me(CH_2)_7CO(CH_2)_8CO_2H$$

Lactarinic acid, $CH_3(CH_2)_{11}CO(CH_2)_4CO_2H$ is related to petroselic acid and to tariric acid, the corresponding 5,6-olefinic and acetylenic acids.[24]

Geronic and **isogeronic acids** are $MeCO(CH_2)_3CMe_2CO_2H$ and

$$MeCOCH_2CMe_2(CH_2)_2CO_2H.[25]$$

C. HYDROXY ALDEHYDE ACIDS

The *-uronic acids* have a carboxyl in place of the primary alcohol group in an aldose. Thus their open chain formulas are $HO_2C(CHOH)_nCHO$ and they exhibit both aldose and acid reactions. They occur widely in plant and animal tissues as complex polyuronides partly as polymers in such materials as the algins of seaweed[26] and the pectins of fruit and partly combined with other materials as in the hemicelluloses and glycoproteins. Because of their wide occurrence in plants and animals they are found in soils.[27] The commonest of these glycuronic acids are D-*glucuronic acid* and D-*galacturonic* acid. D-Glucuronic acid is formed by oxidation in the animal organism. The aldehyde end of the molecule is apparently protected by glucoside formation. Thus when borneol is fed to a dog, borneol glucuronic acid is excreted.[28] Pectin consists largely of polygalacturonic acids.[29]

$$HO_2CCH(CHOH)_3CH \quad HO_2CCHCH(CHOH)_2CH \quad HO_2CCHCH(CHOH)_3CHOH$$

[22] Cason, Prout. *J. Am. Chem. Soc.* **66**, 46 (1944).
[23] *Ann. Rep. Chem. Soc.* (London) 1927, 88.
[24] *ibid.* 1925, 80.
[25] Simonsen. I, p. 92.
[26] Lucas, Stuart. *J. Am. Chem. Soc.* **62**, 1070 (1940).
[27] Shorey, Martin. *J. Am. Chem. Soc.* **52**, 4907 (1930).
[28] Quick. *J. Biol. Chem.* **98**, 537 (1932).
[29] Morell, Link et al. *J. Biol. Chem.* **110**, 719 (1935).

The hexuronic acids are easily decarboxylated by boiling with 12% HCl. Furfural is formed. D-Xylose and L-arabinose may be intermediates from D-glucuronic and D-galacturonic acids, respectively.[30]

D-Mannuronic acid has been made from the alginic acid of certain sea weeds.[31, 32]

Galacturonic acid can be prepared from beet pulp and from citrus fruits.[33]

Ascorbic acid obtained from the adrenal cortex, oranges, cabbages and other materials of high antiscorbutic action (Vitamin C) was formerly regarded as a hexuronic acid.[34] It is actually isomeric with such a substance, being the enediol form of a keto sugar acid instead of an aldehyde sugar acid. Ascorbic acid and its analogs have been the subject of intense study. It has been synthesized and found to have the same Vitamin C action as the natural product.[35, 36]

$$
\begin{array}{c}
\text{HO—C}=\text{C—OH} \\
| \qquad \qquad \diagdown \\
| \qquad \qquad \text{CO} \\
\text{H—C—O} \diagup \\
| \\
\text{HO—C—H} \\
| \\
\text{CH}_2\text{OH}
\end{array}
$$

L-ascorbic acid

The most striking property of ascorbic acid is its easily reversible oxidation. This may well involve the system

$$
\text{H—O—C}=\text{C—O—H} \underset{\text{red.}}{\overset{\text{oxid.}}{\rightleftharpoons}} \text{O}=\text{C—C}=\text{O}
$$

Synthetic Vitamin C is also called *cevitamic* acid (NNR).

XIX. DIBASIC ACIDS

A. SATURATED DIBASIC ACIDS

Oxalic acid, ethane diacid, HO_2CCO_2H, crystallizes as the ortho acid, $(HO)_3CC(OH)_3$. This is another example of the stability of more than one hydroxyl on a single carbon in a molecule containing strongly acid groups. The anhydrous acid can be sublimed. Oxalic acid is a much stronger acid than

[30] Lefèvre, Tollens. *Ber.* **40**, 4517 (1907).
[31] Nelson, Cretcher. *J. Am. Chem. Soc.* **54**, 3409 (1932).
[32] Schoeffel, Link. *J. Biol. Chem.* **95**, 213 (1932).
[33] Isbell. *J. Res. Nat. Bur. Stds.* **33**, 389 (1944). *ibid.* **33**, 401 (1944).
[34] Cox. *Science* **86**, 540 (1937).
[35] Ault et al. *J. Chem. Soc.* **1933**, 1419.
[36] *Ann. Rep. Chem. Soc.* (London) **1933**, 167; **1934**, 177.

its homologs. It naturally differs from them in many ways because it does not have its carboxyl groups attached to a hydrocarbon residue.

Preparation. 1. Heating formates under reduced pressure.

$$2 \ HCO_2Na \rightarrow H_2 + (CO_2Na)_2$$

2. Fusion of carbohydrates (cellulose, saw dust) with mixed NaOH and KOH. Units in the cellulose react as follows:

$$(\overset{|}{\underset{|}{C}HOH)_2} + 2 \ KOH \rightarrow (CO_2K)_2 + 3 \ H_2$$

The reaction mixture is extracted with water, the extract is treated with lime to regenerate the alkalies and form insoluble calcium oxalate which is converted to oxalic acid by H_2SO_4.

3. Oxidation of carbohydrates by nitric acid.

4. Hydrolysis of cyanogen, $(CN)_2$. This is carried out with water, or preferably, with concentrated HCl which gives oxamide first.

5. From CO_2 and hot sodium. This is like a pinacol reduction of a ketone

$$2 \ O{=}C{=}O + 2 \ Na \rightarrow 2 \ O{=}\underset{|}{C}{-}ONa \rightarrow \begin{matrix} OCONa \\ | \\ OCONa \end{matrix}$$

Reactions. 1. Heat gives CO_2 and formic acid and the latter decomposes partly into $CO + H_2O$. It is not possible to make oxalic anhydride.

2. Heating with glycerol gives formic acid or allyl alcohol depending on conditions.

3. Oxidizing agents like $KMnO_4$ convert oxalic acid to CO_2 and water. This is another example of the splitting of a carbon chain by adding two OH groups at a point where adjacent carbons are already partly oxidized.

$$(CO_2H)_2 + [O] + H_2O \rightarrow 2 \ HOCO_2H \rightarrow 2 \ H_2O + 2 \ CO_2$$

It reduces ferric compounds. It is thus valuable in removing iron rust and ink stains by changing them to the more soluble ferrous compounds. This same property is utilized in photography in making platinum prints. While potassium ferric oxalate, $K_3Fe(C_2O_4)_3$, is stable in the dark, it is readily reduced to the ferrous compound by light. Thus it is used on photographic paper which, after exposure, is treated with platinous chloride solution thus precipitating platinum wherever the ferric compound has been reduced by exposure to light. An abnormal reducing power is reported for a sample of oxalic acid treated with insufficient $KMnO_4$ to oxidize it completely.[1]

4. Electrolytic reduction with lead cathodes gives glycolic acid and some glyoxylic acid.

5. Dehydrating agents like concentrated H_2SO_4 give CO_2, CO and H_2O.

[1] *Ann. Rep. Chem. Soc.* (London) **1928**, 78.

6. Acid chlorides ordinarily act as dehydrating agents with oxalic acid. This is presumably because as soon as one OH is replaced the half acid chloride HOCOCOCl reacts internally instead of reacting with more of the reagent used.

$$
\begin{array}{c}
O=C-OH \\
| \\
O=C-Cl
\end{array}
\rightarrow HCl +
\left[
\begin{array}{c}
O=C \\
\diagdown \\
O \\
\diagup \\
O=C
\end{array}
\right]
\rightarrow CO + CO_2
$$

By using a very large excess of PCl_5 both OH groups can be replaced practically simultaneously and a fair yield of oxalyl chloride, $(COCl)_2$, m. 12°, b. 64°, obtained.[2]

7. (a) Alcohols readily give esters. MeO_2CCO_2Me, m. 54°.

(b) The esters can be partially saponified to the ester salts, RO_2CCO_2M.

(c) From the ester salts, free ester acids, ester acid chlorides, etc., can be made.

(d) Oxalic esters react with PCl_5 to give dichloro oxalic esters, RO_2CCCl_2-OR, which react with sodium alcoholates to give tetraalkyl oxalates, $RO_2CC(OR)_3$, esters of the unknown half ortho acid, $HO_2CC(OH)_3$. Tetramethyl and tetraethyl oxalates boil at 76° and 98° at 12 mm. respectively.

8. Oxalic acid yields an amide, a nitrile, etc., by the usual methods. It also forms mixed compounds such as *oxamic acid*, NH_2COCO_2H and its esters, cyanocarbonic esters, $NCCO_2R$, etc.

9. With heavy metal compounds, insoluble oxalates are formed. These are soluble in excess of soluble oxalates to give soluble complexes. Thus chromium gives a soluble potassium oxalochromate, $K_3Cr(C_2O_4)_3$. The physical reality of the complex ion is proved by the existence of this compound in optically active forms (Werner). The six valences of the three oxalate groups can occupy the six coordination points of the central metal in two ways. The products are enantiomorphous. The six coordination points may be regarded as at the points of a regular octahedron with the metal at the center.

10. Oxalic acid gives the expected normal and acid salts, $M_2C_2O_4$ and MHC_2O_4. It also gives a peculiar acid salt of which potassium tetroxalate,

[2] Staudinger. *Ber.* **41**, 3558 (1908).

$KHC_2O_4 . H_2C_2O_4 . 2 H_2O$, is an example. This salt is a valuable standard for bases and oxidizing agents because it can be prepared and kept in a high state of purity.

Oxalic acid is poisonous, presumably, because of the insolubility of calcium oxalate. The toxic nature of ethylene glycol when taken internally may be due to its oxidation to oxalic acid.

Rubeanic acid is dithio-oxamide, $HN = C - SH$. It is used for the spectro-

$$HN = C - SH$$

photometric determination of small amounts of copper.[3]

Malonic acid, propane diacid, $CH_2(CO_2H)_2$, m. 134°, is obtained from its ester which has important synthetic uses. The free acid decomposes when heated to about 160° to give CO_2 and acetic acid. In general, two carboxyls attached to the same carbon form an unstable arrangement.

$$= C(CO_2H)_2 + \text{heat} \rightarrow CO_2 + = CHCO_2H.$$

Substituted malonic acids decompose even more readily to give the corresponding substituted acetic acids, RCH_2CO_2H and $RR'CHCO_2H$. Malonic acid gives all the reactions of carboxylic acids in which either one or both carboxyls may be involved except the formation of an ordinary anhydride. Such an anhydride would have a 4- or 8-membered ring depending on whether one or two molecules took part in its formation. Malonic acid with P_2O_5 gives small yields of *carbon suboxide*, C_3O_2. An excess of methyl Grignard reagent converts carbon suboxide to 2,4,6-triacetylphloroglucinol.[4]

Malonic ester, ethyl malonate, $CH_2(CO_2Et)_2$, b. 198°.

Preparation. Chloroacetic acid is neutralized with sodium carbonate with cooling and is treated with NaCN. The resulting sodium cyanoacetate is "hydrolyzed" and esterified with alcohol and HCl.

$$CN - CH_2CO_2Na + 2 EtOH + 2 HCl \rightarrow CH_2(CO_2Et)_2 + NaCl + NH_4Cl$$

Reactions. 1. Usual ester reactions involving one or both of the ester groups.

2. Replacement of the methylene hydrogens as in acetoacetic ester. Although malonic ester exists practically entirely in the "keto" form, its sodium derivative has the enol structure, $EtOCO - CH = C(ONa)OEt$. Sodium malonic ester reacts with certain halides to give substitution on the methylene carbon. Such substituted malonic esters form sodium derivatives which can react again with halides. It must not be assumed that treatment of malonic ester with one equivalent each of Na and RX will give exclusively a mono-substitution product, $RCH(CO_2Et)_2$. Thus equimolar amounts of Na, malonic ester and benzyl chloride give the following ratios of mono- and dibenzyl malonic esters: in MeOH or EtOH about 2:1; in toluene from 2:1 to 12:1 and

[3] Center, MacIntosh. *Ind. Eng. Chem., Anal. Ed.* **17**, 239 (1945).
[4] Billman, Smith. *J. Am. Chem. Soc.* **61**, 457 (1939).

with no solvent about 1:1.[5] The formation of the disubstituted product is due to the equilibrium:

Na-Malonic ester + R-Malonic ester ⇌ Malonic ester + Na-R-Malonic ester

There is a competition between the two sodium derivatives for the unreacted halide. Fortunately the competition for the sodium is usually in favor of the malonic ester. With halides containing two or more carbons the mono- and di-substituted malonic esters can be separated by fractional distillation. Thus Et- and Et_2-malonic esters b. 201° and 223°. The methyl malonic esters offer a hopeless mixture because malonic esters, the Me- and Me_2-malonic esters b. 198°, 199° and 196°. If a Me and another group, R, are to be introduced into malonic ester, the latter should be introduced first and the R-Malonic ester should be purified before treatment with more Na and MeX. In general, in making RR′-malonic esters, the larger or more sterically hindered group should be introduced first. Instead of adding more Na and the other halide to the reaction mixture it is advisable to isolate and purify the mono-substituted malonic ester. Otherwise the final reaction product will contain R_2-malonic and $R′_2$-malonic esters in addition to the desired product. Such a mixture may be very difficult to separate. In making R_2-malonic esters the two equivalents of Na and RX may be added at once. It is not correct to assume that this procedure gives a disodiomalonic ester. Such a substance would have the *improbable* formula, $ROC(ONa)=C=C(ONa)OR$.

The limitations as to halides which can be used with Na-malonic ester are the same as with Na-acetoacetic ester. The inactivity of phenyl halides is one of the most serious of these because of the importance of phenylethylmalonic ester as an intermediate for phenobarbital. Many reactions have been evolved for making phenylethylmalonic ester for this purpose. Probably the best is the condensation of benzyl cyanide with ethyl carbonate in presence of RONa.[6]

$$PhCH_2CN + CO(OEt)_2 + RONa \rightarrow ROH + EtOH + Na[PhC(CN)CO_2Et]$$

Regardless of whether this compound has the Na attached to O or to N, treatment of it with ethyl bromide followed by acid alcoholysis gives the desired phenylethylmalonic ester.

The mono- and di- substituted malonic esters can be converted to the acids and then heated to give *mono- and di-substituted acetic acids* RCH_2CO_2H, R_2CHCO_2H and $RR′CHCO_2H$. An example of going *up the series* by means of the sodiomalonic ester synthesis is the preparation of the even normal acids and related substances from C_{20} to C_{30} starting with stearic acid. The main changes involved in each addition of C_2 follow:

$$RCO_2H \rightarrow RCH_2OH \rightarrow RCH_2X \rightarrow RCH_2CH(CO_2Et)_2 \rightarrow RCH_2CH_2CO_2H$$

[5] Dunn, Redemann, Lauritsen. *J. Am. Chem. Soc.* 54, 4335 (1932).
[6] Wallingford, Jones. *J. Am. Chem. Soc.* 64, 578 (1942).

3. Another replacement of the methylene hydrogens of malonic acid or its esters involves an aldol condensation with an aldehyde or ketone and a condensing agent such as piperidine or other secondary amine (Knoevenagel).

$$RCHO + H_2C(CO_2H)_2 \rightarrow [RCHOHCH(CO_2H)_2] \rightarrow RCH=C(CO_2H)_2$$

The intermediate "aldol" is unstable because of the reactive α–H. The resulting dibasic acid loses CO_2 readily. These reactions thus become a source for $\alpha\beta$-unsaturated acids.

The alpha-beta unsaturated malonic esters give sodium derivatives if they have at least one H on the gamma carbon. An ordinary 1,3-prototropic change provides a replaceable alpha H.[7]

$$CH_3CH=C(CO_2R)_2 \rightarrow CH_2=CH-C(CO_2R)=C(OR)ONa.$$

Similarly, the condensation product of cyclohexanone and malonic ester gives the sodium derivative of cyclohexenylmalonic ester from which alkylcyclohexenylmalonic esters can be prepared as intermediates for substituted barbituric acids (p. 436).

4. With urea and NaOEt as a condensing agent, the carbethoxyl groups react to give barbituric acid.

5. Sodium malonic ester, in common with all similar compounds such as sodium acetoacetic ester, sodium cyanoacetic ester, sodium nitromethane, sodium phenyl nitromethane, etc., adds to alpha beta unsaturated carbonyl compounds. The addition takes place to the conjugated system.

$$C=C-C=O + NaQ \rightarrow \underset{Q}{C}-C=C-ONa \xrightarrow{H^+} \underset{Q}{C}-C=C-OH \rightleftharpoons \underset{Q}{C}-CH-C=O$$

Since some of the enol present can act as the "acid" the addition of the keto esters can be induced by small amounts of NaOEt. One of the earliest examples of this reaction, although it was not so recognized at the time was the action of sodium malonic ester with ethyl alpha bromoisobutyrate. The expected result was *not* obtained. Instead of dimethylsuccinic acid, the product *obtained* was alpha methylglutaric acid. The alkaline sodium malonic ester first removed HBr from the very reactive tertiary halide to give α-methylacrylic ester, to which more sodium malonic ester then added.

$$Me_2CBrCO_2R \rightarrow CH_2=C(Me)CO_2R \rightarrow$$
$$(RO_2C)_2CH-CH_2CH(Me)CO_2R \rightarrow HO_2CCH_2CH_2CHMeCO_2H$$

Hundreds of cases of this type of addition to $\alpha\beta$-unsaturated compounds have been studied. The formation of a C–C linkage from an O–Na "linkage" is explained as in the action of the sodium derivatives with halides. In order for addition to take place, the Na must leave the rest of the molecule. This reactive fragment or ion can then rearrange before adding to the other end of

[7] Cope, Hoyle, Heyl. *J. Am. Chem. Soc.* 63, 1843–52 (1941).

the conjugated system.

$$\underset{\substack{|\\ C=C}}{ONa} \rightarrow Na^+ + \left[\underset{\substack{|\\ C=C}}{\overset{O\,\ominus}{|}} \longleftrightarrow \underset{\substack{\|\\ C-C}}{\overset{O}{\|}} \ominus \right]$$

Methylmalonic acid, isosuccinic acid, $CH_3CH(CO_2H)_2$.

Preparation. 1. From α-halogenpropionic acid through the α-cyano compound.

2. From sodium malonic ester and methyl iodide or bromide. Dimethylmalonic ester is formed at the same time. These esters cannot be separated because they boil at practically the same temperature. Conversion to the amides and careful crystallization gives pure $MeCH(CONH_2)_2$, m. 208°.[8]

3. It might be expected that ethylidene halides would react with cyanides to give the nitrile of methylmalonic acid, $CH_3CH(CN)_2$. Treatment of CH_3CHBr_2 with KCN followed by hydrolysis gives ordinary succinic acid, $HO_2CCH_2CH_2CO_2H$.

Reactions. Like those of malonic acid except that the loss of CO_2 takes place more readily.

Succinic acid, ethane dicarboxylic acid, butane diacid, $HO_2CCH_2CH_2CO_2H$, m. 183°, has been obtained by the distillation of scrap amber (hence its name) and by fermentation of tartaric acid.

Preparation. 1. By the electrolytic and catalytic reduction of maleic and fumaric acids obtained by the catalytic oxidation of benzene.

$$\underset{\substack{\|\\ CHCO_2H}}{\overset{CHCO_2H}{|}} + 2\,[H] \rightarrow \underset{\substack{|\\ CH_2CO_2H}}{\overset{CH_2CO_2H}{|}}$$

2. By hydrolysis of the dinitrile from ethylene chloride and NaCN.

Reactions. 1. Heat gives succinic anhydride, m. 120°, b. 261°, containing a 5-membered ring. A better preparation is by the action of thionyl chloride.

2. It gives many of the ordinary reactions of carboxylic acids either with one or both of its acid groups.

Ethyl succinate with Na undergoes a double Claisen reaction to give succinylosuccinic ester, succinosuccinic ester, 1,4-dicarbethoxy-2,5-diketocyclohexane, $RO_2C-CH-COCH_2$.

$$\underset{\substack{|\\ CH_2-COCHCO_2R}}{}$$

Ethyl succinate with alkoxides undergoes an aldol type of condensation with ketones to give itaconic acid derivatives.[9] The reaction is quite general:

$$R_2CO + \underset{\substack{|\\ CH_2COOC_2H_5}}{\overset{CH_2COOC_2H_5}{}} \xrightarrow{KOC(CH_3)_3} R_2C=\underset{\substack{|\\ CH_2COOK}}{\overset{CCOOC_2H_5}{}}$$

[8] *Ann. Rep. Chem. Soc.* (London) 1906, 97.
[9] Johnson, Goldman. *J. Am. Chem. Soc.* 66, 1030 (1944).

3. In addition to the expected mono and diamides it gives a cyclic imide on heating the anhydride with NH_3 gas.

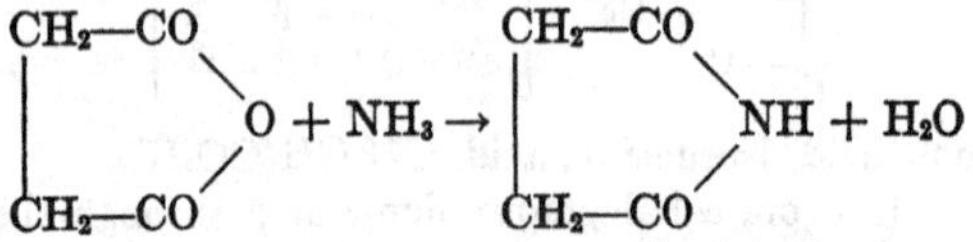

$$CH_2-CO\diagdown \quad CH_2-CO\diagdown$$
$$| \quad\quad O + NH_3 \rightarrow | \quad\quad NH + H_2O$$
$$CH_2-CO\diagup \quad CH_2-CO\diagup$$

Succinimide

On hydrolysis, the ring first opens to form succinamic acid,

$$NH_2COCH_2CH_2CO_2H.$$

Succinimide gives salts with the metal attached to the nitrogen. Chlorination gives N-succinchlorimide, a stable crystalline substance which reacts with water to give HOCl and, consequently has value as a disinfectant (NNR). The Cl attached to nitrogen is a "positive" chlorine.

Succinimide on distillation with zinc dust gives pyrrole. On reduction with sodium and alcohol it gives pyrrolidine (tetrahydropyrrole).

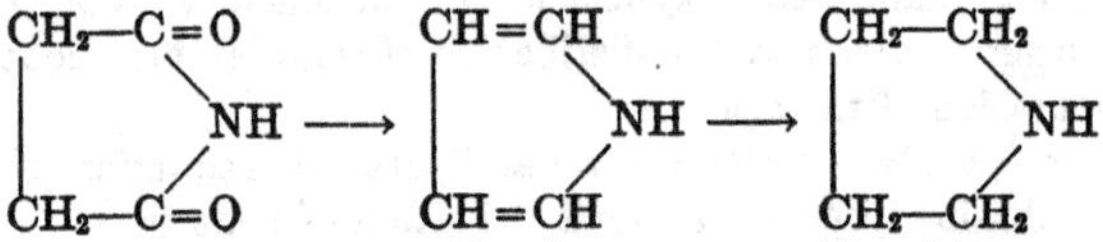

Pyrrole   Pyrrolidine

4. Treatment with PCl_5 gives a small amount of the expected dichloride, $ClCOCH_2CH_2COCl$, but mainly an unsymmetrical dichloride related to butyrolactone.

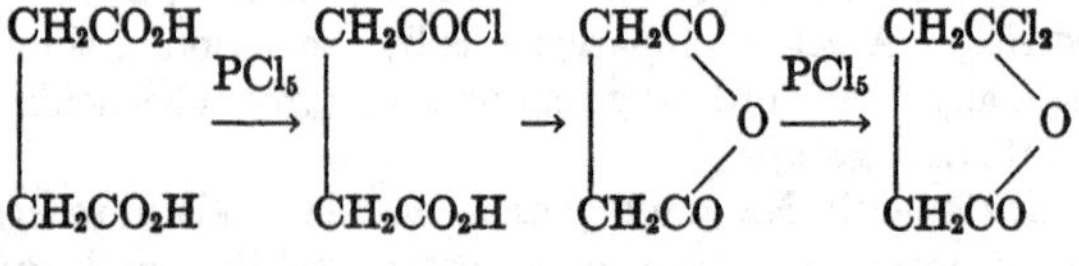

Evidence for this is that the product of PCl_5 with succinic acid reacts with benzene and $AlCl_3$ (Friedel-Crafts) to give

$$C_6H_5COCH_2CH_2COC_6H_5 \quad\quad (C_6H_5)_2C-CH_2-CH_2$$
$$\quad\quad\quad\quad\quad\quad\quad\quad\quad\quad\quad\quad\quad | \quad\quad\quad\quad | \quad 90\%$$
$$\quad\quad\quad 10\% \quad\quad\quad\quad\quad\quad\quad\quad\quad O \quad\text{------}\quad CO$$

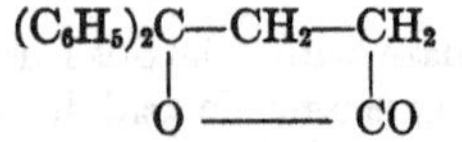

A better yield of the true symmetrical dichloride can be obtained by using a large excess of PCl_5 as in the preparation of oxalyl chloride, or by using $SOCl_2$ and $ZnCl_2$.[10]

[10] Ruggli, Maeder. *Helv. Chim. Acta* **26**, 1476 (1943).

5. The Perkin reaction gives *paraconic acids*,

$$\text{CH}_3\text{CHO} + \underset{\underset{\text{CH}_2\text{CO}_2\text{Na}}{|}}{\text{CH}_2\text{CO}_2\text{Na}} \xrightarrow{\text{Ac}_2\text{O}} \underset{\underset{\text{OH} \quad \text{CH}_2\text{CO}_2\text{Na}}{|}}{\text{CH}_3\text{CH}-\text{CHCO}_2\text{Na}}$$

When this reacts with acid, one of the carboxyl groups is very near the hydroxyl group in space and a lactonic acid results.

$$\text{HO}_2\text{C}-\underset{\underset{\text{CH}_2-\text{CO}}{|}}{\text{CH}}-\overset{\diagdown}{\underset{\diagdown}{\text{CH}}}-\text{CH}_3$$
$$\qquad\qquad\qquad \text{O}$$

Methylparaconic acid
(β-carboxy-γ-methylbutyrolactone)

Distillation converts this mainly into a beta-gamma unsaturated acid (see p. 264). The first step is probably decarboxylation to give valerolactone.

$$\underset{\underset{\text{CH}_2-\text{CO}}{|}}{\text{CH}_2}-\text{CH}-\text{CH}_3 \quad \rightleftharpoons \quad \underset{\underset{\text{CH}_2-\text{CO}}{|}}{\text{CH}_2}-\overset{\oplus}{\text{CH}}-\text{CH}_3 \quad \rightarrow \quad \underset{\underset{\text{CH}_2-\text{CO}_2\text{H}}{|}}{\text{CH}}=\text{CH}-\text{CH}_3$$
γ-Valerolactone   β-γ-unsaturated acid

Two other products obtained in smaller amounts are anhydrides of dibasic acids, homoitaconic and homocitraconic acids (the prefix homo- indicates the next higher homolog with no other change in structure). Writing the formula for methylparaconic acid with the lactone opened, gives:

$$\underset{\underset{\text{CH}_2-\text{CO}_2^\ominus}{|}}{\text{CH}_3\overset{\oplus}{\text{CH}}-\text{CH}-\text{CO}_2\text{H}} \rightleftharpoons \underset{\underset{\text{CH}_2\text{CO}_2^\ominus}{|}}{\text{CH}_3\text{CH}_2-\overset{\oplus}{\text{C}}-\text{CO}_2\text{H}}$$

$$\text{H}_2\text{O} + \underset{\underset{\text{CH}_2-\text{CO}}{|}}{\text{CH}_3\text{CH}}=\text{C}\text{---}\text{CO} \qquad\qquad \text{H}_2\text{O} + \underset{\underset{\text{CH}-\text{CO}}{|}}{\text{CH}_3\text{CH}_2\text{C}}-\text{CO}$$
$$\qquad\qquad\qquad\quad \text{O} \qquad\qquad\qquad\qquad\qquad\qquad\quad \text{O}$$

Homoitaconic anhydride   Homocitraconic anhydride
ethylidenesuccinic anhydride   ethylmaleic anhydride
"Methylitaconic anhydride"   "Methylcitraconic anhydride"

6. Sodium succinate and succinic anhydride react with P_2S_3 to give thiophene.

$$\underset{\underset{\text{CH}_2-\text{CO}_2\text{Na}}{|}}{\text{CH}_2-\text{CO}_2\text{Na}} \xrightarrow{P_2S_3} \underset{\underset{\text{CH}=\text{CH}}{|}}{\text{CH}=\text{CH}} \diagdown{\text{S}}$$

Alkenylsuccinic anhydrides, "ASAA," are available commercially (SPC). They are assigned formulas such as $Me_2C=CH-CH-CO$ These an-

$$O.$$

$$CH_2CO$$

hydrides have important applications in resins, modified drying oils, modified elastomers, plasticizers, detergents, and water repellents.

Methylsuccinic acid, pyrotartaric acid, $HO_2CCH(CH_3)CH_2CO_2H$, was originally obtained by distilling tartaric acid. Pyruvic acid is an intermediate product. The latter gives methylsuccinic acid when heated alone to 170° or at 100° with HCl. An even stranger formation of methylsuccinic acid is by the catalytic hydrogenation of sodium lactate solution under pressure.[11] The loss of a methyl group in the process is hard to explain.

Other methods of preparation. 1. From propylene dibromide through the dicyanide.

2. From acetoacetic ester or malonic ester by means of a methyl halide and a monohalogenacetic ester.

Its reactions are like those of succinic acid. It can be resolved into its optical isomers by means of its salts with the optically active base, strychnine.

sym-Dimethylsuccinic acids, $HO_2CCH(CH_3)CH(CH_3)CO_2H$, exist in two forms, the "para" acid and the "anti" acid. They are both formed in all methods of preparation and are separated by recrystallizing from water, the anti acid being three times as soluble as the para acid at ordinary temperatures. Since these acids have two similar asymmetric carbons, they would be expected to exist in D-, L- and *meso*-forms. The "para" acid (m. 192°) is believed to be a DL-mixture (racemic) while the "anti" acid (m. 122°) may be the *meso* form. All attempts to separate either acid into optically active forms have failed. The corresponding sym-diphenylsuccinic acids have been resolved into optically active forms.[12] Mixtures of the sym-dialkyl succinic acids are obtained from alpha bromopropionic ester (a) with sodium methylacetoacetic ester, (b) with sodium methylmalonic ester, (c) with finely divided silver and (d) with KCN. They are also obtained from sodium methylmalonic ester by treatment with halogens or by electrolysis. Another preparation is by the electrolysis of the ester salt of methylmalonic acid.

$$2\ CH_3-CH-CO_2K \rightarrow 2\left[CH_3-CH-CO_2\right] \rightarrow 2\ CO_2 + CH_3-CH-CH-CH_3$$
$$CO_2R \qquad\qquad CO_2R \qquad\qquad\qquad CO_2R\ CO_2R$$

This is a general method of building up dibasic acids.

unsym-Dimethylsuccinic acid, $HO_2CCH_2C(CH_3)_2CO_2H$.

Preparation. 1. From isobutylene dibromide through the dicyanide.

2. From ethyl beta-chloroisovalerate and KCN.

[11] *Ann. Rep. Chem. Soc.* (London) 1927, 85.
[12] Wren, Still. *J. Chem. Soc.* **107**, 449 (1915).

3. From solid sodium malonic ester and alpha-bromoisobutyric ester in xylene at 190°. If the reaction is carried out in refluxing absolute alcohol, part of the desired product is obtained but a considerable amount of alpha methylglutaric acid is formed by addition of sodium malonic ester to alpha-methylacrylic ester formed by the alkaline action on the bromo ester (p. 380).

4. From mesityl oxide through mesitonic acid.

$$CH_3COCH_2C(CH_3)_2CO_2K + 4\ KOH + 3\ Br_2 \rightarrow$$
$$CHBr_3 + KO_2CCH_2C(CH_3)_2CO_2K + 3\ KBr + 2\ H_2O$$

Glutaric acid, $HO_2C(CH_2)_3CO_2H$, is prepared (1) from trimethylene dibromide through the cyanide, (2) from methylenedimalonic ester or methylenediacetoacetic ester prepared from formaldehyde, (3) from cyclopentanone by oxidation with nitric acid. Glutaric acid resembles succinic acid very closely. It forms a cyclic anhydride and a cyclic imide. Treatment of the silver salts of substituted glutaric acids with I_2 gives substituted butyrolactones.[13]

$$\begin{matrix} CH_2CH_2CO_2Ag \\ | \\ CH_2CO_2Ag \end{matrix} + I_2 \rightarrow 2\ AgI + CO_2 + \begin{matrix} CH_2CH_2O \\ |\qquad | \\ CH_2 \!-\!\! CO \end{matrix}$$

Higher dibasic acids are relatively easily available.

(a) **Adipic acid,** $HO_2C(CH_2)_4CO_2H$, by oxidation of cyclohexanol.[14] Since the development of Nylon, this acid has been of tremendous interest. It forms one component directly and the other indirectly (by conversion to hexamethylene diamine). The original preparation was from benzene to phenol to cyclohexanol which was then oxidized by nitric acid. This step was soon replaced by air oxidation. The latter was then applied to cyclohexane made by hydrogenating benzene. These methods are now being replaced by processes based on agricultural byproducts, particularly oat hulls. The steps involved are oat hulls $\rightarrow$ furfural $\rightarrow$ furan $\rightarrow$ tetrahydrofuran $\rightarrow$ 1,4-dichlorobutane $\rightarrow$ adiponitrile and adipic acid $\rightarrow$ nylon. Treatment of adipic acid with NH_3 at high temperatures gives first adipic diamide and then adipodinitrile. Either of these gives hexamethylene diamine on hydrogenation.

(b) **Pimelic acid,** $HO_2C(CH_2)_5CO_2H$, from pentamethylene dibromide through the cyanide process or from trimethylene dibromide through the malonic ester or acetoacetic ester synthesis. In this process, a cyclic byproduct is obtained. The γ-bromopropylmalonic ester first formed reacts with sodium malonic ester to form free malonic ester and its own sodium derivative. This can react internally to give a cyclobutane derivative.

$$\begin{matrix} & & OR \\ & & \diagup \\ BrCH_2CH_2CH_2C\!=\!C\!-\!ONa \\ & | \\ & CO_2R \end{matrix} \rightarrow \begin{matrix} CH_2\!-\!C(CO_2R)_2 \\ |\qquad\quad | \\ CH_2\!-\!CH_2 \end{matrix} + NaBr$$

[13] *Ann. Rep. Chem. Soc.* (London) **1923**, 91.
[14] "Org. Syntheses."

Pimelic acid is a precursor of biotin in its biosynthesis.

(c) **Suberic acid,** $HO_2C(CH_2)_6CO_2H$, is not readily available. It is formed by the oxidation of cork or castor oil, and by the electrolysis of ethyl potassium glutarate. It is also obtained in the action of trimethylene bromide with magnesium and CO_2. This is a combined Wurtz and Grignard reaction.

If needed in quantity, suberic acid could probably be made by hydrogenating adipic ester with zinc chromite to the $\alpha\omega$-glycol 1,6-hexanediol, converting the latter to the dihalide and treating with NaCN to make suberonitrile.

(d) **Azelaic acid,** $HO_2C(CH_2)_7CO_2H$, is obtained from the oxidation of oleic acid or from pentamethylene dibromide with Na malonic ester or Na acetoacetic ester. The cyclic by-product is obtained in larger amounts than with trimethylene dibromide.

(e) **Sebacic acid,** $HO_2C(CH_2)_8CO_2H$, is obtained by heating a castor oil soap with sodium hydroxide.

$$CH_3(CH_2)_5CHOHCH_2CH=CH(CH_2)_7CO_2Na \xrightarrow{\text{NaOH}}$$
$$NaO_2C(CH_2)_8CO_2Na + C_6H_{13}CHOHCH_3$$
$$\text{capryl alcohol}$$

Sebacic acid was produced commercially in Germany during World War II by treatment of 4,4'-dichlorodibutyl ether with metallic sodium, followed by oxidation of the resulting product with nitric acid.

Many higher acids of this series are known. *Brassilic* and *thapsic acids* are the C_{13} and C_{16} members. The former is obtained by oxidation of erucic and brassidic acids. *Japanic acid,* nonadecane 1,19-dicarboxylic acid, is obtained from Japan wax.[15]

Dibasic acids with an even number of carbons are readily made by Kolbe's electrolytic method from $EtO_2C(CH_2)_nCO_2K$. Other syntheses start with the dibasic acids and involve the following steps.[16]

(a) $\rightarrow -CO_2Et \rightarrow -CH_2OH \rightarrow -CH_2Br$

(b) $-CH_2Br \xrightarrow[\text{ester}]{\text{Na malonic}} -CH_2CH_2CO_2H$ Etc.

(c) $-CH_2MgBr \xrightarrow{ClCH_2OMe} -CH_2CH_2OMe \xrightarrow{HBr} -CH_2CH_2Br$ etc.

When the calcium or, better, the thorium salts of the higher dibasic acids (C_6 to C_{19}) are distilled cyclic ketones result in varying yields.[17] The best yields are obtained with the 5- and 6-ring ketones from adipic and pimelic

---

[15] *Ann. Rep. Chem. Soc.* (London) **1928**, 79.
[16] *ibid.* **1926**, 101.
[17] Ruzicka. *Ann. Rep. Chem. Soc.* (London) **1926**, 112.

acids. As usual, substituting groups force the carboxyl groups nearer in space and favor ring closure. Thus β-Me-adipic acid when heated with CuO gives an 83% yield of 3-Me-cyclopentanone.

The formation of anhydrides from the dibasic acids from C_6 to $_{18}$ has been studied in detail.[18] In addition to rings formed from one and two molecules, higher linear polymers are obtained. Suberic (C_8) and the C_{12} acid give anhydrides containing 18- and 26-member rings of m. 57° and 78° which are stable up to the melting points. The polymeric anhydrides, even with adipic acid are the readily obtained forms. Their polymeric nature may be shown by the action of aniline. Thus the ordinary anhydride of adipic acid when treated with aniline gives adipic mono- and di-anilides and adipic acid while the true monomeric 7-ring anhydride, obtained by vacuum distillation of the polymer, reacts with aniline to give only one product, the monoanilide of adipic acid.

The 15-ring anhydride has a musk odor like the 15-member cyclic ketone and lactone.[19] The effect of alkyl groups on ring closure and stability is again striking. Thus the anhydrides of adipic and β-Me-adipic acids are readily hydrolyzed while those of the Me$_4$-adipic acids are stable to boiling water.[20]

Certain reactions of the derivatives of the dibasic acids differ from those of the monobasic acids because of spatial relations. Thus adipic ester can undergo an internal acetoacetic ester condensation to give cyclopentanone α-carboxylic ester,[21]

$$CH_2CH_2CO$$
$$CH_2 \longrightarrow CHCO_2Et$$

A remarkable reaction is that of the $\alpha\alpha'$-Br$_2$-adipic ester with NaCN to give 1-cyanocyclobutane-1,2-dicarboxylic ester $CH_2C(CN)CO_2Et$, which on hy-

$$CH_2CHCO_2Et$$

drolysis loses CO_2 and give *trans*-cyclobutane 1,2-dicarboxylic acid.[22] This reaction is like that of α-Br-propionic ester with KCN which gives mainly a dimolecular product, $EtO_2CCH(Me)C(Me)(CN)CO_2Et$. An even more surprising reaction of this α-dibromo ester is with secondary amines to give $R_2NCH_2CH_2CO_2Et$ and CH_3COCO_2Et.

The dibasic acids, beginning with glutaric, can be *identified* as the phenacyl or *p*-bromophenacyl esters.[23]

$$(CH_2)_n(CO_2Na)_2 + 2\ BrCH_2COAr \rightarrow (CH_2)_n(CO_2CH_2COAr)_2.$$

[18] Hill, Carothers. *J. Am. Chem. Soc.* 55, 5023 (1933).
[19] Ruzicka. *Bull. soc. chim.* 43, 1145 (1928).
[20] *Ann. Rep. Chem. Soc.* (London) 1927, 84.
[21] Dieckmann. *Ann.* 317, 51 (1901).
[22] Fuson et al. *J. Am. Chem. Soc.* 51, 599, 1536 (1929); 52, 2985 (1930).
[23] Kelly, Kleff. *J. Am. Chem. Soc.* 54, 4444 (1932).

For various values of n the phenacyl esters m. 4, 87°; 5, 72°; 6, 102°; 7, 69°; 8, 80° while the p-Br-phenacyl esters m. 3, 136°; 4, 154°; 5, 136°; 6, 144°; 7, 130°. The usual alternation in melting points is striking.

Ionization constants of dicarboxyl compounds and structure.[24]

B. Unsaturated Dibasic Acids

Methylenemalonic ester, $CH_2 = C(CO_2Et)_2$, is obtained from methylene iodide or bromide with malonic ester and an excess of sodium ethylate. Hydrolysis gives acrylic acid, decarboxylation taking place spontaneously.

Alkylidenemalonic esters, $RCH = C(CO_2Et)_2$ are obtained from aldehydes and malonic ester in the presence of acetic anhydride. The acids are also obtained by the action of aldehydes with malonic acid and a secondary amine. Hydrolysis of the esters and decarboxylation of the acids gives

$$RCH = CHCO_2H.$$

Ethoxymethylenemalonic ester, "EMME," $EtOCH = C(CO_2Et)_2$, is obtained from ethyl orthoformate and malonic ester with Ac_2O and $ZnCl_2$.[25] It is an important intermediate for quinoline derivatives.[26]

Maleic acid, *cis* butenediacid, $H—C—CO_2H$, and **fumaric acid,** *trans*
$$H—C—CO_2H$$

butenediacid, $HO_2C—C—H$, form the most important example of geo-
$$H—C—CO_2H$$

metric isomerism. These acids are both obtained from monohalogen succinic acids and by the dehydration of malic acid (hydroxysuccinic acid). Long heating of the latter at 150° gives fumaric acid. Further heating at 200° gives maleic anhydride. Commercially they are made by the high temperature catalytic oxidation of benzene.[27] They are similarly obtainable from chlorinated hydrocarbons.[28] In Germany during World War II, crotonaldehyde was oxidized using a catalyst containing oxides of Ti, Mo and V, to give a yield of 60 kg. from 100 kg. of the aldehyde.[29]

On reduction maleic and fumaric acids give succinic acid and with halide acids they give the same monohalogen succinic acids. On mild oxidation with $KMnO_4$, maleic acid gives *meso* or internally compensated tartaric acid while fumaric acid gives racemic acid (DL-tartaric acid). Maleic acid, on heating alone or, better, with P_2O_5 or acetyl chloride, gives maleic anhydride. Fumaric

[24] Greenspan. *Chem. Rev.* 1933, I, 339.
[25] Fuson et al. *J. Org. Chem.* 11, 194 (1946).
[26] Price, Roberts. *J. Am. Chem. Soc.* 68, 1204 (1946).
[27] Ellis. "Chemistry of Petroleum Derivatives." Reinhold, 1934. p. 916.
[28] Faith. *Ind. Eng. Chem.* 37, 438 (1945).
[29] Office Publication Board No. 611; 690.

acid, on heating sublimes with a partial change into an anhydride which reacts
with water to give maleic acid.   Long heating, exposure to light, and treatment
with traces of chemical reagents like HCl or I_2 convert maleic to fumaric acid,
which is the more stable form.   The change appears to be related to an
"activation" of the double bond.[30]   Thus a trace of potassium will convert Me
maleate to the fumarate.   The conversion to fumaric acid may also be induced
by chemical reactions in which the acids apparently take no part.   Thus,
neither H_2S nor SO_2 has any effect on maleic acid but the addition of both
reagents converts it to fumaric acid.   Similarly, copper maleate with H_2S
gives fumaric acid.[31]   This effect may be due to "chemical" energy liberated
by the reaction or to some specific effect not now understood.   The reverse
change is brought about by ultraviolet light which converts fumaric acid to
maleic acid, the compound richer in energy.   The best method for this change
is to treat fumaric acid with P_2O_5 or acetyl chloride, distill the maleic anhydride
formed, and treat it with the calculated amount of water.

The properties of the acids are distinctly different.

	Maleic Acid	Fumaric Acid
m.	130°	289° (pressure)
b.	160° dec.	200° subl.
Solubility in cold H_2O	readily soluble	almost insoluble
Taste	disagreeable	acid
Dissociation K $\times 10^{-2}$	1.170	0.093

The isomerism of maleic and fumaric acids is expressed in the space formulas.

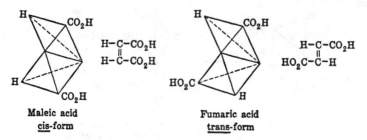

Maleic acid
cis-form

Fumaric acid
trans-form

In the *cis* form, it is possible for the carboxyls to approach each other in
space.   Thus the formation of an anhydride is easy.   Another evidence that
the assignment of these configurations is correct is that the more stable form
should have similar groups as far apart as possible.   The formation of maleic
acid directly by the oxidation of quinone also indicates the *cis* configuration for

[30] *Ann. Rep. Chem. Soc.* (London) 1925, 76.
[31] *Ann. Rep. Chem. Soc.* (London) 1925, 109.

that acid.

The conversion of the two acids into each other is not understood, largely because we know so little about the true nature of a "double bond."

The action of $KMnO_4$ on maleic and fumaric acids is entirely consistent with the space formulas assigned.

Identical
mesotartaric acid

Thus the *cis* form gives the same product, whichever end of the double bond is opened.

Not identical, enantiomorphs
D-tartaric acid   L-tartaric acid

Thus the *trans* form gives D- or L-tartaric acid depending on the end of the double bond opened.

A similar conversion of maleic and fumaric acids to tartaric acids without optical inversion can be produced by treatment with a chlorate and "osmic acid," OsO_4.[32] By this method, fumaric acid gives racemic acid while maleic acid gives mesotartaric acid.

[32] Milas, Terry. *Ann. Rep. Chem. Soc.* (London) 1925, 76.

The addition of HOCl to maleic acid gives only *one product*, a chloromalic acid, m. 145°, while fumaric acid gives in addition an isomeric chloromalic acid, m. 153°. Ease of replacement of Cl by OH in the two acids is in the approximate ratio 10,000:1.[33]

Another fact to show that the formulas used for multiple bonds give only a partial explanation is that acetylene dicarboxylic acid, obtainable from both dibromosuccinic acids, adds halogens and halide acids to give substituted *fumaric* acids instead of the expected *cis* compounds.[34]

Acetylene dicarboxylic acid

Contrary to an obvious prediction, HX is more easily removed from the *trans* position than from the *cis*. Thus chlorofumaric acid loses HCl to give acetylene dicarboxylic acid about fifty times as fast as does chloromaleic acid.[35]

Chloromaleic anhydride, m. 34°, is available commercially (NAD).

Alkyl fumarates are readily obtainable from maleic anhydride, alcohols and acid catalysts. One of their most important reactions is with sodium bisulfite by 1,4-addition to give esters of sodium sulfosuccinic acid,

$$ROCO-CH(SO_3Na)CH_2-CO_2R,$$

a very useful class of detergents (Nacconals) (NAD). Some forty modifications of this type are said to be in use.

An important reaction of maleic anhydride is its functioning as the $\alpha\beta$-unsaturated compound in the Diels-Alder reaction with conjugated dienes. This reaction is valuable synthetically and as a means of forming derivatives of the dienes for identification purposes. Thus butadiene, isoprene, furan, cyclopentadiene, cyclohexadiene, anthracene, etc., react with maleic anhydride to form tetrahydrophthalic anhydride and its substitution products (Diels, Alder).[36]

[33] *Ann. Rep. Chem. Soc.* (London) 1925, 77.
[34] Michael. *J. prakt. Chem.* 52, 289 (1895).
[35] *ibid.* 52, 305 (1895).
[36] Kloetzel. "Organic Reactions," Vol. IV. Wiley, 1948. p. 1.

More complex conjugated butadienes may fail to give such products. Examples of such failure are given by 1,4- and 4,4-disubstituted butadienes and by cis-1,3-pentadiene (piperylene).[37]

Citraconic acid, methylmaleic acid, CH_3-C-CO_2H

$$H-C-CO_2H$$

Mesaconic acid, methylfumaric acid, CH_3-C-CO_2H

$$HO_2C-C-H$$

Itaconic acid, methylenesuccinic acid, $CH_2=C-CO_2H$

$$CH_2-CO_2H$$

These isomeric acids are obtained by heating citric acid. Careful heating gives itaconic anhydride.[38,39] Citric acid first loses H_2O giving aconitic acid, which then loses CO_2.

| | Citric acid | Aconitic acid | Itaconic anhydride |

The itaconic anhydride rearranges partially into citraconic anhydride during the heating.

The amount of the latter in the mixture obtained from citric acid is increased by protracted heating. When itaconic or citraconic acids or anhydrides are boiled for a long time with 10% NaOH, the chief product is sodium mesaconate,

CH_3-C-CO_2Na. Citraconic and mesaconic acids can be converted into

NaO_2C-C-H

each other like maleic and fumaric acids but somewhat less readily. The methyl group facilitates the closing of the anhydride ring. Thus mesaconic acid with acetyl chloride readily gives citraconic anhydride.

The equilibrium between these acids is as follows:

Itaconic acid ⇌	citraconic acid ⇌	mesaconic acid
16%	15%	66%

[37] Robey. *Science* 96, 470 (1942).
[38] "Org. Syntheses."
[39] *Ann. Rep. Chem. Soc.* (London) 1912, 89.

The triad prototropic system is very labile. Thus equilibrium is reached rapidly.

All three acids give methylsuccinic acid on reduction.

Dimethylmaleic acid, pyrocinchoninic acid, a decomposition product of cinchonine, exists only as the anhydride, another example of the effect of methyl groups on ring closure. A remarkable preparation of this substance is from ethylenetetracarboxylic acid which can be made from its ester obtained by boiling a benzene solution of bromomalonic ester with K_2CO_3. Gentle heating gives fumaric acid while very rapid heating at a high temperature gives Me_2-maleic anhydride

$$2\ (HO_2C)_2C = C(CO_2H)_2 \rightarrow 6\ CO_2\ +\ \begin{matrix} MeCCO \\ \parallel \qquad \searrow \\ \qquad \qquad O\ +\ H_2O \\ \parallel \qquad \nearrow \\ MeCCO \end{matrix}$$

Ethylmaleic acid is notable for its easy conversion to methyl itaconic acid.

$$\begin{matrix} MeCH_2{-}CCO_2H \\ \parallel \\ HCCO_2H \end{matrix} \quad \rightarrow \quad \begin{matrix} MeCH = CCO_2H \\ \mid \\ H_2CCO_2H \end{matrix}$$

Methylethylmaleic acid is obtained from hematin.

Diethylmaleic acid, xeronic acid, is obtained as the anhydride by long heating of citraconic anhydride. Its structure is proved by its preparation from $\alpha\alpha$-Br_2-butyric acid and Ag.

Glutaconic acid, $HO_2C - CH = CH - CH_2 - CO_2H$.

Preparation.   1. From citric acid

$$\begin{matrix} CH_2CO_2H \\ \mid \\ C(OH)CO_2H \\ \mid \\ CH_2CO_2H \end{matrix} \xrightarrow{H_2SO_4} \begin{matrix} CH_2CO_2H \\ \mid \\ CO \\ \mid \\ CH_2CO_2H \end{matrix} \rightarrow \begin{matrix} CH_2CO_2H \\ \mid \\ CHOH \\ \mid \\ CH_2CO_2H \end{matrix} \rightarrow \begin{matrix} CHCO_2H \\ \parallel \\ CH \\ \mid \\ CH_2CO_2H \end{matrix}$$

Acetone
dicarboxylic acid

2. From malonic ester and chloroform.

$$2\ CH_2(CO_2R)_2 + CHCl_3 + NaOC_2H_5 \rightarrow (RO_2C)_2C = CH{-}CH(CO_2R)_2$$

$$\downarrow$$

$$HO_2C{-}CH = CH{-}CH_2{-}CO_2H$$

ISOMERISM AND STRUCTURE OF GLUTACONIC ACIDS

Although glutaconic acid belongs to the same type of ethylenic compounds as maleic and fumaric acids, it is known in only one form. This is apparently because of the mobile three carbon prototropic system,

$$-CH = CH - CH_2 - \ \rightleftharpoons\ -CH_2 - CH = CH -$$

The mobility of the H would make impossible a stable *cis*- or *trans*- configuration. The easy formation of glutaconic anhydride by the action of acetyl chloride may indicate the existence of the acid in the *cis*- form or merely the lability of the system by which it can change to the *cis*- form as rapidly as it is removed as anhydride. The anhydride gives a red color with ferric chloride. Consequently it is formulated as having a hydroxyl group.

$$CH_2-CO_2H \qquad CH=C-OH$$
$$| \qquad\qquad\qquad |$$
$$CH \qquad \rightarrow CH \qquad O$$
$$|| \qquad\qquad\qquad ||$$
$$CH-CO_2H \qquad CH-C=O$$

The tendency to form a 6-ring is strengthened by the tendency to conjugation.

The methylene group of glutaconic ester reacts with sodium, sodium ethylate and sodamide like the methylene group in acetoacetic ester and similar compounds. Thus the olefin linkage serves the same role as a carbonyl, a cyanide, or a nitro group.

$$RO_2C-CH=CH-CH_2-CO_2R \rightarrow RO_2C-CH=CH-CH=C-ONa$$

with OR above the final carbon.

Sodium glutaconic ester with methyl halides gives α-methylglutaconic ester. Ozonization shows this to be a mixture of $RO_2C-CH=CH-CHMe-CO_2R$ and $RO_2C-CH_2-CH=CMe-CO_2R$. This does not indicate that the sodium derivative is a mixture but rather that the 3-carbon system shifts:

$$-CH=CH-CH(CH_3)- \rightleftharpoons -CH_2-CH=C(CH_3)-$$

It should be noted that the H atoms at the ends of the triad system in glutaconic acid and its derivatives are *alpha H* atoms and are correspondingly mobile. When the ester is hydrolyzed, the acid is found by ozonolysis to consist entirely of the second form with the methyl on the unsaturated carbon. α-Methylglutaconic acid reacts with acetyl chloride to give a "hydroxy anhydride."

$$CH_2-CO_2H \qquad\qquad CH=C-OH$$
$$| \qquad\qquad\qquad\qquad |$$
$$CH \qquad\qquad\qquad CH \qquad O$$
$$|| \qquad\qquad\qquad\qquad ||$$
$$CH_3-C-CO_2H \qquad CH_3-C—C=O$$

This gives a red coloration with FeCl₃ and can be titrated with dilute bases as a monobasic acid. With concentrated KOH it forms a di-potassium salt which gives "labile" (*cis*) α-methylglutaconic acid. When this is boiled with HCl, the stable (*trans*) acid is formed. The existence of *cis* and *trans* isomers of α-methylglutaconic acid and not of glutaconic acid itself is due to the effect of the methyl group in "freezing" the tautomerism of the three carbon system.

Thus while the system: $-CH=CH-CH_2- \rightleftharpoons -CH_2-CH=CH-$ allows the permanent existence of neither a *cis* nor a *trans* form, the system: $-CH=CH-CH(Me)- \rightleftharpoons -CH_2-CH=C(Me)-$ is sufficiently irreversible in the free acid to allow *cis* and *trans* isomers.

When α-methylglutaconic ester is treated with more sodium and CH_3I, two products are obtained, the αα'-dimethyl (I) and the αα-dimethyl (II) esters.

$$
\begin{array}{c}
\overset{\text{Me}}{\underset{\displaystyle\updownarrow}{\text{RO}_2\text{C}-\text{C}}}=\text{CH}-\text{CH}=\overset{\text{OR}}{\text{C}}-\text{ONa} \rightarrow \text{RO}_2\text{C}-\overset{\text{Me}}{\text{C}}=\text{CH}-\overset{\text{Me}}{\text{CH}}-\text{CO}_2\text{R} \\
\text{(I)}
\end{array}
$$

$$
\begin{array}{c}
\text{NaO}-\overset{\text{OR}}{\text{C}}=\overset{\text{Me}}{\text{C}}-\text{CH}=\text{CH}-\text{CO}_2\text{R} \rightarrow \text{RO}_2\text{C}-\underset{\text{Me}}{\overset{\text{Me}}{\text{C}}}-\text{CH}=\text{CH}-\text{CO}_2\text{R} \\
\text{(II)}
\end{array}
$$

Of the two esters, (I) gives an acid which exists in only one form. The symmetrical arrangement of the two methyl groups gives no favored position for the double bond, the two tautomers being identical. The αα'-dimethyl acid gives a normal anhydride instead of a hydroxy anhydride.

$$
\begin{array}{c}
\text{Me}-\text{CH}-\text{CO} \\
| \qquad\qquad \backslash \\
\text{CH} \qquad\quad \text{O} \\
\| \qquad\quad / \\
\text{Me}-\text{C}-\text{CO}
\end{array}
$$

Ester (II) gives an acid with *cis* and *trans* forms. There is no possibility of a shift of its double bond. Ester (I) with more NaOEt and MeI gives ααα'-Me$_3$-glutaconic ester. The corresponding acid has the same "blocked" arrangement as in (II) and exists in two forms. Ester (II) does not form a sodium derivative. Apparently the grouping $C=C=\overset{\text{OR}}{C}-\text{ONa}$ is impossible.

β-Methylglutaconic acid is made from malonic ester and acetaldehyde.[40]

$$2\,CH_2(CO_2R)_2 + MeCHO \rightarrow MeCH[CH(CO_2R)_2]_2 \xrightarrow{BR_2} \xrightarrow{KOH} \xrightarrow{acid} \xrightarrow{heat}$$

$$HO_2CCH_2-\underset{\text{Me}}{\text{C}}=CHCO_2H$$

The β-methyl acid obtained in this way is the *trans* form. Treatment with acetyl chloride gives an anhydride which, in turn, gives the *cis* form. This, heated with acid, changes to the *trans* form.

[40] Knoevenagel. *Ber.* **31**, 2587 (1898).

Treatment of the β-methyl ester with sodium ethylate and methyl iodide gives $\alpha\beta$-dimethyl, $\alpha\beta\gamma$-trimethyl and $\alpha\alpha\beta$-trimethylglutaconic esters. The first two of these give only one form of acid each. Thus the groupings

$$\overset{\text{Me}}{\underset{|}{}}\overset{\text{Me}}{\underset{|}{}} \qquad \overset{\text{Me}}{\underset{|}{}}\overset{\text{Me}}{\underset{|}{}}\overset{\text{Me}}{\underset{|}{}}$$
$$-C=C-CH_2- \quad \text{and} \quad -C=C-CH-$$

allow a shift of a hydrogen atom. The $\alpha\alpha\beta$-trimethyl acid exists in *cis* and *trans* forms.

Glutaconic ester and both *cis* and *trans* β-methylglutaconic esters add sodium cyanoacetic ester normally.[41]

$$\overset{CH_3}{\underset{|}{}} \qquad \overset{CH_3}{\underset{|}{}}$$
$$RO_2CCH=CCH_2CO_2R \rightarrow RO_2CCH_2-CCH_2CO_2R \rightarrow CH_3C(CH_2CO_2R)_3$$
$$\underset{|}{} $$
$$CH(CN)CO_2R$$

The failure to obtain this addition, together with other unusual relations of the glutaconic acids, formerly encouraged the use of unorthodox formulas for these substances.[42] Later the use of such formulas has been shown to be unnecessary.[43] All the relations of the glutaconic acids and their esters can be interpreted in terms of the ordinary conceptions of tautomerism. The unusual property of this type of substance which was not earlier grasped is the combination in a single molecule, of ordinary keto-enol tautomerism and the tautomerism of the "three-carbon" system as found in allyl carbinols and their halides. Thus in a glutaconic ester, the following equilibria are possible.[44]

$$\overset{H}{\underset{|}{}}$$
$$H-O-C=C-C=C-C=O \rightleftharpoons O=C-C-C=C-C=O \rightleftharpoons$$
$$\overset{H}{\underset{|}{}}$$
$$O=C-C=C-C-C=O \rightleftharpoons O=C-C=C-C=C-O-H$$

These relations are complicated by the existence of *cis* and *trans* isomers in some cases.

Muconic acid is $HO_2C(CH=CH)_2CO_2H$.

Decapentaene 1,10-dicarboxylic acid, dodeca-2,4,6,8,10-pentaene-1,12-diacid, $HO_2C(CH=CH)_5CO_2H$, m. 300° dec., has been made.[45] The last step to the ester is the condensation of oxalic ester with $CH_3(CH=CH)_4CO_2Et$, a good example of the transfer of alpha H properties through a conjugated system (vinylogy).

[41] Kohler, Reid. *J. Am. Chem. Soc.* **47**, 2803 (1925).
[42] Thorpe. *J. Chem. Soc.* **87**, 1669 (1905).
[43] Feist. *Ann.* **428**, 25 (1922).
[44] Kohler, Reid. *J. Am. Chem. Soc.* **47**, 2803 (1925).
[45] Kuhn, Grundmann. *Ber.* **69B**, 1979 (1936).

Crocetin is a highly unsaturated dibasic acid, probably

$$HO_2C(CMe=CHCH=CH)_3CMe=CHCO_2H.[46]$$

Acetylene dicarboxylic acid, butynediacid, $HO_2CC\equiv CCO_2H$, m. 179°, is made from Br_2-succinic acid and alcoholic KOH and from Na_2-acetylide and CO_2.[47] It gives the expected reactions. It easily changes to propargylic acid, $CH\equiv CCO_2H$.

Acetylene dicarboxylic acid adds bromine to give mainly dibromofumaric acid.[48] Thus *trans* addition takes place. This recalls the fact that HBr is much more readily removed from bromofumaric acid than from bromomaleic acid.

Acetylene dicarboxylic acid takes part in Diels-Alder reactions. With butadiene, it gives 3,6-dihydrophthalic acid.

Higher acids $HO_2C(C\equiv C)_nCO_2H$ in which n is 2 and 4 have been made. They are explosive.

C. HYDROXY DIBASIC ACIDS

Tartronic acid, hydroxymalonic acid, $CHOH(CO_2H)_2$, m. 187°, is prepared from bromomalonic acid or from mesoxalic acid, $CO(CO_2H)_2$. It gives the reactions of a secondary alcohol, a dibasic acid and an alpha hydroxy acid. On heating it loses CO_2 and forms glycolide.

Malic acid, hydroxysuccinic acid, $HO_2CCH_2CHOHCO_2H$, is made in large quantities by the hydration of maleic acid from the catalytic oxidation of benzene. It can also be obtained from its calcium salt in "sugar sand," a by-product of the maple sugar industry. Because of its asymmetric molecule, malic acid can exist in optically active forms. The naturally occurring acid shows a peculiar behavior in this respect. A 34 per cent solution at 20° is optically inactive. Dilution gives increasing levo specific rotation while concentration gives increasing dextro rotation. Much study has been devoted to this peculiar effect of dilution on the optical activity of malic acid solutions.[49]

Sodium malate tastes much like NaCl and is used in conditions under which the latter gives undesirable effects as a condiment ("Ekasalt").

Malic acid has been related to the standard of reference, glyceraldehyde, through tartaric acid. Reduction of L_g-(*dextro*)-tartaric acid leads to D_g-(*dextro*)-malic acid, wherein the subscripts refer to the reference standard, glyceraldehyde.

The molecule of active tartaric acid possesses a rotating axis of symmetry

[46] Kuhn, L'Orsa. *Ber.* **64B**, 1732 (1931).
[47] *Ann. Rep. Chem. Soc.* (London) **1926**, 98.
[48] Michael. *J. prakt. Chem.* [2] **52**, 306 (1895).
[49] Bancroft, Davis. *J. Phys. Chem.* **34**, 897 (1930).

(at the tetrahedron junction) and thus reduction of either hydroxyl group will lead to the same product.

Synthetic DL-malic acid is best resolved by crystallization of its cinchonine salts. Cinchonine-D-malate is less soluble than cinchonine-L-malate. When partly racemized malic acid is crystallized as the ammonium molybdate, the form present in excess crystallizes first.[53]

Malic acid gives the reactions which would be expected of its secondary alcohol and two carboxyl groups. Heating gives maleic anhydride and fumaric acid. With urea and H_2SO_4, malic acid gives uracil (I). It first acts as a typical alpha-hydroxyacid to give an aldehyde. This then acts with urea.

$$HO_2CCH_2CHOHCO_2H \rightarrow HO_2CCH_2CHO \rightarrow$$

$$HO_2CCH_2CH = NCONH_2 \rightarrow \begin{array}{c} CH=N-CO \\ | \qquad\quad | \\ CH_2-CONH \end{array} \quad (I)$$

(I) is the diketo tautomer of 2,4-$(OH)_2$-pyrimidine.

Treatment with PCl_5 gives chlorosuccinic acid. When this reaction and the reverse reaction to form malic acid were carried out with optically active materials, one of the most remarkable changes of organic chemistry was discovered. This is an example of the famous *Walden Inversion* (1893).[54]

These changes proved that the reactions involved are not simple metathetical reactions since the entering group does not occupy the *same position* on the asymmetric carbon as that occupied by the group which is removed. Hundreds of researches have led to the conclusion that most metathetical

[53] *Ann. Rep. Chem. Soc.* (London) 1924, 59.
[54] *ibid.* 1911, 60.

reactions in organic chemistry involve an attack on the *back* face of a carbon atom with the expulsion of the replaced group from the *front* (S_N2 reaction).[55] Thus the entering group does not actually replace the other group but takes up a mirror image position in relation to carbon and the replaced group. This *inversion* takes place in all such reactions although it is not detectable except in optically active materials.[56-58] The reasonableness of this *rearward attack* mechanism receives confirmation from the inertness of groups on atoms which cannot conceivably be attacked from the *rear*. Such a case is that of the 1-substitution products of 2.2.2 bicyclo-octane.[59]

In order to attack the *rear* of the 1-carbon, the substituting ion or group would have to get inside the "cage system" formed by the three *cis*-cyclohexane rings.

Citramalic acid is α-hydroxy-α-methylsuccinic acid,

$$HO_2CCH_2C(OH)MeCO_2H,$$

Tartaric acid, dihydroxysuccinic acid, $HO_2CCHOHCHOHCO_2H$, exists in four forms.

(a) L$_g$-(*dextro*)-*Tartaric acid*, (Weinsäure) obtained from argol (crude acid potassium tartrate, cream of tartar) formed as a precipitate in wine vats.

(b) *Racemic acid* (Traubensäure), obtained from the mother liquors of the preparation of L-tartaric acid. It is a definite compound of the D- and L-forms with definite molecules existing even in solution.[60]

(c) *Mesotartaric acid*, obtained along with racemic acid by heating L-tartaric acid with dilute alkalies or organic bases like quinoline. Its acid potassium salt is readily soluble in water.

(d) D$_g$-(*levo*)-*Tartaric acid*, obtained by the resolution of racemic acid by means of an optically active base. Thus, when the cinchonine salts of racemic acid are crystallized, the D-tartrate separates first.

The differences in the properties of the isomeric acids appear in the following table.

[55] Hughes et al. *Nature* **147**, 206 (1941).
[56] Fieser, Fieser. "Organic Chemistry," 1944. p. 279.
[57] Gilman. "Organic Chemistry," 1943. p. 264.
[58] Karrer. "Organic Chemistry," 1946. p. 286.
[59] Bartlett, Cohen. *J. Am. Chem. Soc.* **62**, 1183-1189 (1940).
[60] *Ann. Rep. Chem. Soc.* (London) 1922, 75.

	L-Tartaric	D-Tartaric	Racemic	*meso*-Tartaric
Water of crystallization.............	none	none	+H₂O	+H₂O
Melting point......................	170°	170°	206°	140°
Solubility in water (cold)............	137	137	21	20
Solubility in alcohol (cold)..........	37	37	2	167
Acid K salt, soluble in water.........	0.6	0.6	——	12.5
Dimethyl ester, m.................	61.5°	61.5°	0.85°	54.0°
Diethyl ester, m....................	18.7°	liq.	liq.	——
Ionization constant × 10⁴...........	9.7	9.7	9.7	6.0

The following configurations are assigned to the tartaric acids.

racemic acid

The tartaric acids are best named from their relation to the monoses, natural dextrorotatory tartaric acid being designated L-(*dextro*)-tartaric acid because it is formed by the oxidation of L-threose.

Mesotartaric acid is obtained by the oxidation of both D and L erythrose. The configuration of the monoses are based upon an arbitrary selection for the configuration of D-glyceraldehyde (p. 462).

Ordinary solutions of natural tartaric acid are dextrorotatory but a cold super-saturated solution is levorotatory.

The tartaric acids formed the field in which Pasteur discovered and studied the principles and methods of handling optically active organic compounds. Only modifications of his methods have been introduced since that time.   He discovered the phenomenon of racemization, by which heating an optically active substance alone or with a base inverts the asymmetric arrangement.

Thus the optically active tartaric acids become optically inactive.

The result is a mixture of *meso*-tartaric acid with D- and L-tartaric acids, the latter in equal amounts. Also *meso*-tartaric acid is converted into the same mixture on heating. Heated to a higher temperature, tartaric acid undergoes complex decompositions. The first product, formed at about 160°, is a lactide. At 180°, under reduced pressure, CO, CO_2, formic, acetic, and pyruvic (pyroracemic) acids are formed.[61] Under other conditions, tartaric acid gives as high as a 20% yield of methylsuccinic (pyrotartaric) acid.

Pasteur also discovered and applied *three methods for resolving a racemic material* into its optical isomers.

1. The mechanical separation of enantiomorphic crystals.

While crystallizing various salts of racemic acid, he noted that a solution of sodium ammonium racemate on slow crystallization deposited crystals which were not symmetrical but had *hemihedral* faces. Moreover, the crystals were of two types, related to each other like the right and left hand or as an object and its mirror image. Pasteur sorted these two kinds of crystals and found that a solution of the dextro hemihedral crystals was levorotatory while a solution of the levo hemihedral crystals of the same concentration gave the same angle of dextrorotation. Treatment of the separate crystals with acids gave L-tartaric and D-tartaric acids. The mixing of equal amounts of the two acids in solution gave off heat with the formation of racemic acid. It was later found that the hemihedral crystals separate from a solution of sodium ammonium racemate only below 27°. That is a *transition temperature* below which the Na, NH_4 salts of D- and L-tartaric acids can exist separately and above which they form a compound Na, NH_4-racemate which gives symmetrical crystals. The individual tartrates have the formulas

$$NaNH_4C_4H_4O_6 . 4 H_2O$$

while the true racemate contains only 2 H_2O.

2. By crystallization of compounds of the racemic material with an optically active substance. Thus *l*-cinchonine gives two salts with racemic acid, of which the *l*-cinchonine-*l*-tartrate is less soluble. When repeated recrystallization gives no increase in the levorotation of the salt which separates first, it is assumed to be pure and the *l*-tartaric acid is liberated from the cinchonine salt by careful treatment with an acid or base. If *l*-quinine is used, the *l*-quinine-*d*-tartrate is the less soluble salt.

[61] Chattaway, Ray. *J. Chem. Soc.* **119**, 34 (1921).

3. By biochemical means. Certain lower organisms destroy one optical isomer more rapidly than the other. Thus the *penicillium* of ordinary green mold, grown in ammonium racemate solution, attacks the ammonium *d*-tartrate much more rapidly than the *l*-form. If the process is stopped at the right point practically pure *l*-salt remains.

Rochelle salt is the dextrorotatory sodium potassium tartrate,

$$NaKC_4H_4O_6 . 4 H_2O.$$

It is a laxative. *Seidlitz powders* contain (1) sodium bicarbonate and Rochelle salt and (2) tartaric acid. When dissolved and mixed CO_2 is evolved. Acid potassium tartrate is the acid used with $NaHCO_3$ in tartrate baking powders. It is stable and non-deliquescent.

Diethyl tartrate when pure melts at 18.7°.[62]

Tartaric acid forms complex salts with various heavy metals. Thus ferric salts cannot be precipitated as ferric hydroxide in a solution containing tartrates. The best known of these complexes is that of copper which is found in Fehling's solution, an alkaline cupric solution from which strong reducing agents precipitate red cuprous oxide. The copper is combined with the

hydroxyls,

$$\begin{array}{c} NaO_2CCHO \\ | \diagdown \\ Cu. \\ | \diagup \\ NaO_2CCHO \end{array}$$

Tartar emetic is usually formulated as potassium antimonyl tartrate, $KO_2C(CHOH)_2CO_2SbO \cdot \frac{1}{2} H_2O$. It is difficult to believe, however, that the hydroxyl groups take no part in the formation of the compound. It would seem more reasonable that a ring $\begin{array}{c} -CHOSbOH \\ | | \\ CO\!-\!O \end{array}$, is involved.

Calcium tartrate is insoluble in water. It dissolves in excess of cold KOH or NaOH but reprecipitates on boiling. Presumably the soluble compound is $Ca\begin{array}{c} \diagup OCOCHONa \\ | \\ \diagdown OCOCHONa \end{array}$. Heat hydrolyzes the "alcoholate" groupings.

Tartaric acid with urea and H_2SO_4 gives 2-imidazolone-4-carboxylic acid, $\begin{array}{c} NCONH \\ \| | \\ CH\!-\!CHCO_2H \end{array}$. It thus acts like the half aldehyde of tartronic acid, $OCHCHOHCO_2H$, a result due to the action of H_2SO_4 in converting $-CHOHCO_2H$ to $-CHO + CO + H_2O$.

Trihydroxyglutaric acid, $HO_2C(CHOH)_3CO_2H$, exists in the theoretically possible forms, D-, L-, a racemic and two meso forms. These are related to the

[62] *Ann. Rep. Chem. Soc.* (London) **1922**, 75.

aldopentoses and the pentitols.

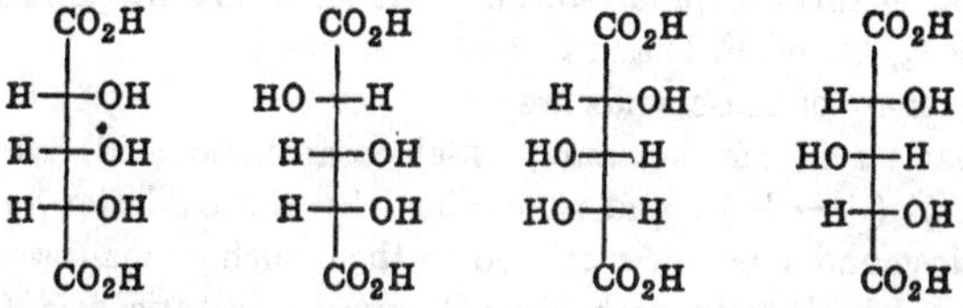

CO₂H	CO₂H	CO₂H	CO₂H

$$\begin{array}{cccc}
\text{CO}_2\text{H} & \text{CO}_2\text{H} & \text{CO}_2\text{H} & \text{CO}_2\text{H} \\
\text{H}-\text{OH} & \text{HO}-\text{H} & \text{H}-\text{OH} & \text{H}-\text{OH} \\
\text{H}-\text{OH} & \text{H}-\text{OH} & \text{HO}-\text{H} & \text{HO}-\text{H} \\
\text{H}-\text{OH} & \text{H}-\text{OH} & \text{HO}-\text{H} & \text{H}-\text{OH} \\
\text{CO}_2\text{H} & \text{CO}_2\text{H} & \text{CO}_2\text{H} & \text{CO}_2\text{H}
\end{array}$$

| *ribo*-trihydroxy-glutaric acid (*meso*) | D− | L− | *xylo*-trihydroxy glutaric acid (*meso*) |

racemic

The first and last are named from their preparation by oxidizing ribose and xylose respectively. The D- and L-forms are obtained from the corresponding arabinose and lyxose.

Tetrahydroxyadipic acid, $HO_2C(CHOH)_4CO_2H$, theoretically can exist in four D-, and four L-, four racemic, and two meso forms. All of these possibilities have been realized in the studies on the aldohexoses. Using the symbols ⊢ and ⊣ to indicate the plane projections $H-C-OH$ and $HO-C-H$ the following pairs of enantiomorphous configurations can be predicted.

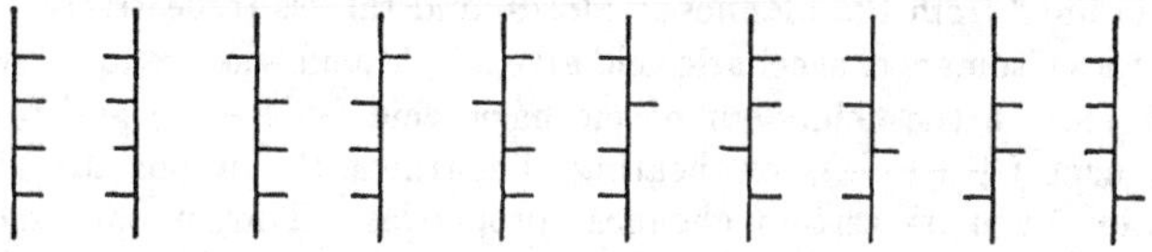

A   B   C   (D)   (E)

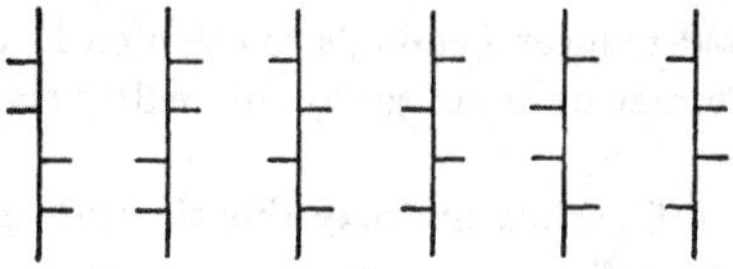

F   G   H

Since the ends of the chain are alike, any one of these projection formulas can be rotated 180° in the plane of the paper without changing the configuration. By this test the members of pair A are *identical* instead of enantiomorphous and the configuration $++++$ represents an internally compensated symmetrical *meso* form. Rotation of the second member of pair B does not make it identical with the first member. Thus B represents a pair of enantiomers. The same is true of C. Rotation of the members of D make them identical with pair C. Similarly pair E is identical with pair B. Pair F is different from any other pair and the members are not identical. G represents another pair of enantiomers. The members of pair H are identical. Thus configurations A and H represent meso forms while B, C, F, and G are racemic forms each consisting of a D- and an enantiomorphic L-form. The following are pairs of *epimers*, A and B, B and H, C and F, C and G.

The commonest of these acids are:

(a) Saccharic acid (Zuckersäure) (D-glucosaccharic acid) which has the configuration C (+ − + +) and is obtained by the oxidation of sucrose, D-glucose, L-gulose and substances related to them such as maltose and starch.[63] Oxidation of saccharic acid with $KMnO_4$ gives L-tartaric acid (− +) while the Hofmann degradation of the diamide gives a dialdehyde which is oxidized to D-tartaric acid (+ −).

(b) **Mucic acid** (Schleimsäure), which is a meso form corresponding to configuration H(+ − − +) is obtained by oxidizing lactose, dulcitol and various natural gums related to galactose and galactonic acid. The wood of the mountain larch is rich in galactan gums which form a commercial raw material for galactonic and mucic acids. Mucic acid is converted by heating in pyridine at 140° to an equilibrium mixture with DL-*talomucic acid*. The two α-carbons have been inverted to give the configurations (− + + +) and (+ − − −). Stronger heating gives *pyromucic acid*, furan-α-carboxylic acid.

(c) **Mannosaccharic, idosaccharic,** and **talomucic acids** exist in D-, L-, and r-forms obtained from the mannoses, idoses, and taloses respectively.

All of these isomers of saccharic acid give adipic acid when reduced with HI.

A difference in the behaviour of saccharic acid (+ + − +) and D-mannosaccharic acid (+ + − −), on heating, illustrates the important effect of configuration even on certain chemical properties. Thus D-mannosaccharic acid forms a dilactone, m. 180–190°, whereas saccharic acid gives a lactonic acid, m. 132°.

Mannosaccharic acid reduces Fehling's solution on heating while saccharic acid does not. The former does not give a difficultly soluble acid potassium salt as does the latter.

Two branched chain 6-carbon hydroxy dibasic acids give examples of the unusual *beta type of lactone*.[64]

$$CH_2(CO_2H)_2 + Me_2CO \xrightarrow{Ac_2O} \begin{array}{c} Me_2C\text{---}CHCO_2H \\ | \quad\quad | \\ O\text{---}CO \end{array}$$

$$HO_2CCMe_2CHBrCO_2H \xrightarrow{Ag_2O} \begin{array}{c} Me_2C \text{---} CHCO_2H \\ | \quad\quad | \\ CO\text{---}O \end{array}$$

The *gem*-dimethyl grouping, as usual, favors ring closure.

[63] *Ann. Rep. Chem. Soc.* (London) 1922, 77.
[64] *Ann. Rep. Chem. Soc.* (London) 1908, 86.

D. Dibasic Ketonic Acids

Mesoxalic acid, HO_2CCOCO_2H, or $HO_2CC(OH)_2CO_2H$, can be made from dibromomalonic acid with $Ba(OH)_2$ or by the hydrolysis of alloxan obtained by the oxidation of uric acid by nitric acid.

It gives ketone and acid reactions. Esters are known of both the true mesoxalic acid and the dihydroxylmalonic acid forms.

Oxaloacetic acid, $HO_2CCOCH_2CO_2H$, is obtained as its ester by condensing oxalic and acetic esters with sodium ethylate. It has similar reactions and uses to malonic ester. The acid can also be made by the oxidation of malic acid with H_2O_2 and $FeSO_4$.[65] Two forms of the acid, m. 184° and 152°, are described as hydroxy fumaric and maleic acids.[66]

$$\begin{array}{ccccc} HO-C-CO_2H & & O=C-CO_2H & & HO-C-CO_2H \\ \| & \rightleftharpoons & | & \rightleftharpoons & \| \\ HO_2C-C-H & & HO_2C-CH_2 & & H-C-CO_2H \end{array}$$

Dihydroxytartaric acid, $HO_2CC(OH)_2C(OH)_2CO_2H$, is obtained from tartaric acid dinitrate and from the action of pyrocatechol with nitrous acid. The first of these processes is an interesting case of intramolecular oxidation. Similar but more complete and violent processes take place when nitroglycerin and related substances are decomposed. The mechanism of the formation from pyrocatechol is not understood.

The sodium salt of dihydroxytartaric acid is only slightly soluble. Heat changes it to sodium tartronate. This involves a rearrangement followed by the loss of CO_2. One of the four hydroxyls may be removed, leaving a carbonium carbon which satisfies itself by attracting an electron pair with its attached CO_2Na from the adjacent carbon.

$$\rightarrow \begin{bmatrix} NaO_2CC(OH)_2 \\ {}^+C(OH)CO_2Na \end{bmatrix} \rightarrow \begin{bmatrix} {}^+C(OH)_2 \\ NaO_2CC(OH)CO_2Na \end{bmatrix} \rightarrow$$

$$\begin{bmatrix} CO_2H \\ C(OH)(CO_2Na)_2 \end{bmatrix} \rightarrow CO_2 + HOCH(CO_2Na)_2$$

Acetone dicarboxylic acid, $HO_2CCH_2COCH_2CO_2H$, is obtained from citric acid by treatment with sulfuric acid, the usual method for removing formic acid from alpha hydroxy acids.

$$(HO_2CCH_2)_2C(OH)CO_2H \rightarrow CO + H_2O + CO(CH_2CO_2H)_2$$

Heat changes it successively to acetoacetic acid and acetone. The reaction cannot be stopped at the intermediate step because acetoacetic acid is less stable to heat than acetone dicarboxylic acid. It has been proposed as a source of CO_2 in baking powders.

[65] Fenton. *J. Chem. Soc.* 69, 546 (1896).
[66] *Ann. Rep. Chem. Soc.* (London) 1912, 85.

αββ-Trimethylketoglutaric acid, Balbiano's acid,

$$HO_2CCHMeCMe_2COCO_2H,$$

is obtained by oxidation of camphor or camphoric acid. It exists in equilibrium with a lactone form. Formerly a cyclopropane form was assumed.[67]

$$
\begin{array}{ccc}
MeCH\!\!-\!\!-\!\!CMe_2 & & MeCH\!\!-\!\!-\!\!CMe_2 \\
| \qquad\quad | & \rightleftharpoons & | \qquad\qquad\quad | \\
CO_2H \quad COCO_2H & & CO\!\!-\!\!O\!\!-\!\!C(OH)CO_2H
\end{array}
$$

The influence of methyl groups on ring closure are shown by the fact that its lower homologs, α-ketoglutaric and β-Me$_2$-ketoglutaric acids exist only in the open chain form while Me$_4$-α-ketoglutaric acid exists only in the cyclic form.[68]

Dihydroxymaleic acid, ketomalic acid,

$$HO_2CC(OH) = C(OH)CO_2H \rightleftharpoons HO_2CCHOHCOCO_2H,$$

is obtained by oxidizing tartaric acid with H_2O_2 and $FeSO_4$.[69] Bromine water converts it to dihydroxytartaric acid. It decomposes on heating with water to 50° to give glycolic aldehyde.[70] Its solution is stabilized by boric acid presumably by the formation of a complex.[71]

2-Heptene-4-one-1,7-diacid, 1-heptene-3-one-1,5-dicarboxylic acid, furonic acid, $HO_2CCH = CHCOCH_2CH_2CO_2H$, has been obtained from furfural through furylacrylic acid. Reduction by HI gives pimelic acid.

XX. POLYBASIC ACIDS

The three chief acids considered here all contain the same carbon skeleton, that of 3-Me-pentane, with the three terminal carbons oxidized to carboxyl groups.

Tricarballylic acid, propane 1,2,3-tricarboxylic acid,

$$HO_2CCH_2CH(CO_2H)CH_2CO_2H,$$

m. 166°, can be obtained (a) from glycerol through the cyanide from 1,2,3-Br$_3$-propane, (b) from citric acid by reduction with HI, and (c) by hydrogenation of aconitic acid obtained by dehydrating citric acid.

Aconitic acid, propene 1,2,3-tricarboxylic acid,

$$HO_2CCH = C(CO_2H)CH_2CO_2H,$$

m. 191° occurs in various plant products and can be made by the cautious dehydration of citric acid. Its structure is proved by its reduction to tri-

[67] *Ann. Rep. Chem. Soc.* (London) 1923, 107; 1928, 81.
[68] *ibid.* 1930, 152.
[69] Fenton. *J. Chem. Soc.* 69, 546 (1896).
[70] *Ann. Rep. Chem. Soc.* (London) 1912, 85.
[71] *ibid.* 1931, 77.

carballylic acid. Since it is doubly an $\alpha\beta$-unsaturated acid this reduction is easy.

Citric acid, hydroxytricarballylic acid, 2-propanol 1,2,3-tricarboxylic acid, $(HO_2CCH_2)_2C(OH)CO_2H$ is prepared (*a*) from the juice of cull lemons through its insoluble calcium salt and (*b*) by the action of various molds on sugar solutions (molasses). It has been synthesized from *sym*-dichloroacetone through the cyanohydrin and cyanide syntheses. Its triethyl ester has been made by the Reformatzky reaction using ethyl bromoacetate and diethyl oxaloacetate with zinc.

Citric acid gives the reactions of a tribasic acid and a tertiary alcohol. On heating, it gives aconitic acid, itaconic anhydride, citraconic anhydride and acetone. With mineral acids it acts as an $\alpha-OH$ acid, losing CO to form acetone dicarboxylic acid. When heated with urea and H_2SO_4 it gives uracil 4-acetic acid.[1]

Magnesium citrate is used as a laxative. Ferric ammonium citrate is used with potassium ferricyanide in blue print paper.

Citric acid reacts with formaldehyde and HCl to give anhydromethylene

$$O$$

citric acid, $(HO_2CCH_2)_2C-\overset{\bigtriangleup}{C}O_2CH_2$, m. 208°. This combines with hexamethylene tetramine to form *helmitol*, a urinary antiseptic (NNR).

Citric acid forms an important link in the biochemical breakdown of carbohydrates.[2]

Various *complex hydroxy polybasic* acids have been found in lichens. *Protolichesteric acid* is formulated as the lactone of

$$n\text{-}C_{13}H_{27}CHOHCH(CO_2H)C(=CH_2)CO_2H;$$

agaracinic acid is $n\text{-}C_{15}H_{31}CH_2(CO_2H)C(OH)(CO_2H)CH_2CO_2H$; and *caperatic* acid is the monomethyl ester of the tribasic acid, *nor-caperatic acid*,

$$n\text{-}C_{14}H_{29}CH(CO_2H)C(OH)(CO_2H)CH_2CO_2H.$$

XXI. CYANOGEN AND ITS DERIVATIVES

A. CYANOGEN AND INORGANIC CYANIDES

Cyanogen, $N\equiv C-C\equiv N$, is related to its derivatives much as is Cl_2. The variety of these is much greater. Cyanogen occurs in coal gas in small traces. It is a poisonous colorless gas (b. $-21°$). It is made by heating the cyanides of mercury or silver and by treating a cupric solution with NaCN. The cupric cyanide first formed is hydrolyzed and the resulting HCN reduces the cupric ions and insoluble cuprous cyanide is precipitated. The latter is a

[1] Hilbert. *J. Am. Chem. Soc.* **54**, 2076 (1932).
[2] Breusch. *Science* **97**, 490 (1843).

component of anti-fouling paints for ships and is available commercially.

$$2 CuSO_4 + 4 NaCN \rightarrow \underline{Cu_2(CN)_2} + (CN)_2 + 2 Na_2SO_4$$

Cyanogen is also formed by the action of P_2O_5 on ammonium oxalate or on oxamide. It is readily soluble in water. Even in the cold it readily hydrolyzes giving a brown precipitate of azulmic acid,[1] $C_4H_5ON_5$, and in the solution oxalic acid, ammonia, HCN, formic acid and urea.

Cyanogen reacts with bases to give cyanides and cyanates (KOCN etc.).

Butadiene and cyanogen react at 400–700° to give 2-cyanopyridine.[2]

Paracyanogen, $(CN)_x$, is a brown powder obtained as a by-product in heating $Hg(CN)_2$. It gives cyanogen at a higher temperature. It is also formed from cyanogen by light.[3]

Oxycyanogen has been prepared in polymeric form, $(OCN)_x$.

Hydrogen cyanide, hydrocyanic acid, prussic acid, b. 26°, may be made from any cyanide with even dilute acids.[4] This is because it is both a weak acid and volatile. Even carbonic acid (moist air) will react with cyanides. It occurs as a glycoside with gentiobiose and benzaldehyde in amygdalin of bitter almonds. Many other cyanogenetic plants are known.[5]

Hydrocyanic acid can also be made from sugar beet residues which contain several per cent of nitrogen as betaine. The steps are as follows:

$$\begin{array}{c} CH_2-N(CH_3)_3 \\ | \qquad | \\ CO-O \end{array} \rightarrow (CH_3)_3N \rightarrow 2 CH_4 + HCN$$

The last step is carried out at 800–1000°.

Hydrogen cyanide is a colorless liquid (m. −14°, b. 26°). It has a peculiar faint odor. Some individuals, however, *cannot detect this characteristic odor*. Minute amounts cause death. Smaller amounts first paralyze the legs. The poisonous effects are not cumulative. Apparently the animal organism is able to destroy HCN at a certain rate without harm. This fact and the low density of the gas makes it less effective as a war gas than toxic materials which are cumulative in effect.

It is completely miscible with water forming an acid one ten-thousandth as strong as acetic acid. The solution in water undergoes hydrolysis to a variety of products. It forms a trimer which is probably aminodicyanomethane.[6]

Hydrocyanic acid is a tautomeric substance giving reactions and derivatives corresponding to two forms

$$H-C\equiv N \;\rightleftharpoons\; H-N=C$$

[1] Jacobsen, Emmerling. *Ber.* 4, 950 (1871).
[2] Janz. *Nature* 162, 28 (1948).
[3] Hogness, Ts'ai. *J. Am. Chem. Soc.* 54, 123 (1932).
[4] "Org. Syntheses."
[5] Briese, Couch. *J. Agr. Res.* 57, 81–108 (1938).
[6] *Ann. Rep. Chem. Soc.* (London) 1922, 69.

This apparent shift of a hydrogen ion or proton from one atom to the next is an example of diad prototropy (Ingold), analogous to the triad prototropy of the keto-enol system. Like that process, it is probably bimolecular in nature.

$$HCN + H^+ \rightleftharpoons [H\overset{+}{C}=NH]^+ \rightleftharpoons H^+ + CNH$$

The HNC form contains "bivalent" carbon.[7] It is perhaps more accurate and certainly more conservative to say that it contains a carbon which behaves like that in carbon monoxide. Both hydrocyanic acid and carbon monoxide exhibit a type of unsaturation different from that shown by the ethylenic compounds. Here the unsaturation is confined to a *single* atom instead of to an adjacent pair. Thus chlorine will add to these substances to give Cl_2CO and $Cl_2C=NH$. Hydrogen chloride presumably forms unstable compounds, formyl chloride (HCOCl) and iminoformyl chloride (HN=CHCl). Thus benzene reacts in the Friedel-Crafts reaction with HCl and CO or HCN to give PhCHO or PhCH=NH.

The conflict in evidence as to the structure of hydrocyanic acid is well illustrated by two methods of preparation.

1. That by the dehydration of ammonium formate or of formamide would indicate the $H-C\equiv N$ structure.

2. That from NH_3 and $CHCl_3$ and alcoholic KOH would indicate the HNC form.

Measurements of the physical properties of hydrocyanic acid show that it exists almost entirely in the HCN form.[8,9]

Hydrocyanic acid may be regarded as an aldehyde of the ammonia system, as an ammonocarbonous acid, and as the analog of CO in the ammonia system.[10]

The analogy between carbon monoxide and hydrocyanic acid may be represented electronically:

$$\left[: \overset{..}{C} : \overset{..}{O} : \right]^0 \qquad \left[: \overset{..}{C} : N : \right] H^+$$

Analogous structures may be assigned to nitrogen and acetylene.

$$\left[: \overset{..}{N} : \overset{..}{N} : \right]^0 \qquad H : C : \overset{..}{C} : H$$

In each case a pair of atoms forms a more or less stable system with ten electrons, the *two* atoms being surrounded by an octet. This peculiar electronic structure is probably related to the remarkable stability of these four substances to high temperatures.

[7] Nef.   *Ann.* **298**, 345 (1897).
[8] Reichel, Strasser.   *Ber.* **64B**, 1997 (1931).
[9] Dadieu.   *Ber.* **64**, 358 (1931).
[10] Franklin.   *Ann. Rep. Chem. Soc.* (London) **1923**, 85.

The *inorganic cyanides* are obtained in a variety of ways.

1. The oldest method is the fusion of "organic" nitrogenous material such as leather scraps with iron and potassium hydroxide or carbonate. Leaching the resulting mass with water gives a beautifully crystalline substance, "yellow prussiate of potash," $K_4Fe(CN)_6$, potassium ferrocyanide. The name "Blutlaugensalz" is derived from its preparation by heating dried blood with lye, the blood supplying the C, N and Fe. When this is treated with an oxidizing agent, usually chlorine, it is converted to the "red prussiate of potash," $K_3Fe(CN)_6$, potassium ferricyanide. The modern way of making potassium ferrocyanide is from the iron oxide which has been used to purify illuminating gas. This contains Prussian blue, ferric ferrocyanide. Boiling with milk of lime gives soluble calcium ferrocyanide which is treated with potassium carbonate.

In the ferro- and ferri-cyanides the Fe atom may be regarded as having two and three CN radicals respectively, thus being ferro*us* and ferr*ic* in the two compounds. In both compounds the "coordination number" of the iron is six.[11] This means that the Fe atom can hold six univalent groups *inside* a complex ion. There is good evidence that these six groups lie four in one plane and one each above and below it.

Tetravalent Ferrocyanide Ion   Trivalent Ferricyanide Ion

These substances do not give iron or cyanide ions in solution. They are practically non-poisonous. Both ferro- and ferri-cyanides exist in alpha and beta forms, stable in basic and in neutral or acid solutions respectively.[12]

The most characteristic reactions of the complex iron cyanides are with iron salts with the formation of characteristic blue precipitates (Prussian Blue and Turnbull's Blue). The formation of Prussian Blue gave the names prussiate and prussic acid. Blue print paper is coated with potassium ferri*cyanide and ferr*ic* ammonium citrate. Exposure to light converts the latter to the ferro*us* salt. Washing with water gives an insoluble blue wherever the light struck.

[11] Werner. *Z. anorg. Chem.* 3, 327 (1893).
[12] Briggs. *J. Chem. Soc.* 99, 1019 (1911).

Heating the complex cyanides with *dilute* sulfuric acid gives hydrocyanic acid.[12a]

$$2\ K_4Fe(CN)_6 + 6\ H_2SO_4 \rightarrow FeK_2Fe(CN)_6 + 6\ HCN + 6\ KHSO_4$$

The potassium ferrous ferrocyanide is not acted on by dilute acid. More concentrated acid gives carbon monoxide due to the hydrolysis of HCN to formic acid and the dehydration of the latter.

$$K_4Fe(CN)_6 + 11\ H_2SO_4 + 6\ H_2O \rightarrow$$
$$6\ CO + FeSO_4 + 6\ NH_4HSO_4 + 4\ KHSO_4$$

The prussiates do not react with dilute solutions of bases, but with hot concentrated solutions they give precipitates of iron hydroxides and cyanides in solution. These reactions are of no value in preparing cyanides because of the large excess of base required.

Many other heavy metals form complex cyanides much like the ferrocyanides and the ferricyanides. The commonest coordination number is 6 but some metals have coordination numbers of 2, 4 and 8. Typical examples are $KCu(CN)_2$, $KAg(CN)_2$, $KAu(CN)_2$, $K_2Cd(CN)_4$, $KAu(CN)_4$, $K_4Co(CN)_6$, $K_4Mo(CN)_8$.

2. When prussiates are heated to high temperatures, they decompose.

$$K_4Fe(CN)_6 \rightarrow 4\ KCN + FeC_2 + N_2$$

The yield of cyanide by this process is increased by heating with metallic sodium.

$$2\ Na + K_4Fe(CN)_6 \rightarrow 4\ KCN + 2\ NaCN + Fe$$

The mixture of cyanides is as useful as a pure cyanide for electroplating and for the cyanide process for extracting gold from low grade ores and tailings.

3. Calcium carbide, calcium cyanamide and sodium carbonate at 500° give NaCN.

4. Calcium cyanamide fused with salt gives a mixture of sodium and calcium cyanides. Treatment with sodium carbonate gives NaCN.

5. Nitrogen reacts with a red hot mixture of coal and soda ash to give NaCN.[13]

$$Na_2CO_3 + 4\ C + N_2 \rightarrow 2\ NaCN + 3\ CO$$

6. Ammonia with red hot coal gives NH_4CN.

7. The present commercial method for making sodium cyanide consists in passing NH_3 through sodium and charcoal at 600–800°. The steps are probably

$$2\ NH_3 \xrightarrow{Na} 2\ NaNH_2 \xrightarrow{C} Na_2NCN \xrightarrow{C} 2\ NaCN$$

[12a] *Ann. Rep. Chem. Soc.* (London) 1923, 85.
[13] Ellis. "Chemistry of Petroleum Derivatives." Reinhold, 1934. p. 293.

8. The increased demand for HCN for the preparation of acrylic esters, methacrylic esters, and acrylonitrile has rendered the ordinary preparations inadequate. The trend is to make HCN by catalytic pyrolysis of a mixture of methane and ammonia.

$$CH_4 + NH_3 \rightarrow HCN + 3 H_2$$

9. Much of the commercial supply of HCN comes from the catalytic dehydration of formamide obtained from methyl formate and NH_3, the former being produced catalytically from CO and methanol (DuPont).

Salts of hydrocyanic acid are often stated to have the isocyanide structure. Mercuric cyanide is remarkable in that it gives no reactions of either mercury or a cyanide. It behaves like a true organic mercurial having the $C-Hg$ linkage.

Hydrocyanic acid adds to aldehydes and ketones to give cyanohydrins. It also adds to $\alpha\beta$-unsaturated compounds. Potassium cyanide speeds this addition much as the sodium derivative of β-keto esters does that to the $\alpha\beta$-unsaturated compounds.

$$RCH{=}CHCOR' + HCN \rightarrow RCHCNCH_2COR'$$

HCN can be added to acetylene to give acrylonitrile. It reacts with ethylene oxide to form ethylene cyanohydrin.

Reduction and hydrolysis convert hydrocyanic acid respectively to methylamine and formic acid. Diazomethane gives mainly methyl cyanide, CH_3CN, with a smaller amount of the isocyanide, CH_3NC.

Hydrocyanic acid can be detected as Prussian blue or as the blood-red ferric thiocyanate. The first is obtained by boiling with NaOH and ferric and ferrous salts and acidifying and the second by evaporating with ammonium polysulfide, dissolving in water and adding a ferric salt.

Hydrocyanic acid, as one of the most deadly poisons, is used as a very effective fumigant especially for orange orchards, holds of ships and for large flour mills. The latter are equipped with permanent piping by means of which the entire mill can be fumigated periodically by simply connecting tanks of liquid HCN to the pipe system. For fumigation, it is available in liquid form and absorbed in porous material (Zyklon). Since its odor in small concentrations is unnoticeable to some individuals, it is often used with *cyanogen chloride*, ClCN, b. 15°, itself a good fumigant and a lachrymator. The latter is prepared from a cyanide solution and chlorine.

Calcium cyanide is so easily hydrolyzed that it cannot be made in the presence of water. It is prepared commercially by the action of anhydrous HCN with calcium carbide. It is a valuable solid fumigant because it acts with the moisture in the air to give HCN.[14]

[14] Young. *Ind. Eng. Chem., Ind. Ed.* 21, 861 (1929).

Soluble cyanides are used for the extraction of gold by dissolving it in the presence of air. The net result is given by the equation

$$4 \text{ Au} + 8 \text{ NaCN} + \text{O}_2 + 2 \text{ H}_2\text{O} \rightarrow 4 \text{ NaAu(CN)}_2 + 4 \text{ NaOH}$$

Such solutions are also used for electroplating with gold and silver. Solutions of $NaAu(CN)_2$ and $NaAg(CN)_2$ are used because of the small concentration of Au^+ and Ag^+ ions which they give. This produces a more compact film of metal.

Sodium nitroprusside, $Na_2[Fe(CN)_5NO]$, formed by the action of nitric acid with potassium ferrocyanide followed by neutralization with Na_2CO_3, gives a violet color with inorganic sulfides. In the complex ion, the Fe has a valence of $+3$ and the NO group is neutral, thus leaving a total valence for the ion of -2.

B. ALKYL CYANIDES

These are also known as *nitriles*, carbonitriles, or acid nitriles, RCN.

Up to C_{13} these are liquids of peculiar but not unpleasant odor and of only slightly toxic properties as compared with the isomeric isonitriles or carbylamines, RNC, which have most disgusting odors and are highly toxic. Acetonitrile, CH_3CN, is soluble in water but the higher members become less soluble.

NITRILES

-cyanide	-nitrile	m. °C	b. °C
Methyl	Acetonitrile	-44	81.5
Ethyl	Propio-	-103	98
Propyl	Butyro-		116
Isopropyl	Isobutyro-		107
Butyl	Valero-		141
Isobutyl	Isovalero-		129
sec-Butyl	Methylethylaceto-		125
t-Butyl	Me₃-aceto- (Pivalo-)	16	106
Amyl	Capro-		163
Hexyl	Oenantho-		184
Heptyl	Caprylo-		199
Octyl	Pelargono-		215
Nonyl	Capric-		236
Decyl	Undeco-		254
Undecyl	Lauro-	4	198 (100 mm.)
Dodecyl	Trideco-		275
Tridecyl	Myristo-	19	226 (100 mm.)
Tetradecyl	Pentadeco-	23	185 (23 mm.)
Pentadecyl	Palmito-	31	251 (100 mm.)
Cetyl, C_{16}	Margaro-	29	208 (10 mm.)
Heptadecyl	Stearo-	41	274 (100 mm.)
Pentacosyl, C_{25}	Ceroto-	58	
Nonacosyl, C_{29}	Mellisso-	70	
Myricyl, C_{31}		75	

Preparation.

1. From inorganic cyanides.

a. KCN or NaCN with alkyl halides in aqueous alcohol solution. As would be expected, the results vary with the halide used. Thus *n*-AmCl gives a 70% yield, *sec*-AmCl a 30% yield and *t*-AmCl none.[15]

b. $Cu_2(CN)_2$, AgCN, $Hg(CN)_2$ or $K_2Hg(CN)_4$ with less reactive or more sensitive halides give cyanides and isocyanides. The latter can be removed by heating with *dilute* acid which has no effect on the former.

c. Inorganic cyanides heated with solid alkali alkyl sulfates.

$$RKSO_4 + KCN \rightarrow RCN + K_2SO_4$$

Again, some isocyanide is formed.

2. From carboxylic acids. The acids are converted to the amides which are then dehydrated.

$$CH_3CO_2H + NH_3 \rightarrow CH_3CO_2NH_4$$

$$CH_3CO_2NH_4 \xrightarrow{\text{Distilled}} CH_3\overset{O}{\overset{\|}{C}}{-}NH_2 + H_2O$$

$$CH_3\overset{O}{\overset{\|}{C}}{-}NH_2 + P_2O_5 \rightarrow CH_3CN + 2 HPO_3$$

This series of reactions shows the relations between the carboxylic acids, the acid amides and the acid nitriles and proves that the latter have the alkyl group attached to the carbon rather than to the nitrogen of the CN group. Higher acids undergo these changes merely by heating in a stream of NH_3. Thus, stearic acid is readily converted to its amide and then to the nitrile. A modification of this preparation is the passage of ammonia and the vapors of a carboxylic acid or its ester over alumina at 500°.

3. A large scale preparation of great promise involves high temperature conversion of olefins and NH_3 to give acetonitrile, propionitrile, butyronitrile, etc. A modification of the process also gives acrylonitrile.

4. From primary amines by treatment with bromine and KOH.

$$RCH_2NH_2 + 2 Br_2 + 4 KOH \rightarrow 4 KBr + 4 H_2O + RCN$$

5. From aldoximes by dehydration with acetic anhydride or thionyl chloride.

$$RCH{=}NOH + Ac_2O \rightarrow RCN + 2 AcOH$$

6. From Grignard reagents with chlorocyanogen, ClCN.

7. From esters with NH_3 passed over hot Al_2O_3 or ThO_2.[15a]

[15] Hass, Marshall. *Ind. Eng. Chem., Ind. Ed.* **23**, 352 (1931).
[15a] Ann. Rep. Chem. Soc. (London) **1918**, 17.

Reactions.

1. Hydrolysis ("saponification"). In basic or acid solution the $C \equiv N$ group is hydrated and then split.

$$RCN + H_2O + NaOH \rightarrow RCO_2Na + NH_3$$
$$RCN + 2 H_2O + HCl \rightarrow RCO_2H + NH_4Cl$$

The hydrolysis of the higher nitriles becomes increasingly difficult because of the inertness of the corresponding amides. Thus $MeEtCHCH_2CN$, treated with 10 volumes of fuming hydrochloric acid for two days, gives only the amide.

2. Milder addition of water to nitriles gives acid amides. This can be accomplished by heating with water at 180° or by treating with hydrogen peroxide and a base at 40°. Some nitriles, on solution in conc. sulfuric acid and dilution with water, give the amides.

$$RC \equiv N + HOH \rightarrow RC\overset{OH}{=}NH \rightarrow RC\overset{O}{-}NH_2$$

Hydrogen sulfide similarly gives thioacid amides, $RCSNH_2$.

3. Dry hydrogen halides form imidohalides. These react with alcohols and mercaptans to give imidoethers and thio-imidoethers. These are stable only as their hydrochlorides. With ammonia, they form amidine hydrochlorides. Hydrolysis gives carboxylic acids. All these compounds resemble ammonium compounds which are stable as salts but not as the free hydroxide. They may be formulated as follows:

$$RC\overset{NH \cdot HX}{\diagdown}X \qquad RC\overset{NH \cdot HX}{\diagdown}OR' \qquad RC\overset{NH \cdot HX}{\diagdown}NH_2$$

4. Aliphatic carboxylic acids add to nitriles to form diacidamides.

$$RCN + RCO_2H \rightarrow (RCO)_2NH$$

This tendency is so great that reactions which would be expected to give γ-cyanopropionic acid actually give succinimide by internal addition. Aromatic carboxylic acids give an apparent metathesis.

$$ArCO_2H + RCN \rightarrow RCO_2H + ArCN$$

Acid anhydrides form tri-acidamides.

$$RCN + (RCO)_2O \rightarrow (RCO)_3N$$

5. Reduction gives primary amines. Side reactions give secondary and tertiary amines. Best results are obtained with alkaline reduction at moderate temperature. Acid reduction favors the hydrolysis of the intermediate reduction product, $RCH = NH$ to give an aldehyde which can condense with

the primary amine to form a Schiff's base, $RCH=NCH_2R$, which produces a *secondary amine* on reduction. Catalytic hydrogenation, at relatively high temperatures, eliminates NH_3 from two molecules of primary amine to give a secondary amine, etc.

6. Grignard reagents give ketones.

$$RCN + R'MgX \rightarrow RR'C=NMgX \rightarrow RR'C=O.$$

7. Sodium or sodamide with alkyl iodides result in alkylation in the alpha position to give secondary and tertiary alkyl cyanides.

$$RCH_2CN \rightarrow RR'CHCN \rightarrow RR'R''CCN$$

Branching of the chain in the reagent and the product slows the reaction and decreases the yield. Considerable work has been done on Na salts of aryl substituted acetonitrile.[16]

8. Treatment with Na or $NaNH_2$ gives dimers and trimers depending on whether a solvent is used. This is a type of aldol condensation and so depends on the presence of α-H.

$$2\ RCH_2CN \rightarrow RCH(CN)C(=NH)CH_2R.$$

Hydrolysis gives a β-keto acid, $RCH_2COCH(R)CO_2H$. Only primary cyanides give the trimers, cyanalkines. These are strongly basic and are formulated as pyrimidines. MeCN gives 2,4-dimethyl-6-aminopyrimidine (I). The first step is a condensation involving the α-H of one molecule and the $-C\equiv N$ group of another, the latter functioning like the $C=O$ group in the aldol condensation. Then the NH group reacts with the CN of a third molecule much as NH_3 reacts with $C=O$. Finally the NH reacts with the CN in the 5-position to close the 6-ring.

The final shift to the conjugated system of double linkages characteristic of the pyrimidines is like the conversion of acetone to its enol form.

Tertiary cyanides cannot be made from KCN and the halide because the alkaline nature of the former causes the removal of HX. The double compound $Hg(CN)_2 \cdot 2\ KCN$ is practically neutral and avoids this difficulty.

[16] Rising, Swartz. *J. Am. Chem. Soc.* **54**, 2021 (1932).

Higher nitriles may be formed from halides and by the action of hypohalites with amines. Thus n-butyl amine gives n-butyronitrile.

$$C_3H_7CH_2NH_2 + 2 Br_2 + 4 NaOH \rightarrow C_3H_7CN + 4 NaBr + 4 H_2O$$

This reaction follows the Hofmann reaction for making amines from acid amides unless the amine formed is volatile enough to escape without reacting with excess of hypohalite solution.

$$RCH_2CONH_2 \xrightarrow[NaOH]{Br_2} RCH_2NH_2 \xrightarrow[NaOH]{Br_2} RC\equiv N$$

Higher nitriles such as lauronitrile and stearonitrile are available commercially from the action of an excess of ammonia on the acids at high temperatures.

$$RCO_2H \rightarrow RCO_2NH_4 \rightarrow RCONH_2 \rightarrow [RC(OH)(NH_2)_2] \rightarrow$$
$$[RC(NH)NH_2] \rightarrow RCN$$

Higher Aliphatic nitriles known as Arnells are useful as plasticizers (Armour).

Nitriles can be *identified* by reaction with phloroglucinol in the presence of $ZnCl_2$, followed by hydrolysis to the ketone, $RCOC_6H_2(OH)_3$. The melting points of such 2,4,6-trihydroxyphenyl alkyl ketones from the first five n-alkyl cyanides are 218, 175, 181, 149, and 120.

Hydroxynitriles are available commercially (ACC). These include **ethylene cyanohydrin,** hydracrylonitrile $HOCH_2CH_2CN$, **glyconitrile,** formaldehyde cyanohydrin, $HOCH_2CN$, **lactonitrile,** acetaldehyde cyanohydrin, $MeCHOHCN$, and **acetone cyanohydrin,** $Me_2C(OH)CN$.

Vinyl cyanide, acrylonitrile, $CH_2{=}CHCN$, b. 78°, is made by dehydrating ethylene cyanohydrin. It is also obtained by the catalytic addition of HCN to acetylene. It is made in large amounts for co-polymerization with butadiene to make Buna N, Perbunan, Hycar OR, Chemigum. It reacts with alpha H atoms in presence of alkaline catalysts by a Michael reaction to introduce cyanoethyl groups, $-CH_2CH_2CN$, in place of the alpha H atoms. Thus acetone could form delta-keto-capronitrile. The difficulty is that several of the alpha H atoms react and also that the alkaline catalysts cause aldol condensations of the acetone. Active H atoms attached to O, N, or S react similarly. Thus H can be replaced by the cyanoethyl group in such widely diverse compounds as phenols, ketones, mercaptans, malonic esters, indene, anthrone, cyclopentadiene, water, glycols, oximes, cellulose, haloforms, nitroparaffins, isobutyraldehyde, mesityl oxide, and crotononitrile.[17, 18] The last two are characteristic of *alpha-beta* unsaturated compounds in this reaction. If possible, isomerization takes place to give alpha H which is replaced by the cyanoethyl group. Thus the products of the last two compounds with acrylonitrile are respectively 2-methyl-3,3-di(cyanoethyl)-1-penten-4-one and

---

[17] Bruson. *J. Am. Chem. Soc.* **64,** 2457 (1942).

[18] Bruson, Niederhauser, Riener, Hester. *J. Am. Chem. Soc.* **67,** 601 (1945).

3,5-dicyano-3-cyanoethyl-1-pentene.   Phenylarsine gives

$$C_6H_5As(CH_2CH_2CN)_2.[19]$$

Benzaldehyde gives products of unknown structure.   The products are usually crystalline compounds obtained in excellent yields.

Another example of the activity of acrylonitrile is its ready addition of thiocyanic acid.   Thus acidification of a mixture of it with aqueous NaSCN gives $NCS-CH_2CH_2-CN$.

Allyl cyanide, $CH_2=CH-CH_2CN$, cannot be made from KCN because the alkaline solution of the latter rearranges the double bond to the $\alpha\beta$-position. The use of cuprous cyanide overcomes this difficulty.[20]

Condensation of ketones with cyanoacetic acid can be used to prepare complex unsaturated nitriles.[21]

Trichloroacetonitrile, Cl_3CCN, is reported to have insecticidal properties.

Nitroacetonitrile, O_2NCH_2CN, is obtained from methazoic acid.

$$O_2NCH_2CH=NOH + SOCl_2 \rightarrow O_2NCH_2CN + SO_2 + 2\ HCl$$

C. ALKYL ISOCYANIDES, ISONITRILES, CARBYLAMINES, RNC

These are liquids of penetrating and disgusting odors and of extreme toxicity.   They boil somewhat lower than the corresponding cyanides.

Preparation.

1. From alkyl iodides and silver cyanide (Gautier 1866).   The first product is a double compound $RNC \cdot AgCN$.   The difference in behaviour of AgCN and KCN in giving mainly isonitriles and nitriles respectively is sometimes expressed by giving the formulas as AgNC and KCN.   The fact is that this difference is not understood.   Electronically the cyanide ion is probably

$$\left[: C : N : \right]^-$$

No information is available as to the factors controlling the attachment of the entering alkyl group to C or to N.   From experience with the alkylation of other metal compounds there is no necessary connection between the attachment of the metal (if other than merely ionic) and the attachment of the alkyl group which is not ionic.   The clue may lie in the double salt formation with AgCN.   Treatment with KCN liberates the isonitrile

$$RNC \cdot AgCN + KCN \rightarrow KAg(CN)_2 + RNC$$

[19] Mann, Cookson. *Nature* **157,** 846 (1946).
[20] "Org. Syntheses."
[21] Shemyakin. *C. A.* **34,** 3676 (1940).

2. Primary amines, with chloroform and a base, give the *carbylamine reaction.*[22, 23]

$$RNH_2 + CHCl_3 + 3\ NaOH \rightarrow RNC + 3\ NaCl + 3\ H_2O$$

This is a delicate test for a primary amine because of the powerful odor of the isocyanides.

Isocyanides are more volatile than the cyanides, MeNC, b. 60°, EtNC, b. 79°.

The *reactions* of the isocyanides depend on a peculiar type of unsaturation located entirely on one atom instead of at two adjacent atoms. Nef concluded that the carbon in the isocyanides is truly bivalent (1899). Gautier, the discoverer of the carbylamines, recognized this peculiarity and formulated them as $R-N=C$. This may be written as $R-N\equiv C$ or

$$R : \overset{..}{N} :: C :$$

1. Halogens add even at low temperatures to give $RN=CX_2$, alkyliminocarbonyl halides. Hydrolysis gives HX, CO_2 and RNH_2 proving the $R-N$ linkage.

2. Heating with sulfur gives mustard oils, $RN=C=S$.

3. Water adds to give alkyl formamides, HCONHR, formed from the initial product $RN=CHOH$. Mineral acids split this product to give formic acid and an amine salt. Hydrogen sulfide gives a thioformamide, HCSNHR.

4. HCl gas adds to give compounds of the type $(RN=CHCl)_2 \cdot HCl$.

5. Acid chlorides add to give alkylimido chlorides of α-ketonic acids, $RCOC(Cl)=NR$.

6. Reduction in alkaline solution or with catalysts gives $RNHCH_3$.

7. Oxidation by HgO gives isocyanates, RNCO.

8. The isocyanides give polymers readily.

9. On heating to 250° they rearrange to cyanides.

XXII. OTHER COMPOUNDS CONTAINING A SINGLE CARBON ATOM AND THEIR DERIVATIVES

These compounds may be classified according to the stage of oxidation in the series.

I	II	III	IV
CH_4 CH_3OH	HCHO	HCO_2H	H_2CO_3

I. As examples of the first stage of oxidation various methyl derivatives have been discussed. These include the alcohol, the amine and related substances, the mercaptan, and the halides.

[22] Hofmann. *Ann.* 146, 109 (1868); *Ber.* 3, 767 (1870).
[23] Nef. *Ann.* 270, 267 (1892).

II. Second stage of oxidation. Formaldehyde, methylene halides, polymers of $(CH_2=NH)$ and (CH_2S), as well as diazomethane have been discussed.

III. Third stage of oxidation. Formic acid, formamide and hydrocyanic acid have already been covered.

1. **Carbon monoxide,** CO, is the classical example of a compound of divalent carbon (Nef).[1] Its similarity to nitrogen in both physical properties and chemical inertness under mild conditions may be shown electronically (Langmuir).

$$: \overset{..}{N} : N : \qquad\qquad : \overset{..}{C} : O :$$

In each case two atoms held by an electron pair are surrounded by an octet of electrons. The catalysts which play so important a role in the chemistry of nitrogen and of carbon monoxide apparently serve to open up this stable arrangement.

Preparation.

(1) On the large scale by the action of steam or CO_2 on hot coal and of methane with steam.

$$H_2O + C \rightleftharpoons CO + H_2 \quad \text{Water gas}$$
$$CO_2 + C \rightleftharpoons 2\,CO$$
$$H_2O + CH_4 \rightarrow CO + 3\,H_2$$

(2) On the laboratory scale by the action of formic acid with H_2SO_4.

Reactions.

(1) With oxygen to give CO_2.

(2) At high temperatures to give CO_2 and C.

(3) With steam and catalysts at high temperatures.

$$CO + H_2O \rightarrow CO_2 + H_2$$

(4) As a high temperature reducing agent in the blast furnace preparation of iron. In this process the finely divided carbon formed by the thermal decomposition of CO plays an important part.

(5) With hydrogen and catalysts, an extraordinary range of products is possible. Probably the oldest reaction is the production of methane.

Methanol Synthesis. This is the chief source of methanol and a variety of higher alcohols (pp. 104, 113, 129, 131).

Fischer-Tropsch Synthesis and related processes. Since 1923, these processes have aroused an increasing amount of world-wide activity.[2-8]

---

[1] Sidgwick. *Chem. Revs.* 9, 77.
[2] Lane, Weil. *Petroleum Refiner* 25, No. 8, 87 (1946); No. 9, 423 (1947); No. 10, 493 (1947).
[3] Storch. *Ind. Eng. Chem.* 37, 340 (1945).
[4] Alden. *Oil Gas J.* 45, No. 27, 79 (1946).
[5] Foster. *Oil Gas J.* 43, No. 15, 99 (1944).
[6] Haensel. *Natl. Petroleum News* 37, R955 (1945).
[7] Herington. *Chemistry & Industry* 1946, 346.
[8] Dewey. Office of Publications Board Rep. 289.

The first practical development was the *Synthol Process*, operated at high pressures and temperatures, which produced a mixture of hydrocarbons and oxygenated compounds valuable as chemical raw materials but of little use as motor fuel. The *Kogasin Process*, operated at lower pressures and temperatures, was more successful. It gave a low octane gasoline which could be re-formed for aviation fuel. Moreover, it gave a good quality of Diesel fuel, lubricating oils and paraffin. The latter was used for making artificial higher fatty acids for soaps and even for edible fats. Large numbers of Kogasin plants played an increasingly important part in the German economy from 1934 to 1944.

At present there are two main types of catalysts for the Fischer-Tropsch reaction. One is composed principally of cobalt and the other type contains iron or iron oxides as the main ingredient. Cobalt catalysts produce principally paraffins and olefins having relatively straight chains. Iron or iron oxide-containing catalysts give hydrocarbon mixtures resembling some of those derived from petroleum, and accordingly this is an important process for making synthetic gasoline, kerosene and Diesel fuel. In addition, a variety of oxygenated compounds are produced. These comprise aliphatic alcohols, ketones, aldehydes and acids.

The *Oxo Process* involves the addition of carbon monoxide to the olefins from the Fischer-Tropsch operations. Hydrogenation of the resulting aldehydes gives primary alcohols. In Germany, the 12–18 carbon olefins have been converted to the next higher alcohols in this way for the production of sodium alkyl sulfonate detergents. The 7–10 carbon olefins similarly give alcohols for the production of phthalic ester plasticizers. The liquid olefins are treated with a cobalt catalyst and water gas at about 200 atm. and 160° to produce aldehydes and alcohols in about 4:1 ratio. The preparation and utilization of the various Fischer-Tropsch catalysts require the utmost care.

Wherever natural gas is readily available, the *synthesis gas* for the above processes is obtained from methane and steam as indicated above because of the great advantages over the *water gas* process.

(6) With acetylene, nickel carbonyl, ethanol, and a trace of hydrogen chloride, carbon monoxide forms ethyl acrylate (Reppe).

(7) Potassium unites with CO to form a compound $C_6(OK)_6$, the potassium compound of hexahydroxy benzene. The first step is like that in the pinacol reduction of acetone and in the formation of the metal ketyls.

$$CO + K \rightarrow \overset{|}{\underset{|}{-C}}-O-K$$

(8) Carbon monoxide unites with many heavy metals to form metal carbonyls, $M(CO)n$. These are volatile liquids which decompose to the metal and CO at high temperatures. The Mond process for extracting nickel depends on the formation, volatilization and thermal decomposition of nickel carbonyl,

$Ni(CO)_4$. Iron carbonyls, $Fe(CO)_5$ and $[Fe(CO)_4]_3$ are known. Carbon monoxide can pass through a red hot iron container by alternate formation and decomposition of iron carbonyls. Iron carbonyl may act as an anti-knock in gasoline. It has never been used successfully.[9]

(9) Carbon monoxide reacts with NaOH under pressure to give sodium formate.

(10) With sodium alcoholates at about 200°, sodium salts of carboxylic acids are formed.

$$RONa + CO \rightarrow RCO_2Na$$

Sodium ethylate with CO at 20° and 80 atm. gives an addition product which reacts with water to give sodium formate and EtOH instead of sodium propionate which is produced at higher temperatures.[10]

(11) With methanol and ethanol and certain catalysts, it gives acetic and propionic acids. Other catalysts give formic esters.

(12) With Grignard reagents, acyloins and olefins are formed.

$$2\,CO + 2\,RMgX \rightarrow 2\,R{-}\underset{\|}{C}{-}OMgX \rightarrow$$

$$\begin{array}{ccc} R{-}\overset{\|}{C}{-}OMgX & R{-}COH & R{-}CHOH \\ R{-}\underset{\|}{C}{-}OMgX & R{-}\underset{\|}{C}{-}OH & R{-}CO \end{array}$$

(13) The olefins formed from RMgX with n carbons contain $2n+1$ carbons, with the double bond next to the middle. Thus ethyl, isopropyl, n-butyl, and isoamyl Grignard reagents give respectively 2-pentene; 2,4-Me_2-2-pentene; 4-nonene, and 2,8-Me_2-4-nonene. The yields are as high as 70%.

(14) Carbon monoxide forms a more stable compound with hemoglobin than does oxygen. This accounts for its toxic nature even in small amounts. Its presence in automobile exhaust gases presents a hazard which may have to be eliminated by its catalytic conversion to CO_2.[11]

2. **Orthoformates,** $HC(OR)_3$, are obtained from chloroform and sodium alcoholates. They are hydrolyzed by acids to give formic acid and ROH. With Grignard reagents they give acetals of aldehydes.

$$HC(OC_2H_5)_3 + RMgX \rightarrow C_2H_5OMgX + RCH(OC_2H_5)_2$$

3. **Formyl chloride** has been made at $-170°$. It undoubtedly exists in equilibrium with CO and HCl.

$$CO + HCl \rightleftharpoons HC{\nearrow}^{Cl} = O$$

[9] Pitesky, Wiebe. *Ind. Eng. Chem.* **37**, 577 (1945).
[10] Scheibler, Frikell. *Ber.* **67B**, 312 (1934).
[11] Frazer. *J. Phys. Chem.* **35**, 405 (1931).

A mixture of CO + HCl gases is used with $AlCl_3$ to introduce the aldehyde group into aromatic compounds.

4. **Formimido chloride** is obtained by the action of dry HCl gas on HCN.

$$HC{\equiv}N + HCl \;\rightleftharpoons\; H{-}\overset{\displaystyle Cl}{\underset{|}{C}}{=}NH$$

It is unstable. With alcohols it gives the corresponding formimido ethers (really esters).

$$HC\overset{\displaystyle Cl}{\underset{|}{}}{=}NH + ROH \rightarrow HC\overset{\displaystyle OR}{\underset{|}{}}{=}NH \cdot HCl$$

Formimido chloride is used in another synthesis for aromatic aldehydes (Gattermann). The condensing agent is anhydrous $ZnCl_2$. The needed reagents are conveniently obtained by treating dry solid zinc cyanide suspended in ether with dry hydrogen chloride gas.[12]

$$Zn(CN)_2 + 4\,HCl \rightarrow ZnCl_2 + 2\,H{-}\overset{\displaystyle Cl}{\underset{|}{C}}{=}NH$$

5. **Fulminic acid** (Knallsäure), carbon monoxide oxime, $H-O-N{=}C$, is unstable in the free state. Mercury fulminate was discovered and is still made by the action of mercury, nitric acid and ethyl alcohol. Mercury fulminate is a detonator for explosives. Mercury fulminate can also be made from sodium nitromethane and mercuric salts.

Silver nitrate crystals cannot be safely washed with alcohol because of the possibility of the formation of silver fulminate which is even more dangerous than the mercury salt.[13]

Reactions of fulminic acid. (1) Polymerization takes place spontaneously even at low temperatures to give a trimer, "meta fulminic acid" which is non-explosive.

(2) Hydrolysis by dilute HCl.

$$H{-}O{-}N{=}C + H_2O \rightarrow \left[HO{-}N{=}\overset{\displaystyle H}{\underset{|}{C}}{-}OH \right] \rightarrow HCO_2H + NH_2OH$$

(3) Concentrated HCl, nitrous acid and other substances add to the bivalent carbon (Nef).

$$H{-}O{-}N{=}C \;\rightarrow\; HO{-}N{=}\overset{\displaystyle H}{\underset{|}{C}}{-}Cl,$$

$$HO{-}N{=}\overset{\displaystyle H}{\underset{|}{C}}{-}NO_2, \qquad HO{-}N{=}CCl_2$$

[12] Adams, Montgomery. *J. Am. Chem. Soc.* 46, 1518 (1914).
[13] Tully. *Chem. Eng. News* 19, 250 (1941).

Fulminic acid.[14]

IV. Fourth state of oxidation, all H atoms of methane replaced by OH, NH$_2$, SH or their equivalent. Since about twenty such compounds or the corresponding derivatives are known, some system of classification is necessary. This might be according to the attachment of the carbon valences to O, N and S atoms. The chlorides would fall under the corresponding $-$OH compounds. In several cases the existence of tautomerism places a given compound in two classes.

DISTRIBUTION OF CARBON VALENCES

(1) *4 to O*	(2) *3 to O, 1 to N*	(3) *2 to O, 2 to N*	(4) *1 to O, 3 to N*
C(OH)$_4$ Orthocarbonic acid	NH$_2$CO$_2$H Carbamic acid	O$=$C(NH$_2$)$_2$ $\rightleftharpoons$ Urea	$\overset{\text{NH}_2}{\underset{\vert}{\text{HO—C}}}=$NH
O$=$C(OH)$_2$ Carbonic acid	NH$_2$COCl	O$=$C$=$N—H $\rightleftharpoons$ Isocyanic acid	H—O—C$\equiv$N Cyanic acid
CO$_2$ CCl$_4$ COCl$_2$			Cl—C$\equiv$N

(5) *4 to N*	(6) *3 to O, 1 to S*	(7) *2 to O, 2 to S*	(8) *1 to O, 3 to S*
HN$=$C(NH$_2$)$_2$ Guanidine	$\overset{\text{OH}}{\underset{\vert}{\text{H—S—C}}}=$O $\rightleftharpoons$ S$=$C(OH)$_2$ Monothiocarbonic acid		
NH$_2$—C$\equiv$N Cyanamide		O$=$C$=$S Carbon oxysulfide	
		O$=$C(SH)$_2$ $\rightleftharpoons$ $\overset{\text{SH}}{\underset{\vert}{\text{HO—C}}}=$S Dithiocarbonic acid Xanthic acid	
		CSCl$_2$	

[14] Ahrens. *Sammlung Chemische-Technischen Vorträge* 1909, 77.

(9)	(10)	(11)	(12)
4 to S	*3 to S, 1 to N*	*2 to S, 2 to N*	*1 to S, 3 to N*

$$S = C(SH)_2$$
Trithiocarbonic acid

$$\begin{array}{c} SH \\ | \\ H_2N\!-\!\!C\!=\!\!S \end{array} \rightleftharpoons HN\!=\!C(SH)_2$$
Dithiocarbamic acid

$$CS_2$$
Carbon disulfide

$$S\!=\!C\!=\!N\!-\!H \rightleftharpoons HS\!-\!C\!\equiv\!N$$
Isothiocyanic acid   Thiocyanic acid

$$S\!=\!C(NH_2)_2 \rightleftharpoons \begin{array}{c} NH_2 \\ | \\ HS\!-\!C\!=\!NH \end{array}$$
Thioruea

(13)	(14)	(15)
2 to O, 1 to N, 1 to S	*1 to O, 1 to N, 2 to S*	*1 to O, 2 to N, 1 to S*

$$\begin{array}{c} SH \\ | \\ O\!=\!C\!-\!NH_2 \end{array} \rightleftharpoons \begin{array}{c} S \\ \| \\ HO\!-\!C\!-\!NH_2 \end{array} \rightleftharpoons \begin{array}{c} SH \\ | \\ HO\!-\!C\!=\!NH \end{array}$$
Monothiocarbamic acid

A. CARBONIC ACID AND ITS DERIVATIVES

1. **Ortho carbonic acid.** $C(OH)_4$, is incapable of existence. Orthocarbonates, $C(OR)_4$, are obtained from sodium alcoholates and chloropicrin.

$$4\ RONa + Cl_3CNO_2 \rightarrow C(OR)_4 + 3\ NaCl + NaNO_2$$

$C(OEt)_4$, b. 159°.

It might be expected that the action of carbon tetrachloride with alcoholates would give orthocarbonates. Such is not the case, however. The production of formates in this reaction was long explained on the basis that the first change is the replacement of one of the halogens by hydrogen due to the "positive" nature of the halogen. This explanation has been eliminated by the work of Kharasch, Alsop, and Mayo[15] on carbon tetraiodide.

Carbon tetraiodide crystallizes from chloroform in a highly evacuated system in small red crystals. Under favorable conditions, including absence of air and light, these do not decompose until heated to 162–165°. In presence of light and air the decomposition point is greatly lowered.

The conclusion is that there is no factual basis for the assumption of a positive iodine atom in carbon tetraiodide. This would be even truer of any such assumption about the bromide and the chloride.

Chloropicrin differs from carbon tetrachloride in having all four groups replaceable by negative alkoxyl groups. The mechanism of this "replacement"

[15] Kharasch et al. *J. Org. Chem.* 2, 76 (1937).

is not understood. Perhaps its first step consists in an addition to the nitro group.

$$\underset{\substack{\|\\ \text{O}}}{\text{Cl}_3\text{C}-\text{N}}\text{O} \rightarrow \underset{\substack{|\\ \text{OR}}}{\underset{\substack{|\\ \text{ONa}}}{\text{Cl}_3\text{C}-\text{N}}}\text{O} \rightarrow \text{NaNO}_2 + \text{Cl}_3\text{C}-\text{OR}$$

Ortho carbonates, on heating with boric anhydride, give ordinary carbonates.

$$\text{C(OR)}_4 + 2\ \text{B}_2\text{O}_3 \rightarrow \text{OC(OR)}_2 + \text{R}_2\text{B}_4\text{O}_7$$

2. **Carbonic acid,** $O = C(OH)_2$, is known only in its salts and esters. Alkyl carbonates are prepared by heating alcohols with chloroformic esters prepared from phosgene.

$$\text{ClCO}_2\text{R} + \text{R'OH} \rightarrow \text{HCl} + \text{RR'CO}_3$$

If different alkyl groups are used the heating must be kept to a minimum to avoid ester interchange

$$\text{RR'CO}_3 + \text{R'OH} \rightarrow \text{R}_2'\text{CO}_3 + \text{ROH}$$

in which ROH represents the more volatile alcohol.

Di-t-butyl carbonate can be prepared from sodium t-butylate and phosgene.[16]

The alkyl carbonates give the ordinary ester reactions but are less active than the corresponding esters of the carboxylic acids. Thus ethyl carbonate is not easily hydrolyzed to alcohol and CO_2.

Dicetyl carbonate is an artificial wax, m. 45° (MCW).

Ethyl carbonate reacts with Grignard reagents to give esters or tertiary alcohols, depending on the proportions of reagents used and the nature of the groups involved.

$$\text{RMgX} + \text{O} = \text{C(OC}_2\text{H}_5)_2 \rightarrow \text{RCO}_2\text{C}_2\text{H}_5 + \text{C}_2\text{H}_5\text{OMgX}$$
$$3\ \text{RMgX} + \text{O} = \text{C(OC}_2\text{H}_5)_2 \rightarrow \text{R}_3\text{COMgX} + 2\ \text{C}_2\text{H}_5\text{OMgX}$$

In the case of groups like naphthyl and tertiary butyl, the reaction is relatively slow and stops at the ester stage.

Alpha H atoms in esters, ketones, and nitriles can be replaced by $-\text{CO}_2\text{R}$ (carbaloxylation) by treatment with alkyl carbonates and sodium alkoxides.[17–19] This reaction gives good preparative methods for monosubstituted malonic esters, beta-ketoesters, and alpha-cyanoesters respectively.

Alkyl carbonates are useful solvents for the metalation and alkylation of malonic esters, beta-ketoesters and alpha-cyanoesters.[20, 21]

[16] Choppin, Rogers. *J. Am. Chem. Soc.* **70**, 2967 (1948).
[17] Wallingford, Homeyer, Jones. *J. Am. Chem. Soc.* **63**, 2056 (1941).
[18] *ibid.* 2252 (1941).
[19] *ibid.* **64**, 576 (1942).
[20] Wallingford, Jones. *J. Am. Chem. Soc.* **64**, 578 (1942).
[21] *ibid.* **64**, 580 (1942).

3. **Carbon dioxide,** carbonic anhydride, CO_2, is involved in many important organic reactions. The most important reaction is its union with water under the influence of sunlight and chlorophyll to give carbohydrates and oxygen. The first step is probably the production of formaldehyde.

$$CO_2 + H_2O \rightarrow HCHO + O_2$$

Starting with this formaldehyde, plants and animals elaborate the countless carbon compounds necessary to their existence. The production of formaldehyde from CO_2 and H_2O and the conversion of the formaldehyde to simple sugars has been achieved by the use of ultraviolet light.[22]

One of the most important sources of carbon dioxide is the action of enzymes in which they decarboxylate organic acids as in the production of ethyl alcohol.

Carbon dioxide enters into a number of ordinary organic reactions. Thus it adds sodium-carbon compounds, Grignard reagents, sodium alcoholates, sodium mercaptides, and ammonia to give RCO_2Na, RCO_2MgX, $ROCO_2Na$, $RSCO_2Na$, $NH_2CO_2NH_4$ and NH_2CONH_2.

The reactions of carbon dioxide are being utilized in making starting materials for chemical and biological studies using the carbon isotopes which are becoming available as the oxides of C^{13} (Sun Oil Company) and radioactive C^{14} (Uranium pile). The reduction of carbon dioxide to CH_3OH by $LiAlH_4$ is one of the most convenient techniques.[23, 24]

4. **Carbon tetrachloride,** as has been seen, does not act as the acid chloride of orthocarbonic acid. First of all, it is unexpectedly unreactive. Secondly, when it does react, it does not give derivatives of carbonic acid. With alcoholic KOH at 100° it gives KCl, CO and $HC(OC_2H_5)_3$ but no trace of K_2CO_3.

5. **Carbonyl chloride,** phosgene, $O=CCl_2$, b. 8°, is very poisonous. On a small scale it is made from carbon tetrachloride and 80 per cent oleum (fuming sulfuric acid containing 80 parts SO_3 added to 100 parts of pure H_2SO_4).

$$CCl_4 + SO_3 \rightarrow COCl_2 + S_2O_5Cl_2, \text{ pyrosulfuryl chloride}$$

Commercially it is made from CO and Cl_2 with a catalyst such as activated carbon.

$$CO + Cl_2 \rightarrow COCl_2$$

Sunlight hastens the reaction. (Hence the name phosgene, Davy).

Phosgene gives acid chloride reactions: (1) with alcohols in the cold only one Cl reacts.

$$ClCOCl + ROH \rightarrow HCl + ClCO_2R$$

The resulting chloroformic esters are themselves acid chlorides.

[22] *Ann. Rep. Chem. Soc.* (London) **1924**, 66.
[23] Nystrom, Yanko, Brown. *J. Am. Chem. Soc.* **70**, 441 (1948).
[24] Calvin et al. "Isotopic Carbon." Wiley, 1949. p. 168.

(2) On warming with alcohols, phosgene gives alkyl carbonates.

$$COCl_2 + 2 ROH \rightarrow 2 HCl + R_2CO_3$$

(3) With NH_3 and primary and secondary amines it gives urea and N-substituted ureas. With primary amines an excess of phosgene gives isocyanates.

$$COCl_2 + RNH_2 \rightarrow (RNH—C=O) \rightarrow HCl + R—N=C=O$$
$$\qquad\qquad\qquad\quad |$$
$$\qquad\qquad\qquad Cl$$

(4) Heated with NH_4Cl, it gives carbamyl chloride.

$$COCl_2 + NH_4Cl \rightarrow NH_2COCl + 2 HCl$$

(5) It reacts in the Friedel-Crafts reaction with aromatic compounds to give acid chlorides and ketones. In one important case, it reacts even without $AlCl_3$. This is the reaction with dimethylaniline to give Michler's ketone, an important intermediate in the making of triphenylmethane dyes.

$$2 Me_2NC_6H_5 + COCl_2 \rightarrow 2 HCl + (Me_2NC_6H_4)_2CO$$

It was this commercial use for phosgene which first made it available as a war gas. Because of its slow reactions with most substances, it was difficult to protect troops against this gas until it was found to react readily, even in small concentrations, with sodium phenolate, C_6H_5ONa, to give diphenyl carbonate.

Phosgene.[25]

In general, the derivatives of carbonic acid react much less rapidly than the corresponding derivatives of the fatty acids. This is true of the following:

$$\left.\begin{array}{l} EtOCO_2Et \\ NH_2CO_2Et \\ NH_2CONH_2 \end{array}\right\} \quad \begin{array}{l} RCO_2Et \\ \\ RCONH_2 \end{array}$$

$$\left.\begin{array}{l} ClCOCl \\ NH_2COCl \\ EtOCOCl \end{array}\right\} \quad RCOCl$$

6. **Chloroformic acid, "chlorocarbonic" acid,** $ClCO_2H$, is known only in its derivatives. Chloroformic esters are made by treating primary or secondary alcohols in the cold with phosgene. Chloroformic esters give a wide range of reactions. With ammonia, primary and secondary amines, amides, imides, and amino acids, they give urethans, alkyl imidodicarboxylates, N-carbethoxy-imidodicarboxylates, allophanates, cyanuric acid, alkyl guanidine-dicarboxylates, and alkyl cyanamidedicarboxylates. Ethyl chloroformate gives the expected reactions with Grignard reagents, with alkyl and allyl

[25] Dyson. *Chem. Rev.* 4, 109.

halides and metals, with sodium sulfide and sodium mercaptides, with metallo derivatives of malonic ester and the like. It is useful in "protecting" phenolic hydroxyl groups by replacing the active H by the carbethoxy group which is reasonably inert to acid reagents but which can be readily removed by bases or by hydrogenolysis.

$$-OH + ClCO_2Et \rightarrow -OCO_2Et$$

B. AMIDES OF CARBONIC ACID

1. **Aminoformic acid, carbamic acid,** NH_2CO_2H, is known only in its derivatives.

(a) *Ammonium carbamate* is formed directly from NH_3 and CO_2.[26]

$$2 NH_3 + CO_2 \rightarrow NH_2CO_2NH_4$$

This is a very soluble white crystalline compound. The carbamates of the alkaline earth metals are soluble. In this they resemble the bicarbonates.

$$NH_2—CO—OCaO—CO—NH_2 \qquad HO—CO—OCaO—CO—OH$$

Ca carbamate   Ca bicarbonate

Ammonium carbamate is hydrolyzed to ammonium carbonate by water at 60°.

(b) Alkyl carbamates are the *urethans*, NH_2CO_2R. They may be prepared (1) by treating chloroformic esters with ammonia or primary or secondary amines (2) by treating carbamyl chloride, NH_2COCl, with an alcohol or, (3) by heating urea to a high temperature with an alcohol,

$$CO(NH_2)_2 + ROH \rightarrow NH_2CO_2R + NH_3$$

(4) by adding alcohols to isocyanic acid or isocyanates.

$$C_6H_5N = C = O + ROH \rightarrow C_6H_5N = C—OH \rightarrow C_6H_5NH—C = O$$
$$\qquad\qquad\qquad\qquad\qquad\qquad\qquad | \qquad\qquad\qquad\qquad | $$
$$\qquad\qquad\qquad\qquad\qquad\qquad\qquad OR \qquad\qquad\qquad\qquad OR$$

Phenyl isocyanate   A phenyl urethan

The phenyl- and naphthylurethans are readily purified crystalline substances very useful in identifying alcohols. The urethans give the reactions of esters and acid amides. They form stable sodium derivatives which react with alkyl and acyl halides to give N-substituted urethans. In this way ethyl chlorocarbonate gives ethyl imidodicarboxylate, $NH(CO_2Et)_2$, m. 50°. The N-alkylated urethans react with nitrous acid and with nitric acid to give N-nitroso and N-nitro compounds. The reduction of these give mono-substituted hydrazines. As has been seen, diazomethane is made by the action of alkali on nitroso-N-methyl urethan, $MeN(NO)CO_2Et$.

Urethan, NH_2CO_2Et, m. 50°, b. 180°, is a hypnotic.

[26] Egan, Potts, Potts. *Ind. Eng. Chem.* **38**, 454 (1946).

2. **Carbamic acid chloride,** carbamyl chloride, NH_2CO-Cl, m. 50°, b. 62°, is obtained by adding dry HCl to cyanic acid and by heating phosgene with ammonium chloride to 400°. It reacts vigorously as an acid chloride even in the Friedel and Crafts reaction. On standing it changes to cyamelide.

3. **Urea, carbamide,** $NH_2-CO-NH_2$, m. 132°, is the chief nitrogenous product of protein metabolism. A human adult excretes about 30 g. per day. It is obtained from urine by evaporation and addition of nitric acid to precipitate the slightly soluble urea nitrate. Historically, the most important synthesis is from a cyanate and an ammonium salt.

$$NH_4-O-C\equiv N \rightarrow NH_2-\overset{\overset{\displaystyle O}{\|}}{C}-NH_2$$

In the laboratory, it is formed in varying yields by the action of ammonia on phosgene, alkyl carbonates, alkyl chloroformates, urethans and carbamyl chloride.

Commercial Syntheses.

(1) From $CO_2 + NH_3$. The ammonium carbamate first formed is heated under pressure.

$$NH_2CO_2NH_4 \rightarrow H_2O + NH_2-CO-NH_2$$

This is like the ordinary formation of an acid amide from its ammonium salt.

(2) By the partial hydrolysis of cyanamide from calcium cyanamide.

$$NH_2-C\equiv N + H_2O \rightarrow NH_2-\underset{\underset{\displaystyle OH}{|}}{C}=NH \rightleftharpoons NH_2-CO-NH_2$$

Formula of Urea. The best evidence indicates it to be a tautomeric substance. An important resonance occurs to stabilize the commonly recognized tautomers.

$$\overset{\displaystyle NH_2}{\underset{\displaystyle NH_2}{\overset{\oplus}{=}}}C-O\ominus, \quad \overset{\displaystyle NH_2}{\underset{\displaystyle NH_2}{|}}C=O \rightleftharpoons \overset{\displaystyle NH_2}{\underset{\displaystyle HN}{|}}=C-OH \rightleftharpoons \overset{\displaystyle NH_3\oplus}{\underset{\displaystyle HN}{|}}=C-O\ominus$$

The last formula has been proposed because of the ease with which urea loses NH_3 to form cyanic acid. Such an assumption is hardly necessary in view of the ease with which substances containing the grouping $C(OH)NH_2$ lose NH_3 to give a $C=O$ group. The reactions are different in many respects from those of ordinary acid amides. Consequently, the OH formula has been favored by some. On the other hand X-ray and crystallographic evidence indicate the carbamide formula (Hendricks 1928). The mass of evidence bearing on the structure of urea includes peculiarities in the following reactions: (a) phosgene + NH_3; (b) urethan, NH_2CO_2Et, with NH_3; (c) effect of heat on alkylated ammonium carbamates; (d) the reaction of urea with nitrous acid;

(e) the formation of nitroso ureas from monosubstituted ureas and nitrous acid instead of primary amines or alcohols,

$$[\text{MeNHCONH}_2 + \text{HNO}_2 \rightarrow \text{MeNHCO}_2\text{H} \rightarrow \text{MeNH}_2 \rightarrow \text{MeOH}]$$

$$\text{MeNHC(OH)NH} + \text{HNO}_2 \rightarrow \text{MeN(NO)C(OH)NH} \quad (90\% \text{ yield})$$

(f) action of Me_2SO_4 to give $\text{HN}=\text{C(OMe)NH}_2$ instead of an $\text{N}-\text{Me}$ compound; (g) the methyl "ether" of isourea is a strong base somewhat like guanidine. These peculiarities are additional evidence for the tautomeric nature of urea.

Reactions. (1) Slow heating removes NH_3 from two molecules to give *biuret*, $\text{NH}_2-\text{CO}-\text{NH}-\text{CO}-\text{NH}_2$. Biuret reacts in basic solution with copper salts to give an intense violet color ("biuret reaction"). Substituted biurets have narcotic effects.[27]

More rapid heating of urea gives cyanic acid which immediately polymerizes.

(2) Acids form salts, for instance, the difficultly soluble nitrate

$$\text{NH}_2\text{CONH}_2 \cdot \text{HNO}_3.$$

The chloride, phosphate and oxalate are also known. It also forms crystalline compounds with various inorganic salts including NaCl and AgNO_3.

(3) It gives salts with metals such as mercury.

(4) Treatment with aqueous acids or alkalies gives NH_3 and CO_2 or their reaction products. An enzyme (*urease*, Sumner) found in soybeans and jackbeans produces the same change at room temperature.

(5) Nitrous acid reacts with urea

$$\text{CO(NH}_2)_2 + 2\,\text{HNO}_2 \rightarrow \text{CO}_2 + 2\,\text{N}_2 + 3\,\text{H}_2\text{O}$$

This reaction never gives the calculated amount of gaseous products. This is not surprising when it is recalled that the "replacement" of $-\text{NH}_2$ by $-\text{OH}$ often involves rearrangements.

(6) Hypohalite solutions or bases and halogens react nearly according to the following equation

$$\text{CO(NH}_2)_2 + 3\,\text{KOBr} + 2\,\text{KOH} \rightarrow \text{K}_2\text{CO}_3 + \text{N}_2 + 3\,\text{KBr} + 3\,\text{H}_2\text{O}$$

In this case a rearrangement (Hofmann) is known to take place.

$$
\begin{array}{ccccc}
\text{NH}_2 & \text{+NH} & \text{NH}_2 & & \\
| & | & | & \xrightarrow{\text{H}_2\text{O}} & \\
\text{C}=\text{O} \rightarrow & \text{C}=\text{O} \rightarrow & \text{NH} & \longrightarrow & \text{NH}_2 + \text{CO}_2 \\
| & | & | & & | \\
\text{NH}_2 & \text{NH}_2 & \text{+C}=\text{O} & & \text{NH}_2
\end{array}
$$

[27] Anderson et al. *Science* 95, 254 (1942).

The hydrazine can be obtained as an insoluble product by adding benzaldehyde. If the reaction is carried out at 0° with only one equivalent of hypohalite, hydrazine is the chief product.

(7) Urea reacts with chlorine and water at 0° to give mono- and di-N-chloroureas.[28]

$$CO(NH_2)_2 + Cl_2 \rightarrow NH_2CONHCl + HCl$$

The chlorine in this compound is "positive." It liberates iodine from iodides. With acids it gives explosive nitrogen chloride.

$$3 \ NH_2CONHCl + 5 \ HCl + 3 \ H_2O \rightarrow 3 \ CO_2 + 5 \ NH_4Cl + NCl_3$$

The dichloro compound, $CO(NHCl)_2$, has similar properties. With NH_3 it gives "para-urazine," which was formerly given the structure $CO(NHNH)_2CO$ but is probably amino-urazole,

NH—CO
| \
           N—NH₂
| /
NH—CO

Both structures agree with the fact of hydrolysis by sulfuric acid to CO_2 and hydrazine sulfate.

(8) Formaldehyde condenses with urea to give *methylol ureas*,

$$NH_2CONHCH_2OH$$

etc. The methylol groups can combine with more urea to form more and more complex molecules until a resin is formed. Urea-formaldehyde resins are among the most important of the plastic industry. Uformite, Plaskon, Beetle, Beckamine.

Wood is hardened by impregnation with methylolurea, $NH_2CONHCH_2OH$.

(9) In the presence of EtONa, urea reacts with malonic esters to give the hundreds of barbituric acid derivatives which have been used or proposed as soporifics (p. 436).

(10) Urea is determined in blood or urine colorimetrically by means of the yellow pigment formed with diacetyl monoxime.[29]

Urea exerts a beneficial effect on slow healing wounds.

Enormous amounts of urea are used in plastics and in fertilizers. It is also used in smokeless powder as a stabilizer to unite with traces of nitrous acid formed by slow decomposition.

[28] Ellis. "Chemistry of Petroleum Derivatives." Reinhold, 1934. p. 502.
[29] Ormsby. *J. Biol. Chem.* 146, 595 (1942).

Formamidoxime is an isomer of urea obtained from hydroxylamine and HCN. It is known as *isuret* or *isouretine*.

$$HCN + NH_2OH \rightarrow HC = NH \rightleftharpoons HC\text{---}NH_2$$

$$\begin{array}{cc} | & \| \\ NHOH & NOH \end{array}$$

Formamide oxime

4. **Alkyl ureas** are known corresponding to the two usual formulas for urea $(NH_2)_2CO \rightleftharpoons HN = C(NH_2)OH$.

Symmetrical di- and tetra-alkylated ureas may be obtained from phosgene or ethyl carbonate with primary and secondary amines respectively. Unsymmetrical products may be obtained by suitable steps.

The properties of the N-substituted ureas resemble those of urea except that the tetrasubstituted products have no H replaceable by the action of HNO_2 or HNO_3.

O-Substituted ureas or "isoureas" are obtained from cyanamide

$$NH_2 - C \equiv N + ROH + HCl \rightarrow NH_2(RO)C = NH.HCl.$$

5. **Ureides,** acyl ureas, are so numerous and important as to deserve a separate classification. As substituted acid amides they can be prepared in a variety of ways starting either with urea and a suitable acid or its derivative such as the chloride, anhydride or ester or with an acid amide and cyanic acid.

Allophanic acid, $NH_2CONHCO_2H$, the ureide of carbonic acid, is known only as the esters obtained from (1) urethans and cyanic acid, (2) chloroformates and urea, and (3) cyanic acid and alcohols. Its amide is biuret, obtained by heating urea. The allophanates of alcohols may be used for identification purposes. Thus the Me, Et and Pr esters melt at 208°, 191° and 160°.

Large numbers of C-substituted **acetylureas** are known. The sedatives *Carbromal* and *Bromural* (NNR) are respectively $Et_2CBrCONHCONH_2$ and the corresponding alpha-bromoisovalerylurea.

Hydantoin is glycolylurea, 2,4-diketotetrahydro-imidiazole, m. 216°. Its tautomeric form is 2,4-(OH)$_2$-imidazole.

$$\begin{array}{cc} H_2C\text{---}NH\text{---}CO & HC\text{---}NH\text{---}COH \\ | \qquad\qquad | & \| \qquad\qquad \| \\ CO\text{------}NH & HOC\text{------}N \end{array} \rightleftharpoons$$

Partial hydrolysis gives *hydantoic acid*, $NH_2CONHCH_2CO_2H$, glycoluric acid, ureido-acetic acid, m. 156° dec. Further hydrolysis gives NH_3, CO_2 and glycine.

The best synthesis of hydantoin is from glycine and potassium isocyanate to give hydantoic acid which undergoes ring closure on boiling with hydrochloric acid.

Many substituted hydantoins have been studied.[30]

Lactylurea is 5-Me-hydantoin. The N-Me isomer, 1-Me-hydantoin, is obtained by the partial hydrolysis of creatinine, 1-Me-2-iminohydantoin. 5,5-Dialkylhydantoins are readily obtained by the action of hypohalite on the diamides of dialkylmalonic acids. The isocyano group formed from the first amido group reacts with the NH_2 of the other amido group to close the 5-ring.

Dilantin (NNR), used in epilepsy, is the sodium salt of 5,5-diphenyl-hydantoin. 5,5-Disubstituted hydantoins having O or S in one of the substituents have anticonvulsant activity.[31]

Hydantoin-3-acetic acid, $NHCONCH_2CO_2H$, is obtained by heating with

$$\overset{|}{C}H_2\!\!-\!\!\overset{|}{C}O$$

excess HCl the product from $COCl_2$ and ethyl aminomalonate. Through its acid chloride, it gives esters and substituted acid amides which are crystalline solids of definite melting points. Thus the first four normal esters of hydantoin-3-acetic acid melt at 91°, 119°, 116° and 95° while the isobutyl, sec-butyl and isoamyl esters melt at 124°, 142° and 100°.

Cyclic ureides, related to dibasic acids, are readily obtained both synthetically and from certain natural products.

Oxalylurea, parabanic acid, 5-ketohydantoin,

$$\begin{array}{cc} O\!=\!CNHC\!=\!O & HOCN\!=\!COH \\ \overset{|}{}\quad\overset{|}{} \rightleftharpoons \overset{||}{}\quad\overset{|}{} \\ NH\!\!-\!\!C\!=\!O & N\!\!-\!\!C\!=\!O \end{array}$$

is prepared from (1) urea and oxalyl chloride, (2) urea, oxalic acid and $POCl_3$, or (3) by oxidation of uric acid with nitric acid. It forms mono and dibasic salts. Water converts the salts into those of *oxaluric acid,*

$$NH_2CONHCOCO_2H.$$

Methylparabanic acid and Me_2-parabanic acid are obtained by oxidation of methyluric acid and caffeine respectively.

Uracil, obtained from malic acid and urea (p. 398), is 2,4-dihydroxypyrimidine.

Barbituric acid, malonylurea, 2,4,6-triketohexahydropyrimidine, 2,4,6-$(OH)_3$-pyrimidine, $CONHCO$, can be made from urea, malonic acid and $POCl_3$.

$$\overset{|}{}\qquad\overset{|}{}$$
$$NHCOCH_2$$

It gives nitro, nitroso (isonitroso), bromo and metallo derivatives. By means of the latter, two alkyl groups can be introduced. Hydrolysis to dialkyl malonic acids proves them to be 5,5-derivatives. Barbituric acid contains

[30] Johnson, Hahn. *J. Am. Chem. Soc.* **39**, 1255 (1917).
[31] Merritt, Putnam, Bywater. *J. Pharmacol. & Exper. Therap.* **84**, 5 (1945).

several possibilities of triad tautomerism. Its ordinary formula will be seen to resemble the keto form of phloroglucinol.

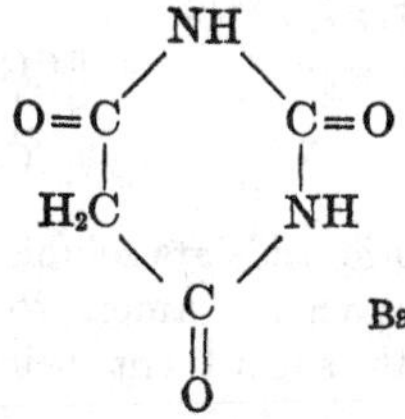

Barbituric acid

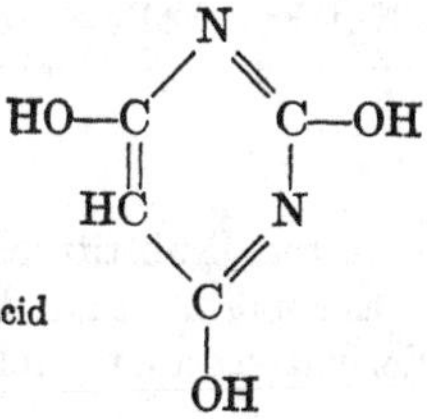

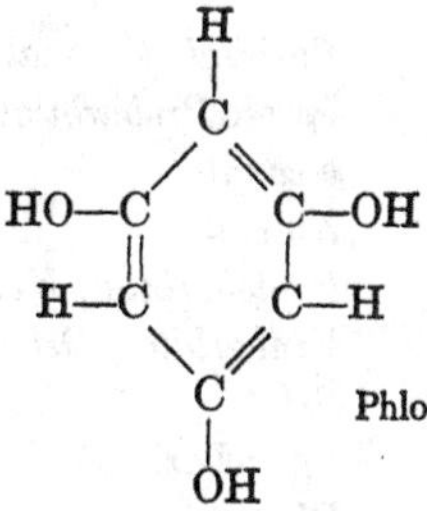

Phloroglucinol

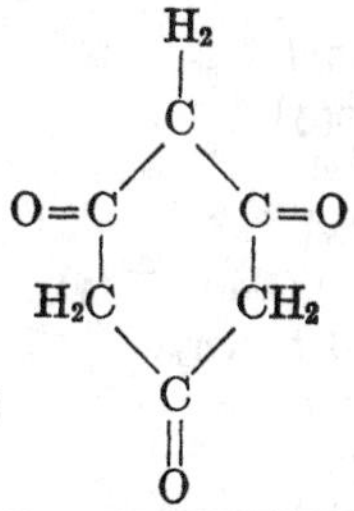

The ease of reaction with nitric acid, nitrous acid, bromine and active metals favor the benzenoid ("phenolic") structure. The pyrimidine ring, a benzene ring with two meta CH groups replaced by N, occurs as such or in tautomeric form in the purines, uric acid and its derivatives.

Barbituric acid is also obtained by condensation of malonic ester and urea in presence of a sodium alkoxide. The primary condensation gives a molecule which can condense internally to give a 6-membered ring.

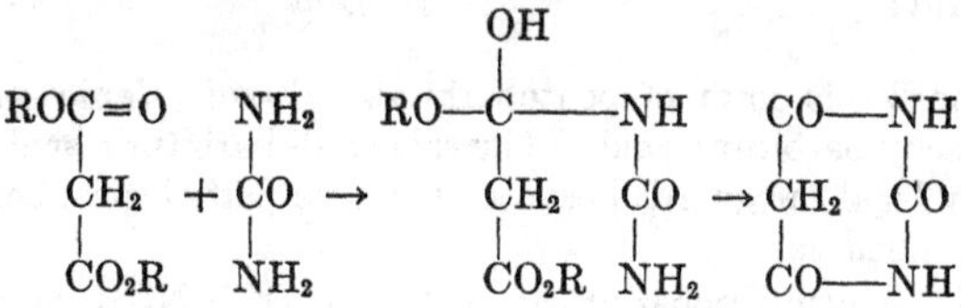

The name of barbituric acid is said to come from its discovery on St. Barbara's Day.

Diethylbarbituric acid, Veronal, Barbital, $Et_2CCO-NH$, is an important

$$\underset{CONHCO}{}$$

soporific made by condensing urea with diethylmalonic ester in presence of NaOEt. Its preparation is typical of the methods used for most of over five hundred 5,5-disubstituted barbituric acids which have soporific power in varying degree and with varying objectionable or dangerous reactions.

$$
\text{Malonic Ester} \xrightarrow{\text{NaOEt}} \xrightarrow{\text{EtBr}} \xrightarrow{\text{NaOEt}} \xrightarrow{\text{EtBr}} \xrightarrow{\text{urea}}
\begin{array}{c}
\text{CO—NH} \\
| \quad\quad | \\
\text{Et}_2\text{C} \quad \text{CO} \\
| \quad\quad | \\
\text{CO—NH}
\end{array}
$$

The sodium salts of the disubstituted barbituric acids are soluble in water and act more rapidly as soporifics than do the free acids. Among the widely used barbituric acid derivatives are the following, the substituents being 5,5- in each case (NNR).

Ethyl, Ethyl	*Barbital, Veronal*
Et, Isopropyl	*Ipral, Probarbital*
Et, *n*-Butyl	*Neonal*
Et, Isoamyl	*Amytal*
Et, 2-Amyl (1-Me-butyl)	*Pentobarbital, Nembutal*
Et, 1-Me-1-butenyl	*Vinbarbital, Delvinal*
Et, *n*-Hexyl	*Ortal*
Et, Phenyl	*Phenobarbital, Luminal*
Et, Δ¹-cyclohexenyl	*Phanodorn, Cyclobarbital*
Methyl, Phenyl	*Rutonal*
n-Propyl, *n*-Propyl	*"Propanal"*
Isopropyl, Allyl	*Alurate, Allonal*
Isopropyl, 2-Br-Allyl	*Nostal (Noctal)*
Isobutyl, Allyl	*Sandoptal*
sec-Butyl, 2-Br-Allyl	*Pernoston (Pernocton)*
2-*sec*-Amyl, Allyl	*Seconal*
Allyl, Allyl	*Dial*

Pentothal is the 2-thio form of pentobarbital. *Evipal (Hexobarbital)* is 1,5-Me₂-5-cyclohexenylbarbituric acid. The sale of all barbituric acid derivatives is controlled by legal safeguards because of serious pathological complications from their continued use.

Violuric acid, 5-isonitrosobarbituric acid, $\begin{array}{c} \text{NHCOC}=\text{NOH} \\ | \quad\quad\quad\quad | \\ \text{CONHCO} \end{array}$, is obtained by action of nitrous acid on barbituric acid or of hydroxylamine with alloxan. Reduction of it or of the 5-nitro compound gives 5-aminobarbituric acid, *uramil,* from which uric acid has been synthesized.

Isovioluric acid, alloxan-6-oxime is known.

Dialuric acid, tartronylurea $\begin{array}{c} \text{NHCOCHOH} \\ | \quad\quad\quad\quad | \\ \text{CONHCO} \end{array}$, is a colorless crystalline compound which readily undergoes oxidation even in air. As $2,4,5,6\text{-}(OH)_4$-pyrimidine, (I), this behavior would be expected. Oxidation gives alloxantin and then alloxan. It is a strong dibasic acid.

Alloxan, mesoxalylurea, NHCOCO, is formed from uric acid or barbituric

$\qquad\qquad\qquad\qquad\qquad$ CONHCO

acid by oxidation. It is strongly acid. With ferrous sulfate it gives an indigo-blue color. Both of these properties correspond to a tauteromerized form of alloxan which is a dihydroxy-quinoid (II) rather than a keto-ureide. It bears the same relation to tartronyl urea that hydroquinone does to quinone.

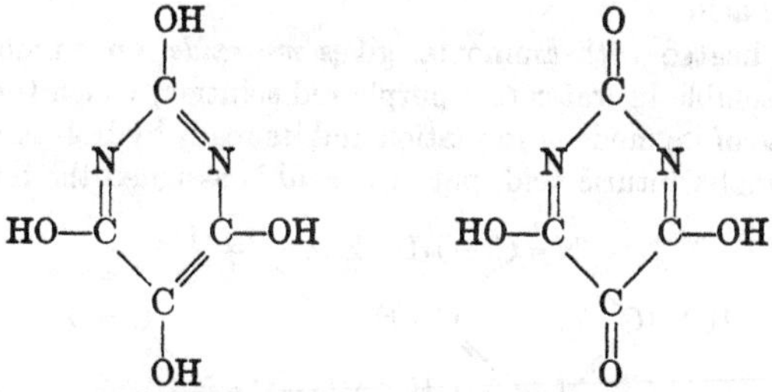

Tartronylurea
I

Mesoxalylurea
II

Alloxantin

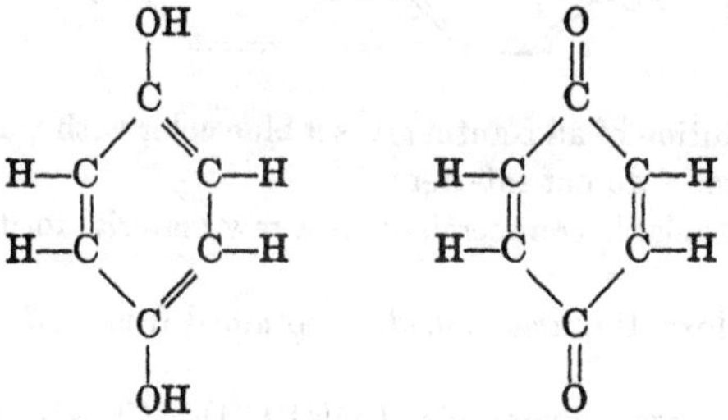

Hydroquinone   Quinone

Quinhydrone

Alloxan combines with $NaHSO_3$. It crystallizes with $4\ H_2O$, one molecule of which is not removed at $100°$ This is presumably combined in the hydrated form of the middle carbonyl group. Reduction converts alloxan to alloxantin and then to tartronyl urea. Alloxan is a good oxidizer, liberating iodine from iodides and converting indigo white to blue. With bases it undergoes internal benzilic acid rearrangement to give an alloxanate,

$$CONHC(OH)CO_2M$$
$$NH{\rule{1.5em}{0.4pt}}CO$$

These reactions resemble those of quinones.

Alloxan is importantly related to riboflavin.[32]

Alloxantin, $C_8H_6O_8N_4$, bears a relation to tartronylurea and mesoxalylurea similar to that which quinhydrone bears to hydroquinone and quinone (p. 526). It can be formed by mixing the two or by mild oxidation of the one or reduction of the other. It was originally obtained by the oxidation of uric acid with nitric acid.

Alloxantin, heated with ammonia, gives *murexide*, an ammonium salt of *purpuric acid*, soluble in water to a purple red solution, which turns blue with bases. Because of its mode of formation and its ready hydrolysis to mesoxalyl- urea and 5-aminobarbituric acid, purpuric acid is assigned the formula

It thus resembles the condensation product of *p*-aminophenol and quinone

An alkaline solution of alloxantin gives a blue color with traces of iron salts. Citrates and tartrates do not interfere.

Alloxantin is available commercially as a raw material for the manufacture of riboflavin.

Tetramethylalloxantin, *amalic acid*, is obtained from caffeine and chlorine water.

Allantoin, 5-carbamidohydantoin, $CONHCHNHCONH_2$, m. 231° dec., is

$$NH{\rule{1.5em}{0.4pt}}CO$$

found in various animal products. It is also formed from uric acid by $KMnO_4$.

Purines and Their Derivatives

Purine is the parent substance for a series of cyclic diureides of which uric acid is the best known. The relation of purine to other important compounds

[32] Banerjee et al. *Science* **101,** 647 (1945).

is shown in the following formulas:

Benzene—b. 80°   Pyrimidine—m. 22°, b. 124°

Cyclopentadiene—b. 41°   Pyrrole—b. 131°   Imidazole—m. 89°, b. 255°

Indene—b. 180°   Indole—m. 52°, b. 254°

Benzimidazole—m. 170°   Purine—m. 217°

All of the rings have conjugated double linkages with corresponding properties.

The common arrangement and numbering of the purine ring system follows:

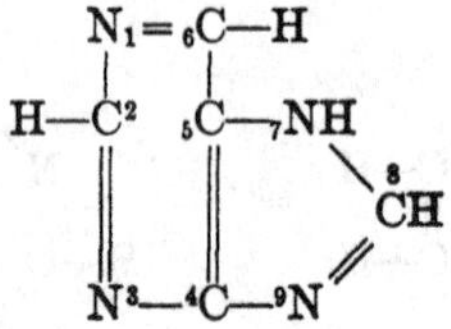

$$N_1 = {}_6C-H$$
$$H-C^2 \quad {}_5C-{}_7NH$$
$$\quad\quad\quad\quad {}^8CH$$
$$N^3-{}^4C-{}^9N$$

This emphasizes the relation to two molecules of urea but conceals the aromatic analogy. Purine is tautomeric with a substance having H in position 9 instead of 7.

Uric acid is usually written as the tri-keto form (A) of 2,6,8-trihydroxy-purine (B) but is actually a tautomeric substance corresponding to both formulas:

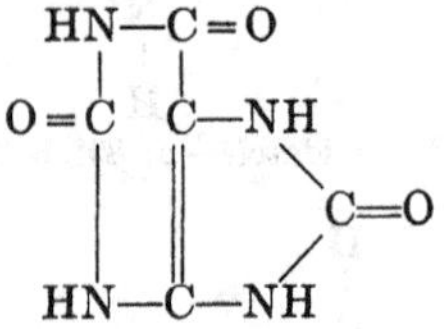

$$HN-C=O \qquad\qquad N=C-OH$$
$$O=C \quad C-NH \qquad HO-C \quad C-N$$
$$\qquad\qquad C=O \qquad\qquad\qquad\qquad C-OH$$
$$HN-C-NH \qquad\qquad N-C-NH$$
$$\quad\quad A \qquad\qquad\qquad\qquad\qquad B$$

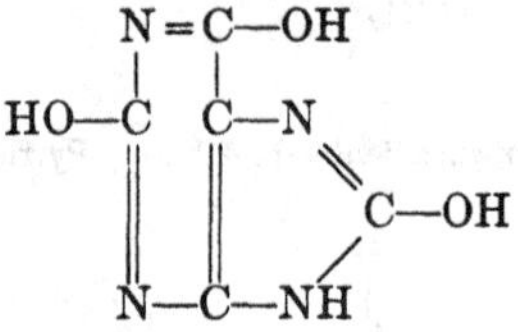

It is worth while to emphasize the phenolic character of uric acid by re-writing its formula

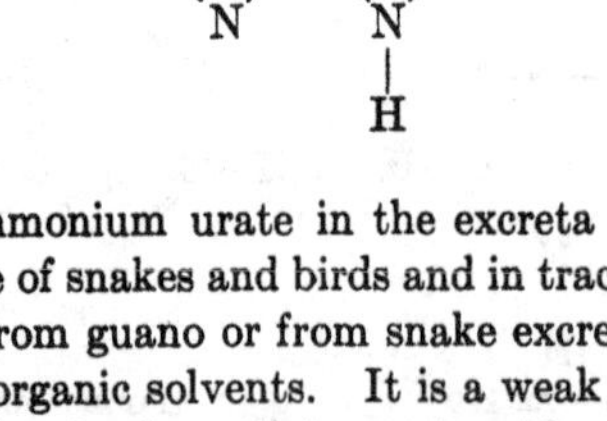

$$OH$$
$$C$$
$$N \quad C-N$$
$$HO-C \quad C \quad C-OH$$
$$N \quad N$$
$$H$$

Uric acid occurs as ammonium urate in the excreta of carnivorous animals, in large amount in those of snakes and birds and in traces in those of mammals. It is usually prepared from guano or from snake excrement. It is practically insoluble in water and organic solvents. It is a weak dibasic acid. The acid grouping is $N=C-OH$ as in the acid form of amides. It is analogous to the carboxyl grouping $O=C-OH$ and to the grouping $C=C-OH$ found in enols and in phenols. The fact that only two H atoms can be replaced by metals is perhaps analogous to the very weak acidic properties of the third H in phosphoric acid. The common salts of uric acid have only one H replaced just as the common salts of phosphoric acid usually have only one or two H atoms replaced. Even alkali salts are only slightly soluble. The acid lithium salt is most soluble (1 part in about 400 of water). This fact was responsible for the popularity of "lithia water" as a supposed solvent for uric acid in various diseases.

The determination of the structure of uric acid formed a most important chapter in early structural chemistry (1863–1883). Looking backward, the essential facts appear as follows:

(1) Careful oxidation gives alloxan and urea. Uric acid is readily oxidized whereas the parent substance, purine, is hard to oxidize. This is much like the easy oxidation of phenols as compared with benzene. This may be added evidence for the pyrimidine structure for uric acid.

(2) It gives a tetramethyl derivative which on hydrolysis gives methylamine as the only nitrogenous product. It is significant that tetramethyluric acid is more readily hydrolyzed and less readily oxidized than uric acid itself. The tetra $N-Me$ compound necessarily has the $O=C-NMe$ grouping which should add water more readily than the $HO-C=N$ grouping which probably predominates in the unmethylated acid. Me_4-uric acid, on careful treatment with dilute bases, loses the No. 2 carbon to give *tetramethylureidin*,

$$\begin{array}{c} MeNHCOC-NMe \\ \| \qquad \qquad \diagdown \\ \qquad \qquad \qquad CO. \\ \| \qquad \qquad \diagup \\ MeNHC-NMe \end{array}$$

Similarly caffeine gives *caffeidin*.

(3) Of two isomeric monomethyl derivatives, one gives alloxan and methylurea while the other gives methylalloxan and urea.

The first two facts indicate the presence of two urea residues and a three carbon chain while the last shows that the molecule is not symmetrical.

The *synthesis* of uric acid has been achieved in many ways. Two of these will be given because of their instructive steps.

A. Synthesis of Behrend and Roosen.

(1) Urea heated with acetoacetic ester gives *methyluracil* which, from its method of formation, is represented as

$$\begin{array}{c} CONHCMe \\ | \qquad \quad \| \\ NHCOCH \end{array}$$

(2) Treatment with nitric acid introduces a nitro group and converts the methyl to carboxyl. Both these changes suggest the properties of an aromatic substance like meta-cresol. On this basis methyluracil can be regarded as

2,4-dihydroxy-6-methylpyrimidine

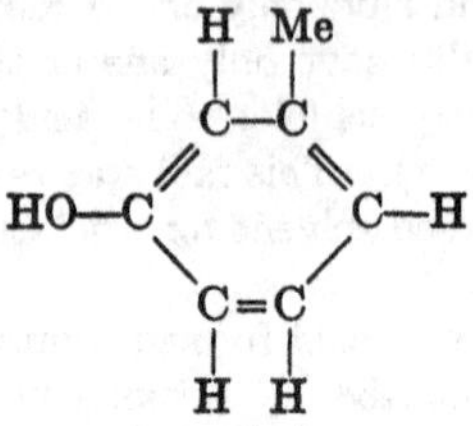

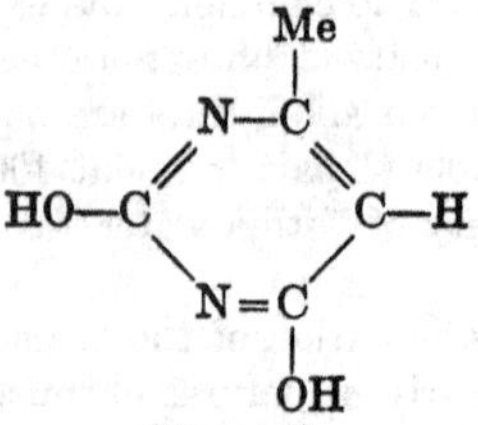

m-cresol

Me-uracil

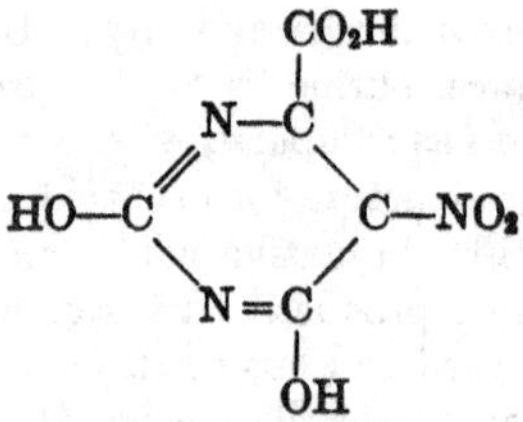

Nitrouracilic acid

(3) Heat eliminates CO_2 leaving nitrouracil.

(4) Reduction and hydrolysis give the corresponding amino and hydroxyl compounds. The latter may be regarded as 5-hydroxy uracil, *isobarbituric acid*, or 2,4,5-trihydroxypyrimidine.

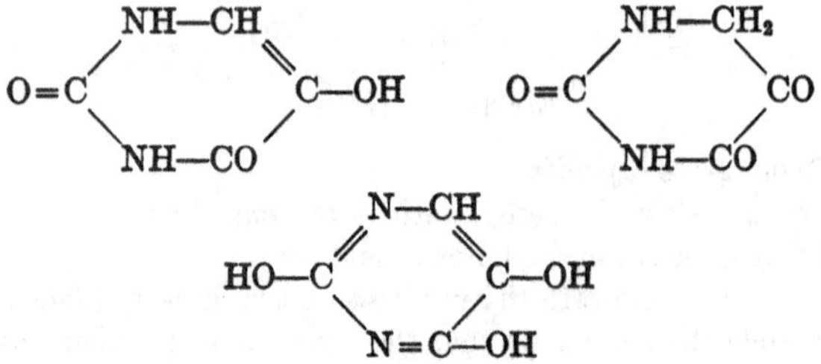

The ready hydrolysis of the NH_2 to OH is reminiscent of the behavior of the groupings:

$$C=C-NH_2 \rightleftharpoons CH-C=NH \xrightarrow{H_2O}$$
$$CH-C(OH)NH_2 \rightarrow CH-C=O \rightleftharpoons C=C-OH$$

(5) Treatment of hydroxyuracil with bromine water gives *isodialuric acid* which corresponds to the following formulas

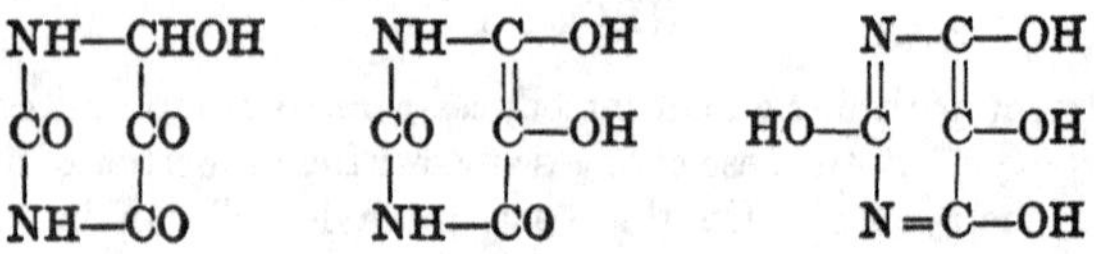

The latter is 2,4,5,6-(OH)$_4$-pyrimidine. The treatment with bromine water thus results in the conversion of $=N-CH=C(OH)-$ to

$$=N-CHOH-C(OH)-.$$

This is like the addition of bromine to an enol, followed by hydrolysis of the resulting alpha bromo compound.

(6) Heating isodialuric acid with urea gives uric acid.

The reason for the ease of this apparently complex reaction is that the urea can add to a carbonyl group existing potentially at any one of the three adjacent carbons Nos. 4, 5 and 6. After the first addition, the carbon in the 5-position to the other N of the urea can present a carbonyl group for 5-ring closure. Regardless of whether the initial addition takes place at carbon 4, 5 or 6, the product is uric acid. The only changes involved are those of tautomerism and of the addition of $-NH_2$ to a carbonyl group, resulting in the imino grouping or its tautomer. Incidentally, if the initial attack is on carbon No. 5, the intermediate product is pseudo uric acid (see below).

B. Synthesis of Bayer as Completed by E. Fischer.

(1) Urea and malonic acid with POCl$_3$ give barbituric acid.

(2) Treatment with nitrous acid gives *violuric acid*, isonitrosobarbituric acid, the 5-oxime of mesoxalylurea, which may also be regarded as a nitroso derivative of the phenolic pyrimidine form of barbituric acid.

(3) Reduction gives *uramil*, 5-aminobarbituric acid.

(4) Treatment with a cyanate and acid changes the NH$_2$ to NHCONH$_2$ forming pseudo uric acid, 5-carbamidobarbituric acid. This involves the usual reaction of cyanic acid with an active H compound.

(5) Dehydration of pseudo uric acid by oxalic acid or, better, by heating with hydrochloric acid gives uric acid. Closure of a 5-ring with either carbon No. 4 or No. 6 gives uric acid.

Reactions of Uric Acid.

With POCl$_3$, uric acid reacts in the OH form giving 2,6-dichloro-8-hydroxypurine and 2,6,8-trichloropurine. The chlorine atoms are very reactive in contrast to those of aromatic chlorides. This is probably due to the grouping $N=C-Cl$ which behaves like the active grouping in acid chlorides, $O=C-Cl$, in contrast to the inactive grouping $C=C-Cl$ in the inactive aromatic and vinyl chlorides. To a certain extent, the replacement of the chlorine atoms can be controlled because of a difference in reactivity in the different positions. This range in activity may be indicated as No. 6 $\gg$ No. 2 $>$ No. 8. Thus HI with trichloropurine gives 2,8-diiodopurine. Reduction of the latter with zinc dust gives purine. With the Cl_3-compound, bases and NH_3 replace the 6-Cl by $-OH$ and $-NH_2$. NaOEt introduces ethoxy groups in positions 6 and 2. Uric acid gives a blue color with tungstic acid and a base.[33]

Important substances related to uric acid are:

> Xanthine, 2,6-$(OH)_2$-purine (A and B below)
> Hypoxanthine, Sarcine, 6-OH-purine
> Guanine, 2-NH_2-6-OH-purine
> Adenine, 6-NH_2-purine

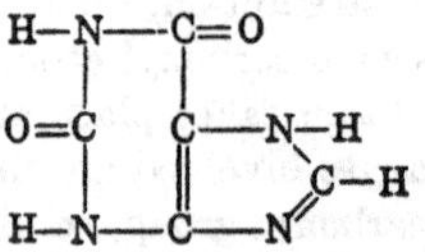

(A) 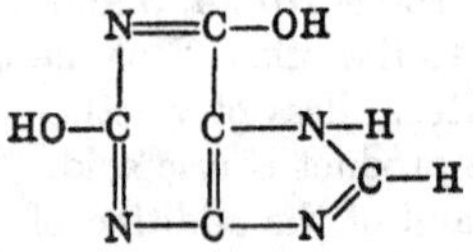   (B)

As in the case of uric acid, the first of these tautomeric forms recalls the relation to urea while the second shows the "phenolic" or "aromatic" acidic form.

The four substances above are obtained by hydrolysis of nucleic acids from cell nuclei.[34] They can be made from uric acid through Cl_3-purine.

The following three purine derivatives are found in plant products such as tea, coffee and cocoa.

> 1,3,7-Me_3-2,6-dioxypurine, caffeine
> 3,7-Me_2-2,6-dioxypurine, theobromine
> 1,3-Me_2-2,6-dioxypurine, theophylline

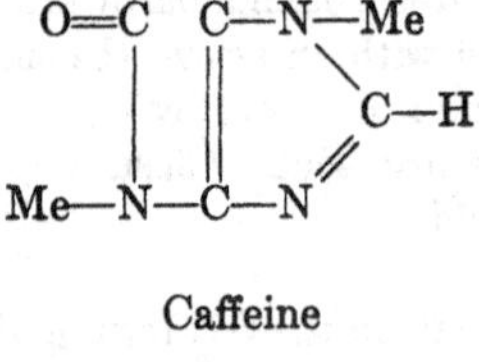

Caffeine

Large amounts of caffeine are made from theobromine, for use in Cola type soft drinks. Synthetic caffeine is now available (Monsanto).

Theophylline is used alone and with ethylene diamine (as "*aminophylline*") for diuresis (NNR). Theophylline is synthesized from 1,3-Me_2-urea and cyanoacetic ester.[35] The systematic names for these substances are less

[33] Frabot. *Ann. Chim. anal.* 9, 371 (1904).
[34] Pryde. "Recent Advances in Biochem." Blakiston, 1931. p. 77.
[35] Traube. *Ber.* 33, 3052 (1900).

confusing if they are regarded as indicating the various methyl derivatives of the *keto* form (C) of 2,6-dihydroxy-purine (D).

$$
\begin{array}{cc}
\text{H–N––C=O} & \text{N=C–OH} \\
\text{O=C} \quad \text{C––N–H} & \text{HO–C} \quad \text{C––N–H} \\
\qquad\qquad \text{C–H} & \qquad\qquad \text{C–H} \\
\text{H–N––C––N} & \text{N––C––N} \\
\text{(C)} & \text{(D)}
\end{array}
$$

The Me$-$N linkage prevents their existence in the tautomeric "aromatic" form *possible* with the H$-$N compounds.

For pyrimidines, purines and nucleic acids see G. 1st, 948–1017; K. 799.

4. Thiourea

Thiocarbamide, m. 172°,

$$(NH_2)_2C=S \rightleftharpoons \overset{NH_2}{NH=C-SH} \rightleftharpoons \overset{\overset{\oplus}{NH_3}}{HN=C-S\ominus}$$

Preparation. (1) By heating ammonium thiocyanate. This process requires a higher temperature than the rearrangement of ammonium cyanate to urea. Also, the reaction is less readily completed.

$$\underset{75\%}{NH_4SCN} \overset{170°}{\rightleftharpoons} NH_3 + HN=C=S \rightleftharpoons \underbrace{\overset{NH_2}{HN=C-SH} \rightleftharpoons (NH_2)_2C=S}_{25\%}$$

(2) By treating cyanamide with ammonium sulfide.

$$NH_2-CN + H_2S \rightarrow NH_2-\overset{SH}{C}=NH \rightleftharpoons (NH_2)_2C=S$$

Thiourea is a much weaker base than urea. It forms stable complex ions with heavy metals. For this reason a solution of thiourea in mineral acid will dissolve even gold.

Reactions. (1) Mercuric oxide removes H$_2$S to give cyanamide.

(2) Careful oxidation by KMnO$_4$ gives urea.

$$(NH_2)_2C=S + [O] \rightarrow (NH_2)_2C=O + S$$

(3) Oxidation with acid KMnO$_4$ solution or with halogens gives a disulfide, as is usual with $-$SH compounds.

$$2\ NH_2-\overset{NH}{\underset{\|}{C}}-SH + [O] \rightarrow NH_2-\overset{NH}{\underset{\|}{C}}-S-S-\overset{NH}{\underset{\|}{C}}-NH_2, \quad \text{formamidine disulfide.}$$

(4) Oxidation by hydrogen peroxide gives a sulfinic acid.

$$\underset{\text{NH}_2}{} \text{NH}_2\text{---}\overset{\overset{\text{NH}}{\|}}{\text{C}}\text{---SH} + 2\,[\text{O}] \rightarrow \text{NH}_2\text{---}\overset{\overset{\text{NH}}{\|}}{\text{C}}\text{---SO}_2\text{H}$$

(5) Alkyl halides react readily to give S-alkyl compounds, as proved by their hydrolysis to mercaptans.

$$(\text{NH}_2)_2\text{C}=\text{S} + \text{RI} \rightarrow (\text{NH}_2)_2\overset{\overset{\text{I}}{|}}{\text{C}}\text{---S---R} \rightarrow \text{HI} + \text{NH}=\overset{\overset{\text{NH}_2}{|}}{\text{C}}\text{---SR}$$

(6) Aldehydes give no action in neutral solution but with a trace of acid form $\text{RCH}=\text{NCSNH}_2$ instantly.

The tendency for thiourea to react in the —SH form is pronounced[36] but the solid apparently exists in the symmetrical form.[37] With diazomethane, thiourea gives $\text{MeSC(NH}_2)\text{NH}$ whereas urea is unchanged.[38]

Substituted thioureas are obtained as follows:

(1) By heating the thiocyanates of primary or secondary amines.

(2) By adding H_2S to a disubstituted cyanamide.

(3) By adding NH_3 or primary or secondary amines to mustard oils (isothiocyanates), RNCS.

$$\text{R---N}=\text{C}=\text{S} + \text{NH}_3 \rightarrow \text{RN}=\overset{\overset{\text{NH}_2}{|}}{\text{C}}\text{---SH} \rightleftharpoons$$

$$\text{RNH---}\overset{\overset{\text{NH}_2}{|}}{\text{C}}=\text{S} \rightleftharpoons \text{RNH---}\overset{\overset{\text{NH}}{\|}}{\text{C}}\text{---SH}$$

(4) By heating primary amines with carbon disulfide.

$$\text{RNH}_2 + \text{CS}_2 \rightarrow \text{RNH---}\overset{\overset{\text{S}}{\|}}{\text{C}}\text{---S---NH}_3\text{R} \rightleftharpoons \text{RNH---}\overset{\overset{\text{S}}{\|}}{\text{C}}\text{---SH} + \text{RNH}_2$$

$$(\text{RNH})_2\text{C(SH)}_2$$

$$\text{H}_2\text{S} + (\text{RNH})_2\text{C}=\text{S}$$

(5) By heating thiophosgene with amines.

$$\text{CSCl}_2 + 2\,\text{RNH}_2 \rightarrow (\text{RNH})_2\text{C}=\text{S}$$
$$+ 2\,\text{RR'NH} \rightarrow (\text{RR'N})_2\text{C}=\text{S}$$

Dithiobiuret, $\text{H}_2\text{NCSNHCSNH}_2$, is available as an intermediate for a wide variety of S and N compounds (ACC).

[36] Werner. *J. Chem. Soc.* 109, 1132 (1916).
[37] Hendricks. *J. Am. Chem. Soc.* 50, 2455 (1928).
[38] *Ann. Rep. Chem. Soc.* (London) 1919, 87.

C. Amidines of Carbonic Acid and Related Compounds

Guanidine, carbamidine, iminourea, $(NH_2)_2C=NH$.

Preparation. (1) Heating ammonium thiocyanate at 180°, presumably with the formation of thiourea and cyanamide as intermediates.

(2) Heating cyanamide with NH_4Cl and ammonia to give guanidine hydrochloride.

(3) Ammonolysis of orthocarbonates.

$$C(OEt)_4 + 3\ NH_3 \rightarrow (NH_2)_2C=NH + 4\ EtOH$$

(4) Ammonolysis of chloropicrin.

$$Cl_3CNO_2 + 7\ NH_3 \rightarrow (NH_2)_2C=NH + 3\ NH_4Cl + N_2 + 2\ H_2O$$

Carbon tetrachloride gives no trace of guanidine on treatment with ammonia.[39]

(5) Ammonolysis of cyanogen chloride.

$$Cl-C\equiv N \rightarrow Cl-\overset{\overset{\displaystyle NH_2}{|}}{C}=NH \rightarrow HN=C=HN \rightarrow (H_2N)_2C=NH$$

Guanidine is a white crystalline hygroscopic substance. It is a *strong* monoacid base, even forming carbonates. It unites with hydrogen ions to form the stable "strong" ion $[C(NH_2)_3]^+$.

Partial hydrolysis gives urea. It can be converted to hydrazine by the following steps.

$$NH=C(NH_2)_2 \rightarrow NH=\overset{\overset{\displaystyle NH_2}{|}}{C}-NH \cdot NO_2 \rightarrow$$

$$NH=\overset{\overset{\displaystyle NH_2}{|}}{C}-NH-NH_2 \rightarrow 2\ NH_3 + CO_2 + NH_2NH_2$$

The best preparation of aminoguanidine is from dicyandiamide.[40] Fusion of the latter with NH_4NO_3 gives guanidinium nitrate, $[(H_2N)_3C][NO_3]$. This, on solution in conc. H_2SO_4 and dilution, gives insoluble nitroguanidine. Electrolytic reduction gives aminoguanidine, isolated as its bicarbonate. The amino compound can be diazotized and isolated as the solid diazonium chloride or hydroxide. These can be coupled with the usual aromatic dye intermediates.[41] The grouping $-N=NNHC(NH)NH_2$ enters the usual positions in relation to the orienting groups.

[39] Nef. *Ann.* **308**, 331 (1899).
[40] Shreve, Carter. *Ind. Eng. Chem.* **36**, 423 (1944).
[41] Shreve, Carter, Willis. *Ind. Eng. Chem.* **36**, 426 (1944).

With nitrous acid, guanidine gives an 80–90% yield of cyanamide.[42] As in the action of nitrous acid with a primary amine, the NH_2 is probably removed with its octet of electrons thus leaving a carbonium ion. This is stabilized by loss of a proton.

$$
\begin{array}{ccc}
\underset{\displaystyle C=NH}{\overset{\displaystyle NH_2}{\diagdown}} & \underset{+}{\overset{\displaystyle NH_2}{C=NH}} & \begin{array}{l} \nearrow\quad HN=C=NH \\ \qquad\quad \Big\updownarrow \\ \searrow\quad H_2NC\equiv N \end{array}
\end{array}
$$

If the carbonium ion coordinated with a molecule of water, the result would be the ordinarily assumed intermediate, isourea, which would certainly not change to cyanamide nearly quantitatively.

Cyanamide with amines gives substituted guanidines. Thus glycine and its N – Me derivative, sarcosine, give

$$H_2NC(NH)NHCH_2CO_2H \quad \text{and} \quad H_2NC(NH)NMe-CH_2CO_2H$$

Glycocyamine   Creatine
Guanidineacetic acid

Creatine is found in muscle and is obtained from beef extract. These substances contain OH and NH_2 in the 1,5-position and are easily dehydrated by heating with acids to give imino-hydantoins. The product from creatine is *creatinine*, $HN=CNHCO$. It is a strong base and forms well-defined salts

$$\underset{MeN\text{——}CH_2}{\vert\qquad\quad\vert}$$

with acids. It also reacts slowly with bases to give salts of creatine by opening of the imide linkage.

Guanylurea, $H_2NCONHC(NH)NH_2$, is available as its sulfate (ACC).

Guanidine, heated with beta-ethoxyacrolein diethyl acetal in absolute ethanolic HCl, gives 2-aminopyrimidine, the intermediate for sulfadiazine, 2-sulfanilamidopyrimidine.

Sulfaguanidine (NNR) is p-$NH_2C_6H_4SO_2N=C(NH_2)_2$.

Tetracene is used in detonator mixtures with lead azide and other initiating explosives. It is 1-guanyl-4-nitrosoaminoguanyltetrazene,

$$H_2N-C(=NH)-NH-NH-N=N-C(=NH)-NH-NH-NO$$

D. Cyanic Acid and Related Compounds

1. **Cyanic acid,** $H-O-C\equiv N$, and isocyanic acid, $H-N=C=O$, have no independent existence. The dry distillation of urea gives cyanuric acid, $H_3C_3O_3N_3$. When this is heated and the vapors are condensed below 0° a volatile liquid of composition HCON is obtained. This is unstable above 0°. Its reactions correspond to the tautomeric system

$$H-O-C\equiv N \;\;\rightleftharpoons\;\; O=C=N-H$$

[42] Bancroft, Belden. *J. Phys. Chem.* **35**, 2684, 2950 (1931).

Cyanic acid is thus an example of triad prototropy. Above 0°, it polymerizes to cyanuric acid and an isomeric substance, cyamelide. Cyamelide can be converted to salts of cyanuric acid by alkalies. The relations between these substances are as follows:

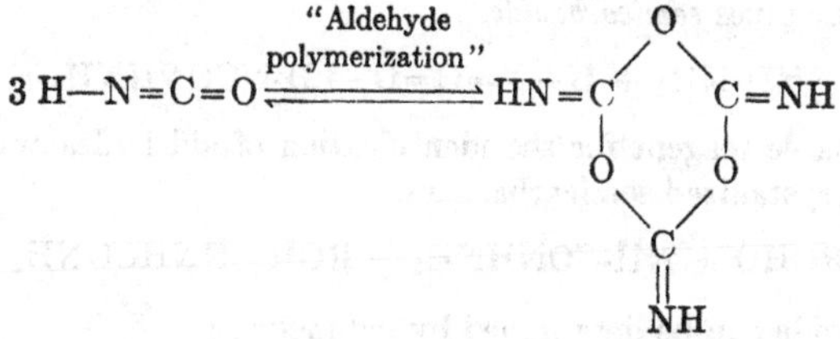

Cyamelide

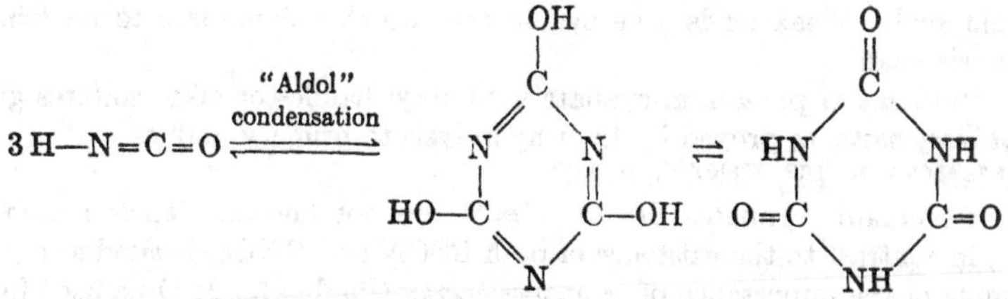

Cyanuric acid

Complex compounds related to cyanuric acid include melon, melam, melem, hydromelonic acid, and cyameluric acid.[43] The last probably has the following interesting structure:[44]

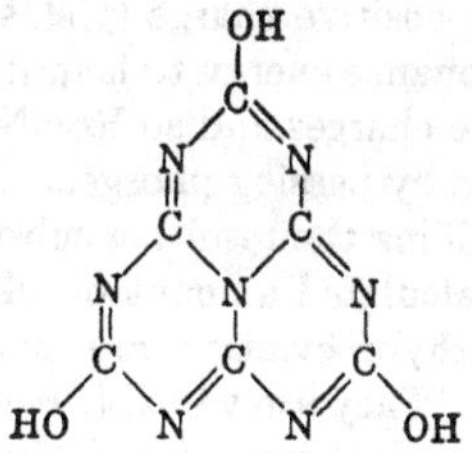

The fact that diazomethane gives the tri-N-methyl compound from cyanuric acid has been given as evidence for the NH formula. This may merely be an example of the easier methylation of NH as compared to OH. The cyanurates are related to the OH formula. O—Hg and N—Hg compounds are known. O-Alkyl compounds are obtained from alkyl iodides and silver cyanurates at low temperatures while the N-alkyl compounds are formed by the polymerization of RNCO or by heating the O-alkyl compounds. The O- and N-compounds are identified by their hydrolysis to ROH and RNH_2.

[43] J. Am. Chem. Soc. 61, 3420 (1939); 62, 842 (1940).
[44] Pauling, Sturdivant. Proc. Natl. Acad. Sci. 23, 615 (1937).

Potassium cyanate is made by oxidizing KCN with PbO or PbO_2, or better, by heating urea with K_2CO_3.

Cyanic acid is useful in replacing an active H by the $-CO-NH_2$ group. This is done by acidifying a mixture of the active H compound and KCNO. Thus hydrazine gives *semicarbazide*.

$$NH_2NH_2 + H-N=C=O \rightarrow H_2NCONHNH_2$$

This is a valuable reagent for the identification of aldehydes and ketones by forming well-crystallized semicarbazones.

$$RCHO + NH_2CONHNH_2 \rightarrow RCH=NNHCONH_2$$

Semicarbazide can be determined by iodimetry.[45]

Inorganic cyanates with strong acids give CO_2 by the easy hydrolysis of cyanic acid. Weak acids give cyanic acid which polymerizes to its trimer, cyanuric acid.

Treatment of potassium cyanate with alkyl halides or alkyl sulfates gives *alkyl isocyanates* as proved by their hydrolysis to primary amines.

MeNCO, b. 45°, EtNCO, b. 60°.

True organic cyanates, $R-O-C\equiv N$, are not known. Their non-existence, in contrast to the existence of both RSCN and RNCS, is cited as a good example of the importance of *resonance energy* (Pauling). If Q be used for O or S, the probable resonating structures include

$$R-Q-C\equiv N, \qquad R-\overset{+}{Q}=C=\bar{N}, \qquad R-\overset{++}{Q}\equiv C-\bar{\bar{N}}.$$

Since O difficultly assumes a positive charge (yields electrons), the first structure is left alone with no resonance energy to help it exist. On the other hand, S can readily assume positive charges and so RSCN is stable.

Isocyanates are also made by passing phosgene over heated primary amine hydrochlorides and then distilling the resulting substituted carbamyl chlorides or treating them with the calculated amount of alkali. In this way phenylisocyanate and alpha-naphthylisocyanate are made from aniline and α-naphthylamine respectively. They are valuable reagents for the identification of alcohols and primary and secondary amines because they form well-crystallized urethans and substituted ureas.

Isocyanates react with active H compounds and with Grignard reagents, the addition taking place on the carbonyl group. A keto shift usually follows as the second step of the reaction. This may be illustrated by the action with water.

$$R-N=C=O + H_2O \rightarrow R-N=C(OH)_2 \rightarrow$$
$$R-NH-\underset{\underset{OH}{|}}{C}=O \rightarrow RNH_2 + CO_2$$

[45] Bartlett. *J. Am. Chem. Soc.* **54**, 2853 (1932).

Similarly, alcohols, ammonia, primary and secondary amines and Grignard reagents give N-alkyl urethans, mono- di- and tri-substituted ureas, and N-substituted acid amides respectively.

Isocyanates are intermediates in the Hofmann and Curtius conversions of RCOOH to RNH_2 (p. 167). They can sometimes be isolated in these reactions.

Alkyl isocyanates polymerize to "esters" of isocyanuric acid. These are really N-substitution products as shown by their hydrolysis to CO_2 and primary amines.

2. **Cyanogen chloride,** chlorocyanogen, ClCN, and BrCN and ICN, are the acid halides of cyanic acid, $HO-C \equiv N$. They are obtained by the direct action of the halogens on a cyanide.

$$NaCN + Cl_2 \rightarrow NaCl + Cl-C \equiv N$$

ClCN, b. 15°, BrCN and ICN, volatile crystals. The cyanogen halides give the ordinary reactions of acid halides with active H compounds like H_2O, alcohols, NH_3 and primary and secondary amines. The products first formed, namely, HOCN, $ROC \equiv N$, $NH_2-C \equiv N$, RNH$-$CN, etc. are capable of reacting further with the active H compounds to give NH_3 imino carbonates, $(RO)_2C = NH \cdot HCl$, guanidine and substituted guanidines.

Cyanogen bromide gives an unusual reaction with tertiary amines. It forms an addition compound which decomposes with the loss of one of the organic groups on heating.

$$R_3N \rightarrow R_3NCN(Br) \rightarrow RBr + R_2N-CN$$

The disubstituted cyanamide can readily be converted to a secondary amine.

$$R_2NCN \xrightarrow{H_2O} R_2N-CO_2H \rightarrow CO_2 + R_2NH$$

With a cyclic amine the ring is opened. Thus N-methylpiperidine gives first $Br(CH_2)_5N(CN)Me$ and then $Br(CH_2)_5NHMe$.

Cyanogen chloride polymerizes on standing to give the acid chloride of cyanuric acid, *cyanuric chloride*, m. 145°, b. 190°. This reacts with alcohols to give true esters of cyanuric acid which hydrolyze to give cyanuric acid and alcohols.

Cyanuric chloride reacts with NH_3 to give the corresponding acid amide, *melamine*. With urea melamine gives increasingly important resins. Their uses include shrink-proofing of wool, heat resistent plastics, ion-exchange, and bonding glues. Melmac, Uformite, Resinene.

If one NH_2 of melamine is hydrolyzed off, *ammeline* results, if two NH_2 groups are replaced by OH, the product is *ammelide*, while replacement of all three NH_2 groups in melamine by OH groups gives cyanuric acid.

3. **Cyanamide,** $NH_2-C\equiv N$.

Preparation. (1) As the calcium salt ("Lime Nitrogen," Kalkstickstoff), from calcium carbide and nitrogen at 1000°.

$$CaC_2 + N_2 \rightarrow CaN-CN + C$$

This product is used as a fertilizer on slightly acid soils because of its gradual conversion to cyanamide, urea, ammonium carbamate and ammonium carbonate.

(2) As the sodium salt in the manufacture of cyanides.

$$2\ NaNH_2 + C \xrightarrow{400°} Na_2N-CN + 2\ H_2$$

The sodium salt may also be obtained by treating calcium cyanamide with Na_2CO_3 solution.

From the sodium salt, free cyanamide may be obtained by means of sulfuric acid.

$$Na_2NCN + 10\ H_2O + H_2SO_4 \rightarrow Na_2SO_4 \cdot 10\ H_2O + NH_2CN$$

The cyanamide is extracted from the crystalline mass with ether or alcohol.

(3) From cyanogen chloride and the calculated amount of ammonia.

$$ClCN + 2\ NH_3 \rightarrow NH_4Cl + NH_2CN$$

(4) By desulfurizing thiourea by heating with HgO.

$$CS(NH_2)_2 + HgO \rightarrow HgS + H_2O + NH_2CN$$

Cyanamide is a colorless soluble crystalline substance, m. 40°. When melted it gives a dimer, "dicyandiamide," cyanoguanidine (I) (ACC),

At higher temperatures it gives a trimer, melamine, the triamide of cyanuric acid (II) (ACC). Cyanamide is a weak base and a weak acid. It exhibits tautomerism.[46]

$$N \equiv C - NH_2 \rightleftharpoons HN = C = NH$$
Basic   Acidic

Cyanamide adds water, ammonia and hydrogen sulfide to give urea, guanidine and thiourea respectively. Its salts react with alkyl halides to give dialkyl cyanamides which are readily converted to secondary amines.

$$Na_2NCN + 2\ RX \rightarrow R_2NCN \rightarrow R_2NCO_2H \rightarrow R_2NH$$

Cyanamide.[47]
Cyanamide, Dicyandiamide, and Melamine.[48]
Calcium cyanamide.[49]

4. **Thiocyanic acid,** rhodanic acid (Rhodanwasserstoffsäure), $H-S-C \equiv N$, and isothiocyanic acid, $H-N=C=S$, are tautomeric. Thiocyanic acid is a crystalline solid, stable below 0° and in dilute solution, in which it shows the properties of a strong acid. At ordinary temperature it polymerizes to a yellow insoluble compound.

Potassium thiocyanate is made by melting KCN with sulfur. Ammonium thiocyanate is obtained as a by-product from the purification of illuminating gas. It is formed from the CS_2 and NH_3 in the gas.

$$CS_2 + 4\ NH_3 \rightarrow NH_4-S-C \equiv N + (NH_4)_2S$$

Organic thiocyanates are obtained from KSCN with alkyl halides or alkyl sulfates. Their structure is proved by their formation from cyanogen chloride and mercaptides. A more conclusive proof is their oxidation by nitric acid to alkyl sulfonic acids. Still another proof is that with sodium malonic ester they give cyanomalonic ester and alkyl disulfides.

MeSCN, b. 133°; EtSCN, b. 142°.

Organic thiocyanates find a wide use as insecticides especially in fly sprays. *Lorol thiocyanate*, mainly dodecyl thiocyanate; *Thanite*, thiocyanoacetic esters of isoborneol and fenchol; *Lethanes*, No. 384, $BuOCH_2CH_2SCN$, No. 60, mainly $C_{11}H_{23}CO_2CH_2CH_2SCN$, No. A70, $O(CH_2CH_2SCN)_2$.

Mustard oils, RNCS, are obtained from primary amines, carbon disulfide and lead nitrate or some similar salt which gives an insoluble sulfide.

$$RNH_2 + CS_2 \rightarrow [RNH-CS_2H] \xrightarrow{Pb(NO_3)_2} PbS + RN=C=S$$

[46] *Ann. Rep. Chem. Soc.* (London) 1915, 85.
[47] *Chem. Eng. News* 18, 759 (1940).
[48] Rochow. *Ind. Eng. Chem.* 18, 1579 (1940).
[49] Ahrens. *Sammlung Chemische-Technischen Vorträge* 1931, 213.

sym-Disubstituted thioureas react with acids to give mustard oils.

$$RNH-CS-NHR + HCl \rightarrow RN=C=S + RNH_3Cl$$

The Raman spectra of mustard oils give no indications of double bonds. On the other hand, alkyl isocyanates show Raman lines for double bonds. Parachor measurements, however, indicate that mustard oils contain true double bonds.[50]

Reactions of Mustard Oils. The carbon-sulfur linkage is very reactive. Water under the influence of acids can give the following reactions:

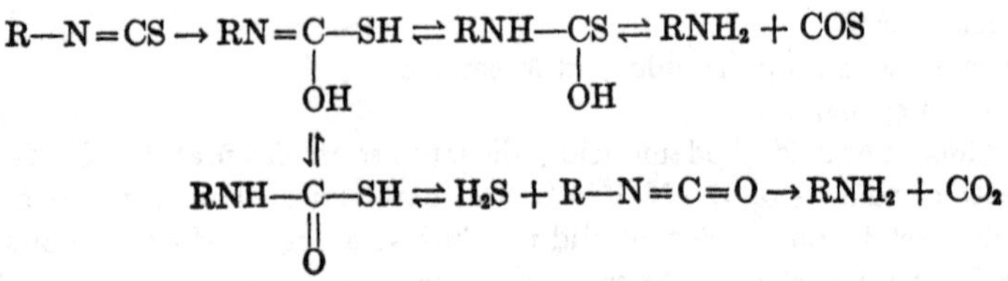

$$R-N=CS \rightarrow RN=C-SH \rightleftharpoons RNH-CS \rightleftharpoons RNH_2 + COS$$
$$\qquad\qquad\quad | \qquad\qquad\qquad |$$
$$\qquad\qquad\quad OH \qquad\qquad\qquad OH$$
$$\qquad\qquad\quad \Updownarrow$$
$$RNH-C-SH \rightleftharpoons H_2S + R-N=C=O \rightarrow RNH_2 + CO_2$$
$$\qquad\quad ||$$
$$\qquad\quad O$$

The products formed depend on the acid used and the conditions.

Anhydrous alcohols, NH_3 and primary and secondary amines give thiourethans, and substituted thioureas.

MeNCS, m. 34°, b. 119°; EtNCS, b. 132°.

Allyl mustard oil occurs as a glucoside, potassium myronate, in mustard seeds. This can be hydrolyzed by acids or by the enzyme myrosin found in mustard seeds.

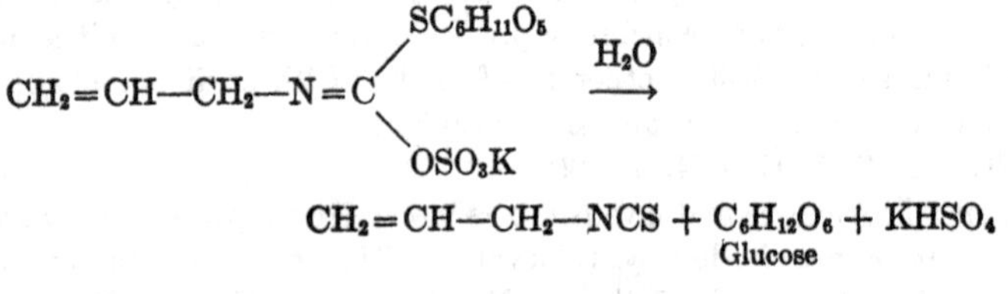

$$CH_2=CH-CH_2-N=C \overset{SC_6H_{11}O_5}{\underset{OSO_3K}{<}} \xrightarrow{H_2O}$$

$$CH_2=CH-CH_2-NCS + C_6H_{12}O_6 + KHSO_4$$
$$\text{Glucose}$$

Many other mustard oils occur in plant products.

Thiocyanogen, $(SCN)_2$, is obtainable from lead or mercury thiocyanate and bromine. It resembles bromine in its action with aromatic compounds and unsaturated compounds.[51-53]

[50] Sidgwick. "Electronic Theory of Valency." Oxford, 1927. p. 127.
[51] Soderback. *Ann.* **443**, 142 (1925).
[52] Bruson, Calvert. *J. Am. Chem. Soc.* **50**, 1735 (1928).
[53] "Org. Reactions," III.

E. THIOCARBONIC ACIDS

1. **Monothiocarbonic acid,** $HS-\overset{\overset{\displaystyle OH}{|}}{C}=O \rightleftharpoons S=C(OH)_2$, is not known in the free state. Derivatives may be prepared as follows:

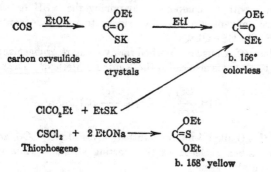

carbon oxysulfide   colorless   b. 156°
crystals   colorless

ClCO$_2$Et + EtSK

CSCl$_2$ + 2 EtONa $\longrightarrow$ $\overset{\displaystyle OEt}{\underset{\displaystyle OEt}{C=S}}$
Thiophosgene

b. 158° yellow

2. **Carbon oxysulfide,** COS, is made (1) from CO and S at red heat, (2) by potassium thiocyanate and 1:1 sulfuric acid

$$H-N=C=S + H_2O \rightarrow H-\underset{\underset{\displaystyle H}{|}}{N}-\underset{\underset{\displaystyle OH}{|}}{C}=S \rightarrow O=C=S + NH_4HSO_4$$

It is a colorless combustible gas with a faint odor, b. $-47°$. It reacts like CS$_2$ with alcoholates, mercaptides, amines and Grignard reagents. It is readily hydrolyzed to CO$_2$ and H$_2$S.

3. **Dithiocarbonic acid,** $O=C(SH)_2 \rightleftharpoons HO-\overset{\overset{\displaystyle SH}{|}}{C}=S$, is known only in its derivatives. The di-S esters are best made from phosgene and mercaptides.

$$COCl_2 + 2 EtSK \rightarrow CO(SEt)_2$$
b. 196°

The S, O derivative is readily obtained from CS$_2$ and alcoholic potash.

$$CS_2 + C_2H_5OH + KOH \rightarrow \overset{\displaystyle OEt}{\underset{\displaystyle SK}{C=S}}$$

Potassium ethyl xanthogenate

This is a stable crystalline product. With copper compounds, *yellow* precipitates are obtained (hence the name). With mineral acids at low temperatures, it gives the free xanthic acid, EtOCS$_2$H. At room temperature this decomposes violently into EtOH + CS_2. The xanthates react with alkyl halides to give the S, O di-esters of dithiocarbonic acid, such as $EtO-CS-SEt$, b. 200°. The two possible methyl ethyl esters are known. They both boil at 184° but can be distinguished by their action with ammonia, one giving MeSH and the other EtSH.

In common with all compounds containing the $-SH$ or $-SM$ (metal) linkage, potassium xanthate is readily oxidized to a disulfide, xanthic disulfide, $EtO-CS-S-S-CS-OEt$.

The Tschugaeff (Chugaev) Method for avoiding rearrangement in converting alcohols to olefins is also called the Xanthate Method. The steps follow:

$$ROH \xrightarrow[\text{KOH}]{\text{CS}_2} ROCS_2K \xrightarrow{\text{MeI}} ROCS_2Me$$

Pyrolysis of this O-alkyl S-methyl xanthate gives MeSH, COS, and the olefin of the same carbon skeleton as the starting alcohol. Thus pinacolyl alcohol gives t-butylethylene.

Thiocarbonyl chloride, thiophosgene, $S=CCl_2$, b. 33°, is made from carbon disulfide (OS). Treatment with Cl_2 and a trace of iodine gives Cl_3CSCl, perchloro methyl mercaptan. Reduction with Fe gives $CSCl_2$, an orange liquid of penetrating odor. Its reactions resemble those of phosgene except that it is much less reactive. Thus it is only slowly decomposed by boiling water. With NH_3 it gives NH_4SCN instead of thiourea or NH_2CSCl. The process may depend on the greater acidic properties of the $-SH$ group as compared to the $-OH$ group formed by the initial addition of NH_3 to the CS and CO groups. Thus successive molecules of NH_3 may act as follows:

$$Cl_2CS \rightarrow Cl_2C(NH_2)SH \rightarrow Cl_2C(NH_2)SNH_4 \rightarrow 2\ NH_4Cl + N \equiv C-SNH_4$$
$$Cl_2CO \rightarrow Cl_2C(NH_2)OH \rightarrow NH_4Cl + ClCONH_2 \rightarrow$$
$$ClC(OH)(NH_2)_2 \rightarrow NH_4Cl + CO(NH_2)_2$$

With primary amines, the reaction is readily stopped at the mustard oil, RNCS. Thiophosgene is polymerized by light.

4. **Thiocarbonic acid, trithiocarbonic acid,** $S=C(SH)_2$, is an unstable red oil. Its soluble salts are readily obtained from alkali sulfides and CS_2. With heavy metal solutions, highly colored precipitates are obtained. The alkali salts react readily with alkyl halides to give esters.

5. Carbon disulfide, CS_2, is made by heating carbon and sulfur in an electric furnace. When purified by distilling over mercuric chloride, its usual disagreeable odor is largely removed. It boils at 47° and has a very low kindling point (below 100°). It is very reactive.

Carbon disulfide can be made by passing methane and sulfur over silica at 400°. The H_2S can be converted to sulfur for reuse.

Reactions. (1) With water, under special conditions it gives COS. Usually the hydrolysis is complete to CO_2 and H_2S.

(2) With chlorine in presence of iodine, a complex series of changes takes place, giving Cl_3CSCl and finally CCl_4 and chlorides of sulfur.

(3) With sulfur monochloride, in the presence of iron as a catalyst it gives carbon tetrachloride.

$$CS_2 + 2 S_2Cl_2 \rightarrow CCl_4 + 6 S$$

The sulfur is converted to S_2Cl_2 and used again. This is the commercial method for making CCl_4.

(4) Reduction by zinc and acid gives trithiomethylene, trimer of thioformaldehyde.

(5) With alkalies and alcohols, it gives xanthates (p. 455). This is the basis of the commercial viscose process for converting cellulose (a poly hydroxyl compound) into rayon. The cellulose is treated with sodium hydroxide and CS_2 to make viscose. This is then forced through minute openings into a dilute acid solution to regenerate the cellulose in fine silky fibers which can be spun into thread and yarn.

$$ROH + CS_2 + KOH \rightarrow ROCS_2K \xrightarrow{\text{acid}} ROCS_2H \rightarrow ROH + CS_2$$

(6) With ammonia it gives the ammonium salt of dithiocarbamic acid.

$$CS_2 \xrightarrow{NH_3} \left[S=\overset{\displaystyle NH_2}{\underset{\displaystyle |}{C}}-SH \right] \rightarrow S=\overset{\displaystyle NH_2}{\underset{\displaystyle |}{C}}-SNH_4$$

(7) With primary and secondary amines it gives amine salts of N-substituted dithiocarbamic acids.

F. Thiocarbamic Acids

1. **Monothiocarbamic acid** exists as O-, N-, and S-alkyl derivatives. The methods of preparation follow:

(a) O-Derivatives

$$EtO-CS-SEt + NH_3 \rightarrow EtO-CS-NH_2 + C_2H_5SH$$

Unsym. diethyl dithiocarbonate   O-Ethyl thiocarbamate m. 58°

(b) S-Derivatives. The S-esters may be made from the O-esters by heating with a trace of alkyl halide.[54]

$$NH_2—CS—OEt \xrightarrow{EtI} \left[NH_2—\overset{\overset{\displaystyle SEt}{|}}{\underset{\underset{\displaystyle I}{|}}{C}}—OEt \right] \rightarrow EtI + NH_2—\overset{\overset{\displaystyle SEt}{|}}{C}=O$$

They may also be obtained by the careful hydrolysis of a thiocyanate with dilute acid.

$$Et—S—C\equiv N + H_2O \rightarrow Et—S—\overset{\overset{\displaystyle NH_2}{|}}{C}=O$$
S-Ethyl thiocarbamate
m. 109°

The S-esters decompose at 150° to give mercaptans and cyanuric acid.

(c) N-Alkyl derivatives are made (1) from carbon oxysulfide and amines and (2) from thiophosgene and amines followed by treatment with alkyl halides or alcoholates.

$$COS + RNH_2 \rightarrow RNH—CO—SNH_3R \xrightarrow{R'I} RNH—CO—SR'$$

$$CSCl_2 + R_2NH \rightarrow R_2N—CS—Cl \xrightarrow{NaOR'} R_2N—CS—OR'$$

The compounds containing the C=S grouping with no free H on the adjacent atoms are orange or red in color and oxidize spontaneously showing phosphorescence. Since the compound RNH−CS−OR' does not show these phenomena its formula probably should be written $RN=\overset{\overset{\displaystyle SH}{|}}{C}—OR'$. This is another indication of the peculiarity of a "double bond" attached to sulfur (pp. 149, 200, 210, 301–2).

2. **Dithiocarbamic acid**, $NH_2—\overset{\overset{\displaystyle SH}{|}}{C}=S \rightleftharpoons NH=C(SH)_2$, is obtained as colorless needles by treating its ammonium salt with HCl. The salt is made from cold alcoholic ammonia and CS_2.

Oxidizing agents such as $FeCl_3$ and the halogens form a disulfide.

$$2\,NH_2—\overset{\overset{\displaystyle S}{||}}{C}—SH + [O] \rightarrow H_2O + NH_2—\overset{\overset{\displaystyle S}{||}}{C}—SS—\overset{\overset{\displaystyle S}{||}}{C}—NH_2$$
Thiuram disulfide

[54] Wheeler, Barnes. *Am. Chem. J.* **22**, 148 (1899).

Alkyl halides give a series of reactions ending with the formation of dithio-carbonate.

$$\overset{\text{RI}}{NH_2-CS-SNH_4} \rightarrow \overset{\text{RI}}{NH_2-CS-SR} \rightarrow$$
Alkyl dithiocarbamate

$$NH=C(SR)_2 \cdot HI \xrightarrow[\text{H}_2\text{O}]{\text{heat}} NH_4I + (RS)_2CO$$
Dialkyl dithio-
carbamate
hydroiodide

Primary and secondary amines can be used in place of ammonia to give the corresponding substituted thiocarbamates. Dimethylamine gives the rubber accelerator, *Tuads*, tetramethyl thiuram disulfide. It is also valuable as an insect repellent in *Arasan*.

$$Me_2NH + CS_2 \rightarrow S=C(NMe_2)SNH_2Me_2 \rightarrow$$

$$S=C(NMe_2)SH \xrightarrow{\text{[O]}} S=C(NMe_2)S-SC(NMe_2)=S$$

Sodium diethyl dithiocarbamate, Et_2NCS_2Na, gives a characteristic golden brown color with very dilute ammoniacal cupric solutions.

Zerlate, a fungicide, is zinc dimethyldithiocarbamate.

Dithane contains the sodium or zinc salt of a bis-dithiocarbamic acid obtained from CS_2 and ethylenediamine, $HS-CS-NHCH_2CH_2NH-CS-SH$.

XXIII. CARBOHYDRATES

The sugars, starches and celluloses form this important group. The original name, carbohydrate, was based on the fact that all of these substances then known had formulas which could be written as $C_m(H_2O)_n$. Even on this basis the name was unsatisfactory because many substances such as formaldehyde and acetic acid could be written as "hydrates" of carbon but were not carbohydrates. It is now known that some sugars deviate from this formula. Thus rhamnose is $C_6H_{12}O_5$. Carbohydrates are also defined as hydroxyaldehydes or hydroxyketones or substances which give those on hydrolysis. This definition is not wholly satisfactory. While glycolic aldehyde, $HOCH_2CHO$, gives the *reactions characteristic of a simple sugar*, namely, (a) reduction of ammoniacal silver solution to give a mirror, (b) reduction of alkaline cupric solution (Fehling's solution) to give a precipitate of cuprous oxide, (c) formation of a yellow-brown color on heating with strong alkalies, (d) the development of a violet color when a solution with alpha naphthol is underlayed with concentrated sulfuric acid, Molisch Test,[1] and (e) the formation of a crystalline osazone (di-phenylhydrazone) when treated with an excess of phenylhydrazine, other hydroxy aldehydes and ke-

[1] Foulger. *J. Biol. Chem.* 92, 345 (1931).

tones fail to give some or all of these characteristic reactions. Simple sugars contain either the equivalent of the $-CHOHCHO$ or $-COCH_2OH$ grouping usually in cyclic combination. Thus simple sugars could be defined as substances which contain an alpha hydroxy aldehyde or an alpha keto primary alcohol. Even this definition fails. Thus ribodesose obtained from thymus nucleic acid[2] contains the $-CH_2CHO$ grouping as does also digitoxose.

<h3 style="text-align:center">CLASSIFICATION OF CARBOHYDRATES</h3>

1. Simple sugars usually contain the grouping $-CHOHCHO$ or $-COCH_2OH$ or their equivalent. They cannot be hydrolyzed to smaller molecules. They are known as *monosaccharides, monosaccharoses, monoses,* or *oses.* They are either aldoses or ketoses. It should be remembered that in these substances the aldehyde or ketone group usually does not exist free but in some union with a hydroxyl group of the same molecule if a 5- or 6-membered ring can be formed within the molecule (glucose) or of another molecule if that is the only way a ring can be formed (glycolic aldehyde).

Oses containing 5 to 10 carbon atoms are indicated as pentoses to decoses. The terms diose, triose and tetrose mean monosaccharides containing 2, 3, or 4 carbon atoms. Glucose is an aldohexose, dihydroxyacetone is a ketotriose.

2. Substances which on hydrolysis give simple sugars. These are called *osides.*

a. Osides which give only *oses* on hydrolysis are *holosides,* more commonly called *polysaccharides, polysaccharoses,* or *polyoses.* Such are the higher sugars like sucrose and the starches and celluloses. Those that have a definitely known number of monosaccharide units are termed oligosaccharides. These include: disaccharides, as sucrose and maltose; trisaccharides, as raffinose; and higher members. The naturally occurring high polymers, as starch and cellulose, have a not definitely ascertained number of monosaccharide units and are designated polysaccharides.

b. Osides which give *oses* and other types of compounds are *heterosides,* usually known as *glucosides* or, more properly, *glycosides.* Amygdalin on hydrolysis gives glucose, HCN and benzaldehyde.

Carbohydrate classification.[3-7]

<h3 style="text-align:center">A. MONOSACCHARIDES</h3>

The simplest sugar, the aldodiose, *glycolic aldehyde,* has already been described. Its reactions are characteristic of the simple sugars except that the smallness of its molecule causes its dimerization in the reaction of hydroxyl

[2] Levene et al. *J. Biol. Chem.* 85, 785 (1930).
[3] Gortner. "Outline of Biochem." Wiley, 1929. p. 487.
[4] Gilman. "Organic Chemistry," 1st Ed., 1938. pp. 1399–1594.
[5] *ibid.* 2nd Ed., 1943. pp. 1532–1719.
[6] Fieser, Fieser. "Organic Chemistry," 1944. pp. 349–80.
[7] Karrer. "Organic Chemistry," 1946. pp. 308–54.

and carbonyl group. It contains no asymmetric carbon atom and, consequently, cannot exist in optically active form as do all natural monoses.

Glyceric aldehyde, glyceraldehyde, propandiolal, $HOCH_2CHOHCHO$, can be obtained in the racemic forms by the following steps.[8]

$$HOCH_2CHOHCH_2OH \rightarrow CH_2=CHCHO \rightarrow$$
$$ClCH_2CH_2CH(OEt)_2 \rightarrow CH_2=CHCH(OEt)_2 \rightarrow$$
$$HOCH_2CHOHCH(OEt)_2 \rightarrow HOCH_2CHOHCHO.$$

It is interesting that it is probably easier to carry out these steps than to oxidize one of the primary alcohol groups in glycerol although that oxidation has been carried out by hydrogen peroxide and $FeSO_4$. When the aldehyde is liberated from its acetal, an oily enolic compound is obtained as a by-product, $HOCH_2C(OH)=CHOH$. Dimolecular glyceraldehyde boiled with pyridine gives as much as 49% dihydroxyacetone, apparently through this enediol which is common to both substances.[9]

Glyceraldehyde exists as a crystalline dimer, m. 138°. In solution this gradually changes to the monomer. ` This behaviour is like that of glycolic aldehyde (p. 331).

It acts like a true aldehyde in giving the Angeli-Rimini hydroxamic acid reaction. In this respect it differs from the higher aldoses which show no true aldehyde group in this reaction due to cyclic hemiacetal formation.

Its oxime when acetylated and treated with ammoniacal silver solution loses HCN. This is a *general method for going from one aldose to the next lower one.*[10] The net result is as follows:

$$HOCH_2CHOHCH=NOH \rightarrow HCN + H_2O + HOCH_2CHO$$

D- and L-Glyceraldehyde have been obtained as syrups. DL-β-Aminolactic aldehyde dimethyl acetal, $NH_2CH_2CHOHCH(OMe)_2$, was treated with *l*-menthylisocyanate, $C_{10}H_{19}NCO$, to give substituted ureas,

$$l\text{-}C_{10}H_{19}NHCONHCH_2CHOHCH(OMe)_2(L)$$

and the L-D isomer. The stereomers were separated and hydrolyzed and the NH_2 group was replaced by OH by means of nitrous acid. Hydrolysis by excess of dilute acid gave the optically active glyceraldehydes.[11]

D-Glyceraldehyde and L-glyceraldehyde have been prepared. The rotations of aqueous solutions of the two gradually decrease by about 50% on standing. Evaporation gives the high rotations again. Thus the change is one of *mutarotation* rather than of *racemization.*

Glyceraldehyde can be slowly fermented by yeast.

[8] "Org. Syntheses."
[9] *Ann. Rep. Chem. Soc.* (London) 1927, 64.
[10] Wohl. *Ber.* 26, 730 (1893).
[11] *Ann. Rep. Chem. Soc.* (London) 1915, 66.

D-Glyceraldehyde is *assigned* the configuration: [12]

All monoses with the same arrangement of the H, OH and CH_2OH on the optically active carbon farthest from the aldehyde or ketone group (the active group) are classed as D- regardless of whether their rotations are actually dextro- or levo-.   The arrangement $H-C-OH$ when the active group is at the top of the formula and in the plane of the paper so that the H and OH are in front of this plane is the D-form, and is indicated as (+) regardless of the actual sign of rotation.   Correspondingly, the arrangement $HO-C-H$ is indicated as (−) regardless of rotation.

Dihydroxyacetone, propandiolone, $HOCH_2COCH_2OH$, is best obtained from glycerol by means of the Sorbose-bacterium.   It crystallizes as a dimer, m. 80°.   The isomerism of this dimer with that of glyceraldehyde is a good example of the effect of symmetry on melting points.

As would be expected, the dimer becomes monomeric on standing in water solution, probably to give the hydrated carbonyl compound.   Although ketones form such compounds less easily than do aldehydes, this molecule gives a possibility of "internal association."

The *gem*-hydroxyls would favor the ring formation on the same principle that cyclopropanone is stable only in the hydrated form, 1,1-$(OH)_2$-cyclopropane. The association suggested depends on forces similar to those acting *intermolecularly* in liquid water and in liquid alcohols.

Dihydroxyacetone oxime, m. 84°.   An excess of phenylhydrazine converts dihydroxyacetone to the osazone of glyceraldehyde.

[12] Rosanoff.   *J. Am. Chem. Soc.* **28**, 114 (1906).

When the semicarbazone of the monoacetate of dihydroxy acetone is boiled with glacial acetic acid, the disemicarbazone of pyruvic aldehyde is produced.[13]

$$\begin{array}{ccc} HOCH_2C=NNHCONH_2 & & CH_3C=NNHCONH_2 \\ | & \rightarrow & | \\ CH_2OCOMe & & CH=NNHCONH_2 \end{array}$$

The interconversions of glyceric aldehyde, dihydroxyacetone, and pyruvic aldehyde are striking examples of the ease with which adjacent carbon atoms can oxidize and reduce each other.

$$\begin{array}{ccc} CH_2OH & CH_2OH & CH_3 \\ | & | & | \\ CHOH & CO & CO \\ | & | & | \\ CHO & CH_2OH & CHO \end{array}$$

A mixture of glyceraldehyde and dihydroxyacetone obtained by the careful oxidation of glycerol is "glycerose."

Dihydroxyacetone was formerly produced commercially as Oxantine.[14-15] It can be fermented.

An interesting homolog of dihydroxyacetone is 1,7-heptanediol-4-one which exists only as a bi-cyclic spiro acetal, *oxetone*. It lacks the hydroxyl groups to make it a ketose.

$$\begin{array}{ccccc} CH_2-CH_2 & & O & CH_2 \\ | & & C & | \\ CH_2 & O & CH_2-CH_2 \end{array}$$

TETROSES, $C_4H_8O_4$

Aldotetroses. In common with the higher aldoses these fail to give certain typical aldehyde reactions such as the hydroxamic acid test of Angeli. They are cyclic hemiacetals.[16]

$$\underset{\overline{\rule{1cm}{0.4pt}\ O\ \rule{1cm}{0.4pt}}}{HOCHCHOHCHOHCH_2}$$

These, because of their similarity to the lactones formed by hydroxy acids, are called *lactols*.

There are four possible configurations of carbons 2 and 3 (counting the active group as 1 in the aldotetroses).

$$\begin{array}{cccc} a & b & c & e \\ + & - & + & - \\ + & - & - & + \end{array}$$

Four aldotetroses are known. As will be discussed under the aldopentoses, each of these can exist in α and β lactol forms.

[13] Bernier, Evans. *J. Am. Chem. Soc.* **60**, 1381 (1938).
[14] Levene, Walti. *J. Biol. Chem.* **78**, 23 (1928).
[15] Spoehr, Strain. *J. Biol. Chem.* **89**, 503 (1930).
[16] Tollens. *Ber.* **15**, 1635 (1882).

D-Erythrose is obtained by the degradation of D-arabic acid by the method of Ruff. This consists in treating the calcium salt with hydrogen peroxide and ferric acetate. The net result is

$$-CHOH-CO_2H + [O] \rightarrow H_2O + CO_2 + -CHO$$

This is an *important general method for going from one aldose to a lower one*. On reduction, D-erythrose gives *meso*-erythritol whose structure also follows from its oxidation to *meso*-tartaric acid. L-Erythrose is obtained similarly from L-arabinose and is similarly related to *meso*-tartaric acid. Since, of the possible configurations of tartaric acid, $++$, $+-$, $-+$, only the first is symmetrical, that must correspond to the internally compensated or meso compound. Thus the configurations a and b must correspond to the two erythroses. Since the $++$ configuration corresponds to that of D-glyceraldehyde, that is assigned to D-erythrose and the other to its enantiomorph, L-erythrose. Following are the processes related to D-erythrose, with the *formulas written as those of aldehydes and acid instead of the true cyclic lactols and lactones*. This procedure is often followed with the sugars but it must be carefully remembered that *the cyclic forms are the actual ones unless the hydroxyl groups are protected by groups such as methyl or acetyl* or unless the carbonyl group is converted to a mercapto grouping, $C(SR)_2$.

D-arabinose   D-arabonic acid   D-erythrose   *meso*-tartaric acid

D-Threose can be obtained by degrading the aldopentose, D-xylose. On reduction it gives D-threitol which on oxidation with nitric acid gives L-tartaric acid. Since D-tartaric acid has the $+-$ configuration the configurations thus determined follow:

D-xylose   D-xylonic acid   D-threose   D-threitol   D-tartaric acid

From the $+$ configuration of its lowest C, comes the classification of D-threitol on the same basis as the monoses. Its specific rotation in water is $+4.3°$ and in alcohol is $-11.5°$.

D-Erythrose and D-threose differ only in the configuration of the No. 2 carbon. One has the configuration $++$ and the other $+-$. Pairs of substances differing in this way are *epimers*. Thus *meso*-tartaric acid is epimeric to both D- and L-tartaric acid. This relation is important because it is possible to change one epimer to the other.

Bromine water converts the aldotetroses to D- and L-erythronic and D- and L-threonic acids.

Ketotetrose. Only one structure is possible, $HOCH_2COCHOHCH_2OH$. Since it contains only one asymmetric carbon, it can exist only in one set of enantiomorphs. Dextrorotatory *erythrulose* is obtained by the action of the Sorbose-bacterium on erythritol (meso). This production of optically active material is characteristic of biochemical reactions involving asymmetric carbon atoms. Ordinary oxidation of a secondary carbinol grouping in *meso*-erythritol would be expected to attack both of these indiscriminately. The attack of one would give D- and of the other, L-erythrulose. The mechanism by which the biochemical "reagents" accomplish one process exclusively is one of the major unsolved problems of chemistry and biology.

On reduction, the dextrorotary erythrulose gives a mixture of two tetritols, erythritol (meso) and L-threitol. On oxidation, the latter gives L-tartaric acid. These relations illustrate the method of deciding on the D- or L-classification for a substance by means of its reactions. The reduction of the carbonyl group gives two possible configurations of the new asymmetric carbon, HCOH and HOCH.

Dextrorotatory erythrulose

$$
\begin{array}{ccc}
& CH_2OH & CO_2H \\
& | & | \\
& HCOH & HCOH \\
& | & \rightarrow \quad | \\
& HCOH & HCOH \\
& | & | \\
& CH_2OH & CO_2H \\
& \text{Erythritol} & \textit{meso}\text{-tartaric acid.}
\end{array}
$$

$$
\begin{array}{ccc}
& CH_2OH & CO_2H \\
& | & | \\
& HCOH & HCOH \\
& | & \rightarrow \quad | \\
& HOCH & HOCH \\
& | & | \\
& CH_2OH & CO_2H \\
& \text{L-Threitol} & \text{L-Tartaric acid.}
\end{array}
$$

$$
\begin{array}{cc}
CH_2OH & CH_2OH \\
| & | \\
HCOH & C=O \\
| & | \\
C=O & HOCH \\
| & | \\
CH_2OH & CH_2OH \\
a. & b.
\end{array}
$$

In order to give these results, the erythrulose must have the configuration *a*. Turning this 180° in the plane of the paper gives *b*, which is identical but has the reactive group at the top of the formula. Thus it is evident that the asymmetric carbon in the dextrorotatory erythrulose is related to that in L-glyceraldehyde and it must therefore be classed as L-*erythrulose*.

Erythrulose, like other ketoses, differs from aldoses in not being attacked by dilute bromine water which converts the aldoses to the corresponding acids.

Pentoses, $C_5H_{10}O_5$, in contrast to the lower monosaccharides, are widely distributed in nature. Treatment with dilute acid converts them to furfural

Aldopentoses have long enough chains so that two types of lactol rings are possible, 6-ring or delta lactols and 5-ring or gamma lactols.

$$\text{HOCH(CHOH)}_3\text{CH}_2 \quad \text{and} \quad \text{HOCH(CHOH)}_2\text{CHCH}_2\text{OH.}$$

Each type of lactol can exist in α and β forms depending on which bond of the carbonyl oxygen opens during ring formation. The possibilities may be illustrated as follows:

Aldehydo form   (*a*)   (*b*)   (*a'*)

Forms *a* and *a'* are identical. They can be seen to represent the enantiomer of *b*. Forms *a* and *b* have the projected configurations HCOH and HOCH or + and −, the plane of the ring being vertical with No. 1 carbon at the top and the ring-O away from the observer. The α form of the cyclic sugar is the one in which the additional asymmetric C formed by ring closure has the HCOH or + arrangement (form *a*, *a'*).

The relation between the alpha and beta forms of the lactols can better be shown by perspective formulas.[17,18] In this representation, the ring lies in the

---

[17] Haworth, Nicholson. *J. Chem. Soc.* **1926,** 1899.
[18] *Ann. Rep. Chem. Soc.* (London) **1929,** 94.

plane of the paper and the atom or group written *below* a C in the ring is *below* the plane of the paper.

alpha form   beta form

These involve a monose in which C No. 2 has the $H-C-OH$ configuration. These relations will be developed further when glucose is considered.

The ordinary 6-ring form (amylene oxide, *pyranose*) is the more stable and less reactive form. The 5-ring form (butylene oxide, *furanose*) is very reactive and can be isolated only in the form of derivatives. The natural valence angle of oxygen is 90°.[19] It has been found that the oxygen angle in tetrahydrofuran is 105° and in tetrahydropyran is 88°.[20] Thus the pyranose ring is less "strained" than the furanose ring.

The alpha and beta forms of the cyclic hemiacetals have been isolated for many of the higher monoses, notably glucose. In solution they are interconvertible and finally form an equilibrium mixture. Since the α and β forms differ sterically they differ in optical activity. Thus if either form or a mixture other than the equilibrium mixture is dissolved, the optical rotation of the solution gradually changes until the equilibrium mixture is reached. This change is *mutarotation* (multirotation, birotation).

A word may be said here as to the physical reality of the conception contained in Emil Fischer's distinction between the HCOH and HOCH or the + and − arrangement of hydroxyl groups in the sugars. The best proof of its reality is its agreement with a multitude of facts in the sugar field discovered long after Fischer's formulation of these principles. The significance of the *sides* on which hydroxyl groups are located becomes doubly important because the ring structure of the sugars keeps these positions definite in relation to each other in a way which would not be possible in open chain compounds in which there is free rotation about each single bond. The ring has the same effect as a double linkage in restricting rotation.

In addition to the active group, the aldopentoses have three asymmetric carbon atoms. The configurations of these could be

+	−	+	−	+	−	−	+
+	−	+	−	−	+	+	−
+	−	−	+	+	−	+	−
a	b	c	e	f	g	h	j

[19] Pauling. *J. Am. Chem. Soc.* **53**, 1367 (1931).
[20] Hibbert, Allen. *J. Am. Chem. Soc.* **54**, 4115 (1932).

Eight optically active aldopentoses are known corresponding to these configurations.

D-Xylose, wood sugar, m. 140–160°, is widely distributed as pentosans (xylans). A possible commercial source is cottonseed hull bran (Acree). On degradation, D-xylose gives D-threose (+ −) and on oxidation it gives D-xylonic acid and then *xylo*-trioxyglutaric acid which is optically inactive and cannot be resolved into optically active isomers. The internal compensation of this acid shows that its third C must have the HOCH or − configuration. Thus D-xylose has configuration (+ − +).

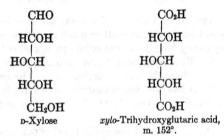

D-Xylose   *xylo*-Trihydroxyglutaric acid,
                              m. 152°.

3,5-Ditritylxylofuranose has been prepared.[21]

D-Xylose is also formed by the degradation of D-gulonic acid. D-Xylose, by the cyanohydrin synthesis, yields D-gulonic acid and D-idonic acid. *This process is valuable for going from one aldose to the next higher one.*[22]

$$-\text{CHO} \rightarrow -\text{CH(OH)CN} \rightarrow -\text{CH(OH)CO}_2\text{H} \rightarrow \text{lactone} \rightarrow -\text{CHOHCHO}$$

The last step is possible because hydroxy lactones are readily reduced to lactols whereas acids are not readily reducible. The process gives a new asymmetric carbon and therefore results in the formation of a pair of epimeric acids (lactones). Since the starting material is optically active the amounts of the two epimers formed are not equal. In this case D-gulonic acid predominates. The oxidation of D-gulonic lactone gives D-tartaric acid (+ −).

Reduction of xylose gives xylitol, an optically inactive non-resolvable substance. The phenylosazone of D-xylose, m. 161°, is different from that of L-arabinose, m. 160°, but is identical with that formed by D-lyxose.

L-Xylose can be obtained from the hexonic acid, L-gulonic acid. It has the configuration (− + −).

Although pentose from wood hydrolysis cannot be fermented by ordinary yeasts, it can be converted to ethanol using a mold of the genus *Fusarium*.[23]

The epimer of D-xylose is D-lyxose (same osazone). Its configuration is therefore (+ − −). The *conversion of an aldose into its epimer* by inverting

[21] McElory. *J. Chem. Soc.* 1946, 100.
[22] Kiliani. *Ber.* 19, 767 (1886).
[23] Nord et al. *Biochem. Z.* 285, 241 (1936).

No. 2 C is an important general process. It consists in oxidizing the aldose to the monobasic acid and heating with pyridine or quinoline to racemize the No. 2 carbon (alpha carbon). The mixture of epimeric acids is then separated by suitable means and the new lactone is reduced to the new aldose by sodium amalgam.

$$
\begin{array}{ccccc}
\text{CHO} & \text{CO}_2\text{H} & & \text{CO}_2\text{H} & \text{CHO} \\
| & | & \text{Pyridine} & | & | \\
\text{HCOH} \rightarrow & \text{HCOH} & \xrightleftharpoons{\qquad\qquad} & \text{HOCH} & \rightarrow \text{lactone} \rightarrow \text{HOCH} \\
| & | & \text{Heat} & | & | \\
\text{D-Xylose} & & & & \text{D-Lyxose}
\end{array}
$$

D-Lyxose, m. 101°, shows mutarotation to a value of $[\alpha]_D = -13.9°$. On reduction it gives D-arabitol (configuration $+ - -$). It gives the same phenyl osazone as D-xylose. Acyclic derivatives of D-lyxose have been prepared, namely, the diethyl mercaptal and its tetraacetate and *aldehydo*-D-lyxose hexaacetate. The tetraacetate of the free *aldehydo* form did not crystallize.[24] L-Lyxose has been prepared. L-Arabinose, "*Arabinose*," m. 160°, is widely distributed as pentosans (arabans) in vegetable gums such as gum Arabic.[25] It is ordinarily prepared by acid hydrolysis of cherry gum or mesquite gum.[26] It is also found in sugar beet residues. L-Arabinose is strongly dextrorotatory and shows mutarotation, the equilibrium value for a 10% solution being $[\alpha]_D = + 105°$. *p*-Bromophenylhydrazone, m. 162°, phenylosazone, m. 160°. This same osazone is formed by L-ribose.

Reduction of L-arabinose gives L-arabitol. Bromine water gives L-arabonic acid. Further oxidation with nitric acid gives L-trihydroxyglutaric acid.

D-Arabinose occurs in certain glucosides and has been made by degrading the oxime of D-glucose,[27] the *general method for forming a lower aldose from a higher one*. The oxime is prepared and treated with acetic anhydride to form the acetylated cyanide. This, with ammoniacal silver solution, gives the acetate of the next lower aldose.

$$
\begin{array}{ccc}
\text{CH}=\text{NOH} & \text{CN} & \\
| & \rightarrow | & \\
\text{CHOH} & \text{CHOCOMe} \rightarrow \text{CHOH(OCOMe)} \rightarrow \text{CHO} \\
| & | & | & |
\end{array}
$$

D-Arabinose, on reduction, gives D-arabitol which is also obtained by the reduction of D-lyxose. Thus D-arabinose and D-lyxose differ only in the relation of the end groups to the three central carbon atoms. Interchanging the active group and the CH$_2$OH group in the formula of one would give the formula of the other. This would not be true of configurations *a*, *b*, *f* and *g* because such

[24] Wolfrom, Moody. *J. Am. Chem. Soc.* **62**, 3465 (1940).
[25] Mason. *Chem. Inds.* **53**, 680 (1943).
[26] Anderson, Crowder. *J. Am. Chem. Soc.* **52**, 3711 (1930).
[27] Wohl. *Ber.* **26**, 730 (1893).

an interchange would make no difference. Since the configuration of D-lyxose is $(+--)$ that of D-arabinose must be $(++-)$.

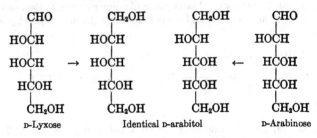

| D-Lyxose | Identical D-arabitol | | D-Arabinose |

DL-Arabinose is probably the pentose found in urine during pentosuria.

DL-Arabinose has been resolved by means of d-amyl mercaptan. The mercaptal of D-arabinose is less soluble in alcohol than that of the L-form. The aldose is regenerated by HgO.

L-Ribose is the epimer of L-arabinose (same osazone). As such it can be obtained by epimerization of the latter by conversion to L-arabonic acid and racemization of the alpha carbon followed by separation and reduction of the lactone to L-ribose $(---)$. On reduction, it gives adonitol, optically inactive by internal compensation $(---)$, m. 102°. The properties of this substance are in sharp contrast to those of the closely related xylitol $(-+-)$. On oxidation L-ribose gives L-ribonic acid and then ribotrihydroxy glutaric acid $(---)$ which forms a hydroxylactonic acid, m. 170°. This is in contrast to xylo-trihydroxyglutaric acid $(-+-)$ which gives no lactonic acid.

D-Ribose is the characteristic sugar of plant nucleic acid (yeast).[28] It is of great importance because of its occurrence in riboflavin. Crystalline beta-Me-D-ribopyranoside has been prepared.[29]

Thyminose, D-2-desoxy-ribose, D-2-ribo-desose, is the sugar of animal nucleic acid.[30]

The configurations of the aldopentoses may be summarized.

D- and L-Arabinose, $++-$ and $--+$.
D- and L-Xylose, $-+-$ and $-+-$.
D- and L-Lyxose, $+--$ and $-++$.
D- and L-Ribose, $+++$ and $---$.

Apiose, $(HOCH_2)_2C(OH)CHOHCHO$, an aldopentose with a branched chain, occurs in the glucoside apiin. Oxidation gives apionic acid which, on reduction with HI and P, gives isovaleric acid, $Me_2CHCH_2CO_2H$. p-Bromophenylosazone of apiose, m. 212°. It does not give furfural with acids.

[28] Levene, Jacobs. Ber. 43, 3147 (1910); 42, 2703 (1909); 41, 2703 (1908).
[29] Jackson, Hudson. J. Am. Chem. Soc. 63, 1229 (1941).
[30] Levene. Am. Rep. Chem. Soc. (London) 1930, 265.

The only ketopentose reported in Nature is L-xylulose which occurs in some human urines. Probably one of the by-products of the condensation of formaldehyde in the presence of $CaCO_3$ to give formose (a hexose) is a ketopentose. This gives the phenylosazone of DL-arabinose.

METHYL ALDOPENTOSES, CH₃CH(CHOH)₃CH

These are found as glucosides and methylpentosans in various plant products.

L-**Rhamnose**, L-manno-methylose, "isodulcite," m. 124°, monohydrate, m. 93°, is found in many glycosides including xanthorhamnin, quercitrin, strophanthin, ouabain and solanin. The freshly prepared water solution of the hydrate is levorotatory but gradually becomes dextrorotatory (mutarotation). At equilibrium in a 10% solution $[\alpha]_D = +8.4°$. The anhydrous rhamnose starts with a dextrorotation of about $+30°$ and drops to the equilibrium value.

Phenylosazone, m. 187°.

Reduction gives rhamnitol, treatment with acids gives methyl furfural, with bromine water gives rhamnonic acid and with nitric acid gives L-trihydroxyglutaric acid, $HO_2C(CHOH)_3CO_2H$ $(--+)$. This settles the configuration of carbons Nos. 2–4 as a group but leaves undecided whether the alpha carbon is $+$ or $-$ and leaves carbon No. 5 uncertain. Both these questions are settled by applications of Hudson's Lactone Rule.[31] This states that, with the projection formulas arranged in the usual way, the lactones having the gamma OH on the right will be dextrorotatory and vice versa. The lactone of rhamnonic acid is levorotatory. Therefore, carbon No. 4 has the $-$ configuration and No. 2 and No. 3 are $++$. On treatment with H_2O_2 and ferric acetate,[32] rhamnonic acid gives a methyltetronic acid whose lactone is levorotatory, thus showing that carbon No. 5 in rhamnose is $-$. The configuration of rhamnose is thus

$$\underset{\begin{array}{cccc} & H & H & OH & OH \end{array}}{CH_3-\overset{\begin{array}{cccc} OH & OH & H & H \end{array}}{C}-C-C-C-CHO.} \quad (--++)$$

Rhamnose can be epimerized through its lactone to give *isorhamnose* with the configuration $--+-$ (the active group on the right as usual). *Isorhamnose,* L-epirhamnose, is identified by its ethyl mercaptal, $-CH(SEt)_2$, m. 98°. Such derivatives obtained from the simple sugars and mercaptans (both aliphatic and aromatic) have proved valuable for identification and to stabilize the

[31] Hudson. *J. Am. Chem. Soc.* **32**, 338 (1910).
[32] Ruff. *Ber.* **31**, 1573 (1898).

aldehyde group. The mercaptal grouping is stable to ordinary reagents but is readily removed by HgO or HgCl₂

$$-CH(SEt)_2 + HgO \rightarrow -CHO + (EtS)_2Hg$$

L-**Fucose** (L-galacto-methylose) is obtained from the methylpentosans of marine algae. D-**Fucose** (rhodeose), **quinovose** (chinovose, iso-rhamnose, iso-rhodeose, D-gluco-methylose), **epi-rhodeose, gulo-methylose, talo-methylose,** and **altro-methylose** are known stereomers of rhamnose.[33]

Digitoxose, $Me(CHOH)_3CH_2CHO$, is a 2,6-didesoxyhexose characteristic of digitalis glycosides.[34]

HEXOSES, $C_6H_{12}O_6$

These are the most important natural sugars. They are most widely distributed in nature both in the free state and in combination as heterosides (glycosides) and holosides ("polymeric" forms like the polysaccharides). They are fermented by various organisms. With acids they give first 5-hydroxymethylfurfural and then levulinic acid, $MeCO(CH_2)_2CO_2H$. There are sixteen theoretically possible aldohexoses and eight 2-ketohexoses. All of these hexoses are actually known. A large portion of the work on the hexoses was done by Emil Fischer and his many students.

ALDOHEXOSES,

$$\overset{\lceil\qquad O\qquad\rceil}{CH_2OHCH(CHOH)_3CHOH} \quad \text{AND} \quad \overset{\lceil\qquad O\qquad\rceil}{CH_2OHCHCHOHCH(CHOH)_2CHOH}$$

The first is the ordinary or stable pyranose form while the second is the reactive *gamma* or furanose form. Each can exist in *alpha* and *beta* forms depending on the configuration of the new asymmetric C formed by ring closure involving the active group (C No. 1).

The number of stereoisomers of a given structure can be calculated by simple formulas. If n is the number of asymmetric carbon atoms, a the number of optically active forms, and m the number of meso forms (optically inactive by internal compensation); then

1. If the molecule contains no two carbons attached to identical groups, $a = 2^n$; $m = 0$.

2. If n is even and the molecule can be divided into two equal halves, $a = 2^{n-1}$; $m = 2^{\left(\frac{n-1}{2}\right)}$.

3. If n is odd and the molecule can be divided into two equal halves by a plane through the middle carbon, $a = 2^{n-1} - 2^{\left(\frac{n-1}{2}\right)}$; $m = 2^{\left(\frac{n-1}{2}\right)}$.

The configurations of the aldohexoses have been determined by the principles already outlined with the lower monoses. Each aldohexose has four

[33] *Ann. Rep. Chem. Soc.* (London) **1929,** 98.
[34] *ibid.* **1930,** 106.

asymmetric carbon atoms besides the active or aldehyde carbon which is responsible for alpha and beta forms. Since each of these carbons is different (the two ends of the molecule are different), there are 2^4 or 16 possible configurations. Arranging the models in the usual way with the active group at the top and the CH_2OH at the bottom with both in the plane of the paper and the H and OH groups in front of this plane and then projecting the formula on paper and finally indicating the HCOH and HOCH arrangements as $+$ and $-$ respectively, we have the sixteen possible combinations:

```
+ -   - +   + -   + -   - +   - +   + -   - +
+ -   + -   - +   + -   - +   - +   - +   + -
+ -   + -   + -   - +   - +   + -   - +   - +
+ -   + -   + -   + -   + -   + -   + -   + -
```

In each pair, the first represents a member of the D-series, related to D-glucose and D-glyceraldehyde and the second the enantiomorph belonging to the L-series. As usual, the configuration of the penultimate C determines whether the compound belongs to the D or the L series.

The methods of assigning these configurations will be indicated only briefly as the principles have already been developed.

D-GLUCOSE, DEXTROSE, GRAPE SUGAR, $(+ + - +)$

This monose is widely distributed in plants. In animals it is a normal constituent of the blood and is related to the development of muscular energy. In diabetes mellitus it occurs in the blood and urine in large amounts. It occurs as the sole constituent of the carbohydrates maltose, cellobiose, trehalose, starch and cellulose and with other monoses in lactose, turanose and sucrose. It is the commonest sugar in natural glycosides (heterosides) and gives the name to these substances.

The discovery of the configuration of glucose was one of Emil Fischer's classic contributions to organic chemistry.[35]

Ordinary dextrose is mainly α-D-glucose, m. 146°, $[\alpha]_D = + 109.6°$. β-D-Glucose can be obtained by crystallizing D-glucose from hot pyridine, m. 148–150°, $[\alpha]_D = + 20.5°$. Both forms show mutarotation, the optical activity of the solutions changing until the value for the equilibrium mixture, $+ 52.7°$, is reached. This corresponds to about 37% α- and 63% β-glucose. The change of one form into the other is catalyzed by hydrogen ions or hydroxyl ions *near the neutral point* but is stopped by as weak a base as pyridine or as weak an acid as cresol.[36] On the other hand, a mixture of these two solvents makes mutarotation about twenty times as fast as in water. The change consists of the

[35] Hudson. *J. Chem. Education* 18, 353 (1941).
[36] Lowry. *J. Chem. Soc.* 127, 1371 (1925).

addition and removal of H^+ as in the bimolecular mechanism of keto-enol isomerization.

The determination of the configuration of the two glucoses depends on a study of the cleavage of their methyl pyranosides with enzymes and with glycol-splitting reagents and on studies of the formation and fissure of their 1,2-epoxy derivatives.

α-D-Glucose, with boric acid, gives a solution of higher conductivity than a similar solution of β-D-glucose and boric acid. The conductivity of the former falls and of the latter rises until the same value is attained. That this is due to the formation of the equilibrium mixture of α and β sugars is shown by the fact that the velocity constants for the changes in conductivity and in optical activity are the same.[37] Thus α-glucose must have two hydroxyl groups in the favorable position while β-glucose does not. In view of the pyranose structure of glucose, the only way to achieve this condition is for the active (aldehydic) asymmetric carbon to have the + or HCOH configuration to match that of the No. 2 carbon. β-D-Glucose correspondingly has the configuration $++-+(-)$, starting with the No. 5 carbon. The No. 5 carbon being in ring combination, no two adjacent OH groups are in the same plane.[38] The same configurations for alpha and beta glucose are obtained by Hudson's Rule (1909).[39]

That the intermediate in the process of mutarotation may be the aldehyde form is indicated by data on the mutarotation in chloroform solution of the penta-acetate of *aldehydo*-galactose.[40]

Various ways of indicating the configuration of the forms of glucose follow.[41]

α-D-Glucopyranose   α-D-Glucofuranose

[37] *Ann. Rep. Chem. Soc.* (London) 1913, 79.
[38] Macpherson, Percival. *J. Chem. Soc.* 1937, 1920.
[39] *Ann. Rep. Chem. Soc.* (London) 1910, 87.
[40] Wolfrom, Morgan. *J. Am. Chem. Soc.* 54, 3390 (1932).
[41] *Ann. Rep. Chem. Soc.* (London) 1929, 94.

While the first formula is configurationally correct it is not in accordance with the space relations although it was used from Emil Fischer's time until 1926 when the second formula was shown to agree with the models.[42] The second formula accurately represents about 99% of *alpha*-glucose. The last, glucofuranose, is a very reactive form which has been isolated only as derivatives.[43] The furanose form is probably of the utmost importance in natural processes. Before they were recognized as furanose forms of the ordinary pyranoses they were called *heteros* or *h*-sugars.[44] The furanose illustrated above is really an alpha form. Reversal of the H and OH on the No. 1 carbon would give the beta form of the sugar. Four isomeric ethyl glucosides have actually been prepared.

Crystallographic evidence shows the presence of the pyranose ring and the *cis* relation of the hydroxyls on carbons Nos. 1 and 2 in alpha D-glucose.[45]

Glucose, in solution, behaves as a tautomeric mixture of the following forms: alpha and beta D-glucopyranose, the hydrated aldehyde form, the 1,2-enediol, alpha and beta D-glucofuranose, and perhaps a minute amount of the free aldehyde form.

Glucose is not as sweet as sucrose. It is made in enormous amounts by the hydrolysis of starch (about one billion pounds per year). Depending on its purity it appears as corn syrup, yellow chip glucose, corn sugar, cerelose (a very pure crystalline monohydrate), etc.

Pure dextrose solution is used intravenously in conditions which lower the natural blood sugar content.

On heating, glucose loses water and becomes *glucosan*, $C_6H_{10}O_5$. A levorotatory isomer, "beta-glucosan" or *levoglucosan* is obtained in fair yields by the vacuum pyrolysis of starch or cellulose.

It has a peculiar bicyclic cage system:

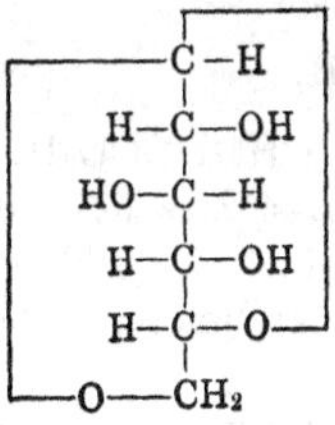

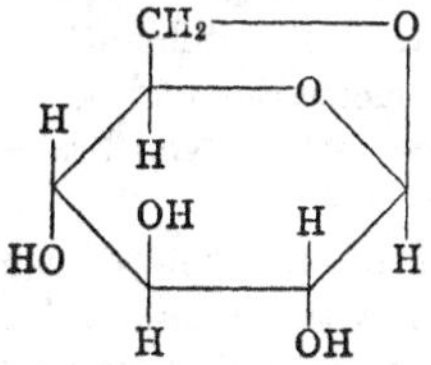

Similar compounds (from D-mannose) have proved invaluable in the synthesis of disaccharides.[46]

[42] Haworth, Nicholson. *J. Chem. Soc.* **1926**, 1899.
[43] Levene. *Chem. Revs.* **5**, 1 (1928).
[44] *Ann. Rep. Chem. Soc.* (London) **1925**, 88.
[45] *ibid.* **1939**, 184.
[46] Hudson et al. *J. Am. Chem. Soc.* **63**, 1447, 1724 (1941); **64**, 1289, 1852, 2435 (1942).

Levoglucosan is also obtained by the action of barium hydroxide on beta glucosides. The alpha forms are inert.[47] This is a good example of trans-elimination in ring closure. The same phenomenon is found in the formation of epoxy sugars.[48]

Periodic acid oxidation of glucosan removes the 3-C as formic acid and converts the 2-C and 4-C to aldehyde groups. Further oxidation with bromine water converts these to carboxyl groups forming the dibasic acid:

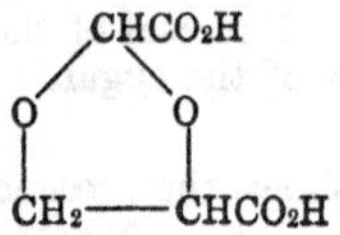

Although this is an acetal, it is stable to hot concentrated hydrochloric acid. This corresponds to the stability of the chloral acetals of glycol and of glycerol. It is the cyclic acetal of glyoxylic acid and glyceric acid.[49]

D-Glucose gives the same osazone, m. 205°, as D-mannose and D-fructose. Thus carbons Nos. 3–5 in these three monoses have the same configuration. Since the degradation of D-glucose gives D-arabinose, the configuration of these carbons is $(+ + -)$. In addition to the methods of Wohl (p. 461) and of Ruff (p. 464) the following method gives excellent yields of arabinose. D-Gluconic anhydride with NH_3 gives the amide which reacts with hypochlorite solution to give D-arabinose.[50] The hydroxyisocyanate, $-CH(OH)NCO$, formed as an intermediate may be assumed to lose HNCO or to be hydrolyzed to $-CH(OH)NH_2$, the NH_3 addition compound of the lower aldose, which readily forms the aldose by reaction with the hypochlorite.

Reduction of D-glucose gives sorbitol, $HOCH_2(CHOH)_4CH_2OH$, m. 111°. Oxidation with bromine water gives D-gluconic acid,

$$HOCH_2(CHOH)_4CO_2H,$$

and with nitric acid gives saccharic acid, $HO_2C(CHOH)_4CO_2H$. The latter gives a characteristic difficultly soluble acid potassium salt. Permanganate oxidation of saccharic acid gives oxalic and L-tartaric acid, configuration $(- +)$. Since carbon No. 3 is $-$ like No. 2 in arabinose, carbon No. 2 in glucose must be $+$.

Glucose is an excellent reducing agent, especially in alkaline solution, and is used as such with ammoniacal silver solution for silvering mirrors and in the vat process for converting insoluble indigo to indigo white.

The oxidation of glucose and the other monoses is a most complex and important problem.[51]

[47] Hudson. *J. Chem. Education* 18, 353 (1941).
[48] *Ann. Rep. Chem. Soc.* (London) 1939, 260, 266.
[49] Jackson, Hudson. *J. Am. Chem. Soc.* 62, 958 (1940).
[50] Weerman. *Ann. Rep. Chem. Soc.* (London) 1915, 74.
[51] Evans. *Ann. Rep. Chem. Soc.* (London) 1929, 281.

In the presence of dilute bases, D-glucose gives an equilibrium mixture with its epimer, D-mannose, and D-fructose, the related ketohexose.[52] The same mixture is obtained from any one of the three sugars. At equilibrium 63.4% glucose, 30.9% fructose and 2.4% mannose are obtained.[53]

The process goes through the enediol, $-C(OH)=CHOH$, common to the three sugars. At the same time, small amounts of nonfermentable products, possibly 3-ketoses or other 2-ketoses, are probably formed.

Stronger bases (8N) give i-lactic acid, α,γ-dihydroxybutyric acid and saccharinic acids.[55-57] These are not to be confused with saccharic acid (Zuckersäure) obtained by oxidizing glucose and gulose. The saccharinic acids (Saccharinsäure) are the results of rearrangements much like the benzilic acid rearrangement. If only H rearranges, the result is a metasaccharinic acid such as $HOCH_2(CHOH)_2CH_2CHOHCO_2H$ while if a carbon rearranges the result is an isosaccharinic acid such as

$$HOCH_2CHOHCH_2C(OH)(CH_2OH)CO_2H$$

Glucose and other sugars dissolve calcium oxide and related substances to form glucosates, etc.

Glucose with an ammoniacal zinc solution gives methyl glyoxaline, indicating splitting by the mild alkali to give glyoxal, MeCOCHO, and formaldehyde.

DERIVATIVES OF D-GLUCOSE

α-Methyl glucoside, Me α-D-glucopyranoside, $HOCH_2CH(CHOH)_3CHOMe$ (configuration $++-++$), m. 166°, is the chief product from the action of dilute methyl alcoholic HCl on glucose. It is strongly dextrorotatory without *mutarotation*. Thus the lactol ring is definitely fixed. It is converted to D-glucose by maltase, an enzyme which splits alpha glucosides. β-Methyl glucoside $(++-+-)$, m. 108°, is made from aqueous glucose and dimethyl sulfate (poison) and alkali, the reagents being added gradually in equivalent amounts. It is levorotatory without mutarotation. It is more easily hydrolyzed than the α-form. Thus treatment of the glucoside with alcoholic HCl gives the α-form

$$\alpha\text{-glucose} + \text{MeOH} \rightleftharpoons \alpha\text{-glucoside}$$
$$\beta\text{-glucose} + \text{MeOH} \rightleftharpoons \beta\text{-glucoside}$$

[52] Lobry de Bruyn, Ekenstein. *Ber.* **28**, 3078 (1895).
[53] Wolfrom, Lewis. *J. Am. Chem. Soc.* **50**, 837 (1928).
[54] Benedict, Dakin, West. *J. Biol. Chem.* **68**, 1 (1926).
[55] Nef. *Ann.* **376**, 1 (1910).
[56] Glattfeld, Woodruff. *J. Am. Chem. Soc.* **49**, 2309 (1927).
[57] Nicolet. *J. Am. Chem. Soc.* **53**, 4458 (1931).

Certain enzymes have the specific ability to hydrolyze *alpha* or *beta* glucosides but not both. Thus maltase from yeast splits only the alpha while emulsin splits only the beta forms.

Large numbers of glucosides have been prepared or isolated from natural products. *Picein* is *p*-hydroxyacetophenone *beta*-D-glucopyranoside. Trichloroethyl *beta*-D-glucoside is found in plants grown in soil containing chloral.[58]

Glucosides of higher alcohols containing 6, 9, 10, and 12 carbons have been made.[59, 60]

A substance of structure analogous to that of the glucosides is the *Cori ester*, typical of biologically important phosphates.[61, 62] It is a D-glucopyranose-1-phosphate containing as the active group $CH-OPO(OH)_2$.

γ-Methyl glucoside is a furanose (butylene oxide ring).[63] It is more properly called methyl *alpha*-D-glucofuranoside. It was first made by Emil Fischer by allowing glucose to stand with 1% methyl alcoholic HCl at room temperature. It is much more reactive to acids and oxidizing agents than the α- and β-glucosides.

Periodic acid oxidation of methyl glucoside eliminates the middle C of the pyranose ring and converts the two carbinol groups next to it to aldehyde groups.[64]

Four **ethyl glucosides** have been prepared. These are known as alpha, beta, gamma and delta. The last two are derivatives of the reactive gamma glucose. They are better called α- and β-glucopyranosides and α- and β-glucofuranosides. Their mps. and specific rotations are 114°, +150°; 73°, −33°; 83°, +98°; 60°, −86°.

Treatment of the methyl glucosides with MeI and Ag_2O[65] or with Me_2SO_4 and NaOH[66] gives tetramethyl methyl glucosides which can be distilled under reduced pressure. Hydrolysis gives Me_4-glucose, b_{20} 182–5°. This has a bitter taste and reduces Fehling's solution on heating but with the formation of only one-fifth as much cuprous oxide as given by glucose itself. The

[58] Miller.   *Contrib. Boyce Thompson Inst.* **9**, 213 (1938); **12**, 465 (1942).
[59] Noller, Rockwell.   *J. Am. Chem. Soc.* **60**, 2076 (1938).
[60] Kreider, Friesen.   *J. Am. Chem. Soc.* **64**, 1482 (1942).
[61] Wolfrom, Pletcher.   *J. Am. Chem. Soc.* **63**, 1050 (1941).
[62] Wolfrom et al.   *J. Am. Chem. Soc.* **64**, 23 (1942).
[63] *Ann. Rep. Chem. Soc.* (London) **1927**, 70.
[64] Jackson, Hudson.   *J. Am. Chem. Soc.* **58**, 378 (1936); **59**, 994 (1937).
[65] Purdie, Irvine.   *J. Chem. Soc.* **83**, 1021 (1903).
[66] Gustus, Lewis.   *J. Am. Chem. Soc.* **49**, 1512 (1927).

presence of the methoxyl group evidently blocks the continued oxidation of the carbohydrate chain. Me_4-glucose can be obtained in α- and β-forms which show mutarotation.

Methylated sugars with acetic anhydride and HBr give the corresponding acetylated bromo-sugars.[67] Thus 2,3,6-Me_3-glucose gives 1-α-Br-tetraacetyl-glucose from which glucose can be obtained by hydrolysis.

The various partly methylated glucoses have been made and are important in relation to the structures of the polysaccharides. The other methylated monoses are similarly important. In all such substances the methoxyl determination of Zeisel is most useful.[68-70]

Pentamethyl-*aldehydo*-glucose, $MeOCH_2(CHOMe)_4CHO$, has been made by methylating diethylmercaptoglucose and then removing the mercaptal groups with $HgCl_2$.[71] This reacts readily with methanol to give the acetal, $-CH(OMe)_2$.

Tetra-allyl α-methyl glucoside has been prepared using allyl bromide in place of MeI.[72]

Alkali salts of glucose and other sugars can be made with the alkali metals in liquid ammonia.[73] These are useful for introducing alkyl and other groups into the sugars.

α- and β-Pentaacetylglucoses, m. 112° and 134°, of the formula

$$\overset{\displaystyle \lceil \underline{\qquad\quad O \qquad\quad} \rceil}{MeOCOCH_2CH(CHOCOMe)_3CHOCOMe}$$

have been prepared. One of the acetyl groups is more easily removed than the rest. The pentaacetate of the true aldehyde formula,

$$MeOCOCH_2(CHOCOMe)_4CHO,$$

has been made by acetylating the mercaptal of glucose, $-CH(SR)_2$, and treating with HgO.[74]

1,2,3,4-tetraacetyl-*beta*-D-glucose, in pyridine with *mesyl chloride* ($MeSO_2Cl$), gives a quantitative yield of the corresponding 6-mesyl derivative, m. 156°.[75]

$$\overset{\displaystyle \lceil \underline{\qquad\quad O \qquad\quad} \rceil}{}$$

Acetobromoglucose, $MeOCOCH_2CH(CHOCOMe)_3CHBr$, m. 89°, is made from the penta-acetate and HBr in glacial acetic acid. Treatment with zinc

[67] Hess, Dziengel. *Ber.* **68B**, 1594 (1935).
[68] Hewitt, Moore. *J. Chem. Soc.* **81**, 318 (1902).
[69] Rigakos. *J. Am. Chem. Soc.* **53**, 3903 (1931).
[70] Zemplen, Bruckner. *Ber.* **64B**, 1852 (1931).
[71] Levene, Meyer. *J. Biol. Chem.* **69**, 175 (1926).
[72] Tomecko, Adams. *J. Am. Chem. Soc.* **45**, 2698 (1923).
[73] Muskat. *J. Am. Chem. Soc.* **56**, 693 (1934).
[74] Wolfrom. *J. Am. Chem. Soc.* **51**, 2188 (1929).
[75] Helferich, Gnüchtel. *Ber.* **71**, 714 (1938).

and acetic acid, followed by hydrolysis, gives D-glucal,

$$HOCH_2CHCHOHCHOHCH=CH.$$
$$|\underline{\hspace{2cm}O\hspace{2cm}}|$$

This is identical with D-mannal. It is a typical glycol.[76]

Hydrochloric and hydrofluoric acids also react with acetylated monoses.[77]

Among the acyl derivatives of the sugars are those obtained with *p*-toluene-sulfonyl chloride. The grouping $-SO_2C_6H_4CH_3-p$. is called the *tosyl* group.

Gold thioglucose, *Solganal-B*, aurothioglucose, is a water soluble compound used in special therapy. It contains the terminal grouping $-CH_2SAu$.[78]

α- and β-forms, m. 160° and 141°, of the *phenylhydrazone* of D-*glucose* can be obtained. These show mutarotation. Thus they are cyclic forms subject to the same type of isomerization as α- and β-glucose,

$$\overline{\hspace{1cm}O\hspace{1cm}}$$
$$HOCH_2CH(CHOH)_3CHNHNHC_6H_5$$

Glucose reacts with one mol of methylphenylhydrazine, $C_6H_5(Me)NNH_2$, but is not oxidized by an excess of the reagent (difference from ketoses). Excess phenylhydrazine gives the well-known osazone formation.

Phenylglucosazone (A), m. 205°.[79] On hydrolysis or, better, on treating with another aldehyde, the phenylhydrazone groups can be removed leaving an alpha keto aldehyde, an *osone*.

$$\qquad\qquad\qquad\qquad\qquad\qquad RCHO$$
(A) $HOCH_2(CHOH)_3C(=NNHC_6H_5)CH=NNHC_6H_5 \xrightarrow{\hspace{1cm}}$
$$\qquad\qquad\qquad\qquad\qquad HOCH_2(CHOH)_3COCHO$$

Reduction of the osones gives ketoses, in this case D-fructose.

Glucose reacts with one or two mols of acetone, adjacent *cis* hydroxyl groups reacting to give isopropylidene derivatives (I). The acetone compounds of the sugars are valuable in blocking certain positions so that others can be methylated.[80] When only one acetone is introduced it reacts in the 1,2-position of the pyranose form. When a second molecule of acetone reacts with glucose, it involves the 5,6-hydroxyls, thus forcing a change to a furanose ring (II).

Similarly carbonates (III) can be formed by chloroformic ester, $ClCO_2Et$, and a base or pyridine.[81] 1,2-Isopropylidene-α-glucofuranose-5,6-carbonate

[76] *Ann. Rep. Chem. Soc.* (London) **1931**, 98.

[77] Brauns, Frush. *Bur. Standards J. Research* **6**, 449 (1931).

[78] "The Formulary." Univ. Mich. Hospital, 1946.

[79] Butler, Cretcher. *J. Am. Chem. Soc.* **51**, 3161 (1929).

[80] *Ann. Rep. Chem. Soc.* (London) **1927**, 74.

[81] *ibid.* **1924**, 71.

can be split by acids to

give the 5,6-carbonate of gamma glucose (IV).[82] This has the characteristic high reactivity of the gamma sugars such as the reduction of a cold permanganate solution and the practically instantaneous union with methanol in the presence of HCl.[83] The carbonates are less readily hydrolyzed than the acetone compounds by dilute acids but are immediately converted to glucose by bases.[84]

Glucose-6-phosphate[85] is an example of the hexose-phosphates which are probably important in fermentation processes.

Glucosamine has NH_2 in place of the α-OH of glucose. It is obtained by the hydrolysis of *chitin*, the bony material of crustacea. It also occurs in *chondroitic acid* in cartilage combined with glucuronic acid, acetic acid and sulfuric acid.

N-Methyl-L-glucosamine is one of the hydrolysis products of streptomycin.

Tetramethylglucoseen-1,2, Me$_4$-glucose-1-ene,

reacts with HCl to form 5-methoxymethyl-2-furaldehyde.[86]

OTHER ALDOHEXOSES

L-Glucose, $(--+-)$, is obtained from L-arabinose by the cyanohydrin synthesis.[87] This gives a mixture of L-mannonic acid and L-gluconic acids. The lactone of the former crystallizes first. The latter is purified as the calcium salt. Reduction of the L-gluconic lactone gives L-glucose. L-Glucose is not fermented by yeast.

[82] *Ann. Rep. Chem. Soc.* (London) **1929**, 96.
[83] Haworth, Maw. *J. Chem. Soc.* **1926**, 1751.
[84] *Ann. Rep. Chem. Soc.* (London) **1930**, 102.
[85] *ibid.* **1931**, 98.
[86] Wolfrom, Wallace, Metcalf. *J. Am. Chem. Soc.* **64**, 265 (1942).
[87] Kiliani. *Ber.* **19**, 767 (1886).

D-**Mannose**, $(++--)$, m. 132°, occurs in glycosides (mannosides) and in polyoses (mannans) in a great variety of natural products. One of the chief of these is the ivory nut from which buttons are made. The waste material from this manufacture is readily hydrolyzed by dilute acid to give mannose.[88]

D-Mannose is readily fermented by yeast. Its phenylhydrazone, m. 205°, is very insoluble in water. In pyridine solution it is dextrorotatory but shows no mutarotation. Excess of phenylhydrazine gives the osazone of D-glucose, m. 205°, levorotatory in pyridine solution.

The configuration of D-mannose as the epimer of D-glucose is shown by its conversion to D-mannonic acid and the inversion of the alpha carbon in the usual way by heating with pyridine, etc. (p. 469).

Whereas aldoses, by the cyanohydrin synthesis, usually give a pair of epimeric acids, D-mannose gives only *one* acid and that in nearly quantitative yield, D-mannoheptonic acid. This is a true *asymmetric synthesis* in which an optically active molecule influences the configuration of a newly formed asymmetric carbon.

A reagent of peculiar selectivity in its reactions among sugars is

$$CH_2(C_6H_4NMeNH_2)_2,$$

unsatisfactorily named diphenylmethane-dimethyl-dihydrazine or bis [(N-α-methylhydrazino)4-phenyl]methane or, better, di(p-α-methylhydrazino-phenyl)methane.[89] It does not act with fructose or the disaccharides. It acts only with aldoses in which carbons Nos. 2, 3 and 4 have at least two adjacent hydroxyls of the same "sign." Thus it reacts with mannose $(++--)$, galactose $(+--+)$, arabinose $(++-)$ and ribose $(+++)$ while it does not act with glucose $(++-+)$ or xylose $(+-+)$. Because it can be used to separate epimers in some cases (glucose and mannose) it has been called von Braun's epimer reagent.

D-Mannose forms a di-acetone compound which shows mutarotation. Thus the active group is free and the acetone residues are combined in the 2,3- and 5,6-positions.

beta-Phenyl-D-mannoside reacts with alkali to give D-mannosan $<1,5>\beta$-$<1,6>$, *levomannosan*, which has been used in the synthesis of epi-cellobiose and cellobiose.[90]

L-**Mannose** is obtainable from L-mannonic lactone which crystallizes first from the mixture with L-gluconic lactone in the Kiliani synthesis from L-arabinose. It is not fermented by yeast.

D-**Galactose**, $(+--+)$, m. 168°, occurs in galactans, in glycosides and in lactose, raffinose and stachyose and in certain cerebrosides. It is best obtained by hydrolyzing lactose and crystallizing out the galactose leaving the glucose in solution. A potential commercial source is found in the galactans

[88] Isbell. *J. Research. Natl. Bur. Standards* 26, 47 (1941).
[89] von Braun. *Ber.* 50, 42 (1917).
[90] Knauf, Hann, Hudson. *J. Am. Chem. Soc.* 63, 1447 (1941).

of the mountain larch.[91] It shows mutarotation. It is more slowly fermented than the other natural monoses. Its methylphenylhydrazone,

$$HOCH_2(CHOH)_4CH = NN(Me)C_6H_5,$$

m. 191°, is only slightly soluble in water. Reduction gives dulcitol, m. 188°, optically inactive by internal compensation. Oxidation gives D-galactonic acid and finally mucic acid, a meso compound. The latter is also obtained from L-galactose. When mucic acid is reduced the product is a racemic mixture of D- and L-galactonic acids, the reduction of one carboxyl giving the former and the other the latter. Further reduction gives DL-galactose. Fermentation destroys the D-form leaving L-galactose.

Galactose is used clinically as an indicator of the degree of liver function.

Various acyclic derivatives of galactose have been made.[92, 93] These are typical of such derivatives of other monoses. They include the following, only groups on the 1-C indicated: D-galactose diethyl mercaptal pentaacetate, $-CH(SEt)_2$;—S-ethyl O-methyl monothioacetal—, $-CH(SEt)OMe$; aldehydo-1-chloro-1-methoxy-D-galactose pentaacetate, $-CH(OMe)Cl$; the corresponding 1-ethylmercapto compound, $-CH(SEt)Cl$; the 1,1-dichloro pentaacetate, $-CHCl_2$; aldehydo-D-galactose heptaacetate, $-CH(OAc)_2$; and the alpha and beta forms of 1-chloro-aldehydo-D-galactose hexaacetate, $-CH(OAc)Cl$.

Studies of the mutarotation of solutions of galactose indicate that they contain the α- and β-pyranose and α- and β-furanose sugars.[94]

Pentaacetyl-aldehydo-galactose can be made from the ethyl mercaptal, $-CH(SEt)_2$. The free aldehyde shows no mutarotation in chloroform. Its crystalline hydrate and its ethyl hemiacetal show mutarotation.[95]

DL-Galactose is obtained by the careful oxidation of dulcitol. Fermentation destroys the D-form. The DL-mixture can be separated by making the d-amylphenylhydrazine[96] by means of $C_6H_5(C_5H_{11})NNH_2$, separating the two compounds DD and DL and removing the hydrazine by treating with benzaldehyde.

L-Idose, $(-+-+)$, m. 156°, is obtained in the usual way from the L-idonic acid which is formed with L-gulonic acid from L-xylose by the cyanohydrin synthesis. D-Gulonic acid can be epimerized to D-idonic acid which can be reduced to D-idose. Reduction and oxidation give iditol, m. 74°, and idosaccharic acid. The latter gives a difficultly soluble copper salt. The idoses are not attacked by yeast.

L-Gulose, m. 156°, (configuration $-+--$), can be made from L-xylose by the cyanohydrin synthesis. It gives sorbitol and saccharic acid. Thus it

[91] Acree. C. A. 25, 5554 (1931).
[92] Wolfrom, Weisblat. J. Am. Chem. Soc. 62, 878, 1149 (1940).
[93] Wolfrom, Brown. 63, 1246 (1941).
[94] Ann. Rep. Chem. Soc. (London) 1929, 92.
[95] Wolfrom. J. Am. Chem. Soc. 53, 2275 (1931).
[96] Neuberg, Federer. Ber. 38, 868 (1905).

484   ALIPHATIC COMPOUNDS

resembles glucose except that the aldehyde and primary carbinol groups are reversed with respect to the four central carbons. This reversal of the ends can be carried out experimentally in the following steps:

D-Glucose → D-gluconic acid → saccharic acid →

$$++-+ \qquad -+-- \qquad -+--$$

D-glucuronic acid → L-gulonic acid → L-gulose

(L-guluronic acid) → (D-gluconic acid) → (D-glucose)

$$-+-- \qquad ++-+ \qquad ++-+$$

D-Talose, m. 188°, $(+---)$, can be made by epimerizing D-galactonic acid, separating and reducing the D-talonic lactone formed. Methylphenyl-hydrazone, m. 154°. It gives talitol, m. 86°, and talomucic acid, m. 158°.

D-Altrose, m. 184°, $(+++-)$, and D-*allose*, m. 184°, $(++++)$, are obtainable from D-ribose $(+++)$ by the Kiliani synthesis. The configurations are settled by the fact that D-altrose gives an optically active dibasic acid, D-talomucic acid, m. 158°, while D-allose gives a meso acid, allomucic acid, m. 170°.

Crystalline *beta*-L-altrose has been obtained from L-ribose.[97]

Hamamelose, $HOCH_2(CHOH)_2C(OH)(CH_2OH)CHO$, a branched chain aldohexose, is found as a glycoside in hamameli tannin.[98]

Aldoses can be identified by conversion to the fully acetylated diethyl mercaptals.[99] The melting points of these derivatives of the common aldoses follow (°C): D-arabinose, 80; L-arabinose, 79–80; D-xylose, 46–8; D-lyxose, 36–7; D-glucose, 45–7; D-galactose, 76–7, 80–1, 90–1; 6-desoxy-L-galactose, (L-fucose), 99–100; D-mannose, 52–3; 6-desoxy-L-mannose (L-rhamnose), 59–61; D-gluco-D-*gulo*-heptose, 99–100; D-gala-L-*gluco*-heptose, 105; D-gala-L-*manno*-heptose, 145–6; D-manno-D-*gala*-heptose, 77; D-gala-L-*gala*-octose, 106.

2-KETOHEXOSES

$$\overline{\quad O \quad}$$
$$CH_2(CHOH)_3C(OH)CH_2OH$$

$$\overline{\quad O \quad}$$
$$HOCH_2CH(CHOH)_2C(OH)CH_2OH$$

The second is the furanose or gamma form. The configurations are determined by reference to those of the aldohexoses which give the same osazones, that is, the aldohexoses which have the same configurations for carbons 3–5. The ketohexoses exist in alpha and beta form and show abnormal or no mutarotation.

D-Fructose, levulose, fruit sugar, $(++-)$, occurs in the sweet juices of fruits and in honey, in sucrose and in inulin. It is best prepared from the

[97] Austin, Humoller. *J. Am. Chem. Soc.* 56, 1153 (1934).
[98] *Ann. Rep. Chem. Soc.* (London) 1930, 106.
[99] Wolfrom, Karabinos. *J. Am. Chem. Soc.* 67, 500 (1945).

latter source as found in the roots of the dahlia and the Jerusalem artichoke.[100]
Invert sugar is the equimolar mixture of D-fructose and glucose obtained by
hydrolyzing sucrose. It is levorotatory since D-fructose has a stronger levo
rotation than D-glucose has dextro.

D-Fructose can be made from D-glucose by the usual reaction for *converting
an aldose into the related ketose*, that is, through the gluosazone which, on treatment with benzaldehyde, gives D-glucosone, $HOCH_2(CHOH)_3COCHO$. This
can be reduced to D-fructose.

D-Fructose reduces Fehling's solution more rapidly than the aldohexoses
but does not react with alkaline iodine solution whereas the aldoses do. It
gives the same osazone as D-glucose and D-mannose. With methylphenylhydrazine it gives both a hydrazone and an osazone (difference from glucose and
other aldoses). Evidently the carbonyl group in the 2-position condenses
readily and No. 1 carbinol group is oxidized even by so mild an agent. On the
other hand, if the $C_6H_5(Me)NN=$ group is on the No. 1 carbon, the No. 2
carbinol group is not oxidized and no osazone can be formed.

The mutarotation of fructose consists in a pyranose-furanose interconversion.[101]

Reduction of D-fructose gives D-mannitol $(++--)$ and sorbitol
$(++-+)$.[102]

It is fermented by yeast. It is sweeter than sucrose. It cannot be used
by diabetics.

Treatment with mild alkaline reagents partly converts D-fructose to
D-glucose and D-mannose.

Methylated fructoses are important in sugar studies.

Tetramethyl-D-fructopyranose
Reacts slowly with HCl + MeOH
No reaction with cold $KMnO_4$
Dextrorotatory

Tetramethyl-γ-fructose
Rapid reaction with
HCl + MeOH and with $KMnO_4$
Levorotatory

The fact that the furanose or gamma form of levulose is found in sucrose,
inulin, etc., is related to its greater reactivity which enables it to unite with
other monose molecules.

Me₄-γ-fructose with dilute acid, changes readily to ω-methoxymethyl-
furfural.[103]

[100] Jackson, McDonald. *Bur. Standards J. Research* 5, 1151 (1930).
[101] Isbell, Pigman. *J. Research Natl. Bur. Standards* 20, 773 (1938).
[102] Creighton. *C. A.* 29, 1725 (1935).
[103] *Ann. Rep. Chem. Soc.* (London) 1927, 71.

D-Fructose gives two isomeric di-acetone compounds which are non-reducing but are furanose derivatives, the combination taking place in the 1,2,4,5- and 2,3,4,5-positions respectively.

D-Allulose (D-psicose) has been isolated as its diacetone derivative from distillery residues.[104] It is an epifructose which gives the same osazone as do D-altrose and D-allose. It is a non-fermentable reducing sugar.

DL-Fructose, alpha acrose, formose, has been obtained artificially by polymerizing through aldol condensation, the following: a. formaldehyde, b. glycollic aldehyde, c. glycerose obtained by careful oxidation of glycerol and consisting of glyceraldehyde and dihydroxyacetone, d. the same mixture obtained from acrolein dibromide, $BrCH_2CHBrCHO$, with barium hydroxide. Several osazones have been obtained from the mixture of sugars formed in this way. The most important is phenyl-α-acrosazone, m. 219°, identical with the osazone obtained from DL-glucose and DL-mannose. The sugar shows the reactions of D-fructose except for its optical inactivity. By the regular monose reactions this synthetic sugar has been converted to D-glucose, D-fructose, D-mannose, D-galactose, D-xylose, etc.

Formose[105] obtained from formaldehyde also contains small amounts of sorbose (β-acrose) and a ketopentose.[106] The formation of carbohydrates in nature probably involves a process much like that of the conversion of formaldehyde to formose.

L-Sorbose, m. 165°, (− + −), occurs in various plant products. It is formed from sorbitol (+ + − +) by the Sorbose-bacterium. It might be thought that this action would give two products, one by the oxidation of the No. 2 carbon and the other by that of the No. 5 carbon, but only the latter is attacked. The Sorbose-bacterium like most biological reagents is highly specific. It can oxidize only a No. 2 carbon in a system $++CH_2OH$ or $--CH_2OH$. That is, it oxidizes a No. 2 carbon in a system where the No. 2 and No. 3 carbons have their OH groups on the same side. Thus it does not oxidize L-iditol (− + − +).

The configurational relationships of L-sorbose are instructive:

D-Glucose   D-Sorbitol   L-Sorbose

[104] Zerban, Sattler. J. Am. Chem. Soc. 64, 1740 (1942).
[105] Butlerow. Ann. 120, 295 (1861).
[106] Ann. Rep. Chem. Soc. (London) 1925, 82.

L-Sorbose on reduction gives sorbitol and L-iditol. It gives the same osazone as L-gulose and L-idose. It is not fermented by yeast.

D-Sorbose is obtained by the action of dilute bases on D-galactose (see under tagatose). Its properties, except its dextrorotation, are like those of its L-enantiomers.

D-Sorbose has also been made from the pentaacetate of its *keto* form which was prepared from the tetraacetate of D-xylonic acid by successive treatment with PCl_5, diazomethane, and acetic acid.[107]

D-Tagatose, m. 124°, (configuration $+ - -$), is formed in small amounts in the action of dilute bases with D-galactose. This is a most complex process involving also D-talose, D-sorbose and *galtose* (possibly a 3-ketohexose).

CHO	CHOH	CHO	CH₂OH
HCOH	COH	HOCH	CHOH
HOCH	HOCH	HOCH	CO
HOCH ⇌	HOCH ⇌	HOCH	HOCH
HCOH	HCOH	HCOH	HCOH
CH₂OH	CH₂OH	CH₂OH	CH₂OH
D-galactose	⇅	D-talose	galtose (?)

CH₂OH	CH₂OH	CH₂OH
CO	C—OH	CO
HOCH ⇌	HOC ⇌	HCOH
HOCH	HOCH	HOCH
HCOH	HCOH	HCOH
CH₂OH	CH₂OH	CH₂OH
D-tagatose		D-sorbose

D-Tagatose gives the same osazone as D-galactose and D-talose.

Higher monosaccharides have been made as high as decoses by the cyanohydrin synthesis usually starting with D-glucose or D-mannose.[108] These higher oses are usually not fermentable. An exception is manno-nonose. *Sedoheptose* is obtained from *Sedum spectabile*, osazone, m. 197°. A mannoketoheptose is obtained from the avocado pear.

Heptoses.[109]

[107] Wolfrom, Olin, Evans. *J. Am. Chem. Soc.* **66**, 204 (1944).
[108] *Ann. Rep. Chem. Soc.* (London) **1922**, 66.
[109] *ibid.* **1934**, 168.

1-Desoxy-*keto*-D-galaheptulose pentaacetate, $CH_3CO(CHOAc)_4CH_2OAc$, has been made from galactonic acid by successive treatment with diazomethane, HCl, NaI, and HI.[110]

Amino sugars and related polyoses are important in various animal products. Glucosamine (chitosamine) has an NH_2 group in place of the alpha hydroxyl of glucose. It exists in α and β forms.[111] Chitin, found in various marine animals, gives glucosamine on acid hydrolysis. A *thiosugar* has been found in the adenine nucleoside of yeast.[112]

B. Polysaccharides, Polysaccharoses, Polyoses, Holosides

These give only simple sugars on hydrolysis. They may be divided into substances which are sweet and readily soluble like ordinary sugar (sucrose), and those which are not, like the starches and celluloses.

Sugar-like polyoses, oligosaccharides. Just as methyl alcohol can unite with glucose to give methyl glucosides other hydroxy compounds can form similar unions with monoses. This includes the monoses themselves. If the second monose forms the glucoside of the first by means of its own *aldehydo* hydroxyl, the resulting biose has no active group and cannot reduce Fehling's solution. Such a case is that of sucrose. If, however, the glycoside is formed through any other OH of the second monose, the active group of the latter is left free and the resulting disaccharide is a reducing sugar like maltose. Such a sugar may also form an osazone and can give mutarotation.

Sucrose and turanose both give glucose and fructose on hydrolysis but turanose reduces Fehling's solution and acts with phenylhydrazine while sucrose does neither.

Disaccharides

The union of two aldoses involving the active groups may be represented as follows, S representing the rest of the sugar molecule:

$$\underset{O}{\overset{H}{S-C}} + (HO)_2\overset{H}{C}-S' \rightarrow S-\overset{H}{C}-O-\overset{H}{C}-S' + H_2O$$
$$\qquad\qquad\qquad\qquad\qquad OH \quad OH$$

Hydrolysis reverses the process to give one or two monosaccharides.

In the above reaction each sugar can act to give its *alpha* or *beta* derivative. Thus two different sugars can unite through their active groups to give four different disaccharides, namely $\alpha\alpha$, $\alpha\beta$, $\beta\alpha$, and $\beta\beta$. If only one sugar is involved, there are only three resultant sugars involving the active groups. These three possibilities are realized in trehalose, isotrehalose and neotrehalose

[110] Wolfrom, Brown, Evans. *J. Am. Chem. Soc.* 65, 1021 (1943).
[111] *Ann. Rep. Chem. Soc.* (London) 1922, 78.
[112] Levene, Sobotka. *J. Biol. Chem.* 65, 551 (1925).

each of which consists of two molecules of D-glucose united through both aldehyde groups. If only one aldehyde group is involved only the α- and β-forms of the glycoside linkage are possible. The free aldehyde group can, however, exist as alpha or beta form.

An interesting substance is the compound,

$$CH_2OHCHOH - O - CH_2OH.$$

m. 148°, isolated from cabbage leaves.[113] It is faintly sweet and is soluble in water. It is not hydrolyzed by acids, does not reduce Fehling's solution and does not form an osazone. These properties are due to the fact that aldehyde groups of the component parts, glycolic aldehyde and formaldehyde, are masked in the "glycoside" union.

Disaccharides, $C_{11}H_{20}O_{10}$. *Vicianose,* on hydrolysis, gives D-glucose and L-arabinose while *primaverose* gives D-glucose and D-xylose.

The disaccharide from hydrolysis of *rutin* is a *gluco-rhamnose* $C_{12}H_{22}O_{11}$.[114]

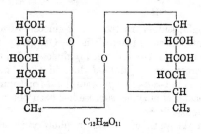

$C_{12}H_{22}O_{11}$

The natural and related members on hydrolysis by acid or by specific enzymes, give mainly glucose either alone or with other hexoses. *Maltose, gentiobiose, cellobiose* and the three *trehaloses* all give glucose alone. *Sucrose* and *turanose* give also D-fructose, while *lactose* and *melibiose* gvie also D-galactose; *apiobiose* gives also apiose. The artificial dissacharide 4-glucosido-mannose gives also mannose.[115]

Sucrose, cane sugar, saccharose, saccharobiose, "sugar," m. 160°, is obtained from the sugar cane and the sugar beet. $[\alpha]_D^{20} = + 66.5°$. It was first made available in granulated form by de Boré in 1794 in New Orleans. It is made in greater tonnage than any other pure organic compound.

Hydrolysis by acid or invertase gives D-glucose (dextrorotatory) and D-fructose (more strongly levorotatory). The rate of the hydrolysis by acid is about 1000 times that of maltose or lactose. The mixture is *invert sugar.* The inversion in the optical activity is used in the quantitative determination of sugar by polarimetry. Sucrose does not reduce Fehling's solution, does not react with phenylhydrazine and does not give mutarotation. Thus it does

[113] Buston, Schryver. *Biochem. J.* 17, 470 (1923).
[114] Zemplen, Gerecs. *Ber.* 68B, 1318 (1935).
[115] Isbell. *Bur. Standards J. Research* 7, 1115 (1931).

not have the active hemiacetal grouping, $-CH(OH)-O-$. Its high rate of hydrolysis is due to the $-O-C-O-C-O-$ grouping as contrasted with the $-O-C-O-C-C-$ linkage in reducing bioses. Vigorous oxidation converts sucrose mainly to oxalic and saccharic acids.

The structure of sucrose has been determined by methylation of the hydroxyl groups and hydrolysis to tetramethyl-D-glucose and 1,3,4,6-tetramethyl-D-fructose.[116] This is different from ordinary tetramethylfructose which is 1,3,4,5. Thus the fructose in sucrose has a furanose ring (gamma) instead of the ordinary pyranose ring. The structure of sucrose is

Sucrose is thus 2-glucosido-γ-fructose or 2-glucosido-fructofuranose. The glucosido group is introduced in place of the $2-OH$ group in gamma fructose. It can also be regarded as 1-(γ-fructosido)-glucose. Hudson gives the full designation of sucrose as 2-[alpha-d-glucosido(1,5)]-beta-d-fructose(2,5). Another possible name is 1-α-d-glucopyranosyl-β-d-fructofuranoside.

By means of a bacterial enzyme, sucrose has been obtained from D-glucose phosphate and D-fructose.[117]

Recent work on the path of carbon in photosynthesis using $C^{14}O_2$ has shown that:

(1) Green algae (*Chlorella*) are able to accumulate reducing power during illumination in the absence of carbon dioxide which can later be used for the dark reduction of carbon dioxide.[118]

(2) After very short photosynthetic experiments, 30 seconds to 90 seconds, the main portion of the newly reduced carbon dioxide is found in the phosphoglyceric acids, triose phosphates and the hexose phosphates. The latter are probably glucose-1-phosphate and fructose-6-phosphate.

(3) The first *free* carbohydrate which appears is sucrose.[119]

Sucrose forms several definite compounds with lime (calcium sucrates or, less correctly, saccharates). These are sometimes used for recovering an additional amount of sucrose from molasses. They can be decomposed by treatment with CO_2.

Triphenylchloromethane, trityl chloride, reacts with *primary* alcohol groups in sugars. Thus sucrose gives a tri-trityl ether.[120]

[116] Avery, Haworth, Hirst. *J. Chem. Soc.* **1927**, 2308.
[117] Hassid et al. *J. Am. Chem. Soc.* **68**, 1465 (1946).
[118] Calvin, Benson. *Science* **105**, 648 (1947).
[119] *ibid.* **109**, 140 (1949).
[120] *Ann. Rep. Chem. Soc.* (London) **1929**, 100.

Although the consumption of sucrose as a food has increased tremendously in the last century, the improvement in the culture of the sugar cane and the sugar beet and the extension of the former culture in tropical countries has given a potential production in excess of its food uses. Industrial uses are actively being sought (Sugar Research Fdn., N.Y.C.).

Turanose, m. 157°, obtainable by partial hydrolysis of melezitose, is 3-*alpha*-D-glucopyranosido-D-fructose.[121] It differs from sucrose in reducing Fehling's solution, and in undergoing mutarotation, indications of the free hemiacetal hydroxyl on the 2-C of the fructose moiety.

Maltose, malt sugar, $C_{12}H_{22}O_{11} \cdot H_2O$, is prepared from starch by diastase. Hydrolysis by acids or maltase gives only d-glucose. It reduces Fehling's solution and forms an osazone. Oxidation gives maltobionic acid which gives D-glucose and D-gluconic acid on hydrolysis. Thus maltose contains the active hemiacetal group $-CH(OH)-O-$. Since maltase, which splits maltose, is found to split only alpha glucosides and not the beta forms, maltose is concluded to be an alpha glucoside. Maltose solution shows an increasing dextro-rotation on standing. Since this is characteristic of the mutarotation of β-glucose, the free glucose lactol ring in maltose is probably of that type. Methylation of maltose followed by hydrolysis gives tetramethylglucose and 2,3,6-trimethylglucose. The latter indicates that the linkage with the second molecule of glucose is through the 4-position since the 5-position is part of the pyranose ring. On the basis of this experimental evidence, maltose is described as 4-α-glucosido-β-glucose and the corresponding formula is

The name indicates that an alpha glucose molecule has been attached through its active OH to the 4-carbon of a beta glucose molecule with elimination of water between the two OH groups. Another descriptive name for maltose is 4-D-glucopyranosyl-α-D-glucopyranoside.

As would be expected from its two primary alcohol groups, maltose forms a di-trityl ether with triphenylmethyl chloride.

Isomaltose is 6-α-D-glucopyranosyl-D-glucopyranose prepared from starch and by synthesis from glucose. It is characterized as a crystalline β-D-octa-acetate.

Cellobiose, cellose, is obtained as its octa-acetate by treatment of cellulose with acetic anhydride and sulfuric acid.[122] Its properties and reactions are

[121] Isbell. *J. Research Natl. Bur. Standards* **26**, 351 (1941).
[122] Skraup König. *Ber.* **34**, 115 (1927).

like those of maltose except that it is less strongly dextrorotatory and is split by emulsin which splits typical beta glucosides but not by maltase. Cellobiose is 4-β-glucosido-β-glucose. It has been synthesized.[123]

Lactose, milk sugar, is prepared from the whey of milk. Hydrolysis gives D-glucose and D-galactose. It contains a free active group as shown by oxidation by Fehling's solution, its action with phenylhydrazine and its mutarotation. Bromine water gives lactobionic acid which, on hydrolysis, becomes D-galactose and D-gluconic acid showing that the free active group is in the glucose part. Such an oxidation to bionic acids by bromine is best carried out in the presence of a benzoate which reacts with the HBr and prevents the breaking of the bionic linking.[124] Since methylation and hydrolysis give Me_4-galactose and 2,3,6-Me_3-glucose, lactose is 4-β-D-galactoside-D-glucose. It exists in α- and β-forms whose rotations fall and rise respectively until the equilibrium value of $+55°$ is reached.

Lactose, with dilute bases, gives a ketose, *lactulose*, which gives D-fructose and D-galactose on hydrolysis. It is 4-β-D-galactosido-α-D-fructose.[125]

Lactose octa-acetate, with $AlCl_3$, gives the chloroacetyl derivative of *neolactose*, a galactosido-D-altrose. This inversion of both the α- and β-carbons is very unusual.[127] This artificial biose differs from most of those occurring in nature by not containing glucose. Cellobiose and lactose have the same skeleton structure as maltose.[127]

Lactose has been synthesized starting with acetone-D-mannosan and acetobromo-D-galactose through epi-lactose.[128]

Lactose.[129]

Gentiobiose, obtained from gentianose, a triose, gives glucose on hydrolysis, is split by emulsin and gives tetramethylglucose and 2,3,4-trimethylglucose on methylation and hydrolysis. It is thus 6-β-glucosido-glucose. The enzymes are true catalysts in that they speed up the reverse process as well as the direct one. Thus emulsin converts D-glucose to gentiobiose. It reduces Fehling's solution and shows mutarotation.

Melibiose is obtained from raffinose. It is 6-α-galactosido-α-glucose and the corresponding -β-glucose. The equilibrium value $[\alpha]_D = + 143°$. The skeletons of gentiobiose[130] and melibiose are

[123] Hudson et al. *J. Am. Chem. Soc.* **64**, 1289 (1942).
[124] *Ann. Rep. Chem. Soc.* (London) 1929, 94.
[125] *ibid.* 1930, 109.
[126] Kunz, Hudson. *J. Am. Chem. Soc.* **48**, 1978 (1926); **48**, 2002 (1926).
[127] *Ann. Rep. Chem. Soc.* (London) 1927, 78.
[128] Haskins, Hann, Hudson. *J. Am. Chem. Soc.* **64**, 1490 (1942).
[129] Whittier. *Chem. Rev.* 1925, 126.
[130] *Ann. Rep. Chem. Soc.* (London) **1927**, 78.

Trehalose, isotrehalose and neotrehalose are glucosido-glucoses containing no free active group. They presumably represent the three possible forms: 1-α-glucosido-α-glucose, 1-α-glucosido-β-glucose and 1-β-glucosido-β-glucose.[131] Trehalose appears as a reserve carbohydrate in many fungi which are unable to form starch.

Rutinose is a biose, β-1-L-rhamnosido-6-D-glucose.[132]

TRISACCHARIDES, $C_{18}H_{32}O_{16}$

Raffinose, melitriose, the most important trisaccharide, is found in small amounts in nature. It is present in cottonseed meal to about 8% and in beet molasses. It is tasteless. Careful hydrolysis by acids or by raffinase gives D-fructose and melibiose while hydrolysis by emulsin gives sucrose and galactose. Thus the two enzymes attack different glucoside linkages. It does not reduce Fehling's solution. Thus there is no free active group. It is probably 1-(γ-fructosido)-6-galactosido-glucose.

Gentianose, with dilute acids or invertase, gives fructose and gentiobiose which can be further hydrolyzed to glucose. It contains no free active group. Thus it is probably 1-fructosido-6-glucosido-glucose.

Melezitose, melecitose, obtained from melezitose honey,[133] gives glucose and turanose on partial hydrolysis. Its resistence to Fehling's solution indicates the absence of hydroxyl on the 2-C. Thus it is probably a 2,3-diglucosidofructose.

Mannotriose, manninotriose, gives 2 molecules of D-galactose and 1 of D-glucose. It reduces Fehling's solution and may be 6-(6-galactosido-galactosido)-glucose,

$$HOCH_2CH(CHOH)_3CH{-}O{-}CH_2CH(CHOH)_3CH{-}O$$
Galactose units

$$-CH_2CH(CHOH)_3CHOH$$
Glucose unit

Stachyose, $C_{24}H_{42}O_{21}$, is a **tetrasaccharide.** It does not reduce Fehling's solution. Careful hydrolysis gives mannotriose and D-fructose. Thus it is possibly 1-fructosido-6-(6-galactosido-galactosido)-glucose.

Verbascose is a **pentasaccharide,** m. 253°, which gives one mol each of D-glucose and D-fructose and three mols of D-galactose on hydrolysis.[134]

Haworth, "Constitution of the Sugars," Arnold and Company, London, 1929.

[131] *ibid.* 1925, 88.
[132] Zemplén, Gerecs. *Ber.* **68B**, 1318 (1935).
[133] Hudson, Sherwood. *J. Am. Chem. Soc.* **42**, 116 (1920).
[134] Murakami. *C. A.* 3691 (1940).

Polysaccharides

These form important sources of reserve carbohydrates in animals and in plants (glycogen, inulin, starch) and material for the cell walls in plants (cellulose). Most of them are of very high molecular complexity and are insoluble. Under certain conditions, they give colloidal solutions. They undergo hydrolysis to sugars with varying degrees of ease. Polysaccharides related to nearly all the types of sugars are known although those derived from glucose are by far the most plentiful.

Starch. The starches obtained from different vegetable sources differ in physical properties but all give D-glucose on complete hydrolysis. Partial hydrolysis of starch gives simpler polysaccharides such as the dextrins. Many ensymes, such as diastase, hydrolyze starch to maltose. Such enzymes are widely distributed in plants and animals.

Starches are mixtures of amylose and amylopectin which are linear and branched (on C-6) polysaccharides respectively.[135]

Corn starch ordinarily contains about 22% amylose.

Amylose gives a characteristic blue color with free iodine.

Starch does not reduce Fehling's solution. With hypobromite it gives maltobionic acid which on hydrolysis gives D-glucose and D-gluconic acid.

Heating starch to about 250° gives *dextrins*, simpler polysaccharides. Acid treatment gives dextrins of different sorts. The products of increasingly vigorous treatment with acid give blue, red, or no color with iodine.

Methylation and hydrolysis of amylopectin (Haworth) give mainly 2,3,6-Me₃-glucose with about 5% of the tetramethyl compound. From this and other data, amylopectin is regarded as a polyglucosidic alpha combination joined by the 1- and 4-positions as in maltose with a branching α-1,6 at about every 24 monomer units.[136]

Amylose probably consists largely of long chains with the following type of maltose linkage.

Starch.[137]

[135] Schoch. Adv. Carb. Chem. 1, 247 (1945).

[136] Wolfrom. Chapter I. *in* Frear Agricultural Chemistry Principles Vol. I, D. Van Nostrand Co. Inc. N. Y. (1950).

[137] Kerr, R. W. (editor), "Chemistry and Industry of Starch." Rev. 2d ed. Academic Press, Inc. New York (1950).

Glycogen, liver starch, is the means used by the animal body to store glucose. It is more soluble in hot water than other polysaccharides. It does not reduce Fehling's solution. With iodine it gives a violet red color. Hydrolysis gives dextrins, maltose and finally D-glucose.

Inulin is a reserve polysaccharide in many plants. It does not reduce Fehling's solution. It is soluble in hot water and is levorotatory. Acids or inulase convert it entirely to D-fructose. Methylation and splitting gives 3,4,6-Me$_3$-γ-fructose. Thus the fructose in inulin is fructofuranose the same as in sucrose. Hydrolysis converts it to the more stable fructopyranose, ordinary D-fructose. The fructose units in inulin are held in glycosidic linkage probably through the 2- and 1-positions of successive units in the chain molecule. Inulin undergoes "depolymerization" to smaller combinations of fructose units with great ease.[140]

Treatment of inulin with nitric acid in chloroform gives a very stable difructose anhydride.[141] Its general skeleton is

A number of isomeric forms have been obtained which differ in part in the ring structures of the components.[142] The stability is due to the dioxan ring.

Levan, laevan, a levulosan like inulin, is produced by certain types of bacteria. It probably has the fructose units held by the 2- and 6-positions.[143]

Cellulose, $(C_6H_{10}O_5)_x$, is the important structural building material of the vegetable kingdom. The amount of it in existence at any one time corresponds to about half the amount of CO_2 in the atmosphere. It is constantly being built up from that source and being broken down to it again by various microorganisms.

Cotton is nearly pure cellulose. A very high grade of alpha cellulose is now obtained from wood by removing all other materials by physical and chemical treatment.

Cellulose is insoluble in most solvents but dissolves in ammoniacal copper solution (Schweitzer's reagent).

Hydrolysis gives D-glucose in almost quantitative yield. Acetolysis of cellulose gives about 40% yield of cellobiose octa-acetate. Cellotriose, m. 238°, cellotetraose, m. 251°, and cellohexaose, m. 266°, have been prepared.[144]

[140] Ann. Rep. Chem. Soc. (London) 1929, 101; 1932, 130.
[141] Irvine. Ann. Rep. Chem. Soc. (London) 1932, 130.
[142] Jackson et al. Bur. Standards J. Research 3, 27 (1929); 5, 1151 (1930); 5, 733 (1930); 6, 709 (1931), McDonald. Adv. Carb. Chem. 2, 265 (1946).
[143] Hibbert et al. Can. J. Research 4, 221 (1931).
[144] Willstätter. Ann. Rep. Chem. Soc. (London) 1931, 104.

Cellulose may be regarded as consisting of long chains of D-glucose molecules linked in glucosidic (beta) combination.[145] These chains orient themselves in the fibers of cellulose. It is experimentally known that there are three hydroxy groups for each glucose unit in cellulose. The nitrates, acetates and xanthogenates of these groups are of the utmost importance in the industries based on nitrocellulose (explosives, lacquers), and in the manufacture of artificial fibers (Rayon, acetate silk, etc.). The ethers of cellulose are also industrially important. The most important of these is *ethyl cellulose* (*Ethocel*). The commercial ethylated material has a low degree of substitution on the three hydroxyl groups of each glucose unit. *Methyl cellulose* is similar, *Methocel*. Cellulose reacts with ethylene oxide to give *Cellosize*. It also forms dibenzylcellulose. *Cellulose acetobutyrate* is more compatible with many resins than is the plain acetate.

Cellulose dinitrate is used with camphor and other plasticizers in *celluloid*. Cellulose diacetate is used in *Lumarith* and in *Tenite*. *Tenite II* contains cellulose having both acetate and propionate groups. Cellulose dipropionate is used in *Fortical*.

Treatment of cellulose with formaldehyde gives *methylene cellulose* which probably contains $-OCH_2O-$ linkages between adjacent C atoms in the pyranose rings and $-OCH_2OCH_2O-$ between adjacent pyranose units.

Cellulose derivatives are increasingly important in the plastics industry.

Methyl chloroacetate gives *carboxymethyl cellulose* CMC which contains $-OCH_2CO_2Me$ groupings. Produced in tonnage quantities it is used to increase the effectiveness of all types of detergents, in textile treatment, and as an additive for oil well drilling mud.[146]

Dimesylcellulose is obtained by means of $MeSO_2Cl$.[147]

The partial acetylation of wood with acetic anhydride and pyridine (vapor reaction) gives a wood resistant to insects and moisture.[148]

Oxycel, polyanhydroglucuronic acid, is an oxidized cellulose that is used to absorb blood in surgical practice since it is aborbed by tissues without damage.

Chemistry of Cellulose, Heuser, Wiley, N. Y. 1944.

Hemicelluloses are complex cell-wall polysaccharides which may be extracted from plant tissues by dilute alkali but not by water. Hot acid hydrolysis splits them into constituent sugar or sugar-acid units.[149]

The polysaccharides have received much attention in recent years and many of the difficult problems in the field including that of their molecular complexity are approaching solution.[150] The formulas suggested for starch

[145] *ibid.* 1924, 88; 1932, 126.
[146] *Chem. Inds.* 62, 733 (1948).
[147] Wolfrom, Sowden, Metcalf. *J. Amer. Chem. Soc.* 63, 1688 (1941).
[148] *Chem. Inds.* 61, 980 (1947).
[149] Yanovsky. *Ind. and Eng. Chem.* 31, 95 (1939).
[150] *Ann. Rep. Chem. Soc.* (London) 1932, 126.

and cellulose do not express the remarkable differences in their physical properties.[151]

Pentosans, arabans, etc. are related to the pentoses much as starch and cellulose are related to glucose. Xylan probably consists of combinations of D-xylose residues held in beta glucosidic linkage like the glucose in cellulose.[152]

Lichenin, moss starch, is found in lichens and seeds. It is a typical hemicellulose or reserve cellulose such as are stored in plants as ready sources of carbohydrate energy. They are readily hydrolyzed by acids and enzymes.[153, 154]

Pectins are condensation polymers of D-galacturonic acid in which the terminal carboxyl groups are esterified with methanol.[155]

Lignin. Notable progress has been made in determining the structure of lignins.[156] By treating vanillin acetate with $AlCl_3$ a product practically like spruce lignin was obtained. The vanillin acetate rearranges, loses water, and polymerizes to give poly-8-methoxydihydrobenzopyrone. The synthetic lignin contains fewer methoxy groups than the natural but $AlCl_3$ is known to remove these groups. In all other properties, such as solubility, appearance, and chemical behavior, the natural and synthetic lignin appear identical.

Lignin.[157, 158]   (Cf. vanillin, p. 683.)

XXIV. AMINO ACIDS

These substances are of the utmost importance since those in which the amino group is alpha to the carboxyl group are the building stones of the proteins of animals and plants. These have been obtained by hydrolysis of proteins and have been built up in the laboratory to form polypeptides which resemble some of the simpler protein compounds.[1]

The amino acids show the reactions of both their characteristic groups. They are thus amphoteric, forming, for example, either hydrochlorides or sodium salts. In the neutral condition they exist as ammonium salts formed between the amino and carboxyl groups, either of the same or different molecules. As ammonium salts, the amino acids are non-volatile, crystalline solids, very soluble in water but practically insoluble in organic liquids.

The place of amino acids in nutrition has been widely studied in recent years.[2] The following are necessary for maximum growth in rats: tryptophan, valine, lysine, phenylalanine, histidine, methionine, leucine, threonine,

[151] Staudinger, Eilers.   *Ber.* **69B**, 819 (1936).
[152] *Ann. Rep. Chem. Soc.* (London) 1929, 105.
[153] *ibid.* 1925, 100.
[154] Hawley, Norman.   *Ind. Eng. Chem.* **24**, 1190 (1932).
[155] Morell, Bauer, Link.   *J. Biol. Chem.* **105**, 1 (1934).
[156] Russel.   *Chem. Inds.* **61**, 612 (1947).
[157] Phillips.   *Chem. Revs.* **14**, 103 (1934).
[158] Hibbert.   *Ann. Rev. Biochem.* **11**, 183 (1942).
[1] Fischer.   *Ber.* **39**, 2893 (1906).
[2] Neuberger.   *Chem. and Ind.* 338 (1946).

isoleucine, and arginine. In addition to these, chicks require glycine and glutamic acid. All ten of these amino acids are now commercially available.

Amino acids form phosphotungstates which are useful in their isolation.

In utilizing the reactions of either of the groups in the amino acids the presence of the other must be remembered. Thus an attempt to make an acid chloride by means of any of the ordinary reagents causes trouble because of the active H atoms of the NH₂ group. If these are "protected" as by conversion to an acetyl derivative, $-NHCOMe$, the carboxyl can be changed to $-COCl$ satisfactorily.

Amino acids.[3, 4, 5]

Preparation of Alpha Amino Acids[6]

1. By hydrolysis of proteins either by heating with dilute acid or by proteolytic enzymes. Methods of separation for the thirty-one amino acids which have been isolated from various proteins are still inadequate.

An important means of separating amino acids from other materials and, to a limited extent, from each other, is supplied by suitably chosen ion-exchange resins.

2. By hydrolysis of alpha amino nitriles obtained by treating aldehydes with KCN and NH₄Cl.[7] This process is assumed to go through the cyanohydrins.

$$RCHO \rightarrow RCH(OH)CN \rightarrow RCH(NH_2)CN \rightarrow RCH(NH_2)CO_2H$$

The amino acids will be represented as in the "free" form although they exist as ammonium salts such as:

$$
\begin{array}{cc}
RCH-NH_3 & \quad NH_3\oplus \\
| \quad\quad | & or \quad RCH \overset{\nearrow}{\underset{\searrow}{}} \\
CO-O & \quad\quad CO_2\ominus
\end{array}
$$

3. By the usual methods for making amines from halogen compounds. The action of NH₃ with alpha halogen acids gives the desired product but also di- and even tri-substitution products of ammonia. On the other hand, the Gabriel phthalimide synthesis is successful with alpha bromo esters (see Method 6).

4. By the reduction of oximes and hydrazone derivatives of alpha keto acids.

5. From aldehydes by condensation with hippuric acid (benzoyl glycine) followed by reduction and hydrolysis

$$RCHO + PhCONHCH_2CO_2H \rightarrow RCH = C(NHCOPh)CO_2H \rightarrow$$
$$RCH_2CH(NHCOPh)CO_2H \rightarrow RCH_2CH(NH_2)CO_2H$$

The reduction step is easy because of the $\alpha\beta$-position of the double bond.

[3] Karrer. "Organic Chemistry," 1946. p. 272.
[4] Fieser, Fieser. "Organic Chemistry." Heath, 1944. p. 404.
[5] Gilman. "Organic Chemistry," 2nd Edition, 1943. p. 1079.
[6] Block. Chem. Revs. 38, 501 (1946).
[7] Strecker. Ann. 75, 27 (1850).

This is the azlactone synthesis.[8,9]

6. Treatment of an alkyl halide of Class 1 (p. 74) with sodium phthalimido-malonic ester, obtained from bromomalonic ester and potassium phthalimide, followed by hydrolysis and heating, has the net effect of adding the grouping $-CH(NH_2)CO_2H$ to the original organic residue

$$RX \rightarrow RC(CO_2Et)_2N(CO)_2C_6H_4 \rightarrow$$
$$C_6H_4(CO_2H)_2 + RC(NH_2)(CO_2H)_2 \rightarrow$$
$$CO_2 + RCH(NH_2)CO_2H.$$

The history of the discovery of amino acids including methionine has been given by Vickery.[10,11]

Reactions of Alpha Amino Acids

A. *Reactions due to the* $-NH_2$ *Group.*

1. Formation of hydrochlorides, $RCH(NH_3Cl)CO_2H$.

2. Alkylation to give $-NHR$ and $-NR_2$. This sometimes involves the formation of a betaine, or internal quaternary ammonium salt,

$$\oplus R_3NCH(R')CO_2\ominus.$$

This, on heating, gives an ester such as $R'CH(NR_2)CO_2R$. The ester alkyl can be removed by acid or alkaline hydrolysis but the $N-R$ linkage is not thus broken.

3. Acylation by acid halides or anhydrides to give $-NHCOMe$ (acetyl), $-NHCOPh$ (benzoyl), $-NHSO_2Tol$ (p-toluenesulfonyl, tosyl), $-NHSO_2CH_3$ (methylsulfonyl, mesyl).

4. With nitrous acid to give N_2 and hydroxy acids and related products. The measurement of the N_2 gas is a method for determining NH_2 groups.[12] Conversion of the NH_2 group to hydroxyl is also produced by various micro-organisms. This replacement of NH_2 by OH apparently does not produce Walden inversion.

Treatment of the *esters* with nitrous acid gives diazo esters, $RC(N_2)CO_2H$, which resemble diazomethane in reactions.

5. Hydriodic acid at 200° removes the amino group leaving unsubstituted acids.

6. Treatment with formaldehyde converts the $-NH_2$ groups to $-NHCH_2OH$ groups. This leaves the carboxyl group free so that it can be titrated with indicators. This is the basis of the "formol titration" of amino acids.

7. They give characteristic crystalline compounds with Reinecke's solution, $H[Cr(NH_3)_2(SCN)_4]$. Thus lysine Reineckate has the formula

$$C_6H_{14}O_2N_2 \cdot 2 \; HR \cdot 4 \; H_2O, \text{ m. } 235° \text{ dec.}$$

[8] "Org. Reactions," III.
[9] Fieser, Fieser. "Organic Chemistry." Heath, 1944. p. 417.
[10] Vickery, Schmidt. *Chem. Revs.* 9, 169 (1931).
[11] Gortner. "Outlines of Biochemistry." Wiley, 1929.
[12] Van Slyke. *J. Biol. Chem.* 12, 275 (1912).

B. Reactions due to the $-CO_2H$ *Group.*

1. Formation of metal salts, $RCH(NH_2)CO_2Na$, etc.

2. Formation of esters, $RCH(NH_2)CO_2R'$, from the acid, alcohol and HCl. This gives the ester hydrochloride which, with mild alkaline treatment, gives the free ester.   These compounds are volatile and have been used in separating amino acids (E. Fischer).

3. Formation of acyl halides, etc.   For this reaction the amino group must be acetylated or benzoylated before treatment with $SOCl_2$, PCl_5, etc., to give $RCH(NHCOMe)COCl$, etc.   After the acid chloride has been used for the desired purpose, the protecting group can be removed by vigorous treatment with HCl to give the hydrochloride of the free NH_2 compound.   Since the chief use of acyl halides of amino acids is in making peptides this process is of limited value because the hydrolysis of the protecting group also breaks the peptide linkage.   Acetylglycylglycine has the formula

$$CH_3CONHCH_2CONHCH_2CO_2H.$$

Thus the protecting group is held by a "peptide linkage" also.   This difficulty has been overcome by Bergmann by using a protecting group which can be removed without hydrolysis (p. 513).

4. Heating with barium hydroxide solution removes CO_2 leaving primary amines.   Similar processes occur in putrefaction.   Thus putrescine,

$$NH_2(CH_2)_4NH_2,$$

is formed from ornithine.

5. Treatment with acetic anhydride and pyridine removes CO_2 and forms an α-acetamidoketone.[13]

$$RCH(NH_2)CO_2H + (MeCO)_2O \rightarrow CO_2 + RCH(NHCOMe)COMe$$

6. Reduction of esters of amino acids with Na and absolute alcohol gives homologs of ethanolamine, $RCH(NH_2)CH_2OH$.

C. Reactions involving both $-NH_2$ *and* $-CO_2H$.

1. Formation of inner salts or dipolar ions (Zwitter ions) as already outlined.

2. Dehydration to give diketopiperazines.

This type of grouping is also formed in the partial hydrolysis of proteins.

3. Amino acids dissolve CuO on heating and give deep blue salts which show the properties of cuprammonium complex salts.   They give no precipitate

[13] *Ann. Rep. Chem. Soc.* (London) 1928, 91.

with bases. They may be formulated as

$$
\begin{array}{c}
\text{RCH—NH}_2 \qquad\quad \text{O——CO} \\
\big| \qquad\qquad\qquad\qquad \big| \\
\qquad\qquad \text{Cu} \\
\big| \qquad\qquad\qquad\qquad \big| \\
\text{CO——O} \qquad\quad \text{NH}_2\text{—CHR}
\end{array}
$$

Dilute $FeCl_3$ solution gives a red color with many amino acids. This is undoubtedly due to a complex formation involving both groups.

4. Various biological processes remove both groups with the formation of primary alcohols.

$$RCH(NH_2)CO_2H + H_2O \rightarrow CO_2 + NH_3 + RCH_2OH.$$

This is the source of the alcohols of fusel oil.

5. Treatment with potassium isocyanate and HCl gives *hydantoins*.

$$
RCH(NH_2)CO_2H \rightarrow RCH(NHCONH_2)CO_2H \rightarrow
\begin{array}{c}
\text{RCH—NH} \\
\big| \qquad\quad \diagdown \\
\qquad\qquad \text{CO} \\
\big| \qquad\quad \diagup \\
\text{CO—NH}
\end{array}
$$

NH_4SCN in acetic anhydride gives *thiohydantoins*.

6. Carbon dioxide in the presence of bases adds to the NH_2 group giving carbamino acids which form stable soluble salts with alkaline earth metals. Thus solutions of the amino acids in barium hydroxide give no precipitate with CO_2 but form soluble salts of the type

$$
\begin{array}{c}
\text{RCH—NH——CO} \\
\big| \qquad\qquad\qquad \big| \\
\text{CO—O—Ba—O}
\end{array}
$$

STEREOCHEMISTRY OF THE AMINO ACIDS

The alpha amino acids, except glycine, contain at least one asymmetric carbon atom. The natural amino acids with the exception of glycine are obtained in optically active forms. With the exception of some D-forms occurring in antibiotics, such as penicillin and gramicidin, all the natural amino acids belong to the L-series as can be shown by their configurational relation to the L-malic acid. The optical activity of the amino acids depends on the specificity of the enzymes which tear down and build up proteins. If they can combine with one configuration they cannot combine with its mirror image. They literally have to be "hand in glove" with their substrates and a right hand cannot fit a left glove.

Introduction and replacement of the NH_2 group in the optically active amino acids give many examples of the Walden inversion. The following changes are illustrative.

NOBr
1. Dextrorotatory (L-) alanine ――――→ L-Br-propionic acid.
NOBr
2. Levorotatory (D-) alanine ――――→ D-Br-propionic acid.

3. Treatment of the optically active bromopropionic acids with NH_3 gives the alanines of the *same* sign of rotation as the starting material. Thus L-alanine can be converted to D-alanine and vice versa by apparently simple metathetical processes. In spite of an enormous amount of work and much speculation, these changes are still not understood.

The racemic synthetic amino acids are readily resolved by means of active bases such as brucine after protecting the NH_2 by the formyl group.[14]

CLASSIFICATION OF AMINO ACIDS

The individual amino acids related to proteins may be classified in various ways. Some have the same number of acid and basic groups while others have an excess of one or the other. While most contain the basic and acid groups in aliphatic chains, with or without aromatic substituents, proline and hydroxy proline have these groups in heterocyclic combination. As will be seen these are closely related to the aliphatic diamines just as the parent pyrrolidine is related to 1,4-diaminobutane.

A. Purely Aliphatic Amino Acids

1. *Simple amino derivatives of the acetic acid series.*

Glycine, glycocoll, aminoacetic acid, $H_2NCH_2CO_2H$, m. 232°, is obtained by hydrolysis of the proteins of glue, gelatin and silk fibroin. Treatment of chloroacetic acid with ammonia gives glycine together with iminodiacetic acid, $NH(CH_2CO_2H)_2$, and nitrilotriacetic acid, $N(CH_2CO_2H)_3$. By modifications of the process, the yield can be raised to 50%.[15] Glycine is obtained by treating Cu, Co, Ni or Zn salts of chloroacetic acid with NH_3. The action of NaCN, NH_4Cl and formaldehyde give $C_9H_{12}N_6$[16] a trimer of methyleneamino-acetonitrile, $CH_2=N-CH_2CN$, which can be converted to glycine by hydrolysis.

Four types of complex copper salts of glycine are known.[17]

A solution of glycine in barium hydroxide gives no precipitate with CO_2 but forms a soluble salt of a carbamino acid.

Nullapon A and *Nullapon B,* used to prevent the precipitation of calcium soaps by hard water, are respectively $N(CH_2CO_2Na)_3$ and

$$[-CH_2N(CH_2CO_2Na)_2]_2.$$

[14] Fischer, Warburg. *Ber.* **38,** 3997 (1905).
[15] Robertson. *J. Am. Chem. Soc.* **49,** 2889 (1927).
[16] Johnson, Rinehart. *J. Am. Chem. Soc.* **46,** 768 (1924).
[17] Borsook, Thimann. *J. Biol. Chem.* **98,** 671 (1932).

Acetylglycine, $CH_3CONHCH_2CO_2H$, is called *aceturic acid*. *Hippuric acid*, benzoylglycine, $C_6H_5CONHCH_2CO_2H$, m. 187°, occurs in urine of the herbivora. It can be made from glycine and benzoic acid or benzoyl chloride. It condenses with aldehydes to give α-benzamido-$\alpha\beta$-unsaturated acids which can be reduced and hydrolyzed to give higher α-amino acids. **Sarcosine,** N-methylglycine, $CH_3NHCH_2CO_2H$, is formed by the decomposition of creatine and caffeine. It occurs in muscle and is found in meat extract. **Betaine,** completely methylated glycine, gives its name to similar internal quaternary ammonium salts of amino acids

$$CH_2\!\!-\!\!CO$$
$$Me_3N\!\!-\!\!\!-\!\!O \quad \text{or} \quad Me_3\overset{\oplus}{N}CH_2\overset{\ominus}{CO_2}$$

Betaines are found in plants probably as end-products of nitrogen metabolism. Betaine itself occurs in sugar beets and can be obtained from beet molasses. It can be made from Me_3N and chloroacetic acid. When heated to 293° it becomes methyl-N-dimethylaminoacetate, $Me_2NCH_2CO_2Me$. At higher temperatures it gives Me_3N.

The physiological activity of *-onium compounds* of the betaine type has been extensively studied by Renshaw and his students.

Alanine, α-aminopropionic acid, $CH_3CH(NH_2)CO_2H$, m. 298°, exists in practically all proteins. It constitutes over 20% of the fibroin of silk. It was first synthesized by Strecker from acetaldehyde. It can be prepared from pyruvic acid, $MeCOCO_2H$, with H_2 and NH_3 in the presence of palladium catalyst. Similar results are obtained when cysteine is used as the catalyst. It is possible that the formation of amino acids in organisms may go through the keto acids.

α-Aminobutyric acid, $C_2H_5CH(NH_2)CO_2H$, is found in very small amounts in a few proteins.

Valine, α-aminoisovaleric acid, $(CH_3)_2CHCH(NH_2)CO_2H$, m. 315°, is found in small amounts in many proteins and to about 5% in edestin and casein. Yeast converts it to isobutyl alcohol. **Norvaline** is α-amino-n-valeric acid, $CH_3CH_2CH_2CH(NH_2)CO_2H$.

Leucine, α-aminoisocaproic acid, $(CH_3)_2CHCH_2CH(NH_2)CO_2H$, m. 295°, is widely distributed among proteins and makes up a large part of such important ones as the globin of hemoglobin, serum albumen, serum globulin, blood fibrin, keratin of horn, hair, etc. and vegetable proteins such as edestin and zein. It is fermented by yeast to isoamyl alcohol. Leucine comprises about 30% of the amino acids in insulin.[18] Isoleucine, α-amino-β-methylvaleric acid, $C_2H_5(CH_3)CHCH(NH_2)CO_2H$, m. 280°, makes up 1–2% of the amino acids of casein. Its molecule contains two asymmetric carbon atoms. The natural material is bitter and is fermented by yeast to optically active amyl alcohol. Mild treatment with barium hydroxide solution racemizes the

[18] Fieser, Fieser. "Organic Chemistry." Heath, 1944. p. 437.

alpha carbon giving *allo-isoleucine* which is sweet and is not attacked by yeast. Allo-isoleucine can also be made from *d*-valeraldehyde by the Strecker synthesis, the new asymmetric carbon being obtained in both stereoisomeric forms. **Norleucine,** caprine, α-amino-*n*-caproic acid, $CH_3(CH_2)_3CH(NH_2)CO_2H$, m. 285°, is found in traces in certain proteins. It is found along with leucine in the proteins from the spinal cords of cattle.

Beta-amino acids can be made from β-halogenated acids and by the addition of ammonia to $\alpha\beta$-unsaturated acids. Heat reverses this process, converting the β-NH$_2$ to $\alpha\beta$-unsaturated acids. *Betaines* of beta amino acids on heating give the R$_3$N salts of $\alpha\beta$-unsaturated acids. Betaines of alpha amino acids decompose on heating to give R$_3$N and $\alpha\beta$-unsaturated acid, thus giving a simple way of removing the amino group by exhaustive methylation.

Gamma- and delta-amino acids can be made from the corresponding halogen acids. On heating they lose water and form gamma lactams analogous to the lactones. The betaines of gamma amino acids, on heating, give R$_3$N and gamma lactones.[19]

Amino acids with more carbons between the active groups are known. These show reactions of both groups. A method of producing the ω-amino caproic and heptylic acids involves the following changes.[20]

$$\text{Benzoylpiperidine} \xrightarrow{PCl_5} C_6H_5CONH(CH_2)_5Cl \xrightarrow[\text{Na malonic ester}]{\text{KCN etc.}} \text{etc.}$$

Reduction of the phenylhydrazones of keto acids gives the corresponding amino acids.[21]

ϵ-Aminocaproic acid when heated gives 20–30% of the 7-membered lactam and 80–70% of a polymer probably containing at least ten molecules of the acid. On the other hand γ-aminobutyric acid and δ-aminovaleric acid, on heating, give the 5- and 6-membered lactams without the formation of any polymers. ζ-Aminoheptoic acid gives only polymeric products and none of the 8-membered lactam.[22] Thus if the distance between the NH$_2$ and CO$_2$H groups is favorable for ring formation, the action of heat is intramolecular. Otherwise, there is an increasing tendency to have the elimination of water take place between different molecules.

2. *Amino derivatives of aliphatic acids containing an additional basic group.*

Ornithine, $\alpha\delta$-diaminovaleric acid, $H_2N(CH_2)_3CH(NH_2)CO_2H$, is obtained as a syrup by the hydrolysis of arginine

$$-CH_2NHC(NH)NH_2 + 2\ H_2O \rightarrow -CH_2NH_2 + CO_2 + 2\ NH_3.$$

[19] *Ann. Rep. Chem. Soc.* (London) 1904, 101.
[20] v. Braun. *Ber.* **40,** 1834 (1907).
[21] Fischer, Groh, Reinhart. *Ann.* **383,** 363 (1911).
[22] Carothers, Berchet. *J. Am. Chem. Soc.* **52,** 5289 (1930).

The extra NH_2 group makes this substance an amine as well as an internal ammonium salt. Putrefaction decarboxylates it to putrescine,

$$H_2N(CH_2)_4NH_2.$$

Proline may be regarded as the inner cyclic imine of ornithine.

The metabolism of ornithine presents interesting chemistry. Birds convert it to *ornithuric acid*, $PhCONH(CH_2)_3CH(NHCOPh)CO_2H$. Mammals produce more complicated changes, giving in succession:

$$citrulline, \ H_2NCONH(CH_2)_3CH(NH_2)CO_2H;$$

$$arginine, \ HN=C(NH_2)NH(CH_2)_3CH(NH_2)CO_2H;$$

$$glycocyamine, \ HN=C(NH_2)NHCH_2CO_2H;$$

and

$$creatine, \ HN=C(NH_2)N(CH_3)CH_2CO_2H.$$

The next to the last step involves reaction with glycine with the regeneration of ornithine. The last step is a case of *transmethylation* by methionine or choline.

Lysine, $\alpha\epsilon$-diaminocaproic acid, $H_2N(CH_2)_4CH(NH_2)CO_2H$, m. 224°. The presence of the free amino group lowers the m.p. 60° as compared with norleucine. Lysine occurs in all the important classes of proteins except the prolamines. It occurs to about 10% in serum albumen and in fibrin from clotted blood. Decarboxylation by the bacteria of putrefaction gives cadaverine, $H_2N(CH_2)_5NH_2$. Lysine is an essential for nutrition.

Arginine, α-amino-δ-guanidylvaleric acid,

$$H_2NC(NH)NH(CH_2)_3CH(NH_2)CO_2H,$$

m. 238° dec., is best prepared from gelatine.[23] It is obtained as a by-product in the manufacture of pearl essence. It is a strong enough base to give a stable carbonate. Clupeine and salmine, proteins of the sperm of herring and salmon, contain arginine to the extent of two-thirds of their amino acid residues. In smaller amounts it occurs in nearly all proteins. It apparently contributes their basic properties. Partial hydrolysis converts arginine to citrulline, α-amino-δ-carbamidovaleric acid.[24] In the same way that a 1,4-diamine can lose NH_3 to form a cyclic imine, a 1,4-amino-guanidino compound such as arginine might lose a molecule of guanidine to give *proline*, pyrrolidine-α-carboxylic acid. The parallel occurrence of arginine and proline in most proteins may indicate such a relation.

Suberyl arginine is a constituent of the bufotoxins from toad venom.

Creatine, N-methylguanidylacetic acid, $H_2NC(NH)N(CH_3)CH_2CO_2H$, is present in muscle and, as creatine phosphoric acid, is intimately connected with

[23] Cox. *J. Biol. Chem.* **78**, 475 (1928).
[24] *ibid.* **123**, 3 (1938).

muscular activity. An inner anhydride, *creatinine*, is found in plants and in beef extract.

$$\begin{array}{c} \text{NH-CO} \\ \text{HN=C} \diagup \qquad | \\ \diagdown \\ \text{MeN-CH}_2 \end{array}$$

The hydrolysis of creatine gives sarcosine, $MeNHCH_2CO_2H$.

3. *Amino derivatives of dibasic acids.*

Aspartic acid, aminosuccinic acid, $HO_2CCH_2CH(NH_2)CO_2H$, m. 251°, an acid as well as an ammonium salt, is widely distributed among proteins. Its synthesis from malic acid through chlorosuccinic acid is easy. With optically active materials, these changes have given many important examples of the Walden inversion. **Asparagin,** $HO_2CCH(NH_2)CH_2CONH_2$, the half amide of aspartic acid, is found in asparagus.

Glutamic acid, α-aminoglutaric acid, $HO_2C(CH_2)_2CH(NH_2)CO_2H$, m. 206°, is found in most proteins. It gives acid properties because of its free carboxyl group. It is the most plentiful single component of the hydrolytic products of casein, amounting to 15–20%. It forms over 40% of the gliadin from wheat gluten. Sodium glutamate (Ajinomoto, Aji, Chuyu, Gluta, Ve-tze-sin) is used as a condiment in Japan and China to give a meatlike taste to vegetable foods.[25] It can be tasted in 1:3000 dilution whereas the limits for sucrose and NaCl are 1:200 and 1:400. It is prepared commercially from the Steffens waste of beet sugar manufacture.

Glutamic acid constitutes about 20% of the amino acids of insulin. Yeast converts glutamic acid to succinic acid, a product found in small amounts in fermentation residues. **Glutamine,** $HO_2CCH(NH_2)CH_2CH_2CONH_2$, is the half amide of glutamic acid. Insulin is believed to contain glutamine combined with cystine.[26]

4. *Hydroxy aliphatic amino acids.*

Serine, α-amino-β-hydroxypropionic acid, $HOCH_2CH(NH_2)CO_2H$, m. 228° dec., is obtained by acid hydrolysis of silk gum, which consists mainly of the protein *sericin*.[27] It occurs in small amounts in other proteins. With nitrous acid it gives glyceric acid. It can be synthesized from hippuric ester and formic ester, the latter reacting as an aldehyde with the alpha H in the former.

[25] Han. *Ind. Eng. Chem.* 21, 984 (1929).
[26] Jensen, Wintersteiner. *J. Biol. Chem.* 97, 93 (1932).
[27] Shelton, Johnson. *J. Am. Chem. Soc.* 47, 412 (1925).

A better synthesis starts with alpha, beta-dibromopropionic ester.[28]

$$\text{BrCH}_2\text{CHBrCO}_2\text{R} \xrightarrow{\text{NaOMe}} \text{CH}_2{=}\text{CBrCO}_2\text{R} \xrightarrow{\text{NaOMe}}$$

$$\text{MeOCH}_2\text{CHBrCO}_2\text{R} \xrightarrow{\text{NH}_3} \text{MeOCH}_2\text{CHNH}_2\text{CO}_2\text{R} \xrightarrow{\text{HBr etc.}}$$

The expected preparation from beta-OH-alpha-Br-propionic acid with ammonia under pressure, gives α-OH-β-NH$_2$-propionic acid (*isoserine*).[30] The changes involved are probably as follows:

$$\begin{array}{cccc} \text{CH}_2 & \text{CH}_2\text{OH} & \text{CH}_2 & \text{CH}_2\text{NH}_2 \\ \| & | & | \diagdown & | \\ & & & \text{O} \\ \text{CH} & \text{CHBr} & \text{CH} \diagup & \text{CHOH} \\ | & | & | & | \\ \text{CO}_2\text{H} & \text{CO}_2\text{H} & \text{CO}_2\text{H} & \text{CO}_2\text{H} \end{array}$$

Threonine, α-amino-β-hydroxybutyric acid.[30] Its synthesis utilizes the addition of a mercuric salt to an olefinic linkage.[31] Crotonic ester is treated with mercuric acetate in methanol to form *alpha*-acetoxymercuri-*beta*-methoxybutyric ester. Treatment with KBr and Br$_2$ replaces the mercury by Br. Hydrolysis, treatment with NH$_3$ and then with fuming HBr to convert the methoxyl to hydroxyl gives the hydrobromide of threonine.

Natural threonine has been converted to dextrorotatory L-*alpha*-aminobutyric acid and to levorotatory D-lactic acid, thus proving its configuration.

$$\begin{array}{c} \text{CO}_2\text{H} \\ | \\ \text{H}_2\text{N}{-}\text{C}{-}\text{H} \\ | \\ \text{H}{-}\text{C}{-}\text{OH} \\ | \\ \text{CH}_3 \end{array}$$

Hydroxyglutamic acid, hydroxylysine, and **hydroxyvaline** have been reported in small amounts from protein hydrolysis. **Diaminotrihydroxydodecanoic acid** and **diaminodihydroxysuberic acid** have been obtained from casein.

5. *Aliphatic amino acids containing sulfur.*

Cysteine, α-amino-β-mercaptopropionic acid, HSCH$_2$CH(NH$_2$)CO$_2$H, is readily oxidized to its disulfide *cystine*. The oxidation-reduction system $-$S$-$S$-$ and $-$SH represented by these two compounds is of the utmost

[28] "Org. Syntheses"; Wood, Du Vigneaud. *J. Biol. Chem.* **134**, 413 (1940).
[29] *Ann. Rep. Chem. Soc.* (London) **1907**, 104.
[30] Meyer, Rose. *J. Biol. Chem.* **115**, 721 (1936).
[31] Carter. *J. Biol. Chem.* **112**, 769 (1936).

importance in biological processes. The oxidation of cysteine is catalyzed by even minute amounts of metals like Fe, Cu, and Mn. The sulfhydryl group (mercapto group) in cysteine has important growth stimulating properties.[32] Other mercapto compounds such as thioglycolic acid, $HSCH_2CO_2H$, thioglucose, and thiocresols have similar effects when applied to seedlings and to wounds in animals. Cysteine hydrochloride solution buffered with sodium borate is recommended as an aid to the healing of wounds (Squibb).

While cysteine has not been found in proteins it is the active part of the tripeptide glutathione which is widely distributed in cells.

Cysteine cannot be considered separately from cystine. The two probably form a biological unit involved in all processes of oxidation and growth. Naturally the material from dead tissues is found mainly in the oxidized form, cystine.

Homocysteine is α-amino-γ-mercaptobutyric acid.[33]

Cystine, bis-[α-aminopropionic acid]-β-disulfide,

$$[-S-CH_2CH(NH_2)CO_2H]_2,$$

m. 260° dec., occurs plentifully in hair, wool and horn. This occurrence is probably related to the growth promoting effect of cysteine which on oxidation gives cystine, the ordinary change from a sulfhydryl compound to a disulfide. It is similarly found in hemoglobin and in insulin, the hormone of the pancreas. In both these proteins it is intimately related to oxidizing function. The activity of the sulfur in cystine and related proteins presents a complex problem. Thus while boiling with 0.1 N, sodium carbonate liberates 2.8% of the sulfur in cystine as H_2S, the same treatment of dialanylcystine and dialanylcystine anhydride give 18.6% and 91.8% respectively.[34]

Cystine with excess iodine solution gives *cysteic acid,*

$$HO_3SCH_2CH(NH_2)CO_2H.$$

Cystine is essential for nutrition.

Homocystine is the first oxidation product of homocysteine.

Djenkolic acid, SS'-methylene-bis-cysteine, SS'-methylene-bis-(α-amino-β-mercaptopropionic) acid,[35] $CH_2(SCH_2CHNH_2CO_2H)_2$.

Methionine, α-amino-γ-methylmercaptobutyric acid,

$$CH_3SCH_2CH_2CH(NH_2)CO_2H,$$

is found in casein in small amounts and in ovalbumin and edestin to the extent of 4.5 and 2%. It has been synthesized in several ways.[36] One preparation is by the phthalimidomalonic ester synthesis from 1-chloro-2-methylmer-

[32] Hammett, Hammett. *Protoplasma* 19, 161 (1933).
[33] Butz, Du Vigneaud. *J. Biol. Chem.* 99, 135 (1932).
[34] Sheppard, Hudson. *Ind. Eng. Chem., Anal. Chem.* 2, 73 (1930).
[35] van Veen, Hyman. *Rec. trav. chim.* 54, 493 (1933).
[36] *Ann. Rep. Chem. Soc.* (London) 1931, 234.

captoethane, $ClCH_2CH_2SMe$, obtainable from ethylene chloride and $NaSMe$. Synthetic methionine has been resolved by means of the brucine salts. The L-isomer is identical with the natural material. Methionine may act like cystine in metabolism. The body requires choline for the synthesis of methionine from homocysteine. This has been regarded as evidence that the organism is unable to synthesize methyl groups and that the synthesis takes place through a transmethylation process.[37]

Although *taurine*, aminoethyl sulfonic acid, $H_2NCH_2CH_2SO_3H$, does not belong in this series it may be mentioned here because of its occurrence in bile and its possible relation to methionine from which it might be obtained by oxidation and decarboxylation. It is readily synthesized by standard reactions.

Pantothenic acid, N-(3,3-dimethyl-2,4-dihydroxybutyryl)-β-aminopropionic acid, $HOCH_2CMe_2CHOHCO-NHCH_2CH_2CO_2H$, the important growth factor in yeast, has been synthesized.[38]

B. α-Aminopropionic Acid (Alanine) with Cyclic Substituents in the Beta Position

1. *With purely aromatic substituents.*

Phenylalanine, α-amino-β-phenylpropionic acid, $C_6H_5CH_2CH(NH_2)CO_2H$, m. 280° ±, is found in all classes of proteins except the protamines. It gives the xanthoproteic reaction with nitric acid.

Tyrosine, *p*-hydroxyphenylalanine, α-amino-β-*p*-hydroxyphenylpropionic acid, p-$HOC_6H_4CH_2CH(NH_2)CO_2H$, m. 318°, occurs in all types of proteins except the protamines. It gives the xanthoproteic test with HNO_3. With Millon's reagent (mercuric nitrate in nitric acid containing nitrous acid), it gives a red brown precipitate. With diazotized sulfanilic acid (benzene *p*-diazonium sulfonate) it gives a red color due to formation of an azo dye (Pauly reaction). Decarboxylation gives *tyramine*, β-*p*-hydroxyphenylethyl amine. 3-Amino-*tyrosine* is effective against staph. aureus. The ethyl ester of tyrosine is used as an anti-oxidant in edible fats. N,N-Dimethyltyramine is *hordenine*.

Surinamine, ratanhine, is N-methyltyrosine,

$$p\text{-}HOC_6H_4CH_2CH(NHCH_3)CO_2H.$$

3,4-Dihydroxyphenylalanine, $HO\underset{OH}{\underset{\diagup}{\bigcirc}}CH_2CH(NH_2)CO_2H$, is found in only a few proteins.

[37] du Vigneaud et al. *J. Biol. Chem.* **131**, 57 (1939).
[38] Fieser, Fieser. "Organic Chemistry." Heath, 1944. pp. 276, 1013.

Iodogorgoic acid, 3,5-diiodotyrosine,

$$HO\langle \rangle CH_2CH(NH_2)CO_2H,$$

has been obtained from the protein of the thyroid gland.[39]

Thyroxine, (NNR), 3,5-diiodo-4-(3,5-diiodo-4-hydroxyphenoxy)-phenyl-alanine, α-amino-β-4-(2,6,3′,5′-tetraiodo-4′-hydroxydiphenyloxidyl)-propionic acid,

$$HO\langle \rangle O\langle \rangle CH_2CH(NH_2)CO_2H,$$

was first isolated from the thyroid gland by Kendall.   Harington, in Barger's laboratory, proved its structure and synthesized it.[40]

2. With heterocyclic substituents.

Tryptophan, β-indolylalanine, α-amino-β-3-indolylpropionic acid, m. 290° ±.

It is one of the essential amino acids.   Zein (corn) and gliadin (wheat) contain no tryptophan and will not support life.   It can be synthesized from β-aldehydo-indole and hippuric acid.   It is more readily prepared by an interesting modification of the malonic ester synthesis.   Ethyl acetamino-malonate, solid powdered NaOH, and toluene are heated with *gramine,* 3-dimethylaminomethylindole, skatyldimethylamine.   The latter is readily prepared by the Mannich reaction of indole with formaldehyde and di-methylamine.

The betaine of tryptophan, *hypophorine,* is found in some seeds.

With concentrated nitric acid, tryptophan gives a yellow color which changes to orange with ammonia (xanthoproteic reaction).   With glyoxylic acid and conc. sulfuric acid, it gives a blue violet color.

Tryptophan gives rise to the growth-hormone, hetero-auxin, indole-3-acetic acid.[41]

Histidine, β-4-imidazolylalanine, α-amino-β-4-imidazolylpropionic acid, N ——— C—CH₂CH(NH₂)CO₂H, m. 235°, is obtained from many proteins.

[39] *Ann. Rep. Chem. Soc.* (London) 1931, 234.
[40] *ibid.* 1926, 234; 1928, 261; 1930, 274; 1931, 234.
[41] Kögl.   *Ann. Rep. Chem. Soc.* (London) **1935,** 426.

It has been synthesized by the azlactone method from 4-aldehydoglyoxaline and hippuric acid.[42]   It gives the Pauly reaction with diazotized sulfanilic acid. Putrefactive organisms decarboxylate it to *histamine*, β-4-imidazolylethylamine, a substance of high physiological activity.   It is related to asthma and similar pathological conditions.   Anti-histamine drugs are becoming increasingly important.   Histidine is essential for life.

The *betaine of histidine* is found in certain edible mushrooms.

Ergothioneine (thiozine), the betaine of 2-mercaptohistidine, is found in ergot and in blood.[43]

C. Pyrrolidinecarboxylic Acids

As has been indicated, these may be related genetically to arginine which has nitrogen substitutions in the 1,4-position in a saturated ring and is consequently closely related to these 1,4-cyclic imines.

Proline, pyrrolidine-α-carboxylic acid, CH_2—$CHCO_2H$, m. 220°, is unique

among the amino acids in being readily soluble in alcohol.   It is found in all classes of proteins.   Methylation gives the betaine, *stachydrine*.   **Hydroxyproline**, 3-hydroxypyrrolidine-2-carboxylic acid, m. 270° ±, is obtained from few proteins and in only minute amounts except in gelatin in which it is reported to the extent of 15%.   The position of the hydroxyl group is not definitely settled.   If hydroxyproline is related to a hypothetical hydroxyarginine, the 3-position seems more probable than the 4-position which is usually assigned to the hydroxyl.

Amino Acids Formed from Individual Proteins

The methods of separation and identification of amino acids are still so inadequate that the following figures are merely indications of the order of magnitude of the occurrence of various amino acids and of the fact that the compositions of different proteins vary over wide ranges.   In most cases the numbers given are averages of those reported by various authorities.

Some amino acids can be identified by their N-mesyl derivatives.   Thus the glycine compound $CH_3SO_2NHCH_2CO_2H$ melts at 174°.   The corresponding ethyl ester, m. 42.5°.   It can be vacuum distilled.   N-mesyl-DL-alanine, m. 80°.

POLYPEPTIDES

These contain two or more amino acids held by the amido linkage —NH—CO— or —N=C(OH)—.   They are obtained by cautious hydrolysis of proteins and by synthetic means.[44]

[42] Fieser, Fieser.   "Organic Chemistry."   Heath, 1944.   p. 418.
[43] Eagles.   *J. Am. Chem. Soc.* **50**, 1386 (1928).
[44] Gortner.   "Outlines of Biochemistry."   Wiley, 1929.   pp. 309–21.

APPROXIMATE PERCENTAGES OF THE COMMONER AMINO ACIDS IN REPRESENTATIVE PROTEINS

	Alanine	Arginine	Aspartic acid	Cystine	Glutamic acid	Glycine	Histidine	Hydroxyproline	Leucine	Lysine	Methionine	Phenylalanine	Proline	Serine	Tryptophan	Tyrosine	Valine	Ammonia
Protamines																		
Salmine	4	89											4	3			1.6	
Histones																		
Globin		5	4	0.3	1.7		11	1	29	4		5	3	0.6	3	2		
Albumins																		
Serum	4	5	4	7	8		4	1	30	11		4	4	0.6	1	5		
Egg	2	3	2	0.3	9		1		8	3		4	2		2	1	2	
Globulins																		
Animal																		
Serum	2	4	2.5	4	8	3	2		19	7		2.5	2.5		4	7	1	2
Fibrin	4	7	2	1	10	3	3		15	10		2.5	4		5	7	1	
Miosin	4	5	0.5		14	0.5	3		8	3		2.5	3			2	6	
Vegetable																		
Edestin	4	16	4	0.2	14	4	4	2	14	4	4	2.5	2	0.8	1	2	2	2
Prolamines																		
Gliadin	3.5	3	0.6	1.8	44		2		7		2	2	13			1	2	5
Zein	3.8	1.5	2		31		0.8		25			8	9			5		4
Scleroproteins																		
Keratin	1.5	3	1	8	7	2	0.5		10	0.5		1	3	1		3	2	
Fibroin (silk)	23	1.5				40	1		1.5	1		1	1	2		10	2	
Gelatin	5	8	2		4	20	0.5	15	5	4		1	7	0.4				0.4
Phosphoproteides																		
Casein	1.5	3	1	0.2	15		4		8	7	0.2	3	7	0.5		4	7	

The simplest dipeptide, glycylglycine, $H_2NCH_2CONHCH_2CO_2H$, can be made in a variety of ways which illustrate the methods of linking amino acids together:

1. By action of bases on the bimolecular anhydride, diketopiperazine.

2. By action of NH_3 on the product of the action of chloroacetyl chloride with the amino group of glycine. For the preparation of higher peptides, the chloroacetyl glycine can be converted to its acid chloride and treated with another amino acid, etc. The final step is the replacement of the chlorine in the chloroacetyl group by means of ammonia.

A natural tripeptide, *glutathione*, glutamylcysteinylglycine,[45]

$$HO_2CCH(NH_2)CH_2CH_2CONHCH(CH_2SH)CONHCH_2CO_2H$$

is a constituent of practically all cells. It is intimately involved in cell growth.

The polypeptides containing a large number of amino acid units have many properties resembling the simpler proteins.

E. Fischer and E. Abderhalden have built up polypeptides containing 18 and 19 amino acid residues respectively. Unfortunately these differ from natural proteins in many important ways besides being smaller molecules. They contain:

1. Only the simplest amino acids.

2. No free amino, carboxyl or guanidyl groups which characterize the important proteins. The methods used in their preparation had distinct limitations including the difficulty of removing the protecting group from the NH_2 without hydrolyzing peptide linkages and even racemizing the asymmetric alpha carbon. These difficulties have been overcome by Bergmann and Zervas who have achieved two advances which open up tremendous new possibilities in the study of polypeptides, proteins and the related enzymes.

1. A method of protecting the amino group and removing the protecting group without using any hydrolytic process, thus avoiding the breaking of the peptide linkages.

2. A method of degrading a polypeptide by removing one amino acid residue at a time and identifying it, instead of using the old method of vigorous hydrolysis which gave all the amino acid molecules with no clue as to the order in which they were linked.

A most inspiring aspect of these two spectacular advances in one of the most difficult fields of organic chemistry is that they are based on relatively simple and long known reactions.

Bergmann's method of protecting the amino group is by the carbobenzoxy group, introduced by $C_6H_5CH_2OCOCl$, which is readily formed from phosgene and benzyl alcohol. After the acid chloride or acid azide has been formed and has been combined with the desired amino acid the protecting group is removed

[45] *Ann. Rep. Chem. Soc.* (London) 1930, 254.

by catalytic *hydrogenation* with palladium black.

$$-NHCO_2CH_2C_6H_5 \rightarrow -NH_2 + CO_2 + C_6H_5CH_3$$

This involves no attack on any other linkage in the peptide. Optically active amino acids and peptides protected by the carbobenzoxy group are remarkably stable to racemizing influences. The catalytic reduction fails with cystine compounds. In the case of guanidine derivatives the strongly basic group is masked by nitration. The nitroguanidine compound then behaves like an ordinary amino acid. During the catalytic reduction the nitro group is removed.

$$O_2NNHC(NH)NH- \rightarrow NH_2C(NH)NH-$$

By the improved methods, dipeptides of the natural amino acids (levorotatory) have been made as follows: glutamyl- and lysyl-glutamic acids, lysyl-glycine, lysyl-histidine, lysyl- and tyrosyl-aspartic acids, glutamyl- and tyrosyl-tyrosine, and glycyl-arginine.

Bergmann's method of degrading the polypeptides is even more brilliant.[46] The free amino group is first blocked by treatment with phenyl isocyanate to give a phenylureide.

$$H_2N- \rightarrow C_6H_5NHCONH-$$

The free carboxyl group of the protected peptide is methylated by diazomethane. Treatment of the ester with hydrazine gives the hydrazide, which, with nitrous acid, gives the azide. Treatment of the azide with benzyl alcohol gives the benzyl-urethan by rearrangement. Having two $-NH-$ groups on the same carbon atom, this gives an aldehyde. Thus controlled hydrolysis and catalytic hydrogenation of the urethan give toluene and CO_2 (from the protecting group), ammonia and an aldehyde (from the terminal amino acid of the polypeptide) and the phenylureide of a polypeptide containing one less amino acid residue than the starting material. The essential steps follow:

$$Q-CONHCHRCO_2H \rightarrow -CO_2Me \rightarrow -CONHNH_2 \rightarrow -CON_3 \rightarrow$$
$$(Q-CONHCHR-NCO) \rightarrow Q-CONHCHRNHCO_2CH_2Ph \rightarrow$$
$$PhCH_3 + CO_2 + NH_3 + RCHO + Q-CONH_2$$

Identification of the aldehyde gives the structure of the terminal acid of the polypeptide as $RCHNH_2CO_2H$. Treatment of $QCONH_2$ with hydrazine and then with nitrous acid gives an azide which can be rearranged as above and hydrolyzed and hydrogenated to give an aldehyde $R'CHO$ corresponding to the second amino acid in the original polypeptide. This process can be continued until only a single amino acid residue remains as the amide of the phenylureide, $PhNHCONHCHR''CONH_2$. Vigorous hydrolysis gives $R''CHNH_2CO_2H$. By this method the isomeric dipeptides, alanylleucine and

[46] Bergmann. *Science* 79, 439 (1934).

leucylalanine could be distinguished. The first aldehyde obtained in the two cases would be respectively isovaleraldehyde and acetaldehyde. If the first aldehyde obtained in this polypeptide degradation proved to be the half aldehyde of succinic acid or of malonic acid, the end component would be indicated respectively as glutamic or aspartic acid.

Splitting of Peptides by Enzymes

While Emil Fischer used the enzymes as tools in studying the proteins, later workers, starting with R. Willstätter[47] and continuing with Waldschmidt-Leitz and his students and collaborators such as A. K. Balls,[48] have used the peptides as tools in studying enzymes. By this means it has been possible to diagnose the results of the fractionations of enzymes by adsorbents. Thus erepsin (from intestines) contains several enzymes, among others a dipeptidase which carries on the last step of the splitting of a protein and an amino-polypeptidase which attacks higher peptides by first attaching itself to the amino group. Trypsin (from pancreas) gives a carboxylpolypeptidase which is specific for peptides of the proper degree of acidity. Just as papainase can combine with isoelectric (neutral) protein molecules and split them there is a dipeptidase which can split neutral "dipeptides" like chloroacetyl-o-nitro-aniline (Balls). Apparently the enzyme, in this case, combines directly with the peptide linkage.

The use of some of the earlier polypeptides cast little light on the action of the enzymes because of their inertness to these enzymes. The work of Bergmann and Zervas in making available peptides more nearly resembling the natural proteins has opened new possibilities in this field (1933–6). By studying the action of dipeptidase on dipeptides of all significant types they find:

1. The nine dipeptides listed on p. 514 are split. These all contain levo-rotatory amino acids with the exception of the glycyl group in the last one which is, of course, symmetrical. They also contain what may be called a normal peptide linkage, $H_2N - CHR - CONH - CHR' - CO_2H$. This includes, between the free amino and carboxyl groups an alpha H, a $- CONH -$ grouping and another alpha H.

2. Dipeptidase fails to react with

a. β-L-Asparagyl-L-tyrosine which has an extra CH_2 between the free NH_2 and CO_2H groups.

b. Substances like glycylsarcosine and L-alanyl-L-proline which have no free imido H.

c. Substances like L-alanyl-α-aminoisobutyric acid and aminoisobutyryl-glycine in which one or other of the two alpha H atoms is lacking.

d. Dipeptides containing one or both amino acids of the dextrorotatory instead of the natural levorotatory form. These facts are summarized in

[47] Willstätter et al. *Z. physiol. Chem.* 142, 14 (1925).
[48] Balls, Köhler. *Ber.* 64B, 34 (1931).

diagrammatic illustration. The enzyme is pictured as fitting spatially the

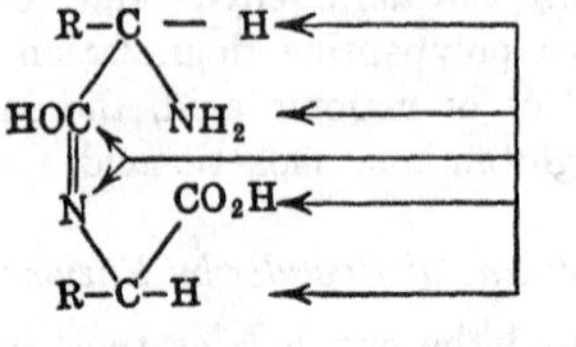

Dipeptide	Enzyme

peptide according to Emil Fischer's lock and key analogy. The hand and glove analogy is probably more accurate because a key has a *plane of symmetry* whereas an enzyme apparently has none.

The necessary contact in space between the enzyme and the peptide would fail:

1. If there were no imino H and the $-N=C(OH)-$ grouping were thus impossible. In that case the NH_2 and CO_2H groups could not approach each other closely enough to unite with the enzyme. The *trans* form of the above structure cannot be attacked for similar reasons. It can change to the *cis* form, however.

$$N=C(OH) \rightleftharpoons NH-CO \rightleftharpoons N=C(OH)$$

trans	*cis*

2. If another atom lay between the NH_2 and CO_2H and thus changed the size of the ring.

3. If one or other of the two alpha H atoms is missing or is interchanged in space with R or R'.

PROTEINS

These are found universally distributed in larger or smaller amounts in both animal and vegetable organisms. They are complex combinations of relatively simple amino acids. The form of combination of these units is only partly known. While they undoubtedly include polypeptide linkages there are also other more labile linkages.[49, 50]

The molecular weights of most proteins are 34,000 or higher (Svedberg).[51] They are colloidal in character. Consequently their solutions offer most complex problems.[52] Protein solutions are optically active (levorotatory).

The structure of proteins has been considered in detail in extended mathematical and X-ray studies.[53]

[49] Gortner. "Outlines of Biochemistry." Wiley, 1929. p. 318.
[50] Klarmann. *Chem. Rev.* 1927, 51.
[51] Svedberg, Nichols. *J. Am. Chem. Soc.* 48, 3081 (1926).
[52] Sörenson. *Compt. rend. Lab. Carlsberg* 12, 1 (1917).
[53] Wrinch. *Science* 88, 149 (1938).

Fibrous protein-like molecules with molecular weights above 1,000,000 have been synthesized from the anhydrides of N-carboxy-α-amino acids by a chain reaction, initiated by water and continuing through the alternate loss of CO_2 and rupture of the anhydride linkages.[54]

Plants are able to synthesize their proteins from inorganic material. Animals cannot do this. In the processes of metabolism, proteins are hydrolyzed by enzymes to give simpler products including amino acids. These are synthesized to give the specific proteins needed by the species involved. The animal body is unable to build up certain amino acids such as cystine, histidine, lysine, and tryptophan.

Combinations of proteins with non-protein materials form the *proteides* such as hemoglobin of blood, nucleoproteins of the cell nuclei and mucins of the mucous membranes. The proteides are also called *conjugated proteins*.

The proteins usually contain about the following percentages: carbon, 52; hydrogen, 7; oxygen, 22; nitrogen, 18; sulfur, 0.2. Many contain other elements such as phosphorus, iron and iodine. Some of the simplest contain no sulfur.

Acidic and basic properties of proteins. Isoelectric point.

The amino acids of which proteins are composed are of three types as to acid or basic properties.

1. Truly amphoteric substances having an equal number of $-NH_2$ and $-CO_2H$ groups as in glycine and cystine.

2. Substances with more basic tendencies because of an extra $-NH_2$ group as in arginine and lysine.

3. Those with acidic tendencies because of an extra carboxyl group as in glutamic and aspartic acids.

Depending on their amino acid content, proteins have varying acidic or basic properties. Thus zein forms salts only with bases and on hydrolysis gives no basic amino acids. The protamines, however, are strongly basic and yield an excess of basic amino acids on hydrolysis.

A thorough study of the addition of gaseous HCl to all types of organic nitrogen compounds[55] has led to the conclusion that a true peptide linkage, $-CONH-$, should add HCl stoichiometrically. It is found that 30% and 92% of the nitrogen in gliadin and edestin add HCl in this way while zein does not add HCl at all, thus indicating that it may not contain true peptide linkages.

When the hydrogen ion concentration (pH) of a protein solution is changed by adding HCl and NaOH, the following equilibria are involved:

$$\overset{\oplus}{H_3}N-Q-CO_2H \underset{OH\ominus}{\overset{H\oplus}{\rightleftharpoons}} \overset{\oplus}{H_3}N-Q-CO_2^{\ominus} \underset{H\oplus}{\overset{OH\ominus}{\rightleftharpoons}} H_2N-Q-CO_2^{\ominus}$$

[54] Woodward, Schramm. *J. Am. Chem. Soc.* 69, 1553 (1947).
[55] Bancroft, Barnett. *Proc. Natl. Acad. Sci.* 16, 118 (1930).

The pH at which the dipolar ions are at a maximum is the *isoelectric point* of the protein. At this point the solution shows minimum conductivity, osmotic pressure and viscosity. At this pH the protein shows the least swelling with water and does not undergo cataphoresis. That is, the colloidal particles move toward neither electrode. Proteins coagulate best and contain the least amount of inorganic matter at their isoelectric points.

The isoelectric points (in pH units, neutrality = 7, with acids below and bases above this value) are as follows: glutenin, 4.5; gelatin, 4.7; egg albumin, 4.8; serum albumin, 5.4; edestin, 5.7; oxyhemoglobin, 6.8; gliadin, 9.2; clupeine, 12.1.

Classification of the Proteins

1. **Protamines.** These are the simplest proteins and contain no sulfur. They are strongly basic, have low molecular weights (about 3000, cf. p. 520), are readily soluble in water and do not coagulate on heating. They occur in the sperm of fish: clupeine in herring, cypreine in carp, salmine in salmon and trout, and sturine in sturgeon. These apparently all contain diamino and monoamino acids in the ratio 2:1.[56] Thus the probable molecular proportions of amino acids in clupeine is arginine 10, proline 1, alanine 1, serine 2, valine 1, and in salmine it is arginine 14, proline 3, serine 3, valine 1.

2. **Histones** resemble the protamines but are more complicated and contain sulfur. They are obtained from hemoglobin, from the white blood corpuscles, from nucleoproteides and from the sperm of animals. They are coagulated by heat but the coagulum is soluble in dilute acid. Their solutions give precipitates with solutions of many other proteins.

3. **Albumins** are amphoteric with no excess of acid or basic properties. They are soluble in water and are coagulated by heat. The albumins of egg, serum and milk are characteristic. Leucosin is a vegetable albumin from wheat. They are difficultly precipitated by salt solutions. Hydrolysis gives no glycine. Myogen forms 80% of muscle protein.

4. **Globulins** are weakly acidic. They are insoluble in water and soluble in dilute salt solutions but are precipitated by higher salt concentrations. This definition is highly incomplete because different salts and the same salts at different concentrations have markedly different effects.[57] Heat coagulates the globulins.

Animal globulins are found in the body fluids while plant globulins occur as solids in seeds, etc. *Myosin* of muscle is a typical globulin. Its coagulation produces *rigor mortis*. Edestin is found in hemp seeds and ovoglobulin in egg yolk.

5. **Prolamines** (gliadins) are vegetable proteins soluble in 70–80% alcohol but insoluble in absolute alcohol and in water. Their hydrolysis gives much proline, but no glycine, lysine or tryptophan. Thus they cannot support life

[56] Waldschmidt, Leitz. *Monatsh.* 66, 357 (1935).
[57] Gortner. "Outlines of Biochemistry." Wiley, 1929. p. 365.

without other proteins. Examples are *gliadin* from wheat and other cereals, *hordein* from barley and *zein* from corn. Milk also gives an alcohol-soluble prolamine.

6. **Glutelins** are insoluble in neutral solvents but dissolve in very dilute acids and bases. Examples are glutenin from wheat and oryzein from rice.

7. **Scleroproteins,** (albuminoids) of skin and related structures, are insoluble in water and solvents in general. They contain much sulfur and resist hydrolysis especially by the ordinary proteolytic enzymes. Examples are collagen (glue and gelatine), keratin (horn, wool, etc.), fibroin (silk), and elastin (animal ligaments).

Classification of the Proteides (Conjugated Proteins)

The first three classes contain phosphoric acid in various combinations:

1. **Phosphoproteides** are weak polybasic acids. They are not coagulated by heat. On hydrolysis with HCl in the presence of pepsin they give proteoses and a complex phosphoric acid derivative, *paranucleic acid,* which is different from animal and plant nucleic acids in that it contains no xanthines, pyrimidines nor pentoses.

Examples are *vitelline* of egg yolk and *casein* of milk.

2. **Nucleoproteides** from cell nuclei contain proteins combined with nucleic acid. The latter is of two kinds typified by plant nucleic acid from yeast cells and animal nucleic acid from the pancreas. The nucleoproteides are soluble in bases and insoluble in acids.

3. **Lecithoproteides** are soluble in such solvents as ether. They include complex compounds of phosphoric acid (lecithins, phosphatides, phospholides) and occur in cytoplasm and cell membranes.

4. **Chromoproteides** are combinations such as hemoglobin and its oxidation product oxyhemoglobin. The action of glacial acetic acid and NaCl gives the hydrochloride of hematin, the red pigment combined with the globin. Other examples of chromoproteides are hemocyanin of the blood of marine animals, and *phycoerythrin* and *phycocyan* of seaweeds. Black hair, wool, etc., contain a protein combined with *melanin.*

5. **Glucoproteides,** glycoproteides or mucins, contain proteins combined with carbohydrate derivatives such as glucosamine and glucuronic acid. They include the proteins of the various mucous fluids of animals and probably part of egg white. The mucins are acid and insoluble in neutral solution but soluble in slightly basic solution.

6. **Lipoproteins** (Mathews) contain as the prosthetic (additional) group a higher fatty acid.

Classification of Protein Cleavage Products (Derived Proteins)

1. **Proteans.** Insoluble in water. First products of cleavage.

2. **Metaproteins,** products formed by further action of acids, or enzymes. Soluble in dilute acids and bases. Insoluble in neutral solvents.

3. **Proteoses,** albuminoses, soluble hydrolysis products of proteins. Not coagulated by heat. Precipitated by excess solid ammonium sulfate.

4. **Peptones,** hydrolysis products. Soluble, not coagulated by heat, not precipitated by ammonium sulfate.

5. **Peptides,** definite compounds consisting of two or more amino acids united through the $-NHCO-$ or $-N=C(OH)-$ linkage.

The proteolytic enzymes give different degrees of cleavage: pepsin gives proteoses and peptones; trypsin gives proteoses, peptones and amino acids; and erepsin changes peptones to amino acids and NH_3.

Ninhydrin (triketohydrindene) gives a blue color with amino acids, polypeptides, and peptones.

Molecular Weights of Proteins

From his studies of the rates of sedimentation in the ultracentrifuge, Svedberg concludes that the molecular weights of typical proteins are about as follows: egg albumin, 34,000; hemoglobin, 66,000; casein, 98,000; gelatin, 150,000; and the blood pigments of certain lower marine animals, as high as a million. Values even approaching five million have been indicated.[58] The order of magnitude of these values is confirmed by purely chemical means. Thus hemoglobin with its 0.4% of Fe would have a molecular weight of 14,000 if its molecule contained only *one atom* of iron, which is rather improbable. If there is *only one cystine* unit in hemoglobin the molecular weight is 16,000. The value of 75,000 to 100,000 for casein, obtained with the ultracentrifuge taken in conjunction with the percentages of sulfur, phosphorus, cystine, tryptophan, tyrosine, and histidine of 0.78, 0.86, 0.49, 1.24, 5.55 and 1.78 respectively, gives the value 98,000 cited above.[59]

Visual purple is a protein of about 250,000 molecular weight. It has about ten chromophores per molecule.[60] Thus there is one chromophore per Svedberg unit.

Protein Reactions

These are very complex but a few are of general importance.

1. Many proteins form soluble alkali salts. A typical example is casein, which has at least five potassium salts containing from one to sixteen potassium atoms per molecule of casein.

2. Proteins form salts with acids. Varying amounts of HCl will unite with proteins. This salt formation is not limited by the number of free $-NH_2$ groups.[61]

3. Free amino groups in protein liberate nitrogen when treated with nitrous acid.[62] The amount of NH_2 groups varies widely, the percentage of the

[58] *Ann. Rep. Chem. Soc.* (London) 1931, 231.
[59] Carpenter. *J. Am. Chem. Soc.* 53, 1812 (1931).
[60] Broda, Goodeve, Lythgoe. *J. Physiol.* 98, 397 (1940).
[61] Bancroft, Barnett. *Proc. Natl. Acad. Sci.* 16, 118 (1930).
[62] Van Slyke. *J. Biol. Chem.* 12, 275 (1912).

total nitrogen existing as NH_2 groups being: zein, 0; gliadin, 1.1; edestin, 1.8; gelatin, 3.1; casein, 5.5.

4. Heating with bases in the presence of a trace of copper salt gives a red violet color. This is called the *biuret reaction* because it was first noted with biuret, $NH_2CONHCONH_2$. It is also given by other nitrogen compounds but not by amino acids. Thus the completeness of protein hydrolysis can be detected by the failure of the biuret test.

5. Nitric acid gives a yellow color which turns orange with ammonia (xanthoproteic reaction). When a solution of a protein is underlayed with nitric acid, the protein is precipitated in a white ring at the junction of the two liquids (Heller's test for albumin in urine).

6. Millon's reaction. Mercury dissolved in twice its weight of conc. nitric acid and diluted with two volumes of water gives a red color when heated with a solid protein.

7. Potassium ferrocyanide and acetic acid give hydroferrocyanic acid which gives precipitates with many proteins.

8. Salts of heavy metals such as Cu, Hg and Pb give precipitates.

9. Reagents for alkaloids give precipitates with most proteins.

Proteins.[63]

NYLON

The nylon polymers from which plastics, filaments and fibers are made resemble the proteins in having very large molecules consisting of small units united by peptide linkages. They are polyamides of the general formula $(-NH-CO-Q-CO-NH-Q'-)_n$ in which Q and Q' may be practically any bivalent organic grouping. In ordinary Nylon, Q is a tetramethylene chain and Q' is a hexamethylene chain. Nylon grew out of the purely scientific interest of W. H. Carothers in the properties of high molecular weight materials formed from reactions of bifunctional compounds such as dibasic acids and glycols, ω-hydroxy acids, ω-amino acids and the like. Compounds were chosen with enough carbon atoms between the functional groups to prevent intramolecular ring closure. Adipic acid and hexamethylene diamine formed an ideal pair. It was early observed that if fibers drawn from a melt were pulled suddenly they had little tensile strength. If, however, they were cautiously drawn to three or four times their original length, the resulting filament had remarkable strength. By this process the random arrangement of the long molecules becomes essentially parallel. This original discovery has been adapted to the present manufacturing process.

Ordinary nylon is Polymer 66, so called because of the six C atoms in each of its intermediates. Adipic acid is converted to hexamethylene diamine and the latter is mixed with a molecular amount of adipic acid in water solution to give Nylon-salt. Evaporation and heating of this salt under rigidly con-

[63] Fieser, Fieser. "Organic Chemistry." Heath, 1944. pp. 426–47; 498–511.

trolled conditions gives the polymer from which nylon is made in its multitudinous forms.

The start of nylon manufacture in Japan may presage the end of the natural silk industry.   The thousands of acres devoted to mulberry trees may be freed for food crops in the same way that the enormous acreage devoted to the growing of madder and indigo in the 19th century was released for food production by the development of synthetic alizarin and indigo.   Just as there are many useful modifications of the natural dyes, there will be many kinds of nylon.

PART II

ALICYCLIC OR POLYMETHYLENE COMPOUNDS

I. GENERAL

A. OCCURRENCE

The alicyclic compounds, as the name indicates, are cyclic compounds having aliphatic properties. Many with 6- and 5-membered rings occur very widely in nature in various plant products and in crude petroleums. Most petroleums with the exception of Appalachian (Pennsylvania grade) and particularly Michigan crudes contain large amounts of naphthenes or complicated alicyclic compounds.

The following alicyclic hydrocarbons have been isolated from a midcontinent crude[1] (listed in order of increasing b.p.): cyclopentane, methylcyclopentane, cyclohexane, 1,1-Me_2-cyclopentane, t-1,3-Me_2-cyclopentane, t-1,2-Me_2-cyclopentane, methylcyclohexane, ethylcyclopentane, 1,1,3-Me_3-cyclopentane, c,t,c-1,2,3-Me_3-cyclopentane, 1,1,2-Me_3-cyclopentane, 1,1,-Me_2-cyclohexane, c-1,3-Me_2-cyclohexane, n-propyl-cyclopentane, ethylcyclohexane, 1,1,3-Me_3-cyclohexane, c,t,c-1,2,4-Me_3 cyclohexane. The terpenes and camphors of plants are mainly alicyclic in nature.

Many alicyclic compounds have been synthesized by applying the ordinary synthetic processes in such a way that they take place *intramolecularly* with the formation of rings instead of *intermolecularly* to form larger molecules. Up to 1926 rings of 3 to 9 carbons had been made.

B. BAYER STRAIN THEORY

Bayer in 1885 first noted that the angle subtended by the corners and center of a regular tetrahedron, namely about 109.5°, lies between the value for the angles in the regular pentagon and hexagon, namely 108° and 120°. On this was based the Bayer Strain Theory that 5- and 6-membered rings form most readily and are most stable. Qualitatively this conception has been most helpful. That other factors are involved is shown by the greater occurrence of 6-membered rings as compared with 5-membered rings although the latter have practically the "natural" angle for a tetrahedral carbon atom. Influences which tend to give the normal angle between the carbon valences give easier formation and greater stability. Thus the *spiro compound A* containing a 6- and a 3-ring with one carbon in common is more easily made and is more

[1] Rossini. *Chem. Eng. News* **25**, 230 (1947).

stable than the cyclopropane compound B.[2]

The increased angle in the 6-ring (120°) tends to force together the other valences of the spiro carbon and thus stabilize the 3-ring. Thus A is not opened by conc. HCl at 240° whereas B is opened by 5% acid at 200°. Further support to this reasoning is given by the fact that compound C has the same stability as B.[3] The 5-ring does not change the natural angle.

Large strainless rings[4] have been found in nature and have been synthesized.[5–7] These have been made with as many as 32 carbon atoms. In these compounds the atoms of the ring do not lie in one plane. Thus the natural angle of the carbon atoms is not strained to equal the angle of a large plane polygon.

The most common phenomenon involving the Bayer strain conception is the fact that suitably reactive groups in the 1,5 or 1,6 position tend to react *intramolecularly* with ring closure.

The presence of substituent groups greatly changes the ease of closing rings and the stability of the rings.[8]

C. Reactions for Forming Alicyclic Compounds

These are mainly familiar aliphatic reactions. Among these are the Wurtz reaction, acetoacetic ester and malonic ester syntheses with dihalides, and *intramolecular* examples of the aldol and acetoacetic ester condensations and pinacol reductions. Probably the oldest and most general reaction is the production of a cyclic ketone by heating the salt of a dibasic acid.

[2] *Ann. Rep. Chem. Soc.* (London) 1915, 109.

[3] *ibid.* 1921, 94.

[4] Sachse-Mohr. *Ann. Rep. Chem. Soc.* (London) 1924, 92.

[5] Ruzicka et al. *Helv. Chim. Acta* 9, 230, 249, 389, 499, 715, 1008 (1926); 10, 695 (1927); 11, 496, 670, 686, 1159, 1174 (1928); 13, 1152 (1930); 15, 1459 (1932); 16, 487, 493 (1933); 17, 78 (1934); *Bull. soc. chim. Belg.* 41, 565 (1932); *Chem. Weekblad* 25, 614 (1928); *C.A.* 20, 2333 (1926); 22, 2755 (1928); 23, 1648 (1929); 25, 3012 (1931).

[6] Hill, Carothers. *J. Am. Chem. Soc.* 55, 2023, 5031, 5039 (1933).

[7] Carothers, Hill. *J. Am. Chem. Soc.* 55, 5043 (1933).

[8] *Ann. Rep. Chem. Soc.* (London) 1922, 126; 1925, 129.

D. METHODS FOR OPENING ALICYCLIC RINGS TO GIVE ALIPHATIC COMPOUNDS

These also involve familiar reactions such as oxidation of olefinic or ketonic compounds. The 3- and 4-membered ring compounds are exceptional in that they can be converted to propane and butane derivatives by catalytic hydrogenation and to propylene and butylene derivatives by heating with catalysts like Al_2O_3.

E. CHANGE IN SIZE OF RINGS

Just as rearrangements in the aliphatic series involve the shift of a $C-C$ linkage, so in the alicyclic series such rearrangements result in the formation of large or smaller rings.

1. Treatment of aliphatic amines with nitrous acid produces rearranged products in practically all cases.

$$Me_3CCH_2NH_2 \rightarrow Me_2C(OH)CH_2Me$$

Similarly alicyclic amines give rearranged products with different rings as well as the normal products of replacement of NH_2 by OH.

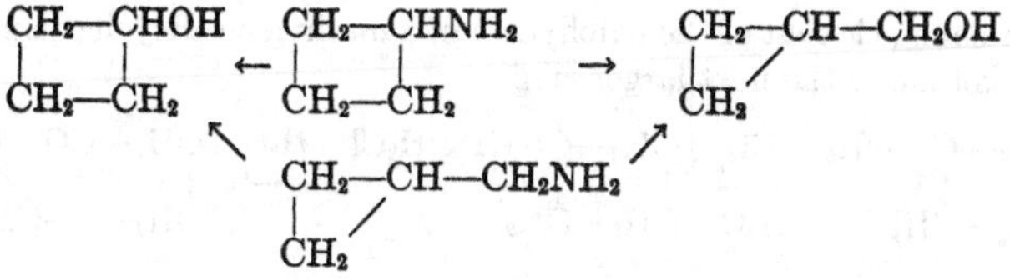

2. Dehydration of alcohols often gives rearrangements.

$$Me_3CCHOHMe \rightarrow Me_2C=CMe_2$$

Similarly

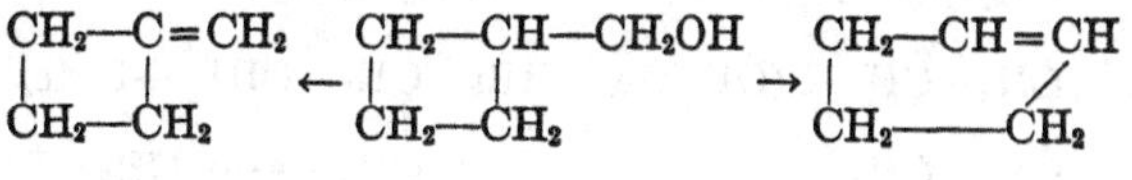

An analogous narrowing of the ring has not been produced with alicyclic alcohols. The dehydration of borneol to give camphene instead of bornylene involves a similar conversion of one ring into another.

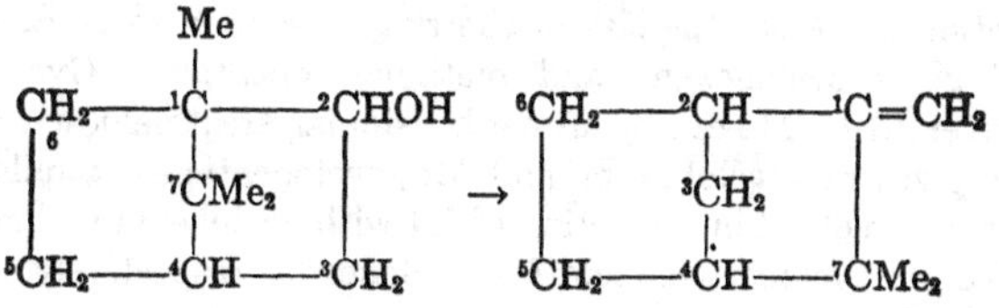

The breaking of the linkage 1:6 and the establishing of the linkage 2:6 converts the 6-ring into a 5-, and one of the 5-rings into a 6-.

To obtain a narrowing of the ring by dehydration, a grouping such as 2,2-Me$_2$-cycloheptanol would be necessary. This should give olefins with the skeleton of isopropylcyclohexane.

3. Probably the oldest authentic rearrangement is that of pinacol to pinacolone.

$$Me_2C(OH)C(OH)Me_2 \rightarrow Me_3CCOMe$$

In the alicyclic series the pinacolone rearrangement produces both narrowing and widening of the ring.

$$\begin{array}{cc} CH_2-CH_2-C(OH)Me & CH_2-CH_2-C(Me)COMe \\ | \qquad\qquad | & \rightarrow \quad | \qquad\qquad | \\ CH_2-CH_2-C(OH)Me & CH_2-\!\!-\!\!-\!\!-CH_2 \end{array}$$

$$\begin{array}{cc} CH_2-CH_2 & \\ | \qquad\qquad \searrow & CH_2-CH_2-CO \\ | \qquad\qquad\quad C(OH)-C(OH)Me_2 \rightarrow \quad | \qquad\qquad\quad | \\ | \qquad\qquad \nearrow & CH_2-CH_2-CMe_2 \\ CH_2-CH_2 & \end{array}$$

A similar change is that of the halohydrin of a methylene alicyclic compound to give the ketone of the next larger ring.

$$\begin{array}{ccccc} CH_2-C=CH_2 & Cl_2 & CH_2-C(OH)CH_2Cl & H_2O & CH_2-CO-CH_2 \\ | \qquad\quad | & \xrightarrow{\ \ } & | \qquad\qquad | & \xrightarrow{\ \ } & | \qquad\qquad \nearrow \\ CH_2-CH_2 & H_2O & CH_2-CH_2 & & CH_2-\!\!-\!\!-CH_2 \end{array}$$

4. Both narrowing and widening of the ring have been observed in addition to the normal reactions in supposedly metathetical processes.

$$\begin{array}{ccc} CH_2-CH_2-CHI & AgNO_2 & CH_2-C(NO_2)CH_3 \\ | \qquad\qquad \nearrow & \xrightarrow{\ \ } & | \qquad\qquad | \\ CH_2-\!\!-\!\!-CH_2 & & CH_2-CH_2 \end{array}$$

$$\begin{array}{ccc} CH_2-CH-C(OH)Me_2 & HBr & CH_2-CHBr-CMe_2 \\ | \qquad\quad | & \xrightarrow{\ \ } & | \qquad\qquad \nearrow \\ CH_2-CH_2 & & CH_2-\!\!-\!\!-CH_2 \end{array}$$

Cyclopropylcarbinol with HBr gives the normal product and some cyclobutyl bromide. Likewise cyclobutylcarbinol gives largely cyclopentyl bromide. No entirely analogous reactions are known in the aliphatic series.

5. Reduction decreases the size of the ring in some cases. Benzene with HI at 300° gives cyclohexane and methylcyclopentane. Cyclohexane is partially converted to Me-cyclopentane by similar treatment with HI while cycloheptane gives Me-cyclohexane and Me$_2$-cyclopentane. Similar changes are undoubtedly involved in the action of HI with cyclohexyl and cycloheptyl iodides to give Me-cyclopentane and Me-cyclohexane respectively.

1,3-Cycloheptadiene with H_2 and Ni gives cycloheptane or Me-cyclohexane depending on the temperature. An analogous change takes place with Me_3-ethylene and H_2 in a dark electric discharge to give Me_4-methane.[9]

$$\begin{array}{c} -CH \\ \diagdown \\ C \rightarrow \\ \diagup \\ -CH_2 \end{array} \quad \begin{array}{c} -CH-CH_3 \\ | \\ -CH_2 \end{array} \qquad \begin{array}{c} Me-C=CH-Me \\ | \\ Me \end{array} \rightarrow \begin{array}{c} Me \\ | \\ Me-C-CH_3 \\ | \\ Me \end{array}$$

Cyclohexane can be isomerized to Me-cyclopentane by an alumina catalyst.[10]

6. The oxidation of cyclobutanol is reported to give cyclopropane carboxylic acid. No analogous oxidation is known in the aliphatic·series.

7. α-Halogencycloketones with bases give carboxylic acids of the next lower ring.

$$\begin{array}{c} CH_2-CH_2-CHX \\ | | \\ CH_2-CH_2-CO \end{array} \xrightarrow{ KOH } \begin{array}{c} CH_2-CH_2-CH-CO_2K \\ | \diagup \\ CH_2 CH_2 \end{array}$$

8. The addition of HCl to pinene to give bornyl chloride involves a change of a 4- to a 5-membered ring.

9. Diazomethane with cyclohexanone gives cycloheptanone.[11] This is analogous to the action with an acid chloride.

$$RCOCl + CH_2N_2 \rightarrow N_2 + RCOCH_2Cl$$

$$\begin{array}{c} CH_2-CH_2-CH_2 \\ | | \\ CH_2-CH_2-CO \end{array} + CH_2N_2 \rightarrow N_2 + \begin{array}{c} CH_2-CH_2-CH_2-CH_2 \\ | \diagup \\ CH_2-CH_2 CO \end{array}$$

II. ISOMERISM OF ALICYCLIC COMPOUNDS

This is more complex than that of the aliphatic compounds. It may be illustrated by the cyclic isomers, C_6H_{12}. As will be recalled, this formula represents 13 olefins with 4 possible in *cis* and *trans* forms and one in D- and L-forms, a total of 18 isomers of all kinds. It also represents polymethylene compounds of which the following are representative:

1. Cyclohexane. Theoretically this should exist in two stereoisomeric forms which can readily be shown by models. Since the natural angle of the carbon valences, 109.5°, varies considerably from that of a regular hexagon, 120°, it is not possible for the six carbon atoms to lie in one plane. A model shows that atoms 2,3,5,6 lie in one plane while atoms 1,4 can lie both in one other plane (*cis*) or each in a separate plane, one above and one below the plane

[9] Menghini, Sorgato. *Gazz. chim. ital.* 62, 621 (1932).
[10] Ipatief, Dovgelevich. *J. Russ. Phys. Chem. Soc.* 43, 1431 (1911).
[11] Mosettig, Burger. *J. Am. Chem. Soc.* 52, 3456 (1930).

of the four middle atoms (*trans*). Actually only one form of cyclohexane is
known and this has been shown to be almost wholly the *trans* form.[1, 2] It has
not been possible to isolate the *cis* form because of the ease with which it
changes to the *trans*.

The *cis* and *trans* forms are sometimes called the "*boat*" and "*seat*" forms.

2. Methylcyclopentane offers no possibilities for isomerism.

3. Dimethylcyclobutanes can exist as 1,1-, 1,2-, and 1,3-isomers. The
four carbon ring occupies one plane. In the 1,1-isomer one methyl is above
and the other below this plane. The 1,2-compound can have both methyls in
one plane (*cis*) or one above and one below the plane of the ring (*trans*).

A plane following the dotted line perpendicular to the plane of the ring is a
plane of symmetry for the *cis* but not for the *trans* form. The latter has no
plane of symmetry. It can, therefore, exist in enantiomorphic forms (D- and L-).

The 1,3-isomer can also exist in *cis* and *trans* forms. In this case a plane
through the 1,3-corners and at right angles to the plane of the ring includes

[1] Aston, Szasz, Fink. *J. Am. Chem. Soc.* **65**, 1135 (1943).

[2] Beckett, Pitzer, Spitzer. *J. Am. Chem. Soc.* **69**, 2488 (1947).

both methyl groups and is a plane of symmetry. Thus no enantiomorphism is possible for either 1,3-isomer.

4. Ethylcyclobutane can have only one form.

5. Trimethylcyclopropanes can exist as 1,1,2- and 1,2,3-isomers. The first cannot exist in *cis* and *trans* forms but can exist in D- and L-forms because the molecule is asymmetric.

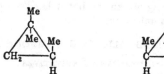

On the other hand the 1,2,3-isomer exists in *cis* and *trans* forms but optical isomers are impossible because both forms have at least one plane of symmetry through one corner bisecting the opposite side, making enantiomorphs impossible.

6. Methylethylcyclopropanes can exist as 1,1- and 1,2-isomers. The latter has *cis* and *trans* isomers which can also exist in D- and L-forms.

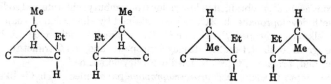

7. Propylcyclopropanes permit branching in the side chain giving the normal and iso form. Thus C_6H_{12} represents 28 alicyclic isomers, structural and stereo (geometric and optical).

All of the asymmetric molecules noted in the optically isomeric alicyclic compounds cited above contain at least one asymmetric carbon (attached to four different groups). The alicyclic series provides an example of an asymmetric optically active molecule containing *no asymmetric atom*, 1,4-methylcyclohexylideneacetic acid, which has been resolved into its active forms.[3]

$$
\underset{H}{\overset{Me}{C}}\diagdown \quad \underset{CH_2-CH_2}{\overset{CH_2-CH_2}{}} \diagup C=C \underset{CO_2H}{\overset{H}{\diagdown}}
$$

[3] Perkin, Pope, Wallach. *J. Chem. Soc.* 95, 1789 (1909).

Models show that a plane including the Me and H bisects the ring but is not symmetrical to the H and carboxyl, while the plane of the ring includes the H and carboxyl but is not symmetrical to the Me and H. Thus the molecule has no plane of symmetry in spite of the fact that no one of the atoms is attached to four different groups.

Unsaturated alicyclic compounds. Numerous substances containing one or more double bonds in the ring are known but it has been possible to obtain triple bonds only in large strainless rings.[4]

III. INDIVIDUAL ALICYCLIC COMPOUNDS

A. Cyclopropane Derivatives

Cyclopropane, trimethylene, $CH_2 \overset{\displaystyle CH_2}{\diagup \diagdown} CH_2$, m. $-127°$, b. $-34°$, is made from trimethylene halides and sodium or zinc. Because of its value as an inhalation anesthetic, attempts are being made to cheapen its preparation.[1] At high temperatures alone, or with catalysts (Pt, etc.) at lower temperatures, it isomerizes to propylene. With H_2 and Ni at 80° it gives propane. Br_2 and HBr react even in the cold to form trimethylene dibromide and n-propyl bromide respectively. Strangely, chlorine opens the ring but also forms *cyclopropyl* chloride, b. 43°. Alkyl substituted cyclopropanes usually undergo ring opening more easily even than does the parent substance. Cyclopropane acts slowly with $KMnO_4$, thus differing from propylene.

Homologs of cyclopropane are made from 1,3-dibromides and zinc. Me-cyclopropane, b. 5°, is obtained in low yields from isobutyl chloride and sodium.[2] 1,1-dimethylcyclopropane, b. 21°, can be obtained by the action of sodium or sodium alkyls on neopentyl chloride.[3] The action of sodium on neohexyl chloride gives both 1,1-methylethylcyclopropane and 1,1,2-trimethylcyclopropane, and, probably, traces of 1,1-dimethylcyclobutane.[4]

All attempts to make *methylenecyclopropane* have failed. This is like the failure to prepare pure cyclopropanone (p. 531).

Cyclopropyl alcohol, $(CH_2)_2-CHOH$, b. 100–103° has been reported from the reaction of ethylmagnesium chloride and epichlorohydrin.[5] The presence of metallic halides, such as $FeCl_3$, was found necessary for the reaction. This type of reaction has been explained as an oxidation-reduction.[6]

$$Cl-C-C\overset{\displaystyle O}{\diagup \diagdown}C + C_2H_5MgCl \rightarrow C\overset{\displaystyle C}{\diagup \diagdown}C-OH + C_2H_4 + MgCl_2$$

[4] Ruzicka, Hürbin, Boekenoogen. *Helv. Chim. Acta* 16, 498 (1933).
[1] Hass et al. *Ind. Eng. Chem.* 28, 1778 (1936).
[2] Conden, Smith. *J. Am. Chem. Soc.* 69, 956 (1947).
[3] Whitmore et al. *J. Am. Chem. Soc.* 61, 1616 (1939); 63, 125 (1941); 64, 1783 (1942).
[4] Chapman Ph.D. Thesis, Penn. State College 1950.
[5] Magrane, Cottle. *J. Am. Chem. Soc.* 64, 484 (1942); Stahl, Cottle. *J. Am. Chem. Soc.* 65, 1782 (1943).
[6] Kharasch, Fields. *J. Am. Chem. Soc.* 63, 2316 (1941).

Cyclopropanol forms some alcohol derivatives, is changed by heat to propan-aldehyde and dimerizes to 2-methyl-2-penten-1-al over anhydrous potassium carbonate.

A great variety of other methods have failed to give it.[7] Glycerol-1,3-dibromohydrin with Zn gives allyl alcohol.[8] The corresponding esters with Na give allyl esters.[9]

Cyclopropylamine, $(CH_2)_2CHNH_2$, b. 50°, is made by the Hofmann reaction on the acid amide obtained from the acid chloride and NH_3 or from the cyanide and H_2O_2. With nitrous acid, it gives allyl alcohol instead of cyclopropyl alcohol.[10]

$$CH_2 - CHNH_2 + HNO_2 \rightarrow N_2 + H_2O + CH - CH_2OH$$

The change is analogous to the action of nitrous acid on neopentylamine except that the transfer of electrons within the electronically deficient cyclopropyl residue breaks a $C-C$ linkage without forming a new one. The resulting allyl fragment combines with hydroxyl ion to give allyl alcohol. Cyclopropylamine can be completely methylated. Its quaternary hydroxide decomposes to give cyclopropene, b. $-35°$, a substance very reactive with oxygen and bromine and readily subject to polymerization.[11,12]

$$\underset{CH_2-CH_2}{CH-NH_2} \rightarrow \underset{CH_2CH_2}{CH-NMe_3(OH)} \rightarrow Me_3N + \underset{CHCH_2}{CH}$$

Cyclopropanone has not been prepared in the free state in spite of many attempts.[13] Apparently a doubly linked oxygen attached to a cyclopropane ring involves great strain. The reaction of ketene with diazomethane gives cyclobutanone instead of the expected product

$$H_2C=C=O + 2 N_2CH_2 \rightarrow \underset{CH_2-CH_2}{CH_2-C=O}$$

With excess ketene and a little water or methanol the product is the hydrate or hemiacetal, $(CH_2)_2C(OH)_2$ or $(CH_2)_2C(OH)OMe$, in which the strain of the doubly linked oxygen is avoided. The crystalline hydrate gradually changes

[7] Lipp, Buchkremer, Seeles. *Ann.* **499**, 1 (1932).
[8] Aschan. *Ber.* **23**, 1833 (1890).
[9] Sattler. *J. Am. Chem. Soc.* **54**, 830 (1932).
[10] Kishner. *Chem. Zentr.* 1905, I, 1703.
[11] Demjanow, Dojarenko. *Ber.* **56**, 2200 (1923).
[12] Schlatter. *J. Am. Chem. Soc.* **63**, 1733 (1933).
[13] Lipp, Buchkremer, Seeles. *Ann.* **499**, 1 (1932).

to propionic acid. This is not an ordinary rearrangement but probably a splitting of the cyclopropane ring as in the action of HBr.

$$CH_2\underset{CH_2}{\overset{C(OH)_2}{\diagup}} \xrightarrow{H^+} [CH_3CH_2C(OH)_2]^+ \rightarrow H^+ + CH_3CH_2CO_2H$$

The hemiacetal, when shaken with a dilute base, evolves heat and changes to methyl propionate. The stability of the hydrate as compared with the free ketone recalls the great stability of 1,1-Cl$_2$-cyclopropane.[14]

Cyclopropylcarbinol, $(CH_2)_2CHCH_2OH$, can be made by reducing the corresponding ester. The corresponding bromide, not obtainable from the alcohol, is best made by von Braun's substituted benzamide method.[15,16] The corresponding amine is made by reducing cyclopropyl cyanide obtained from γ-bromobutyronitrile and a solid base.

$$BrCH_2CH_2CH_2CN \xrightarrow{KOH} (CH_2)_2CHCN \xrightarrow{[4H]} (CH_2)_2CHCH_2NH_2$$

$$(CH_2)_2CHCH_2NH_2 \xrightarrow{PhCOCl\ PBr_5} \longrightarrow (CH_2)_2CHCH_2Br + PhCN$$

Cyclopropyl methyl ketone, b. 112°, is made from ethylene dibromide, acetoacetic ester and an excess of NaOEt. The first product,

$$BrCH_2CH_2CH(COMe)CO_2Et,$$

can react with sodioacetoacetic ester to give

$$EtO_2C(MeCO)CHCH_2CH_2CH(COMe)CO_2Et,$$

or with NaOEt to give a Na derivative which can reach internally to give $CH_2\text{—}C(COMe)CO_2Et$ which can be split like other acetoacetic esters to give
$\diagdown$
CH_2
either cyclopropyl methyl ketone or cyclopropane carboxylic acid. The ketone is split by HBr to give $Br(CH_2)_3COMe$.

Cyclopropanecarboxylic Acids

These will be discussed rather fully because they illustrate principles important in this series.

1. The **monocarboxylic acid,** ethyleneacetic acid, m. 17°, b. 181°, is made by heating the 1,1-dicarboxylic acid. It is also obtained from trimethylene

[14] Gustavson. *J. prakt. Chem.* (2) **42**, 495 (1890).
[15] Arvin, Adams. *J. Am. Chem. Soc.* **50**, 1983 (1928).
[16] v. Braun. *Ann.* **445**, 201 (1925).

glycol, through the dibromide, and the bromocyanide.

$$Br(CH_2)_3CN \xrightarrow{\text{KOH}} \underset{\underset{CH_2}{|}}{CH_2{-}CHCN} \to (CH_2)_2CHCO_2H$$

The removal of HX takes place better in liquid NH_3-ether solution with $NaNH_2$.[17] The carboxyl group stabilizes the ring. Thus bromine replaces the alpha H instead of opening the ring. Fuming HBr acts only at 175° to give γ-Br-butyric acid. The acid chloride and Et ester b. 119° and 133°.

2. The **1,1-dicarboxylic acid,** ethylenemalonic acid, vinaconic acid, m. 140°, is made from ethylene bromide, malonic ester and excess NaOEt. The by-product is ethylene dimalonic ester, $(RO_2C)_2CHCH_2CH_2CH(CO_2R)_2$. Heat converts the dibasic acid to cyclopropanecarboxylic acid.

3. The **1,2-dicarboxylic acids** are obtained by decarboxylating the 1,1,2-tricarboxylic acid made from malonic ester, $\alpha\beta$-Br$_2$-propionic ester and NaOEt. The *cis* or *maleinoid* form of the 1,2-acid is obtained as its anhydride, m. 59°, by heating the 1,1,2-acid above 200° or by treating it with acetyl chloride. The anhydride unites with H_2O at 140° to give the 1,2-acid, m. 139°. It corresponds not only to maleic acid with a ring in place of a double bond but also to mesotartaric acid because it contains two asymmetric carbon atoms which compensate each other. When the *cis*-1,2-acid or its anhydride is heated a short time at 240° with fused KOH, cooled and acidified, the *trans* or *fumaroid* 1,2-acid, m. 175°, is obtained. This corresponds to racemic acid and can be resolved through the brucine salts into the D- and L-*trans*-cyclopropane-1,2-dicarboxylic acids, m. 175°, $[\alpha]^D = \pm 84.8°$. The silver salt of the *trans* acid, on heating, gives the anhydride of the *cis* acid. Another preparation of the *trans* acid is from acrylic ester and diazoacetic ester through pyrazoline-3,5-dicarboxylic ester.

$$CH_2{=}CHCO_2R + N_2CHCO_2R \xrightarrow{50°}$$

$$\underset{\underset{RO_2CCHN=N}{|\qquad\quad|}}{CH_2{-}CHCO_2R} \xrightarrow{180°} \underset{RO_2CCH{-}CHCO_2R}{\overset{CH_2}{\diagup\ \diagdown}}$$

It is notable that the 1,2-di-carbethoxycyclopropane is stable to HBr whereas the 1,1-compound is split by this reagent.[18]

The ease of closing the cyclopropane ring depends on the valence angles as determined by the volumes of the attached groups. Thus the calculated

[17] Cloke et al. *J. Am. Chem. Soc.* **53,** 2791 (1931).
[18] *Ann. Rep. Chem. Soc.* (London) 1914, 117.

angles in the open-chain compounds are as follows:

$$CH_2 \diagup \begin{array}{c} CH_2CO_2R \\ \\ CH_2CO_2R \end{array} \qquad MeCH \diagup \begin{array}{c} CH_2CO_2R \\ \\ CH_2CO_2R \end{array} \qquad Me_2C \diagup \begin{array}{c} CH_2CO_2R \\ \\ CH_2CO_2R \end{array}$$
$$115° \qquad\qquad 112° \qquad\qquad 109°$$

As the angle decreases the ease of 3-ring formation increases.[19] A good example of this effect is found in the action of 1,3-dibromides with sodiomalonic ester. Whereas $Br(CH_2)_3Br$ gives the expected cyclobutane compound, $Me_2C(CHBrCO_2R)_2$ (I) gives a cyclopropane derivative and $\alpha\alpha'$-Br$_2$-glutaric ester gives both types of products.[20] In (I) the arrangement of the groups evidently favors 3-ring closure.

$$(I) \rightarrow \begin{array}{c} RO_2C-CH-CH(CO_2R)_2 \\ \diagup \\ Me_2C-CHBr-CO_2R \end{array} \rightarrow \begin{array}{c} RO_2C-C-CH(CO_2R)_2 \\ \diagup \; | \\ Me_2C-CH-CO_2R \end{array}$$

4. The **cis-1,2,3-tricarboxylic acid,** m. 153°, is obtained by heating the tetracarboxylic acid made from malonic ester, NaOEt and dibromosuccinic ester. The *trans*-1,2,3-acid, m. 220°, is obtained by heating pyrazoline-3,4,5-tricarboxylic esters obtained from diazoacetic ester and fumaric ester.

5. The **1,1,2,2-tetracarboxylic acid,** m. 200° dec., is obtained by the action of bromine on the disodio derivative of propane-1,1,3,3-tetracarboxylic ester made from methylene chloride and sodiomalonic ester. The first action of the bromine is to substitute on one of the alpha carbons. The resulting bromo compound then reacts internally to give the 3-ring.

$$\begin{array}{c} RO_2C \qquad\qquad CO_2R \\ \quad| \qquad\qquad\quad | \\ NaO(RO)C=C-CH_2-C=C(OR)ONa \rightarrow \end{array}$$

$$\begin{array}{c} CO_2R \qquad\qquad\qquad\qquad CH_2 \\ \quad| \qquad\qquad\qquad\qquad \diagup \quad \diagdown \\ (RO_2C)_2CBrCH_2=C(OR)ONa \rightarrow (RO_2C)_2C \rule{1.5cm}{0.4pt} C(CO_2R)_2 \end{array}$$

The **cis-1,2,3-trans-1-acid,** m. 100° dec., is made from malonic ester, NaOEt and Br$_2$-succinic ester. Heat converts it to the *cis-1,2,3-acid*. The *cis-1,2-trans-1,3-acid*, m. 198°, is made from malonic ester, NaOEt and bromomaleic ester. In this case the malonic ester adds to the $\alpha\beta$-unsaturated ester first. Then the Br acts intramolecularly with the sodium.

$$\begin{array}{c} RO_2CCBr \qquad RO_2CCHBr \qquad\qquad NaOEt \qquad RO_2CCH \\ \quad\| \qquad \rightarrow \qquad | \qquad\qquad \xrightarrow{\qquad\qquad} \qquad | \diagdown \\ RO_2CCH \qquad RO_2CCH-CH(CO_2R)_2 \qquad\qquad RO_2CCH-C(CO_2R)_2 \end{array}$$

The acid loses CO_2 to give the trans-1,2,3-acid.

[19] *Ann. Rep. Chem. Soc.* (London) **1922,** 115.
[20] *ibid.* **1925,** 126.

The ring-chain tautomeric formula formerly assigned to Balbiano's acid, obtained by the oxidation of camphoric acid, was $1,3,3$-Me_3-2-OH-cyclopropane-$1,2$-dicarboxylic acid.[21]

$$Me_2C—CH(Me)CO_2H \rightleftharpoons \left[\begin{array}{c} Me_2C—C(Me)CO_2H \\ | \quad / \\ C(OH)CO_2H \end{array} \right]$$

$$\begin{array}{c} | \qquad | \\ CO—CO_2H \end{array}$$

It has since been shown that the tautomeric form is a lactone of the hydrated ketoacid.[22]

Many *complex cyclopropane derivatives* have been made by processes such as the addition of malonic ester to benzalacetophenone, bromination and removal of HBr from the $\alpha\gamma$-positions.[23]

$$PhCH=CHCOPh \rightarrow PhCH—CH_2COPh \xrightarrow{Br_2} \xrightarrow{Base} PhCH—CHCOPh$$

$$\begin{array}{c} | \\ CH(CO_2R)_2 \end{array} \qquad\qquad \begin{array}{c} | \quad / \\ C(CO_2R)_2 \end{array}$$

The actual existence of a 3-membered ring is proven in a conclusive manner by the splitting of such a compound at the three different sides of the triangle: (1) reduction gives $PhCOCH_2CHPhCH(CO_2R)_2$, (2) NH_3 gives

$$PhCH=C(COPh)CH(CO_2R)_2,$$

and (3) HBr in glacial acetic acid gives $PhCOCH_2C(CO_2H)_2CHBrPh$, which loses water, CO_2 and HBr to give

$$PhCH=CCH=CPh$$

$$\begin{array}{c} | \qquad | \\ CO——O \end{array}$$

Another preparation for *complex cyclopropane derivatives* is by the decomposition of pyrazoline compounds formed from $\alpha\beta$-unsaturated carbonyl compounds with diazomethane or diazoacetic ester. These have been illustrated in the preparation of some of the cyclopropane carboxylic acids.

Natural cyclopropane derivatives are the thujenes and carenes which have internal and external 3-membered rings fused to the 6-membered terpene rings. Similar compounds have been synthesized. The stability of the cyclopropane ring is surprisingly illustrated by the oxidation of such substances containing CO or $C=C$ in the 6-ring to give $1,1$-Me_2-cyclopropane-$2,3$-dicarboxylic acid, *caronic acid*, from carone, its homolog, *homocaronic acid*, having $-CH_2CO_2H$ in place of one CO_2H from carene,[24] and α-*tanacetogendicarboxylic acid*, 1-isopropyl-2-carboxycyclopropane-1-acetic acid, from thujone. One of the

[21] *Ann. Rep. Chem. Soc.* (London) **1922**, 114.
[22] *ibid.* **1928**, 81.
[23] Kohler et al. *J. Am. Chem. Soc.* **39**, 1404, 1699, 2405 (1917).
[24] *Ann. Rep. Chem. Soc.* (London) **1923**, 99; **1928**, 81.

acids from the insecticide pyrethrum is pyrethronic acid.[25]

$$Me_2C—CHCO_2H$$
$$|\qquad|$$
$$CHCH=CHEt$$

Oxidation gives a glycol and then *trans*-caronic acid.

Derivatives of *bicyclobutane* and *tricyclobutane* have been prepared.[26]

$$MeCH_2CHO + 4HCHO \rightarrow MeC(CH_2OH)_3 \rightarrow MeC(CH_2CO_2R)_3 \rightarrow$$

$$MeC(CHBrCO_2R)_2(CH_2CO_2R) \xrightarrow[\text{heat}]{\text{KOH}} \xrightarrow{\text{acid}} MeC—CCO_2H$$

$$\begin{matrix} CHCO_2H \\ | \\ MeC—CCO_2H \\ | \\ CHCO_2H \end{matrix}$$

The product 1-Me-bicyclobutan-2,3,4-tricarboxylic acid exists in three forms, m. 154°, 165° and 193°. The first is removed by heating the mixture and extracting the anhydro acid of the 154° isomer with ether. Treatment with acetic anhydride converts the 165° isomer to the anhydro compound which is removed by ether. The residue is the 193° isomer which does not form an anhydro acid. Bromination of the three acids proceeds very differently: the 154° acid giving a stable bromo acid; the 165° acid giving a lactonic acid free of bromine; and the 193° acid giving methyltricyclobutane tricarboxylic acid (p. 537). In harmony with their reactions the acids are assigned the following configurations.

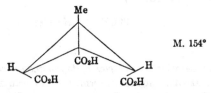

M. 154°

A study with atomic models shows that the 2,3,4-carbons, the 3-carboxyl and the two H atoms are in one plane while the 2,4-carboxyls lie near each other slightly below this plane. Anhydride formation should be easy. The mono-bromide gives no opportunity for reaction with a carboxyl or the other H.

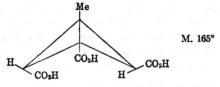

M. 165°

Anhydride formation would be difficult. The α-Br is near a carboxyl group.

[25] *ibid.* 1923, 98.
[26] *ibid.* 1920, 87.

The carboxyls and carbons 2,3,4 are in one plane.

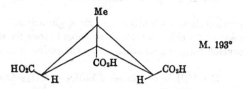

M. 193°

No anhydride would be expected. The α-H atoms are near each other so that the α-Br can react with the other α-H to close the third ring. The carboxyls are in one plane. The degree of strain in this *cage system* of four cyclopropane rings as indicated by models is much less than might be expected.

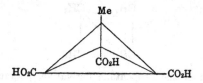

B. Cyclobutane Derivatives

$$\begin{array}{c} CH_2CH_2 \\ | \quad | \\ CH_2CH_2 \end{array}$$

The preparation of **cyclobutane**, CH_2CH_2, b. 15°, involves familiar reactions.

1. Trimethylene bromide, $Br(CH_2)_3Br$, with malonic ester and NaOEt gives cyclobutan-1,1-dicarboxylic ester.

2. Hydrolysis and heat give the monocarboxylic acid.

3. Conversion to the amide and application of the Hofmann reaction give cyclobutylamine, b. 81°.

4. Exhaustive methylation of the amine gives cyclobutene.

5. Reduction by Ni and H_2 at 100° gives cyclobutane.

Cyclobutane is stable to cold HI and to KMnO₄. Catalytic hydrogenation at 120° give *n*-butane. Thus the 4-membered ring is more stable than that of cyclopropane.

Cyclobutene shows the ordinary olefin reactions. Attempts to make cyclobutadiene give butadiene or acetylene.[27,28]

Cyclobutanol, $(CH_2)_3CHOH$, b. 123°, along with cyclopropylcarbinol is obtained from the corresponding amine and nitrous acid.

Cyclobutylcarbinol, $(CH_2)_3CHCH_2OH$, is obtained by the catalytic reduction of the acid chloride. Its dehydration gives cyclopentene and *methylenecyclobutane*, $(CH_2)_3C=CH_2$, b. 41°. This substance in presence of acids or on

[27] Willstätter, v. Schmaedel. *Ber.* **38**, 1992 (1905).
[28] Baker. *J. Chem. Soc.* 1945, 258.

heating with alumina[29] gives some *methylcyclobutene*, $\begin{matrix} CH=CMe \\ | \quad\quad | \\ CH_2-CH_2 \end{matrix}$, b. 37°.

Methylenecyclobutane with nitric acid gives glutaric acid. This is related to the conversion of its chlorohydrin to cyclopentanone by warming with water. Methylenecyclobutane is also obtained by the action of zinc on pentaerythrityl bromide

$$C(CH_2Br)_4 + 2\ Zn \rightarrow 2\ ZnBr_2 + \begin{matrix} CH_2-C=CH_2 \\ | \quad\quad | \\ CH_2-CH_2 \end{matrix}$$

Some Me-cyclobutene is formed at the same time. After many years of study of this peculiar reaction some of the originally expected product spiropentane, b. 39°, has been isolated.[30]

The conversion of the neopentane derivative, pentaerythrityl bromide, through the cyclobutane derivative to cyclopentanone involves rearrangements, ring formation and ring widening which can be shown as follows with the * indicating a carbon with only six electrons.[31]

The initial reaction may be the formation of 1,1-di-bromomethylcyclopropane, followed by similar rearrangement with ring widening.[32]

[29] Dojarenko. *Ber.* **59**, 2933 (1926).
[30] Murray, Stevenson. *J. Am. Chem. Soc.* **66**, 812 (1944).
[31] Whitmore. *J. Am. Chem. Soc.* **54**, 3247 (1932).
[32] *Ann. Rep. Chem. Soc.* (London) **1923**, 116.

Cyclobutanone, $(CH_2)_3CO$, b. 100°, is made by the Hofmann reaction on α-bromocyclobutanecarboxylic acid amide. Oxidation with nitric acid gives succinic acid. Heating the Ca salt of glutaric acid gives no trace of cyclobutanone. On the other hand Ca $\beta\beta$-Me$_2$-glutarate apparently gives the ketone which rearranges to mesityl oxide.[33]

$$Me_2C(CH_2CO_2)_2Ca \rightarrow Me_2C(CH_2)_2CO \rightarrow Me_2C=CHCOCH_3$$

The effect of the *gem*-Me$_2$ groups in ring closure is notable.[34]

Substituted cyclobutandiones, $RCH(CO)_2CHR$, have been prepared and found to be different from the dimers of the ketenes $RCH=C=O$.[35]

Cyclobutyl methyl ketone, $(CH_2)_3CHCOCH_3$, b. 134°, is made from the acid chloride and Me$_2$Zn.

Cyclobutanecarboxylic Acids

The **monocarboxylic acid,** $(CH_2)_3CHCO_2H$, b. 194°, is stable to HBr. HI at 200° gives *n*-valeric acid. Br$_2$ gives α-substitution. Alkaline KMnO$_4$ forms oxalic acid.

The **1,1-dicarboxylic acid,** m. 155°, loses CO$_2$ above 200°. It is not attacked by cold Br$_2$ or HBr.

The **cis-1,2-dicarboxylic acid,** m. 138°, is made by heating the 1,1,2,2-tetracarboxylic acid in a sealed tube with water at 200°. At 300° it forms an anhydride.

The **trans-1,2-dicarboxylic acid,** m. 131°, is obtained by heating the *cis* acid with hydrochloric acid. It is said to react readily with acetyl chloride to give the anhydride of the *cis* acid, m. 75°. This conversion of a *trans* acid into the *cis*-anhydride is also observed in the truxillic acids (p. 541). The *trans*-acid can be made by hydrolyzing and heating the product of the action of NaCN on $\alpha\alpha'$-Br$_2$-adipic ester.[36, 37]

$$
\begin{array}{c}
CH_2\!-\!CHBrCO_2R \\
| \quad\quad | \\
CH_2\!-\!CHBrCO_2R
\end{array}
\rightarrow
\left[
\begin{array}{c}
CH_2\!-\!CH(CN)CO_2R \\
| \quad\quad | \\
CH_2\!-\!CHBrCO_2R
\end{array}
\right]
\rightarrow
$$

$$
\begin{array}{c}
CH_2\!-\!C(CN)CO_2R \\
| \quad\quad | \\
CH_2\!-\!CHCO_2R
\end{array}
\rightarrow
\begin{array}{c}
CH_2\!-\!CHCO_2H \\
| \quad\quad | \\
CH_2\!-\!CHCO_2H
\end{array}
$$

The introduction of the CN renders the adjacent α-H reactive enough to close the ring by action with the other Br.

The **cis-1,3-dicarboxylic acid,** m. 136°, is obtained by heating the 1,1,3,3-tetracarboxylic acid.

Norpinic acid, obtained by oxidizing pinene, is 2,2-Me$_2$-cyclobutan-1,3-dicarboxylic acid. It exists in *cis* and *trans* forms. The synthesis of this acid

[33] *Ann. Rep. Chem. Soc.* (London) 1921, 96.
[34] Ingold. *J. Chem. Soc.* **119,** 305 (1921).
[35] *Ann. Rep. Chem. Soc.* (London) 1917, 108.
[36] Fuson, Kreimeier, Nimmo. *J. Am. Chem. Soc.* **52,** 4074 (1930).
[37] Bode. *Ber.* **67B,** 332 (1934).

has offered extreme difficulties.   Apparently the *gem*-Me$_2$ prevents the closing of the 4-ring in most reactions.   The difficulty was overcome by using a cyclic imide as starting material for the ring closure.[38]

$$\text{Me}_2\text{C}\underset{\text{CH(CN)—CO}}{\overset{\text{CH(CN)—CO}}{\diagdown \hspace{-0.5em} \diagup}}\text{NH} \xrightarrow[\text{NaOMe}]{\text{CH}_2\text{I}_2}$$

$$\text{Me}_2\text{C}\underset{\text{C(CN)—CO}}{\overset{\text{C(CN)—CO}}{\diagdown \hspace{-0.5em} \diagup}}\text{CH}_2 \diagdown \hspace{-0.5em} \diagup \text{NH} \rightarrow \text{Me}_2\text{C}\underset{\text{C(CO}_2\text{H)}_2}{\overset{\text{C(CO}_2\text{H)}_2}{\diagdown \hspace{-0.5em} \diagup}}\text{CH}_2$$

Caryophyllenic acid, from the oxidation of caryophyllene, is 1,2,2-Me$_3$-cyclobutane-1,3-dicarboxylic acid.

The **trans-1,3-dicarboxylic acid,** m. 171°, is obtained in small yield by the action of NaOEt on α-Cl-propionic ester.

The **1,1,2,2-tetracarboxylic acid,** m. 150° dec., is made by the action of Br$_2$ on the disodio derivative of ethylene dimalonic ester.

$$\begin{matrix} \text{CH}_2\text{—CH(CO}_2\text{R)}_2 \\ | \\ \text{CH}_2\text{—CH(CO}_2\text{R)}_2 \end{matrix} \xrightarrow{\text{NaOEt}} \xrightarrow{\text{Br}_2} \begin{matrix} \text{CH}_2\text{—C(CO}_2\text{R)}_2 \\ | \quad\quad | \\ \text{CH}_2\text{—C(CO}_2\text{R)}_2 \end{matrix}$$

The **1,1,3,3-tetracarboxylic acid** is made by the polymerization of methylene malonic ester

$$\begin{matrix} (\text{RO}_2\text{C})_2\text{C}\!=\!\text{CH}_2 \\ \\ \text{CH}_2\!=\!\text{C(CO}_2\text{R)}_2 \end{matrix} \rightarrow \begin{matrix} (\text{RO}_2\text{C})_2\text{C}\text{——}\text{CH}_2 \\ | \quad\quad\quad | \\ \text{CH}_2\text{—C(CO}_2\text{R)}_2 \end{matrix}$$

The **truxillic acids** are stereoisomeric 2,4-diphenylcyclobutane-1,3-dicarboxylic acids while the *isotruxillic* or *truxinic* acids are the corresponding 3,4-diphenyl-1,2-dicarboxylic acids.   Some of their isomers occur in coca alkaloids.

A suspension of trans-cinnamic acid in water irradiated for two months gives a 50–75% conversion to β-truxinic acid

$$\text{C}_6\text{H}_5 \quad\quad \begin{matrix} \text{C}_6\text{H}_5 \\ | \\ \diagup\hspace{-0.5em}\square\hspace{-0.5em}\diagdown \text{COOH} \\ | \\ \text{COOH} \end{matrix}$$

Only when the cinnamic acid is slowly recrystallized and subsequently irradiated is the α-truxillic acid formed and then in low yields.   The best way to prepare the truxillic acid is to dimerize cinnamaldienmalonic acid with

[38] Kerr.   *J. Am. Chem. Soc.* **51,** 614 (1929).

subsequent bromination, dehydrobromination and finally oxidation.[39]

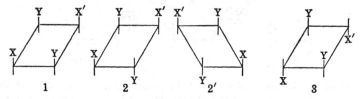

The formation of benzil, PhCOCOPh, by oxidation of truxinic acids shows their phenyl groups to be in the 1,2-position. Thus the truxillic acids are formed from cinnamic acid as follows.

$$PhCH=CHCO_2H \qquad PhCH—CHCO_2H$$
$$\rightarrow$$
$$HO_2CCH=CHPh \qquad HO_2CCH—CHPh$$

Five stereoisomeric truxillic acids are known.[40] They are designated as alpha (α-), gamma (γ-), peri or eta (η-), epi, and epsilon (ϵ-). The alpha acid is found in coca leaves and is formed from cinnamic acid. On heating, the alpha acid gives anhydrides which can be converted by water to the gamma- and peri-acids. The peri-acid, with hot HCl, gives the epi-acid which reacts with hot acetic anhydride to give the anhydride of the ϵ-acid. It is to be noted that the γ-, peri- and ϵ-acids readily form anhydrides which can be converted back to the *same* acids while the α- and epi-acids do not give anhydrides of their own but of isomeric acids. Another important bit of evidence is that the α- and γ-acids give anilic acids, containing $-CO_2H$ and $-CONHPh$, which can be resolved into optically active forms while the corresponding compounds of the other three acids cannot be resolved.

The possible configurations of the truxillic acids and of their anilic acids are as follows, indicating the carboxyl as X, the $-CONHPh$ as X', and the phenyl as Y and H as the vacant end of a vertical line.

[39] Bernstein, Quimby. *J. Am. Chem. Soc.* 65, 1845 (1943).
[40] *Ann. Rep. Chem. Soc.* (London) 1926, 128.

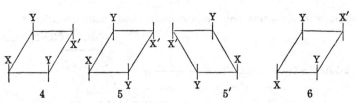

1,3 and 4 are identical with their mirror images and could not exist in optically active forms. 2,2′ and 5,5′ are pairs of enantiomorphs. Thus the α- and γ-acids must correspond to formulas 2 and 5 but, of these, only the γ-acid forms its own anhydride. Therefore the γ-acid has configuration 2 which has the carboxyls on the same side of the plane. The α-acid corresponds to 5 in which the carboxyls are *trans* to each other as are also the phenyls. The most stable acid which forms its own anhydride is the ϵ-acid which corresponds to 3. The relatively stable acid formed by heating with HCl but not capable of forming its own anhydride is the epi-acid corresponding to 4. This leaves 1 for the peri-acid which forms an anhydride but is unstable to hot HCl.

The changes relating the five acids thus are as follows:

The ready "Walden inversion" of the groups under the influence of heat alone or with reagents is worthy of thought. It is not explained by any of our present conceptions of the stability of the C−C bond.

The stereoisomers of the theoretically possible *truxinic acids* may be represented as follows using the horizontal line to represent the edge of the cyclobutane plane and long and short lines to represent the phenyl and carboxyl groups respectively.

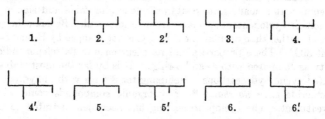

Configurations 2,2'; 4,4'; 5,5' and 6,6' represent four pairs of enantiomorphic dibasic acids whereas 1 and 3 is each identical with its mirror image. Acids 1 and 3 each could give a pair of enantiomorphic anilic acids. Thus 1 could form

$$\begin{array}{ccc}
& & \\
\text{Ph} \quad \text{CONHPh} & \text{and} & \text{CONHPh} \quad \text{Ph} \\
\text{Ph} \quad \text{HO}_2\text{C} & & \text{CO}_2\text{H} \quad \text{Ph}
\end{array}$$

The situation is considerably complicated by the fact that *each* of the other eight acids could give a pair of anilic acids which are not enantiomorphic. Acid 2 could give

$$\begin{array}{ccc}
\text{Ph} \quad \text{CO}_2\text{H} & & \text{Ph} \\
\text{Ph} & \text{and} & \text{HO}_2\text{C} \quad \text{CONPh} \quad \text{Ph} \\
\text{CONHPh} & &
\end{array}$$

Four truxinic acids are known. They are designated as the beta (β-), delta (δ-), zeta (ζ-), and neo-acids. To these have been tentatively assigned the configurations 3, 5(5'), 6(6') and 2(2') respectively.

Before the truxillic acids were solved the five forms of cyclobutan-1,3-dicarboxylic-2,4-diacetic acids were isolated and identified.[41]

[41] Ingold, Perren, Thorpe. *J. Chem. Soc.* 121, 1765 (1922).

C. Cyclopentane Derivatives

Cyclopentane, $\begin{array}{c}CH_2CH_2 \\ | \qquad >CH_2 \\ CH_2CH_2\end{array}$ b. 50.5°, has been found in various naphthenic petroleums. It is made from pentamethylene dibromide and zinc or by the reduction of cyclopentanone made from adipic acid. Bromine in sunlight replaces H rather than opening the ring. It is not changed by H_2 and catalysts even at 300°. These properties and its occurrence in petroleum indicate its stability as compared with 4- and 3-rings. It is by far the most stable of the saturated cyclic hydrocarbons. Refluxing for 8 days with anhydrous aluminum chloride gave *no change*.[42] Few organic compounds could withstand this treatment. The cyclopentane ring has been demonstrated to be nonplanar.[43]

Methylcyclopentane, $(CH_2)_4CHCH_3$, b. 71°, occurs in various petroleums.[44] It is readily formed by the isomerization of cyclohexane with aluminum chloride.[45] Hydrogen and Ni at 460° also convert cyclohexane to methylcyclopentane. It is more readily oxidized than cyclopentane, presumably because of the tertiary H. Heating with dil. nitric acid replaces this H by a nitro group.

The *dimethylcyclopentanes* are all known including the *cis* and *trans* forms of the 1,2- and 1,3-isomers. Their synthesis is best achieved from the cyclopentanones, which are obtained from the decarboxylation of the substituted adipic acids, made in turn from the cresols and xylenols.

The boiling points of the isomers are 1,1- 81.8°, *t*-1,3- 90.8°, *c*-1,3- app. 92°, *t*-1,2- 91.9°, *c*-1,2- 99.2°.

The *trimethylcyclopentanes* have four different structural arrangements—1,1,2-, 1,1,3-, 1,2,3- and 1,2,4-. All isomers are known except the *c,c,c*-1,2,4-trimethylcyclopentane.

The 1,1,3-isomer, b. 104.9, can be made from 1,1,3-Me$_3$-cyclohexanol. It has also been made by high pressure polymerization of isobutylene.

1,1,2-trimethylcyclopentane, b. 113.7, can be made from camphor through the following steps: camphor → camphoric acid → camphoric anhydride → isolauronolic acid → isolaurolene (1,1,2-Me$_3$-2-cyclopentene) → 1,1,2-Me$_3$-cyclopentane. The 1,2,3- and 1,2,4-isomers are best made by the following series of reactions: 1,2,4-xylenol → 3,4-dimethylcyclohexanol → 2,3-dimethyladipic acid (1 part) and 3,4-dimethyladipic acid (2 parts) → *t*-2,3-dimethylcyclopentanone (I), b. 152°, + *t*- and *c*-3,4-dimethylcyclopentanone (II), b. 158° and 164°; (I) → 1,2,3-Me$_3$-cyclopentanol → 1,2,3-Me$_3$-cyclopentene → 1,2,3 Me$_3$-cyclopentanes, *c,t,c*- b. 110°, *c,c,t*- b. 117°, *c,c,c*- b. 123°; (II) →

[42] Cox. *Bull. soc. chim.* **37** (4), 1549 (1925).
[43] Aston, Fink, Schumann. *J. Am. Chem. Soc.* **65**, 341 (1943).
[44] Hicks-Bruun, Brunn. *Bur. Standards J. Research* **7**, 799 (1931).
[45] Stevenson, Beeck. *J. Am. Chem. Soc.* **70**, 2890 (1948).

1,3,4-Me$_3$-cyclopentanol → 1,3,4-Me$_3$-cyclopentene → 1,2,4-Me$_3$-cyclopentanes, c,t,c- b. 109°, c,c,t- b. 116°.

Cyclopentene, b. 45°, is made from the bromide in the usual way and shows ordinary olefin reactions. The plant growth hormones, Auxin A and B[46] contain a 3,5-di-sec-butylcyclopentene ring with an additional group in the 2-position. For Auxin A the group is $-$CHOHCH$_2$(CHOH)$_2$CO$_2$H, and for Auxin B,[47] $-$CHOHCH$_2$COCH$_2$CO$_2$H.

Methyl cyclopentenes are made by dehydrating the corresponding alcohols. 1-Methylcyclopentene, b. 75.2°, is the only isomer obtained from methylcyclopentanol. 3- and 4-Methylcyclopentenes, b. 65.1 and 65.7, are produced in approximately equal amounts by alumina dehydration of 3-methylcyclopentanol. Acid dehydration gives some of the 1-methylcyclopentene. Pure 3-methylcyclopentene is best prepared by adding HCl to cyclopentadiene and coupling the 3-chlorocyclopentene with methyl Grignard reagent.[48]

Methylenecyclopentane, b. 75.3, has been made from the pyrolysis of the xanthate of cyclopentylcarbinol. Dehydration of the primary alcohol gives primarily the 1-methylcyclopentene, with some cyclohexene.

Laurolene, 1,2,3-Me$_3$-cyclopentene, b. 120°±, is obtained by the decarboxylation of laurolenic acid by distillation. It has also been synthesized from $\alpha\alpha'$-Me$_2$-cyclopentanone.[49]

Cyclopentadiene, $\begin{array}{c} \text{CH}=\text{CH} \\ | \qquad\qquad \text{CH}_2 \\ \text{CH}=\text{CH} \end{array}$ b. 41°, is obtained from the low boiling

portions of crude benzene from coal tar. The CH$_2$ group attached to the ends of the conjugated system shows the reactivity of an α-CH$_2$. Metallic K replaces one H. Aldehydes and ketones with alkaline condensing agents give

fulvenes, $\begin{array}{c} \text{CH}=\text{CH} \qquad\quad \text{R} \\ | \qquad\qquad \text{C}=\text{C} \\ \text{CH}=\text{CH} \qquad\quad \text{R}' \end{array}$, in which R and R' may be H, alkyl or aryl.

The color of the fulvenes increases with the size of R and R', becoming red with aromatic substitution. The fulvenes add alkali metals and form dimers.

$$\begin{array}{c} \text{CH}=\text{CH} \qquad\qquad\qquad\qquad\qquad \text{CH}=\text{CH} \\ | \qquad\qquad\qquad\qquad\qquad\qquad\qquad\qquad | \\ \qquad\qquad \text{CNa}-\text{CH}_2\text{CH}_2-\text{CNa} \\ \text{CH}=\text{CH} \qquad\qquad\qquad\qquad\qquad \text{CH}=\text{CH} \end{array}$$

Phenyl fulvene is the basis of a colometric determination of cyclopentadiene.[50]

[46] *Ann. Rep. Chem. Soc.* (London) **1934, 358.**

[47] *ibid.* **1934,** 216.

[48] Crane, Boord, Henne. *J. Am. Chem. Soc.* **67,** 1237 (1945).

[49] Noyes, Kyriakides. *J. Am. Chem. Soc.* **32,** 1064 (1910).

[50] Uhrig, Lynch, Becker. *Ind. Eng. Chem., Anal. Ed.* **18,** 550 (1946).

Like all conjugated dienes cyclopentadiene polymerizes readily. At room temperatures it forms di-cyclopentadiene, b. 170°. Distillation gives some monomer and some higher polymers. Various formulas have been assigned to the polymers.[51]

$$
\begin{array}{l}
\text{HC} \overset{\text{CH}_2}{\diagup} \text{CH—CH} \quad \overset{\text{CH}_2}{\diagup} \text{CH} \\
\text{HC} \text{——CH —CH——CH}
\end{array}
\qquad
\qquad
\begin{array}{l}
\text{HC} \overset{\text{CH}}{\diagup} \text{CH——CH} \\
\text{HC} \overset{\text{CH}_2}{\diagdown} \text{CH} \quad \text{CH} \\
\quad\quad \text{CH} \quad \text{CH}_2
\end{array}
$$

The last has been proved correct for the dimer.[52] The polymerization takes place in the same way as does that of isoprene. The process is essentially the same as that of the Diels-Alder reaction.

Di-cyclopentadiene exists in two stereoisomeric forms. Because of the *p*-methylene bridge the cyclohexane ring is necessarily in the *cis* or "boat" form with the 2,3,5,6-carbons in one plane and the 1,4-carbons above it. In the *cis* dimer, the 5-ring is on the same side of the plane as the endomethylene group while in the *trans*-dimer it is on the opposite side. Two trimers are known, each related to one of the dimers and formed from it by addition of a molecule of cyclopentadiene to the double bond in the 6-ring to give another 6-ring with an endomethylene group.

$$
\begin{array}{ccc}
& \text{CH} \quad\quad \text{CH} & \\
\text{CH} \overset{}{\diagup} \text{CH} \overset{}{\diagup} \text{CH} & \text{CH——CH} \\
\| \quad \text{CH}_2 \quad | \quad \text{CH}_2 \quad | & \| \\
\text{CH} \quad\quad \text{CH} \quad\quad \text{CH} & \text{CH} \\
& \text{CH} \quad \text{CH} \quad \text{CH}_2 &
\end{array}
$$

In much the same way that cyclopentadiene polymerizes, it can act as the diene in the Diels-Alder reaction with $\alpha\beta$-unsaturated carbonyl compounds. The products again are cyclohexane derivatives having a bridged or *endo*-methylene group.

$$
\begin{array}{ccc}
\text{—CH} =\text{CH—CO—} & & \text{—CH—CH—CO—} \\
\text{CH —CH}_2\text{— CH} & \rightarrow & \text{CH —CH}_2\text{— CH} \\
\quad \text{CH—CH} & & \quad \text{CH}=\text{CH}
\end{array}
$$

Chlordane, an important insecticide, is a viscous, colorless odorless liquid, b. 175°C. at 2 mm. It consists of about 60% 1,2,4,5,6,7,8,8-octachloro-4,7-methano-3a,4,7,7a-tetrahydroindane and related compounds equally toxic. It is prepared by the Diels-Alder addition of hexachlorocyclopentadiene to cyclopentadiene followed by the addition of chlorine to the double bond in the

[51] Egloff, Bollman, Levinson. *J. Phys. Chem.* **35**, 3489 (1931).
[52] *Ann. Rep. Chem. Soc.* (London) **1931**, 87.

1,2 position.[53] It has been successful in combating cockroaches, ants, and locust.[54]

Cyclopentadiene has been used in the synthesis of di-, tri-, and tetra-cyclopentanes.[55]

Methylcyclopentadienes are obtained in the cracking of petroleum naphtha. They form dimers much like cyclopentadiene.[56]

Pentaphenylcyclopentadienyl, m. 260°, is probably the most stable free radical. It forms violet crystals which show no tendency to associate.[57] Its high degree of stability is probably related to its extraordinary possibilities for resonance.

$$
\begin{array}{c}
\text{Ph—C}\text{———}\text{C—Ph}\\
\ \ \ \|\qquad\qquad\|\\
\text{Ph—C}\qquad\text{C—Ph}\\
\diagdown\ \ \diagup\\
\text{C}\\
\diagup\ \diagdown\\
\text{Ph}
\end{array}
$$

Cyclopentanol, cyclopentyl alcohol, $(CH_2)_4CHOH$, b. 140°, is obtained by reducing the ketone. It is an entirely ordinary secondary alcohol. It gives halides which react normally. The unsaturated halide, 3-chlorocyclopentene, acts as a cyclic allyl halide. It has been used to synthesize chaulmoogric acid.[58]

Cyclopentandiol-1,2 is obtained from the olefin, cis- m. 47°, trans- m. 10°. Dehydration gives the ketone, cyclopentanone.

Cyclopentanone, adipone, $(CH_2)_4CO$, b. 130°, is made

1. By distilling the calcium salt of adipic acid, obtained by the oxidation of cyclohexanol, prepared in turn by the catalytic hydrogenation of phenol. These steps give a method for going from a 6-carbon aromatic ring to the 5-carbon aliphatic ring. Another method is to heat a mixture of adipic acid and $BaCO_3$ at 300°.

2. By internal acetoacetic ester condensation of adipic ester.

$$
\begin{array}{l}
CH_2CH_2CO_2R\\
|\\
CH_2CH_2CO_2R
\end{array}
\xrightarrow{\text{NaOEt}}
\begin{array}{l}
CH_2CH_2CO\\
|\qquad\diagup\\
CH_2CHCO_2R
\end{array}
\rightarrow
\begin{array}{l}
CH_2CH_2CO\\
|\qquad\diagup\\
CH_2CHCO_2H
\end{array}
\rightarrow (CH_2)_4CO
$$

[53] Kerns, Ingle, Metcalf. *J. Econ. Entom.* **38**, 661 (1945).

[54] Wellman. *Chem. Ind.* **62**, 914 (1948).

[55] Goheen. *J. Am. Chem. Soc.* **63**, 744 (1941).

[56] Edson, Powell, Fisher. *Ind. Eng. Chem.* **40**, 1526 (1948).

[57] Ziegler. *Ann. Rep. Chem. Soc.* (London) 1928, 156.

[58] Perkin, Cruz. *J. Am. Chem. Soc.* **49**, 1070 (1927).

Cyclopentanone gives all ketone reactions both of the CO and of the $\alpha-CH_2$ groups with great ease. The folding back of the α-carbons by the ring closure seems to favor such activity.

$$\begin{array}{c} CH\text{---}CH_2 \\ | \quad\quad | \end{array}$$

Methylcyclopentenone, $CH_3C\text{---}COCH_2$, b. 157°, is obtained along with the C_5 and C_6 cyclic ketones in the products of wood distillation. Alkylcyclopentenones are obtained from the action of P_2O_5 on lactones of the type

$$R-CH_2-\overline{CR'-CH_2-CH_2-COO}.^{[59]}$$

The active principles of pyrethrum flowers contain the cyclopentane nucleus.[60]

Four main constituents have been identified, pyrethrins I and II (formula A) and cinerins I and II (formula B).

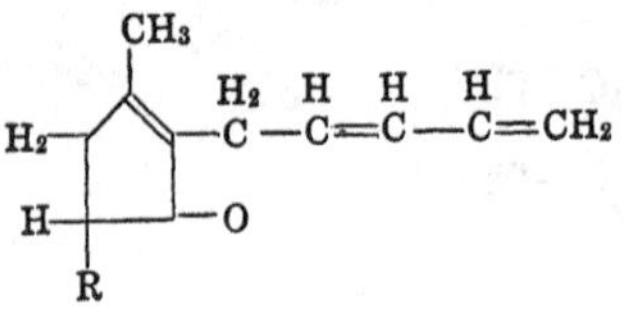

A 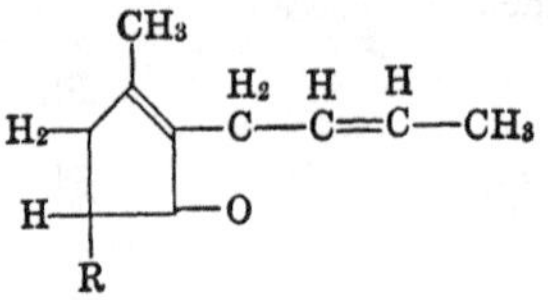   B

For pyrethrin I and cinerin I

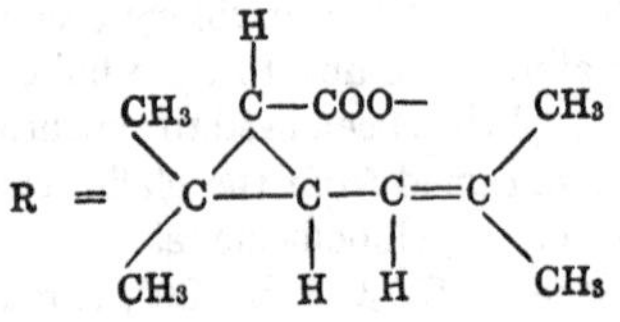

For pyrethrin II and cinerin II

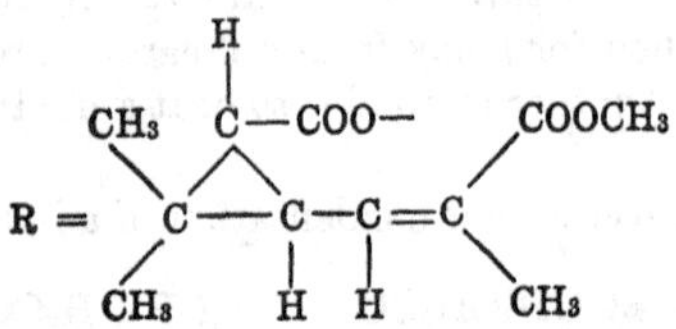

Many compounds have been found which increase the insecticidal activity of pyretheum (synergists). The most promising of these are piperonyl butoxide and piperonyl cyclonene (see piperonal) (Pyrenones). These compounded insecticides can be used with complete freedom where food is handled.[61]

---

[59] Frank et al. *J. Am. Chem. Soc.* **70**, 1379 (1948).

[60] LaForge, Barthel. *J. Org. Chem.* **1**, 38 (1936); **12**, 199 (1947).

[61] Mail. *Chem. Ind.* **61**, 218 (1947); **62**, 418 (1948).

1,2,3-Triketotetramethylcyclopentane, $\underset{\underset{CMe_2-CMe_2}{|}}{COCOCO}$ is blue. It forms a colorless hydrate.[62]

Pentaketocyclopentane, cyclopentanpentone, *leuconic acid,* exists as the tetrahydrate. The water cannot be removed without decomposing the compound. It probably exists with only one free ketone group.

$$\begin{array}{c} (HO)_2C-C(OH)_2 \\ |\qquad\qquad \backslash \\ |\qquad\qquad CO \\ |\qquad\qquad / \\ (HO)_2C-C(OH)_2 \end{array}$$

Leuconic acid is made by treating triquinoyl, the hydrate of hexaketocyclohexane, with a base to give a benzilic acid rearrangement, acidifying and heating and oxidizing the *croconic acid,* the enediol of tetraketohydroxycyclopentane. Leuconic acid gives a pentoxime and condenses with two mols of an *o*-diamine showing it to have two pairs of adjacent carbonyls as in diacetyl and similar compounds. As would be expected, the accumulated carbonyl groups give instability toward bases. Simply boiling with sodium carbonate solution gives oxalate and mesoxalate.

Xanthogallol is 1,2,3,3-tetrabromo-4,5-diketocyclopentane, resulting from the action of bromine water on tribromopyrogallol.[63] Treatment with a base and then with bromine water converts xanthogallol to oxalic acid and Br_5-acetone.

Methylated cyclopentanones can be made from the various methyladipic acids, many of which can be made from the cresols and xylenols (Reilly) with reduction followed by nitric acid oxidation. The presence of the alkyl groups favors ring closure. 3,3,4-Me_3-cyclopentanone has been found in acetone oil. Isoprene and 2,3-dimethylbutadiene react with CO under high pressure and temperature to give 3-methyl and 3,4-dimethylcyclopentanone.[64]

Cyclopentane carboxylic acids of greater or less complexity exist in the naphthenic acids of certain petroleums. The *monocarboxylic acid,* b. 215°, is obtained by heating the 1,1-dicarboxylic acid made from malonic ester and 1,4-Br_2-butane. It can also be made by CO_2 with cyclopentyl Grignard reagent which acts entirely normally. *The 1,2-dicarboxylic acid* is made from the tetracarboxylic ester obtained by the action of bromine on the disodio derivative of 1,3-trimethylene dimalonic ester. It exists as *cis,* m. 140°, and *trans,* m. 160°, acids. Heat converts both to the *cis* anhydride while heating with HCl converts the *cis* to the *trans* form.

Three *1,2,3-tricarboxylic acids* are obtained from $\alpha\alpha'$-Br_2-adipic esters, malonic ester and NaOEt. The *cis* form and the *cis trans cis* form are meso

[62] Shoppee. *J. Chem. Soc.* **1936**, 269.
[63] *Ann. Rep. Chem. Soc.* (London) 1917, 108.
[64] Raasch, Theobald. *C.A.* **42**, 2987 (1948).

while the *cis cis trans* form is racemic and can be resolved by means of the brucine salts.   The first and last readily give anhydrides.[65]

The naphthenic acids are obtained as sludges from the refining of petroleum with sulfuric acid and caustic.   The naphthenic acid content of petroleum varies from .1% (Pennsylvania) to 3% (California & Russia).   The potential supply of these acids is more than 200,000 tons annually.   In the United States the greater part of the naphthenic acids used are converted into metallic naphthenates for use as paint driers.   The copper salts are used to keep sand bags from rotting.

The acids include both mono and bicyclic types, and range from 120° C. (12 mm.) to 300° C. (<.1 mm.).   Cyclopentyl derivatives with the carboxyl group in the side chain are present in the greatest proportion.   Among the acids of this type identified are cyclopentyl-, 3-methylcyclopentyl-, 2,3-dimethylcyclopentylacetic, and 2-methylcyclopentylpropionic.

Apocamphoric acid, 2,2-Me_2-cyclopentane-1,3-dicarboxylic acid, and *camphoric acid,* its homolog with an extra methyl group in the 1-position, are important oxidation products in the terpene series.   Camphoric acid contains two asymmetric carbons and exists in four optically active forms.   The specific rotations are all about 48°.   The *d*- and *l*-camphoric acids, m. 187°, form anhydrides and are thus *cis* forms while the *d*- and *l*-isocamphoric acids, m. 171°, are *trans* forms.   By the action of Br_2 and sodium carbonate on camphoric acid, *laurolenic acid,* 1,2,3-Me_3-2-cyclopentene-1-carboxylic acid, is obtained.

Br-camphoric acid   Laurolenic acid

Reduction of camphoric anhydride changes one of the CO groups to CH_2 giving the lactone, *campholide.*   This reacts with KCN to give the nitrile of homocamphoric acid.   Distillation of its Ca salt gives camphor.

Homocamphoric acid   Camphor

[65] *Ann. Rep. Chem. Soc.* (London) 1921, 87; 1922, 77.

Isocampholic acid is 2,2,3-Me₃-cyclopentane-1-acetic acid.

Chaulmoogric acid, m. 68°, *hydnocarpic*, alepric and aleprylic *acid* are

$$CH = CH$$

ω-cyclopentenyl acids, | $CH(CH_2)_nCO_2H$, in which n is 12, 10, 8 and

$$CH_2-CH_2$$

6 respectively.[66, 67] Related acids containing the saturated cyclopentyl group have been made.[68]

α-**Campholytic acid,** 2,3,3-Me₃-cyclopentene-4-carboxylic acid, is believed to exist in intra-annular tautomerism with 1,2,2-Me₃-3-carboxy-bicyclo(2.1.0)-pentane.[69] (See pp. 567–8.)

D. Cyclohexane Derivatives

These are very widely distributed. They occur in the naphthenic portions of crude petroleums, in terpenes, camphors and many other natural products. Artificially they have been prepared in great numbers by reactions which take place intramolecularly in the 1,6-position. This includes reactions of two molecules having reactive 1,3-positions. First, two molecules combine and then the 1,6-process closes the ring.

The relation of *true hexamethylene and true benzene derivatives* is very close. Sometimes the transition from one to the other is spontaneous as in the tautomeric system in which phloroglucinol, 1,3,5-trihydroxybenzene, is in keto-enol equilibrium with 1,3,5-triketocyclohexane. Many aromatic compounds can be reduced by nascent hydrogen from Na and alcohol or from sodium amalgam. These fall in two general classes represented by the dibasic acids and the meta di- and tri-hydroxy compounds. The former contain a grouping like that in the aliphatic αβ-unsaturated ketonic compounds which are similarly reduced and the latter exist in equilibrium with ketones which are

[66] Stanley, Adams. *J. Am. Chem. Soc.* 51, 1515 (1929).

[67] Cole, Cardoso. *J. Am. Chem. Soc.* 61, 2349 (1939).

[68] Yohe, Adams. *J. Am. Chem. Soc.* 50, 1503 (1928).

[69] *Ann. Rep. Chem. Soc.* (London) 1922, 114.

readily reduced.

$$\text{O}=\text{C}\underset{\text{HO}}{\overset{\text{HO}}{\diagup}}\underset{\text{}}{\bigcirc}\underset{}{\overset{\text{OH}}{\diagdown}}\text{C}=\text{O} \qquad \text{OC}\underset{\text{CH}_2-\text{CO}}{\overset{\text{CH}_2-\text{CO}}{\diagdown}}\text{CH}_2$$

Other aromatic compounds can be converted to the corresponding hydro-
aromatic or cyclohexane derivatives by catalytic hydrogenation in presence of
Ni, Pt, Pd etc. Conversely, various methods of oxidation and dehydrogenation will change cyclohexane derivatives to aromatic compounds. The
interconversion of quinone and hydroquinone is one of the easiest and the best
known of these processes.

$$\text{HO}-\text{C}\overset{\text{CH}-\text{CH}}{\underset{\text{CH}=\text{CH}}{\diamond}}\text{C}-\text{OH} \quad \overset{\text{Oxid.}}{\underset{\text{Red.}}{\rightleftharpoons}} \quad \text{O}=\text{C}\overset{\text{CH}=\text{CH}}{\underset{\text{CH}=\text{CH}}{\diamond}}\text{C}=\text{O}$$

Dehydrogenation by means of selenium has been of the greatest value in
converting a multitude of naturally occurring cyclohexane compounds to the
more easily identified aromatic compounds. The "excess" hydrogen is removed as hydrogen selenide. In some cases sulfur can be used similarly but it
usually has the disadvantage of further reacting to form aromatic sulfur
compounds, whereas the less reactive selenium has less of this tendency. Such
methods of conversion to aromatic compounds have been valuable in the study
of the constitution of the sterols, the bile acids, the sex hormones, toad and
arrow poisons and various alkaloids.

Any series of changes which gives three double bonds in a 6-ring takes
place with special ease. Thus carvone changes readily to carvacrol under the
influence of acid.

$$\text{Me}-\text{C}\overset{\text{CH}-\text{CH}_2}{\underset{\text{CO}-\text{CH}_2}{\diamond}}\text{CH}-\underset{\text{Me}}{\overset{}{\text{C}}}=\text{CH}_2 \rightarrow \text{Me}-\text{C}\overset{\text{CH}-\text{CH}}{\underset{\text{HO}-\text{C}=\text{CH}}{\diamond}}\text{C}-\overset{\text{H}}{\underset{}{\text{CMe}_2}}$$

Acid causes the migration of the double bond from the isopropenyl group to the
ring where the enolization of the keto group supplies the third double bond to
give aromatic character to the ring.

The isomerism of cyclohexane compounds follows the usual principles. The
disubstitution products are 1,1-(*gem*); 1,2-(*ortho*); 1,3-(*meta*); and 1,4-(*para*).
The stereochemistry of mononuclear cyclohexane compounds is simplified by
the existence of the ring in only one form (p. 527–8). Existence of *cis* and
trans rings occurs in compounds containing two or more cyclohexane rings.

X-Ray studies of *trans*-1,4-Br$_2$-cyclohexane indicate that the ring is of the *trans* or *seat* type.[70]

The *gem*-disubstituted cyclohexanes have one group above and the other below the plane of the ring. The other isomers can have *cis* and *trans* forms. If the two groups are alike the *cis* forms are all incapable of existence in optically active forms but the *trans* forms of the *ortho* and *meta* compounds have no plane of symmetry and can exist in enantiomorphic optical isomers. The *trans para* compounds are symmetrical.

Saturated Hydrocarbons of the Cyclohexane Series

Cyclohexane, hexamethylene, hexahydrobenzene, $(CH_2)_6$, m. 6.5°, b. 81°, occurs in various petroleums.[71] It is made from benzene by hydrogenation with Ni at about 200° or with platinum black in glacial acetic acid at 20°. It can also be obtained from many of its derivatives by ordinary reactions. In most of its reactions it resembles the paraffins. Bromine substitutes in the light. Anhydrous aluminum chloride at 150° gives no change. Hot permanganate gives adipic acid. Fuming sulfuric acid converts it to benzene sulfonic acid. These reactions show it to be less stable and more reactive than cyclopentane. At temperatures above 200°, Pt or Pd catalysts give dehydrogenation to form benzene in good yields. On the other hand, Ni at 280° gives benzene and *methane*.[72] Cyclohexane with HI or with Al_2O_3 at high temperatures rearranges to Me-cyclopentane.[73]

Methylcyclohexanes are made by the hydrogenation of toluene, the xylenes, etc. The mono compound, b. 103°, and the 1,3-Me$_2$-compound occur in naphthenic petroleums. The Me$_2$-cyclohexanes have been made in all seven theoretically possible forms. They all boil near 120°. The 1,3-dimethylcyclohexanes are reversed in their boiling points, the *cis* boiling lower than the *trans*.[74] The 1,2,4- and 1,3,5-Me$_3$-compounds have been obtained in *cis* and *trans* forms, b. about 140°. The homologs of cyclohexane with longer side chains behave differently with aluminum chloride than do the corresponding benzene derivatives. Thus Et-cyclohexane gives a mixture of Me$_2$-cyclohexanes while Et-benzene gives mainly Et$_2$-benzene and benzene. Similarly *n*-Pr- and *n*-Bu-cyclohexanes give Me$_3$- and Me$_4$-cyclohexanes respectively.[75]

p-Methylisopropylcyclohexane, *p-menthane*, b. 170°, is the parent substance to which many of the important terpenes and terpene alcohols are related. It occurs, to a small extent, in the extract from the longleaf pine stump and can be prepared by hydrogenating *p*-cymene (*p*-Me-isoPr-benzene) in which case

[70] Hassel. *Z. Elektrochem.* **37**, 540 (1931).

[71] Brunn, Hicks-Brunn. *Bur. Standards J. Research* **7**, 607 (1931).

[72] Egloff. "The Reactions of Pure Hydrocarbons." Reinhold Publishing Co., 1937.

[73] Ipatieff, Dovgelevich. *J. Russ. Phys. Chem. Soc.* **43**, 1413 (1911).

[74] Pitzer, Beckett. *J. Am. Chem. Soc.* **69**, 977 (1947).

[75] Grignard, Stratford. *Compt. rend.* **178**, 2149 (1924).

both the *cis* and *trans* isomers are formed.

p-Menthane

The meta isomer is related to a few unusual terpenes (sylvestrenes).

The *aliphatic* hemiterpenes, C_5H_8, and monoterpenes, $C_{10}H_{16}$, such as isoprene and myrcene and their derivatives such as geraniol and linalool have already been discussed.

Unsaturated Hydrocarbons of the Cyclohexane Series

Mono-olefins. *Cyclohexene*, b. 83°, is prepared in the usual way from the alcohol or the halide.   It shows all the usual olefin reactions except that it is slightly more reactive than the open chain $RCH = CHR'$ type, presumably because the ring facilitates the "opening" of the double bond.   Acetyl chloride and $AlCl_3$, at low temperatures, convert it to 1-chloro-2-acetylcyclohexane. At higher temperatures a peculiar reduction takes place to give methyl cyclohexyl ketone.[76]   Cyclohexene has been proposed as a standard to measure the auto-oxidation of cracked gasolines and the efficiency of inhibitors against oxidation.   Three of the four mono-olefins of methylcyclohexane are obtained by dehydration of the hexahydrocresols, 1-Me- b. 111°, 3-Me- b. 102°, and 4-Me- b. 103°.   The best preparation of methylenecyclohexane, b. 103°, is the pyrolysis of cyclohexylcarbinyl acetate.[77]   The latter has a semicyclic double bond.   It readily isomerizes to give 1-Me-cyclohexene.

α- and *β*-cyclogeraniolenes, 1,1,3-Me₃-3 (and 2)-cyclohexene are obtained in the ratio 3:1 by the action of sulfuric acid with geraniolene,

$$Me_2C = CHCH_2CH = CMe_2$$

and its isomers.[78]

A most important process for making complex cyclohexene derivatives in great variety is the Diels-Alder reaction (1928) between a conjugated diene and an *αβ*-unsaturated carbonyl compound.   The simplest possible case is between acrolein and butadiene to give 1,2,3,6-tetrahydrobenzaldehyde.   The net result can be explained on the basis that the *αβ*-double bond *opens and*

[76] Nenitzescu, Cioranescu.   *Ber.* **69B**, 1820 (1936).

[77] Arnold.   *J. Am. Chem. Soc.* **70**, 2590 (1948).

[78] Simonsen.   "The Terpenes," Vol. 1, 2nd Ed.   Cambridge, 1947.   p. 107.

adds to the conjugated diolefin in the 1,4-position, forming a new double bond in the 2,3-position.

$$CH=CH_2 \quad CH_2 \qquad CH-CH_2-CH_2$$
$$| \qquad\quad + \quad || \qquad \rightarrow \quad ||$$
$$CH=CH_2 \quad CH-CHO \qquad CH-CH_2-CH-CHO$$

The conjugated diolefin group can be supplied by ordinary hydrocarbons or by compounds such as furan and N-Me-pyrrole. In the latter cases the resulting cyclohexene ring has an O or N bridge in the 1,4-position. The $\alpha\beta$-unsaturation can be supplied by aldehydes, ketones, acids, and even by substances like benzoquinone. The latter can react with one or two molecules of diolefin. The commonest $\alpha\beta$-compound used is maleic anhydride. Hydrogenation or dehydrogenation converts the products to derivatives of cyclohexane or of benzene.

The six possible *p-menthenes* are known. They are named as follows:

1- 2- 3- 1(7)- 4(8)- 8-

The latter is also designated as 8(9)- *p*-menthene. The fourth isomer can be called 7- without ambiguity. The most common *menthene* is the 3 isomer, b. 168°, obtained from menthol (3-hydroxy-*p*-menthane) by dehydration or through the xanthate or the chloride. It exists in *d-*, *l-* and inactive forms. Oxidation gives β-Me-adipic acid. The menthenes are not found in nature as are the menthadienes but they may be prepared by selective hydrogenation of the latter. The most readily prepared is the *1*-isomer (carvomenthene).

Diolefins. Mixtures of 1,3- and 1,4-cyclohexadiene (1,2- and 1,4-dihydrobenzenes) are obtained by elimination reactions from both 1,2- and 1,4-disubstituted cyclohexanes. Oxidation gives oxalic, succinic and malonic acids. The 1,3-isomer gives a stable dimer, 1,4-endoethylene-1,4,5,6,9,10-H_6-naphthalene (I) by the ordinary process for the dimerization of a conjugated diene.[79]

[79] *Ann. Rep. Chem. Soc.* (London) **1932**, 112.

1,3-Cyclohexadiene can be obtained by heating quinitol and anhydrous MgSO$_4$.[80] Butadiene undergoes the Diels-Alder reaction with itself to give the dimer, 4-vinylcyclohexene. This is a by-product of the butadiene synthesis. The p-menthadienes (p-terpadienes) occur widely in plant products. They are usually characterized by conversion to the hydrohalides, tetrabromides, or nitrosyl chlorides. The common forms follow:

| 1,3-
α-Terpinene
b. 175° | 1,4-
γ-Terpinene
b. 183° | 1,5-
α-Phelladrene
b. 175° | 1(7),2-
β-Phelladrene
b. 171° |

| 1,4(8)-
Terpinolene
b. 188° | 1,8(9)-
Limonene
Dipentene (dl)
b. 176° | 1(7),3-
β-Terpinene
b. 174° | 2,8(9)-
Isolimonene
b. 173° |

The addition of two isoprene molecules to give a p-menthadiene may be a special case of polymerization[81] under the influence of H$^+$. The addition of the H$^+$ to a conjugated diene may be expressed electronically as follows:

$$\ddot{C} :: C : C :: \ddot{C} \xrightarrow{\text{H}^+} H : \ddot{C} : C : C :: \ddot{C} \rightarrow H : \ddot{C} : C :: C : \ddot{C}$$

$$CH_2 = CH - CH = CH_2 + H^+ \rightarrow CH_3 \underset{+}{C}HCH = CH_2 \rightarrow CH_3 - CH = CH - CH_2{}^+$$

in which the $^+$ indicates the carbon with only 6 electrons which is in a condition to unite with a negative ion or to add to one carbon of an olefinic linkage, leaving the other with only 6 electrons.

$$CH_2 = C(Me) - CH = CH_2 \xrightarrow{\text{H}^+} CH_3 - \overset{+}{C}(Me) - CH = CH_2 \rightarrow$$
$$CH_3 - C(Me) = CH - CH_2{}^+$$

This fragment can then add to another molecule of isoprene *exactly as the H$^+$ added to the first*. The same electronic shift would give

$$CH_3C(Me) = CH - CH_2 - CH_2 - C(Me) = CH - CH_2{}^+$$

[80] Zelinskii, Denisenko, Eventova. *Compt. rend. acad. sci. U.R.S.S.* 1, 313 (1935); *C.A.* 29, 4435 (1935).

[81] Whitmore. *Ind. Eng. Chem.* 26, 94 (1934).

The electronically deficient carbon is now in the 1,6-position to a double bond and adds to it to give a p-menthane skeleton with a double bond carbon 1 and the 8-carbon with only 6 electrons. This can be stabilized by loss of H^+ from carbon 9 or 4.

Addition and loss of H^+ could give all the isomeric dienes. The above mechanism explains the synthesis of terpenes by the acid polymerization of isoprene but it does not explain the ability of individual plants to make specific menthadienes. Thus citrous fruits produce d-limonene while coniferous trees produce the l-form. The closely related terpinolene occurs in coriander oil. Isoprene has not been established as a plant constituent. The close relation of this series to the aromatic series is shown by the ready conversion of the p-menthadienes to p-cymene by dehydrogenation.

Moslene is 2,4- or 2,5-menthadiene.[82]

Sylvestrene, 1,8(9)-m-menthadiene, was formerly believed to be a plant product. It is now known to be formed by the cleavage of the cyclopropane ring of either 3- or 4-carene.

The menthadienes or monocyclic terpenes give the reactions of two double bonds forming *tetrabromides, dihydrochlorides, dihydrobromides,* di-*nitrosochlorides* (with NOCl or with EtONO, acetic acid and HCl), *nitrosites* (by addition of NO and NO_2 from N_2O_3) and *nitrosates* (by addition of 2 NO_2). Most of these are crystalline compounds with definite melting points. The *nitrolamines* (from amines and nitroso chlorides) are also valuable for characterization purposes. The nitroso chlorides have the blue color of monomolecular nitroso compounds only when addition takes place in the 4,8-position. Such products are volatile with steam. In other cases the nitroso compound changes to the colorless bimolecular form or to the isonitroso compound, $CH-NO \rightleftharpoons C=NOH$.

Menthadienes can be converted to terpene dimer of unknown structure by acid polymerization. This dimerization is usually accompanied by dehydrogenation.

Triolefins. 1,3,5-p-Menthatriene is p-cymene. 2-Me-2,6,8-p-menthatriene has been obtained from the reaction product of MeMgX with carvone, 2-keto-6,8-p-menthadiene.

Various *semi-benzenes,* containing two double bonds in the ring and a third

[82] *Ann. Rep. Chem. Soc.* (London) 1921, 89.

as a semi-cyclic double bond, have been obtained.[83]   In presence of acid these readily rearrange to aromatic compounds.

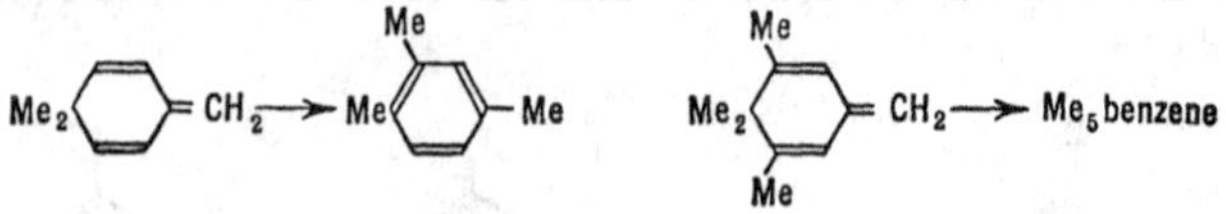

Alcohols

Cyclohexanol, hexahydrophenol, hexalin, m. 20°, b. 161°, is made by the catalytic hydrogenation of phenol with Ni at about 170°.   It is produced commercially as a solvent and as an intermediate in the manufacture of specific acids.   It gives all the usual secondary alcohol reactions.   Gentle oxidation gives the ketone while more vigorous action forms adipic acid $(CH_2)_4(CO_2H)_2$.

A mixture of cyclohexanols can be obtained by the hydrogenolysis of wood waste or similar lignin-containing materials using copper chromium oxide catalyst and high temperature and pressure.

Quinitol, 1,4-cyclohexandiol, obtained from hydroquinone, exists as *cis*, m. 102°, and *trans*, m. 139°.   These taste sweet.   Oxidation gives quinone. The *o*- and *m*-isomers are also known.   The 1,2,3-triol from pyrogallol exists in three stereo forms, m. 108°, 124° and 148°.   The 1,3,5-triol, *phloroglucitol*, m. 184°, is made from phloroglucinol by Na_xHg reduction.

Quercitol, $CH_2(CHOH)_5$, pentahydroxycyclohexane, m. 234°, $[\alpha]^D = +24°$, occurs in plant products.   Reduction with HI gives various products including pyrogallol, phenol and benzene.   Oxidation gives malonic acid, mucic acid (configuration $+--+$) and L-trihydroxyglutaric acid $(+--)$. The possible configurations for quercitol follow:

(*a*) a form with the 5 hydroxyls *above* the plane of the ring.   The other possibilities will have 1 or 2 of the hydroxyls *below* this plane.   With the CH_2 group as No. 6 these possibilities will be (*b,c,d*) with the 1,2, *or* 3-OH below; (*e,f,g,h*) with the 1-OH *and* the 2,3,4 or 5-OH *below;* (*i,j*) with the 2-OH *and* the 3 *or* 4-OH *below.*   A study of the models shows that only configuration (*g*) with the 2,3,5-hydroxyls on one side of the plane and the 1,4-hydroxyls on the other can be optically active and contain the configurations in mucic and L-(OH)$_3$-glutaric acid.

Inositol, hexahydroxycyclohexane, $(CHOH)_6$, occurs widely in nature either free or in combination.   It may be formed by the condensation of 6 molecules of formaldehyde.   The final change would be the condensation of CH in the 6-position with CO in the 1-position to close the ring.

The possible stereoisomeric forms of inositol are: (a) with all groups *above* the plane of the ring, (b) with *one* group *below* the plane; (c,d,e) with *two*

[83] *ibid.* 1922, 96.

groups *below* in *ortho*, *meta* or *para* position to each other, and (*f*,g,h) with *three* groups below in *vicinal* (1,2,3), *symmetrical* (1,3,5), or *unsymmetrical* (1,2,4) arrangement in respect to each other. Of these all have a plane of symmetry except the last which is the only one capable of optical activity. Thus the naturally occurring optically active inositols have configuration (h) in which the 1,2,4-hydroxyls are on one side of the ring and the 3,5,6- are on the other. It is notable that this corresponds to the configuration of active quercitol.

Meso-inositol, m. 225°, occurs free in muscle and in a variety of other animal and plant products. It is essential for the growth of yeast. It is widely distributed in plants as *phytin*, an acid hexaphosphoric ester of inositol which occurs as salts of Ca, Mg, K, etc. Because of the formation of allomucic acid (+ + + +) by the oxidation of meso-inositol and because its monophosphoric ester cannot be resolved into optically active material, the configuration (b) is adopted. Configurations (d) to (h) could not give the + + + + configuration of allomucic acid. Configuration (c) would give unsymmetrical monophosphoric esters. *d-* and *l-Inositols* have been made respectively by the hydrolysis of the monomethyl ethers *pinitol* and *quebrachitol*. Other methyl ethers are *bornesitol* and *dambonitol*.

Cocositol is an optical isomer of inositol.[84]

p-**Menthanols.** The most important, *menthol*, 3-hydroxy-*p*-menthane, m. 43°, b. 216°, $[\alpha]^D = - 49.7°$, occurs in peppermint oil. Hydrogenation of thymol, 3-OH-*p*-cymene, gives *dl*-menthol and *dl*-neomenthol. The three asymmetric carbons in menthol make possible 8 optical isomers all of which are known.[85] Although menthol may be synthesized by the above reduction, the problem of separating the isomer has made these procedures unsuitable for commercial exploitation. Cyclization of citronellal with acid catalysts to isopulegol followed by reduction gives a simpler mixture of isomers.[86] The presence of other geometric forms in menthol impart bitterness and impair its physiological activity.

l-**Menthylamine,** 3-amino-*p*-menthane, is obtained along with three optical isomers by the reduction of the oxime of *l*-menthone. It is a strong base and is very useful in resolving *dl*-acids.

Carvomenthol, 2-OH-*p*-menthane, is obtained by vigorous reduction of carvone.

p-**Menthadiols.** The most important of these is 1,8-*terpin*, 1,8-*p*-menthadiol, obtained as its hydrate by the treatment of turpentine with dilute acid. The *cis*-form of 1,8-terpin, m. 104°, is obtained from the hydrate on standing over sulfuric acid or on heating. Heating with acid gives 1,8-cineole or eucalyptole, m. 1°, b. 174°, which is found in many ethereal oils and 1,4-cineole, m. −46°, b. 172°. Monocyclic terpenes and terpene alcohols are by-products

[84] Manske. *Can. J. Research* **19B**, 34 (1941).
[85] Simonsen. "The Terpenes," Vol. 1. Cambridge, 1947. p. 230 ff.
[86] Glass. *C. A.* **32**, 5160 (1938); Terwilliger. *C. A.* **32**, 5160 (1938).

of this reaction.

1,8-Cineole   1,4-Cineole

Cineoles form oxonium salts even with such weakly acid substances as phenol. *trans*-1,8-Terpin is known, m. 157°. It does not form a hydrate.

Terpin hydrate, m. 121°, is produced commercially by hydration of turpentine with dilute acid. Its principal use is in pharmaceuticals. The melting point of the hydrate is not surprising. For instance, Me_5-ethanol, m. 17°, forms a hydrate, m. 83°.

1,8-Terpin hydrate has been made by the action of 3 MeMgX on cyclohexanone-4-carboxylic ester and by ring closure of geraniol or nerol,

$$Me_2C = CH(CH_2)_2C(Me) = CHCH_2OH,$$

with acid. Since nerol reacts more rapidly it is assigned the *cis* structure.

1,4-Terpin, 1,4-*p*-menthadiol, has been made in *cis* and *trans* forms. It gives 1,4-cineole, b. 174°.

p-Menthenols. Many of these occur in nature or have been produced artificially.

	Position of OH	Position of double bond	m.	b.	$[\alpha]_D$
α-Terpineol	8	1	36°	219°	+98.5°
β-Terpineol	1	8	33°	210°	
γ-Terpineol	1	4(8)	70°		
1-Terpineol	1	3		210°	
4-Terpineol	4	1		212°	+25.4°
Piperitol	3	1		103°±(19 mm.)	
Dihydrocarveol	2	8			
Isopulegol	3	8			

Although the terpineols occur widely in ethereal oils, the chief source is the longleaf pine stump. The terpineols are presumably formed in the stump by slow hydration of the terpene hydrocarbon originally present. Pine oil is a mixture of terpene alcohol with α-terpineol predominating. Synthetic pine oils are made by the acid-catalyzed hydration of turpentine and by the dehydration of terpin hydrate. The stepwise oxidation of α-terpineol is as follows:

| | Homoterpenylic acid | Terpenylic acid | Terebic acid |

Terpenylic acid is the lactone of 2-OH-isobutane-1,1-diacetic acid and *terebic acid* is the γ-lactone of 1-carboxy-2-OH-isobutane-1-acetic acid. An *aldehyde* of this series, found in nature, is *perilla aldehyde* having CHO in place of 7-Me of limonene, 1,8-menthadiene. Its anti-oxime (Peryllartine) is the sweetest substance known, being some 2000 times as sweet as sucrose. The taste-principle of saffron, *picrocrocine*, is a glucoside of 1-aldehydo-2,6,6-Me$_3$-4-OH-cyclohexene.[87]

Ketones

Cyclohexanone, b. 156°, is best prepared from the alcohol by catalytic dehydrogenation with Cu at 250°. It gives all the usual ketone reactions. It condenses with 2 PhCHO to give dibenzalcyclohexanone (I) m. 118°. The yellow color is due to the accumulation of conjugated double bonds (cf. phenyl fulvenes).

Cyclohexanone contains about 8% of the enol form. It does not give a color with FeCl$_3$, however. Caro's acid opens the ketone ring giving the lactone of ω-hydroxycaproic acid. This reaction is general for cyclic ketones. In sunlight, the aqueous alcoholic solution of cyclohexanone gives caproic acid or CH$_2$=CH(CH$_2$)$_3$CHO. The first change is like the similar splitting of acetone to give methane and acetic acid. The second would correspond to a splitting of diethyl ketone to give ethylene and propionaldehyde. With an excess of CH$_2$O and lime, cyclohexanone gives 2,2,6,6-tetramethylol-1-cyclohexanol.[88]

Isophorone, b. 215°, m. −8°, is readily prepared by the basic condensation of acetone. It is the most powerful ketone solvent for nitrocelluose known.

[87] *Ann. Rev. Biochem.* 460 (1934).
[88] *Ann. Rep. Chem. Soc.* (London) 1923, 92.

Diketones

The 1,3-isomer, *dihydroresorcinol*, m. 105°, is made by reducing resorcinol with Na_xHg in solution kept neutral by CO_2 or with H_2 and Ni in alkaline solution.[89] This is the behavior to be expected of an $\alpha\beta$-unsaturated ketone.

$$
\begin{array}{c}
\text{OH} \\
| \\
\text{C} \\
\diagup \quad \diagdown \\
\text{HC} \qquad \text{CH} \\
| \qquad \quad || \\
\text{HC} \qquad \text{C—OH} \\
\diagdown \quad \diagup \\
\text{CH}
\end{array}
\rightleftharpoons
\begin{array}{c}
\text{O} \\
|| \\
\text{C} \\
\diagup \quad \diagdown \\
\text{HC} \qquad \text{CH}_2 \\
|| \qquad \quad | \\
\text{HC} \qquad \text{C=O} \\
\diagdown \quad \diagup \\
\text{CH}_2
\end{array}
\xrightarrow{[2\,H]}
$$

$$
\begin{array}{c}
\text{CO} \\
\diagup \quad \diagdown \\
\text{H}_2\text{C} \qquad \text{CH}_2 \\
| \qquad \quad | \\
\text{H}_2\text{C} \qquad \text{CO} \\
\diagdown \quad \diagup \\
\text{CH}_2
\end{array}
\rightleftharpoons
\begin{array}{c}
\text{CO} \\
\diagup \quad \diagdown \\
\text{H}_2\text{C} \qquad \text{CH} \\
| \qquad \quad || \\
\text{H}_2\text{C} \qquad \text{C—OH} \\
\diagdown \quad \diagup \\
\text{CH}_2
\end{array}
$$

The *1,4-isomer*, m. 78°, can be made from its dicarboxylic ester formed from succinylosuccinic ester by acid ring closure.

$$2\,RO_2CCH_2CH_2CO_2R \xrightarrow{\text{NaOEt}}$$

$$RO_2CCH_2CH_2COCH(CO_2R)CH_2CO_2R \rightarrow$$

$$
\begin{array}{c}
RO_2C\text{—}CHCOCH_2 \\
| \qquad\qquad | \\
CH_2COCH\text{—}CO_2R
\end{array}
\rightarrow
\begin{array}{c}
CH_2COCH_2 \\
| \qquad\quad | \\
CH_2COCH_2
\end{array}
$$

The 1,2-isomer has not been obtained.

Dimethyldihydroresorcinol, 5,5-Me$_2$-1,3-cyclohexanedione, methone, dimedone, $CH_2(COCH_2)_2C(CH_3)_2$, is made from mesityl oxide, malonic ester and NaOEt. First, malonic ester adds to the $\alpha\beta$-double bond giving $MeCOCH_2C(Me_2)CH(CO_2R)_2$. The α-Me is 1:6 to a carboxyl and undergoes an internal acetoacetic ester condensation with it. Treatment of the resulting cyclic carboxylic ester with KOH and HCl gives methone, an important reagent for identifying aldehydes by condensation of the carbonyl of the latter with the methylene groups of two methone molecules to give

$$RCH[CH(COCH_2)_2CMe_2]_2.$$

[89] "Org. Syntheses."

The *2,2-Br₂-compound* is a special brominating agent which gives no HBr. It reacts with dimethylaniline as follows:

$$C_6H_5NMe_2 + Br_2C(COCH_2)_2CMe_2 \rightarrow BrC_6H_4NMe_2 + BrC\underset{\diagdown COCH_2}{\overset{\diagup C\text{—}CH_2}{\diagup}}\overset{OH}{\diagdown}CMe_2$$

The resulting enol can be converted back to the dibromide for reuse. The mechanism is like that for enolization except that a positive Br is expelled from the α-C instead of a positive H.[90]

$$O=\underset{|}{\overset{|}{C}}\text{—}\underset{|}{\overset{|}{C}}\text{—Br} \xrightarrow{H^+} \left[H\text{—}O\text{—}\underset{|}{\overset{+}{C}}\text{—}\underset{|}{\overset{|}{C}}\text{—Br} \right] \rightarrow Br^+ + H\text{—}O\text{—}\underset{|}{\overset{|}{C}}=\underset{|}{\overset{|}{C}}$$

Phloroglucinol, in its keto form, is *1,3,5-cyclohexatrione*. *Angustione* is a triketone, 4,4,6-Me₃-6-acetyl-1,3-cyclohexanedione. It gives the typical reactions of a β-diketone, ferric chloride reaction, copper compound, loss of acetyl with alkali, etc.[91]

Hexaketocyclohexane, cyclohexanehexone, *triquinoyl*, (CO)₆, is obtained in hydrated form by oxidizing hexahydroxybenzene. With alkali, it undergoes the benzilic acid rearrangement to give 1-OH-tetraketocyclopentane-1-carboxylic acid.

Menthone, b. 207°, and **carvomenthone** are the 3- and 2-keto-*p*-menthanes corresponding to menthol and its isomer.

Menthone chemistry.[92]

Unsaturated Ketones

Resorcinol in its ketonic form acts as *1-cyclohexen-3,5-dione*. Ring closure gives 1-acetyl-4-Me-cyclohexene from MeCO(CH₂)₃CH(Me)CH₂CHO, obtained from citronellal and O₃ (S I 61). The first step is an internal aldol condensation between the aldehyde group and the alpha H in the 6-position.

Pulegone, b. 224°, is 3-keto-4(8)-*p*-menthene. Reduction gives menthol. Heating with water at 120° gives acetone and 2-Me-cyclohexanone in the same way that citral gives acetaldehyde and methylheptenone. *Isopulegone* has an 8(9) double bond and is related to isopulegol obtained by ring closure of citronellal CH₂=C(Me)(CH₂)₃CH(Me)CH₂CHO. Treatment of citronellal with acetic anhydride gives *isopulegol acetate*, a ring formation typical of the

---

[90] Robinson. *Ann. Rep. Chem. Soc.* (London) 1922, 102.

[91] *Ann. Rep. Chem. Soc.* (London) **1930**, 166.

[92] Read. *Chem. Revs.* **7**, 1 (1930).

terpene series.

This change involves an *intramolecular* aldol condensation of an H which is alpha to an ethylenic linkage. Such an aldol condensation would be most improbable *intermolecularly* because the more reactive H alpha to carbonyl would react by preference.

Piperitone, 1-*p*-menthen-3-one, b. 236°, occurs in certain eucalyptic oils. This $\alpha\beta$-unsaturated ketone condenses with benzaldehyde to give a benzal compound involving the Me, thus illustrating the activity of H alpha to a double bond which is $\alpha\beta$ to a carbonyl group as in $CH-C=C-C=O$[93, 94].

Carvone, 1,8-*p*-menthadien-6-one, b. 231°, is found widely in plants. On heating with acids, it gives carvacrol, 2-OH-*p*-cymene, thus showing its carbon skeleton and the position of the oxygen. The 1-double bond is $\alpha\beta$ to the carbonyl group, and consequently easily reduced. Oxidation of the *dihydro-carvone* thus obtained with $KMnO_4$ gives 4-acetyl-1-methyl-2-ketocyclohexane, thus showing the position of the 8-double bond. In carvone, the latter can be saturated by adding HBr and replacing the Br by H to give carvotanacetone, 1-*p*-menthen-6-one, which gives pyruvic acid and isopropyl succinic acid on oxidation thus proving the position of the 1-double bond.

If HBr is added to carvone at 8(9) and then removed by alkali, the product is the 7-membered ring *eucarvone*, 2,6,6-Me₃-2,4-cycloheptadien-1-one. Surprisingly, if the 1-double bond is first reduced and HBr then added at 8(9), its removal does not widen the 6-ring but closes a 3-ring at the 3-position giving carone. This is like the formation of cyclopropyl cyanide from γ-Br-propyl cyanide (p. 532).

Eucarvone   Carone

A diketone (I) or its mono-enol exists in the *p*-menthane series in *diosphenol* or *Buchocamphor*, m. 83°. The 2,3-position of the oxygen atoms is shown by its preparation by the action of dilute bases on the $\alpha\alpha$-Br₂-compounds of both

[93] Simonsen. "The Terpenes," Vol. I, 1934. p. 361.
[94] *Ann. Rep. Chem. Soc.* (London) 1921, 90.

menthone and carvomenthone.

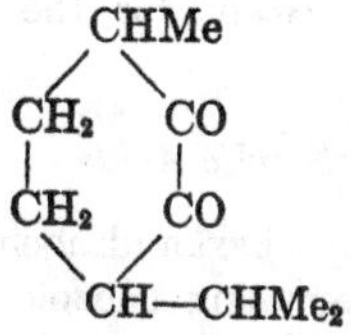

(I)

Such a compound could exist in two mono-enol and one di-enol form. Since it forms only one monoxime, it apparently exists in only one mono-enol form or else one of the two mono-enols reacts much more rapidly with hydroxylamine. Another possibility would be that the carbonyl *ortho* to the methyl group reacts more rapidly than that *ortho* to the isopropyl.

The artificial violet perfumes, α- and β-*ionones*, belong to this series. They are obtained by an ordinary aldol condensation of citral with acetone to give pseudo-ionone, followed by an acid ring closure. The conversion of the 1,5-diene to a 6-ring by acid is an internal *polymerization*[95] initiated by H+.

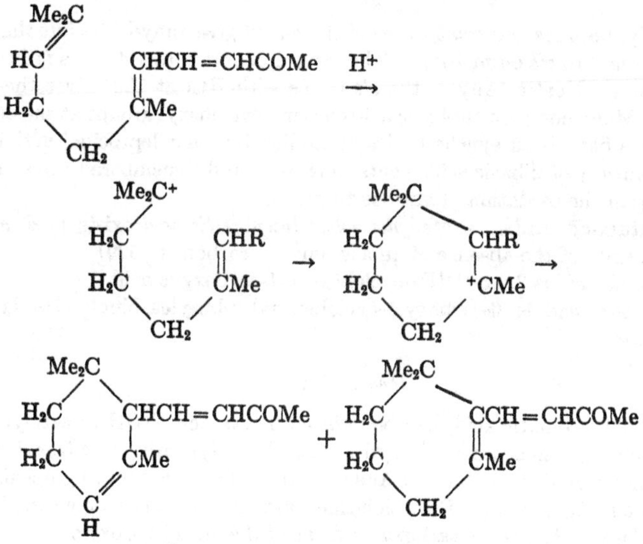

α-Ionone   β-Ionone

The isomers are separated by means of the $NaHSO_3$ compounds of which the α- is less soluble in NaCl solution. The previously accepted cycloheptene structure for the irones has been shown recently to be wrong. α-Irone, 4-(2,5,6,6-tetramethyl-2-cyclohexen-1-yl)-3-buten-2-one, and β-irone, 4-(2,5,6,6-

[95] Whitmore. *Ind. Eng. Chem.* **26**, 94 (1934).

tetramethyl-1-cyclohexen-1-yl)-3-buten-2-one, have the same structure as the corresponding α- and β-ionones except for the additional methyl in the 5-position on the ring.[96]

Carboxylic Acids

Cyclohexanemonocarboxylic acid, hexahydrobenzoic acid, $(CH_2)_5CHCO_2H$, m. 31°, b. 233°, is obtained by reducing benzoic acid by Na and EtOH or catalytically. It is better prepared by the action of NaOBr on methyl cyclohexyl ketone obtained from acetyl chloride, cyclohexene and $AlCl_3$. The dicarboxylic acids have been extensively studied.[97-99] They are obtainable from the aromatic dibasic acids by reduction with Na or catalytically.

	Cis m.	*Trans* m.	*Trans* $[\alpha]^D$
H₆-Phthalic acid.................	192°	215°	18.5°
H₆-Isophthalic acid...............	163°	148°	23.4°
H₆-Terephthalic acid..............	161°	200°	0

Strangely, both *cis* and *trans* forms of the *o*-acid give anhydrides but the *trans* form changes to *cis* on melting. The *trans*-1,3- and both 1,4-acids fail to give anhydrides. Heating any of the *cis* forms with HCl at 180° gives the *trans* forms. Many complex aliphatic acids having cyclohexyl groups at the ends of the chains have been synthesized and studied for their leprocidal activity.[100] A wide variety of dibasic acids containing 3-, 4- and 5-membered rings are obtained from the oxidation of various terpenes.[101]

Unsaturated acids. *p-Methylcyclohexylideneacetic acid* exists in *d*- and *l*-forms in spite of the absence of an asymmetric carbon (p. 529).

Shikimic acid is 3,4,6-(OH)₃-cyclohexene-1-carboxylic acid.

Sedanolic acid is 6-carboxy-1-cyclohexenyl-*n*-butylcarbinol. Its lactone is *sedanolide*.

Hydroxy Acids

Although aromatic acids can be reduced by Na and EtOH or catalytically and although phenols are easily hydrogenated catalytically, *o*-hydroxybenzoic acid, salicylic acid, with Na and AmOH gives pimelic acid by ring opening.

Quinic acid, 1,3,4,5-(OH)₄-cyclohexanecarboxylic acid is widely distributed among plants. It has the carboxyl and one of the meta hydroxyls on the same side of the plane of the ring since a stable lactone is formed. The *m'*- and

[96] *Helv. Chim. Acta* 30, 1807 (1947).

[97] Baeyer. *Ber.* 19, 1797 (1886); *Ann.* 245, 103 (1888); 258, 145 (1890); 269, 145 (1892).

[98] Baeyer, Villiger. *Ann.* 276, 255 (1893).

[99] *Ann. Rep. Chem. Soc.* (London) 1905, 119.

[100] Adams, Stanley, Stearns. *J. Am. Chem. Soc.* 50, 1475 (1928).

[101] Simonsen. "The Terpenes," Vols. I and II, 1931.

p-hydroxyls are also *cis* to each other since they condense with acetone to give a 5-ring acetal which is not possible with *trans-o*-hydroxyls. Methylation of the acetone compound followed by hydrolysis of the acetal gives a *m*-methoxy-$m'p$-dihydroxy acid which does not form a lactone. Thus the m'-OH is *trans* to the carboxyl and quinic acid has the carboxyl and one meta hydroxyl on one side of the ring while the other three hydroxyls are on the other.

An acid of the cyclohexane series containing a 1,4-ether linkage is *cantharidin*, m. 218°, (I) from Spanish flies, the anhydride of 1,2-Me$_2$-1,2-dicarboxy-3,6-endoxycyclohexane. A related substance (II) can be made by the Diels-Alder reaction between maleic anhydride and furan, the latter supplying the diene system.

```
    CH    Me                    CH
H2C      C—CO              HC       CHCO
       O        O                O        O
H2C      C—CO              HC       CHCO
    CH    Me                    CH
        (I)                        (II)
```

E. Bicyclic Terpenes and Camphors

These important natural substances may be compared to p-menthane. The more important parent hydrocarbons are

```
        Me                          Me
CH2—CH —CH2                  CH —CH —CH2
CH2—CH —CH2                  CH2—C ——CH2
     CHMe2                        CHMe2
   p-Menthane              Sabinane (Thujane)
                                  (I)

     Me                    Me                      Me
CH2—CH —CH2          CH —CH —CH2          CH2—C ——CH2
 Me2C     |             CMe2  |              CMe2  |
CH2—CH—CH            CH2—CH—CH2          CH2—CH—CH2
   Carane               Pinane              Camphane
    (II)                 (III)                (IV)
```

The systematic names are (I) 1-isopropyl-4-Me-bicyclo[3.1.0]hexane; (II) 2,2,5-Me$_3$-bicyclo[4.1.0]heptane; (III) 2,2,4-Me$_3$-bicyclo[3.1.1]heptane, and (IV) 1,7,7-Me$_3$-bicyclo[2.2.1]heptane. The numbers in parenthesis indicate how many carbon atoms are in the three bridges in the bicyclic system. The largest possible ring is used in numbering substituents, using as No. 1 the bridgehead which will give the smallest numbers for the substituents.

The numbering of the positions in the ordinary system is different from that of the related menthanes (p. 554). The *gem*-Me₂ system is 7,8,9 and the single Me is 10. The numbering of the 6-rings differs in the four series and will be given under the individual members.

Sabinane Group

Sabinane (thujane), b. 157°, is obtained by hydrogenating the naturally occurring unsaturated derivatives. Its 6-ring is numbered with the isopropyl at 1, the Me at 4, and the internal cyclopropane ring including carbons 1,5, and 6.

Sabinene, 4(10)-thujene, 1-isopropyl-4-methylenebicyclo[3.1.0]hexane, b. 164°, occurs widely in essential oils.

Oxidation removes the 10-carbon leaving *sabina ketone* which on further oxidation, gives α-tanacetogen dicarboxylic acid, 1-isopropyl-2-carboxy-cyclopropane-1-acetic acid. Sabinene, with acids, undergoes addition to the double bond and opening of the cyclopropane ring. HCl gives 1,4-Cl₂-menthane and dilute sulfuric acid gives the corresponding di-tertiary alcohol, 1,4-terpin.

α- and β-*Thujene* have their double bonds at 3 and 2.

Sabinol is 3-OH-sabinene.

α- and β-*Thujone* (*tanacetone*) both have a 3-carbonyl group and differ only in the configuration of the 4-carbon. Heat opens the ring between carbons 1 and 5 replacing it by a 4-double bond thus producing *carvotanacetone* (p. 564). Sulfuric acid opens the ring between carbons 1 and 6 giving *isothujone*, 2,3-Me₂-4-isopropyl-2-cyclopenten-1-one. Reduction gives *thujyl alcohol*. Careful oxidation breaks the 6-ring between the CO and the tertiary C to give α-*thujaketo acid*, 2-acetyl-1-isopropyl-cyclopropane-1-acetic acid. This acid has been reconverted to thujone.[102]

Carane Group

Carane is rare, b. 50° at 9 mm. Its 6-ring is numbered with Me at 4 and the external cyclopropane ring including carbons 1,6 and 7, the latter bearing the *gem*-Me₂. The 5-ketone, *carone*, b. 210°, is made by the action of KOH on 8-Br-2-menthone (p. 564). HBr with carone reverses the process. Heat opens the 3-ring in carone between carbons 6 and 7 giving *carvenone*, 3-menthen-2-one.

Carenes are widely distributed in small amounts.[103] The commonest has the double bond at 3(4).

[102] Ruzicka. *Ann. Rev. Biochem.* **3**, 460 (1934).
[103] *Ann. Rep. Chem. Soc.* (London) **1925**, 131.

Pinane Group

Members of this group are named as derivatives of norpinane (CA), bicyclo(3.1.1)heptane, which is unknown. Its formula and numbering (CA) are as follows:

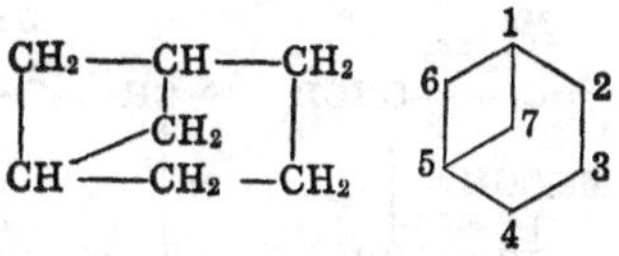

It should be noted that the following positions are identical: 1 and 5, 2 and 4, 6 and 7. The cage structure contains two 6-rings, 123456 and 123457 and a 4-ring 1657. The stability of the latter depends on its relation to the two 6-rings of which it is a part.

In this system, pinane (p. 567) is 2,6,6-Me_3-norpinane (CA).

Pinane, b. 166°, is obtained by catalytic hydrogenation of pinene which gives a *cis-trans* mixture.

Two pinenes are known. Common pinene or α-*pinene*, 2,6,6-Me_3-2-norpinene, b. 156°, is the chief constituent of turpentine. Its formula is written in a variety of ways including—

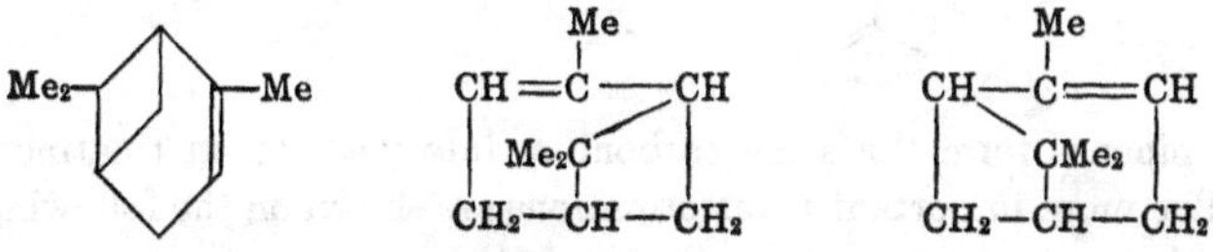

Regardless of the planar representation, the molecule may be visualized as sitting on a puckered 4-ring having carbons 6 and 7 lower than 1 and 5. Carbons 2,3,4 form a plane at right angles to the imaginary line connecting 6 and 7. α-Pinene is found in various turpentines in *d-*, *l-* and *racemic* forms. *d*-α-Pinene has $[\alpha]_D^{20} = +53.0$.

Nopinene, β-pinene, b. 164° differs from ordinary pinene in having a methylene double bond.

It is 2-methylene-6,6-Me_2-norpinane. It occurs in smaller amounts in turpentine.

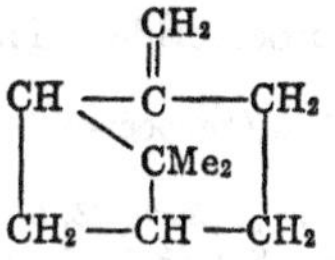

Nopinene

The pinenes are very reactive.

Pyrolysis of α-pinene gives a great variety of products including isoprene and aromatic hydrocarbons from benzene to methylanthracene.[104, 105]

α-Pinene, with air and moisture, gives pinol hydrate (sobrerol) which yields pinol with acids.

Pinol hydrate   Pinol

Such a breaking of a 4-ring by oxygen, followed by closure of an ether ring *without affecting the double bond* is most remarkable. Thermal isomerization of α-pinene gives dipentene, allo-ocimene, and α- and β-pyronenes.[106]

Both pinenes form the same carbonium intermediate on treatment with acids. The more important transformations are shown on the following page. Many of these reactions are reversible (p. 571).

Oxidation of α-pinene gives successively α-pinonic acid[107] (2,2-Me$_2$-3-acetylcyclobutane-1-acetic acid), pinic acid, (2,2-Me$_2$-3-carboxycyclobutane-1-acetic acid), and norpinic acid (2,2-Me$_2$-cyclobutane-1,3-dicarboxylic acid). Oxidation by SeO$_2$ or by air using higher fatty acid salts of Co or Mn gives *verbenone* (III below).

The addition of dry HCl to α-pinene is of great practical importance as the first step in the first commercial synthesis of camphor and of great theoretical interest as illustrating a type of rearrangement very common in the bicyclic terpene series. The process is entirely analogous to that by which *t*-butyl-ethylene gives Me$_3$-ethylene hydrochloride. The changes may be formulated

[104] Hurd. "The Pyrolysis of Carbon Compounds." Chemical Catalog Co., New York, 1929.
[105] Davis, Goldblatt, Palkin. *Ind. Eng. Chem.* 38, 53 (1946).
[106] Goldblatt, Palkin. *J. Am. Chem. Soc.* 63, 3517 (1941).
[107] Egloff et al. "Isomerization of Pure Hydrocarbons," 1942. p. 139.

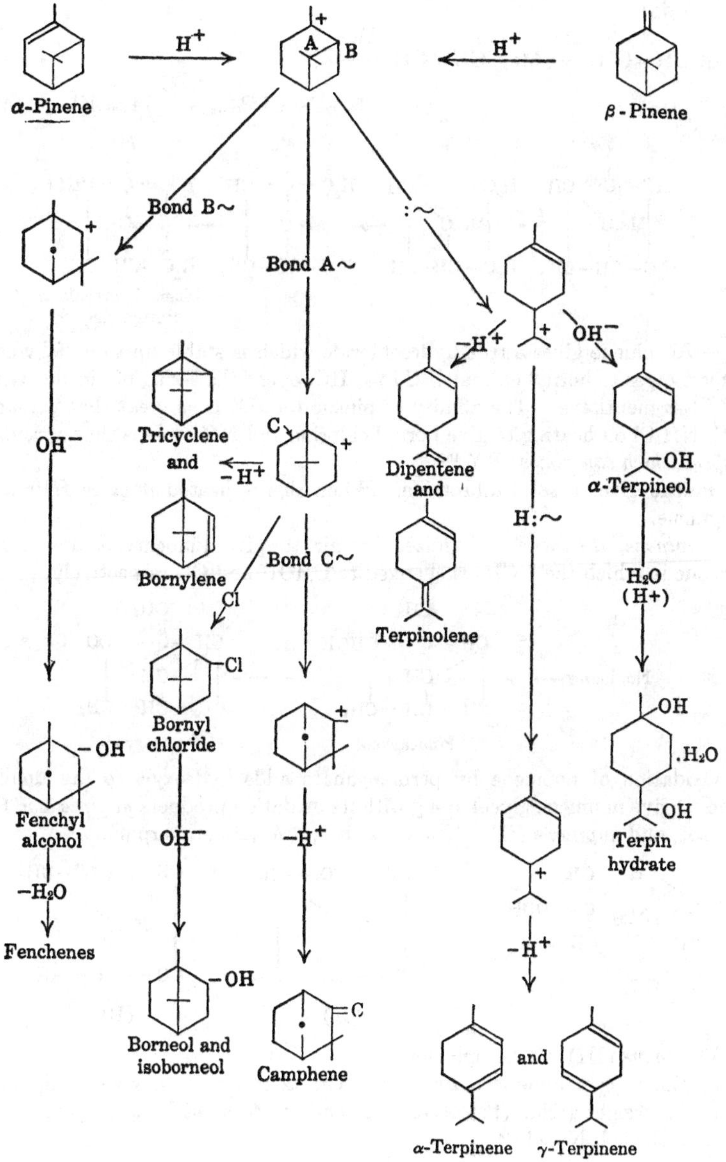
H+
A
B
H+
α-Pinene
β-Pinene
Bond B~
:~
Bond A~
−H+
OH−
OH−
Tricyclene
and
C
−H+
Dipentene
and
−OH
α-Terpineol
H:~
Bornylene
Bond C~
Cl
H₂O
(H+)
Terpinolene
Cl
OH
Bornyl
chloride
− OH
.H₂O
Fenchyl
alcohol
OH−
−H+
−OH
Terpin
hydrate
−H₂O
−H+
Fenchenes
−OH
=C
Borneol and
isoborneol
Camphene
and
α-Terpinene
γ-Terpinene

as follows:

$$Me_3CCH=CH_2 \xrightarrow{\ H^+\ } Me_3\overset{+}{C}CH\text{—}CH_3 \xrightarrow{\ Me\sim\ }$$

$$\overset{+}{Me_2C}\text{—}CHMe_2 \xrightarrow{\ Cl\ominus\ } Me_2ClC\text{—}CHMe_2$$

"Pinene hydrochloride"
Bornyl chloride

At $-70°$ pinene gives a true hydrochloride which is stable up to $-10°$ where it rearranges to bornyl chloride. Moist HCl opens the 4-ring of pinene giving 1,8-Cl_2-p-menthane. The affinity of pinene for HX is so great that it reacts with NH_4Cl on heating to give bornyl chloride and NH_3. It is thus a *neutral reagent which can remove HX*.[108]

Pinene gives a solid dibromide. When this is heated it gives HBr and p-cymene.

Nopinene, β-Pinene is oxidized by air to give pinocarveol and pinocarvone in which the 3-CH_2 is changed to CHOH and CO respectively.

Pinocarveol   Pinocarvone

Oxidation of nopinene by permanganate adds hydroxyls to the double bond to give nopinene glycol along with its oxidation products *nopinic acid* (I) m. 128°, and *nopinone* (II). The latter is 6,6-Me_2-2-ketonorpinane.

(I)   (II)   (III)

Verbenone (III) is 4-ketopinene.

Nopinene is polymerized by Friedel-Crafts catalysts to give hard, high molecular weight resins (Piccolyte). It adds to formaldehyde to give a C_{11} terpenic alcohol, Nopol.[109]

[108] *Ann. Rep. Chem. Soc.* (London) 1920, 81.
[109] *Chem. Inds.* 63, 808 (1948).

Camphane Series

Norcamphane, m. 87°, bicyclo[2.2.1]heptane, 1,4-endomethylenecyclohexane, norbornylane, is readily made by reduction of β-norbornyl chloride or bromide with sodium and alcohol.[110] It has a peculiar damp almost stupefying odor. Evidently the camphor odor characteristic of neo carbon groupings is absent.

Camphane, (p. 567), m. 159°, b. 162°, is obtained by Na and alcohol reduction of bornyl chloride from pinene and HCl. It has the ordinary reactions of 5- and 6-membered alicyclic compounds. The combination of a *cis*-6-ring with the 1,4-positions linked by the Me_2C group forming two 5-rings with three atoms in common is free from strain.

Just as camphane has a carbon bridge between the 1 and 4 carbons, it is possible to have a two carbon bridge between these positions without introducing a strain into the molecule. Bicyclo(2.2.2)octane, m. 170°, has been prepared from the corresponding ketone by heating its semicarbazone with NaOEt.[111, 112]

Such an arrangement really consists of three *cis*-cyclohexane rings having the 1 and 4 carbons in common.

Unsaturated Hydrocarbons

Norbornylene (I), subl. 52–4°, is obtained by dehydration of norborneol by P_2O_5 or from norbornyl bromide and quinoline. It is bicyclo[2.2.1]-2-heptene.[113] No rearrangement is possible as in the corresponding reactions of borneol and bornyl halides (see below). Actually this is because the "normal" and rearranged products are identical.

(I)

Bornylene, m. 114°, b. 150°, cannot be made from borneol by ordinary dehydrating agents as these produce camphene by rearrangement. Conversion

[110] Kommpa, Beckmann. *Ann.* **512,** 172 (1934).
[111] Alder et al. *Ann.* **514,** 1 (1934).
[112] Kommpa. *Ber.* **68B,** 1267 (1935).
[113] Kommpa, Beckmann. *Ann.* **512,** 172 (1934).

of borneol to the Na derivative, treatment with CS_2, methylation and heating at 200° gives bornylene.[114]

Me   Me
CH₂–C—CHOCS₂Me   CH₂–C —CH
    CMe₂   CMe₂
CH₂–CH—CH₂   ⟶   CH₂–CH—CH
                                    Bornylene

Bornylene gives ordinary olefin reactions.

Camphene is reported in various forms, m. 49°, b. 159°. It is obtained by dehydrating the borneols with acid reagents, by removing HX from bornyl halides, or by direct isomerization of pinene. A rearrangement of the ordinary kind (p. 120) is involved.[115, 116] The first addition product gives the unstable carbonium ion II in which the electronically deficient 2-C attracts the spatially near electron pair of the 6-C thus closing a new 5-ring and leaving the 1-C deficient. This deficiency is overcome by attraction of a pair from the 7-C, with the expulsion or removal of a proton leaving the 1,7-double bond of camphene.

        Me   ⁷Me
         C   ⁺C¹
⟶   CH₂ CMe₂CH+   ⁶CH₂⁸CMe₂ CH   ⁶CH₂–CH—C=CH₂
    CH₂–CH—CH₂   ⟶   ⁵CH₂⁴CH–³CH₂   ⟶   ³CH₃
        (II)   (III)   ⁵CH₂–CH—CMe₂
                                                  Camphene

Camphene adds acids to give bornyl and isobornyl derivatives by a reverse of this rearrangement.[117, 118] The first step is the addition of a proton from the acid to the 7-C with the formation of carbonium ion III. Under ordinary conditions this reverts to II which forms the more stable secondary alcohol and its derivatives. On the other hand, use of milder conditions give the tertiary derivatives related to III. Addition of dry HCl at low temperatures gives unstable *camphene hydrochloride* without rearrangement. Dilute alkali gives *"camphene hydrate"* a tertiary alcohol which can be dehydrated to give camphene of the same optical activity as the original. A stereoisomer is made from camphenilone and MeMgX.[119]

[114] Tschugaef. *J. Russ. Phys. Chem.* **36**, 1034 (1904).

[115] Meerwein. *Ann.* **405**, 129 (1914).

[116] Wagner, Brykner. *Ber.* **33**, 2121 (1900).

[117] Meerwein et al. *Chem. Ztg.* **44**, 693 (1920); *Ber.* **53B**, 1815 (1920); **55B**, 2500 (1922); *Ann.* **435**, 190 (1923).

[118] Hückel. "Theoretische Grundlagen der Organischen Chemie," Vol. II, 1931. p. 215.

[119] *Ann. Rep. Chem. Soc.* (London) **1916**, 117.

Oxidation of camphene in acid gives camphor. Permanganate oxidation gives a glycol which on treatment with acid gives *camphenilanic aldehyde.* The same aldehyde is obtained from camphene and chromyl chloride. Oxidation gives the corresponding acid and then, by loss of the carboxyl, *camphenilone,* 2,2-Me_2-3,6-endomethylenecyclohexanone. This is also formed by the ozonolysis of camphene although the chief product results from further oxidation to form the lactone of *hydroxycamphenilic acid.*

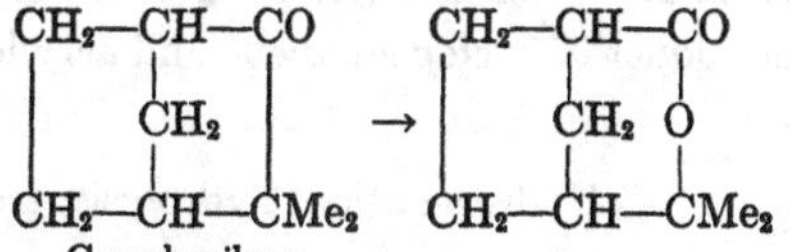
CH₂—CH—CO
CH₂
CH₂—CH—CMe₂
CH₂—CH—CO
CH₂ O
CH₂—CH—CMe₂
Camphenilone

Oxidation of camphene with alkaline permanganate gives *camphenic acid,* α-3-carboxycyclopentylisobutyric acid.

Toxaphene, an important insecticide, is a yellow, waxy solid melting at 65° to 90°. It is a chlorinated camphene containing 67 to 69% Cl—an average formula of $C_{10}H_{10}Cl_8$. Toxaphene has a density of 1.65 at 20° C. and is highly soluble in common organic solvents.[120]

Endocamphene, 3,3-Me_2-4,7-endomethylenecycloheptene, has been made.[121]

α-Fenchene is obtained by dehydrating fenchyl alcohol with rearrangement.[122]

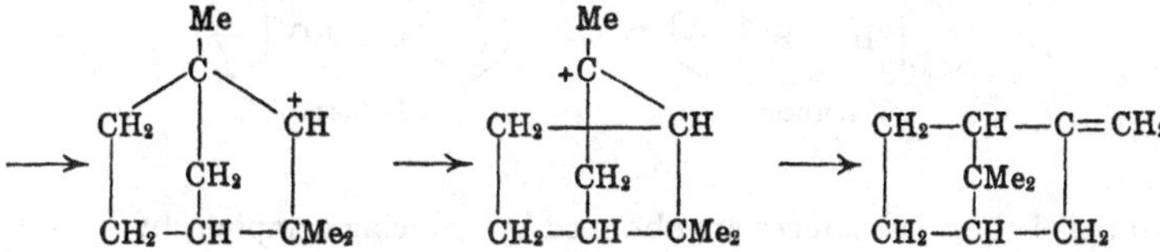
Me
C
CH₂
+
CH
CH₂
CH₂—CH—CMe₂
Me
+C
CH₂
CH
CH₂
CH₂—CH—CMe₂
CH₂—CH—C=CH₂
CMe₂
CH₂—CH—CH₂
α-Fenchene

β-Fenchene, with the Me_2 and 5-H_2 groups interchanged, is obtained from isofenchyl alcohol by a similar rearrangement. This alcohol, by the xanthate method (no rearrangement), gives a *fenchylene,* 3,4,4-Me_3-3,6-endomethylenecyclohexene. *Santene,* 1,2-Me_2-3,6-endomethylenecyclohexene, is the chief dehydration product of *camphenilol.* The lesser product is the unrearranged *camphenilene* with the double bond in the 1,6-position.

[120] Parker, Beacher. *Delaware Agr. Expt. Sta.—Bull.* **264**, *Tech. Bull.* **36**, 26 (1947); *Chem. Inds.* **60**, 76 (1947).

[121] *Ann. Rep. Chem. Soc.* (London) 1927, 120.

[122] *ibid.* 1917, 110.

Alcohols

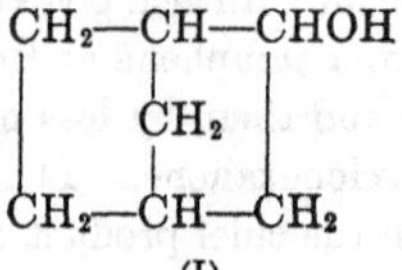

$$\begin{array}{ccc} CH_2 & CH & CHOH \\ & CH_2 & \\ CH_2 & CH & CH_2 \end{array}$$

(I)

β-**Norborneol** (I), b. 177°, can be made by a series of standard reactions from the Diels-Alder adduct of cyclopentadiene and acrylic ester.

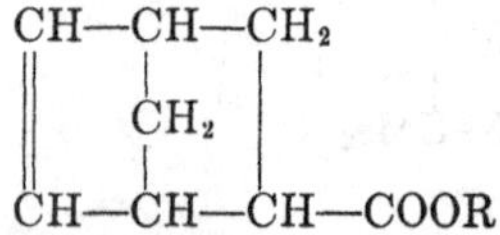

$$\begin{array}{ccc} CH & CH & CH_2 \\ \parallel & CH_2 & \\ CH & CH & CH-COOR \end{array}$$ Hydrogenation, hydrolysis, treatment with PCl_3,

and with NaN_3 gives the isocyanate. This on heating with HCl gives the hydrochloride of *β-aminonorcamphane*. This reacts with acetic acid and $NaNO_2$ to give (I). The last reaction gives no rearrangement of the bicyclic ring system because the normal and rearranged products are identical. Contrast the action of aliphatic amines (p. 172).

Borneol (II), m. 208°, b. 213°, and **isoborneol**, m. and b. 216°, each exist in *d*- and *l*-forms. They are stereoisomers, borneol having its OH on the opposite side of the molecule from the *gem*-Me₂ group and isoborneol having its OH near that group.

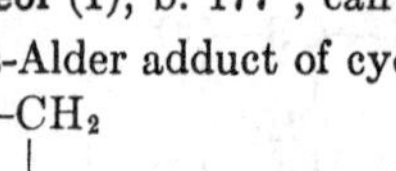

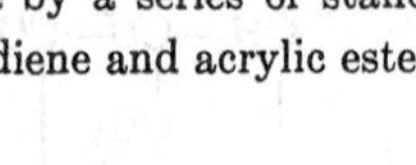

dl-Borneol   *dl*-Isoborneol
(II)

Mixtures of these substances are obtained by reducing camphor, by hydrating pinene and camphene and from the bornyl and isobornyl halides. Isoborneol is more soluble in most solvents than is borneol.

Dehydration of borneol and isoborneol give the rearranged camphene instead of bornylene.

d-**Borneol** or Borneo camphor is found in many ethereal oils. The *l*-form or Ngai camphor is much rarer in nature.

Bornyl chloride, m. 133°, b. 207–8, is made from pinene and HCl. It was the first "artificial camphor." With alkalies it gives camphene. *Bornyl iodide* with alkalies gives bornylene as well as camphene.

Fenchyl alcohol, fenchol, is an isomer of borneol with a 3-*gem*-Me₂ group and a 1,4-CH₂ bridge. It is found among the products of the reaction of acids with pinene. *Isofenchyl alcohol* differs by having the OH in the 6-position. It is obtained by the reduction of isofenchone.

Camphenilol, 2,2-Me_2-3,6-endomethylenecyclohexanol, the next lower homolog of fenchyl alcohol, without the 10-Me, is made by reducing camphenilone, obtained by ozonolysis of camphene.

Tertiary methyl borneol (I) and tertiary methyl fenchol (II) have been made by the action of MeMgX on camphor and fenchone respectively. Dehydration of these two alcohols gives the same product, α-Me-camphene (III). Thus the first undergoes rearrangement while the second does not.[123]

Hydration of III gives 4-Me-isoborneol (IV) which on dehydration gives mainly β-Me-camphene (V) and some of the α-Me compound (III)

Both III and V, on hydration, give IV. The changes involved include rearrangements which are classified under three names, Wagner (change of ring), Meerwein (shift of methyl) and Nametkin (shift of Me and change of ring). Actually the three changes involve the same fundamental rearrangement as that found in the aliphatic series in the pinacolic and retropinacolic rearrangements and the unnamed rearrangement of a methyl group in certain olefins during hydration. In every case the occasion for the rearrangement is the formation of a carbon with only six electrons which can thus attract a pair of electrons (and their attached group) from an adjacent carbon. As to which of the electron pairs (and attached groups) will be attracted, data are lacking to make prediction possible.

[123] *Ann. Rep. Chem. Soc.* (London) 1932, 148.

The electronically deficient carbon (indicated by $^+$) may be formed by *removing* a hydroxyl with the complete octet of electrons held by the oxygen.

1. $Me_2C(OH)C(OH)Me_2 \rightarrow Me_2\overset{+}{C}C(OH)Me_2 \rightarrow$

$$Me_3C\overset{+}{C}(OH)Me \rightarrow Me_3CCOMe$$

2. $Me_3CCHOHMe \rightarrow$

$$Me_3C\overset{+}{C}HMe \rightarrow H^+ + Me_3CCH=CH_2 \ (3\%)$$

$$Me_2\overset{+}{C}C\overset{\downarrow}{H}Me_2 \rightarrow H^+ + Me_2C=CMe_2 + CH_2=C(Me)CHMe_2$$

The electronically deficient carbon may also be formed by *adding* hydrogen ion to a double bond

$$: \overset{..}{C} :: \overset{..}{C} : \xrightarrow{\ H^+\ } : \overset{..}{C} : \underset{\underset{H}{|}}{\overset{..}{C}} :$$

3. $Me_3CCH=CH_2 + H^+ \rightarrow Me_3C\overset{+}{C}H-CH_3 \rightarrow$

$$Me_2\overset{+}{C}CHMe_2 \xrightarrow{\ OH^-\ } Me_2C(OH)CHMe_2$$

In each of these reactions the essential change has been the shift of Me to the electronically deficient carbon.

4. Dehydration of (I) to give (III)

(I) $\rightarrow$ (I')   (III') $\rightarrow$ H$^+$ +   (III)

The C$^+$ in I' attracts the electron pair holding carbons 1 and 6 and this transfers the attachment of carbon 6 to carbon 2 forming a new 5-ring. The C$^+$ in III' is stabilized by attracting an electron pair from its Me group and expelling the H held by that pair. Evidently this process takes place by preference over the apparently similar one by which the C$^+$ in I' would have expelled an H from its methyl group or from the neighboring CH$_2$ group without rearrangement. Apparently such a change would have about the same chance as the formation of *t*-Bu-ethylene from *t*-BuMe-carbinol. Evidently, of the rings only the last is stable in the presence of H ions.

5. Dehydration of (II) to give (III).

$$(II) \longrightarrow \quad (II') \quad \longrightarrow \quad H^+ \; + \quad (III)$$

$$(II') \qquad\qquad (III)$$

6. Hydration of (III) to give (IV).

$$(III) \xrightarrow{\;H^+\;} \quad (II') \xrightarrow[\sim]{Me} \quad \longrightarrow \quad \xrightarrow{\;OH^-\;} (IV)$$

$$(II')$$

7. Dehydration of (IV) to give (V) and (III).

$$(IV) \longrightarrow \quad (IV') \rightleftharpoons \quad \rightleftharpoons \quad (V)$$

$$(IV') \qquad\qquad\qquad (V)$$

$$\Big\downarrow Me\sim$$

$$(II') \rightleftharpoons (III)$$

8. Hydrations of III and V to give IV take place by the reversal of the steps in 7.

Ketones

Camphor (I) is the 2-ketone of camphane. *d-Camphor*, Japan camphor, m. 179°, b. 209°, is the common variety. The *l*-form is Matricaria camphor. Large amounts of camphor are used in celluloid, pyralin and similar products from nitrocellulose. In addition to the natural product, camphor is synthesized from pinene obtained from turpentine.

$$\text{Pinene} \xrightarrow[\text{earth}]{\text{acid}} \text{camphene} \xrightarrow[\text{acid}]{\text{formic}} \text{isobornyl formate} \xrightarrow{\text{steam}}$$

$$\text{isoborneol} \xrightarrow[\text{dehydrogenation}]{\text{catalytic}} \text{camphor.}$$

Camphor, with P_2O_5, gives p-cymene, and with I_2, gives carvacrol (2-OH-cymene).   The bridged ring structure of camphor follows from its oxidation to camphoric acid, 1,2,2-Me$_3$-cyclopentane-1,3-dicarboxylic acid; *camphanic acid* (the lactone of the corresponding 3-OH acid), *camphoronic acid* (2,2,3-Me$_3$-3-carboxyglutaric acid), and finally to Me$_3$-succinic acid.

dl-Camphor

Camphor has been synthesized starting with oxalic and $\beta\beta$-Me$_2$-glutaric esters going through camphoric and homocamphoric acids.[124]   Camphor reacts like ordinary ketones.   Reduction gives the stereoisomeric secondary alcohols, borneol and isoborneol.   With nitrous acid it gives an isonitroso compound, the $C=NOH$ group being in the 3-position, α- to the CO.   Hydrolysis gives 3-ketocamphor or *camphorquinone.*   The *oxime* of camphor, m. 119°, b. 250° dec., on dehydration forms a cyanide which is hydrolyzed to *campholenic acid,* 2,3,3-Me$_3$-cyclopentene-4-acetic acid.   The dichloride of camphor obtained by means of PCl$_5$ at low temperature is the α-dichloride, 2,2-Cl$_2$-camphane, m. 148°.   This readily rearranges to the β-dichloride, 2,6-Cl$_2$-camphane, m. 178°.[125]

α-**Bromocamphor,** (3-Br), m. 77°, b. 274°, gives the ordinary reactions of a halogenated ketone but in addition gives a fair yield of a Grignard reagent with Mg.   With CO$_2$, this gives camphor, 3,3'-dicamphor and camphor-3-carboxylic acid. β-*Bromocamphor,* (10-Br, ω-Br), m. 79°, is obtained by heating the corresponding sulfonyl bromide.   It reacts normally with hydroxylamine, whereas the α-compound does not. π-*Bromocamphor,* (9-Br), m. 93°, is similarly made from the $-SO_2Br$ compound.   The Br is very inactive, not being changed by AgNO$_3$.

Sulfonic Acids of Camphor

Camphor α-sulfonic acid, substituted in the 3-position, is made from camphor and ClSO$_3$Me.   The methyl ester is soluble in bases and is believed to be an enol form.

Camphor β-sulfonic acid (Reychler acid), substituted in the 10-Me, is formed by sulfuric acid in acetic anhydride.   The d-α-Br-β-sulfonic acid is valuable for separating optically active materials.

Camphor π-sulfonic acid, substituted in one of the *gem*-Me$_2$ (8 or 9), is made by the action of oleum or of chlorosulfonic acid.   Both the d- and l-acid have been largely used for resolving racemic materials.   The α-Br-π-*sulfonic acid* is similarly made and used.[125a]

[124] Komppa. *Ber.* **36**, 4332 (1903); *Ann.* **370**, 225 (1909).

[125] Meerwein, Wortmann. *Ann.* **435**, 190 (1923).

[125a] Cornubert, R., "Le Camphre et ses Derives." Paris, 1933. pp. 424.

Epicamphor or β-**camphor** in which the CO and α-CH$_2$ groups are interchanged is known.[126]

Isocamphor is merely an unrelated isomer of camphor, 1-acetyl-3-isopropylcyclopentene.[126] It is also called *isopinolone*.

Apocamphor, fenchocamphorone, is the lower homolog of camphor without the 7-Me group. Oxidation gives apocamphoric acid, 2,2-Me$_2$-cyclopentane-1,3-dicarboxylic acid.

Homocamphor, having an additional CH$_2$ between the CO and C$_4$, has been synthesized.[127]

Substituted camphors with alkyl or aryl on C$_4$ are obtained by treating camphor with a Grignard reagent, dehydrating to give an α-substituted camphene, hydrating to a 4-substituted isoborneol and oxidizing to the desired product.

Fenchone, m. 6°, b. 193°, is an isomer of camphor with the *gem*-Me$_2$ attached in the 3-position and a 1,4-CH$_2$ bridge. The lower symmetry of the molecule is reflected in its lower melting point. It is found in *d*- and *l*-forms in nature. It is more resistant to oxidation than camphor. This is understandable from a comparison of the environment of the ketone group in the two compounds.

$$\begin{array}{cc} & \text{C} & & \text{C} \quad\quad \text{C} \\ & | & & | \quad\quad\quad | \\ \text{C—C—CO—CH}_2\text{—} & & \text{C—C—CO—C—C} \\ & | & & | \quad\quad\quad | \\ & \text{C} & & \text{C} \quad\quad \text{C} \end{array}$$

Fenchone has been synthesized using levulinic ester, bromoacetic ester and MeI.[128] A substance having the structure of fenchone with the lone methyl changed to carboxyl is *camphenonic* acid (III) obtained by distillation of camphenic acid (II) made by alkaline permanganate oxidation of camphene (I).[129]

$$\begin{array}{ccccc} \text{CH}_2\text{—CH—C}=\text{CH}_2 & & \text{CH}_2\text{—CH—CO}_2\text{H} & & \overset{\textstyle\text{CO}_2\text{H}}{\text{CH}_2\text{—C—CO}} \\ |\quad\quad\;\; | & & |\quad\quad\;\; | & & |\quad\quad\;\; | \\ \;\;\text{CH}_2 & \rightarrow & \;\;\text{CH}_2 & \rightarrow & \;\;\text{CH}_2 \\ |\quad\quad\;\; | & & |\quad\quad\;\; | & & |\quad\quad\;\; | \\ \text{CH}_2\text{—CH—CMe}_2 & & \text{CH}_2\text{—CH—C(Me}_2)\text{CO}_2\text{H} & & \text{CH}_2\text{—CH—CMe}_2 \\ \text{(I)} & & \text{(II)} & & \text{(III)} \end{array}$$

The first step involves an unusual rearrangement.

[126] *Ann. Rep. Chem. Soc.* (London) 1912, 151.
[127] Lipp. *Ann. Rep. Chem. Soc.* (London) 1920, 91.
[128] Ruzicka. *Ann. Rep. Chem. Soc.* (London) 1917, 109.
[129] *Ann. Rep. Chem. Soc.* (London) 1916, 117,

Isofenchone has the *gem*-Me₂ attached to the 5-position.   Oxidation gives isofenchocamphoric acid, 2,2,4-Me₃-cyclopentane-1,4-dicarboxylic acids, its α-OH acid and finally α,α,α′,α′-Me₄-glutaric acid.

F. Tricyclic Terpene Derivatives

Tricyclene, m. 68°, b. 153°, an isomer of the bicyclic terpenes ($C_{10}H_{16}$), may be regarded as a camphane with the 2,6-positions joined to form a cyclopropane ring with Carbon No. 1.   It may be represented in various ways.

Tricyclene   Bornylene   Camphene

Breaking the cyclopropane ring at the 2,6-linkage and at the 1,2-linkage (or 1,6-) would give bornylene and camphene respectively.   Before it was obtained as a definite compound, tricyclene was assumed to be the intermediate in the interconversion of the borneols and camphenes and related terpene rearrangements.   This function is impossible as tricyclene is stable under conditions which produce many of these changes.   Tricyclene is obtained in small amounts as a by-product in many terpene reactions.   It is prepared by the action of HgO on the hydrazone of camphor.   In systematic nomenclature, tricyclene is 1,7,7-Me₃-tricyclo[2.2.1.0²·⁶]heptane.   The first three numbers in parenthesis give the numbers of carbon atoms in the three bridges connecting the two common atoms in the bicyclic system while the fourth number gives the number of carbon atoms in the fourth bridge and its superscripts show the carbons linked by this bridge.   The numbering corresponds to that of the largest ring in the tricyclo system.

Tricyclenic acid, m. 148°, b. 264°, has a carboxyl group in place of methyl on the 1-C of tricyclene.   It is prepared from ω-nitrocamphene by treatment with sulfuric acid at low temperature.   The ω-nitrocamphene is prepared by treating isocamphane with nitric acid under pressure.

Teresantalic acid from sandalwood oil has a carboxyl in place of one of the *gem*-Me₂ in tricyclene.   Heat decomposes its Ca salt to give *santene,* 1,2-Me₂-3,6-endomethylenecyclohexene.   Its ester can be reduced to *teresantalol* which can be oxidized to the aldehyde and then converted to tricyclene by heating the semicarbazone with NaOEt.[130]

Adamantane, $C_{10}H_{16}$, m. 268° (sealed tube), obtained from Hodonin crude petroleum from Moravia, is one of the most interesting organic compounds

¹³⁰ *Ann. Rep. Chem. Soc.* (London) **1927,** 121.

found in nature.[131] It crystallizes in the cubic system, has an odor suggesting camphor and turpentine, and is stable to chromic acid, permanganate and nitric acid. The structure of adamantane follows from its analysis, molecular weight, crystal form and melting point. The latter is especially significant. Of the nearly 200 other known substances having the formula $C_{10}H_{16}$ all are liquids except camphene, bornylene and tricyclene with m. 53°, 98° and 68° respectively. The melting point of 268° indicates a remarkable symmetry which is satisfied by the arrangement of the carbon atoms in diamond. This is confirmed by X-ray studies of the hydrocarbon. Thus adamantane is *tricyclo [3.3.1.1³·⁷] decane*. Its model shows *four trans cyclohexane rings*. If a unit of 10 carbon atoms could be taken out of diamond and supplied with 16 hydrogen atoms the result would be a molecule of adamantane. This most interesting hydrocarbon has been synthesized.[132]

G. Higher Terpenes and their Derivatives

Sesquiterpenes

These fall in five groups characterized by the following molecular refractions: Open chains with 4 double bonds, 69.6; monocyclic with 3 double bonds, 67.9; bicyclic with 2 double bonds, 66.1; tricyclic with 1 double bond, 64.4; and tetracyclic with no double bond, 62.7. With conjugated double bonds the value is even higher. Thus it is 68.4 for zingiberene.

The cyclic sesquiterpenes have 6-membered rings of the terpene type with aliphatic terpene side chains. The structures are those to be expected from the polymerization of three isoprene units with one or two ring closures.[133]

Bisabolene (I) and *zingiberene* (II) are monocyclic, *cadinene* (III), *isozingiberene* (IV), *α-selinene* (V), *β-selinene* (VI), and the related tertiary alcohols, *α-eudesmol* (VII) and *β-eudesmol* (VIII) are bicyclic, while *copaene* (IX) is tricyclic.[133a, 134, 135, 136]. Although considerable work has been done on the elucidation of the structures of the sesquiterpenes, some doubt still remains

[131] Landa, Macháček. *Collection Czechoslov. Chem. Communications* 5, 1 (1933).

[132] Prelog, Seiwert. *Ber.* **74B**, 1644 (1941).

[133] Bogert. *Chem. Revs.* 10, pp. 265–94.

[133a] Semmler and Stenzel, *Ber.*, **47**, 2555 (1914).

[134] *Ann. Rep. Chem. Soc.* (London) 1929, 146; **1932**, 153.

[135] Campbell, Soffer. *J. Am. Chem. Soc.* 64, 417 (1942).

[136] Soffer *et al*, *J. Am. Chem. Soc.*, 66, 1520 (1944).

that the double bond positions have been correctly assigned in all of those studied.

(I)   (II)   (III)

(IV)   (V)   (VI)

(VII)   (VIII)   (IX)

(X)

Machilol, cryptomeradol, and certain other sesquiterpene alcohols are identical with eudesmol. Acids convert *farnesol* and *nerolidol* to (I) and its isomers and then to (III). A different internal polymerization of the open chain sesquiterpenes gives (V). The structures of (III) and (V) and related bicyclic compounds follows from their dehydrogenation by S or catalysts to give *cadalene*, 1,6-Me$_2$-4-isopropylnaphthalene and *eudaline*, 1-Me-7-isopropyl-naphthalene, respectively. Substances like selinene and eudesmol lose the *angular methyl group* on dehydrogenation.

Santonin is a lactonic ketone related to α-selinene (V)

It can be converted to 1-Me-7-Et-naphthalene.[137] *Artemisin* is 7-OH-santonin, helenin contains *alantolactone* and isomeric substances also related to santonin.[138, 139]

Tricyclic sesquiterpenes related to (III) and (V) are known. *Copaene* (IX) from copaiba balsam reacts with HCl to give the same di-hydrochloride as does (III). α-Santalene gives teresantalic acid on oxidation and has structure (X) (p. 584). Less known members of the tricyclic series are *cedrene* and *gurjenene*. Ketones and hydroxy ketones of the sesquiterpenes are known.[140] Thus *eremophilone* is the 3-keto compound of (V).[141] Other ketones are *atlantone* and *turmerone* (1933).

Sesquiterpenreihe. L. Ruzicka. Berlin 1928.

Di- and Tri-Terpenes

Many *di-* and *tri-terpenes* are known. Just as the terpenes are based on the benzene nucleus and the sesquiterpenes on the naphthalene nucleus, the diterpenes are related to the phenanthrene nucleus.[142] The nature of the cyclic nucleus in the tetracyclic triterpenes is not known. *Camphorene*, $C_{20}H_{32}$, has a cyclohexene ring with side chains of aliphatic terpene type (XI). It can be regarded as the product of the polymerization of two myrcene molecules. Further cyclization by acid gives a hydroanthracene derivative which can be oxidized to benzene-1,2,4,5-tetracarboxylic acid.[143] *Vitamin* A, $C_{20}H_{30}O$ (XII), can be regarded as the primary alcohol of a monocyclic diterpene with extra unsaturation giving complete conjugation of the 5 double bonds.[144]

(XI) (XII)

The resin acids, the chief component of rosin, are diterpene carboxylic acids. The eight individual resin acids which occur in rosin have one of two parent structures: i.e. abietic-type or primaric-type.[145]

[137] Ruzicka. *Ann. Rev. Biochem.* **3**, 459 (1934).
[138] Kon. *Ann. Rep. Chem. Soc.* (London) **1932**, 156.
[139] *Ann. Rev. Biochem.* 462 (1934).
[140] *Ann. Rep. Chem. Soc.* (London) **1932**, 155.
[141] Ruzicka. *Ann. Rev. Biochem.* **3**, 463 (1934).
[142] *Ann. Rep Chem. Soc.* (London) **1924**, 101.
[143] *ibid.* **1927**, 125.
[144] Karrer. *Ann. Rev. Biochem.* 464 (1934).
[145] Harris, Sanderson. *J. Am. Chem. Soc.* **70**, 334, 2079 (1948).

The abietic type acids of all degrees of saturation are readily dehydrogenated to retene (1-Me-7-isopropyl-phenanthrene). The pimaric-type acids are readily dehydrogenated to pimanthrene (1,7-Me₂-phenanthrene).

Abietic-Type Resin Acids

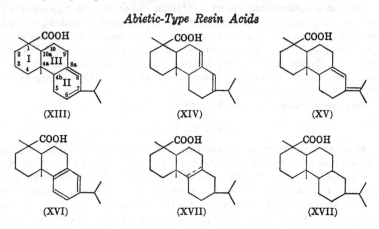

(XIII)   (XIV)   (XV)

(XVI)   (XVII)   (XVII)

Pimaric-Type Resin Acids

(XIX)   (XX)

Levopimaric acid (XIII), $[\alpha]_D^{24}$ − 276°, m. 163–7°, is isomerized almost completely to abietic by acid or heat. This resin acid is the only one which has the *cis*-configuration of the two double bonds, and, therefore, the only one which reacts readily with dienophiles such as maleic anhydride. In the reaction of rosin with maleic anhydride, levopimaric is present in small amounts in equilibrium with the other two double bond abietic-type acids, and, as it reacts, the equilibrium is shifted to consume all of the double bond abietic-type acids to give the levopimaric acid—maleic anhydride adduct, m. 226–7°.

Abietic acid (XIV), $[\alpha]_D^{24}$ − 106°, m. 173–7°, is the chief constituent of commercial rosins which accounts for rosin sometimes being called abietic acid. It is highly susceptible to heat and acid isomerization, and to oxidation.

Neoabietic acid (XV), $[\alpha]_D^{24}$ + 159°, m. 167–9°, is the last of the two double bond abietic-type acids found in natural rosins. Its behavior is similar to that of abietic and levoabietic.

Saturation of one or both of the double bonds or stabilization by aromatization gives a series of less reactive abietic-type acids.

Dehydroabietic acid (XVI), $[\alpha]_D^{24} + 63°$, m. 177–9°, is the only resin acid which contains a partially aromatic nucleus, thus its range of chemical reactivity is greatly increased. It is formed by removal of one mole of hydrogen from the double bond acids.

Dihydroabietic acids (XVII) are formed in the first step of stabilizing abietic-type acids by hydrogenation. In the hydrogenation of one double bond an asymmetric carbon results, thus two isomers. The individual isomers have not been isolated due to their extreme similarity in properties. All isomers give the same lactone, m. 131–2°, on treatment with cold H_2SO_4.

Tetrahydroabietic acids (XVIII) are fully saturated and will undergo only reaction of the carboxyl group. The nucleus contains two asymmetric carbon atoms and, therefore, a mixture of isomers. None of the isomers has been isolated.

Dextropimaric acid (XIX), $[\alpha]_D^{24} + 75°$, m. 215–7°, and **Isodextropimaric acid** (XX), $[\alpha]_D^{24}$ 0°, m. 159–162°, differ only in spatial configuration of the methyl and vinyl groups at the 7-carbon. They are relatively stable to heat and acid isomerization and to oxidation, presumably because the double bonds are not conjugated.

The resin acids are readily esterified with methanol, glycerol, and pentaerythritol (Abalyn, ester gum, Pentayln). Reduction of the acids or the methyl esters gives the primary alcohol (hydroabietyl alcohol) one of the few high molecular weight alcohols readily available. The carboxyl group can also be converted to a primary amine (Rosin Amine D) through the nitrile.

Agathic acid is a bicyclic diterpene acid which gives a tricyclic acid, *isoagathic acid*, on cyclization with formic acid. With Se the latter gives 1,7-Me₂-phenanthrene.[146] A related di-tertiary alcohol is *sclareol*.[147] *Manool* is a bicyclic diterpene alcohol with two double bonds which gives the same "trihydrochloride" as sclareol.[148]

The *sapogenines* (saponines) are related to the *triterpenes*. *Digitonine* has a structure much like that of the sterols.[149] It precipitates sterols having the 3-OH in the *trans* position but not their epimers. Various cyclic triterpenes have been obtained from the open-chain compound squalene.[150] *Elemolic acid*, $C_{30}H_{48}O_{31}$, is a related hydroxy acid. The *amyrenes* and *betuline* are triterpene alcohols. *Ursolic*, *suma-resinolic*, *sia-resinolic* and *oleanolic* acids as well as *hederagenine* and *gypsogenine* are triterpene derivatives. Their dehydrogenation with Se has produced such substances as 1,2,3,4-Me₂-benzene, 2,7-Me₂-

[146] *Ann. Rep. Chem. Soc.* (London) 1932, 162.
[147] *ibid.* 163.
[148] Hosking. *Ber.* 69B, 780 (1936).
[149] Tschesche. *Angew. Chem.* 48, 569 (1935).
[150] *Ann. Rep. Chem. Soc.* (London) 1927, 127.

naphthalene, 1,2,7-Me₃-naphthalene (sapotalene), hydroxysapotalene, poly-methyldinaphthyls and Me₃-picene.[151]

Carotenoids

They are the highly unsaturated fat-soluble coloring matters found in many plant and animal products. The color is due to large numbers of con-jugated double bonds. These compounds are *terpenic* in that they can be formulated as constructed from isoprene residues. The two chief hydrocarbons of this series, *lycopene* and *carotene*, have the formula $C_{40}H_{56}$ and so are tetra-terpenes containing 8 isoprene units but lacking 8 H atoms.

Lycopene (lycopin), the red coloring matter of tomatoes, adds 13 H_2 on catalytic hydrogenation. In dry oxygen it absorbs 11 atoms of O and becomes colorless. It thus contains no rings. Ozonation and various oxidation procedures give acetone, acetaldehyde, acetic acid, levulic acid, succinic acid, methylheptenone and more complex products. Its formula (Karrer) is

$$Me_2C = CHCH_2CH_2C(Me)(= CHCH = CHCMe)_2 = CHCH =$$
$$CHCH = (CMeCH = CHCH =)_2C(Me)CH_2CH_2CH = CMe_2$$

The molecule is symmetrical and is made up of two units having the same chain as phytol and farnesol but having alternate double and single bonds starting at the middle and omitting only the double bonds which would be in 4,5-position to each end. Thus, if built up from reactions of isoprene or related units, the process could not go from one end of the molecule to the other as in the ordinary linear polymerization of isoprene. It might start at the center from an isomer of myrcene obtained by the "reverse" addition of two molecules of isoprene.

Such a molecule could add an isoprene molecule in the ordinary way at *each* end to give a lower analog of squalene containing a double bond in the central group of four atoms

The addition of one and two more isoprene units at *each* end in the usual way would give the skeletons of squalene (with 2 too few H) and of lycopene (with

[151] *Ann. Rev. Biochem.* 469 (1934).

8 too many H). The mechanism of the hydrogenation and dehydrogenation of these substances in nature is still a mystery.

Lycopene, in common with the other carotenoids, is readily oxidized by air and gives deep blue colors with conc. sulfuric acid, $SbCl_3$, etc. These colors are of analytical value.

The *carotenes*, $C_{40}H_{56}$, isomeric with lycopene, are among the most widely distributed natural coloring matters. *α-Carotene*, m. 188°, is dextrorotatory while its isomer, β-carotene, m. 184°, is optically inactive. The structure of β-carotene (Karrer) depends on its absorption of 11 H_2 and its oxidation to the α,α-Me$_2$-derivatives of adipic, glutaric, succinic and malonic acids, together with large quantities of acetic acid. The formation of the first of these (geronic acid) indicates the presence of β-ionone rings. β-Carotene is thus the product which would be obtained by closing a 6-ring at each end of the lycopene molecule

$$\left[\begin{array}{l} CH_2CMe_2C{-}CH{=}CHCMe{=}CHCH{=}CHCMe{=}CHCH{=} \\ CH_2CH_2{-}CMe \end{array} \right]_2$$

β-Carotene

β-Carotene is convertible by the animal organism to Vitamin A and is consequently called Provitamin A. The change involves splitting the molecule in half and changing the $-CH=$ group to a primary alcohol group $-CH_2OH$. α-Carotene gives geronic and isogeronic (β,β-Me$_2$-adipic) acids on ozonolysis. Thus it differs from β-carotene by having one α-ionone ring, $CH_2CMe_2CH{-}$.

$$CH_2CH{=}CMe$$

A third isomer, γ-carotene, occurs with α and β. (Usual composition of carotene is 15% α, 85% β, and trace γ.) γ-Carotene is like β-carotene at one end and like lycopene at the other: m. 178°.

β-Carotene with hydrogen peroxide gives the aldehyde corresponding to Vitamin A; reduction of this gives the vitamin.[152]

Zeaxanthin, $C_{40}H_{56}O_2$, the coloring matter of maize, is a dihydroxyl derivative of β-carotene. Its oxidation gives α,α-Me$_2$-succinic acid but no higher acid. Thus the rings at the ends of the molecule are 2,6,6-Me$_3$-4-OH-cyclohexenyl. *Xanthophyll* (lutein) of green leaves is an isomer of zeaxanthin related to α-carotene. *Crocetin*, $C_{20}H_{24}O_4$, is a dibasic acid with 7 conjugated olefinic linkages and 4 Me groups,

$$({=}CHCH{=}CMeCH{=}CHCH{=}CMeCO_2H)_2.$$

Treatment with $TiCl_3$ gives *dihydrocrocetin*, the addition being said to be at the ends of the long conjugated system. The dihydro compound is much more

[152] *Ann. Rev. Biochem.* (1947).

easily oxidized than crocetin. *Bixin* is the monomethyl ester of *norbixin* (I), a C_{24}-dibasic acid related to crocetin

$$(=CHCH=CMeCH=CHCH=CMeCH=CHCO_2H)_2. \qquad (I)$$

It thus differs from the latter by having no Me groups on the α-carbons.

Relation of the carotenoids to each other. All those of known structure contain the central grouping, $=CH-CH=CH-CH=$, between tertiary carbon atoms. Zeaxanthin and xanthophyll may be regarded as products of mild oxidation of the carotenes which in turn could be formed by cyclization of each end of the lycopene molecule. Oxidation of lycopene by removing *each end* as far as the second double bond would leave norbixin while oxidation of the carotenes with removal of both 6-rings by breaking the adjacent double bonds would give crocetin. The mechanism for such highly selective oxidations is related to the possibility of the existence of such unusual molecules as those of the polyenes. In many respects conjugated double bonds differ radically from the same number of isolated double bonds. While Thiele's idea of partial valences by which a system of conjugated double bonds involves something like a uniform flux of valence forces along the chain should not be taken too literally, it does represent an approximation of the facts.

$$-CH=CH-CH=CH \rightleftharpoons -CH=CH=CH=CH-$$

Thus lycopin may exist in space in some such arrangement as follows:

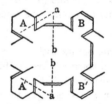

The four rings may be regarded as of chelate nature. When A and A' change to true rings by elimination of the end double bonds, β-carotene is formed. A different position of one double bond in one ring gives α-carotene. Oxidation at a and b would give norbixin and crocetin respectively. These might contain the chelate rings B and B'.

Vitamins and Hormones. Heilbron. Volume II. Carotenoids. Bogert, in Gilman. Vol. II, 1st Ed.

IV. CHOLANE SERIES

The *cholane* nucleus contains a hydrogenated phenanthrene with a 3-carbon chain attached in the 1,2-position of the phenanthrene to give a 5-ring, two angular methyl groups and a 2-*s*-amyl group. It has been prepared from cholesterol and cholanic acid.[1] The special system of numbering is used with the large numbers of related natural substances.

This grouping and its modifications occur in the bile acids, the sterols, toad poisons, heart stimulants of the digitalis and arrow poison type such as strophanthin and ouabain, the sex hormones and in the cortical substances.

Fieser and Fieser, Natural Products Related to Phenanthrene, A.C.S. Monograph No. 70, 1936. (Revised 1949).

A. Bile Acids (Gallensäuren)

These are saturated hydroxy acids having the cholane nucleus. They occur in the animal organism combined with acids as in glycocholic and taurocholic acids related to glycine and taurine respectively. They can be reduced to a parent carboxylic acid, *cholaic acid (cholanic acid)* having the end (24) Me in the 17-side chain converted to CO_2H. Cholanic acid is related to coprostanol (coprosterol).[2] Thus the four rings are united *cis, trans, trans*.

Cholanic acid has been converted to norcholanic, bisnorcholanic, etiocholanic and etiobilianic acids by successive removals of carbon atoms from the side chain and finally by opening Ring IV.[3]

The common bile acids are hydroxyl derivatives of cholanic acid: *lichocholic acid*, 3-OH; *deoxycholic, hyodeoxycholic* and *chenodeoxycholic acids*, 3,12-; 3,6- and 3,7-(OH)$_2$ respectively; *cholic* and *β-phocaecholic acids*, 3,7,12- and 3,7,23-(OH)$_3$ respectively.

The hydroxyl groups in the bile acids offer points of attack which have finally clarified not only their own structures but that of the sterols and related products. Some of the degradation products of deoxycholic acid are deoxybilianic, choloidanic, pyrocholoidanic, prosolanellic, solanellic, pyrosolanellic and norsolanellic or biloidanic acids.[4]

Choleic acids are complex coordination compounds of deoxycholic (3,12-dihydroxycholanic) acid with various other substances such as fatty acids and even aromatic substances such as phenol and naphthalene. The natural

[1] *Ann. Rep. Chem. Soc.* (London) 1927, 130.
[2] *Ann. Rep. Chem. Soc.* (London) 1927, 130.
[3] *ibid.*
[4] *Ann. Rev. Biochem.* 92 (1934).

choleic acid in bile contains 8 molecules of the bile acid combined with 1 molecule of stearic, oleic, or palmitic acid.[5-7]

B. STEROLS

These are widely distributed in plants (phytosterols) and in animals (zoosterols). The animal sterols are hydroxyl derivatives of omega-isopropyl cholane (p. 591). Thus the 17-side chain is $-CH(Me)(CH_2)_3CHMe_2$. The bile acids probably have their origin in the oxidation of the sterols, one point of oxidation being at the *tert*-H of the isopropyl group thus leading to the elimination of that group.

$$-CH_2CHMe_2 \rightarrow -CH_2C(OH)Me_2 \rightarrow -CO_2H + Me_2CO$$

The most studied example of the zoosterols, *cholesterol*, $C_{27}H_{45}$—OH, $[\alpha]^D = -39.5°$, m. 151°, first found in gall stones, has been assigned the following structure.[8,9]

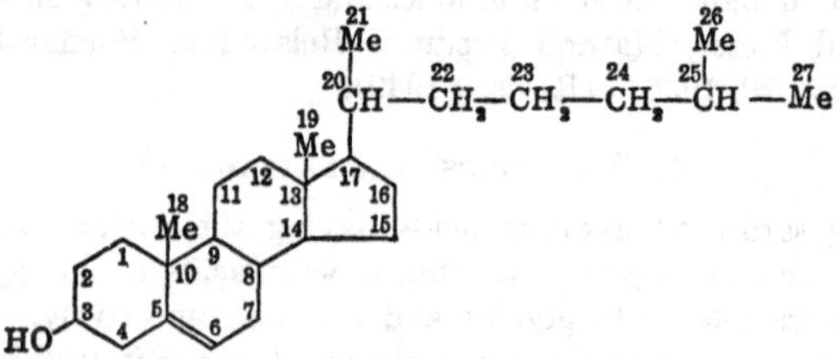

It is thus 3-OH-Δ5-ω-isopropyl-cholene.

Vigorous dehydrogenation with Se gives chrysene, the process thus eliminating the two angular methyls, opening the 5-ring to give a 6-ring at the expense of the 19-Me and removing the side chain. *Mild hydrogenation* saturates the 5-double bond giving dihydrocholesterol or *cholestanol* (β-cholestanol). Replacement of the OH of cholesterol by H gives the parent hydrocarbon *cholestane*, $C_{27}H_{48}$. X-ray measurements[10] and surface film studies both indicate that the sterols have relatively *flat* molecules. Thus the 9,8-linkage must be *trans*, that is, 7- and 10-carbons must be on opposite sides of the plane of carbons 5,6,8,9,12,13 and hydrogens 8 and 9 are on opposite sides of the molecule. In cholestane, rings I and II are also *trans*, the 5-H and 10-Me being on opposite sides of the plane of the molecule. Rings III and IV are believed to be *trans* in all the sterols as well as other cholane derivatives. Cholestane can be converted to a stereoisomer of cholanic acid (iso-, allo- or hyo-cholanic acid). The configurations of the saturated alcohols related to cholesterol have been determined as follows:

[5] Wieland, Sorge. *Z. physiol. Chem.* **97**, 1 (1916).
[6] Sobotka, Goldberg. *Biochem. J.* **26**, 555 (1932).
[7] Sabotka. *Chem. Rev.* **15**, 135 (1940).
[8] *Ann. Rep. Chem. Soc.* (London) **1933**, 198.
[9] *Ann. Rev. Biochem.* 97 (1934).
[10] Bernal et al. *Trans. Roy. Soc.* (London) **239A**, 135 (1940).

Cholestanol, β-cholestanol, *cis, trans, trans, trans.* The *cis* refers to the configuration of the 3-OH group with respect to the 10-Me. The three ring fusions are all *trans.* The β- corresponds to the fact that cholesterol is precipitated by digitonin. Compounds of opposite configuration at carbon No. 3 are not so precipitated and are called α-.

*epi-***Cholestanol,** α-cholestanol, *trans, trans, trans, trans,* is the epimer of cholestanol, differing only in the configuration at carbon No. 3. It is not precipitated by digitonin.

Coprostanol, coprosterol, *pseudo*-cholestanol, *cis, cis, trans, trans,* is found in feces. It can be made by hydrogenating *allo*-cholesterol. It belongs to the β-series, precipitated by digitonin.

*epi-***Coprostanol,** *trans, cis, trans, trans.*

Allocholesterol and *epi-***allocholesterol,** both with $\Delta 4$ instead of $\Delta 5$ as in cholesterol, are obtained as a mixture by aluminum isopropylate reduction of cholestene-4-one-3 made from cholesterol in the following steps.

1. Protection of the 5-double bond by addition of bromine.
2. Oxidation of the 3-CHOH group to carbonyl.
3. Treatment with zinc and acetic acid to remove the bromine and shift the double bond to the 4-position, in conjugation with the ketone group.

The important phytosterols include *sitosterol* (24-Et-cholesterol), *stigmasterol* (24-Et-$\Delta 22$-dehydrocholesterol), *spinasterol*, and *ergosterol*. The last has the same structure as cholesterol but with an additional double bond in the 7-position and the 17-side chain as $-\text{CHMeCH}=\text{CHCHMeCHMe}_2$. It is thus 24-Me-$\Delta 5,7,22$-cholestatrien-3-ol. It gives color reactions with Cl_3-acetic acid.[11] Ultraviolet irradiation converts ergosterol to substances of antirachitic or Vitamin D activity (*lumisterol, tachysterol, calciferol* or Vitamin D$_2$, *suprasterol-1* and *-2,* and *toxesterol*). For this change it is necessary to have the second ring in the sterol with double bonds at positions 5 and 7. Irradiation then opens this ring in the 9,10-position as follows:

The product is Vitamin D$_2$ (calciferol).

Vitamin D$_3$ or activated 7-dehydrocholesterol has the same structure with the 17-side chain of cholesterol.[12, 13] Another product of the irradiation of ergosterol is tachysterol believed to be formed by the opening of bond 9,10 with the formation of a double bond at 9,11.[14, 15]

[11] Rosenheim, Callow. *Biochem. J.* **23**, 47 (1931).
[12] *Ann. Rep. Chem. Soc.* (London) 1933, 210.
[13] *Ann. Rev. Biochem.* 101 (1934).
[14] Grandmann. *Z. Physiol. Chem.* **252**, 151 (1938).
[15] *Biol. Rev.* 135 (1940).

C. CARDIAC AGLUCONES

These are closely related to the bile acids and the sterols but contain angular OH groups and a complex lactone system in the 17-position. They are obtained by hydrolysis of glucosides found in many plants which have been used as sources of heart stimulants and arrow poisons. The best known aglucones contain 23 C atoms. They are *strophanthidin, digitoxigenin, gitoxigenin, digoxigenin, periplogenin, sarmentogenin, uzarigenin,* and *ouabegenin. Scillaridin A* has 25 C atoms.[16] The most studied of these is *strophanthidin.* To it is assigned the formula.[17]

The analogs of strophanthidin have only minor modifications of its molecule. Thus *periplogenin* has the 18-Me instead of an aldehyde group. *Digitoxigenin* differs from periplogenin by not having the 5-OH. *Digoxigenin* is like periplogenin but has the OH on C_{12} instead of C_5.

D. TOAD POISONS

These contain, besides nitrogenous compounds (the bufotenins), cholane derivatives, the *bufagins* and *bufotoxins*.[18,19] The bufagins are unsaturated polyhydroxy cholane derivatives related to the cardiac aglucones. Hydrolysis gives formic or acetic acid. This may be due to an ester group or to some splitting of the molecule to give formaldehyde and acetaldehyde. Bufotoxins are compounds of the bufagins with suberyl arginine.

Bufotalin has a five-carbon unsaturated lactone ring.[20]

[16] *Ann. Rep. Chem. Soc.* (London) **1934**, 218.
[17] *Ann. Rev. Biochem.* 164 (1946).
[18] Wieland et al. *Ann.* 524, 203 (1936).
[19] Jensen. *J. Am. Chem. Soc.* 59, 767 (1937).
[20] *Ann. Rev. Biochem.* 137 (1942).

E. STEROIDAL SAPOGENINS

These have the full cholesterol carbon skeleton[21] with a spiroketal side-chain involving a 16,27-dihydroxy-22-keto system.[22] They are obtained by hydrolysis of glycosides found in plants belonging to the families *Liliaceae, Dioscoreaceae, Scrophulariaceae* and *Amaryllidaceae*. There are thirty members of this group, all of which have been chemically interrelated. The most widely occurring are *diosgenin, smilagenin, sarsasapogenin, hecogenin, manogenin, gitogenin, neodiosgenin* and *kryptogenin*. The most important is *diosgenin*. To it is assigned the formula

Its relation with cholesterol is shown by Clemmensen reduction, followed by PBr₃, KOAc on the 3-Br and Zn-HCl treatments.[23]

The sapogenins are readily degraded to compounds of the hormone series.[24]

F. SEX HORMONES

These are related to the cholane and sterol series in containing the hydrogenated cyclopentenylphenanthrene nucleus. The chief female and male hormones are *estrone* (theelin) (I) and *testosterone* (II)

[21] Tschesche, Hagedorn. *Ber.* **68**, 1412 (1935).
[22] Marker et al. *J. Am. Chem. Soc.* **69**, 2167 (1947).
[23] Marker, Turner. *J. Am. Chem. Soc.* **63**, 767 (1941).
[24] Marker. *J. Am. Chem. Soc.* **62**, 3350 (1940).

The phenolic properties of theelin make possible its commercial extraction from pregnancy urine. *Estriol* differs from estrone in having OH groups at 16 and 17. (For numbering see p. 592.) Dehydration, with rearrangement of H, converts it to estrone. *Equilenin* differs from *estrone* in having rings A and B benzenoid. Its total synthesis has been accomplished.[25] *Androsterone* differs from testosterone in having its first ring saturated and in having hydroxyl at 3 and ketonic oxygen at 17. The corpus luteum hormone, *progesterone*, differs from testosterone by having MeCO- in place of the OH at 17. Reduction of its two ketone groups to secondary alcohol groups and hydrogenation of its double bond converts progesterone to a mixture, *pregnandiol* and *allo-pregnandiol*. The *allo-* indicates that the H added at 5 is *trans* with respect to the 10-Me. This usage must not be confused with that used in the case of allocholesterol (p. 593) to indicate a different position of the double bond. Replacement of the MeCHOH-group in *allo*-pregnandiol by ketonic oxygen gives androsterone.

Most of the possible reduction products of progesterone have been isolated from pregnancy urine. Nonpregnancy female urine contains octahydrotheelin.[26]

G. ADRENAL SUBSTANCES

These are closely related to *allo-pregnandiol* and *progesterone* and may contain additional oxygenated positions at 11, 17 and 21. The more important substances having cortical activity are *desoxycorticosterone* (I), *dehydrocorticosterone* (II) and *17-hydroxydehydrocorticosterone* (cortisone).

Partial syntheses of I and II from bile acids have been accomplished.[27]

The Steroids, W. H. Strain, in Gilman vol. II, 2nd Ed.

[25] Bachman, Cole, Wilds. *J. Am. Chem. Soc.* 61, 974 (1939); 62, 824 (1940).

[26] Marker et al. *J. Am. Chem. Soc.* 59, 2291 (1937); 60, 1725 (1938); 61, 588 (1939); 62, 518, 898 (1940).

[27] Reichstein. *Helv. Chim. Acta* 20, 1164 (1937); 26, 747 (1943).